AL 509676

336 De8h

1274/06

QUANTITATIVE FINANCE
AND
RISK MANAGEMENT

A Physicist's Approach

QUANTITATIVE FINANCE AND RISK MANAGEMENT

A Physicist's Approach

Jan W Dash

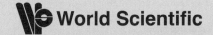
World Scientific

NEW JERSEY · LONDON · SINGAPORE · BEIJING · SHANGHAI · HONG KONG · TAIPEI · CHENNAI

Published by

World Scientific Publishing Co. Pte. Ltd.
5 Toh Tuck Link, Singapore 596224
USA office: Suite 202, 1060 Main Street, River Edge, NJ 07661
UK office: 57 Shelton Street, Covent Garden, London WC2H 9HE

British Library Cataloguing-in-Publication Data
A catalogue record for this book is available from the British Library.

First published 2004
Reprinted 2005

QUANTITATIVE FINANCE AND RISK MANAGEMENT
A Physicist's Approach

Copyright © 2004 by World Scientific Publishing Co. Pte. Ltd.

All rights reserved. This book, or parts thereof, may not be reproduced in any form or by any means, electronic or mechanical, including photocopying, recording or any information storage and retrieval system now known or to be invented, without written permission from the Publisher.

For photocopying of material in this volume, please pay a copying fee through the Copyright Clearance Center, Inc., 222 Rosewood Drive, Danvers, MA 01923, USA. In this case permission to photocopy is not required from the publisher.

ISBN 981-238-712-9

This book is printed on acid-free paper.

Printed in Singapore by Mainland Press

Dedication

I dedicate this book to my father and mother, Edward and Honore Dash. They inspired learning and curiosity, and advised never to take anything for granted.

Table of Contents

ACKNOWLEDGEMENTS xix

PART I: INTRODUCTION, OVERVIEW, AND EXERCISE 1

1. Introduction and Outline 3
   Who/ How/What, "Tech. Index", Messages, Personal Note 3
   Summary Outline: Book Contents 5

2. Overview (Tech. Index 1/10) 7
   Objectives of Quantitative Finance and Risk Management 7
   Tools of Quantitative Finance and Risk Management 9
   The Traditional Areas of Risk Management 11
   When Will We Ever See Real-Time Color Movies of Risk? 13
   Many People Participate in Risk Management 13
   Quants in Quantitative Finance and Risk Management 15
   References 17

3. An Exercise (Tech. Index 1/10) 19
   Part #1: Data, Statistics, and Reporting Using a Spreadsheet 19
   Part #2: Repeat Part #1 Using Programming 22
   Part #3: A Few Quick and Tricky Hypothetical Questions 23
   Messages and Advice 24
   References 24

PART II: RISK LAB (NUTS AND BOLTS OF RISK MANAGEMENT) 25

4. Equity Options (Tech. Index 3/10) 27
   Pricing and Hedging One Option 27
   American Options 30
   Basket Options and Index Options 31
   Other Types of Equity Options; Exotics 33
   Portfolio Risk (Introduction) 33
   Scenario Analysis (Introduction) 33
   References 34

5.	FX Options (Tech. Index 4/10)	35
	FX Forwards and Options	35
	Some Practical Details for FX Options	38
	Hedging FX Options with Greeks: Details and Ambiguities	39
	FX Volatility Skew and/or Smile	41
	Pricing Barrier Options with Skew	45
	Double Barrier Option: Practical Example	47
	The "Two-Country Paradox"	48
	Quanto Options and Correlations	50
	FX Options in the presence of Stochastic Interest Rates	51
	Numerical Codes, Closed Form Sanity Checks, and Intuition	51
	References	52
6.	Equity Volatility Skew (Tech. Index 6/10)	53
	Put-Call Parity: Theory and Violations	54
	The Volatility Surface	55
	Dealing with Skew	55
	Perturbative Skew and Barrier Options	56
	Static Replication	58
	Stochastic Volatility	60
	Local Volatility and Skew	62
	The Skew-Implied Probability Distribution	63
	Local vs. Implied Volatility Skew; Derman's Rules of Thumb	63
	Option Replication with Gadgets	65
	Intuitive Models and Different Volatility Regimes	68
	The Macro-Micro Model and Equity Volatility Regimes	69
	Jump Diffusion Processes	69
	Appendix A: Algorithm for "Perturbative Skew" Approach	69
	Appendix B: A Technical Issue for Stochastic Volatility	71
	References	72
7.	Forward Curves (Tech. Index 4/10)	73
	Market Input Rates	73
	Construction of the Forward-Rate Curve	76
	References	83
8.	Interest-Rate Swaps (Tech. Index 3/10)	85
	Swaps: Pricing and Risk	85
	Interest Rate Swaps: Pricing and Risk Details	91
	Counterparty Credit Risk and Swaps	107
	References	109
9.	Bonds: An Overview (Tech. Index 2/10)	111
	Types of Bonds	111
	Bond Issuance	115

Table of Contents   ix

       Bond Trading 116
       Bond Math 118
       References 121

10. Interest-Rate Caps (Tech. Index 4/10) 123
       Introduction to Caps 123
       The Black Caplet Formula 125
       Non-USD Caps 127
       Relations between Caps, Floors, and Swaps 127
       Hedging Delta and Gamma for Libor Caps 128
       Hedging Volatility and Vega Ladders 129
       Matrices of Cap Prices 131
       Prime Caps and a Vega Trap 131
       CMT Rates and Volatility Dependence of CMT Products 132
       References 136

11. Interest-Rate Swaptions (Tech. Index 5/10) 137
       European Swaptions 137
       Bermuda/American Swaption Pricing 141
       Delta and Vega Risk: Move Inputs or Forwards? 143
       Swaptions and Corporate Liability Risk Management 144
       Practical Example: A Deal Involving a Swaption. 146
       Miscellaneous Swaption Topics 148
       References 150

12. Portfolios and Scenarios (Tech. Index 3/10) 151
       Introduction to Portfolio Risk Using Scenario Analysis 151
       Definitions of Portfolios 151
       Definitions of Scenarios 153
       Many Portfolios and Scenarios 155
       A Scenario Simulator 157
       Risk Analyses and Presentations 157

PART III: EXOTICS, DEALS, AND CASE STUDIES 159

13. A Complex CVR Option (Tech. Index 5/10) 161
       The M&A Scenario 161
       CVR Starting Point: A Put Spread 162
       CVR Extension Options and Other Complications 162
       The Arbs and the Mispricing of the CVR Option 164
       A Simplified CVR: Two Put Spreads with Extension Logic 165
       Non-Academic Corporate Decision for Option Extension 167
       The CVR Option Pricing 169
       Analytic CVR Pricing Methodology 173

	Some Practical Aspects of CVR Pricing and Hedging 176
	The CVR Buyback 180
	A Legal Event Related to the CVR 180
	References 180

14. Two More Case Studies (Tech. Index 5/10) 183
    Case Study: DECS and Synthetic Convertibles 183
    D_{123} : The Complex DEC Synthetic Convertible 188
    Case Study: Equity Call with Variable Strike and Expiration 193
    References 199

15. More Exotics and Risk (Tech. Index 5/10) 201
    Contingent Caps 201
    Digital Options: Pricing and Hedging 205
    Historical Simulations and Hedging 207
    Yield-Curve Shape and Principle-Component Options 209
    Principal-Component Risk Measures (Tilt Delta etc.) 210
    Hybrid 2-Dimensional Barrier Options—Examples 211
    Reload Options 214
    References 217

16. A Pot Pourri of Deals (Tech. Index 5/10) 219
    TIPS (Treasury Inflation Protected Securities) 219
    Municipal Derivatives, Muni Issuance, Derivative Hedging 221
    Difference Option on an Equity Index and a Basket of Stocks 224
    Resettable Options: Cliquets 226
    Power Options 230
    Path-Dependent Options and Monte Carlo Simulation 231
    Periodic Caps 231
    ARM Caps 231
    Index-Amortizing Swaps 232
    A Hypothetical Repo + Options Deal 236
    Convertible Issuance Risk 239
    References 241

17. Single Barrier Options (Tech. Index 6/10) 243
    Knock-Out Options 245
    The Semi-Group Property including a Barrier 247
    Calculating Barrier Options 248
    Knock-In Options 249
    Useful Integrals for Barrier Options 251
    Complicated Barrier Options and Numerical Techniques 252
    A Useful Discrete Barrier Approximation 252
    "Potential Theory" for General Sets of Single Barriers 253

	Barrier Options with Time-Dependent Drifts and Volatilities 255
	References 256
18.	Double Barrier Options (Tech. Index 7/10) 257
	Double Barrier Solution with an Infinite Set of Images 258
	Double Barrier Option Pricing 260
	Rebates for Double Barrier Options 262
	References 263
19.	Hybrid 2-D Barrier Options (Tech. Index 7/10) 265
	Pricing the Barrier 2-Dimension Hybrid Options 267
	Useful Integrals for 2D Barrier Options 268
	References 269
20.	Average-Rate Options (Tech. Index 8/10) 271
	Arithmetic Average Rate Options in General Gaussian Models 272
	Results for Average-Rate Options in the MRG Model 276
	Simple Harmonic Oscillator Derivation for Average Options 277
	Thermodynamic Identity Derivation for Average Options 278
	Average Options with Log-Normal Rate Dynamics 278
	References 280

PART IV: QUANTITATIVE RISK MANAGEMENT 281

21.	Fat Tail Volatility (Tech. Index 5/10) 283
	Gaussian Behavior and Deviations from Gaussian 283
	Outliers and Fat Tails 284
	Use of the Equivalent Gaussian Fat-Tail Volatility 287
	Practical Considerations for the Fat-Tail Parameters 288
	References 294
22.	Correlation Matrix Formalism; the $\mathcal{N}$-Sphere (Tech. Index 8/10) ... 295
	The Importance and Difficulty of Correlation Risk 295
	One Correlation in Two Dimensions 296
	Two Correlations in Three Dimensions; the Azimuthal Angle 297
	Correlations in Four Dimensions 300
	Correlations in Five and Higher Dimensions 301
	Spherical Representation of the Cholesky Decomposition 303
	Numerical Considerations for the $\mathcal{N}$-Sphere 304
	References 305
23.	Stressed Correlations and Random Matrices (Tech. Index 5/10) 307
	Correlation Stress Scenarios Using Data 307

	Stressed Random Correlation Matrices 313
	Random Correlation Matrices Using Historical Data 313
	Stochastic Correlation Matrices Using the $\mathcal{N}$-sphere 314
24.	Optimally Stressed PD Correlation Matrices (Tech. Index 7/10) 319
	Least-Squares Fitting for the Optimal PD Stressed Matrix 321
	Numerical Considerations for Optimal PD Stressed Matrix 322
	Example of Optimal PD Fit to a NPD Stressed Matrix 323
	SVD Algorithm for the Starting PD Correlation Matrix 325
	PD Stressed Correlations by Walking through the Matrix 328
	References 328
25.	Models for Correlation Dynamics, Uncertainties (Tech. Index 6/10) 329
	"Just Make the Correlations Zero" Model; Three Versions 329
	The Macro-Micro Model for Quasi-Random Correlations 331
	Correlation Dependence on Volatility 335
	Implied, Current, and Historical Correlations for Baskets 338
26.	Plain-Vanilla VAR (Tech. Index 4/10) 341
	Quadratic Plain-Vanilla VAR and CVARs 344
	Monte-Carlo VAR 346
	Backtesting 347
	Monte-Carlo CVARs and the CVAR Volatility 347
	Confidence Levels for Individual Variables in VAR 350
	References 351
27.	Improved/Enhanced/Stressed VAR (Tech. Index 5/10) 353
	Improved Plain-Vanilla VAR *(IPV-VAR)* 353
	Enhanced/Stressed VAR *(ES-VAR)* 357
	Other VAR Topics 365
	References 368
28.	VAR, CVAR, CVAR Volatility Formalism (Tech. Index 7/10) 369
	Set-up and Overview of the Formal VAR Results 369
	Calculation of the Generating Function 371
	VAR, the CVARs, and the CVAR Volatilities 374
	Effective Number of SD for Underlying Variables 376
	Extension to Multiple Time Steps using Path Integrals 378
29.	VAR and CVAR for Two Variables (Tech. Index 5/10) 381
	The CVAR Volatility with Two Variables 381
	Geometry for Risk Ellipse, VAR Line, CVAR, CVAR Vol 382

Table of Contents

30. Corporate-Level VAR (Tech. Index 3/10) 387
    Aggregation, Desks, and Business Units 387
    Desk CVARs and Correlations between Desk Risks 389
    Aged Inventory and Illiquidity 391

31. Issuer Credit Risk (Tech. Index 5/10) 393
    Transition/Default Probability Matrices 394
    Calculation of Issuer Risk—Generic Case 399
    Example of Issuer Credit Risk Calculation 403
    Issuer Credit Risk and Market Risk: Separation via Spreads ... 406
    Separating Market and Credit Risk without Double Counting 407
    A Unified Credit + Market Risk Model 410
    References 413

32. Model Risk Overview (Tech. Index 3/10) 415
    Summary of Model Risk 415
    Model Risk and Risk Management 416
    Time Scales and Models 416
    Long-Term Macro Component with Quasi-Random Behavior ... 417
    Liquidity Model Limitations 417
    Which Model Should We Use? 418
    Model Risk, Model Reserves, and Bid-Offer Spreads 418
    Model Quality Assurance 419
    Models and Parameters 419
    References 420

33. Model Quality Assurance (Tech. Index 4/10) 421
    Model Quality Assurance Goals, Activities, and Procedures 421
    Model QA: Sample Documentation 424
    User Section of Model QA Documentation 425
    Quantitative Section of Model QA Documentation 425
    Systems Section of Model QA Documentation 428
    References 430

34. Systems Issues Overview (Tech. Index 2/10) 431
    Advice and a Message to Non-Technical Managers 431
    What are the "Three-Fives Systems Criteria"? 431
    What is the Fundamental Theorem of Systems? 432
    What are Some Systems Traps and Risks? 432
    The Birth and Development of a System 433
    Systems in Mergers and Startups 435
    Vendor Systems 436
    New Paradigms in Systems and Parallel Processing 437
    Languages for Models: Fortran 90, C^{++}, C, and Others ... 438
    What's the "Systems Solution"? 440

| | Are Software Development Problems Unique to Wall Street? .... 440 |
| | References.. 440 |

35. Strategic Computing (Tech. Index 3/10) 441
    Introduction and Background.. 442
    Illustration of Parallel Processing for Finance................................ 442
    Some Aspects of Parallel Processing 443
    Technology, Strategy and Change... 446
    References.. 447

36. Qualitative Overview of Data Issues (Tech. Index 2/10)................... 449
    Data Consistency 449
    Data Reliability 450
    Data Completeness... 450
    Data Vendors 450
    Historical Data Problems and Data Groups.................................... 451
    Preparation of the Data 451
    Bad Data Points and Other Data Traps... 451

37. Correlations and Data (Tech. Index 5/10) 453
    Fluctuations and Uncertainties in Measured Correlations 453
    Time Windowing 454
    Correlations, the Number of Data Points, and Variables 456
    Intrinsic and Windowing Uncertainties: Example 458
    Two Miscellaneous Aspects of Data and Correlations 460
    References.. 460

38. Wishart's Theorem and Fisher's Transform (Tech. Index 9/10)....... 461
    Warm Up: The Distribution for a Volatility Estimate 462
    The Wishart Distribution... 464
    The Probability Function for One Estimated Correlation 465
    Fisher's Transform and the Correlation Probability Function 466
    The Wishart Distribution Using Fourier Transforms...................... 468
    References.. 473

39. Economic Capital (Tech. Index 4/10).. 475
    Basic Idea of Economic Capital 475
    The Classification of Risk Components of Economic Capital.......... 479
    Exposures for Economic Capital: What Should They Be? 480
    Attacks on Economic Capital at High CL 480
    Allocation: Standalone, CVAR, or Other? 481
    The Cost of Economic Capital 483
    An Economic-Capital Utility Function.. 484
    Sharpe Ratios 484
    Revisiting Expected Losses; the Importance of Time Scales 485

Table of Contents xv

    Cost Cutting and Economic Capital 487
    Traditional Measures of Capital, Sharpe Ratios, Allocation 487
    References 488

40.   Unused-Limit Risk (Tech. Index 6/10) 489
    General Aspects of Risk Limits 489
    The Unused Limit Risk Model: Overview 491
    Unused Limit Economic Capital for Issuer Credit Risk 497

PART V: PATH INTEGRALS, GREEN FUNCTIONS, AND OPTIONS 499

41.   Path Integrals and Options: Overview (Tech. Index 4/10) 501

42.   Path Integrals and Options I: Introduction (Tech. Index 7/10) 505
    Introduction to Path Integrals 506
    Path-Integral Warm-up: The Black Scholes Model 509
    Connection of Path Integral with the Stochastic Equations 521
    Dividends and Jumps with Path Integrals 523
    Discrete Bermuda Options 530
    American Options 537
    Appendix A: Girsanov's Theorem and Path Integrals 538
    Appendix B: No-Arbitrage, Hedging and Path Integrals 541
    Appendix C: Perturbation Theory, Local Volatility, Skew 546
    Figure Captions for this Chapter 546
    References 556

43.   Path Integrals and Options II: Interest-Rates (Tech. Index 8/10) 559
    I. Path Integrals: Review 561
    II. The Green Function; Discretized Gaussian Models 562
    III. The Continuous-Time Gaussian Limit 566
    IV. Mean-Reverting Gaussian Models 569
    V. The Most General Model with Memory 574
    VI. Wrap-Up for this Chapter 578
    Appendix A: MRG Formalism, Stochastic Equations, Etc. 579
    Appendix B: Rate-Dependent Volatility Models 586
    Appendix C: The General Gaussian Model With Memory 589
    Figure Captions for This Chapter 591
    References 594

44.   Path Integrals and Options III: Numerical (Tech. Index 6/10) 597
    Path Integrals and Common Numerical Methods 598
    Basic Numerical Procedure using Path Integrals 600
    The Castresana-Hogan Path-Integral Discretization 603

Some Numerical Topics Related to Path Integrals 608
A Few Aspects of Numerical Errors 614
Some Miscellaneous Approximation Methods 618
References 624

45. Path Integrals and Options IV: Multiple Factors
(Tech. Index 9/10) 625
Calculating Options with Multidimensional Path Integrals 628
Principal-Component Path Integrals 629
References 630

46. The Reggeon Field Theory, Fat Tails, Chaos (Tech. Index 10/10) ... 631
Introduction to the Reggeon Field Theory (RFT) 631
Summary of the RFT in Physics 632
Aspects of Applications of the RFT to Finance 637
References 638

PART VI: THE MACRO-MICRO MODEL (A RESEARCH TOPIC) 639

47. The Macro-Micro Model: Overview (Tech. Index 4/10) 641
Explicit Time Scales Separating Dynamical Regions 641
I. The Macro-Micro Yield-Curve Model 642
II. Further Developments in the Macro-Micro Model 646
III. A Function Toolkit 647
References 648

48. A Multivariate Yield-Curve Lognormal Model (Tech. Index 6/10) . 649
Summary of this Chapter 649
The Problem of Kinks in Yield Curves for Models 650
I. Introduction to this Chapter 650
IIA. Statistical Probes, Data, Quasi-Equilibrium Drift 653
IIB. Yield-Curve Kinks: Bête Noire of Yield Curve Models 655
III. EOF / Principal Component Analysis 656
IV. Simpler Lognormal Model with Three Variates 658
V. Wrap-Up and Preview of the Next Chapters 659
Appendix A: Definitions and Stochastic Equations 659
Appendix B: EOF or Principal-Component Formalism 662
Figures: Multivariate Lognormal Yield-Curve Model 667
References 680

49. Strong Mean-Reverting Multifactor YC Model (Tech. Index 7/10) . 681
Summary of this Chapter 681
I. Introduction to this Chapter 682
II. Cluster Decomposition Analysis and the SMRG Model 685

Table of Contents xvii

    III. Other Statistical Tests and the SMRG Model............................ 691
    IV. Principal Components (EOF) and the SMRG Model 694
    V. Wrap-Up for this Chapter.. 694
    Appendix A: Definitions and Stochastic Equations......................... 695
    Appendix B: The Cluster-Decomposition Analysis (CDA)............. 697
    Figures: Strong Mean-Reverting Multifactor Yield-Curve Model ... 701
    References... 715

50.   The Macro-Micro Yield-Curve Model (Tech. Index 5/10).............. 717
    Summary of this Chapter 717
    I. Introduction to this Chapter 718
    Prototype: Prime (Macro) and Libor (Macro + Micro).................... 720
    II. Details of the Macro-Micro Yield-Curve Model 721
    III. Wrap-Up of this Chapter... 724
    Appendix A. No Arbitrage and Yield-Curve Dynamics.................. 725
    Figures: Macro-Micro Model.. 726
    References... 730

51.   Macro-Micro Model: Further Developments (Tech. Index 6/10) 731
    Summary of This Chapter 731
    The Macro-Micro Model for the FX and Equity Markets 733
    Macro-Micro-Related Models in the Economics Literature 735
    Related Models for Interest Rates in the Literature 735
    Related Models for FX in the Literature.. 736
    Formal Developments in the Macro-Micro Model 737
    No Arbitrage and the Macro-Micro Model: Formal Aspects........... 739
    Hedging, Forward Prices, No Arbitrage, Options (Equities) 741
    Satisfying the Interest-Rate Term-Structure Constraints 744
    Other Developments in the Macro-Micro Model 745
    Derman's Equity Regimes and the Macro-Micro Model................. 745
    Seigel's Nonequilibrium Dynamics and the MM Model................. 745
    Macroeconomics and Fat Tails (Currency Crises)........................... 746
    Some Remarks on Chaos and the Macro-Micro Model................... 747
    Technical Analysis and the Macro-Micro Model 749
    The Macro-Micro Model and Interest-Rate Data 1950-1996 750
    Data, Models, and Rate Distribution Histograms 751
    Negative Forwards in Multivariate Zero-Rate Simulations 752
    References... 753

52.   A Function Toolkit (Tech. Index 6/10) 755
    Time Thresholds; Time and Frequency; Oscillations 756
    Summary of Desirable Properties of Toolkit Functions 757
    Construction of the Toolkit Functions... 757
    Relation of the Function Toolkit to Other Approaches.................... 762
    Example of Standard Micro "Noise" Plus Macro "Signal" 764

The Total Macro: Quasi-Random Trends + Toolkit Cycles 767
Short-Time Micro Regime, Trading, and the Function Toolkit 768
Appendix: Wavelets, Completeness, and the Function Toolkit 769
References 771

Index 773

Errata 783

Acknowledgements

First, I owe a big debt of gratitude to Andy Davidson, Santa Federico, and Les Seigel, all super quants, for their support. The work of two colleagues contributed to this book: Alan Beilis (who collaborated with me on the Macro-Micro model), and Juan Castresana (who implemented numerical path-integral discretization). I thank them and the other members of the quant groups I managed over the years for their dedication and hard work. Many other colleagues helped and taught me, including quants, systems people, traders, managers, salespeople, and risk managers. I thank them all. Some specific acknowledgements are in the text.

Rich Lee, extraordinary FX systems person killed on 9/11, is remembered.

I thank Global Risk Management, Salomon Smith Barney, Citigroup for a nine-month leave of absence during 2002-03 when part of this book was written.

The Centre de Physique Théorique (CNRS Marseille, France) granted a leave of absence during my first few years on the Street, for which I am grateful.

The editors at World Scientific have been very helpful.

I especially thank my family for their encouragement, including my daughter Sarah and son David. I could not have done any of this without the constant understanding and love from my wife Lynn.

PART I

INTRODUCTION, OVERVIEW, AND EXERCISE

1. Introduction and Outline

Who/ How/What, "Tech. Index", Messages, Personal Note

1. For Whom is This Book Written?

This book is primarily for PhD scientists and engineers who want to learn about quantitative finance, and for graduate students in finance programs[1]. Practicing quantitative analysts ("quants") and research workers will find topics of interest. There are even essays with no equations for non-technical managers.

2. How Can This Book Benefit You?

This book will enable you to gain an understanding of practical and theoretical quantitative finance and risk management.

3. What is In This Book?

The book is a combination of a practical "how it's done" book, a textbook, and a research book. It contains techniques and results for quantitative problems with which I have dealt in the trenches for over fifteen years as a quant on Wall Street. Each topic is treated as a unit, sometimes drilling way down. Related topics are presented parallel, because that is how the real world works. An informal style is used to convey a picture of reality. There are even some stories.

4. What is the "Tech. Index"? What Finance Background is Needed?

The "Tech. Index" for each chapter is a relative index for this book lying between 1-10 and indicating mathematical sophistication. The average index is 5. An index 1-3 requires almost no math, while 8-10 requires a PhD and maybe more. No background in finance is assumed, but some would definitely be helpful.

---

[1] **History:** The book is an outgrowth of my tutorial on Risk Management given annually for five successive years (1996-2000) at the Conference on Intelligence in Financial Engineering (CIFEr), organized jointly by the IEEE and IAFE. The attendees comprised roughly 50% quantitative analysts holding jobs in finance and 50% PhD scientists or engineers interested in quantitative finance.

5. How Should You Read This Book? What is in the Footnotes?

You can choose topics that interest you. Chapters are self-contained. The footnotes add depth and commentary; they are useful sidebars.

6. Message to Non-Technical Managers

Parts of this book will help you get a better understanding of quantitative issues. Important chapters have discussions of systems, models, and data. Skip sections with equations (or maybe read chapters with the Tech. Index up to 3).

7. Message to Students

You will learn quantitative techniques better if you work through derivations on your own, including performing calculations, programming and reflection. The mathematician George Polya gave some good advice: "The best way to learn anything is to discover it by yourself". Bon voyage.

8. Message to PhD Scientists and Engineers

While the presentation is aimed at being self-contained, financial products are extensive. Reading a finance textbook in parallel would be a good idea.

9. Message to Professors

Part of the book could be used in a PhD finance course (Tech. Index up to 8), or for MBAs (Tech. Index up to 5). Topics you may find of interest include: (1) Feynman path integrals and Green functions for options, (2) The Macro-Micro model with explicit time scales connecting to both macroeconomics and finance, (3) Optimally stressed correlation matrices, (4) Enhanced/Stressed VAR.

10. A Personal Note

This book is largely based on my own work and/or first-hand experience. It is in part retrospective, looking back over trails traversed and sometimes blazed. Some results are in 1988-89 CNRS preprints when I was on leave from the CNRS as the head of the Quantitative Analysis Group at Merrill Lynch, in my 1993 SIAM Conference talk, and in my CIFEr tutorials. Footnotes entitled "*History*" contain dates when my calculations were done over the years, along with recollections[2].

---

[2] **History:** To translate dates, my positions were VP Manager at Merrill Lynch (1987-89); Director at Eurobrokers (1989-90), Director at Fuji Capital Markets Corp. (1990-93), VP at Citibank (1993), and Director at Smith Barney/Salomon Smith Barney/Citigroup (1993-2003). I managed PhD Quantitative Analysis Groups at Merrill, FCMC, and at SB/SSB/Citigroup through various mergers.

Summary Outline: Book Contents

The book consists of six divisions.

I. Qualitative Overview of Risk

A qualitative overview of risk is presented, plus an instructive and amusing exercise emphasizing communication.

II. Risk Lab for Derivatives (Nuts and Bolts of Risk Management)

The "Risk Lab" first examines equity and FX options, including skew. Then interest rate curves, swaps, bonds, caps, and swaptions are discussed. Practical risk management including portfolio aggregation is discussed, along with static and time-dependent scenario analyses.

This is standard textbook material, and directly relevant for basic quantitative work.

III. Exotics, Deals, and Case Studies

Topics include barriers, double barriers, hybrids, average options, the Viacom CVR, DECs, contingent caps, yield-curve options, reloads, index-amortizing swaps, and various other exotics and products.

By now, this is mostly standard material. The techniques presented in the case studies are generally useful, and would be applicable in other situations.

IV. Quantitative Risk Management

Topics include optimally stressed positive-definite correlation matrices, fat-tail volatility, Plain/Stressed/Enhanced VAR, CVAR uncertainty, credit issuer risk, model issues and quality assurance, systems issues and strategic computing, data issues, the Wishart Theorem, economic capital, and unused-limits risk. This is the largest of the six divisions of the book.

Much of this material is standard, although there are various improvements and innovations.

V. Path Integrals, Green Functions, and Options

Feynman path integrals provide an explicit and straightforward method for evaluating financial products, e.g. options. The simplicity of the path integral technique avoids mathematical obscurity. My original applications of path integrals and Green functions to options are presented, including pedagogical examples, mean-reverting Gaussian dynamics, memory effects, multiple variables, and two related straightforward proofs of Girsanov's theorem. Consistency with the stochastic equations is emphasized. Numerical aspects are treated, including the Castresana-Hogan path-integral discretization. Critical exponents and the nonlinear-diffusion Reggeon Field Theory are briefly discussed.

The results by now are all known. The presentation is not standard.

VI. The Macro-Micro Model (A Research Topic)

The Macro-Micro model, developed initially with A. Beilis, originated through an examination of models capable of reproducing yield-curve dynamical behavior – in a word, producing yield curve movements that look like real data. The model contains separate mechanisms for long-term and short-term behaviors of rates, with explicit time scales. The model is connected in principle with macroeconomics through quasi-random quasi-equilibrium paths, and it is connected with financial models through strong mean-reverting dynamics for fluctuations due to trading. Further applications of the Macro-Micro model to the FX and equities markets are also presented, along with recent formal developments. Option pricing and no-arbitrage in the Macro-Micro framework are discussed. Finally a "function toolkit", possibly useful for business cycles and/or trading, is presented.

I believe that these topics will form a fruitful area for further research and collaborations.

2. Overview (Tech. Index 1/10)

In this overview, we look at some general aspects of quantitative finance and risk management. There is also some advice that may be useful. A reminder: the footnotes in this book have interesting information. They function as sidebars, complementing the text[1].

Objectives of Quantitative Finance and Risk Management

The general goal of quantitative finance and risk management is to quantify the behavior of financial instruments today and under different possible environments in the future. This implies that we have some mathematical or empirical procedure of determining values of the instruments as well as their changes in value under various circumstances. While the road is long, and while there has been substantial progress, for many reasons this goal is only partially achievable in the end and must be tempered with good judgment. Especially problematic are the rare extreme events, which are difficult to characterize, but where most of the risk lies.

Why is Quantitative Finance a Science?
Outwardly, the quantitative nature of modern finance and risk management seems like a science. There are models that contain theoretical postulates and proceed along mathematical lines to produce equations valuing financial instruments. There are "experiments" which consist of looking at the market to determine values of financial instruments, and which provide input to the theory. Finally, there are computer systems, which keep track of all the instruments and tie everything together.

Why is Quantitative Finance not a Science?
In science there is real theory in the sense of Newton's laws (F = ma) backed by a large collection of experiments with high practical predictive power and known

---

[1] **Why Read the Footnotes?** Robert Karplus, the physicist who taught the graduate course in electromagnetism at Berkeley, said once that the most interesting part of a book is often in the footnotes. The footnotes are an integral part of this book.

domains of applicability (for Newton's laws, this means objects not too small and not moving too fast).

In contrast, financial theoretical "postulates", when examined closely, turn out to involve assumptions, which are at best only partially justifiable in the real world. The financial analogs to scientific "experiments" obtained by looking at the market are of limited value. Market information may be quite good, in which case not much theory is needed. If the market information is not very good, the finance theory is relatively unconstrained. Finance computer systems are always incomplete and behind schedule (this is a theorem).

Quantitative Finance is Not Science but Phenomenology

The situation characterizing quantitative finance is really what physicists call "phenomenology". Even if we could know the "Newton laws of finance", the real world of finance is so complex that the consequences of these laws could not be evaluated with any precision. Instead, there are financial models and statistical arguments that are only partially constrained by the real world, and with unknown domains of applicability, except that they often break when the market conditions change in an extreme fashion. The main reason for this fragility is that human psychology and macroeconomics are fundamentally involved. The worst cases for risk management, such as the onset of collective panic or the potential consequences of a deep recession, are impossible to quantify uniquely—extra assumptions tempered by judgment must be made.

What About Uncertainties in the Risk Itself?

A characteristic showing why risk management is not science deals with the lack of quantification of the uncertainties in risk calculations and estimates. Uncertainty or error analysis is always done in scientific experiments. It is preferable to call this activity "uncertainty" analysis because "error" tends to conjure up human error. While human error should not be underestimated, the main problem in finance often lies with uncertainties and incompleteness in the models and/or the data. Risk measurement is standard, but the *uncertainty* in the risk itself is usually ignored.

We will discuss one example in determining the uncertainty in risk when we discuss the uncertainties in the components of risk (CVARs) that lead to a given total risk (VAR) at a given statistical level. We hope that such measures of uncertainty will become more common in risk management.

In finance, there is too often an unscientific accounting-type mentality. Some people do not understand why uncertainties should exist at all, tend to become ill tempered when confronted with them, and only reluctantly accept their existence. The situation is made worse by the meaningless precision often used by risk managers and quants to quote risk results. Quantities that may have uncertainties of a factor of two are quoted to many decimal places. False confidence, misuse and misunderstanding can and does occur. A fruitless activity is attempting to explain why one result is different from another result under somewhat different circumstances, when the numerical uncertainties in all these results are unknown and potentially greater than the differences being examined.

Tools of Quantitative Finance and Risk Management

The main tools of quantitative finance and risk management are the models for valuing financial instruments and the computer systems containing the data corresponding to these instruments, along with recipes for generating future alternative possible financial environments and the ability to produce reports corresponding to the changes of the portfolios of instruments under the different environments, including statistical analyses. Risk managers then examine these reports, and corrective measures or new strategies are conceived.

The Greeks

The common first risk measures are the "Greeks". These are the various low-order derivatives of the security prices with respect to the relevant underlying variables. The derivatives are performed either analytically or numerically. The Greeks include delta and gamma (first and second derivatives with respect to the underlying interest rate, stock price etc.), vega[2] (first derivative with respect to volatility), and others (mortgage prepayments, etc.). The Greeks are accurate enough for small moves of the underlying, i.e. day-to-day risk management.

Hedges

Hedges are securities that offset risk of other securities. Knowledge of the hedges is critical for trading risk management. Say we have a position or a portfolio with value C depending on one or several variables $\{x_\alpha\}$ (e.g. interest rates, FX rates, an equity index, prepayments, gold, ...). Say we want a hedge H depending on possibly different variables $\{y_\beta\}$. Naturally, the trader will not hedge out the whole risk, because to do so he would have to sell exactly what he buys (back-to-back). Therefore, there will be a decision, consistent with the limits for the desk, to hedge out only part of the risk. Hedging risk can go wrong in a number of ways. Generically, the following considerations need to be taken into account:

---

[2] **Vega the Greek?** Who thought up this name? Vega does happen to be the 5th brightest star, but this is irrelevant.

1. The hedge variables $\{y_\beta\}$ may not be the same as the portfolio variables $\{x_\alpha\}$, although reasonably correlated. However, historically reasonable hedges can and do break down in periods of market instability.

2. Some of the hedge variables may have little to do with the portfolio variables, and so introduce a good deal of extra risk on their own.

3. The hedge may be too costly to implement, not be available, etc.

Scenarios ("What-if", Historical, Statistical)

In order to assess the severity of loss to large moves, scenario or statistical analyses are employed. A "what-if" scenario analysis will postulate, during a given future time frame, a set of changes of financial variables. A historical scenario analysis will take these changes from selected historical periods. A statistical analysis will use data, also from selected historical periods, and categorize the anomalous large moves ("fat tails") statistically. Especially important, though because of technical difficulties often overlooked, are changes in the correlations. We will deal with these issues in the book at length.

Usually the scenarios are treated using a simplistic time-dependence, a quasi-static assumption. That is, a jump forward in time by a certain period is assumed, and at the end of this period, the changes in the variables are postulated to exist. The jump forward in time is generally zero ("immediate changes") or a short time period (e.g. 10 days for a standard definition of "Value at Risk"). This can be improved by choosing different time periods corresponding to the liquidity characteristics of different products (short periods for liquid products easy to sell, longer periods for illiquid products hard to sell).

Usually, the risk is determined for a portfolio at a given point in time. Scenarios can also involve assumptions about the future changes in the portfolios. For example, under stressed market conditions and losses, it might be postulated that a given business unit would sell a certain fraction of inventory, consistent with business objectives. Extra penalties can be assessed for selling into hostile markets. These require estimation of the action of other institutions, volumes, etc. The worst is an attitude similar to "I don't care what you think your buggy whip is worth, I won't pay that much" that leads to the bottom falling out of a market.

Monte-Carlo Simulation

A more sophisticated risk approach uses a Monte Carlo simulator, which generates possible "worlds" marching forward in time. Either a mathematical formula can be given to generate the possible worlds, or successive scenarios can be chosen with subjective probabilities. Such calculations have more assumptions (this is bad) but lead to more detailed information and avoid some of the crude approximations of the static analyses (this is good).

For the most part, complete Monte-Carlo simulations (except for small portfolios or large portfolios on a limited basis) are a topic for future risk management. Implementation requires a parallel-processing systems effort, as described in Ch. 35.

Data and Risk

Knowledge of the historical data changes in different time frames plays a large role in assessing risk, especially anomalously large moves as well as the magnitudes of moves at different statistical levels. It is also important to know the economic or market forces that existed at the time to get a subjective handle of the probability that such moves could reoccur. While "the past is no guarantee of the future", the truth is nonetheless that the past is the only past we have, and we cannot ignore the past.

Problematic Topics of Risk: Models, Systems, Data

A summary of topics treated in more detail in other parts of the book includes:

1. *Models:* (Model Risk; Time Scales; Mean Reversion; Jumps and Nonlinear Diffusion; Long-Term Macro Quasi Random Behavior; Model Limitations; Which Model; Psychological Attitudes; Model Quality Assurance; Parameters).

2. *Systems:* (What is a System?; Calculators; Traps; Communication; Birth and Development; Prototyping; Who's in Control?; Mergers and Startups; Vendors; New Paradigms; Systems Solutions)

3. *Data:* (Consistency, Reliability, Completeness, Vendors)

The Traditional Areas of Risk Management

Risk management is traditionally separated into Market Risk and Credit Risk. There is a growing concern with Operation Risk.

Market Risk

Market risk is the risk due to the fluctuation of market variables. Market risk is separated out into functional business areas, including Interest Rates, Equities, FX, Emerging Markets, Commodities, etc. Further subdivisions include cash products (bonds, stocks), derivatives (plain vanilla, exotics), structured products (MBS, ABS), etc. Individual desks correspond to further detail (e.g. the desk for mortgage derivatives, or the high-yield bond desk). Each area will have its own risk management expertise requirements. We will spend a lot of time in this book discussing market risk.

A corporate-level measure of market risk is called VAR (Value at Risk). We will discuss various levels of sophistication of VAR, ending with a quite sophisticated measure that in this book is called Enhanced/Stressed VAR.

Credit Risk

Credit risk includes several areas, including traditional credit risk assessment of corporations, credit issuer risk, and counterparty risk.

Traditional credit risk assessment by rating agencies involves financial statistics (balance sheet, cash flows), comparisons, and in-depth analysis.

Credit issuer risk is the risk due to defaults and downgrades of issuers of securities, notably bonds. We will discuss issuer risk in some detail in this book.

Counterparty risk is due to nonperformance of counterparties to transactions. We will only mention counterparty risk briefly.

Unified Market and Credit Risk Management?

Market risk and credit risk are correlated. In times of bad markets, credit risk increases. Conversely, if credit risk is increasing because of economic weakness, the markets will not be bullish. Moreover, there are double counting issues. For example, market risk for credit products is determined using spread fluctuation data. There are technical market spread components and potential default spread components. We will see in Ch. 31 that spread movements can be distinguished for particular definitions of market and credit risk. However, it would be better if market and credit risk management were integrated. Unfortunately, the languages spoken by the two departments are largely disjoint and there can be legacy structural issues that hamper communication and integration.

Operational Risk

Operational risk deals with the risk of everything else, losses due to the "1001 Risks". One presentation tried to get the topics of operational risk on one slide. The slide contained such small font that it appeared black. Operational risk can be thought of as "Quantifying Murphy's Law" with large entropy of possibilities that can go wrong. Included here are human error, rogue traders, fraud, legal risk, organizational risk, system risk etc. Model risk could be regarded as operational (it has to go somewhere). The recent accounting and analyst scandals would also be classified as operational risk. The worst part about a major loss from operational risk is that it is always new and unexpected.

When Will We Ever See Real-Time Color Movies of Risk?

Soon after starting work on the Street in the mid 80's, I had a vision of real-time risk management, with movies in color of risk changing with the market and with the portfolio transactions. I'm still waiting. Drop me an e-mail if you see it.

Many People Participate in Risk Management

In a general sense, a vast number of people are involved in risk management. These include traders, risk managers (both at the desk level and at the corporate level), systems programmers, managers, regulators, etc in addition to the quants. All play important roles. Commonly, risk management is thought of in terms of a corporate risk management department, but it is more general.

Systems Programmers
Systems play a large role in the ability to carry out successful risk management. Systems programmers naturally need expertise in traditional computer science areas: code development, databases, etc. It is often overlooked but it is advantageous from many points of view if programmers understand what is going on from a finance and math point of view.

Traders
Traders need to understand intimately the risks of the products they trade. Sometimes traders are quants on the desk. Risk reports are designed by analysts or traders and coded by the programmers. There are however innumerable exceptions—for example, a trader writing her own models on spreadsheets and generating local desk risk reports from them. Traders use the models while exercising market judgment to gauge risk.

Risk Managers
Desk and corporate risk managers need some quantitative ability and must possess a great deal of practical experience. Risk managers also have a responsibility to understand the risks of business decisions and strategies (e.g. customer-based or proprietary trading, new products, etc).

Corporate Risk Management
Corporate risk management aggregates and analyzes portfolio risk, and analyzes deals with unusual risk. Corporate risk management also performs limit oversight for the business units. The risk results are summarized for upper management in presentations. A corporate-level assessment of risk is extremely difficult because

of the large number of desks and products. Collecting the data can be a monumental task. Inconsistent risk definitions between desks and other issues complicate the task.

Two Structures for Corporate Risk Management

There are two common structures for corporate risk management. The diagram illustrates the alternatives of the two-tier or one-tier risk management structure:

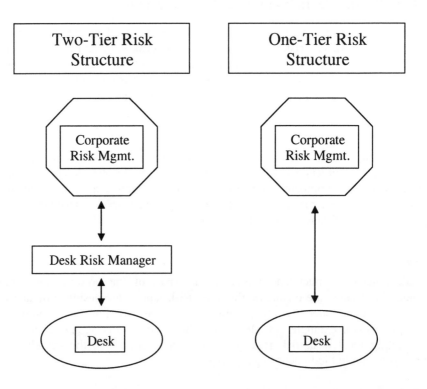

In the one-tier risk structure, the corporate risk management department is in direct contact with traders on a given desk. In this structure, the corporate risk manager follows the day-to-day trading risk details as well as participating in the other activities of corporate risk management. In the two-tier risk structure there is an intermediate desk or business risk manager. Here, the desk risk manager interacts with the traders. The desk risk manager then summarizes or emphasizes unusual risk to the corporate risk department[3].

---

[3] **Risk Management Structure:** The paradigm adopted depends on the risk-management philosophies of the trading desks and of corporate risk management. Different structures may apply to different desks. The two-tier solution requires a division of responsibilities. There are advantages and disadvantages for each of the structures.

Quants in Quantitative Finance and Risk Management

First, what is a "Quant"? This is a common (though not pejorative) term mostly applied to PhDs in science, engineering, or math doing various quantitative jobs on Wall Street [4], also called The Street.

Jobs for Quants

We start with jobs involving models. Risk is measured using models. Here the standard paradigm is that models are developed by PhD quants writing their own code, while systems programmers develop the systems into which models are inserted.

A quant writing a model to handle the risks for a new product needs to understand the details of the financial instruments he/she is modeling, the theoretical context, the various parameters that become part of the model, the numerical code implementing the model etc. The numerical instabilities of the models need to be assessed and understood.

Many other jobs for quants exist besides writing or coding models. They include risk management, computer work, database work, becoming a trader, etc.

Looking for a Job

If you are serious about pursuing a career, try to find people in the field and talk to them. Networking is generally the best way for finding a job. Headhunters can be useful, but be aware that they probably have many resumes just as impeccable as yours. At this late date, there are many experienced quants out there. If you get to the interview stage, find out as much as possible about what work the group actually does. You have to want to do the job, and be willing to give 110%. Enthusiasm counts.

On the Job: What's the Product?

The product on the job depends on the situation. Changing conditions can and do lead to changing requirements. Flexibility is important. Don't be afraid to make a suggestion – you may have a good idea. An essential piece of advice is to "Solve problems and don't cause problems".

Creativity and the 80-20 Rule

Creative thinking and prioritized problem-solving abilities are key attributes for a quant, along with the skill to apply the "80-20" rule (get 80% of the way there with 20% of the effort) in a reasonable time.

---

[4] **"Wall Street":** This means any financial institution, not just the street in New York.

Communication Skills

Communication is critical. Decisions often must (or at least should) be made with input involving quantitative analyses. Specifically the skills needed to write clear concise memos or to give quick-targeted oral presentations in non-quantitative terms while still getting the ideas and consequences across are very important and should not be overlooked[5].

Message to PhD Scientists and Engineers Who Want to Become Quants

For model building and risk management, you need to know how to program fluently in at least one language (C, C++, or Fortran) [6]. No exceptions. Fluent means that you have had years of experience and you do not make trivial mistakes. Prototyping is important and extremely useful. Prototyping can be done with spreadsheets (also Visual Basic), or with packages like Mathematica, PV Wave, Matlab, etc. However, prototyping is not a replacement for serious compiled code. Knowledge of other aspects of computer science can also be useful (GUIs, databases and SQL, hardware, networking, operating systems, compilers, Internet, etc).

For background, in addition to this book (!), read at least one finance text and some review articles or talks[i]. Conferences can be useful. Try to learn as much as possible about the jargon. Become acquainted to the extent possible with data and get a feel for numerical fluctuations. Be able to use the numerical algorithms for modeling, including Monte Carlo and diffusion equation discretization solvers. Learn about analytic models. Learn about risk.

Message to Quants Who Want to Become Quant-Group Managers

If you learn too much about quantitative analysis, finance and systems—and if you can manage people—you may wind up as the manager of a Quant Group. You now have to work out the mix between managing responsibilities and continuing your work as a quant.

Managing quants can be rather like a description of the Israeli Philharmonic Orchestra when it was founded: the orchestra was said to be hard to conduct because all the players thought they should be soloists. There are good books about managing people[ii], and there are in-house and external courses. My advice is to be genuine, work with and alongside your quants, understand the details,

---

[5] **Exercise:** Please note the practical and amusing but really dead serious exercise in the next chapter. Communication skills are a major part of this exercise.

[6] **Language Wars:** It is amazing how heated discussions on computer languages resemble fights over religious dogma. It is easiest to go along with the crowd, whatever that means. See Ch. 34.

understand the difficulties, gain the group's respect, set achievable goals that are appreciated by the management, and generally be a leader.

Depending on the situation, you can have the option of providing innovative thinking and leadership while working hard and hands-on. You need to have the strength to work independently. You need to continue to give an effort of 110%. You need to deliver the product, but be very careful about definitions. Try never to use pronouns, and especially not the pronoun "it"[7].

You will broaden your horizons, meeting smart, friendly experts who can teach you a lot, interacting with management, and experiencing the sociology of the many tribes in finance, all speaking different languages. Always assume that you can learn something.

Sometimes you will need courage. While most people are up-front and helpful, you will encounter a variety of sharks. You may also need to cope at various times with adversity, possibly including: misunderstanding, tribalism, secrecy, Byzantine power politics, 500-pound gorillas, wars, lack of quantitative competence, sluggish bureaucracy, myopia, dogmatism, interference, bizarre irrationality, nitpicking, hasty generalizations, arbitrary decisions, pomposity, and unimaginative people who like to Play Death to new ideas. However, while they do occur, these negative features are exceptions, not the rule.

All in, it is fascinating, challenging, and even fun. Bon voyage.

References

[i] **Finance: Sample References**
Hull, J. C., *Options, Futures, and Other Derivative Securities*. Prentice Hall, 1989.
Willmott, P., Dewynne, J., and Howison, S., *Option Pricing – Mathematical models and computation*. Oxford Financial Press, 1993.
Dash, J. W., *Derivatives in Corporate Risk Management*. Talk, World Bank, Finance Professionals Forum, 1996.

[ii] **Management**
Lefton, R.E., Buzzotta, V.R., and Sherberg, M., *Improving Productivity Through People Skills – Dimensional Management Strategies*. Ballinger Publishing Co (Harper & Row), 1980.

---

[7] **The Dangerous "It" Word and Too Many Pronouns:** The pronoun "it" is probably the most dangerous word in the English language, leading to all sorts of misunderstandings and friction. More generally, people speak with "too many pronouns". What you refer to by "it" is in all probability not exactly what your interlocutor is thinking, and the two concepts may not be on the same planet. You can be burned by "it".

3. An Exercise (Tech. Index 1/10)

This practical and amusing (but dead-serious) exercise will give you some glimmer in what it can be like carrying out a few activities in practical finance regarding a little data, analysis, systems, communication, and management issues. The exercise is illustrative without being technical. There are important lessons, the most important being communication. The idea is not just to read the exercise and chuckle, but actually try to do it.

Remember, the footnotes provide a running commentary and extension of the topics in the main text. Footnotes are actually sidebars and form an integral part of the book.

Part #1: Data, Statistics, and Reporting Using a Spreadsheet

Step 1: Data Collection
Find the 3-month cash Libor rate and the interest rates corresponding to the prices of the first twelve Euro-dollar 3-month futures[1]. Keep track of each of these thirteen rates every day[2] for two weeks[3] using the spreadsheet program Excel[4,i] and note the rate changes each day.

---

[1] **Libor and ED Futures**: There are a number of different interest rates used for different purposes. You need to spend some time learning about the conventions and the language. Libor is probably the most important to understand first. The "cash" or "spot" 3-month USD Libor rate is the interest rate for deposits of US dollars in banks in London starting now and lasting for 3 months. The related Eurodollar (ED) futures give the market "expected" values of USD Libor at certain times called "IMM dates" in the future. ED futures are labeled by MAR, JUN, SEP and DEC with different years, e.g. SEP04. The interest rate in %/year corresponding to a future is [100 - price of the future].

[2] **"The Fundamental Theorem of Data"**: Collecting and maintaining reliable data is one of the Black Holes in finance (this statement is a Theorem). What you are being asked to do here is to get a tiny bit of first-hand experience of how painful this process really is. Notice for example that you weren't told where to find the data.

[3] **Time**: Two weeks is 10 business days. Time in finance is sometimes measured in weird units. For example, one year can be 360 days (used for Libor and ED futures).

[4] **Spreadsheets**: If you don't know much about spreadsheets, regardless of what you know about programming or quantitative packages, this seemingly trivial exercise is

Step 2: Statistics and Reporting

At the end of the 2 weeks, calculate the average and standard deviation of the nine changes of the first futures rate, expressed in bp/yr[5]. Type exactly what you did on the top of the spreadsheet with the dates and label everything[6]. Next, use the Excel Wizard to draw a graph of these changes, label the graph clearly including the units, and print it out. Also, print out the spreadsheet as a report[7].

Step 3: Bis for Steps 1,2

Repeat the above two steps for the 10 daily differences of the first future rate minus the cash Libor rate[8].

Step 4: Correlations

Calculate the correlation of the daily changes of cash Libor with respect to the daily changes of the first futures rate[9]. Next, define the "Return" from time t to time t + 1 by this daily change of the rate divided by the rate at time t. Calculate the correlation of the cash Libor returns with respect to the first future rate

---

already a potential problem for you. Spreadsheets are ubiquitous. For quants, spreadsheets are useful for prototyping. The alternative can be a restriction in your employment possibilities. Excel is the de-facto standard spreadsheet.

[5] **Basis Point bp:** A basis point is %/100 or 0.0001 in decimal. Time-differences of rates (and also spreads, i.e. differences between different rates at the same time) are commonly quoted in bp/yr. Usually the /yr unit is left off.

[6] **Spreadsheet Labeling and Organization:** Clear spreadsheet formatting is key to help prevent errors and confusion that easily arise, especially in large spreadsheets. This is not to mention confusion for yourself if you come back in 6 months and try to understand what you did. One handy tip is to use colors with bold type for important quantities (e.g. input numbers green, intermediate results yellow, output results blue). Unlabelled spreadsheets create misunderstandings.

[7] **Graphs and Reports:** Graphs and reports are ubiquitous in risk management. Reports that are clear to people apart from the creator of the report are sadly not always the norm.

[8] **Why do Another Calculation?** There are many reported quantities. This one measures Libor curve risk. This step gets you initiated to repetitive work that can be part of the job.

[9] **Correlations:** Correlations are critical in risk management. We will spend a lot of time in the book discussing correlations, including how to stress them consistently. By now you should have figured out that a spreadsheet has built-in canned functions to do correlation calculations and many others.

returns. Now look at how different the result is for the correlation using the rate returns from the correlation using the rate changes[10].

Step 5: Written Communication

Write a two-paragraph memo about what you did, clearly, carefully and neatly enough so you could turn it in to your old English professor[11].

Step 6: Verbal Communication

Staple your nice spreadsheet report and graph to the memo, and hand over the package to somebody. Tell her what you did in no more than 3 minutes, and ask her to spend no more than 3 minutes looking at the material. Ask her to feed back what she understood[12].

Step 7: Celebrate

Go have a beer[13].

Analysis of Part #1

You were just walked through a soup-to-nuts exercise. Each step corresponds to a common activity. This included written and verbal communication..

---

[10] **Definitions:** The correlations for differences and the correlations for returns are not exactly the same. Many risk measures have different possible definition conventions. You may have to dig deep to find out which definitions are being used in a report.

[11] **Written Communication, Management, and Goat Language:** This part is important. You shouldn't skip it because if you can't write what you did clearly, you may not be paid as much. Assuming you do not report to another quant, your potential Wall Street manager will not speak your language and is probably neither willing nor capable of learning it. The communication of even a summary of technical information or its significance is often hard. On the other hand, some managers have excellent intuition, understand the thrust of a technical argument quickly, and make valuable suggestions for improvement. You will be lucky if you report to such a person. You should make memos as clear and simple as possible without sacrificing the message. A wise manager, Gary Goldberg, gave some good advice for quants to use simple "Goat Language". Good luck.

[12] **Verbal Communication and Management:** This part is important and difficult. Again, you shouldn't skip it because if you can't clearly describe orally what you did, you won't be paid as much. By the way, your manager may only speed-read your memos. Face-to-face communication may be the only way you can transmit your message. You will probably not be given much time for the meeting. Hit the important points. Be prepared for a possibly arcane experience. You have to try to learn how to adjust.

[13] **Beer:** This is not pointless, and gives some idea of the sociology. Still, this activity is not as common as some people might imagine. People work hard and go home.

Communication skills are essential and many quants perform them badly, to their detriment. *The worst error consists of using the pronoun IT* [14]. Hopefully you used Microsoft Word or Word Perfect (not the *vi* editor in UNIX) to write your short report so that it looks like a professional document that a manager will take seriously. Appearance counts[15].

For you PhDs who feel insulted by the trivial technical aspects of this exercise, be aware that most quant work on the Street is not academic. You may well have sideline activities like those described above – though the work can be more difficult, fast moving, and challenging than you might imagine.

Part #2: Repeat Part #1 Using Programming

Instead of the spreadsheet, write a program in your favorite computer language along with file inputs to perform the same steps as in Exercise Part #1. Document your source code[16] by clearly writing at the top what it is you are doing, in good English with complete sentences. If you skipped the memo and verbal communication, it's time to bite the bullet [17]. Print out your report and graph[18].

---

[14] **Second Warning: The Most Dangerous Word in the English Language is "It":** Again, the probability is 100% that every person will have a different definition of the word "it" for any given reference. Besides the confusion generated, you can get severely burned if the management thinks you are saying something or promising something other than what you are intending.

[15] **Appearance of Documents and Presentations:** There are people who have greatly improved their careers producing easy-to-read documents and PowerPoint color presentations. Upper management is usually NOT interested in the details and IS interested in getting summary information quickly and painlessly. You can learn from these people.

[16] **Source Code Comments:** Good programming practice, remember? I once had to try to make sense out of some complex mortgage code that had no comments at all. The remark of the programmer was that the code was completely obvious. *&*(%$#.

[17] **Written and Verbal Communication:** No, you can't skip this activity.

[18] **Graphs and Programmers:** Can you (ahem!) produce graphs using your compiled code? It is hard to describe the frustration with systems groups that as a matter of "principle" only work with compiled code and hate spreadsheets, but have trouble producing reasonable reports and graphs.

Analysis of Exercise, Part #2

You have just carried out the same exercise in "production mode", as opposed to the spreadsheet "prototype mode".

Part #3: A Few Quick and Tricky Hypothetical Questions

Question 1: System Risk

What would you estimate to be the amount of data such that programming would be preferred over spreadsheets? [19] Under which situations would you recommend replacing the files with a database? Next, suppose you are ordered to take over either the spreadsheet or the source code from somebody who has left the company and has documented nothing[20]. Now which approach would you favor?

Question 2: Should we do this Deal or not?

Say you have to estimate the risk for a Backflip Libor option lasting for 1 year. What time period of historical data would you recommend in order to get a handle on the potential risk of this animal, and when will you have the answer?[21]

Question 3: Market Risk

Based on the ridiculously small 10-day sample, if your boss came to you right now, what would you say would be a reasonable measure for Libor risk?[22, 23].

---

[19] **Programs vs. Spreadsheets:** Portfolios can have hundreds of correlated variables and thousands of deals. On the other hand, you may need to provide an answer by 2 p.m. for a risk analysis depending on a few variables for a deal perhaps about to go live.

[20] **Personnel Risk:** The situation described is not academic – it happens all the time. By the way, did *you* document *your* code?

[21] **The Backflip Option and the Time Crunch:** You have never heard of the Backflip Option; the name is fictitious. In practice, you may not have time to analyze a complicated option in detail or even get the precise definition much in advance. You need to get used to the pace – you're not going to publish a journal article. In fact, the desk wants an answer by 2 pm. So based on the information, what are you going to do?

[22] **Management Communication Will Not Go Away:** Your boss knows nothing about statistics other than having a hazy impression of the basic concepts. Do you think that he/she understood what you said? It is critical for your career that the management understands what you are doing and why it is important. The challenge for you is that their eyes may glaze over after two minutes of explanation. The particular issue of Libor risk in the example will not come up because industrial-strength databases and quants have solved it, but the communication issue is always there.

Messages and Advice

To Computer Programmers Who Skipped Exercise Part #1

Now is the time to go back and do Exercise Part #1 in Excel that you may have skipped, no matter how impure you find it and no matter how much you hate Microsoft. You will be regarded as more useful and valued by a business unit if you ALSO know how to use a spreadsheet for quick calculations, reports, and prototyping. Please try writing the memo and describing your work verbally – these skills will really be useful for you. You will be definitely be more effective by learning something about finance, even if you are a programmer.

To Those Who Can't Program and Skipped Exercise Part #2

So maybe you are on your way to sales or investment banking. Still, this is a good time to learn at least the rudiments of a compiled computer language if you don't already know one. Even if you never have to program in your future career on the Street, the chances are high that you will be interacting in some fashion with computer people. The more you know about systems and the way the technology guys think, the better you will be able to communicate with them, understand what the problems are, and get done what you want. Otherwise, computer land can turn out to be a frustrating black-box experience.

References

[i] **Excel**
Campbell, M., *Using Excel.* Que Corporation, 1986.
Jackson, M. and Staunton, M., *Advanced modelling in finance using Excel and VBA.* John Wiley & Sons, 2001.

---

[23] **Simple Procedures, Accuracy, and Communication:** The use of simple procedures is a double-edged sword. An important potential advantage is better communication to non-technical management. The downside is that management may neither understand nor remember the limitations of a simple procedure. Try to get a handle on the uncertainty. If the approximation is reasonable, communicate up front that you are using a simple but reasonable approximation, possibly with the aim of improving it as priorities permit.

PART II

RISK LAB (NUTS AND BOLTS OF RISK MANAGEMENT)

4. Equity Options (Tech. Index 3/10)

In this chapter, we begin the analysis of standard or "plain-vanilla" equity options[1,i]. Similar remarks will hold for other options, e.g. foreign exchange (FX) options. Just for balance, we treat some topics in the FX options chapter that are directly applicable to equity options and vice-versa. We will treat the subject qualitatively, reserving the formalism for later chapters (see especially Ch. 42).

Pricing and Hedging One Option

In order to present the material in the way you might encounter it on a system, a spreadsheet format is shown. Following each section are some quick comments[2]. We begin with the deal definition, which specifies the "kinematics". This information is put into the official database of the firm's "books and records".

Deal Definition

1	Deal ID	ABC123
2	Option type	Call
3	Strike $E	$100
4	Calculation date	6/25/02
5	Settlement date	6/27/02
6	Expiration date	6/25/04
7	Payment date	6/27/04
8	Principal	$1 MM
9	Divide option by spot?	No
10	Divide option by strike?	No

Comments: Deal Definition

---

[1] **Acknowledgements:** I thank many traders, especially Larry Rubin and Alan Nathan, for helpful conversations on practical aspects of equity options.

[2] **Comments for System:** If you ever had to deal with a derivatives pricing system, you know that the definitions of the various quantities on the screen are sometimes not totally clear. If comments are put into help screens, time is saved and some errors can be prevented for the uninitiated user.

1   For database ID purposes
2   Call Option: Investor has right to buy something
3   Option is at the money if strike = stock price
4   Date the calculation is done (e.g. today)
5   Payment for the option is made at settlement date
6   A European option has one expiration date
7   Any option payout, if it exists, is paid at this date
8   Normalization factor
9   If Yes, delta will be changed
10 Another convention

The parameter inputs are shown next. These describe the "dynamics" of the deal related to the option model. The *types* of parameters form part of the model specification. The *numerical values* of the parameters are *not* part of the model[3].

Parameters
1	Spot Price $Spot	$100
2	Interest Rate r	5 %
3	Discounting Spread s	0 bp
4	Dividend Yield y	1 %
5	Implied Volatility σ	15 %
6	Type of Volatility	Lognormal
7	Skew Correction dσ	0 %

Comments: Parameters
1 "Spot" means current (today's) price
2 Need to specify type of rate, units
3 Some models have two interest rates differing by a spread
4 Alternatively to dividend yield, can specify discrete dividends
5 Implied Volatility produces the option price using model formula
6 "Lognormal" = model assumption of Gaussian behavior for log price changes
7 Volatility Skew: Adjustment designed to fit certain option data

The results of the model are shown next. First is the option price.

Price
1	Option value $C	$12.225

Comments: Price
1   Discounted expected value of payout cash flow

---

[3] **Definition of a Model:** The inclusion of the types of parameters as a part of the model is not an idle formality. The numerical algorithm only specifies part of the story.

Equity Options

Either analytic or numerical solution to some accuracy

Next, the hedging details of the option are presented. Notice that there is a variety of conventions for the hedging quantities. For example, "gamma" can be defined in practice in many different ways, both analytically and numerically.

Hedging

1	Delta Δ	0.6698
2	Δ _Fixed_Dividend	0.6828
3	$Delta $ Δ	$66.983
4	Strike Delta	-0.5476
5	Gamma γ	0.01641
6	$Vega	$0.496
7	$Rho	$1.073
8	$Phi	-$1.294
9	$Theta	

Comments: Hedging
1 Standard Δ = $dC/$dSpot, may be numerical derivative
2 Delta with fixed $Dividends (y = $Dividends/$Spot)
3 $Spot * Delta, the third convention used
4 Strike Delta $dC/$dE is approximately = - Δ
5 Gamma γ = $d²C/$dSpot²; many possible conventions
6 Change $dC for dσ = 1% (sometimes 1bp is used)
7 Change $dC for dr = 1% (sometimes 1bp is used)
8 Change $dC for dy = 1% (sometimes 1bp is used)
9 Change $dC for dt = 1 day (see footnote)

Did you notice theta didn't show up? You found a bug! Great! See the footnote[4].

The hedging of one equity option takes place using Taylor's theorem. We get

$$dC = \Delta \cdot dS + \tfrac{1}{2}\gamma \cdot (dS)^2 + \text{Vega} \cdot d\sigma + \text{Rho} \cdot dr + \text{Phi} \cdot dy + \text{Theta} \cdot dt + ...$$

(4.1)

---

[4] **Good Work, You Found the Bug:** Now that you've discovered the bug, you get to have the experience of dealing with it. Maybe theta has been calculated and just doesn't appear (GUI bug), in which case you will be talking with systems personnel. They are hassled, and will probably put your request at the end of a long priority list. You may now have to go to meetings to monitor the status of the system. Maybe there is a calculation bug – something wasn't defined etc. In that case, perhaps you will get to debug the code to calculate theta. This means you will decipher a lot of code written by other people, some long gone, with less than sterling documentation. Aren't you glad you found the bug?

Higher order terms can be important especially for options near expiration and/or large moves in the underlying. However, the traditional characterization of an option and most risk reporting is done with these hedging parameters.

American Options

For risk management, we need to know the risk as a function of future time. For European options, the expiration date is unique. For American options,[ii] which can be exercised at any time, there is no definite expiration time[5]. American options must be valued with numerical codes, discretizing the diffusion equation (corresponding to an assumed random movement of the stock price, along with a drift term specified by "no-arbitrage" constraints). The hedges $(\Delta, \gamma...)$ for an American option need to be determined numerically. The question remains as to how to distribute the risk as a function of future time. This can be done with the probabilities of exercise at different times. To see how that works, consider the illustration below.

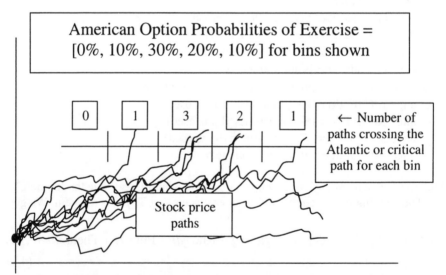

The probabilities of exercise at different future times can be found by a path counting technique[6]. The "critical path" (or as I like to call it, the "Atlantic path"), corresponds to the stock price as a function of time at which American

---

[5] **Other Types of Exercise:** A Bermuda option can be exercised only on certain dates. Sometimes there is a lockout period during which exercise is not allowed. These types of options are very common with interest rates. See Ch. 43.

[6] **Path Integrals and Options:** See Ch. 41-45.

option exercise occurs. Once the critical path is determined, the fraction of paths crossing the critical path in each interval in time can be found, for example by Monte Carlo simulation. This fraction defines $\mathcal{P}_{Exercise}(t)dt$, namely the probability of exercise at time t in dt. The above picture has several bins in time for ten paths with a stylized (constant) critical path.

The determination of the actual critical path is done using a back-chaining procedure from last exercise time along with a dictum of maximizing profit—if at a given time, option exercise is more profitable than the expected return holding the option, the option is exercised.

Given the probabilities of exercise the hedges can be distributed in time as $\Delta(t) = \Delta \cdot \mathcal{P}_{Exercise}(t)$ for $t < T$ (maturity). The remainder of Δ is put into $\Delta(T)$. This procedure, while accurate, is cumbersome. Physically, the risk is mostly at the last exercise time if the option is far out of the money (OTM), and at the first possible exercise time if the option is far in the money (ITM)[7].

Basket Options and Index Options

We have used the word "stock" as if it were a single stock. However the most common stock options correspond to stock index options, e.g. options on the S&P 500 index. The index B is a linear combination, or "basket", of stocks $\{S_\alpha\}$ with certain weights $\{w_\alpha\}$, viz $B(t) = \sum w_\alpha S_\alpha(t)$. Hence, an index stock option is actually an option on a basket.

The first and still most common theoretical assumption made for stock price movements is "lognormal"—namely relative changes or "returns" $\mathcal{R}_\alpha(t) = dS_\alpha(t)/S_\alpha(t)$ of a stock price are distributed normally. In small (formally infinitesimal) time interval dt starting at time t and ending at time $t + dt$, we write $dS_\alpha/S_\alpha = \mu_\alpha dt + \sigma_\alpha dz_\alpha$. Here dz_α is a Gaussian random variable with width $\sqrt{dt}$. The drift μ_α, "volatility" σ_α, and dz_α are all functions of time[8]. A little arithmetic shows that $dB(t)/B(t) \neq \sum w_\alpha \mathcal{R}_\alpha(t)$, i.e. the return of the index is not the weighted sum of the returns of the components. For that reason, the return of the basket is not lognormal if the components are lognormal. Instead, we find that

---

[7] **OTM, ITM, ATM Notation:** OTM = out-of-the-money, ITM = in-the-money, ATM = at-the-money. These abbreviations are used throughout the book.

[8] **Skew:** In addition, the volatility can depend on the stock price; this is called "skew". We will discuss skew at some length in Ch. 5,6.

$$dB(t)/B(t) = \sum w_\alpha S_\alpha(t)[\mu_\alpha(t)dt + \sigma_\alpha(t)dz_\alpha(t)] / \sum w_\alpha S_\alpha(t) \quad (4.2)$$

We suppose $\{dz_\alpha\}$ are correlated[9] according to $\langle dz_\alpha dz_\beta \rangle = \rho_{\alpha\beta}\sigma_\alpha\sigma_\beta dt$, and suppose that the basket and stock values do not vary too much from their values at some specific time t_0. Denoting $\langle XY \rangle_c = \langle XY \rangle - \langle X \rangle\langle Y \rangle$ we obtain

$$\sigma_B^2 dt \equiv \left\langle [dB(t)/B(t)]^2 \right\rangle_c \approx \sum_{\alpha\beta} w_\alpha w_\beta \rho_{\alpha\beta} S_\alpha(t_0) S_\beta(t_0) \sigma_\alpha \sigma_\beta dt \Big/ [B(t_0)]^2$$

(4.3)

With these assumptions, the lognormal assumption with the basket volatility σ_B is then used for baskets. The basic reason is that, even though somewhat inconsistent, anything else would be infeasible[10]. Sometimes, again rather inconsistently, a time-dependent basket volatility using the above formula is employed replacing the time t_0 with an arbitrary time t.

In practice, the basket volatility may use the above formula for small custom basket deals. For index options—e.g. the S&P 500—the index volatility $\sigma_{S\&P}$ is just parameterized and no attempt is used to build it up from the components. This is because the detailed information simply does not exist.

Basket Options in Disguise: Swaptions

Although we have been discussing equity basket options, an exactly similar set of steps is used to derive the volatility for an interest-rate swap break-even rate. There the stock prices are replaced by "forward rates". The swap is a weighted "basket" of forward rates minus a swap rate, and the "break-even" swap rate is weighted "basket" of forward rates. The nominal model for forward-rate fluctuations is again lognormal, and the break-even rate is (again somewhat inconsistently) assumed lognormal. We will examine swaptions in Ch. 11.

---

[9] **Correlation:** The average is statistical. Mathematically an infinite set of Gaussian random numbers is supposed to be used for the average, but of course, this never happens in practice. We discuss correlations thoroughly in Ch. 22-25, 37.

[10] **Living With Inconsistency:** This is one of the reasons why it is not too profitable to try to axiomatize option theory with a lot of unnecessary mathematical rigor.

Other Types of Equity Options; Exotics

There are a large number of types of equity options, many exotic. "Exotic" basically means something that cannot be priced using a Black-Scholes formula with simple parameters. We will deal with a representative set of exotics, but it would require a heavy volume to categorize the full panoply of equity options.

Portfolio Risk (Introduction)

Once we have the future time distribution of hedges for each deal, we can add up the risks for all deals in a portfolio. Then we know in principle what the portfolio risk looks like as a function of future time t as seen from the valuation date today t_0. Now the future risk for each deal changes as the calendar time t_0 moves forward. Moreover, a portfolio of deals changes as a function of time as new deals are done, old deals are exercised or expire, hedges are modified etc. For this reason, the risk must be monitored regularly.

The periodicity of the risk reporting depends on the user. For the trading desks, the risk reports are run daily (usually overnight). For corporate reporting purposes, the time scales for risk reporting are much coarser, for example quarterly. The risks of particular deals may be monitored by the corporate risk group on an as-needed basis.

Scenario Analysis (Introduction)

It should be clear that the use of the Greeks will not work well if large moves in the stock price occur. For example, a short[11] put option with a large notional that is far OTM[7] might contribute negligibly to the risk. However, if the market suddenly tanks[12], this option can suddenly become ATM with large risk. Conversely, an apparent large risk may not mean much. For example, a short put option ATM near expiration with large negative gamma may have little real downside risk[13].

For this reason, scenario analyses assuming various types of large moves are analyzed from time to time. These scenarios can be one of three types:

Scenario Type 1. "Ad-hoc" scenarios (e.g. suppose the stock price drops 59%)

---

[11] **Long, Short:** "Long" means the option was bought; "short" means the option was sold.

[12] **Jargon:** The phrase "the market tanks" is a technical term that means the market is quickly going to pot.

[13] **Homework:** Do you believe this statement?

Scenario Type 2. Historical scenarios (e.g. the worst case stock price drop = 37%)
Scenario Type 3. Statistical scenarios (e.g. 99.9% CL stock price drop = 42%) [14]

We will look at scenario risk analysis for portfolios in some detail in Ch. 12.

References

[i] **Some Classic Options References**
Black, F. and Scholes, M. , *The Pricing of Options and Corporate Liabilities.* J. Political Economy, 3, 1973. Pp. 637-654.
Merton, R. C., *Theory of Rational Option Pricing.* Bell J. Econ. Management Science,Vol. 4, 1973. Pp. 141-183.

[ii] **American Equity Options**
Jarrow, R. and Rudd, A., *Option Pricing.* Richard D. Irwin, Inc., 1983
Wilmott, Pl, DeWynne, J., Howison, S., *Option Pricing – Mathematical models and computation.* Oxford Financial Press, 1993.
Dash, J. W., *Path Integrals and Options – I.* Centre de Physique Théorique, CNRS Marseille preprint CPT-88/PE. 2206, 1988.

---

[14] **Statistical Scenarios:** The reader may wonder how a statistical scenario based on historical information can be worse than the worst-case historical scenario. This is because the statistical scenario typically will be defined using a Gaussian (bell-curve) fit to the tails of the distribution – e.g. at a 99% CL – and then extrapolated to the desired confidence level. We discuss such subjects in Ch. 21.

5. FX Options (Tech. Index 4/10)

In this chapter we discuss foreign-exchange FX derivatives. We start with FX forwards and simple FX options. We give some practical details for FX options, including hedging with Greeks. We introduce volatility skew (or smile). We give some examples of pricing exotic barrier FX options. We present the "two-country paradox". We discuss quanto options and correlations, FX options in the presence of stochastic interest rates, and comment on numerical codes and sanity checks.

FX Forwards and Options

Consider the picture for the idea of "interest-rate parity" for FX forwards.

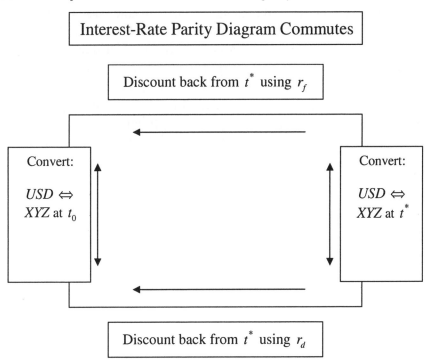

Say we are in the US with US dollars (USD). We can convert the USD to XYZ currency at time t_0, put the XYZ amount in the XYZ bank at foreign rate r_f and finally convert back at t^* from XYZ to USD. The result should be the same as if we kept the USD in the USD bank at domestic rate r_d. Equivalently, we can take the proceeds at t^* in each currency and discount back to t_0 at the respective rates to compare the results at t_0. The diagram commutes; we can go around it either way.

We use the following notation relative to the USD.

$$\eta = \frac{\#Units(XYZ)}{OneUSD}$$

$$\xi = \frac{\#USD}{OneUnit(XYZ)} \tag{5.1}$$

I used to live in France, and even though the Euro now exists I still think in French francs and recall definitions using $\eta \approx 5 \cdot FF/USD, \xi \approx 0.2 \cdot USD/FF$ [1]. Market quotes can exist using either η or ξ. There are old traditions, e.g. ξ for GBP and for the currencies of the British Empire. Sometimes η is called a "European quote" and ξ an "American quote". My definition for ξ is the same as in Andersen's book [i] and is like any price, e.g. $\xi = \$0.50$/apple. Other conventions exist: "GBP-USD = 1.50" or "GBP/USD = 1.50" instead of $\xi = 1.50 \cdot USD/GBP$ here. Cross FX rates UVW/XYZ are treated in exactly the same fashion. You need to make sure you know the convention being used by somebody or some model.

Pricing FX Forwards and FX European Options

Call the spot FX rate[2] ξ_0 and call the domestic and foreign interest rates $r_d = r_d(\tau^*)$, $r_f = r_f(\tau^*)$ for a time interval τ^*. The forward FX rate ξ_{fwd} derived from the interest-rate parity argument given above is

---

[1] **Old Habits Die Hard:** Prices in French stores are quoted in FF side by side with Euros. Moreover prices for land as recently as 1980 were quoted not in FF but in "anciens francs", each worth 1/100 FF.

[2] **Notation:** Interest rates are exhibited as continuously compounded.

FX Options

$$\xi_{fwd} = \exp\left[\left(r_d - r_f\right)\tau^*\right]\xi_0 \qquad (5.2)$$

Note that the continuously compounded foreign interest rate acts like a "dividend yield" in the equivalent formula for equity options. The "FX swap" is the forward - spot difference $\xi_{swap} = \xi_{fwd} - \xi_0$, multiplied by some conventional normalizing power of 10.

Models for FX options[1] are similar to those used for equity options. One difficulty is conceptual; it is important to label things and keep track of the units.

Assuming lognormal behavior of $\xi(t)$ and using the forward ξ_{fwd}, the usual formalism goes through, as first derived by Garman and Kohlhagen. The option prices and deltas today t_0, up to an overall normalization, are:[3]

$$C_{call,put}\left(\xi_0,t_0\right) = \exp\left[-r_d\tau^*\right]\left[\pm\xi_{fwd}N\left(\pm d_{1\xi}\right) \mp E_\xi N\left(\pm d_{2\xi}\right)\right] \qquad (5.3)$$

$$\Delta_{call,put}\left(\xi_0,t_0\right) = \pm\exp\left[-r_f\tau^*\right]N\left(\pm d_{1\xi}\right) \qquad (5.4)$$

The normalization is important, and will be discussed below. In Eqn. (5.4) $\Delta(\xi_0,t_0) = \partial C/\partial \xi_0$ is the textbook definition, although other definitions exist. Also,

$$d_{1\xi}, d_{2\xi} = \frac{1}{\sigma\sqrt{\tau^*}}\left[\ln\left(\xi_{fwd}/E_\xi\right) \pm \sigma^2\tau^*/2\right] \qquad (5.5)$$

These formulae are for a call or put on the same currency XYZ, or mathematically just the Black-Scholes formula for a call or put on the variable ξ. They need further normalization, discussed below. Also the volatility $\sigma(\tau^*)$ is appropriate for the expiration at time $t^* = t_0 + \tau^*$.

Now a call option on *USD* physically is the same as a put option on *XYZ*. This is because if *USD* appreciates with respect to *XYZ* then *XYZ* depreciates with respect to *USD*. It is important to note that for an option on USD, we are in XYZ-land. Then the natural option variable is η and the "domestic" and "foreign" rates are reversed [4]. Written as a function of η, we have

---

[3] **Options Formalism:** We will get a lot of math later; for now we just quote the formulae. Here, N(x) is the cumulative normal distribution with N(x) + N(-x) = 1.

[4] **Lognormal Behavior for ξ, η and a Preview of the Two-Country Paradox:** Because $\eta = 1/\xi$, lognormal behavior of ξ implies lognormal behavior of η. However making this

$$\eta_{fwd} = \exp\left[\left(r_f - r_d\right)\tau^*\right]\eta_0 \qquad (5.6)$$

$$C_{call,put}\left(\eta_0,t_0\right) = \exp\left[-r_f\tau^*\right]\left[\pm\eta_{fwd}N\left(\pm d_{1\eta}\right) \mp E_\eta N\left(\pm d_{2\eta}\right)\right] \qquad (5.7)$$

$$\Delta_{call,put}\left(\eta_0,t_0\right) = \pm\exp\left[-r_d\tau^*\right]N\left(\pm d_{1\eta}\right) \qquad (5.8)$$

$$d_{1\eta}, d_{2\eta} = \frac{1}{\sigma\sqrt{\tau^*}}\left[\ln\left(\eta_{fwd}/E_\eta\right) \pm \sigma^2\tau^*/2\right] \qquad (5.9)$$

Some Practical Details for FX Options

There are several ways to quote the results for options and a number of different conventions for reporting the Greeks. There are also modifications of the option formula to take account of the specific features of the FX options market.

Normalization of FX Options

The overall normalization factor for the option formula is important. We need to express the option price in units of a definite currency and get rid of the FX currency ratio. The same option can be reported in four ways:

Method 1. Call on XYZ using the variable ξ. Normalization to get USD units: Divide by spot ξ_0, multiply by USD notional $\mathcal{P}_{USD}$.

Method 2. Call on XYZ using the variable ξ. Normalization to get USD units: Either multiply by XYZ notional $\mathcal{P}_{XYZ}$ or divide by strike E_ξ and multiply by USD notional $\mathcal{P}_{USD} = E_\xi \mathcal{P}_{XYZ}$.

Method 3. Put on USD using the variable η. Normalization to get XYZ units: Divide by spot η_0, multiply by XYZ notional $\mathcal{P}_{XYZ}$.

Method 4. Put on USD using the variable η. Normalization to get XYZ units: Either multiply by USD notional $\mathcal{P}_{USD}$ or divide by strike E_η and multiply by XYZ notional $\mathcal{P}_{XYZ} = E_\eta \mathcal{P}_{USD}$.

While all this may seem trivial (not to mention annoying), the output from some model will lead to confusion until you know the convention. The existence

---

change of variable modifies the drifts. This leads to a conundrum, discussed below, the "two-country paradox". The paradox does not show up for ordinary options.

FX Options

of the different conventions is because the deal may be booked either in the US or in XYZ-land.

For example, suppose XYZ = MXN (Mexican peso). If X lives in Mexico and buys a USD put, X pays in MXN. From X's point of view, a put on something (e.g. P_{USD} dollars) is the right to sell that thing for a fixed number $P_{MXN} = E_\eta P_{USD}$ of pesos, so X naturally would use the definition #4.

However, say the broker-dealer BD (who sells the USD put to X) is in the US. If the option pays off, BD will have to give P_{MXN} pesos to X in exchange for P_{USD} dollars. Then BD must purchase P_{MXN} pesos at expiration t^* with a number $P^*_{USD} = P_{MXN}/\eta^*$ of dollars, where η^* is the exchange rate at t^*. Hence from BD's point of view, the put is normalized at time t by $1/\eta(t)$, or today t_0 by $1/\eta_0$ (with spot e.g. $\eta_0 = 10 \cdot MXN/USD$). Hence the BD uses definition #3.

Hedging FX Options with Greeks: Details and Ambiguities

The Greeks are used for option risk management. It is important to understand that the Greeks do not have unique definitions. Although this information is in this section on FX options, the same remarks in this section apply to equity options. For FX, there are ambiguities depending on the normalization, as above.

Delta

Because the spot changes day-by-day (indeed minute by minute), if the option is divided by the spot there is a change in the formula for the delta hedging of the option. Besides the variation of normalization factors $1/\eta_0$, a modified delta can be defined by multiplying delta by η_0. The bottom line is that there are several possible definitions of delta. When trying to understand FX option risk reports (usually just labeled cryptically with words like "delta") you obviously need to know the normalization conventions[5].

For plain vanilla options, ordinary differentiation can be used. More complex options require numerical code, possibly including "skewed" volatilities, and possibly with boundaries (e.g. barriers) or complicated boundary conditions (e.g. American options). Then numerical difference procedures must be used to get delta and gamma. Sometimes symmetric differences are used, and sometimes

---

[5] **On Getting the Conventions:** This may not be so easy to do. Here is a hypothetical scenario. The guy who wrote the code for this risk report left the company last month. He was so busy producing other customized reports for the new head trader on the desk that he had no time to document anything (which he regarded as a waste of time anyway). Of course since I just made up that scenario, you will never see anything like it.

only one-directional moves are used. A one-directional move can become problematic for a barrier option when spot gets near the barrier.

The amount of move in spot used for the numerical difference is another hidden variable. A move up (e.g. 1%) might cross a barrier, leading to a very different result than with a smaller move not crossing the barrier.

Gamma

There are at least nine different ways to define gamma, some with very different numerical values. Gamma may or may not be defined to include varying the $1/\eta_0$ normalization factor, and may contain other factors of η_0.

Vega

Vega can be defined by a continuous derivative, by numerical differences with volatility changes of 1%, by numerical differences with volatility changes of 10 bp (sometimes followed by scaling the result up by 10), and by differences using a percentage of the volatility. Differences in procedure can be noticeable when the volatility is low and/or the option is away from ATM.

Rho and Phi

Rho (sensitivity to r_d) and phi (sensitivity to r_f) can be defined by continuous derivatives, numerical differences with interest-rate changes of varying amounts, (1% or less, sometimes followed by scaling the result back up to an equivalent 1% move). Differences in procedure can be noticeable when rates are high, which happens for emerging markets.

Theta

Theta is essentially a poor man's simulator. There are many ways to define theta. Possibilities include moving the calculation date one day forward or one business day forward, moving the settlement day one day forward (or not moving the settlement date), leaving the spot rate fixed or moving the spot rate forward, using the same volatilities and interest rates or reinterpolating the volatilities and interest rates corresponding to the new dates. Sometimes a continuous derivative is used and the result has to be rescaled to get equivalent one-day differences.

Higher Greeks (Speed, Charm, Color)

Because the usual Greeks just represent the lowest order terms in the Taylor series of the option, some higher-order terms can be monitored. These include[ii] *Speed, Charm,* and *Color. Speed* is the derivative of gamma with respect to the underlying. *Charm* is the derivative of delta with respect to time. *Color* is the derivative of gamma with respect to time[6].

Other Option Details

Other conventions include different time periods for the diffusion, for carry, and for discounting. Here is a picture of how FX options actually work.

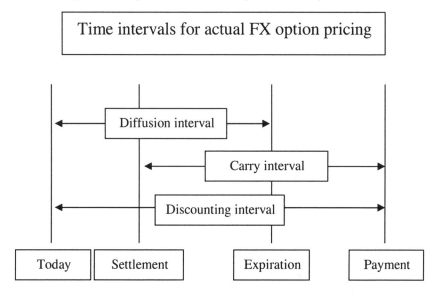

The payment for the option premium (at settlement) is due some time after the deal is done or calculated today. Any payment from the option payoff takes place some time after expiration. These times are usually 2 business days. The diffusion takes place from today to the option expiration. The carry (i.e. the term from the interest rates r_d, r_f in the drift) takes place from settlement to the option payoff date. The discounting is from the option payoff date back to today. The extra discounting from settlement back to today is called "Tom Next". Other details include the usual plethora of conventions for interest rates and volatilities.

FX Volatility Skew and/or Smile

The skew dependence of volatility is generally thought of as a monotonic dependence on the option strike, viz $\sigma(E)$. The strike dependence is needed in

---

[6] **Envy?** Maybe these names (charm, color) just prove that some finance guys really wanted to be high-energy physicists.

order to reproduce European option values using the standard Black-Scholes formula. A more complicated non-monotonic dependence is often seen in FX, called a "smile".[7] Here is a freehand picture illustrating the idea:

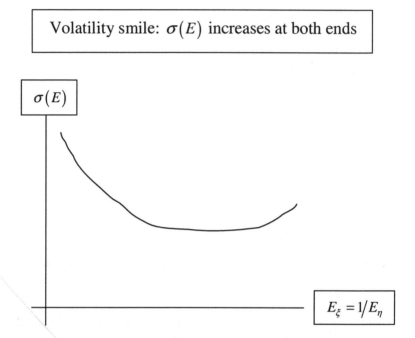

Physical Motivation for the FX Volatility Smile

The smile behavior for FX can be qualitatively understood using a "fear" idea. For illustration, say spot is $\xi_0 = 1.50 \cdot USD/GBP$ or $\eta_0 \approx 0.67 \cdot GBP/USD$. Consider an OTM GBP put for a low strike $E_\xi^{Low} = 1.40 \cdot USD$ that pays off at expiration if $\xi^* < E_\xi^{Low}$. The fear that the GBP might depreciate this much can induce more investors to buy these puts. The extra demand raises the put price. Thus we get a premium on this volatility $\sigma(E_\xi^{Low})$ relative to the ATM vol, resulting in volatility skew.

Now, consider an OTM call on the GBP for a high strike (say $E_\xi^{High} = 1.60 \cdot USD/GBP$) that pays off if $\xi^* > E_\xi^{High}$. This option, on the other hand, is an OTM put on USD with a low strike in the inverse variable

---

[7] **More About Skew:** A lot more information about skew for equity options is in the next chapter.

FX Options

$E_\eta^{Low} \approx 0.71 \cdot GBP/USD$ that pays off if $\eta^* < E_\eta^{Low}$. The fear that the USD might depreciate this much can also place a premium on this volatility $\sigma(E_\eta^{Low})$. Since these two options are the same thing, $\sigma(E_\eta^{Low}) = \sigma(E_\xi^{High})$.

Thus, vol premia for both low and high strikes can exist; this is the smile. Because the fear intensities are generally not equal, the currencies generally being of different strengths, the smile will not generally be symmetric.

General Skew, Smile Behavior for FX

A more general phenomenological parameterization of the complexity in the strike dependence of the volatility includes both monotonic skew and non-monotonic smile terms, and the mix is time-dependent. A study along these lines was reported for dollar-yen [iii].

Fixed Delta Strangles, Risk Reversals, and Hedging

A strangle is the sum of a call and put, and a risk reversal is the difference of a call and put. The FX option vol skew/smile is often quoted using "25-delta" strangles and risk reversals[8]. Since 50-delta would be ATM and 0 delta would be completely OTM, the 25-delta is "halfway" OTM. A call $C_{call}(\sigma_{call}, E_{call})$ and put $C_{put}(\sigma_{put}, E_{put})$ are used with $\Delta_{call} = -\Delta_{put} = 0.25$ for a given expiration time t^*. We need to find the strikes to produce this 25 delta condition, remembering that $\sigma_{call}(E_{call})$, $\sigma_{put}(E_{put})$ are functions of the strike. Other delta values are also used, e.g. 10-delta.

Let us examine the fixed delta conditions a little more. The relation between the strikes of the call and the put for a given value of delta is related to the forward.

When the deltas are equal and opposite as for the 25-delta condition, we have

$$d_{1,call}(\sigma_{call}, E_{call}) = -d_{1,put}(\sigma_{put}, E_{put}) \quad (5.10)$$

To get an idea of what this implies, consider the situation without skew where this relation simplifies. Up to a volatility term, the forward is related to the product of the call strike and the put strike. From equations (5.4), (5.5) we get

---

[8] **Risk Reversal Convention:** The price of a risk reversal is the difference of the call and put prices on the USD. The market is also quoted in terms of the difference in the implied volatilities of the USD call and put. Again, remember that a USD call is an XYZ put so a positive risk reversal means a bearish market on the XYZ currency. Also, a risk reversal has other names: combo, collar, cylinder, ...

$$E_{put}E_{call} = \xi_{fwd}^2 \exp(\sigma^2 \tau^*) \tag{5.11}$$

This means that if we own a call on GBP with 25-delta, ignoring skew, we can hedge it locally with a put on GBP having the same delta (and opposite sign) provided we choose the put strike E_{put} according to the above relation.

Implied Probabilities of FX Rates and Option "Predictions"

Formally, the second derivative of the standard option formula with respect to the strike is the Green function, including the discount factor[9]. The implied Green function or probability distribution function (pdf), $G_{\text{Implied pdf}}$, is defined by inserting the market prices for options $C_{\text{Market Data}}(E)$ expiring at some time t^* with different strikes. The second derivative is approximated numerically. With x either $\ln \xi$ or $\ln \eta$, we get the relation

$$G_{\text{Implied pdf}}\left[x_0, t_0; x^* = \ln(E), t^*\right] = E \frac{\partial^2 C_{\text{Market Data}}(E)}{\partial E^2} \tag{5.12}$$

Looking at the height of the implied pdf at some level, the option market's "prediction" today about the probability that the FX rate will find itself at a that level at time t^* in the future can be numerically ascertained.

As mentioned, the physical basis of this "prediction" is just that the fear factor leading purchasers to buy XYZ puts at elevated prices or vols produces an elevated implied pdf value for future XYZ currency depreciation. If these fears are realized, the prediction will come "true". While the statement that "I am afraid XYZ will drop and will pay extra for put insurance, therefore my prediction is that XYZ will drop" may seem like a tautology, the implied pdf does produce a quantitative evaluation of the effect.

Here is a picture of the idea for a lognormal pdf and a modified implied pdf including skew effects [10]. The fat tails coming from the increased volatility for low values of the underlying increase the implied pdf values at low values, with respect to the unskewed lognormal pdf.

---

[9] **Green Functions:** We will look at the math later; for right now just try to get the idea.

[10] **Plot:** Actually this plot comes from the S+P 500 index data in Ch. 4, and it serves to illustrate the idea generally. The implied pdf illustrated is just a lognormal pdf with the skewed vol put in at each value of the underlying and the total renormalized to 1.

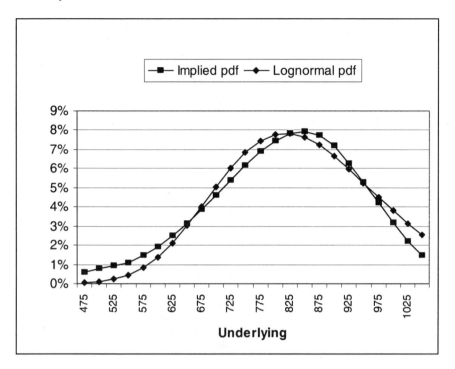

If there is substantial real market pressure on the XYZ currency due to selling XYZ spot accompanied by feedback of bearish purchases of XYZ puts, the option market "prediction" will indeed come "true". However the amount of such depreciation really relies on complex factors including transaction volumes, macroeconomic information, and local political conditions instead of theoretical peaks in the implied option probability distribution. To say the least, the impact of these real factors on the FX market is not easy to evaluate successfully and consistently over time [11].

Still, option pricing including skew information is desirable since information from the option market is included, including the fear-factor distortion.

Pricing Barrier Options with Skew

Barrier options are options that change their character if the underlying variable hits the barrier value. Barrier options are discussed in detail later in the book. Skew effects can be important for barrier option pricing. Consider an up and out

---

[11] **Trading:** Hey, that's why good traders make the big bucks, right?

call option[12] on XYZ currency as a function of ξ where the barrier is K_ξ. For simplicity ignore smile and just assume skew. As ξ increases along a path going near the barrier, the XYZ currency is appreciating and the local volatility decreases. This lowers the probability of knockout compared to the classic situation with a constant volatility independent of the ξ level.

The drawing below gives the idea:

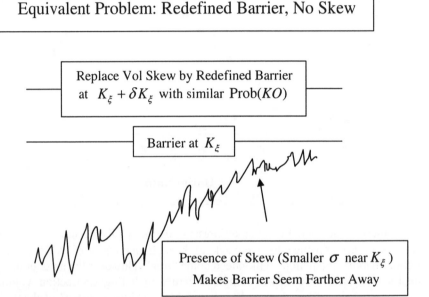

From the point of view of a drunk gremlin staggering along such a path defined by a given set of appropriate random numbers[13], the barrier at K_ξ is "hard to reach" in his skewed world. Equivalently to an observer, the gremlin gets less drunk as it approaches the barrier and staggers less, making it less likely to cross. We can replace this gremlin by a new gremlin in an unskewed world if we also replace the barrier K_ξ with a new barrier $K_\xi^{UnskewedWorld} = K_\xi + \delta K_\xi$. This is done such that the unskewed-world gremlin on a path with the same

---

[12] **Up-Out Option:** This means that if the underlying goes above the barrier at any time before option expiration, the call option ceases to exist.

[13] **Diffusion and Drunks:** The classic model for diffusion is a drunk staggering out of the bar. The drunk performs a random walk about the forward path. Mean reversion can be modeled by attaching one end of a spring to one leg and the other end to a rod along the forward path. A jump can be modeled by having a big gust of wind knock the drunk one way or the other. Long-term drifts can be modeled by sloping the terrain appropriately.

FX Options

random numbers (but with the higher unskewed volatility) has the same probability of knockout with respect to this new barrier $K_\xi^{UnskewedWorld}$ as does the skewed-world gremlin on his path with respect to the original barrier K_ξ

Although we cannot find one new barrier to force equivalence on all paths, we can do it on the average given the average barrier crossing or KO time τ_{KO}. An academic formula for τ_{KO} exists. The approximate change in the barrier is [14]

$$\delta K_\xi = \xi_0 \left[\sigma(\xi_0) - \sigma(K_\xi) \right] \sqrt{\tau_{KO}} \tag{5.13}$$

Qualitative "back-of-the-envelope" ideas of this sort can be used to develop intuition, or possibly for quick estimations on the desk. Another use is to provide sanity checks for model code that is being developed, for debugging.

Double Barrier Option: Practical Example

Double barrier options are more complicated than single barrier options. Double barrier options have two barriers—one above and one below the starting FX rate. The simplest case is a double knock-out option, where the option ceases to exist if either barrier is crossed at any time to expiration.

The formalism for double-barrier options is in Ch. 18. Here, we just give a practical example. Double barrier options are often used in FX. To give an idea, here is a 2-barrier USD call example, complete with hedging information, presented in a spreadsheet-like fashion that you might see in a system [15, 16].

Quantity	Value
Double KO call on USD	79,653 DM
Standard call (no barriers)	249,846 DM
Forward FX rate	1.52163 DM/USD
Probability of knock out	44.3 %

---

[14] **History:** This particular heuristic idea of moving the barrier to account approximately for skew was developed by me in 1995.

[15] **Parameters:** This DEM/USD example was in 1996 before the Euro. Spot = 1.5303, lower barrier = 1.45, upper barrier = 1.60. Strike = 1.52, vol = 8% (no skew), DEM rate = 3.2366% ctn/365, USD rate = 5.4913% ctn/365, expiration = 92 days, notional = $10MM, no rebate. Dates: Settlement = Valuation, Payout = Expiration, no Tom-next.

[16] **Numerical Accuracy Reporting:** There is no way that the numerical accuracy of options models quotes should be believed to many decimals. You might like to construct your own code and check the numerical accuracy of my pricer results above.

Delta = Spot * dC/dSpot	486,369 DM
Gamma = 1% * Spot * d²C/dSpot²	-561,443 DM
Vega (Vol up 1%)	-15, 529 DM
Theta (1 day)	797 DM
Rho (DM rate up 100 bp)	2,979 DM
Phi (100 bp)	-3,343 DM

The probability that knockout occurs from either boundary is around 44 %. This probability decreases as the barriers are moved away. For example, if the lower barrier were 1.0, the probability of knock out drops by a factor of two. If one of the barriers is removed, the double-barrier option becomes a single barrier option. For example, the double-barrier KO option with the lower barrier dropped substantially turns into an up-out single barrier option.

The fact that $\gamma < 0$ means that, as spot η_0 moves up, the presence of the upper barrier lowers the option price and lowers delta. Eventually we will get $\Delta < 0$ and the option price will go to zero as η_0 approaches the upper barrier.

The negative vega is also related to the barriers, because as the volatility increases, the probability of knock-out also increases. If the vol goes up 1%, the probability of knockout becomes 55%.

The increases of 100 bp in the interest rates are very large for the 3-month period of the option. Still the option value changes only by around +-3,000 DM. This indicates the relative insensitivity of the FX option to the interest rates.

Rebates

A "rebate" may be part of the contract, providing for a fixed amount to the option holder in case a barrier is crossed. This rebate may either be paid at expiration or paid when the barrier is crossed ("at touch"). Rebates exist for single and double barrier options. A rebate amounts to an extra knock-in digital option.

The "Two-Country Paradox"

The two-country paradox is generated by a logical argument[iv]. We have two equivalent notations for expressing foreign exchange related by $\eta = 1/\xi$. It cannot possibly make any difference if we announce (ignoring commissions etc.) that 1 USD buys 10 MXN ($\eta = 10 \cdot MXN/USD$) or that 1 MXN buys 0.10 USD ($\xi = 0.10 \cdot USD/MXN$). For this reason, we should be able to make the change of variable $\eta = 1/\xi$ in the mathematical expression for any financial FX instrument without any physical change in the result. This, however, is not generally the case. The drift after the change of variable is inconsistent with interest-rate parity. Consistency with interest-rate parity for both variables

FX Options

requires a separate normalization for the drift for each variable. These separate normalizations are inconsistent with the direct change of variables $\eta = 1/\xi$. This inconsistency is the paradox [17].

Mathematically, we can express the paradox noting that for any Gaussian random variable W we have $\langle \exp(W) \rangle = \exp\{\langle W \rangle + 1/2[\langle W^2 \rangle - \langle W \rangle^2]\}$. We first set $W_1 = \ln \xi$ and then set $W_2 = \ln \eta = -W_1$, both assumed to be Gaussian with constant volatility σ. We get at time t, after time interval τ, the result

$$\langle \xi(t) \rangle \langle \eta(t) \rangle = \exp[\sigma^2 \tau] \tag{5.14}$$

This however is inconsistent with interest-rate parity (IRP) for which the right-hand-side would be 1, not $\exp[\sigma^2 \tau]$:

$$\langle \xi(t) \rangle_{IRP} = \xi_{fwd} = \xi_0 \exp[(r_d - r_f)\tau]$$
$$\langle \eta(t) \rangle_{IRP} = \eta_{fwd} = \eta_0 \exp[(r_f - r_d)\tau] = 1/\langle \xi(t) \rangle_{IRP} \tag{5.15}$$

Since IRP is fundamental, it is necessary to define the drift for each variable separately, consistent with IRP, and inconsistently with $\eta = 1/\xi$. It further adds to the confusion that the change of variables $\eta = 1/\xi$ in the standard FX option formula expressed in terms of ξ does in fact lead to the standard option formula expressed in terms of η. This is due to a fortuitous cancellation. The cancellation does not occur for digital options or rebates for example.

It is not clear to me whether realizable arbitrage exists for this paradox. However, consider one system S_ξ that, for bookkeeping purposes,[18] expresses a given portfolio in terms of ξ. Consider another system S_η that chooses to express the same portfolio in terms of η. Say that the two systems used Monte Carlo simulators, one in $\ln \xi$ and the other in $\ln \eta$, with the same random number generator and the same starting seed, along with an extra minus sign for changes in $\ln \eta$. For the system S_ξ, a point $\xi(t)$ on a given path is equivalent to

---

[17] **Martingale Disguise:** Mathematicians will announce that the requirement should be that the drift is determined for each variable using its own martingale condition. This just says that the drift of each variable should be determined by interest rate parity. The inconsistency with the change of variables η = 1/ξ remains.

[18] **Use of one single FX Convention:** A good example of what I am talking about is in Riehl & Rodriguez, p. 43, problem 2.12 answer (Ref).

$\eta(t) = 1/\xi(t)$. However, this point $\eta(t)$ would not be attained on the corresponding η path for the system S_η. Again, this is because of the mismatch in the drift for $\ln \eta$ using the change of variable relative to interest-rate parity. For this reason, exercise conditions, barrier conditions, rebates, etc. would not match up between the two simulators, and the results would differ. Naturally, if the two systems were in two different firms, no one would notice.

The FX market is the only case of which I am aware that has the potential for this paradox. Consider, for example, interest rate term-structure constraints in the case that a physical rate $r(t)$ is derived from a change of variable from some other variable $y(t)$. The most common case is the lognormal change of variable, $y = \ln r$, and others are possible. The term structure constraints from zero-coupon bonds are used to determine the drifts of the physical rates $r(t)$. Using the change of variables, the drifts of the variables $y(t)$ are then determined. There is no separate normalization for the drift of $y(t)$.

Quanto Options and Correlations

Stock options with FX characteristics exist, called quanto options. The basic idea is that a stock is purchased in one country with its local currency (e.g. JPY) and the stock hedge in JPY is managed from a desk in another country with a different currency (e.g. USD). The USD investor X who buys a quanto pays an extra premium and is contractually insured against FX risk. The FX risk appears in the quanto option price. The critical issue here is the correlation $\rho_{Stock,FX}$ between the movements of the stock price and the movements of the FX rate[19]. It can be proved rather generally that the quanto effect can be taken into account by modifying the stock-price drift[20]. Equivalently, for standard quantos, the stock dividend yield $q_{DivYield}^{Stock}$ can be changed to an "effective" stock dividend yield $q_{DivYield}^{Eff}$ according to the prescription

---

[19] **Correlation Sign Convention – Watch Out:** The sign of the correlation needs attention. The example corresponds to the correlation between relative moves of dS/S and dη/η. The correlation between dS/S and dξ/ξ has the opposite sign.

[20] **Quanto Options:** The idea is to use a two-factor lognormal stock + FX model and integrate out the FX degrees of freedom to obtain an effective one-factor stock model. This can be done if the payoff only depends on the stock price locally (e.g. in JPY). The n-factor model formalism is discussed in Ch. 45. We leave the application of this formalism to this section as an exercise for the reader.

FX Options

$$q_{DivYield}^{Eff} = q_{DivYield}^{Stock} - \rho_{Stock,FX} \sigma_{Stock} \sigma_{FX} \qquad (5.16)$$

In Ch. 22-25 and Ch. 37, we discuss correlation risk in detail. A major problem with quanto options revolves around the instability of the correlations over the time scales of interest for hedging and reporting.

FX Options in the presence of Stochastic Interest Rates

We have exhibited the standard formalism of FX options in the presence of a deterministic term structure of interest rates. This means that the domestic and foreign interest rates r_d, r_f are numbers, dependent on the time to expiration, but are not random. The incorporation of stochastic interest rates implies a three-factor model (ξ or η and r_d, r_f), as was recognized early on[21]. The numerical analysis and technology are involved. Most of the risk of FX options is due to the FX rate. Some idea of the contribution of the interest rate fluctuations can be obtained from the rho and phi sensitivities, along with simple scenarios along outlier paths for the interest rates.

Numerical Codes, Closed Form Sanity Checks, and Intuition

Exotic FX options, options with FX components, bonds with FX-related payouts, etc. form a rich tapestry that go beyond the scope of this book. Along with American options and options including skew, no closed form solution may exist. Techniques of numerical codes for diffusion are then used. These include:

- Binomial or trinomial discretization
- Monte Carlo simulation
- Finite difference or finite element analysis

Descriptions of these numerical methods are well documented, and we refer the interested reader to the literature [v].

The choice of the type of numerical approximation depends on the analysis. Barriers are more difficult to represent in some schemes than others, for example. Monte Carlo simulators, while perhaps the most flexible, need careful attention to hedging parameter stability.

When developing numerical code, it is always important to have the ability to look at approximations based on closed-form solutions or simple modifications of

---

[21] **History:** For example, the three-factor FX option formalism is in my notes from 1989.

them. It is difficult to interpret numerical code output physically, and physical intuition is extremely valuable. This is because:

- Intuition is behind the language traders, salespeople, and risk managers use and it's a good idea to be able to communicate physically why your code is producing the results it is.
- Sanity checks provided by closed form solutions are useful in code development. In particular, if your code output differs by say 30% from a closed-form approximation, you are much more confident that there are no bugs in your code than if the @#*$&^% black box code puts out a result different from the approximate sanity check by say an order of magnitude.

References

[i] **Standard FX Options and Market References**
Garman, M. B. and Kohlhagen, S. W., *Foreign Currency Option Values*. Journal of International Money and Finance, Vol 2, 1983. Pp. 231-237
Andersen, T. J., *Currency and Interest-Rate Hedging – A User's Guide to Options, Futures, Swaps, and Forward Contracts*. New York Institute of Finance, Prentice-Hall, 1987.
Derosa, D. F., *Options on Foreign Exchange*. Probus Publishing Co., 1992.
Riehl, H., Rodriguez, R. M., *Foreign Exchange & Money Markets – Managing Foreign and Domestic Currency Options*. McGraw-Hill Book Co, 1983.

[ii] **Higher-Order Greeks**
Simons, H., *The Color of Money*. Futures, 1/96, Pp. 40-41.

[iii] **General Skew, Smile Parametrization**
Brown, G. and Randall, C., *If the Skew Fits ....* Risk Magazine, April 1999. P. 62.

[iv] **The "Two-Country Paradox"**
Garman, M. B., *Exotic Options: Their Pricing, Risk Management, and Financial Engineering*. RISK seminar, 1993.
Taleb, N., *Dynamic Hedging – Managing Vanilla and Exotic Options*. John Wiley & Sons, 1997. Pp. 433-437.

[v] **Numerical Techniques Textbook**
Wilmott, Pl, Dewynne, J., Howison, S., *Option Pricing – Mathematical models and computation*. Oxford Financial Press, 1993.

6. Equity Volatility Skew (Tech. Index 6/10)

In this chapter, we consider volatility skew for equity options. We also include some formalism and skew models. Volatility skew refers to the strike dependence of the volatility. For example, some S+P 500 index option volatility data as a function of the strike E using the Black-Scholes model are shown below[1]:

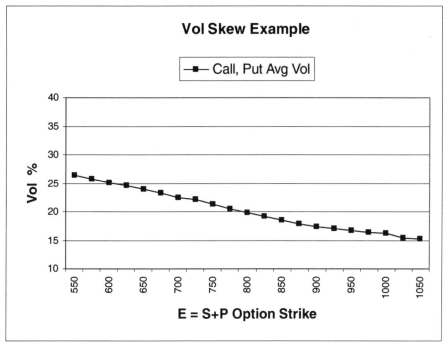

What this graph indicates is that if a single volatility is put into the usual stock-option Black-Scholes formula, then in order to reproduce the market option prices this volatility decreases as a function of the strike of the option. Physically, this condition seems imply that *"fear is greater than greed"*. Thus, a premium for OTM puts with low strikes to protect the downside exists [fear], relative to

---

[1] **Data:** Data are the averaged call and put midpoint DEC 1997 S&P index option vols on 5/14/97 when spot was 837.54. Nothing special happened on that day. We thank Citigroup for the use of these data.

OTM calls at high strikes to participate in the upside [greed]. Also, for that reason, skew increases in importance after severe market perturbations[i].

Put-Call Parity: Theory and Violations

Call and put option payoffs at option expiration satisfy the obvious relation[2] $\left(S^* - E\right)_+ - \left(E - S^*\right)_+ = \left(S^* - E\right)$. The "Put-Call Parity" statement for European options both with strike E is a direct consequence. It says that the call-put European option price difference today is the volatility-independent forward,

$$C_{call} - C_{put} = e^{-r\tau^*}\left(S_{fwd} - E\right) \qquad (6.1)$$

Therefore, call and put options theoretically have the same volatility. This is only approximately true. There is a strike skew effect, shown below[3]:

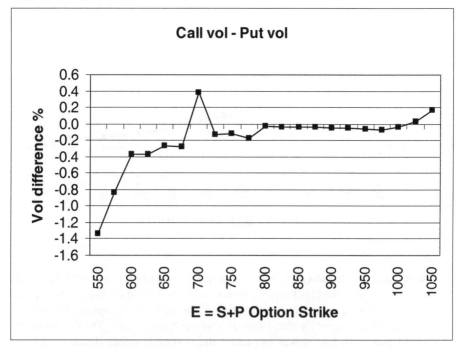

---

[2] **Notation:** $(S^* - E)_+ = S^* - E$ if $S^* > E$ and 0 otherwise.

[3] **What's the blip at E = 700?** Put vols were anomalously low at this point, even below put vols at 720. Should we take the market at face value and put in the spike, or smooth it out? You be the judge. Just another example of Dealing With the Data.

Skew with Less Liquid Options

The complexities with skew are heightened for less liquid options. The simplest exotics are S+P barrier options. Even these relatively simple exotics are highly illiquid and little skew information is directly available. It is therefore common practice to model the skew, using a model for barrier options consistent with the skew information from the standard options market[4]. Options on individual stocks usually do not have enough information even to construct a volatility surface (see below). Skew from the S+P index might be added on to the single-stock vol, with consequent vol basis risk between the S&P and the stock. Skew uncertainties are compounded for options on baskets of individual stocks.

Therefore, skew is only well defined for non-exotic index options. However, the word "stock" will be used in this section instead of "index" for clarity.

The Volatility Surface

Before proceeding further with skew, we can consider different expiration times t^*. If we put all volatility information on the same graph and fill in the gaps, we would get a surface $V(E, t^*)$. This is called the "volatility surface" [5].

Dealing with Skew

There are six methods to handle skew.

Skew Method 1. Perturbative Skew: Use a standard model with a prescription for skew modification of the standard model.

---

[4] **Philosophy of Skew:** We may feel that we are discovering something fundamental about the stock-price process from the S+P standard option market, in which case we would naturally want to use this same process to price all S+P options, including barrier options. A slightly less religious position, though equally pragmatic and leading to the same activities, would simply state that it at least gives more confidence to have a model that prices simple options correctly before tackling complicated options.

[5] **The Volatility Surface:** The volatility surface is not easy to parameterize. First, the data for different maturities do not cover the same ranges in strikes (short-dated options have a more restricted range of strikes than long-dated options). Second, away from the money the options are more illiquid, the bid-ask spreads increase, and the vols generally have uncertainties. These uncertainties are partly due to the decreased option vega or price sensitivity to the volatility away from the money. Third, the market for some parameters may be one-sided (on this date there were no bids for DEC calls below 550, and no bids for DEC puts below 475). Fourth, technical supply/demand effects appear that are not constant with respect to (E, t*). The vol surface therefore sometimes has anomalies, and these anomalies can cause difficulties with the option models.

Skew Method 2. Static Replication: Use a combination of standard options that approximately replicates a boundary condition, e.g. at a barrier.

Skew Method 3. Stochastic Volatility: Assume that the volatility is itself stochastic with an assumed process including a "vol of vol" noise term.

Skew Method 4. Local Volatility Function: Make the volatility in the stock diffusion process at time t a function of the stock price $\sigma_{SkewFit}\left[S(t),t\right]$ such that the S+P option prices are fit, replicating the skew.

Skew Method 5. Intuitive Models: These are called "sticky strike" or "sticky delta / sticky moneyness"; they are sometimes used to describe the stock-price dependence of the skew as time marches on.

Skew Method 6. Jump Diffusion: These models parametrize jump processes that are included along with Brownian motion. The jumps modify the effective volatility.

Perturbative Skew and Barrier Options

A possible perturbative skew approach with an up-and-out (UO) call option is illustrated. The idea[6] is to start with the standard barrier option model. Then a skew correction is added perturbatively such that the boundary condition, terminal condition, and limiting properties are maintained.

Some requirements for a reasonable skew approach are:

- The option vanishes as the stock price approaches the boundary $\hat{S}_0 \to K$
- The standard call, with the correct volatility, is recovered as the boundary is removed ($K \to \infty$)
- The terminal payoff condition at expiration time t^*_{mat} is not modified[7].

A perturbative approach to include skew (the "T decomposition formula") similar in spirit but different in detail, was suggested by Taleb [ii].

The skew correction in this approach is generated by the skew of a "replicating portfolio". At expiration, this portfolio replicates the barrier option. In a rough sense, the portfolio also approximates the barrier option before expiration. The portfolio has a simple form (two calls and a digital option). Only vols at the strike E and barrier K are needed.

---

[6] **History:** This particular perturbative skew approach was developed by me in 1997.

[7] **Notation:** The subscript "mat", short for option maturity, is there because other intermediate times will show up below.

We consider an up-out UO call. Other single barrier options can also be treated in the same way or using sum rules [8]. Assume that the UO call has strike E, barrier K, and time to expiration or maturity τ^*_{mat}. At expiration, t^*_{mat} the payoff (intrinsic value) looks like this:

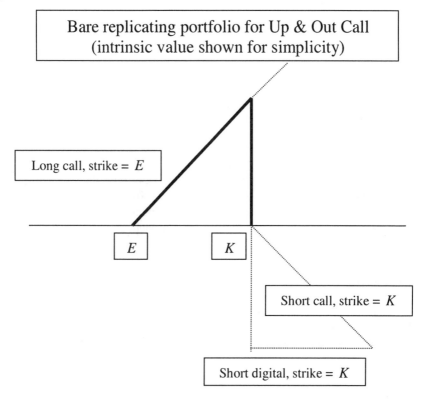

Again, the idea is that the skew of this portfolio is used in order to generate the skew of the barrier option. The procedure is in App. A.

---

[8] **Skew for other barrier options:** Other single barrier options can be obtained using sum rules. For example, the down and out (DO) call can be obtained from the DO put and the DO forward. Up and in (UI) options can be obtained from the standard options and the DO options. Double-barrier option skew can be obtained approximately by using a perturbative approach modifying the standard double-barrier closed-form solutions by skew corrections involving a limited number of images (e.g. one image for each barrier), along with sum rules.

Some Numerical Results for Barrier Options with Skew

While naturally not identical, the results of this model are qualitatively similar to other approaches that have skew. For example, an up-out S+P call option[9] was analyzed with the following results:

UO Call Price with No Skew (zeroth order approximation):

$$C_{academic}^{\text{Strike vol, No Skew}} = 55.8, \quad C_{academic}^{\text{Spot vol, No Skew}} = 59.3$$

UO Call Price with Different Skew Models[10]:

$$C_{\text{Perturbative Skew}} = 61.2, \quad C_{\text{MC Simulation}}^{\text{Local Volatility}} = 61 \pm 0.5, \quad C_{\text{Static Replication}} = 60.8$$

It is seen that the zeroth order approximation is better with the spot vol than the strike vol. The results including skew using three different model approaches are similar. We have included results from a MC local volatility simulation (see below), and from Derman's static replication approach, to which we now turn.

Static Replication

Static replication[iii] is a clever way of approximating a complicated path-independent option by a replicating portfolio consisting of simple options. This set replicates the payoff and boundary conditions of the original option. Because payoff and boundary conditions uniquely determine any option once the stochastic equation for the underlying variable is given, the original option can be replaced by the replicating portfolio. Derman noticed that this can be achieved as a function of time, so once the replicating portfolio is chosen, the same portfolio remains a replicating portfolio provided that the parameters in the stochastic equations do not change.

---

[9] **Barrier option example - parameters:** Strike 800, barrier 1,000, spot 940, time to expiration 0.3 yrs, rebate 64 paid at touch, strike vol 35%, spot vol 27.5%, barrier vol 24.5%, interest rate 6 % (ctn365), dividend yield 1.7% (ctn365). Parameters rounded off.

[10] **Calculation Details for the Perturbative Skew Approach:** The academic model (including rebate) with spot vol was used for the 0th order approximation. Multiplicative skew corrections with averaging over knockout times were used. The rebate was included in the digital option for the bare replicating portfolio used to construct the skew. A call-spread approximation was used for the digital. I thank Tom Gladd for assistance in the calculations.

Equity Volatility Skew

The inclusion of skew using this approach is carried out by putting skew into each of the simple options in the replicating portfolio. Because the options are simple, the job of finding their skew is in principle relatively straightforward.

In general, the replicating portfolio has an infinite number of options; in practice, this is approximated by a finite portfolio. Therefore, the achieved replication is only approximate.

We next give a summary of Derman's static replication method using path integrals and Green functions for a continuous-barrier up-out European call option[11]. Although the allowed region for paths contributing to this barrier option is below the barrier, the idea is to replace the existence of the barrier by a replicating set of OTM call options that have no payoff for any path below the barrier. The OTM call option payoffs above the barrier are propagated back using the standard Green function without the barrier. That is, an equivalent problem with no barrier is used to solve the original problem with the barrier. This is similar to using images; here the images can be thought of as the set of OTM call options replicating the zero boundary condition along the barrier.

The UO call $C^{(UO\_Call)}(t^*, E)$ is equivalent to an ordinary call $C(t^*, E)$, both expiring at t^* with strike E, plus the replicating portfolio $V_{replicating}$ for the barrier at K. An OTM call $C_\ell = C(t_\ell, E_\ell)$ in $V_{replicating}$ has strike $E_\ell \geq K$, expires at $t_\ell < t^*$ with payoff C_ℓ^*, and has weight w_ℓ. The weights $\{w_\ell\}$ are chosen to enforce the zero boundary condition. At time t_j, $V_{replicating}$ consists of those $\{C_\ell\}$ with $t_\ell > t_j$ that have not expired. Exhibiting the variable $x_j = \ln S_j$, which *is required to be below the barrier* $x_j < \ln K$,

$$V_{replicating}(x_j, t_j) = \sum_{t_\ell > t_j} w_\ell C_\ell(x_j, t_j) \tag{6.2}$$

The standard Green function (including discounting) for propagation between times t_j and $t_\ell > t_j$ in the *absence* of the barrier is written as $G(x_j, t_j; x_\ell, t_\ell)$. Then the option $C_\ell(x_j, t_j)$ is given by the expectation

$$C_\ell(x_j, t_j) = \int_{-\infty}^{\infty} dx_\ell \cdot G(x_j, t_j; x_\ell, t_\ell) \cdot C_\ell^* \tag{6.3}$$

---

[11] **Path Integrals and Barrier Options:** See Ch. 17, 18 for barrier option formalism, path integrals and Green functions in Ch. 41-45. The standard Green function for a constant volatility has a Gaussian or normal form in the x variables. Skewed vol input is used. You do not have to know anything more to follow the logic in this section.

Notice that the region of integration is over *all* values of x_ℓ, but the integrand is nonzero only for $x_\ell > \ln E_\ell \geq \ln K$. We now determine the weights $\{w_\ell\}$ recursively by setting $V_{replicating}\left(x_j = \ln K, t_j\right) = 0$ at each t_j.

This concludes the discussion of static replication. We now turn to the stochastic volatility method for including skew.

Stochastic Volatility

Stochastic volatility is a natural idea. First, casual examination of a given historical time series over different time periods for a given window size usually produces noticeably different standard deviations, dependent on the window. These differences in standard deviations cannot be explained by the trivial observation that the window size is finite, and that some statistical finite-sample noise must be present in the standard deviation.

Because the option volatility is (philosophically) supposed to be the market perception of the future underlying standard deviation, it is logical to postulate that standard-deviation instabilities will show up as instabilities in the option volatility. These volatility instabilities can be modeled in a stochastic volatility framework using a "volatility of volatility" or "vol of vol".

Instabilities in the option volatilities are also evident from the time dependence of the implied vol for a given option. The implied vol generally jumps around from one day to the next, as market supply/demand conditions change. If the distribution of implied vols is plotted and the standard deviation measured, we have an approximation to the vol of vol for that option.

Stochastic volatility is often assumed to exist for risk management purposes. For scenario analysis, VAR[12] and other risk measures, the vol of vol is important because it represents an extra vega risk that can contribute significantly.

A simple and pedagogically accessible model of stochastic volatility is now presented[13]. More sophisticated frameworks were developed by Hull and White, and by Heston where a stochastic process for the time dependence of the volatility is postulated. A comprehensive review was given by Carr[iv]. The model presented here is equivalent to assuming that movements in the local volatility in all intervals $(t, t+dt)$ are equal, with this uniform fluctuation chosen at random. That is, the local volatility fluctuations are parallel as a function of time. This

---

[12] **VAR:** VAR is an acronym for Value at Risk, and is discussed in Chapters 26-30.

[13] **History:** The stochastic volatility model in the text was constructed by me in 1986 after Andy Davidson pointed out the stochastic nature of historical and implied volatilities for T-bond futures. The original idea, also due to Davidson, was to find an average volatility with more stable properties.

Equity Volatility Skew

model is transparent and sophisticated enough to exhibit the idea of stochastic volatility.

We postulate a probability distribution function (pdf) $\mathcal{P}[\sigma]$ for volatility σ. For example, this volatility pdf could be a Gaussian distribution centered around some "renormalized" volatility σ_R with vol-of-vol width Δ_σ, between limits for $\sigma \in (\sigma_R - \Lambda_L, \sigma_R + \Lambda_U)$, for some cutoff parameters Λ_L, Λ_U. We write

$$\mathcal{P}[\sigma] = \mathcal{N} \exp\left[-\frac{1}{2\Delta_\sigma^2}(\sigma - \sigma_R)^2\right] \tag{6.4}$$

Here the normalization $\mathcal{N} = N(\Delta_\sigma, \sigma_R; \Lambda_L, \Lambda_U)$ is such that $\int \mathcal{P}[\sigma] d\sigma = 1$. Consider a European call option with strike E with the volatility σ dependence made explicit (we include the discounting factor in the Green function):

$$C(x_0, t_0; \sigma; E) = \int_{\ln E}^{\infty} dx^* G_0(x^* - x_0, t^* - t_0; \sigma)\left[\exp(x^*) - E\right]_+ \tag{6.5}$$

This just produces the usual Black-Scholes (BS) formula. If we now assert that the volatility takes value σ with probability $\mathcal{P}[\sigma]$ in $d\sigma$, we can define the volatility-probability-averaged call option $C^{Avg-\sigma}$ as

$$C^{Avg-\sigma}(x_0, t_0; E) = \int_{\sigma_R - \Lambda_L}^{\sigma_R + \Lambda_U} d\sigma \mathcal{P}[\sigma] \cdot C(x_0, t_0; \sigma; E) \tag{6.6}$$

Clearly, this $C^{Avg-\sigma}$ has a different strike dependence than for the volatility unaveraged result. If we now try to interpret $C^{Avg-\sigma}$ as being given by the BS formula, we will need to compensate for the volatility averaging by placing a strike-dependent skew volatility $\sigma_{Skew}(E)$ into the Green function in the BS formula such that we get the same numerical result for $C^{Avg-\sigma}$ as above, viz:

$$C^{Avg-\sigma}(x_0, t_0; E) = \int_{\ln E}^{\infty} dx^* G_0\left[x^* - x_0, t^* - t_0; \sigma_{Skew}(E)\right]\left[\exp(x^*) - E\right]_+ \tag{6.7}$$

In this manner, stochastic volatility generates a skew dependence $\sigma_{Skew}(E)$.

We could also define a volatility-averaged Green function $G^{Avg-\sigma}$ as

$$G^{Avg-\sigma}\left(x^* - x_0, t^* - t_0\right) = \int_{\sigma_R - \Lambda_L}^{\sigma_R + \Lambda_U} d\sigma \mathcal{P}[\sigma] \cdot G_0\left(x^* - x_0, t^* - t_0; \sigma\right) \quad (6.8)$$

So also

$$C^{Avg-\sigma}\left(x_0, t_0; E\right) = \int_{\ln E}^{\infty} dx^* G^{Avg-\sigma}\left(x^* - x_0, t^* - t_0\right)\left[\exp(x^*) - E\right]_+ \quad (6.9)$$

Note although the above two different integrals produce the same $C^{Avg-\sigma}$, the integrands are not pointwise equal. Some technical comments are in App. B.
This ends the discussion of stochastic volatility. We turn to local volatility.

Local Volatility and Skew

Local volatility is the volatility in the stochastic equation for the stock-price relative changes $[S(t+dt) - S(t)]/S(t)$ in the time interval $(t, t+dt)$. The technology of local volatility methods was pioneered by Derman et. al. [v]. It seems that these methods are now preferred in the industry for incorporating skew for equity options.

We can use path integrals[11] to discuss local volatility. The path integral is well formulated for a local volatility $\sigma_{S,t}$ determining the infinitesimal stock price diffusion at time t, which can depend on the stock price $\sigma_{S,t} \equiv \sigma[S(t), t]$. This is because each stochastic equation in $(t, t+dt)$ that depends on the local volatility $\sigma_{S,t}$ is inserted as a constraint into the path integral.

If the each integral in the path integral at time t_l is discretized into N_S points and there are N_t forward times in the time partition, then there are $N_S N_t$ total points in the discretization, and by construction, there are the same number $N_S N_t$ of local volatilities. Explicitly, at each expiration time t_l ($l = 1...N_t$) we specify N_S European option prices with different strikes $\{E_\varsigma(t_l)\}$ ($\varsigma = 1...N_S$). We can then numerically invert the equations determining the option prices $\{C[E_\varsigma(t_l)]\}$ for the local volatilities $\{\sigma(S_\varsigma, t_l)\}$ in the diffusion process. It is not necessary that the strikes E_ς and the discretized lattice prices S_ς be equal.

Equity Volatility Skew

Analytic Results in Perturbation Theory

Analytic results can be obtained for local volatility in perturbation theory[vi]. The formalism is advanced and requires path integrals (c.f. Ch. 42, Appendix D).

The Skew-Implied Probability Distribution

The discounted expected integral for a European option with strike E_ς and exercise time t_I can be differentiated with respect to the strike of the option. As discussed in Ch. 5, if we thus differentiate twice, it is easy to see that the Green function is obtained with final co-ordinate $x_I = \ln\left(E_\varsigma(t_I)\right)$. Specifically, (exhibiting only strike, exercise labels), and including discounting, we get

$$E_\varsigma(t_I)\frac{\partial^2 C\left[E_\varsigma(t_I)\right]}{\partial\left[E_\varsigma(t_I)\right]^2} = G_0\left[x_I = \ln\left(E_\varsigma(t_I)\right)\right] \quad (6.10)$$

While this type of relation has been known forever, the more recent application is to turn the statement around backwards to *derive* an empirical Green function $G_{empirical}$ or probability distribution function pdf from the strike curvature of the option prices, as determined from market data. This $G_{empirical}$ is often plotted to exhibit visually the deviation from the standard lognormal pdf. Naturally, numerical issues arise in finding the option strike curvature since the options are only known at discrete points, market anomalies (supply/demand) exist, etc. Finally, Dupire and Derman have shown that a diffusion equation is satisfied by the Green function written as a function of the strike.

Local vs. Implied Volatility Skew; Derman's Rules of Thumb

An implied volatility, roughly speaking, is an average over local volatilities of the diffusion process all the way to expiration. This average is specified for an option of a given strike as the (single) number that has to be inserted in a Black-Scholes formula for the volatility to get the market price. The local volatilities depend on the stock price and the time, and they specify the diffusion process over small time steps. For this reason, the local volatility dependence on the underlying price and the implied volatility dependence on the strike are not equal.

The following picture shows implied and local volatilities, illustrated with two paths.

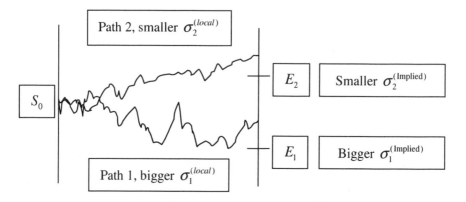

Derman's First Rule of Thumb

Derman's "1st Rule of Thumb" says that local volatility varies with the stock price about twice as rapidly as the implied volatility varies with the strike[v]. The implied volatility is an average of the local volatilities along the paths contributing to the option payoff. For a rough approximation, use ½ the sum of the initial and final local volatilities. Then if the implied vol increases, the final local volatility has to increase by a factor 2 to cancel the averaging ½.

Consider two call options with the strikes indicated. Paths like path #1 contribute to $C_{call}(E_1)$ with bigger implied vol, but they do not contribute to $C_{call}(E_2)$ with smaller implied volatility. Therefore, paths like path #1 must have a bigger local volatility. Similarly, paths like path #2 must have a smaller local volatility. So, as the stock price drops the local vol increases. Similarly, as the stock price goes up the local vol decreases. We could have done the analysis with put options. The bigger local vol for lower stock prices also reproduces the put options at lower strikes made more expensive by the "fear" factor.

We need to consider what happens to the skew of an option as real time moves forward [14]. Now the dependence of the local volatility $\sigma[S(t),t]$ on the stock price at forward times—viewed from today—can be determined today by the set of options today. Because the local volatilities determine the diffusion process completely, the behavior of the implied volatility for a given option is then theoretically determined as real calendar time moves forward. Thus, if on Jan. 1 we determine the local vols from option data, then the values of implied

---

[14] **Crystal-Ball Analogy:** This is similar to imagining the behavior of an option along future worlds specified by future Wall Street Journals, viewed through a crystal ball today.

vols on Feb. 1 are predicted once a given stock level on Feb. 1 is specified. Theoretically, at any point along a given stock price path, we can predict the value of any option and thus any implied volatility.

Derman's Second Rule of Thumb
As time changes, the stock price will change. Derman's "2nd Rule of Thumb" says that the change in implied volatility of a given option for a change in stock price is approximately the change in implied volatility for a change in strike. This is reasonable since the price and the strike are in a sense dual variables.

Derman's Third Rule of Thumb
The hedging of options is affected by skew. Because the skew slope is negative, the local volatility decreases as the stock price moves up. Therefore, in that case, a call option increases less than it would in the absence of skew. Thus, Δ_{call} is reduced by skew. This reduction is the negative change in C_{call} produced by the negative change in volatility $\delta\sigma_{call}$ for the given increase δS. This is Derman's "3rd Rule of Thumb"; the reduction in Δ_{call} is

$$\delta\Delta_{call} = \frac{\delta C_{call}}{\delta\sigma_{call}} \frac{\delta\sigma_{call}}{\delta S} \approx \frac{\delta C_{call}}{\delta\sigma_{call}} \frac{\delta\sigma_{call}}{\delta E} \approx - \text{Vega} * |\text{Skew Slope}| \quad (6.11)$$

Option Replication with Gadgets

Derman and co-workers observed [vii] a clever identity for an option expiring at some $t^* = t_\ell$ in terms of a weighted integral of options with different strikes expiring at $t^* - dt = t_{\ell-1}$. Derman called a "gadget" the difference of the option and the weighted integral and proposed that the identity be used as a method for hedging an option with a set of other options. Because the hedging can be done locally in time, the methodology hedges local volatility. Of course, the utility of gadgets relies on the existence of gadgets in the market. The idea is to express an option that expires at some time t_ℓ with strike E_ℓ as a weighted integral over options that expire at an earlier time $t_{\ell-1}$ with different strikes $\left\{ E_{\ell-1}^{(\alpha)} \right\}$.

Consider the following figure.

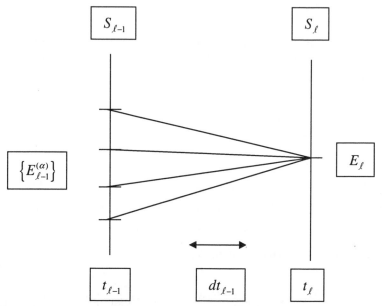

Here α looks like a discrete label (and for the trinomial model $\alpha = 1, 2, 3$) but will actually be continuous for our derivation, i.e. we will integrate over the variable $E_{\ell-1}$. We use the subscript ℓ because Derman's idea is to use gadgets at any time in the future.

Note that this is not "static replication". As we saw above, static replication is concerned with the determination of a set of options that can be used to replicate a boundary condition, e.g. a barrier for an up and out call.

Derman used a trinomial model to illustrate the gadgets. We can derive the gadget identity using path integrals and Green functions. We need to evaluate the various options. Since we use lognormal dynamics, we set $y_{\ell-1} = \ln E_{\ell-1}$, $x_{\ell-1} = \ln S_{\ell-1}$, etc. Denoting C as a call option, the gadget definition is

$$Gadget(S_0, t_0) \equiv C(E_\ell, t_\ell) - \int_{-\infty}^{\infty} dy_{\ell-1} G^{(R)}(y_\ell, y_{\ell-1}; dt_{\ell-1}) C(E_{\ell-1}, t_{\ell-1})$$

(6.12)

Here the weight function $G^{(R)}(y_\ell, y_{\ell-1}; dt_{\ell-1})$ is written in terms of the strikes, and will be determined to make the gadget value equal to zero (so that the gadget looks like a hedged portfolio). It will turn out that $G^{(R)}(y_\ell, y_{\ell-1}; dt_{\ell-1})$ is closely related to the usual Green function $G_{\ell-1,\ell}$ written in terms of the stock variables for the last propagation between $t_{\ell-1}$ and t_ℓ.

The usual Green function $G_{0,\ell-1}$ propagating between the current time and the time $t_{\ell-1}$ is common to all the options. We get

$$Gadget(S_0, t_0) = \int_{-\infty}^{\infty} dx_{\ell-1} G_{0,\ell-1} \cdot \{I - II\} \tag{6.13}$$

Here the integrals I and II are produced by the two terms in Eqn. (6.12)

$$I = \int_{-\infty}^{\infty} dx_\ell G_{\ell-1,\ell} \left(e^{x_\ell} - e^{y_\ell}\right)_+ \tag{6.14}$$

$$II = \int_{-\infty}^{\infty} dy_{\ell-1} G^{(R)}(y_\ell, y_{\ell-1}; dt_{\ell-1}) \left(e^{x_{\ell-1}} - e^{y_{\ell-1}}\right)_+ \tag{6.15}$$

We want the equality of the integrals $I = II$, making the gadget $= 0$. Evidently, we want a Gaussian in the strike variables for $G^{(R)}(y_\ell, y_{\ell-1}; dt_{\ell-1})$ to make it look like $G_{\ell-1,\ell}$, up to some normalization factor. We also have a different variable of integration. We make the change of the dummy variable $y_{\ell-1} = -x_\ell + x_{\ell-1} + y_\ell$. After a little algebra, we obtain the identity $I = II$, provided we set

$$G^{(E)}(y_\ell, y_{\ell-1}; dt_{\ell-1}) = e^{-r_{\ell-1} dt_{\ell-1}} \frac{E_\ell}{E_{\ell-1}} \cdot$$

$$\cdot \frac{1}{(2\pi\sigma^2_{\ell-1} dt_{\ell-1})^{1/2}} \exp\left[-\frac{[y_\ell - y_{\ell-1} - (r_{\ell-1} - \sigma^2_{\ell-1}/2) dt_{\ell-1}]^2}{2\sigma^2_{\ell-1} dt_{\ell-1}}\right] \tag{6.16}$$

Because $y_\ell - y_{\ell-1} = x_\ell - x_{\ell-1}$ from the change of variable, we find

$$G^{(R)}(y_\ell, y_{\ell-1}; dt_{\ell-1}) = e^{-r_{\ell-1} dt_{\ell-1}} \frac{E_\ell}{E_{\ell-1}} G_{\ell-1,\ell} \tag{6.17}$$

Noting that $dy_{\ell-1} = dE_{\ell-1}/E_{\ell-1}$ and $dx_\ell = dS_\ell/S_\ell$ we can cast this equation into a notation closer to Derman's. This ends the discussion of local volatility.

Intuitive Models and Different Volatility Regimes

The heavy artillery of the local volatility described above contrasts with two simple intuitive models for the movement of the implied volatilities with changes in the stock price. These are called the "sticky-strike" model and the "sticky-delta" or the closely related "sticky moneyness" model.

Derman advocates the use of the different volatility models under different regimes of realized stock volatility and he has analyzed the different time periods defining these regimes from S+P volatility data [i].

Derman suggests that the local tree model is appropriate when the realized volatility takes a sudden jump and the market moves downward rapidly.

Sticky Strike, Sticky Delta, and Sticky Moneyness Models

The *"sticky-strike"* model assumes that the implied volatility for a given strike remains unchanged regardless of the change in the stock price. For this reason, $\Delta(E)$ for a given option of strike E does not contain the extra term $\delta\Delta$ discussed above. Derman proposes this as a reasonable model in case the market is trending sideways without change in realized volatility, so that keeping the implied vols for existing options unchanged is a good match.

The *"sticky-delta"* model assumes that the ATM volatility remains unchanged regardless of the change in the stock price. The "sticky-delta" therefore refers to the ATM delta, i.e. to an unchanged Δ_{ATM} of the ATM option. The delta $\Delta(E)$ for a given option of strike E has an extra term $\delta\Delta$ with the opposite sign from the $\delta\Delta$ of the local volatility model.

The *"sticky-moneyness"* rule is related to the sticky-delta rule. Moneyness is defined as the ratio E/S. Thus an ATM option has moneyness = 1, so constant ATM vol means constant moneyness vol, at least near moneyness = 1.

Derman suggests that the sticky-delta or sticky-moneyness rules are appropriate for hedging when the market is trending with a constant realized volatility. The constant realized volatility about the trending average corresponds to the choice of constant implied volatility at and near to the ATM point.

This ends the discussion of the intuitive models and the use of the various volatility models in different regimes of stock price movement.

The Macro-Micro Model and Equity Volatility Regimes

It is interesting to note that the various regimes as characterized by Derman (trending, range-bound, jumpy) seem to correspond to the characteristics of the Macro-Micro model discussed in Ch. 47-51, as applied to equities. In the Macro-Micro model, we have quasi-random drifts or slopes corresponding to trending (with positive or negative drifts) and range-bound behavior (small drift), all with limited volatility. Jumps can exist, playing a role in the micro component. Skew in the Macro-Micro model can be accommodated using the stock price in the probability distribution functions for each of the components.

Jump Diffusion Processes

Andersen and Andreasen have examined the effects of jump diffusion processes on generating volatility smile effects. We refer the reader to their paper[viii] and the references therein.

Appendix A: Algorithm for "Perturbative Skew" Approach

The steps in the perturbative model described in the text follow. Although the procedure might seem complicated, it is easy to implement numerically:

Step 1. Start: Use the standard "academic" UO barrier call formula[15] $C_{academic}^{(E,K)}\left(\hat{S}_0; \tau_{mat}^*; \hat{\sigma}_0\right)$ as the zeroth order approximation. The volatility $\hat{\sigma}_0$ is chosen phenomenologically to minimize the skew correction (the ATM τ_{mat}^* vol works well). Here $\hat{S}_0$ is the current (spot) price.

Step 2. Construct the "bare replicating" BR portfolio[16] *for the payoff of the UO call at some intermediate time* t^*, $C_{BareRep}^{(E,K)}\left[\hat{S}_0; \tau^*; \sigma_E\left(t^*\right), \sigma_K\left(t^*\right)\right]$. This BR portfolio will *not* be used to approximate the barrier option but its skew dependence *will* be used to construct the skew correction perturbatively by an add-on to the standard UO barrier formula in Step 1. $C_{BareRep}^{(E,K)}$ with skew has three

---

[15] **Barrier Option Academic Formalism:** We will discuss this in Chapter 17.

[16] **Derman's replicating portfolio:** The "replicating portfolio" here is *not* the same as Derman's replicating portfolio. Derman's replicating portfolio is exact but the composition of his replicating portfolio is complicated to construct because it relies on a back-chaining algorithm.

pieces: long call [strike E, vol $\sigma_E\left(t^*\right)$]; short call [strike K, vol $\sigma_K\left(t^*\right)$]; short digital option[17] [strike K, height $K - E$]. We first choose t^* as the average knockout time t_{AvgKO}; this maintains the boundary condition as $\hat{S}_0 \to K$, the terminal condition as $\hat{t}_0 \to t^*_{mat}$, and the limit $K \to \infty$ (boundary is removed).

Step 3. Construct the no-skew BR portfolio, i.e. without skew for the vols, $C^{(E,K)}_{\text{BareRep}}\left[\hat{S}_0; \tau^*; \sigma_0\left(t^*\right), \sigma_0\left(t^*\right)\right]$ where $\sigma_0\left(t^*_{mat}\right) = \hat{\sigma}_0$.

Step 4. Subtract the skewed and unskewed BR portfolios in (2) and (3) above to get the first approximation for the skew correction, $\delta C^{(E,K)}_{Skew}$.

Step 5. Probability-weight $\delta C^{(E,K)}_{Skew}$ by $\mathcal{P}^{(K)}_{K.O.}\left(\hat{S}_0; \tau^*; \hat{\sigma}_0\right)$, the probability[18] that knock out occurs at t^*, and then integrate over t^*. In order to maintain the knockout boundary condition, a subtraction[19] $\delta C^{(E,K)}_{Subt}$ first has to be made to $\delta C^{(E,K)}_{Skew}$, "renormalizing" it to $\delta C^{(E,K)}_{RenSkew} = \delta C^{(E,K)}_{Skew} - \delta C^{(E,K)}_{Subt}$. The skew correction including knock-out probability weighting is then of the form

$$\delta C^{(E,K)}_{FinalSkewCorrection} = \int_0^{t^*_{mat}} \mathcal{P}^{(K)}_{K.O.}\left(\hat{S}_0; \tau^*; \hat{\sigma}_0\right) \delta C^{(E,K)}_{RenSkew}\left(\tau^*; \ldots\right) d\tau^* \quad (6.18)$$

The result for the barrier option including skew is

$$C^{(UO\_Call)}_{WithSkew} = C^{(E,K)}_{academic} + \delta C^{(E,K)}_{FinalSkewCorrection} \quad (6.19)$$

---

[17] **Common Digital Option Approximation:** The digital option can be approximated by a call spread (the difference of call options) having strikes K, K+ δK with δK a small amount. Strictly speaking, the skew for these two options should also be included. This approximation is the way digital options are actually hedged.

[18] **Smearing out the Skew Correction with the Knockout Probability:** We use the academic formula for the knockout probability. We need to renormalize this probability to one. This is because we are merely smearing out the skew correction derived with the average knockout time. Note that then the replicating portfolio theoretically becomes an infinite number of options.

[19] **The Subtraction:** The subtraction $\delta C_{Subt}^{(E,K)}$ has the same form as $\delta C_{Skew}^{(E,K)}$ with the current stock price replaced by K and the nominal volatility $\sigma_0(t^*)$ replaced by $\sigma_K(t^*)$.

Equity Volatility Skew

An alternative multiplicative prescription for skew exists[20]. The idea is to scale the skew using a reasonable base, e.g. the BR portfolio as defined in Step 3, $C_{BareRep}^{(E,K)}[Step(3)]$. Define a "skew factor" $\mathcal{F}_{SkewFactor}$ including knockout probability weighting as

$$\mathcal{F}_{SkewFactor} = \int_0^{t_{mat}^*} \mathcal{P}_{K.O.}^{(K)}\left(\hat{S}_0;\tau^*;\hat{\sigma}_0\right) \delta C_{RenSkew}^{(E,K)} / C_{BareRep}^{(E,K)}[Step(3)] d\tau^* \quad (6.20)$$

This produces the multiplicative prescription for the skew perturbation

$$C_{WithSkew}^{(UO\_Call)} = C_{academic}^{(E,K)}\left[1 + \mathcal{F}_{SkewFactor}\right] \quad (6.21)$$

Appendix B: A Technical Issue for Stochastic Volatility

This appendix has some comments on large volatility contributions for the stochastic volatility model in the text, which we believe present a technical issue for stochastic volatility models in general.

If we let the limits of integration $(\sigma_R - \Lambda_L, \sigma_R + \Lambda_U)$ for the volatility become $\sigma \in (0, \infty)$, we can get a closed form solution for $G^{Avg-\sigma}$ in terms of an infinite series of modified Bessel functions of the third kind by expanding the exponential containing terms linear in σ and using the identity (for any ν with $\text{Re}(a), \text{Re}(b)$ positive) [ix]

$$\int_0^\infty \sigma^{2\nu-1} \exp\left(-\frac{a}{\sigma^2} - b\sigma^2\right) d\sigma = \left(\frac{a}{b}\right)^{\nu/2} K_\nu\left(2\sqrt{ab}\right) \quad (6.22)$$

However, interchange of the multiple $\int d\sigma \int dx^*$ integrals is problematic if the volatility becomes infinite (even if large volatilities are suppressed), because the Gaussian spatial damping for large x^* is not uniform. It makes a difference whether we let $x^* \to \infty$ for fixed σ or we let $\sigma \to \infty$ at fixed x^*, or we take a scaled limit as both variables become infinite. The non-uniformity in the two variables x^* and σ just described would seem to pose a problem for stochastic volatility models that do not have some lattice cutoff for the volatility.

---

[20] **Acknowledgement:** I thank Cal Johnson for suggesting the multiplicative idea, for discussions on Derman's work, and for many other conversations.

References

[i] **Phenomenological Skew Analysis**
Derman, E., Regimes of Volatility – *Some Observations on the Variation of S&P 500 Implied Volatilities*. Goldman Sachs working paper, 1999.

[ii] **Perturbative Approach to Skew**
Taleb, Nassim *Dynamic Hedging – Managing Vanilla and Exotic Options*. John Wiley & Sons, 1997. See p. 335.

[iii] **Static Options Replication:**
Derman, E., Ergener, D., Kani, I., *Static Options Replication*. Goldman Sachs, May 1994.

[iv] **Stochastic Volatility**
Dash, Jan, *A Path Integral Approach to Volatility Averaging in the Black-Scholes Model*, Merrill Lynch working paper, unpublished, 1986.
Hull, J. C. and White, A., *The Pricing of Options on Assets with Stochastic Volatilites*. Journal of Finance, 42 (1987), pp. 281-300.
Heston, S., *A Closed-Form Solution for Options with Stochastic Volatility with applications to Bond and Currency Options*. Review Financial Studies Vol 6, 1993.
Carr, P., *A Survey of Preference-Free Option Valuation with Stochastic Volatility*. Bank of America Securities working paper, 2000, and references therein.

[v] **Local Volatility Skew Methods**
Derman, E., Kani, I., Chriss, N., *Implied Trinomial Trees of the Volatility Smile*. The Journal of Derivatives, Summer 1996, p. 7.
Derman, E., Kani, I., Zou, J., *The Local Volatility Surface: Unlocking the Information in Index Option Prices*. Financial Analysts Journal, July/August 1996, p. 25.

[vi] **Perturbation Theory and Local Volatility**
Dash, J. W., *Path Integrals and Options*. Invited talk, Mathematics of Finance section, SIAM Conference, Philadelphia, July 1993.

[vii] **Replication and Gadgets**
Kani, I., Derman, E., and Kamal, M., *Trading and Hedging Local Volatility*. Goldman Sachs working paper, 1996.

[viii] **Jump Diffusion and Smile**
Andersen, L. and Andreasen, J., *Jump-Diffusion Processes: Volatility Smile Fitting and Numerical Methods for Pricing*. General Re Financial Products working paper, 1999.

[ix] **Bessel Functions**
Erdélyi, A. et. al., *Tables of Integral Transforms*, McGraw-Hill, 1954. Vol 1, p. 146, #29.
Gradshteyn, I. S. and Ryzhik, I. M., *Table of Integrals, Series, and Products*. Academic Press, 1980. See p. 340, #9.

7. Forward Curves (Tech. Index 4/10)

In this chapter, we discuss the construction of the forward-rate curves needed especially for pricing interest-rate derivatives. We begin with a discussion of the input rates to the forward curve construction models, and then discuss the mechanics.

Market Input Rates

Fixed-income securities pricing and hedging rely on forward-rate curves. We now consider the input market rates for the construction of the forward-rate curve in a generic derivative system. The input data needed to construct the curve are described below. The input data are typically obtained either from vendor sources or from the desk in the case of a broker-dealer BD[1].

The choice of which input rates are used depends to some extent on the algorithm chosen to construct the yield curve. The yield curve is used to construct the set of discount factors employed for discounting future cash flows[2]. The input rates used also depend on the currency, since different instruments are available in different currencies[3]. The rates include various types. For the US market, these are cash rates, ED futures, Libor swaps, and US Treasuries.

---

[1] **The Close of Business:** Rates of course vary during the course of the trading day. Official pricing to be put into the books and records will generally use rates from the close of business (COB). It is best if these COB rates are saved automatically to a database so that later analysis is facilitated.

[2] **Future and Forward Jargon:** Do not mix up the word "future" in a phrase like "future cash flow" (cash to be paid some time hence) with "a future" (a contract to buy/sell something at a fixed price at some date hence). Also, do not mix up the word "forward" in a phrase like "forward time period" with "a forward", which is "a future" modified by a small "convexity" correction, as we shall discuss later.

[3] **Yield Curves for Different Currencies:** The main currencies for the largest swap markets and their notations are USD ($US), JPY (Japanese Yen), EUR (Euro), and GBP (Great Britain Pound) or STG (Sterling). Financial futures and swaps are available at different maturities for these currencies. The use of *liquid* instruments is in principle important for reliable pricing. For other currencies including emerging markets there is less choice, and there may only be a few instruments available to construct the yield curve, liquid or not.

Cash Rates

Cash rates are short-term "money-market" rates[i], usually involving the cash Libor interest rates[4] paid for deposit times of various lengths out to 1 year for deposits in London, e.g. US Libor (USD "Eurodollar" deposits), JPY Libor (JPY "Euroyen" deposits) etc.

Futures

For USD, the Eurodollar (ED) futures, based on 3-month ED deposits, are used [ii]. ED futures are cash-settled (i.e. no delivery of a security as for treasury futures). The price of a future P_f is between 0 and 100, and corresponds to a future rate $r_f = 100 - P_f$. The ED futures have a notional of \$1MM and correspond to a 3M forward time period[5], so a 1bp change in rates[6] corresponds to a price change[7] $\$dP_f = \$25$. This is the change in the interest for a \$1MM deposit over ¼ year[8], if rates were to change by 0.0001/yr. Using Excel spreadsheet notation[9] we have $\$dP_f = \$10 \wedge 6 * 0.25\,yr * 0.0001/yr = \$25$ per bp change in rate. The

---

[4] **Libor and Hibor:** The acronym Libor stands for "London Inter Bank Offer Rate". Libor rates in different currencies sometimes have different names (e.g. HIBOR for Hong Kong dollar).

[5] **Quantity and Time Notation:** Abbreviations for quantity [MM = million; B = billion; M, K = thousand]. Abbreviations for time [Y or yr = year; M = month; D = day].

[6] **Basis Points, Units and Some Advice:** Small changes in rates are measured in "basis points" or bp. Numerically one bp = 0.0001. Actually there is an implicit time unit, since interest rates are usually quoted in amounts paid per year (e.g. r = 4%/yr) so the relevant "one basis point" change in an annual rate r would be dr = 1bp/yr = 0.0001/yr. Usually the symbol bp is used without the units. While it sounds trivial, dropping the units easily leads to confusion. *The potential quant is highly advised to put all units in all equations, at least for himself/herself.*

[7] **Sensitivities of Futures in other Currencies:** The price changes for futures for 1 bp/yr change in rates for other currencies have other conventions, but they are all determined by the same sort of equation as presented in the text.

[8] **Money Market Day-Count Conventions**: The meaning of "one year" changes from currency to currency for these futures. For USD, there are considered to be 360 days/yr but for GBP there are 365 days/yr. These "money-market day count" conventions are not the same as the conventions for bonds. And you thought that it was stupid to have different units for the measurement of electricity!

[9] **Arithmetic Notation:** Finance people often denote multiplication by a star * and powers by a caret ^ used in Excel spreadsheets, so 4*4*4 = 4^3 = 64.

Forward Curves

value of the future increases if rates decrease. A given ED future will have its value determined at a definite IMM fixing or reset date[10].

Libor Swaps

A Libor swap is an exchange of fixed payments determined by a "fixed rate" for floating payments based on US Libor values at various times in the future. A "swap rate" is that fixed rate which makes the swap value zero, or indifferent to the choice of fixed or floating payments. USD Libor swap rates are available with high precision. They serve as constraints, since swaps priced by the swap model using the model yield curve have to agree with the input swap rates. Historically the swap rates are expressed as spreads to treasuries[11]. We look at swap pricing in Ch. 8. The USD swap rates based on Libor are now the most liquid instruments available in general, even more than treasuries, and the swap rates are now tending to become considered as the fundamental rates to which other rates are compared. Swap rates in other currencies (e.g. Euro, GBP, JPY) provide similar constraints for input to pricing deals in those currencies.

Treasuries

For the USD curve, these are US-government notes and bonds. For other currencies, the available government securities are used. US treasury securities are auctioned from time to time. The last-auctioned ("on-the-run") securities are those that are the most liquid. For the Libor curve, treasuries are really only used for representing the swap rates as swap spreads to treasuries.

Treasuries are used directly to construct the Constant Maturity Treasury (CMT) curve that is used to price CMT deals. The CMT curve takes account of the various yields and results from a model-dependent interpolation[12] to get a

---

[10] **IMM Fixing/Reset dates and Settlement dates for Futures:** The (IMM = International Monetary Market) "fixing" or "reset" dates, at which the values of the ED futures are determined, are the 3rd Mondays of March, June, September, and December in future years (abbreviated e.g. as MAR04 for March, 2004). Payment (settlement) is made 2 days later, a tradition started by snail mail. This is a minor but annoying complication, since the risk of an instrument depending on a Libor future is zero after that rate is set, so rate uncertainty or diffusion only goes up to Monday, but an extra 2 days of discounting for the cash flow should be included up to Wednesday. Conventions are different for different currencies. Finally, there have recently been introduced ED futures other than 3M (e.g. 1M).

[11] **Bid, Offer, Mid Swap Spreads:** There is a bid and offer side to the swap spreads. The bid might be 40 bp/yr and the offer 44 bp/yr. This means that a potential fixed rate payer is ready to pay 40 bp/yr and a potential fixed rate receiver wants to receive 44 bp/yr (above the corresponding treasury rate). The "mid" swap spread is the average (42 bp/yr).

[12] **Off the Run Govies and CMT:** "Off-the-run" securities are. securities from previous auctions. They may be included in the algorithm giving the CMT rates through interpolations. However, off-the-run securities are less liquid because many bonds have

nominal value for a treasury yield[iii] at any maturity. Note that of course securities do not have their original time to maturity since they were sold so the "10 year treasury" will be plotted at somewhat less than 10 years[13].

Construction of the Forward-Rate Curve

We now discuss the construction of the US Libor forward-rate curve given the input rates discussed above. Actually, the interest-rate curve can be expressed in a number of different conventions.

The forward rate $f^{(T)}(t_0;\Delta T)$ is the interest rate we obtain in a contract made today t_0 for a hypothetical deposit starting at time T and ending at time $T+\Delta T$. Although we could define the forward rate for any T, in practice the forward rates are only modeled at a given discrete set of starting times $\{T_j\}$, and with a fixed ΔT. The set of rates thus obtained is the forward-rate curve used in pricing. A summary of the kinematics is pictured next:

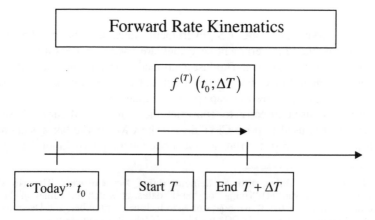

The notation is as follows:
- Today's forward rate curve is the set of $f^{(T)}(t_0;\Delta T)$ for different T,

---

already been placed in portfolios where they are held to term. Therefore, these get less weight in the fit. By the way, government securities are known as "Govies".

[13] **Sliding Down the Yield Curve:** Because securities that have already been issued do not have their original time to maturity, the jargon is that the securities have "slid down the yield curve" from the point where they were issued. The picture is that the yield curve is positively sloped (yields for bonds of longer maturity are greater than bonds of shorter maturity), which is usually – but by no means always – the case.

- t_0 = "today", the date of the input data and also the contract date for the hypothetical $(T, T + \Delta T)$ deposit,
- T = Start of the deposit time period
- $T + \Delta T$ = End of the deposit time period

General Forward Rates

A generalization[iv] of the forward rate specifies a future time t with $t_0 < t < T$, at which time we determine what interest we will get paid for a deposit starting at T and ending at $T + \Delta T$, still using the input data from t_0 today[14]. This interest rate we can call $f^{(T)}(t;t_0;\Delta T)$. We need this generalization if we simulate the forward rate moving in time t, always corresponding to a deposit from T to $T + \Delta T$, starting from today's forward rates determined by today's data[15]. The set of such forward rates at time t for different T forms the forward-rate curve at time t. The kinematics are illustrated below:

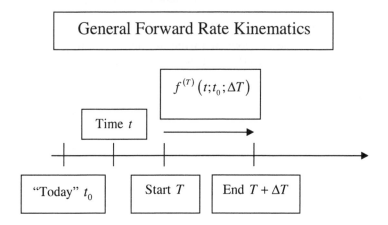

---

[14] **Forward-Rate Example**: For example, take a 3-month deposit of money from $T = 3/1/04$ to $T + \Delta T = 6/1/04$. We may want to know what rate we can expect to contract as of $t = 9/1/03$ for this deposit, using input data from today $t_0 = 6/1/03$. We will wind up with a set of the possible values of this particular rate, for example using a Monte Carlo simulator of the diffusion process for the forward rate.

[15] **HJM**: The most complete formulation of this idea is due to Heath, Jarrow, and Morton (HJM). See the references.

The Forward Rates and ED Futures

We need today's forward rate curve $\{f^{(T)}(t_0; \Delta T)\}$ for different maturities $\{T\}$. In practice[16], we can choose the type of forward rate $\Delta T = 3M$. This will allow us to use data from today's ED futures, which are 3-month rates, for constraints. Say T is one of the IMM fixing dates t_{IMM}. Then the forward and ED future deposit start-times coincide, so the rates must (one would think) coincide. We get the reasonable-looking constraint:

$$f^{(T=t_{IMM})}(t_0; 3M) \cong f_{EDfuture}^{(t_{IMM})}(t_0) \qquad (7.1)$$

The Forward vs. Future Correction

The above equation is not quite correct. There is a difference between the ED future rate and the forward rate even when the deposit start times $T = t_{IMM}$ coincide[17]. This correction can be modeled, and it typically turns out to behave roughly as the square of the time difference $(T - t_0)^2$. The correction is added to the future's price or subtracted from the future rate to get the forward rate. For example, the mean-reverting Gaussian rate model produces [v, 18]

$$f^{(T)}(t_0; 3M) = f_{future}^{(T)}(t_0) - \frac{\sigma^2}{2\omega^2}\left[1 - e^{-\omega(T-t_0)}\right]^2 \qquad (7.2)$$

---

[16] **Meaning of 3M:** 3M means 3 months. With the US money-market convention that one year has 360 days, 3M means 90 days.

[17] **Futures and Margin Account Fluctuations:** Consider a buyer X of ED futures. He must have a "margin account" that must maintain sufficient funds covering changes in the value of the futures. Every day, this account is "marked to market", meaning that funds can be withdrawn by X from the account if the future price goes up (rates down), and funds must be added by X to the account if the future price goes down (rates up). This changes the economics of buying futures. The fluctuations in the margin account thus clearly increase with increased rate fluctuations (called rate volatility). The buyer of a forward rate contract does not have a margin account.

[18] **Theoretical Calculation of the Forward vs. Futures Correction:** The correction was calculated in my 1989 path integral II paper (Ref). It arises from straightforward integration. The "classical path" rate $r^{(CL)}$ about which rate fluctuations occur is the ED future rate in the instantaneous limit $\Delta T = \varepsilon \rightarrow 0$. For small mean reversion ω, the difference between the future and forward rate is quadratic in $T - t_0$. The parameter σ is the Gaussian rate volatility (diffusion constant). See Ch. 43 on path integrals.

Forward Curves

The correction can also be backed out numerically from implementing overall consistency of the forward rate curve in the futures region with other data[19]. The results are roughly consistent with the model for appropriate choices of σ, ω.

Use of Swaps in Generating the Curve

In general, a swap-pricing model is used along with the determination of the forward rates (we discuss the specifics of swap pricing below). Each fixed rate $E_a^{(model)}$ for the corresponding model par swap with zero value $S_a = 0$ (obtained from a putative set of forward rates) is compared with data $E_a^{(data)}$. The forward rates are than chosen such that each $E_a^{(model)} = E_a^{(data)}$, up to some small numerical error.

Cash Rates and the Front Part of the Curve

The initial part of the forward rate curve is determined by cash. The forward rate for a deposit starting today and ending in 3M is just defined as the 3M cash rate; this starts the forward rate curve. There are some consistency issues[20].

The Break or Changeover Point

In practice, a famous problem occurs at a break point or changeover point. Below this point, the forward rates are obtained with futures (with the correction) and above this point, swaps are used [21]. The forward rates tend to be quite

---

[19] **Numerical Value of the Futures vs. Forward Correction:** Numerically the corrections are roughly ("ball-park" as they say) < 1 bp for the front contracts, but become substantial - on the order of 10 bp at for contracts settling in 5 years. Since the bid-offer spreads of plain vanilla swaps are only a few basis points, these corrections are significant.

[20] **Use of Cash Rates in Generating the Curve:** The cash rates other than the 3M rate can also be used as inputs for generating the forward rates, especially for short-term effects (e.g. 1 week). Generally, the cash rates do not influence the curve further out than a few months much since there is a good overall consistency in the market between cash and futures. One chronic nuisance is that the combination of a 3M cash rate with a 3M forward rate starting in 3M should in principle just be the 6M cash rate. Ask your friendly curve constructor quant if he really gets that one right.

[21] **Specification of the Changeover or Break Point Between Futures and Swaps:** The choice of where this changeover of the use of the futures and the swaps depends on the algorithm. It can also depend on the traders who may change it around. Usually it is between 2 and 4 years for USD. For other currencies that have fewer futures, the changeover point is earlier.

discontinuous for T around this point. Other discontinuities can also occur. These discontinuities can lead to substantial effects in pricing some derivatives.

Curve-Smoothing Algorithms

Smoothing algorithms can be used to smooth out discontinuities in the curve generated by the above procedure[22]. Philosophically it is unclear whether the curve really should be made as smooth as possible. The curve acts as if it has a stored "potential energy". In practice, if you make the curve drop in one place it tends to bulge up in another place, somewhat like sitting on a large balloon. The smoothing algorithms can also result in smooth but noticeably large oscillations. These oscillations affect pricing of those securities that are not in the constraint set (i.e., pricing still occurs with all the given constraints realized).

Example of a Forward Curve

Here is an illustrative picture.

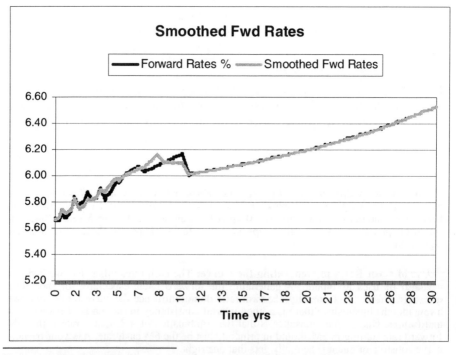

---

[22] **How to Be Smooth:** Cubic splines are a popular choice for smoothing a curve with discontinuities generated by an algorithm. Another is to calculate independently the forward-rate curves generated by futures and by swaps. These are then added together in an overlap region with weights that are chosen as some function of T. Other smoothing choices exist, some sophisticated and proprietary, which I cannot discuss here.

Forward Curves

The graph is of a forward 3M Libor curve without smoothing and with a simple smoothing algorithm[23]. The unsmoothed curve has regions of somewhat zigzag behavior. The smoothed curve has some oscillations[24].

Model Risk and Curve Construction

It should be recognized that the construction of the forward-rate curve is not well posed in a mathematical sense. This is because there are an infinite number of forward rates $\{f^{(T)}(t_0;\Delta T)\}$ for all $\{T\}$, or at least a large number at different $\{T_j\}$, but only a relatively small set of data points for constraints. Therefore different algorithms can and do result in different curves.

Later in this book, we will thoroughly discuss model risk. The construction of the forward curve is a good example where different but perfectly sane arguments can lead to different results. The differences in turn affect pricing and hedging of derivatives differently; this is part of model risk.

Rate Units and Conventions

One of the annoying features of interest-rate products is the plethora of units or rate conventions. We need to know how to translate between these conventions. Perhaps a system uses one set of units in calculating, or a salesman wants to quote a result in a certain convention.

Let r_1 be the convention for a rate. This means there are assumed to be N_{days_1} days per year (e.g. 360, 365), a frequency f_1 of compounding (1 Annual, 2 Semiannual, 4 Quarterly, 12 Monthly), and the number n_1 of compounding periods in a year (numerically $n_1 = f_1$). For example, $r_1 = 0.05/yr$ with convention SA365 means $f_1 = 2$, $N_{days\_1} = 365$. If time differences are measured in calendar days, the convention is called "actual". The relation between the rate

---

[23] **Illustration Only:** The curve is meant to be illustrative and was produced by some software I wrote. The data are not current. The smoothing algorithm is a simple technique invented for the book. Your algorithm will no doubt be much better. See the next footnote.

[24] **Which Curve is Better: Smoothed or Unsmoothed?** Say for argument that you have this great whiz-bang smoothing technique but the rest of the world uses unsmoothed rates (or "inferior" smoothing techniques). Then you will be off the market – which is after all determined by the rest of the world - for some OTC products that are sensitive to the regions of differences of the smoothed and unsmoothed rates. Conversely, if the world smoothes the rates and you don't, you will also be off the market. Again, the smoothing algorithm, when used, is not unique. Have fun.

conventions is given by the requirement that, independent of convention, the same physical interest is produced in one year with $Dt_{1\,yr}$ calendar days, viz

$$\left(1 + \frac{r_1 \cdot Dt_{1\,yr}}{f_1 \cdot N_{days\_1}}\right)^{n_1} = \left(1 + \frac{r_2 \cdot Dt_{1\,yr}}{f_2 \cdot N_{days\_2}}\right)^{n_2} \tag{7.3}$$

The logic is as follows. First, assume $Dt_{1\,yr} = N_{days\_1}$ and take $f_1 = 1$. For an initial amount $\$N$, at the end of the year of N_{days_1} days, by definition the r_1 convention produces interest $r_1 \$N$. If the frequency $f_1 = 2$, at the end of $N_{days\_1}/2$ days the r_1 convention produces interest $r_1 \$N/2$. This interest is reinvested for the remaining half of the N_{days_1} days, resulting in compounding. Similarly, at the end of N_{days_2} days, with $f_2 = 1$ the r_2 convention produces interest $r_2 \$N$. Hence at the end of N_{days_1} the r_2 convention produces interest $\frac{N_{days\_1}}{N_{days\_2}} r_2 \cdot \$N$. Generalizing the logic gives the result. Equivalently, we have

$$r_2 = \frac{N_{days\_2}}{N_{days\_1}} f_2 \left[-1 + \left(1 + \frac{r_1}{f_1}\right)^{f_1/f_2}\right] \tag{7.4}$$

For example, if the r_2 convention is Q360 with $f_2 = 4$, $N_{days\_2} = 360$, we get $r_2 = 0.0490/yr$. This is ten basis points less than for the r_1 convention. Such an amount is definitely significant, being larger than the bid-ask spread for many swaps in the market.

See the footnote for another convention[25] called 30/360.

Compounding Rates

Compounding rates is related to rate conventions, but the emphasis is different. Instead of coming up with two representations of the same rate, we wish to generate a composite rate from two other rates. For example, suppose that we

---

[25] **30/360 Day Count:** This assumes that there are 30 days per month and 360 days per year. So the number of days between two calendar dates apart by (N_{years} N_{months} N_{days}) is calculated as $360*N_{years} + 30*N_{months} + N_{days}$. Corrections are made as follows: If the first date is on the 31st, reset it to the 30th. If the first date is Feb. 28 and it's not a leap year or if the first date is Feb. 29, reset it to the "30th of February". If the second date is on the 31st, and first date is on the 30th or 31st, reset the second date to the 30th.

Forward Curves

have two neighboring 3M forward rates. We can find an equivalent 6M forward rate that generates the same interest as applying the first 3M forward rate and reinvesting the interest in compounding with the second 3M forward rate.

Suppose that $r_a = f^{(T=t_a)}(t_0; 3M)$ is the 3M forward rate starting at time t_a and ending at $t_m = t_a + 3M$, while $r_m = f^{(T=t_m)}(t_0; 3M)$ is the 3M forward rate starting at t_m and ending at $t_b = t_m + 3M$. The equivalent 6M forward rate $r_{ab} = f^{(T=t_a)}(t_0; 6M)$ starting at t_a and ending at $t_b = t_a + 6M$ (using $N_{days} = 360$ for rates, with time differences "actual") is given by

$$\left[1 + r_{ab}\left(\frac{t_b - t_a}{360}\right)\right] = \left[1 + r_a\left(\frac{t_m - t_a}{360}\right)\right]\left[1 + r_b\left(\frac{t_b - t_m}{360}\right)\right] \quad (7.5)$$

Rate Interpolation

Often we need to interpolate rates. A curve-generating algorithm may, for example, generate 3M forward rates at IMM dates. However, we need 3M rates at arbitrary times. Usually, simple linear interpolation suffices. So, if f_{l-1} is the IMM forward rate starting at t_{l-1}, and f_l is the IMM forward rate starting at t_l, then the 3M forward rate f_a starting at an intermediate time $t_{l-1} < t_a < t_l$, and reducing to the appropriate limits at the end points, is given by

$$f_a = f_{l-1}\left(\frac{t_l - t_a}{t_l - t_{l-1}}\right) + f_l\left(\frac{t_a - t_{l-1}}{t_l - t_{l-1}}\right) \quad (7.6)$$

References

[i] **Money Market**
Stigum, M., *The Money Market*, 3rd Edition. Business One Irwin, 1990.

[ii] **Futures**
Fink, R. and Feduniak, R., *Futures Trading*. NY Institute of Finance, Prentice-Hall, 1988. P. 452.
Rothstein, N. (Editor) and Little, J., *The Handbook of Financial Futures*. McGraw-Hill, 1984.
McKinzie, J. and Schap, K., *Hedging Financial Instruments*. Probus, 1988.

[iii] Yield Calculations
Allen, S. and Kleinstein, A., *Valuing Fixed-Income Investments and Derivative Securities*. NY Institute of Finance, Simon & Schuster, 1991.

[iv] HJM Forward Rate Dynamics
Heath, D., Jarrow, R. and Morton, A., *Contingent Claim Valuation with a Random Evolution of Interest Rates*. The Review of Futures Markets 9 (1), 1990; *Bond Pricing and the Term Structure of Interest Rates: A New Methodology*. Econometrica 60, 1 (1992).

[v] Modeling the Forward - Future Correction
Dash, J., *Path Integrals and Options II: One Factor Term Structure Models*. CNRS – CPT Marseille, 1989. Eqn. 4.11.

8. Interest-Rate Swaps (Tech. Index 3/10)

Swaps: Pricing and Risk

In this chapter, we begin a detailed discussion of pricing and risk management of interest rate derivatives[1,2]. We start with interest rate swaps[i]. The simplest version of an interest-rate swap is the exchange of fixed-rate interest payments for floating-rate interest payments. For example, corporation ABC might receive the fixed rate E and pay a floating rate. Here is a picture of a plain-vanilla swap[3].

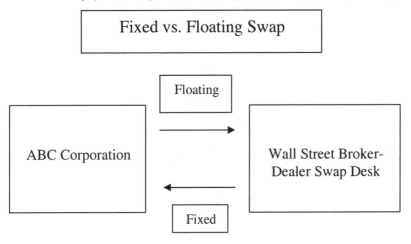

---

[1] **Acknowledgements to Traders, Brokers, Sales:** I thank brokers at Eurobrokers and traders at FCMC for breaking me into real-world aspects of interest-rate derivatives. I also thank many traders at Citibank, Smith Barney, Salomon Smith Barney, and Citigroup through the various mergers, for helpful interactions. I thank sales people for informative conversations. Much of the presentation of the practical real-world Risk Lab part of the book comes from what I learned from these people.

[2] **History:** I got direct exposure to practical fixed-income derivatives at Eurobrokers, and as the Middle-Office risk manager for FCMC. This work was based on pricing and risk-management software that I designed and wrote.

[3] **Swap Picture:** Diagrams are used to clarify deals. With complicated transactions involving various counterparties and many swap legs, diagrams are essential.

The most common floating rate is Libor. Other rates are Prime and CMT, which we discuss Ch. 10 on interest-rate caps.

Why is this risk management for ABC? ABC Corporation may have issued fixed rate debt but prefers, for its own reasons (for example asset-liability matching), to pay floating rate debt. Therefore, ABC exchanges the interest rate payments. Alternatively, in the case that ABC issued floating rate debt but prefers to pay fixed, the swap would go the other way. From the point of view of the broker-dealer BD, the swap in the picture is a "pay-fixed" swap. The BD swap desk will hedge the swap in a manner that we will consider in some detail.

Of course, ABC may not really be performing risk management. Maybe ABC thinks that rates will decrease, due to its own analysis. Now usually the forward curve is upward sloping, implying that the markets "expectation" is that rates will increase. Thus, ABC is "betting against the forward curve". Sometimes this has worked in the past for corporations, and other times it has failed.

We next show a plain-vanilla USD fixed-float swap in a spreadsheet-like format, maybe similar to a system you might encounter, along with comments[4]. We begin with the deal definition. Again, these quantities give the deal "kinematics" that would be stored by the books and records of the firm.

Deal Definition
1	ID	ABC Corp. 0001
2	Fixed Rate	6.35%
3	Floating Rate	3 months
4	Deal date	3/19/96
5	Start date	3/27/96
6	End Date	3/27/01
7	Notional	$100 MM
8	Currency	USD
9	Notional Schedule?	No
10	Floating Rate Spread	0
11	Floating rate type	USD Libor
12	Cancelable?	No
13	Payments in Arrears?	Yes

Comments: Deal Definition
1 Deal identification that will be put into the database.
2 Conventions for rate units will be specified (30/360, act/360 etc.)
3 Time between successive specifications ("fixings") of the floating rate
4 Date the deal is being done. This is an old deal.
5 First date specified in the contract for some sort of action.

---

[4] **Systems Appearance:** These range from actual spreadsheets to GUIs of whatever eclectic esthetics the systems designer chooses (naturally with feedback from the traders who have to use it). Probably the appearance would be much more attractive than what I use here for illustration. The comments would not appear; they are for the reader.

6   Last date specified in the contract. This deal has expired.
7   Normalization for calculating cash flows.
8   Payments in different currencies lead to "cross currency swaps".
9   "No" means that the normalization is the same for each date.
10 Means no additional interest paid above the floating rate.
11 Other interest rates are possible but less common.
12 If cancelable, options called "swaptions" are involved.
13 Payments are made 3 months after floating rates are determined.

Next, we give the input parameters for pricing the swap. These parameters are the forward-rate curve, where the forwards correspond to the particular floating rate on which the swap is based (3-month Libor is the most common). Normally this curve would be produced in another part of the system and read into the module used for pricing and hedging. Therefore, we have:

Parameters
1   Forward Rate Curve

Comments: Parameters
1   See previous chapter for discussion

The graph shows the forward rate curve used for the swap in this particular case along with the break-even (BE) rate of the swap.

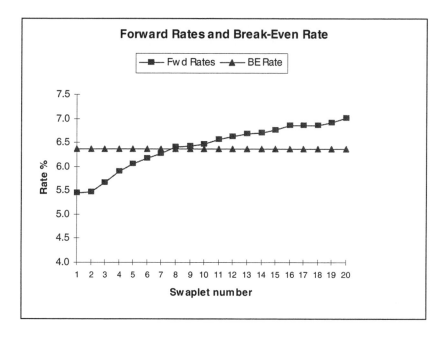

The BE rate is the average of all 20 rates with discounting included (see the section at the end for the math). The "swaplets" are the components of the swap and are discussed below.

We now give the representative pricing and overview of hedging results for this swap[5].

Pricing and Overview of Hedging
1 Swap value $000 $40.5
2 BE Swap Rate   6.359%
3 Delta total   1714.5
4 Gamma total   -87.2

Comments: Pricing and Overview of Hedging
1 This swap was practically "at the money" with very little value.
2 Equivalent fixed rate that would have led to zero swap value.
3 The initial hedge would be to buy 1714 ED futures contracts ($43K/bp).
4 Gamma is expressed as 100 * change in number of contracts/bp.

This swap is viewed from the point of view of the counterparty paying the fixed rate, at the deal date. Since the fixed rate (6.35 %) was slightly below the BE rate (6.359 %), the swap had a small positive value. A swap has a small convexity, i.e. second derivative with respect to rates producing gamma (γ), due to the discount factors being nonlinear in the rates. However, gamma cannot be hedged with Eurodollar (ED) futures since ED futures have no convexity (a change in rates by 1bp changes any ED future by $25 without any discounting correction). Therefore, gamma would need to be hedged with other instruments. Alternative reporting specifications of Δ and γ in $US are also used, multiplying by $25/contract/bp.

Looking Inside a Swap: Swaplets

The details of the swap involve the breakdown into "swaplets" as shown below.

Swaplet	Fwd Rate	Start	End	$Value $000	Delta	100 Gamma
1	5.438	3/27/96	6/27/96	$ (230.00)	101	-4.7
2	5.476	6/27/96	9/27/96	$ (217.00)	100	-4.7
3	5.667	9/27/96	12/27/96	$ (165.00)	97	-4.6
4	5.898	12/27/96	3/27/97	$ (107.00)	95	-4.5
5	6.058	3/27/97	6/27/97	$ (69.00)	95	-4.6

---

[5] **Where Did this Swap Example Come From?** This swap is the same as in my first 1996 CIFER tutorial. While rates have dropped dramatically since 1996, the principles are general and the results for current swaps will be similar. I wrote the swap pricer.

Interest-Rate Swaps

Swaplet	Fwd Rate	Start	End	$Value $000	Delta	100 Gamma
6	6.162	6/27/97	9/29/97	$ (45.00)	96	-4.7
7	6.277	9/29/97	12/29/97	$ (17.00)	91	-4.5
8	6.402	12/29/97	3/27/98	$ 11.00	86	-4.3
9	6.424	3/27/98	6/29/98	$ 17.00	91	-4.6
10	6.464	6/29/98	9/28/98	$ 25.00	86	-4.4
11	6.559	9/28/98	12/28/98	$ 44.00	85	-4.4
12	6.629	12/28/98	3/29/99	$ 58.00	83	-4.3
13	6.681	3/29/99	6/28/99	$ 68.00	81	-4.3
14	6.706	6/28/99	9/27/99	$ 72.00	80	-4.2
15	6.762	9/27/99	12/27/99	$ 82.00	78	-4.2
16	6.849	12/27/99	3/27/00	$ 98.00	77	-4.1
17	6.85	3/27/00	6/27/00	$ 97.00	76	-4.1
18	6.855	6/27/00	9/27/00	$ 96.00	75	-4.1
19	6.906	9/27/00	12/27/00	$ 103.00	72	-3.9
20	7.005	12/27/00	3/27/01	$ 118.00	70	-3.8

Swaplet Composition of a Swap

There are 20 swaplets in a 5-year swap. The swaplet hedges are individually given in equivalent numbers of ED futures contracts. These result from the sensitivities to forward rates for which a given swaplet depends. The biggest sensitivity is to the forward rate in the same line in the table as the swaplet, but there are other sensitivities due to the discount factors.

Here is a picture of one swaplet inside a swap.

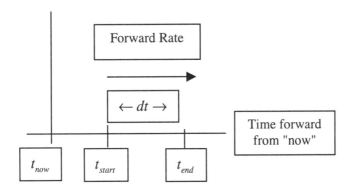

Diagram for one swaplet

Simple Scenario Analysis for a Swap

Now we consider the change in the characteristics of a swap under a hypothetical simple scenario. The scenario is that all forward rates $f^{(T)}$ regardless of maturity are raised by 10 bp keeping time fixed, i.e. $df^{(T)}_{Scenario} = 10bp$. Sometimes rates can change suddenly by this magnitude, e.g. in less than 1 day. We could also envision a time-dependent scenario, although the simplest and most common procedure is to separate out the time dependence by moving time forward while keeping rates fixed. The change in the forward rates (10 bp) is assumed equal for all forward rates—a "parallel shift". Other scenarios, with the various forward rate changes taken as unequal, give the yield-curve shape risk.

The revalued or reval ("new") results are shown below for the value of the swap and the Greeks along with the initial ("old") results, and the changes.

Scenario Reval	New	Old	Change
1. Swap value $000	$468.1	$40.5	$427.5
2. Delta Hedge (contracts)	1,705.8	1,714.5	(8.7)
3. Gamma (100 * Change in # Contracts)	-86.9	-87.2	0.3
4. Break-even Swap Rate	6.46%	6.36%	0.10%

Comments
1. Pay fixed swap makes money because the received floating rates increased
2. Fewer contracts needed to hedge the swap
3. Very little change in gamma
4. Break-even rate for the swap after rates increase also goes up 10 bp

The "Math Calc" for Price Changes Using the Greeks

We now compare the approximate results using the Greeks to the exact results obtained through scenario revals. The delta contribution gives almost all the change in the value of a swap:

$$\$dS_{Delta} = \Delta_{contracts} \cdot \frac{\$25}{bp \cdot contract} \cdot df^{(10bp)}_{Scenario} \qquad (8.1)$$

Inserting the old $\Delta = 1714.5$ contracts gives $\$dS_{Delta} = \$428.6K$. The convexity due to gamma is negative because the discount factors decrease with increasing rates. We have

$$\$dS_{Gamma} = \gamma_{dcontracts/bp} \cdot \frac{\$25}{bp \cdot contract} \cdot \frac{1}{2} \left[df^{(10bp)}_{Scenario} \right]^2 \qquad (8.2)$$

Inserting $\gamma = -0.87$ change in contracts per bp gives the small result $\$dS_{Gamma} \approx -\$1.1K$. The sum gives the "Math Calc" change $\$dS_{MathCalc}$,

$$\$dS_{MathCalc} \equiv \$dS_{Delta} + \$dS_{Gamma} \tag{8.3}$$

The result for $\$dS_{MathCalc}$ is equal to the scenario revaluation difference to this accuracy, $\$dS_{Scenario} = \$427.5K$. Actually, this result is "too accurate" because a real-world hedge would just involve an even multiple of 100, i.e. 1700 contracts.

Now we need to check the change in delta $d\Delta_{contracts}$ due to gamma. We should have to a good approximation

$$d\Delta_{contracts} = \gamma_{dcontracts/bp} \cdot df_{Scenario}^{(10bp)} \tag{8.4}$$

Plugging in gamma as $\gamma = -0.87$ gives the change in the number of contracts as $d\Delta_{contracts} = -8.7$, in agreement with the reval result.

Interest Rate Swaps: Pricing and Risk Details

In this section, we follow up the introduction to swaps with some more detail. Consider a broker-dealer swap desk BD that transacts a new swap with a customer ABC. Typically, the new swap risk to BD will be hedged immediately. This new deal also may be considered in the context of a portfolio of deals already held by BD. The actual hedge put on by BD for the new deal may depend on what happens to be in its book, as well as the risk appetite and view (if any) of BD regarding the market, perhaps leading to a specific strategy. Regardless, no hedging strategy will be exact, and residual risks will remain under any circumstance. These residual risks will then go into the portfolio risk and change the overall risk of the desk[6].

In order to be concrete, we consider the hedging of an isolated swap in USD for fixed payments vs. floating payments[7]. Swaps involving more than one currency also exist[ii, 8].

---

[6] **Limits for Risk:** The desk risk may be controlled by limits set by an independent risk management group. This involves intensive negotiations and periodic reviews. See Ch. 40.

[7] **Basis Swaps:** There are also swaps with floating payments in one rate vs. floating payments in another rate. These are called "basis swaps". The hedging then depends on the correlations between the two floating rates.

Hedges

The following instruments are available as possible hedges:

- Short-term money-market cash instruments
- Eurodollar (ED) futures
- FRAs
- Other swaps
- Treasuries
- Treasury futures

The swap can be divided up into constituent swaplets, as we have seen. These swaplets have individually different cash flow specifications with definite rules for determining the reset values of the rate on which the swap is based, the cash flow dates[9], etc. according to the contract[iii] between BD and ABC[10]. The following complications theoretically enter into the hedging of the swap.

Hedging with Cash Instruments

The swap may have an uncertain cash flow before the first Eurodollar future IMM date[11]. This leads to "spot risk", and can be hedged with the short-term

---

[8] **Cross Currency Swaps:** If payments are made in more than one currency, we have a "cross currency swap". Extra complications involving FX hedging then occur. One distinction with single-currency swaps is that the notionals for a cross-currency swap (in the different currencies) are exchanged at the beginning and then reversed at the end of the cross-currency swap. If there is only one currency, notionals are not exchanged, only interest payments. Cross-currency swaps can get very complicated as cash flows are converted into several currencies successively in the same deal. See Beidleman (Ref.)

[9] **Payments in Arrears:** The various swap cash-flow dates are generally "in arrears", which means that a cash payment is made some time after the rate determining that cash payment is made (or "reset" or "fixed"). An extra discount factor for the extra time period from the reset time to the cash-flow time is included. If a payment is not "in arrears", the payment is said to be "up front".

[10] **The All-Important "Term Sheet" and Why You Need It:** The legal contract for the deal will be preceded (before the deal is transacted) by a "term sheet" containing the details of the proposed transaction. These term sheets may or may not be generic, and they can change up to the last minute depending on the transaction dynamics between the desk BD and the customer ABC. In early discussions, the term sheet may only be schematic. The quant should always get a term sheet – hopefully the latest one, even if indicative or provisional, and even if the salesman argues that he doesn't have it yet – before wasting time modeling, pricing, and hedging using wrong assumptions regarding the nature of the deal. Having said this, a fair amount of back-and-forth quantitative analysis often occurs for a given deal if there are non-generic aspects.

[11] **Initial Cash flow in the Swap:** New swaps can also have an initial cash flow that is set in the contract. A swap already in the books may have a cash flow that has been reset but not yet paid. There is a risk due to the uncertainty of the discounting from the analysis

money market cash instruments. Sometimes this is not done, and the spot risk will be approximately hedged with the nearest ED future[12].

Hedging with ED Futures

The cash flows originating in the forward time period where the ED futures are relatively liquid can be hedged with ED futures[13]. In general, the cash flows for the swap will occur at various times that will not coincide with the ED future IMM dates[14]. The following picture should clarify things.

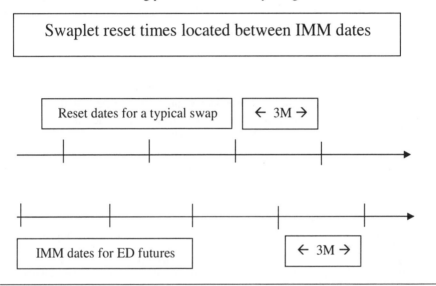

date until the payment date for the entire cash flow. However, if the swap is sold, the cash to be paid from the reset to the transaction date is sometimes set on an "accrual" basis not including discounting.

[12] **The End-Year Effect:** There is an "end-year" effect that happens at the end of the year which leads to anomalies in the money market and which needs to be hedged separately. For the year 2000, this effect was very large.

[13] **Hedging with ED Futures:** The liquidity of the ED futures is greatest for times within the front few years, and the implementation of ED future transaction is more difficult further out in time. In addition, futures are generally transacted in groups of 100 (anything smaller is called an "odd lot" and is inconvenient to transact). Often an ED hedge will be transacted first quickly using the front few contracts for the whole swap risk and then some time later the hedge will be distributed among hedge instruments with other transactions. Various residual risks corresponding to the inexactness of the hedging for each of the points raised above will occur.

[14] **IMM Swaps:** Some swaps, generally short term swaps, do have fixing dates at the ED future IMM dates (these are imaginatively called IMM swaps). IMM swaps are very quickly transacted and are popular with short-term swap desks.

An interpolation of the risks in time then needs to be performed. In general, this is just done proportionately to the time intervals from a given swaplet cash flow to the IMM dates before and after that cash flow.

For a mneumonic, see the footnote[15].

Hedging with FRAs

An FRA[iv] (forward rate agreement) is the same as a one-period swap, i.e. a swaplet[16]. FRAs have the advantage that the periods over which they exist are measured in months from today, not at IMM dates. For this reason the dates can be tailored to match those of a new swap.

Hedging with Other Swaps

If the ABC swap happens to be a swap depending on other rates[17] (e.g. CP, Prime, Muni, CMT etc.) then a Libor swap can be transacted in the opposite direction. The residual risks here involve "basis risk" due to the difference in the behavior of the rate from Libor. Date or time risk also occurs since the cash flow dates of the two swaps will generally not be identical. There is also normalization risk since the ABC swap notional or principal[18] may not match the hedge swap

---

[15] **ED Futures and Swaps Hedging Mnemonic:** Put your hands together with interlocking fingers. Your right hand intersects at the swap-reset times and your left hand intersects at the IMM dates. You need to interpolate the risks of each of the swap-reset dates of each right-hand finger with the neighboring two futures of the neighboring fingers on your left hand.

[16] **FRAs:** FRA = Forward Rate Agreement. Either say the letters "F.R.A." or say the acronym "fra" pronounced to rhyme with "bra". FRAs are short-term money market instruments (see Stigum, Ref.). The floating side of an FRA is a forward rate starting at $\tau = p$ months and ending at $T = q$ months. FRAs are represented as pxq (e.g. 1x4, 3x6, 6x12). For 3x6, say "Threes Sixes". FRAs are cash settled proportional to the difference of the floating rate and the contract rate. There is a consistency between the FRA market and the futures market.

[17] **Other Rates:** Examples include CP = Commercial Paper, Prime = Prime rate set by banks, CMT = Constant Maturity Treasury, Muni = A municipal index rate. These "basis" rates have their own swap rates often expressed as a "spread" or difference with respect to Libor swap rates (except for Muni rates which use ratios with respect to Libor rates depending on tax considerations). Some of these swaps (e.g. CP) involve complicated rules regarding averaging the rates over various date periods.

[18] **Notional and Principal:** "Notional" and "principal" mean the same thing. The notional is a normalization factor that multiplies the overall expression for the swaplets comprising the swap. Sometimes the swap market is characterized by the total notional in all transactions.

Interest-Rate Swaps

principal. See the ISDA[19] reference[v] for data on notionals[20]. If the ABC swap is an amortizing swap, (the principal is different for the different swaplet cash flows), further complications exist[21].

Hedging with Treasuries

U.S. treasuries (which are very liquid) can also be used as hedges. Again there is basis risk since treasury rates are generally not the same as the rate on which the swap is based, date risk corresponding to mismatched cash flows, normalization risk, etc. An exception is for CMT swaps, which have a natural hedge in treasuries. The market for "repo" becomes involved [22,23].

---

[19] **ISDA:** This is the International Swaps and Derivatives Association, Inc. ISDA describes itself as the global trade association representing leading participants in the privately negotiated [fixed income] derivatives industry. ISDA was chartered in 1985 and today has over 550 members around the world. ISDA Master Agreements are always used for contracts for plain-vanilla deals (see the Documentation Euromoney Book Ref). You can find out more at www.isda.org.

[20] **The Total Swap Notional and Why You Don't Care:** A recent figure for the total outstanding notional for interest rate swaps, interest rate options and currency swaps is over $50T (Trillion) according to *ISDA News* (Ref.). "Outstanding" means all deals already done that have not expired or closed. This initially somewhat scary number has very little to do with the real risk in swaps which is many orders of magnitude less. For example we saw above that a $100MM notional swap has a risk of around $40K per bp change of rates to the BD if the swap is unhedged. The BD will generally hedge this risk down to a very small fraction of $40K. Finally, swaps on the other side (receive fixed vs. pay fixed) will behave in the opposite fashion. Forget about the $50T.

[21] **Notional or Principal Schedules for Amortizing Swaps:** Amortizing swaps are often specified by the customer ABC according to a "schedule" of the different notionals for the different cash flows, corresponding to specific ABC needs. For example, the first cash flow could have $100MM notional and the second cash flow $90MM notional, etc. Naturally, ABC will have to pay an extra amount to BD for such custom treatment, but part of this will be eaten up by the residual risk forced on the BD.

[22] **Repo:** The repo market is an art unto itself. The simplest version is overnight repo, where a dealer sells securities to an investor who has a temporary surplus of cash, agreeing to buy them back the next day. This amounts to the BD paying an interest rate (repo) to finance the securities. The arrangement can be made such that the BD keeps the coupon accrual. The complexities of repo (including "specials") enter the hedging considerations. We give an illustrative example of a repo deal in Ch. 16.

[23] **Acknowledgement:** I thank Ed Watson for discussions on repo and many other topics.

Hedging with Treasury Futures

Treasury futures involve complications, including the "delivery option" for the "cheapest to deliver"[24]. The inclusion of these complications into risk systems is standard.

Example of Swap Hedging

We already considered the basic idea in a previous section. Again consider the above 5-year swap, non-amortizing with $100MM notional, based on 3-month Libor with payments in arrears, where the BD pays a fixed rate to the customer ABC. The risk to BD from the ABC swap is that rates decrease. This is because then BD would then receive less money from ABC determined by the decreased floating Libor rate that ABC pays to BD. Therefore the pay-fixed swap of the BD must be hedged with instruments that increase in value as rates decrease. These include for example buying bonds or buying ED contracts or both[25,26].

As we saw, the total delta or DV01[27] from a swap model for this swap is short 1714 equivalent ED contracts[28]. This means that if ED futures were to constitute

---

[24] **Bond Futures and the CTD:** A bond future (as opposed to ED futures) requires delivery of a bond to the holder of the bond future from a party that is short the bond future. However, this does not mean delivery of a definite bond, but rather of any bond chosen at liberty from a set of bonds specified for that particular bond future. One of these possible deliverable bonds (the "cheapest to deliver or CTD") is economically the best choice for the holder of the short position, who gets to choose. The CTD bond (with today's rates) thus determines the characteristics of the bond future today. However, since the future behavior of rates is uncertain, another different bond may wind up being the cheapest to deliver when it actually becomes time to deliver a bond. This uncertainty shows up as a correction to the bond future's price and its sensitivity to interest rates. Other complexities exist for bond futures.

[25] **"Short and Long the Market" for Swaps:** The pay-fixed swap makes the BD "short the market" and the pay-fixed swap loses money as rates decrease. In order to hedge the pay-fixed swap, the BD must go "long the market", buying instruments that make money as rates decrease. Generally, going long the market is in the same direction as going long (buying) bonds, whose prices increase as rates drop.

[26] **More Swap Jargon:** Say that the bid swap spread is 40 bp/yr and the offer 44 bp/yr. This means that a potential fixed rate payer is ready to pay 40 bp/yr and a potential fixed rate receiver wants to receive 44 bp/yr (above the corresponding treasury rate). A "bid side swap" for a BD means that the BD pays the fixed rate. Since the BD in that case receives floating rate payments, which increase when the swap spread increases for fixed treasury rate, the BD who pays fixed is said to be "long the spread" and "long the swap".

[27] **DV01 Warning:** Careful. Some desks use the convention that DV01 or delta corresponds to a rate move up, and some desks use the convention that DV01 corresponds to a rate move down. This complicates aggregation of risk between desks. The story gets worse. See the following footnote.

[28] **More on Different DV01 Conventions and Risk Aggregation Problems:** Other conventions for expressing delta or DV01 exist. One convention uses the notional for a

the total hedge for the swap, the BD would have to buy ("go long") 1714 ED futures. For a one basis point (1/100 of 1%) decrease in Libor, one ED future increases in value by $25, so the risk to BD for the ABC swap is a loss of 1714*$25 or around $43,000 for each bp of overall (parallel) rate increase[29]. There is also a small correction from second-order derivative "gamma" terms[30].

The "Delta Ladder", or Bucketing: Breaking Up the Hedge

The "ladder" or set of "buckets" is a breakup of the total hedge in maturity steps. For example, we can use 3-month "buckets". The units are equivalent ED future contracts. An approximate hedge for this front part of the swap could consist of buying 100 of the first three "strips"[31]. That is, the approximate hedge for the first

---

given bond or swap (e.g. 10 year). That is, the risk is expressed in terms of the number of 10-year treasuries it would take to hedge the overall risk. Sometimes the risk is expressed in terms of zero-coupon rate movements. Sometimes the risk is expressed in terms of a mathematical rate present in a model which is used to generate the dynamics, but which actually has no physical meaning. The difference between the reported DV01 using these different conventions can easily be on the order of a few percent. Finally, sometimes the magnitude of the rate move will be different between desks (e.g. 1 bp, 10 bp, 50 bp) and thus may include some convexity. This is often done as a compromise involving numerical stability issues of the models dealing with options that are also in the portfolio. These conventions will sometimes but not always be marked on the desk risk reports. All this can generate confusion when aggregating risk. Similar anarchy unfortunately exists for other risks as well. Sometimes this whole problem is ignored, or goes unrecognized.

[29] **The Movements of Rates is Roughly Parallel:** This risk is expressed in the simplest case for all forward rates determining the swap increasing together by one bp. While different forward rates by no means move by the same amount in the real world, nonetheless in practice, the approximation of this "parallel shift" covers much of the risk and is the measure commonly used for a first-order approximation. The extent to which the forward rates do not move in parallel will present risks insofar as the individual swaplet deltas are not individually hedged. One way to look at rate movements is using principal components, which we treat in Ch. 45 and in Ch 48 (Appendix B).

[30] **Gamma for a Swap:** A Libor swap has small gamma (second derivative) risk, so most of the risk is just due to the first derivative or delta (or DV01). The gamma risk is expressed by specifying the change in the number of equivalent contracts per bp increase in rates. For this swap, this number is about -0.9 for the entire swap. As we shall see below, the details of bucketing gamma in forward time is complicated since gamma is really a matrix of mixed second partial derivatives corresponding to the different forward rates comprising the swap. Thus, there is really no exact gamma ladder. However, an effective approximate gamma ladder will be constructed.

[31] **More Handy ED Futures Jargon:** A "strip" is a set of four neighboring ED futures contracts corresponding to one year. These strips are given names. The "Front Four" are first contracts (ED1 – ED4). The "Reds" are the next four (ED5 – ED8), then the "Greens", the "Blues", the "Golds", etc. Moreover, the months corresponding to expiration are abbreviated (e.g. December is called DEC pronounced "Deece"). The names are convenient shorthand for the traders.

3 years could consist of buying 100 contracts of each of the first 12 contracts ED1—ED12. The "spot" risk up to the first contract would need to be hedged separately, as would the tail risk from 3—5 years of 561 equivalent contracts (see discussion below). Residual risks would be put into the portfolio.

A graph of the first 3 years of the delta ladder hedge needed for the swap is shown below.

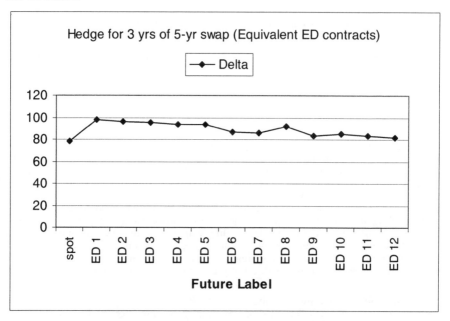

Hedging Long-Dated Swaps

The hedging of a long-dated swap is achieved by first choosing hedging instruments and then minimizing the risk. The risk can be broken up into buckets, but in general, it will not be possible to make the risk zero in every bucket unless the buckets are suitably chosen large enough.

Back Chaining Hedging Procedure: Example

Consider, for example, an amortizing 15-year Libor swap. Aside from back-to-back offsetting of this swap with a similar 15-year swap, there is no single natural hedge. A general strategy is to employ a mixture of the hedging instruments above. For illustrative purposes, we use five, ten, and thirty-year treasuries along with the first 2 years of ED futures. The buckets are thus 0-2 years, 2-5 years, 5-10 years, and 10-30 years.

The logic is shown in the next figure.

Interest-Rate Swaps

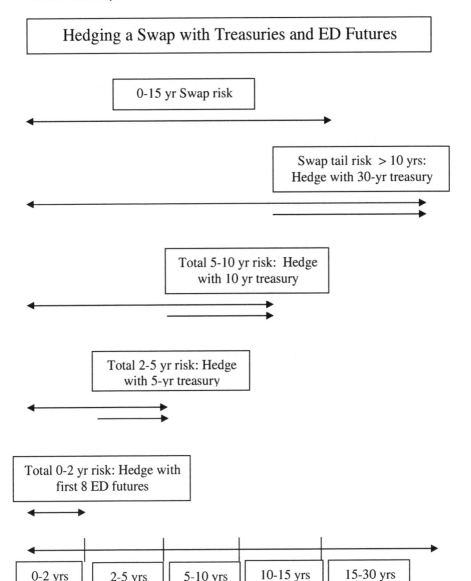

The hedge components that hedge the deal in the various buckets can be found using a computer minimization routine. The result amounts to a back-chaining algorithm. In practice, a simpler hedge could be executed (e.g. a rough hedge only with 10-year treasuries) and later adjusted to a complicated hedge.

Algorithm

The algorithm indicated in the drawing above is as follows:

- The 10-15 year "tail" risk of the 15-year swap $\Delta_{\text{Swap}}^{(10-15\,yr)}$ is hedged by an appropriate number of 30-year treasuries. This is done by choosing the 10-30 year treasury risk[32] $\Delta_{30\text{-yr Tsy}}^{(10-30\,yr)} = -\Delta_{\text{Swap}}^{(10-15\,yr)}$. Thus, the tail risk of the swap past 10 years is hedged. The detailed risk is not hedged however; the risk report will show the total [swap + 30-year treasury hedge] having canceling risks between 10-15 years and 15-30 years.

- The combined 5-10 year risk $\Delta_{\text{Swap}+30\text{-yr Tsy}}^{(5-10\,yr)}$ of the swap and 30-year treasuries is hedged by 10-year treasuries with $\Delta_{10\text{-yr Tsy}}^{(5-10\,yr)} = -\Delta_{\text{Swap}+30\text{-yr Tsy}}^{(5-10\,yr)}$.

- The combined 2-5 year risk of the swap and the 10, 30-year treasury hedges $\Delta_{\text{Swap}+10,\,30\text{-yr Tsy}}^{(2-5\,yr)}$ is hedged by 5-year treasuries, $\Delta_{5\text{-yr Tsy}}^{(2-5\,yr)} = -\Delta_{\text{Swap}+10,\,30\text{-yr Tsy}}^{(2-5\,yr)}$.

- The combined 0-2 year risk of the ABC swap and all treasury hedges $\Delta_{\text{Swap}+5,10,\,30\text{-yr Tsy}}^{(0-2\,yr)}$ is hedged with ED futures $\Delta_{\text{ED Futures}}^{(0-2\,yr)} = -\Delta_{\text{Swap}+5,10,\,30\text{-yr Tsy}}^{(0-2\,yr)}$.

Changes in the Hedges with Time

The total DV01 or delta risk of the ABC swap plus the treasuries plus the ED futures is zero at the time the above back chaining algorithm is implemented (modulo odd lots or fractional residual effects). However all this was valid only at one time. As time progresses, changes to the hedges need to be made. These include events like these:

- A payment is made or received for the swap. At this time, a corresponding hedge must be removed because the swap risk has changed.

- An IMM date is past. Before this happens the trader will "roll over" the front contract which is about to expire with the second contract about to become the front contract.

Swap Pricing Math and Risk

Here is a summary of the math for an interest rate swap. The value $\$S$ of the swap "today" is given by the sum of swaplet values $\$S_l$,

---

[32] **How to Break up a Treasury to Implement the Hedging Back-chaining Algorithm:** The bucketed risk of a treasury bond is determined in the same way as the bucketed risk of the fixed leg of a swap. The forward rates determine the discount factors that in turn determine the bond price. The discount factors are simply differentiated with respect to the forward rates. The amounts of the various treasuries in the hedges are simply obtained by determining the notional values corresponding to the integrated risks in the intervals needed and specified in the back-chaining procedure.

Interest-Rate Swaps

$$\$S = \sum_{l=0}^{N_{swap}} \$S_l = \sum_{l=0}^{N_{swap}} \$N_l \cdot \left(\hat{f}^{(T_l)} - E \right) \cdot dt_{swap} \cdot P^{(T_l)}_{Arrears} \qquad (8.5)$$

Here $\hat{f}^{(T_l)}$ is the forward rate of the type specified by the contract[33], for the given maturity (or fixing or reset) date. In this section we use a caret ^ to emphasize that the forward rates are determined by today's data. The fixed rate E may be defined to include a spread. The time interval dt_{swap} is specified by the contract (e.g. 3 months) as the time between successive T_l. The notionals or principals $\$N_l$ are usually constant, but as noted above may not be constant depending on the client's requirements. If, as is common, the notionals $\$N_l$ decrease with time, the swap is called an amortizing swap.

Each discount factor $P^{(T_l)}_{Arrears}$ is the zero-coupon bond that discounts back from the time that the cash flow occurs to today. Normally this cash flow is "in arrears", a time dt_{swap} later than the maturity or reset date T_l. Note that $P^{(T_l)}_{Arrears}$ depends on all the forward Libor rates $\left\{ \hat{f}^{(T_j)} \right\}$ up to the date T_l.

The sum runs over all terms specified in the contract from the 0th term (which usually involves accrued interest) to the last term N_{swap}. For a single-currency swap, the exchange of the notional is not done.

The Swap Break-Even Rate

The BE rate R_{BE} is given by setting the value of the swap to zero, and is the average of the forward rates with discounting included, viz

$$R_{BE} = \frac{\sum_{l=0}^{N_{swap}} \$N_l \cdot \hat{f}^{(T_l)} \cdot dt_{swap} \cdot P^{(T_l)}_{Arrears}}{\sum_{l=0}^{N_{swap}} \$N_l \cdot dt_{swap} \cdot P^{(T_l)}_{Arrears}} \qquad (8.6)$$

---

[33] **More on Swap Specification Details:** This includes rate specification, including the type (Libor etc.), the day-count convention (money market, 30/360 etc.), holidays when payments cannot be made, and a number of other items.

Given the BE rate, the value of the swap is determined. We define the quantity $\$X = \sum_{l=0}^{N_{swap}} \$N_l \cdot dt_{swap} \cdot P_{Arrears}^{(T_l)}$, and we write

$$\$S = (R_{BE} - E) \cdot \$X \qquad (8.7)$$

In this form, it is clear that the value of the swap is (up to normalization) given by the difference between the break-even rate and the fixed rate.

Swap Risk, the Delta Ladder, and the Gamma Matrix

In this section, we consider the swap risk in detail. In particular, we will consider different definitions of the delta ladder. We also provide a correct treatment of gamma using the gamma matrix and we will see why a simple gamma ladder (usually defined) is not viable. Our conclusions are more general than just for swaps; in particular, they apply to swaptions as well.

Swap Risk

In order to break down the swap risk, we need to consider the individual risk for each of the forward rates. The delta ladder is the collection $\{\Delta_l\}$ whose exact definition will be considered below. Gamma is the derivative of delta. This means that gamma is really a matrix with elements $\gamma_{ll'}$ since Δ_l can be differentiated with the forward rate with $l' \neq l$.

The change $dS_{MathCalc}$ in the value of the swap using the delta ladder and the gamma matrix including the normalization $\$N_{ED} = \$25/(bp \cdot contract)$ is the usual Taylor series result to second order,

$$\$dS_{MathCalc} = \$N_{ED} \left[\sum_{l=0}^{N_{swap}} \Delta_l \cdot \delta f_l + \frac{1}{2} \sum_{l,l'=0}^{N_{swap}} \gamma_{ll'} \cdot \delta f_l \delta f_{l'} \right] \qquad (8.8)$$

Here, δf_l is the assumed change in $\hat{f}^{(T_l)}$. Note for parallel shifts, i.e. a constant value $df_{Scenario}$ for all δf_l, we get the expression we had previously, namely

$$\$dS_{MathCalc} = \$N_{ED} \left[\Delta \cdot df_{Scenario} + \frac{1}{2} \gamma \cdot (df_{Scenario})^2 \right] \qquad (8.9)$$

Interest-Rate Swaps

Here, the total delta is $\Delta = \sum_{l=0}^{N_{swap}} \Delta_l$ and the total gamma is $\gamma = \sum_{l,l'=0}^{N_{swap}} \gamma_{ll'}$.

The Delta Ladder

We now consider the delta ladder $\{\Delta_l\}$. We will see that there are several possible definitions. First, the variation of the discount factors, while complicated, contributes as a rule of thumb only around 10% to delta. So to get an approximate but reasonable estimate of delta, you can ignore the discount factors. In this approximation, delta Δ_l (in ED contracts) for the lth forward rate is obtained by two steps: (1): Differentiate the swaplet $\$S_l$ with respect to the appropriate forward rate $\hat{f}^{(T_l)}$ for that term, and (2): Divide the result by the normalization $\$N_{ED} = \$25/(bp \cdot contract)$.

There are two ways that the variation of the discount factors $\{P^{(T_l)}_{Arrears}\}$ can be included. We note first that the discount factor can be written in terms of the forward rates as

$$P^{(T_l)}_{Arrears} = \prod_{l'=0}^{l} \frac{1}{\left(1 + \hat{f}_{l'} \cdot dt_{l'}\right)} \tag{8.10}$$

The two methods of including the discount-factor variation are then:

Method #1. The delta Δ_l is assumed to contain the total variation of the whole swap price $\$S$ (i.e. all swaplets) with respect to the lth forward rate. Noting that the swaplet index has to be at least l in order to have a dependence on $\hat{f}^{(T_l)}$ we get

$$\Delta_l^{(Method\ \#1)} = \frac{1}{\$N_{ED}} \frac{\partial \$S}{\partial \hat{f}^{(T_l)}} = \frac{1}{\$N_{ED}} \sum_{l'=l}^{N_{swap}} \frac{\partial \$S_{l'}}{\partial \hat{f}^{(T_l)}} \tag{8.11}$$

Method #2. The delta Δ_l is assumed to contain the total variation of the individual *swaplet* price $\$S_l$ with respect to *all* forward rates. The forward rate index only goes up to l for nonzero sensitivity of $\$S_l$. We get

$$\Delta_l^{(Method\,\#2)} = \frac{1}{\$N_{ED}} \sum_{l'=0}^{l} \frac{\partial \$S_l}{\partial \hat{f}^{(T_{l'})}} \qquad (8.12)$$

The first method will produce Δ_l that can easily be interpolated using futures contracts from the IMM dates, since the total swap dependence on a given rate is specified. However, the delta for the lth swaplet is then not given by Δ_l. In the second method, the delta for the lth swaplet is given by Δ_l. It can be useful for risk management to look at both of these methods.

A numerical IMM delta ladder can be produced by varying each forward rate $\hat{f}_m^{(IMM)}$ at the corresponding IMM date $t_m^{(IMM)}$, then reconstructing the forward rate curve and recalculating the swap[34].

A further complication is that for Libor other than 3-month (e.g. 6-month Libor or 1-month Libor), an appropriate change of variables has to be made to compare to the 3-month IMM ladder.

Non-Parallel Shifts, the Gamma Matrix, and the Sick Gamma Ladder

Unfortunately, many systems are incapable of dealing with gamma matrices, and incorrectly calculate only the diagonal elements γ_{ll}. A simple gamma ladder $\{\gamma_l\}$ uses only the diagonal elements, $\gamma_l = \gamma_{ll}$. This diagonal gamma ladder is however inaccurate, as we shall now see[35].

Let us (following these systems) incorrectly identify the total gamma with the diagonal elements only, γ_{Diag_Only}, i.e.

$$\gamma_{Diag\_Only} = \sum_{l=0}^{N_{swap}} \gamma_{ll} \qquad (8.13)$$

This procedure clearly only works if the off-diagonal gamma matrix elements are small compared to the diagonal terms. This can produce misleading results with non-parallel yield-curve shifts. How can we see this? Imagine a forward rate

---

[34] **So Which Delta Ladder does *Your* Report Show?** In the text, we have exhibited several possibilities. Maybe the guy who programmed the risk system has just left the firm to become a junior trader. Did he document the formula that he programmed? Hmm?

[35] **Sick Gamma Ladders and Risk Reports:** Good traders know that the diagonal gamma ladder is not useful for math risk calculations. However, there are risk reports that obstinately show diagonal gamma ladders anyway. It is not clear how many people understand the problems with gamma ladders. In order to cure the diagonal gamma ladder, off-diagonal gamma matrix corrections have to be included (see the next section).

Interest-Rate Swaps

shift scenario that goes up by 10 bp in the first bucket and then down by 10 bp in the second bucket. To a good approximation, the BE rate is unchanged and therefore the swap value is unchanged. However, a math calculation done with Δ_1, γ_{11} for the first bucket and Δ_2, γ_{22} for the second bucket leads to a cancellation of the delta terms and an addition of the gamma terms, implying the incorrect conclusion that the swap value should change.

Generally, in order to perform risk management with changes in yield-curve shape, some systems can typically only perform reval computations without the possibility of analytic calculations.

An Approximate Factorization Method for Incorporating the Gamma Matrix in the Gamma Ladder

Here is a trick to get the off-diagonal gamma matrix elements that enables reasonably accurate approximate results for non-parallel yield-curve moves to be calculated analytically[36]. It is important to emphasize that this approximation *works on a deal-by-deal basis only*. It is given by

$$|\gamma_{ll'}| \approx \sqrt{|\gamma_{ll}\gamma_{l'l'}|} \qquad (8.14)$$

So, for a given swap, the off diagonal gamma matrix elements are approximately given by the square root of the geometric average of the two diagonal matrix elements. In order to motivate this, note that $\gamma_{ll'}$ involves the variation of the discount factors. The relatively simple form of the discount factors then leads after some algebra to the above approximate relation. We also need to insert the sign. We assume that the gamma matrix has elements all of the same sign. For a swap, this sign is equal to the sign of the notional.

Using this approximation, we find, now for arbitrary parallel or non-parallel shifts in the forward rate curve, for a given deal,

$$\$dS_{MathCalc} = \$N_{ED} \left\{ \sum_{l=0}^{N_{swap}} \Delta_l \cdot \delta f_l + \frac{1}{2} \left[\sum_{l=0}^{N_{swap}} \sqrt{|\gamma_{ll}|} \cdot \delta f_l \right]^2 \cdot \text{sgn}(\$N) \right\} \qquad (8.15)$$

Now notice that, as in the above scenario, if one rate is moved up and its neighbor rate is moved down, the gamma term (now correctly) is approximately zero since the rate changes roughly cancel out in the sum, as desired.

In this way, it is possible to define a modified set of diagonal gamma matrix elements that incorporate the effects of the off-diagonal terms. In that way, an

---

[36] **History:** This approximation for the gamma matrix was discovered by me in 1991. It also turns out to work reasonably well for swaptions, even American swaptions.

effective gamma ladder $\left\{\gamma_l^{eff}\right\}$ including the effects of the off-diagonal terms can be resurrected. The numerical results for gamma ladders in this book have been corrected to include off-diagonal gamma matrix effects.

A Non-Amortizing Swap is an FRN and a Non-Callable Bond

If $\hat{f}^{(T_l)}$ is Libor and the notionals are constant, the swap can be rewritten. Formally adding and subtracting a fictitious notional payment at the end of the swap, and breaking apart the sums, we get the result for a pay-fixed swap written in terms of an FRN and a non-callable bond:

$$\text{Swap}_{\text{Non-Amortizing, Pay-Fixed}} = \text{FRN} - \text{Bond} \qquad (8.16)$$

Here FRN is a floating rate note containing the floating-rate part of the swap along with a notional payment at the end. It is easy to see that, at reset points, the FRN is at par. Between two reset points, the FRN is only slightly different from par (see below). Therefore, in this case, the risk of the swap can be obtained from the risk of a non-callable bond[37] with a coupon equal to the fixed rate of the swap and discounted with Libor. For a bond with a coupon including a credit spread with respect to Libor, the equation could be turned around to read Bond = FRN - $\text{Swap}_{\text{Non-Amortizing, Pay-Fixed}}$, provided the fixed rate in the swap includes the credit spread.

FRNs are at or Close to Par

We close this section by considering FRNs briefly. At the first reset date t_0 the definition of the FRN is (leaving off the notional and with rates in decimal),

$$FRN^{[N]}(t_0) = \sum_{l=0}^{N} \hat{f}_l dt_l \cdot P_{Arrears}^{(T_l)} + 1 \cdot P_{Arrears}^{(T_N)} \qquad (8.17)$$

Here, the finite time interval between payments is called $dt_l = t_{l+1} - t_l$. The discount factor written in terms of the forward rates in money-market convention is

---

[37] **Noncallable Bonds:** These bonds, also called "bullet bonds " are guaranteed, in the absence of issuer default, to pay coupons to maturity. Callable bonds contain embedded options giving the issuer the right, at any one of a set of defined times, to pay a certain amount to the bondholder and cancel the rest of the bond.

$$P_{Arrears}^{(T_l)} = \prod_{l'=0}^{l} \frac{1}{\left(1 + \hat{f}_{l'}.dt_{l'}\right)} \quad (8.18)$$

It is simple then to see the recursion $FRN^{[N]}(t_0) = FRN^{[N-1]}(t_0)$. On the other hand, at $N = 0$ (one cash flow $\hat{f}_0 dt_0$ paid at t_1) we immediately get $FRN^{[0]}(t_0) = 1$. Hence, at reset, the FRN is at par.

Note that if the value date is before the first reset, there is an extra discount factor. Similarly, if the value date is between the first reset and the second reset at t_1 the FRN is not at par, but will be close to par.

The reader should carefully note that the rewriting of a swap as an FRN and a bond is *not* valid for amortizing swaps, where the swaplets have different notionals. Further, in a later chapter we will examine index-amortizing swaps. These are amortizing swaps with complicated knock-out features. Misleading results can be obtained using an incorrect FRN - Bond association of such more complicated instruments.

Counterparty Credit Risk and Swaps

A counterparty of a deal is the just other party to the deal, for example the other side of a swap. We will look at the risk from the point of view of a broker-dealer BD. The counterparty will be called ABC. The counterparty risk for BD is the risk that the counterparty ABC defaults on some condition of the deal.

Counterparty risk can be calculated using multi-step Monte-Carlo (MC) simulations of underlying variables, along with models for the securities at future times [vi]. The MC simulations move forward in time.

Potential counterparty default events are built into the simulation. For a given MC path p_α, the potential default events cause losses at future time t for any security "in the money" (ITM) to BD at time t, i.e., for which the counterparty ABC owes money to BD. The simulator retains all ITM cash flows along each path p_α. Other cash flows are ignored. The potential losses are then tabulated for different MC paths at different future times, and various statistics (average loss, loss at some CL, etc.) are calculated. For example, we can calculate the counterparty risk at a 99% CL at time t by picking out the 100th worst potential loss in a simulation of 10,000 paths at that time.

For illustrative purposes, counterparty risk can be obtained rather easily for scenarios, e.g. at the 99% CL scenario envelope of interest rates in the future generated by a model of interest-rate diffusion.

We cannot enter into all the complexities, and merely give an example.

Illustrative Counterparty Credit Risk Example for a Swap

Consider a new deal, a pay-fixed swap from the point of view of the broker-dealer BD with counterparty ABC, using a scenario. The notional is $1 MM, the maturity is four years with semi-annual payments, starting today at par. The scenario assumed is at a 99% CL for the break-even rate with a lognormal volatility of 0.2. The rate scenario starts at $r_0 = 5\%$ and increases to around 13% after four years, viz $r(4yrs) = 5\% * \exp(2.33 * 0.2 * \sqrt{4}) = 12.7\%$. The swap value starts increasing with time, since BD receives the floating rates assumed to increase with time under the scenario. Ultimately at maturity, the value goes to zero because the swap disappears. Hence, there is a maximum point for the forward swap value on this rate path, occurring at around 1.5 years.

The positive swap value represents counterparty risk to BD. This is because if the counterparty ABC defaults on the terms of the swap, BD will lose the positive value of the remaining swap payments owed to BD (relative to the smaller value of the fixed-rate payments owed to ABC by BD).

Here is a picture with semiannual forward swap values, discounted back to today. The picture would look more complicated (a saw-tooth behavior) if the risk were plotted at intermediate times. As an approximation, we use the time-averaged swap value for the given scenario. This is shown by the constant line.

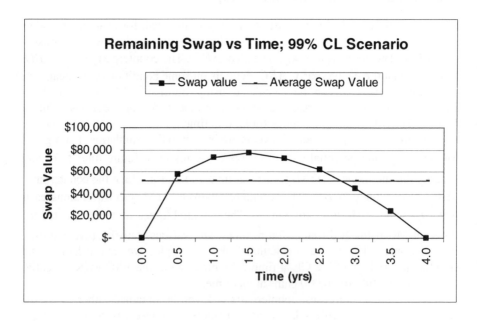

Counterparty Credit Risk for General Portfolios

To deal with firm-wide counterparty risk, a big MC simulator is run for the possibly thousands of deals in the various portfolios for different business units. The calculations generally have to be done over a long time, corresponding to the possible times that the counterparty can default in the future, until the ends of the deals. Appropriate netting logic has to be incorporated for clearly offsetting positions. Such counterparty risk calculations can be a huge enterprise, both for computation and for the collection of the data. On a firm-wide basis, because securities depend on many types of variables, the simulator has to include multivariate statistics for the generation of the paths.

Models for the securities are often very computationally intensive for such simulations. Approximate models can sometimes be used (e.g. a swap for a deep-in-the-money Bermuda swaption) corresponding to a large market move that, coupled with counterparty default, leads to a large loss.

References

[i] **Swaps**
Beidleman, C. R., *Interest Rate Swaps*. Business One Irwin.

[ii] **Cross-Currency Swaps**
Beidleman, C. (Editor), *Cross Currency Swaps*. Richard D. Irwin, Inc. 1992.

[iii] **Swaps Contracts, Documentation**
Gooch, A. and Klein, L., *Documentation for Derivatives*. Euromoney Books, 1993.

[iv] **FRAs**
Stigum, M., *The Money Market*. Business One Irwin, 1990.

[v] **Total Swap Notional**
International Swaps and Derivatives Association, *ISDA News, Market Survey*. p.3. December, 2001.

[vi] **Counterparty Credit Risk**
Picoult, E., *Measuring Counterparty Credit Exposure, Credit Risk and Economic Capital on a Portfolio Basis*. Talk, ISDA, 1/7/98.
Counterparty Risk Management Policy Group, *Improving Counterparty Risk Management Practices*. The Bond Market Association, 1999.

9. Bonds: An Overview (Tech. Index 2/10)

Bonds are debt instruments of many different sorts issued to raise money. Issuers of bonds include corporations, governments, and agencies. In this book, an issuer is labeled as ABC and an investor as X. Bonds are obligations of ABC to pay back to X the borrowed money (called the "notional" or "par" amount) at the maturity date of the bond and in some cases earlier. In addition, the bonds have coupons (so called because the investor used to clip off "coupons" as pieces of paper to get paid).

The world of bonds is an extremely complicated zoo; few people are expert in more than one sector[1]. There are many good finance books and reference articles on each market, to which we refer the interested reader[i]. The outline of the rest of this chapter is:

- Types of Bonds
- Bond Issuance
- Bond Trading
- Bond Math

Types of Bonds

Bonds are classified in several ways. A short quick list follows. In the next section, we discuss some aspects of specific issuers.

Fixed-Rate Bonds

The most common type of coupon is the same for each payment in time, defining a fixed-rate bond. The amount of the coupon is determined by the credit rating of the issuer[2], along with conditions in the market at the time of issuance. Although

---

[1] **Bond Land:** There is no way that the world of bonds can be described in detail while keeping this book portable. The reader is referred to any bookstore with a finance section where you can spend hundreds of dollars to learn the details.

[2] **Credit Ratings:** Credit rating agencies (e.g. S&P and Moody's) rate the credit of issuers through a highly specialized procedure. Naturally, there are different credit notations. Investment grade credits are defined as BBB or higher (S&P), and Baa or higher (Moody's). Non-investment grade (also called high-yield or junk) is defined as BB or lower (S&P), and Ba or lower (Moody's). The credit ratings are regularly reviewed

the coupon does not change, the price of the bond can change with the market through the yield, defined below.

Floating-Rate Bonds

Some coupons change with time as determined by a rule, involving a changing rate or floating rate index. There are many such indices. The most common is Libor. Typically, a floating-rate bond will be issued at Libor plus a spread (e.g. Libor + 1%, or Libor + 100bp). Other floating-rate instruments are short-term money market notes based on, for example commercial paper (CP) [ii]. Floating rate bonds stay near par, as discussed in the previous chapter.

Zero-Coupon Bonds

If no coupon exists, the bond unimaginatively is called a zero-coupon (ZC or 0C) bond. We call $P_{ZC}^{(T)}$ the price of a ZC bond with maturity date T. Clearly, since he gets no coupon payments, the investor X will pay much less than the notional for one of these bonds (X buys the bond at a "discount"). Sometimes issuers do not want to pay coupons for cash-flow reasons. It is possible to decompose or "strip" a set of coupon bonds into zero-coupon bonds. There are government bond traders that look at and try to cash in on the (small) mispricings using this technique.

Callable Bonds and Puttable Bonds

Many bonds are "callable". This means that at certain dates the issuer ABC can, for prices, which are known up-front, demand that investors X sell back the bond. A few bonds are "puttable". This means that at certain dates, investors X can make the issuer ABC buy back bonds for pre-determined prices. Generally, puttable bonds are also callable. The call and put features mean that the bonds contain embedded options. The quantitative analysis of such bonds is complicated; we shall examine some of the features in this book. Some derivative products, notably swaptions, are connected with these bonds.

Deterministic Coupon Changes: Step-Up Bonds

Some bonds have coupons that change according to a rule. If the coupons increase with time, the bond is called a "step-up". Step-ups appeal to bearish investors who believe rates will increase and want increasing coupons that keep pace without having to sell existing bonds and buy new ones. Usually step-ups

---

and changed if deemed necessary by the rating agencies. Traders have their own view on credit that may or may not agree with the rating agencies, especially on a short-term basis when a rating seems suspicious. Agency credit ratings generally change over a longer time frame, except when an issuer is clearly in difficulty.

are callable, so the investor loses price appreciation in bull markets when rates drop and the bonds are called[iii].

Bonds Depending on Other Markets

Some bonds have coupons that depend on other markets besides interest rates. For example, equity-linked bonds pay more if a given stock or index increases. Some structured notes pay a lower coupon in return for a potential equity gain. Other possibilities include bonds whose coupon depends on a commodity (e.g. gold), or on the value of an FX exchange rate (e.g. Japanese Yen vs. USD), etc.

Convertible Bonds

Convertible bonds ("converts") combine both interest-rate coupons and potential equity[iv, 3]. Converts can be exchanged for (or "converted into") equity. Converts have a wide array of complex side conditions. Lower-credit issuers often use convertibles. Simple versions called PERCS, DECS, etc. also exist. We will look at some possibilities in Ch. 14.

Mortgage-Backed Securities

Mortgage-backed securities (MBS) are the repackaging of homeowner mortgages of various sorts into securities. MBS have non-deterministic coupons. This is because the coupons progressively disappear as homeowners increasingly prepay their mortgages, thus removing the underlying collateral. The determination of expected prepayments along with their uncertainties is extremely complicated and involves many variables[4]. Complex mortgage products called CMOs

---

[3] **Convertible Bonds:** Convertible bonds have a lower coupon than ordinary bonds of the same credit. To compensate the investor receiving this low coupon, convertibles contain the implicit option to be converted to shares of stock under certain conditions. A simple but potentially misleading picture is to think of a convertible as an ordinary bond plus a stock option. This is because the conversion option value requires analysis on each possible stock price path at each time in the future.

[4] **Mortgage Prepayment Modeling:** The subject of prepayment modeling is best thought of to first approximation as a complicated mix of phenomenological wizardry, accompanied by fitting large amounts of data. The goal of a prepayment model is a parameterization of the historical behavior of people regarding the financing of their most valuable asset (their homes), and human behavior is not easy to model. Moreover, when pricing mortgage products, the parameters of a prepayment model are often changed from historical values to "implied values", thus producing "implied prepayments". These implied parameters are chosen such that the pricing model fits the market prices for the mortgage products.

(collateralized mortgage obligations) generally have many "tranches" with different characteristics and complicated payment logic[5].

This book does not cover MBS. There are many excellent references, to which we refer the reader [v].

Asset-Backed Securities

Asset-Backed Securities[vi] (ABS) are the repackaging of anticipated cash flows from different sorts of assets into securities. These include auto loans, credit cards, home equity loans, equipment leases, etc. Theoretically, any potential set of cash flows with enough certainty could be repackaged.

Munis

Municipal or "muni" bonds are issued in a large variety of types, short and long term, with different funding goals by local and state entities, depending on their capital needs. The interest is generally exempt from federal tax and may be exempt from state and local tax[6].

Non USD Bonds

Bonds can be issued in different currencies besides the US dollar (USD), for example British pound GBP, Euro EUR, Japanese yen JPY, etc. The choice of the currency depends on the markets and the appetites of investors in various countries to buy the bonds[vii].

Guaranteed bonds; Bradys

Some bonds have guarantees (e.g. Brady bonds[viii] for emerging-market countries that have partial guarantees on some of their coupons). Some mortgage and muni bonds are also guaranteed. The investor pays an extra premium to cover the guarantee.

---

[5] **CMO Logic:** There are entire systems to analyze the cash-flow rules for the different tranches in CMOs. This must be done painstakingly from the contracts. Prepayments affect different tranches in very different ways. Rules include the relative amounts of interest or principal and the ordering of the cash-flow payments into the various tranches.

[6] **Muni Bond Land:** This is a complex zoo. Muni bonds include General Obligation (GO) Bonds, Revenue Bonds (housing, utility, health care transportation, industrial), Municipal Notes (TANs, RANs, GANs), etc. Some bonds called private activity bonds are not tax exempt but are subject to the alternative minimum tax AMT.

Loans

Banks make loans to clients. Bonds are generally riskier than loans. This is due to the generally longer maturity of bonds relative to the loans that the banks are willing to make, and due to the often-lower credit rating of the bond issuers relative to those corporations to which banks are willing to lend. Loans can also be repackaged into securities.

Bond Issuance

We now give a quick overall idea of issuance. A broker-dealer BD will participate in facilitating the issuance of various types of fixed income securities. Here are the data for U.S. bond issuance for the first half of 2002 along with a little commentary to give some flavor[7]. The specifics naturally change with time.

Bond Issuance Data (First 6 months of 2002)

Type of Issuance	Amount Issued ($B)
U.S. Treasury	$250 B
Federal Agencies	$450 B
Municipal	$200 B
Corporate	$400 B
Asset-Backed	$225 B
Mortgage-related	$1,000 B
Commercial paper	$1,325 B

Treasury Issuance

U.S. Government Treasury issuance depends on tax receipts and thus on the strength of the economy, the debt ceiling, government spending, and projected budget deficits.

Agency Issuance

The bulk of the agency issuance[8] is from the FHLB, Freddie Mac, and Fannie Mae. Note that this issuance is debt—i.e. bonds funding the agencies.

---

[7] **Data Source:** *Research Quarterly*, The Bond Market Association report, August 2002. Issuance data are rounded off to the nearest $25B (billion) and trading data to the nearest $10B. Data are for the first half of 2002. Some of the issuance commentary in the text is from this source.

[8] **Federal Agencies Issuing Debt:** FHLB is the Federal Home Loan Bank, Freddie Mac is FHLMC (Federal Home Loan Mortgage Corporation), and Fannie Mae is FNMA

Muni Issuance

Muni issuance in 2002 was at record levels following recent turbulence in the equity markets along with recent low rate levels. The refunding of bonds (from higher to lower coupons at current low interest rate levels) was up along with new issuances.

Corporate Issuance

Corporate issuance is sector dependent (e.g. telecommunications, manufacturing) and highly dependent on credit. Most corporate issuance is investment grade. High-yield issuance is an order of magnitude less than investment grade issuance. Convertible bond issuance is a small fraction of the total, and it has been decreasing. Overall, issuance in 2002 was down partly because some corporations were hurt by scandals.

Asset-Backed Issuance

ABS issuance depends on investor demand, which in turn depends on relative yields and perceived risk versus other bond markets.

Mortgage Issuance

Most mortgage issuance is through the agencies (FHLMC and FNMA, with GNMA somewhat less); there is also a small non-agency or "private label" component. Issuance of these mortgage "products" depends on the collateral of homeowner mortgages. The increasing prepayment of old mortgages and concurrent increased refinancing by homeowners in the current low rate environment led to more MBS/CMO issuance in 2002.

CP Issuance

CP issuance was negatively affected by concerns over issuer credit quality.

Bond Trading

The secondary market is the market for trading bonds after they are issued. A broker-dealer will have trading desks in the various bond markets, with different traders specializing in different narrow sectors. Each market is highly specialized.
To give an idea, here are the trading volume data[7] for the first half of 2002.

(Federal National Mortgage Association). Other smaller agencies issue smaller amounts of bonds. These include Sallie Mae (dealing with student loans), the Farm Credit System FCS and the Tennessee Valley Authority TVA. Ginnie Mae or GNMA (Government National Mortgage Association) issues mortgage-backed securities, but not debt.

Secondary Trading Volume Data (First half of 2002)

Market	Daily trading volume ($B)
U.S. Treasury	$350 B/day
Federal Agencies	$80 B/day
Municipal	$10 B/day
Corporate	$20 B/day
Mortgage-related	$140 B/day

Trading, Flight to Quality, Diversification, Convergence, and Asteroids

A flight to quality occurs when investors take refuge in treasuries from risky assets in stressful markets. People generally try to escape risk by *diversification*, buying instruments in different sectors or different markets. There are also *convergence plays or trades* in which you "just have" to make money when two different instruments eventually "must" become identical.

A stressed market, if severe enough, can bring on collective panic[9]. We might pictorially call the onset of a particularly crazy stressed market as due to an *"asteroid"*. Diversification strategies can be roiled in an asteroid-panic environment. This is because, no matter what the instruments, investors want to shed risk. Selling pressure increases dramatically as more and more people give up strategies and pursue the flight to quality. All instruments except the most risk-free (generally treasuries) drop in value, and diversification fails.

Traders can be hit badly in stressed markets along with investors, losing buckets of money. To a first rough (but not misleading) approximation, traders cleverly buy relatively cheap but potentially risky products and sell treasuries for hedging. When the asteroid hits they lose twice—once because their risky assets (that cannot find buyers) drop in value, and again because their short positions in treasuries drop in value. Convergence plays can also get demolished before the theoretical convergence can take place. Before being disbanded by management, the sophisticated Salomon Arb Group was a victim of such phenomena. Other victims of the same 1998 disaster included a variety of hedge funds, most notably LTCM [ix].

---

[9] **Phase Transitions and Collective Panic:** Theoretically, the flight to quality can be thought of as a *phase transition*, where we go from a highly disordered state (buyers and sellers of various securities in comparable numbers) to a highly ordered state (basically only sellers with buyers waiting). One possible framework is the critical 2nd-order phase transition in the Reggeon Field Theory, discussed in Ch. 46. Other possibilities can form a rich area for research.

Bond Math

In this section, we give some basics on "bond math". The reader should be warned that the practical details are messy.

Discount Factors

Discount factors are essential to understand because they describe, once a cash flow is determined in the future, how much the cash flow is worth today. The set of zero-coupon bonds $\left\{P_{ZC}^{(T)}\right\}$ are the set of discount factors. The zero-coupon bond $P_{ZC}^{(T)}$ equals today's value of a future cash flow of amount $1 to be paid by issuer ABC at a time T in the future. Conversely, $1/P_{ZC}^{(T)}$ tells you how much $1 invested today will be worth at time T if invested in an interest-bearing account with rate typical of the ABC coupon. Theoretically, $\left\{P_{ZC}^{(T)}\right\}$ arise from stripping coupon bonds of different maturities. Since in general there are only a discrete (and sometimes small) number of bonds for a given issuer, aggregation is used to get discount factors for a given credit rating in a given sector. Interpolation schemes must be adopted in order to calculate the discount factors for arbitrary maturities between the known ZC bond maturity dates.

Yields

Bond prices can be recast equivalently as "yields". The yield y is a common rate to be used in all discount factors for all coupons in order to produce the given price of the bond[10]. A quick mnemonic is "yield up, price down". What this means is that if rates go up and the coupons of newly-issued bonds increase, investors will pay less for an already-issued bond because these pay a lower coupon.

Take a coupon bond paying f coupons per year, with the jth coupon to be paid at date T_j, a time interval τ_j years from today. There are "day-count

---

[10] **Other Yield Conventions:** There is also a "nominal yield". This is the annual coupon and a "current yield", which is the annual coupon divided by the price in decimal. If the bond is callable, the coupons up to the first call date are employed to give the "yield to call" YTC. The "yield to maturity" YTM is the definition in the text. The "yield to worst" YTW is the minimum of YTM and YTC. For a premium bond called at par, YTW = YTC and for a discount bond YTW = YTM. For munis, by regulation the YTW must be the yield quoted to clients.

Bonds: An Overview

conventions" for what constitutes a year[11]. The yield has the attribute of the frequency f compounding, and the discount factor for that cash flow is $P_{ZC}^{(T_j)} = [1+y/f]^{-f\tau_j}$. Note as $f \to \infty$ we get continuous compounding, i.e. $P_{ZC}^{(T_j)} \to \exp(-y\tau_j)$. The price $B_{ABC}^{(T)}$ of the ABC bond of maturity T, a time interval τ from today, with annual coupon c_{ABC}, is the sum of all the discounted cash flows, i.e.

$$B_{ABC}^{(T)}(y) = \sum_{j=0}^{N} \frac{c_{ABC}}{f}[1+y/f]^{-f\tau_j} + 100[1+y/f]^{-f\tau} \quad (9.1)$$

The $j = 0$ accrued interest term is omitted for the clean price, and is present for the "dirty price" with the discount factor omitted[12]. The $j = 1$ term has the first full coupon. The last N^{th} coupon is paid at maturity. In real calculations, there are a myriad of details[13].

Duration and Convexity

The duration D and convexity C of a bond with price B are defined as:

$$D = -\frac{1}{B}\frac{\partial B}{\partial y}, \quad C = \frac{1}{B}\frac{\partial^2 B}{\partial y^2} \quad (9.2)$$

The duration, defined with the minus sign, is positive. Note that the duration, up to a factor, is the weighted average time of payments[14]. To second order, we get the relation

---

[11] **30/360 Day Count:** This assumes that there are 30 days per month and 360 days per year, as described in the preceding chapter.

[12] **Accrued Interest:** Accrued interest is calculated from and including the last interest payment date, up to but not including the settlement date of the trade. Settlement (regular way) is t + 3 days for corporates and t + 1 day for US governments, where t is the trade date. Accrued interest is paid to the seller of the bond.

[13] **Bond Conventions:** These can get very messy and depend on the contract details of the bonds. In particular, bonds have a variety of conventions determining the actual interest payments. To get a complete description of the complexities, consult The Bloomberg (hit the GOVT button and then type DES and HELP). You will get around 45 pages listing 600 conventions for calculating interest for government bonds in different countries. This is cool.

[14] **Other Durations:** Macaulay duration is defined to include a factor (1+y/f).

$$\delta B = \left[-D \delta y + \frac{1}{2} C (\delta y)^2 \right] \cdot B \qquad (9.3)$$

The DV01 is the change in the bond price δB for a one bp/yr change in yield, viz $\delta y = 10^{-4}/yr$. This includes all changes in the bond price, including any changes due to embedded options in the bond if they exist. The DV01 is generally defined numerically, since call features in bonds and other complexities cannot be described analytically. The conventions for the actual number of basis points moved vary. If the move δy is too small, the numerical algorithms can become unstable and/or produce unphysical numerical jumps[15]. If the move is too big, large second order convexity effects enter. For example we can move $\delta y = 50 bp/yr$. Then the change δB is scaled back down by δy to get the DV01.

Spreads

Universally, the world of bonds is described by their spreads. Differences between a bond's yield and the yield at the same maturity for a given benchmark (e.g. government or Libor) define that bond's spread. Typically, spreads are quoted for a given credit and a given sector (e.g. BBB US Industrials).

The spread contains all information about the price, given the benchmark. Therefore, all the effects determining the bond price enter in the spread. These include the perception of risk due to potential default and credit downgrades, technical supply/demand factors, market psychology, and generally all the information used by bond traders[16].

---

[15] **Numerical Instabilities and DV01:** Numerical code involves discretization and if the yield change is too small, the code gets two nearly equal prices. The price error may be relatively small, but the difference between the nearly equal prices is small enough to be very sensitive to these price errors, sometimes unfortunately resulting in instabilities in DV01. If this happens, increasing the shift δy can help. All this is hard to explain to some people.

[16] **Spreads and Implied Probabilities of Default:** Some analysts assume that bond spreads are entirely due to the possibility of default. Actually, the logic is turned around to get "implied probabilities of default" $p_{imp\_default}$ from the spreads. This is done using bond spreads in the discount factors along with logic that eliminates bonds that happen to default in the future based on the probability $p_{imp\_default}$. Then $p_{imp\_default}$ is varied until the bond price is obtained. However, historical statistics give actual default probabilities that are quite different from the theoretical probabilities $p_{imp\_default}$.

Option-Adjusted Spreads (OAS)

The option-adjusted spread (OAS) for bonds with embedded options is defined as a calculated spread added to the benchmark curve, such that the bond model price is the same as the market price. The benchmark curve is the curve off which spreads are defined, e.g. US Treasury. The model needs to use whatever logic is needed to include the embedded options. The model numerical algorithm can be a formula, a numerical calculation using a discretized lattice, a Monte-Carlo simulation, etc. Given the benchmark curve, the OAS is therefore a translation of the bond price, including the effect of the options.

Callable bonds, mortgage products, etc. are commonly quoted using OAS. Duration and convexity are often defined such that the OAS is held constant.

General, Specific, and Idiosyncratic Risks

Risk can be classified in successive degrees of refinement. If an average risk is taken over many bonds, such as a large bond index[17], the risk is called "general". As refinements are made, narrowing the focus to an average over bonds in a sector, the risk becomes more "specific". If the details of a specific issuer are specified, the risk is highly specific. The difference between specific and general risks is also a form of "idiosyncratic risk". The problem in getting idiosyncratic risk essentially is that the market prices are not known for all bonds. Hence, various approximations have to be made.

Matrix Pricing and Factor Models

There are various approximate pricing methods. One is called "matrix pricing" where known prices of some bonds in a given sector are marked on a matrix of coupon vs. maturity, and interpolation is used to get other prices.

"Factor models" assign components of bond spreads to various issuer credit and sector characteristics etc., and thereby arrive at an approximate theoretical price for a bond.

References

[i] **Bonds (General)**
Sharpe, W. and Alexander, G., *Investments*, 4th Ed. Prentice Hall, 1990. See Ch.12-14.
Fabozzi, F. and Zarb, F., *Handbook of Financial Markets*, 2nd Ed.. Dow Jones-Irwin 1986.
Fabozzi, F. , Fabozzi, T. D., *Bond Markets, Analysis and Strategies*. Prentice Hall 1989.
Fabozzi, F., *Bond and Mortgage Markets.*, Probus Publishing, 1989.

[17] **Bond Indices:** A standard general bond index used as a benchmark is the Lehmann Brothers Aggregate Bond Index. Sector bond indices for corporates, munis, etc. exist.

Allen, S. L. and Kleinstein, A. D., *Valuing Fixed-Income Investments and Derivative Securities.* New York Institute of Finance, 1991.
Series 7, *General Securities NYSE/NASD Registered Representative Study Manual,* Securities Training Corp., 1999.

ii Money Market
Stigum, M., *The Money Market.* Business One Irwin, 1990.

iii Step-Up Bonds
Derivatives Week Editors, *Leaning Curves Vol II.* Institutional Investors, 1995. See p. 78.

iv Convertibles
Zubulake, L., *Convertible Securities Worldwide.* John Wiley & Sons, Inc. 1991.

v Mortgage-Backed Securities
Bartlett, W., *Mortgage-Backed Securities.* NY Institute of Finance, 1989.
Davidson, A., Herskovitz, M., *Mortgage-Backed Securities.* Probus Publishing Co., 1994.
The Mortgage-Backed Securities Workbook. Mc-Graw Hill, 1996.
Davidson, A., Ho, T. and Lim, Y., *Collateralized Mortgage Obligations.* Probus Publishing Co., 1994.
Fabozzi, F. (Ed), *The Handbook of Mortgage-Backed Securities.* Probus Publishing Co.
Hayre, L. (Ed), *Guide to mortgage-backed and asset-backed securities.* Salomon Smith Barney, John Wiley & Sons, 2001.

vi Asset Backed Securities
Pavel, C. Securitization: *The Analysis and Development of the Loan-Based/Asset-Backed Securities Markets,* Probus Publishing, 1989.

vii Non USD Bonds
Bowe, M., *Eurobonds.* Dow Jones-Irwin 1988.
Viner, A., *Inside Japanese Financial Markets.* Dow Jones-Irwin 1988.

viii Brady Bonds
Govett, H., *Brady Bonds – Past, Present and Future.* www.bradynet.com/e24.htm, 1996.
Molano, W. T., *From Bad Debts to Healthy Securities? The Theory and Financial Techniques of the Brady Plan.* www.bradynet.com/n025.htm, 1996.
Graicap FI Research, *Introduction to Brady Bonds.* www.bradynet.com/e52.htm, 1997.

ix Flight to Quality and LTCM
Lowenstein, R., *When Genius Failed.* Random House, 2000.
Bloomberg, L. P., *Meriwether Apologizes for LTCM Collapse, Cites Flaws,* 8/21/00; *Long-Term Capital CEO Meriwether's Letter to Investors,* 9/2/98; *Long-Term Capital's Problems Loom over Markets,* 10/5/98; *Long-Term Capital Management's Investments,* 10/9/98.
New York Times, *At Long-Term Capital, a Victory of Markets over Minds,* 10/11/98.

10. Interest-Rate Caps (Tech. Index 4/10)

Introduction to Caps

In this chapter, we discuss interest-rate caps. Caps provide insurance against rising interest rates by paying off if rates go up enough. The picture below gives the idea for one piece of a cap, called a caplet.

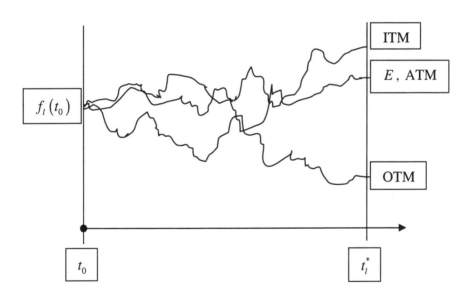

Diffusion of forward rate f_l starts with value $f_l(t_0)$ at time t_0, and ends at maturity t_l^*. The strike of the corresponding caplet is E. Paths in/at/out of the money are notated as ITM/ATM/OTM, respectively.

We first discuss standard Libor caps, and then go on to discuss Prime caps and CMT caps. Other rates (e.g. CP, Muni) are also used[i]. We present the standard market model that uses lognormal forward rates. The new ingredient for options is, of course, volatility.

The picture above illustrates a particular Libor forward rate $f_l(t)$ diffusing from the present time t_0 to its maturity t_l^* with some volatility σ_l. The set of volatilities $\{\sigma_l\}$ is called a "term structure of volatilities" for the forward rates.

While many possible interest-rate processes have been suggested, volatilities as traded in the market are quoted in lognormal-rate language. This means that $\sigma_l^2 dt = \left\langle \left\{ [f_l(t+dt) - f_l(t)]/f_l(t) \right\}^2 \right\rangle_c$, averaged over stochastic variability, in time dt. In addition, σ_l could depend on t. Volatilities for other processes have different definitions and different units. Sometimes model volatilities are used for hidden variables before a transformation is made into physical rates[1].

At the Money Options and the Trader on a Stick

In the above figure, we show the strike E of a hypothetical call option on the forward rate f_l along with three paths, ending in the money (ITM), out of the money (OTM), and at the money (ATM).

We might irreverently call the ATM path the "frozen trader-on-a-stick path"[2].

As we have seen, a swap is a collection of swaplets. In the same way, a cap is a collection or a basket of caplets. The lth caplet is priced using its own volatility σ_l. This volatility is placed in the lognormal process of the forward rate corresponding to the caplet option expiration date t_l^*. From the lognormal

---

[1] **Different Models, Trading, Reporting:** Sometimes one model will be used for trading and another (usually simpler) model for reporting. This leads to inconsistencies in the way the desk views its risk (using the trading model) and the way the risk is reported to corporate risk management (using the reporting model). Sociologically, the desk does not care much about the reporting model, while risk management is often concerned about consistency. Discussions backed up by authority may be needed for resolution.

[2] **Frozen-Trader-on-a-Stick Near Expiration ATM:** This path winds up very near at the strike at expiration. Near expiration, the hypothetical trader sits frozen, since he does not know whether to lift the hedges as needed for OTM where $\Delta = 0$ or keep the hedges as needed for ITM where $\Delta = 1$. The quant would say that near expiration and at the money, gamma (the change of delta for small changes in rates) becomes very large. The trader would not be amused.

assumption for $f_l(t)$, it follows that the Black model with the caplet volatility is applicable[3]. The cap value is the sum of caplet values, $\$C_{cap} = \sum_l \$C_l$.

The Black Caplet Formula

The classic Black formula for the lth caplet value evaluated at time t_0 is[4]

$$\$C_l\left[\hat{f}^{(t_l^*)}, t_0\right] = \$\mathcal{N}_l\left[\hat{f}^{(t_l^*)} N(d_{l+}) - EN(d_{l-})\right] \cdot dt_{cap} \cdot P_{Arrears}^{(t_l^*)} \qquad (10.1)$$

Here $\hat{f}^{(t_l^*)} = f_l(t_0)$, where we use carets ^ indicating that the forward rates are determined today. The time interval to the caplet maturity is τ_l. $\$\mathcal{N}_l$ is the lth notional, in cursive to avoid confusion with the normal functions $N(d_{l\pm})$. Also $P_{Arrears}^{(t_l^*)}$ is the zero coupon bond price for discounting in arrears (payment made after reset), dt_{cap} is the time interval between resets (1/4 year for 3 month Libor), E is the rate strike (which can also depend on the caplet), and

$$d_{l\pm} = \left[\ln\left(\hat{f}^{(t_l^*)}/E\right) \pm \frac{1}{2}\sigma_l^2 \tau_l\right] / \left[\sigma_l \sqrt{\tau_l}\right] \qquad (10.2)$$

Floors and Floorlets
Interest-rate floors are collections of floorlets. Floorlets are put options on rates and have corresponding formulae within the context of the Black model.

---

[3] **Black Formula for Caplets:** This is the Street-standard model. To motivate it, recall that forward rates and futures are linearly related, up to a convexity correction. Ignoring margin accounts, there is no cost for futures. Futures are not assets since they cost nothing to buy. Therefore, the no-arbitrage drift of changes in forward rates is close to zero, producing the Black formula. Equivalently, we can use the Black-Scholes formula with a "dividend" yield exactly canceling a "risk-free rate".

[4] **Homework:** We spare the reader the details of the mathematics at this stage of the book. However, the reader may want to derive the caplet and floorlet Black formulae.

Cap and Caplet Implied Volatilities

A *cap* implied volatility (or vol for short), $\sigma_{cap}^{(impl)}$, is a *single* volatility to be used in each Black formula for every caplet in that particular cap. The cap implied vol $\sigma_{cap}^{(impl)}$ just serves as a proxy for that cap price. Caps of different maturities will have different cap implied vols.

Brokers quote implied cap vols using the Black model. The quotes are generally a composite of the deals done in the market that day. Not all maturities in general will have traded. This is especially true for longer maturities, and for caps on rates other than Libor, which are less liquid. The cap vols will also generally be quoted for at or near the money options, since these are the most liquid. Model pricing and extrapolation techniques are used for options away from the money.

Cap and caplet implied vols are not the same. *Caplet* implied vols are the market-determined values $\left\{\sigma_l^{(impl)}\right\}$ of the volatilities $\left\{\sigma_l\right\}$ for the individual forward rates. The same caplet implied vol is used for each cap containing that caplet.

A procedure to obtain a term structure of volatility for the caplets with different volatilities $\left\{\sigma_l^{(impl)}\right\}$ is employed, varying the caplet volatilities until the market cap prices are obtained. In order to do this a recursive procedure can be used starting with the shortest maturity cap to get the volatility of the shortest maturity forward rates, and then subsequently progressing outward in maturity to longer maturities. Here is a graph of an illustrative caplet vol term structure for a 5-year cap on 3M Libor, taken from my CIFEr tutorial:

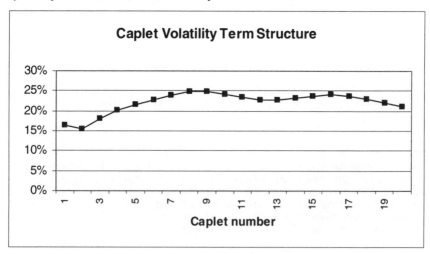

Analytic Formula for Cap Implied Vol
A useful analytic approximation for the implied volatility of a cap involves averaging, with the weights being the caplet values[5]. It reads

$$\sigma_{cap}^{(impl)} \approx \frac{\sum_l \sigma_l^{(impl)} \cdot \$C_l}{\$C_{cap}} \qquad (10.3)$$

Non-USD Caps

In different currencies, the implied volatilities for a given maturity will be different. Insurance costs against rising interest rates in, say, Germany are not the same as insurance costs against rising US rates. This follows since the macroeconomic situations determining the long-term behaviors as well as the technical factors determining the short-term behaviors are in general different for Germany and the US[6].

Relations between Caps, Floors, and Swaps

Put-Call Parity for Libor
A put-call parity relation says that the price of a cap minus the price of a floor with the same "kinematics" (resets, notionals, etc.) is the price of a plain-vanilla pay-fixed swap[7]. This is the analog of put-call parity for equity or FX options. In this case it reads

---

[5] **Implied Cap Vol Formula:** This approximation works reasonably well even with different notionals for the caplets. The physical motivation is that the cap vol is controlled by the dominant caplet vols, but only if the caplet values (that depend on the rates and strike) are significant. Hence, both caplet vols and caplet values enter. The formula is only applicable for one deal at a time, and in particular does *not* hold for portfolios of caps including long and short positions.

[6] **Non-USD Rate Caps:** Conventions differ. For example in the US, new caps generally start with the first reset in the future (e.g. in 3 months to be paid at 6 months). However, in some currencies the convention is that new caps start with the first reset determined now (with the rate known), and with this known cash flow paid (e.g.) at 3 months.

[7] **More Homework:** You'll remember the put-call relation better if you derive it yourself than if you just read about it. Hint: $N(x) + N(-x) = 1$.

$$C_{\text{Cap}} - C_{\text{Floor}} = S_{\text{Pay-fixed Swap}} \qquad (10.4)$$

Now a Libor swap has no volatility dependence. Hence, the implied volatilities of caps and floors should be identical, provided the deal kinematics are the same. Illiquidity and technical supply-demand factors enter into the real world, breaking the theoretical dictum. So in practice, floor implied vols are not exactly the same as cap implied vols.

Limiting Relations

As interest rates rise significantly, a cap becomes deep ITM, while the floor with the same kinematics becomes deep OTM and worthless. Therefore, a cap approaches a pay-fixed swap as interest rates rise[8].

Hedging Delta and Gamma for Libor Caps

We have already looked at Libor interest rate swaps. The interest rate risk measured by Δ and γ for interest rate caps is treated in the same way as for swaps. The same comments regarding the Δ ladder and the gamma matrix, along with the hedging instruments, hold here.

The delta hedging of caps is done with a combination of futures, swaps and bonds (treasuries). Futures can in general only be transacted cheaply in batches of 100 and are mostly liquid only for short maturities—especially in non-USD currencies. Swap transactions can be costly, and bonds are subject to repo rate risk. Long-dated transactions are the most problematic. For these reasons, exact delta hedging is impossible in practice.

As explained in the previous chapter for swaps, ladder or bucket risk reports are generated giving the hedging mismatches as a function of maturity. These reports can take several forms, equivalent in content, but emphasizing different points of view. Examples include ladders in forward rates, ladders in swap rates, ladders in zero-coupon bonds etc. Some traders and managers get used to looking at certain reports and prefer those. Therefore, it is useful to become familiar with all the reports.

---

[8] **A Story: Swaps, Caps, and Objects in a C^{++} System:** Once upon a time there was an industrial-strength C^{++} object-oriented vendor derivatives system, whose object model specified two objects for each derivative. One object was attached to each leg of a fixed/floating swap. There seemed to be only one object for a cap, so the cap was assigned one object, and cap's dollar price was used as the other object. When it was pointed out that deep ITM a cap, assigned a single object, turns into a swap, which required two objects, it was clear that the object specification was inconsistent. The system developers were startled. Naturally, it was much too late to do anything about it. What lessons can we draw from this story? The answer is not at the back of the book.

Hedging Volatility and Vega Ladders

We naturally have vega ladders to describe the details of the volatility dependence of interest-rate products. Each caplet has its own vega, defined as the sensitivity of the caplet value to its own volatility. These vegas correspond to the maturities of the caplets. An illustrative vega ladder for a 5-year forward cap:

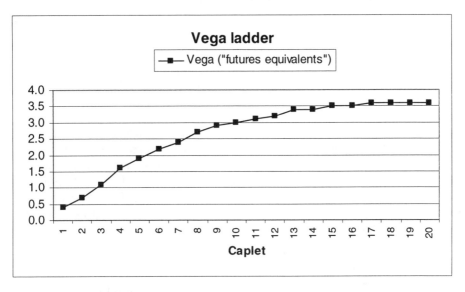

The vega normalization here is in "futures equivalents", defined as one future equivalent = $2500, for a change in vol of 1%. If a caplet volatility σ_l changes by $\delta\sigma_l$ in %, then the caplet value $\$C_l$ will change by

$$\$dC_l = \left(\frac{\$2500}{\%}\right) \cdot Vega_l^{(\text{Fut.Equiv})} \cdot \delta\sigma_l\,(\%) \tag{10.5}$$

For example, if the total vega for this cap is 53.6, and if the volatility goes up by 1% for all caplets, the change in the value of the cap is $134,000.

By linear interpolation, these caplet vegas corresponding to the caplet expiration dates $\{t_l^*\}$ can be mapped onto the IMM dates $\{t_m^{(\text{IMM})}\}$. The use of the IMM dates has an advantage at the short end, because pit options can be used for hedging purposes. Pit options in the US are options on Eurodollar futures, and have IMM maturities. They are described like caplets and floorlets[9,ii]. There are no natural hedges farther out in maturity.

Sometimes swaptions (options on entire swaps) are used to hedge caps (baskets of options on single forward rates)[10]. We look at swaptions in Ch. 11.

Implied Vols, Realized Vols, and Hedging Strategies

The implied caplet volatilities $\sigma_j^{(impl)}$ and the realized rate volatilities $\sigma_j^{(realized)}$ (i.e. the actual volatility of the $f_j(t)$ rate as time progresses) are related. This relation however is not perfect and it exhibits instabilities. It is sometimes said that an implied volatility gives the "market expectation into the future" of the to-be-realized volatility of the forward rate.

However, implied option volatilities are driven by supply and demand. Moreover, implied vols possess some strike dependence (skew). Such complexities for implied vols do not have much to do with realized volatilities.

The hedging and desk strategy for interest-rate options requires considerable empirical skill. In the canonical example, a trader tries to buy volatility cheap relative to his projections of implied or realized vol levels. In addition, because models are simplified, model hedges may be followed only approximately by experienced traders.

Bid/Ask Vols and Illiquidity

Bid vols σ_{bid} and ask vols σ_{ask} correspond to buying and selling price levels. The mid vol σ_{mid} is the (bid, ask) average. In order for these vols to be well defined, the caps actually have to trade in the market. The bid-ask vol spread

---

[9] **Pit Options:** A put on a ED future is comparable to a caplet, which is a call on a forward rate. A pit option is also described by Black formula with some normalization differences. In particular, a pit option does not have the dt_{cap} factor. This is because a future has no units, whereas a rate has units 1/time. In addition, pit options are American options, which gives an extra premium for the possibility of early exercise. Nonetheless, the European option approximation is reasonable for risk purposes. For details, see Hull's book (Ref).

[10] **The Dangers of Cross-Volatility Hedging:** Trying to hedge caps with swaptions can backfire when forward rates become decoupled from swap rates. For example, as we saw in the last chapter, the forward rates can change in such a way that the swap rates do not change. There are well-known examples of blowups in hedging procedures that run across volatility types. For this reason, even if volatility risk is aggregated for corporate risk reporting, it is best to keep track of each type of volatility risk on the desk separately.

Interest-Rate Caps 131

$\sigma_{spread} = \sigma_{ask} - \sigma_{bid}$ will be well determined for liquid caps that trade often and not well determined for illiquid caps[11].

For reporting purposes, different philosophies can be adopted. Vol can be marked at the mid level and a reserve taken for the amount projected to unwind the position. The reserve can be built into the reporting directly by, for example, marking the long vol positions to the bid vol (thus under valuing with respect to the mid vol).

Matrices of Cap Prices

For convenience, matrices of cap prices can be constructed with different strike levels and maturities. Separate matrices will exist for bid, ask and mid vols. The model is used for the interpolation, since only a small fraction of the caps in the matrix will be quoted in the market on a given day. Naturally, similar remarks hold for floors.

Prime Caps and a Vega Trap

The Prime rate is the interest rate that US banks charge to their most creditworthy customers. The Prime rate generally changes across the banking industry, depending on macroeconomic conditions, when a major bank decides to change it. Prime caps provide insurance against increases in the Prime rate. Now because the Prime rate is only changed sporadically, it has a behavior that does not look like diffusion at all. Rather the Prime rate has a step-like behavior where it is fixed for a relatively long time on each step[12]. Each step can last for macroscopic times—e.g. months—during which time the Prime rate is unchanged.

Nonetheless, Prime caps are priced assuming a Prime rate diffusion process with a Prime volatility. Model assumptions are used to get the Prime volatility

---

[11] **Bid/Ask Vol Illiquidity Problems:** For illiquid vol products, it can happen that only one side of the market exists. For example, dealers may only be selling a given product to clients but not trading the product with other dealers. In this case the ask vol is known but the bid vol is not known (vols are from the point of view of the dealer). Hence, the mid vol is not known either. In that case, some assumptions have to be used. If suddenly the dealer has to buy back this vol because of strategy changes or whatever, the real bid-ask spread may be nowhere near the assumption.

[12] **The Prime Rate and the Macro-Micro Model:** The constancy of the Prime rate for macroscopic times was the prototype for the Macro-Micro model for a macro-economically determined rate in the absence of trading. That is, even though derivative instruments on the Prime rate exist (swaps, caps) and their market prices fluctuate, the Prime rate itself does not fluctuate. The Macro-Micro model is described in Ch. 47-51.

σ_{Prime}. Typically σ_{Prime} is related to the Libor vol σ_{Libor} for equivalent maturity. There are two common methods for pricing Prime caps. Both use the Black model with lognormal volatility, but with different parameters. These are[13]:

1. Use $\sigma_{\text{Prime}} = \dfrac{R_{Libor}}{R_{\text{Prime}}}\sigma_{Libor}$ where $R_{\text{Prime}} = R_{Libor} + s_{\text{Prime}\_Libor}$ is the Prime rate given by Libor plus the Prime-Libor spread. This assumes that the Gaussian vols σ_G^{Prime} and σ_G^{Libor} given by changes in Prime and Libor rates, using the formula $\sigma_G^2 dt = \left\langle \left[R(t+dt) - R(t) \right]^2 \right\rangle_c$, are equal, and therefore the lognormal vols are multiplied by the rates. The strike E_{Prime} is as specified in the Prime cap.

2. Use the Libor vol σ_{Libor} and Libor rate R_{Libor} along with an equivalent Libor strike obtained by subtracting the Prime-Libor spread from the Prime strike, $E_{Libor}^{Equiv} = E_{\text{Prime}} - s_{\text{Prime}\_Libor}$. Lowering the strike has a similar effect to raising the rate. However, the vol prescription is different than in method 1.

There is a potential trap related to Prime vega that depends on the method used. Commonly what is done is to quote vega by "changing input vols by 1%" and finding the change in the option value. In both methods, the input vol is Libor. However, in method 1, the change in Prime vol will be less than 1%. If, just by a change in the semantics, we redefined "input vol" to mean the vol used as input to the Black model, then we would change the Prime vol by 1%. Therefore, it is clear that an innocuous-sounding change in the definition leads to a change in the risk quoted for the option[14].

CMT Rates and Volatility Dependence of CMT Products

Caps and swaps are also written on CMT (Constant-Maturity Treasury) rates and on CMS (Constant-Maturity Swap) rates[15, iii]. A CMT rate has a definite maturity

---

[13] **Models, Rigor, and Clients:** These seemingly crude approximations for Prime caps may offend the rigorously minded quant. Be aware that on the desk, empirical models are sometimes employed regardless of "theory". There may be no alternative. What would be *your* theory of Prime rate dynamics? By the way, you need an indicative Prime cap price right after lunch, because that is when the salesman is going to call up the client.

[14] **Definitional Traps for Risk:** Prime vega ambiguity is not the only example of why it is important to know the details of how the risk is defined. Once you find out, it is a good idea to document things and then periodically monitor the situation in case something changes.

[15] **CMT Rates and CMT Derivatives:** The CMT rates are obtained by fitting the treasury curve, and are contained in a weekly Federal Reserve Bank "H15" report.

τ (e.g. $\tau = 5$ for 5-yr CMT). We need the maturity-τ CMT forward rate at time t. There is an extra complexity, namely a state dependence is present for the various possible values of a forward CMT rate at a given time[16].

A major consequence or complication is that the CMT rates are volatility dependent. The forward CMT rate is a composite rate that depends on the diffusion probability of getting to a given value at time t. For example, if a short-rate model is used for the underlying dynamics[17], this diffusion probability depends on the short-rate volatility. The diagram gives the setup:

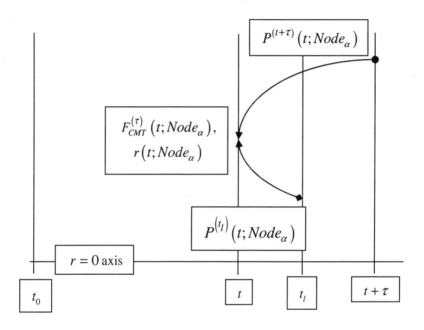

Derivatives written on CMT rates are used by, for example, insurance companies that have products such as SPDAs that have payouts based on CMT, for example the 5-yr CMT. If this rate goes up, the company loses money. A CMT cap can protect against this risk. CMT products are also used as hedging vehicles for mortgage-related activities, since mortgage rates are correlated with treasury rates. CMT caps are illiquid and broker quotes can remain unchanged for long periods of time.

---

[16] **Simplified CMT Models:** Sometimes the complexity of the node dependence of CMT rates is ignored. See Hogan et al (ref).

[17] **Short-rate Formalism:** The interested reader might consult Ch. 43 for some details.

The forward CMT rate (of maturity τ, at time t, at node α) is called $F_{CMT}^{(\tau)}(t; Node_\alpha)$. The nodes arise from a discretization of rates at each time for implementing numerical algorithms.

The CMT rate $F_{CMT}^{(\tau)}(t; Node_\alpha)$ and the short rate $r(t; Node_\alpha)$ are shown. The discount factor $P^{(t+\tau)}(t; Node_\alpha)$ from maturity date $t + \tau$ back to time t at node α is also shown along the discount factor $P^{(t_l)}(t; Node_\alpha)$ corresponding to an intermediate date t_l. CMT rates can be built up from short rates using the same short-rate model used to price other derivatives, for consistency.

This forward CMT rate is equal to the coupon that is obtained from setting the corresponding forward treasury coupon bond to par at the forward node α, namely $B_{Tsy}^{(\tau)}(t; Node_\alpha) = 100$. The bond $B_{Tsy}^{(\tau)}(t; Node_\alpha)$ depends on the state at time t because the forward discount factors used to construct this bond depend on the state. There are other treasury-market related corrections[18].

Explicitly, the discount factors[19] $P^{(t_l)}(t; Node_\alpha)$ back from times t_l to time t at specific node α are used to construct the bond[20] with coupon $C_{Tsy}^{(\tau)}$,

$$B_{Tsy}^{(\tau)}(t; Node_\alpha) = \sum_{t_l=t}^{t+\tau} C_{Tsy}^{(\tau)} \cdot P^{(t_l)}(t; Node_\alpha) + 100 \cdot P^{(t+\tau)}(t; Node_\alpha) \quad (10.6)$$

At par (i.e. 100) for the bond, $C_{Tsy}^{(\tau)}$ by definition is $F_{CMT}^{(\tau)}(t; Node_\alpha)$, i.e.

$$F_{CMT}^{(\tau)}(t; Node_\alpha) = 100 \cdot \left[1 - P^{(t+\tau)}(t; Node_\alpha)\right] \bigg/ \sum_{t_l=t}^{t+\tau} P^{(t_l)}(t; Node_\alpha) \quad (10.7)$$

Given a model for the discount factors, in this way we obtain the CMT rates.

---

[18] **Repo and Auction Complexities:** Because CMT rates are connected with the treasury market, some complexities of repo (including forward repo curves) and auction effects enter CMT calculations.

[19] **Discount Factors:** For CMT we use the treasury discount factors, for CMS Libor we would use Libor discount factors.

[20] **Kinematics:** In practice for CMT and CMS, we may need to put in extra factors (e.g. a factor ½ for semi-annual coupons), the rate convention (e.g. 30/360, money market), etc.

Interest-Rate Caps 135

CMS Rates

To get a CMS rate we would set the appropriate forward Libor swap to par (zero value) at each node at time t and follow the same procedure.

A CMT Swap is a Volatility Product

A CMT swap is considerably more complicated than the plain-vanilla swaps we considered in the previous chapter because of the volatility dependence of the CMT rate. We need to compute the rates $F_{CMT}^{(\tau)}(t; Node_\alpha)$ at the different reset times of the swaplets, and for the different nodes at those reset times. The swaplet contribution at a given node is proportional to the difference of $F_{CMT}^{(\tau)}(t; Node_\alpha)$ and the other rate in the swap. This other rate can be fixed at some value E or can be a floating rate like Libor. In the latter case, the swap is called a CMT-Libor basis swap.

Because of the volatility dependence of the CMT rate, CMT swaps have nonzero vega. Even though there is no optionality written in the contract, because CMT swaps have vega they could be put in an "option" book.

CMT Caps

We now consider CMT caps[21]. We also need the index l specifying which CMT caplet we are talking about[22], so we write $F_{l,CMT}^{(\tau)}(t; Node_\alpha)$. We numerically determine, at the lth CMT caplet maturity t_l^* and at the node α at t_l^*, if $F_{l,CMT}^{(\tau)}(t_l^*; Node_\alpha) > E$. If so, the caplet gets a contribution proportional to the difference. The contributions from all nodes at t_l^* are added up to get the lth CMT caplet value. The CMT cap value is the sum of the CMT caplets.

Note that because a CMT swap is volatility dependent, CMT caps and floors have different volatility dependencies.

---

[21] **Lognormal CMT Rate Dynamics Comments:** The CMT (or CMS) rates can be derived as composite rates from an underlying process as we illustrate here. Alternatively, these composite rates can be stochastically modeled directly. It should be noted that it is inconsistent to model a composite rate as lognormal and at the same time model the "elementary" rates contained in the composite rate as lognormal. This is because the any function – even a sum – of lognormal processes is at best only approximately lognormal. Naturally, this does not stop the market from quoting the composite CMT volatilities as lognormal.

[22] **CMT Rate Notation:** Ugly as the notation looks, there is actually another attribute, namely the date t_0 at which data are used to construct the CMT forward curve.

Numerical Considerations in Calculating CMT Derivatives

The numerical calculations for CMT products are highly numerically intensive. This is because of the dependence on the discretization nodes of the quantities needed for the pricing. Most of the time using brute-force numerical code is spent calculating the zero-coupon bonds at each node. Once we have the zero coupon bonds at a node we immediately get the CMT rate at that node, as we saw above.

We can speed up the calculations. The basic idea is to use a fast analytic method based on a mean-reverting Gaussian process for calculating the CMT rate, including the volatility correction, at any future time and for any given future short-term rate. Explicitly we can use the mean-reverting Gaussian short-rate model for an approximation to the discount factors $P^{(t_l)}(t; Node_\alpha)$. The explicit expression is given in the discussion of this model in Ch. 43.

The actual values of the future short-term rates come from a separate code, for example a lognormal short-rate process code[23]. Hence no negative rates actually appear[24].

References

[i] **Definitions**
Downes, J., Goodman, J. E., *Dictionary of Finance and Investment Terms*. Barron's Educational Series, Inc., 1987.

[ii] **Pit Options**
Hull, J. C., *Options, Futures, and other Derivative Securities*. Prentice-Hall, 1993.

[iii] **CMT Products**
Hogan, M., Kelly, J., Paquette, L., *Constant Maturity Swaps*. Citibank working paper, 1993.
Smithson, C., *ABC of CMT*. Risk Magazine Vol. 8, p. 30, Sept. 1995.
O'Neal, M., *CMT-Based Derivatives*. Learning Curves, Derivatives Week. Institutional Investor, Inc. 1994. See p. 117.

---

[23] **History:** I got the idea of mixing analytic and numerical methods to speed up the calculations for CMT products in 1994. The speedup over the brute-force lognormal numerical code was an order of magnitude.

[24] **Gaussian Models and Negative Rates:** The analytic Gaussian model gives a reasonably good numerical approximation for discount factors, and thus for CMT rates. There is some negative short-rate contribution to discount factors in Gaussian models. *However*, the short-rate grid in the CMT rate calculation is lognormal, and no negative rates appear in the grid. The short rate = 0 axis is in the figure to emphasize this point.

11. Interest-Rate Swaptions (Tech. Index 5/10)

We described interest-rate caps in the last chapter. A cap is a collection or basket of options (caplets), each written on an individual forward rate. A swaption, on the other hand, is one option written on a collection or a basket of forward rates, namely all the forward rates in a given forward swap[1]. The fact that the swap option is written on a composite object means that correlations between the individual forward rates are critical for swaptions.

Swaptions are European if there is only one exercise date, Bermudan if there are several possible exercise dates, and American if exercisable at any time. The forward swap into which the swaption exercises can be either a pay-fixed or a receive-fixed swap[2].

Swaptions are usually based on Libor. Swaptions on other rates also exist.

European Swaptions

A European swaption is an option that, at the swaption exercise date, gives the right to the swaption owner X to enter into a forward swap. There are two numbers that characterize a European swaption. The first is the time interval τ^* from the value date t_0 to the swaption exercise date t^*. The second is the time interval τ_{swap} from start date t_{Start} of the forward swap to the maturity date T_{Mat} of the forward swap[3].

The picture gives the idea for European swaptions:

---

[1] **Forward Swap:** A forward swap is a swap that starts sometime in the future, at t_{start}, and ends at T_{mat}, in the notation of the diagram on the next page.

[2] **Names for Swaptions:** If the swap from the point of view of the broker dealer is a receive-fixed swap, the swaption is called a receiver's swaption or a call swaption. If the swap from the point of view of the broker dealer is a pay-fixed swap, the swaption is called a payer's swaption or a put swaption.

[3] **Jargon:** The two numbers τ^* and τ_{Swap} characterize the swaption by saying "In τ^* for τ_{Swap}" or "τ^* by τ_{Swap}". So if $\tau^* = 3$ years and $\tau_{Swap} = 5$ years the swaption would be called "In 3 for 5" or "3 by 5" swaption. Sometimes the total time from now until the end of the swap is used, i.e. $\tau^* + \tau_{Swap}$ is substituted, so the swaption would be called "In 3 for 8" or "3 by 8". You need to check the convention used locally.

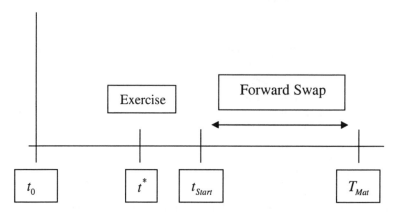

The market-standard formula for a European swaption price $\$C_{swt}$ is the Black formula with a constant volatility. Swaption volatilities are quoted in this language in the market. The result is obtained by assuming that the forward swap breakeven (BE) rate R_{BE} is lognormal[4, i]. We have, for a pay-fixed swaption with strike rate E,

$$\$C_{swt} = \$X \left\{ R_{BE} N(d_+) - E N(d_-) \right\} \tag{11.1}$$

Here,

$$d_\pm = \frac{1}{\sigma \sqrt{\tau^*}} \left[\ln(R_{BE}/E) \pm \frac{1}{2} \sigma^2 \tau^* \right] \tag{11.2}$$

The kinematic factor $\$X$ of notionals and discount factors corresponds to the forward swap,

---

[4] **Black Swaption Formula and Consistency Issues:** Jamshidian (ref. i) has proved that the Black formula is an exact result for European swaptions under certain assumptions. Physically the BE rate corresponds to a swap with zero value which costs nothing. Hence the BE rate is not an asset, and therefore the swaption is given by the Black formula. However there are consistency difficulties. Note that if the break-even rate is lognormal, individual forward rates cannot be exactly lognormal. Also, note that the break-even rates of all swaps cannot all consistently be lognormal. For example, the BE rates of the three swaps between successive time intervals (t_1, t_2), (t_2, t_3), and the total time interval (t_1, t_3) are related and cannot be independently lognormal.

$$\$X = \sum_{l \in FwdSwap}^{N_{swap}} \$N_l \cdot dt_{swap} \cdot P_{Arrears}^{(T_l)} \qquad (11.3)$$

The discount factors serve to discount the in-arrears cash-flow payoffs in the forward swap back to the value date, today t_0.

Note that deep in the money $R_{BE} \gg E$, the pay-fixed swaption approaches the pay-fixed forward swap. This means that exercise becomes highly probable if the swaption holder X can exercise into a swap with a large positive swap value, to X's benefit.

Put-Call Parity and European Swaptions

We have a relation between options that follows simply from the fact that the sum of all probabilities of all paths starting at a given point has to be one. This becomes the statement of put-call parity for European options[5]. The probability sum rule is translated into the statement about the normal functions $N(x) + N(-x) = 1$. We illustrate the result for European swaptions:

$$C_{\text{Payers Swaption}} - C_{\text{Receivers Swaption}} = S_{\text{Pay-Fixed Swap}} \qquad (11.4)$$

This equation also holds for the hedges (Δ, γ, vega).

European Swaption Volatility

In general, the volatility for a composite rate (the BE rate) will be less than the volatilities for individual rates. This is because correlations are less than perfect and some diversification occurs. The situation can be thought of similarly to the volatility for a basket of equities. The volatility of the basket of equities is lowered because some stock prices can go up while others go down. In a similar fashion, some forward rates can go up while others go down, keeping the BE rate relatively constant or less volatile.

In particular, if $\rho_{ll'}$ is the correlation between forward rate returns, if all dynamics are lognormal, and if we ignore the (relatively small) rate dependence of the discount factors, then the lognormal BE volatility σ_{BE} is approximately

$$\sigma_{BE}^2 \approx \sum_{l,l'} \zeta_l \zeta_{l'} \sigma_l \sigma_{l'} \rho_{ll'} \qquad (11.5)$$

---

[5] **Put-Call Parity and American Options:** Because of the complex nature of exercise, although the sum rule for probabilities of course remains, no simple relation exists for American or Bermuda options.

Here, a sort of "component weight" ζ_l is given in terms of the ratio of the time-averages of the lth forward rate in the forward swap and the BE rate. This time average is indicated by brackets $\langle ... \rangle$. It is to be taken over the same time window as that defining the volatility and correlations. To get this formula, factorization approximations were used for ratios of averages and for products of averages. Explicitly,

$$\zeta_l = \frac{\langle f_l \rangle \cdot \$N_l \cdot dt_{swap} \cdot P_{Arrears}^{(T_l)}}{\langle R_{BE} \rangle \cdot \$X} \tag{11.6}$$

Hedging European Swaptions: Delta and Gamma (The Matrix)

European swaption Δ risk is hedged using the same techniques as previously described using ladders. Note that the Δ ladder will only be significant in the period of the forward swap. Before the start of the forward swap, small residual Δ effects are present from the discount factors.

European swaption γ risk needs to take into account the fact that gamma is a matrix $\gamma_{ll'}$. The approximate factorization relation $|\gamma_{ll'}| \approx \sqrt{|\gamma_{ll}\gamma_{l'l'}|}$ mentioned in the section on swaps works reasonably well for swaptions to include the effects of the off-diagonal terms. We emphasize again that this only holds deal-by-deal, i.e. for each swaption separately, not for a portfolio.

A Paradox, a Paradox, a Most Ingenious Vega Paradox[6]

European swaption vega risk is tricky and involves a little paradox. On the one hand, if the swaption is exercised all volatility dependence disappears. This is because the swap obtained by exercising the swaption has no volatility dependence. This would imply that the vega should be concentrated in the vega bucket containing the swaption exercise time t^*. On the other hand, the forward rates f_l individually making up the forward swap associated with the swaption live at times t_l after t^*. Hence this argues that the sensitivities to the individual forward rate volatilities σ_l should be spread out in the buckets at times t_l after t^*. The paradox therefore is how to construct the vega risk report[7].

---

[6] **Musical Reference:** Listen to the trio of Ruth, Frederic, and the Pirate King in *The Pirates of Penzance*, Gilbert and Sullivan (No. 19). The paradox in the operetta has to do with whether birthdays should appear or disappear, relative to Frederic's ability to exercise an option to leave the pirates and subsequently exterminate them.

[7] **Vega Paradox:** There is no good way out of this paradox. Different desks report vega using either of the two methods presented in the text. If the procedure spreading out vega

Bermuda/American Swaption Pricing

A Bermuda swaption has a discrete exercise schedule (usually every 6 months after a lockout period during which no calls are allowed). Swaptions and callable bonds are closely related. A 10NC3 bond means a 10-year bond that is callable after 3 years. The call option embedded in the bond corresponds to a Bermuda swaption, which can be exercised after 3 years. The exercise prices usually vary with time in a schedule. For a non-amortizing swap, we have seen that we can rewrite a receiver's swap as a long bond and a short FRN. Thus on a given exercise date, the schedule might specify that ABC would need to pay 102 for the bond originally at par (100).

American swaptions, which can be exercised at any time (perhaps after a lockout period), are less common than Bermudas largely because of the close relation of Bermuda swaptions with callable bonds.

Bermudas/Americans are priced with a dynamical stochastic rate model, using the same backward back-chaining algorithms used to price callable bonds. For example, a short-rate diffusion process can be used[8]. A description is given later in the book (c.f. Ch. 44).

Determination of the Local Forward Term-Structure of Volatility

In order to determine the local forward term structure of volatility for the Bermuda algorithm, we proceed using the normalization to the European swaption market. This involves the pricing of a number of European swaptions using the Bermuda algorithm, and varying the local short-rate forward volatility in the Bermuda algorithm until agreement with a grid of European swaption prices is obtained. This procedure has to be watched carefully, as it can become numerically unstable. In particular, cutoffs need to be imposed such that the local volatilities do not become negative, unphysically small, etc. The underlying cause of this sort of problem is that the European swaption market prices are not completely internally consistent, viewed from the Bermuda model perspective.

Bermuda/American Swaption Hedging

Vega for a Bermuda/American swaption book is hedged using other volatility products. Because the algorithms are normalized to the European swaption

---

over the period of the forward swap is used, the risk system must be clever enough to drop all the risk at the swaption exercise date. Far from exercise, it seems more reasonable to use the spread-out vega method, but close to exercise, it is probably better to use the concentrated-vega method.

[8] **Other Rate Processes for Bermuda Swaption Pricing and Job Possibilities:** There is a variety of short-rate processes available. The most common "Street-Standard" is lognormal. Multifactor models can also be used. Transformations of hidden variables into physical short rates can be employed. There would seem to be employment possibilities for quants here that could last for years.

market, and because in some sense Bermuda options are a mélange of European options shielded from each other by the complex exercise logic, European swaptions are commonly used as approximate hedges. These hedges must be modified as time progresses[9].

Hedging depends on the shape of the yield curve and the shape changes. This is because as the yield curve shape changes, the probability of early exercise changes, so the European hedges need modification.

Vega for deep ITM swaptions is small and hard to evaluate. Oscillations in vega for discrete codes can be observed as a function of the amount of volatility change $\delta\sigma$. These oscillations are typical of numerical noise.

A significant potential problem is that the Bermuda (and American) swaptions are quite illiquid. Therefore in a stressed market, a Bermuca swaption seller can suffer substantial losses that cannot be hedged away.

Swaption Delta Hedging and Numerical Noise

Delta Δ and gamma γ are hedged using numerical differences derived from the algorithms. The magnitude of the change of the curve δr to get these differences needs to be chosen carefully in order to avoid spurious numerical instabilities for the discretized algorithm; in particular δr cannot be too small, certainly more than 1bp[10]. At least 10 bp should be used, and 50 bp has been used.

The Δ ladder for Bermuda/American swaptions is tricky, especially at the short end. Occasional instabilities can be observed where neighboring delta buckets have canceling large fluctuations, indicating small overall risk but spurious differential calendar risk[11]. These instabilities are worse with deals near exercise. This is partially because then only a few nodes of the algorithm can be used to determine the swaption. Instabilities are also magnified by anomalies in

---

[9] **Acknowledgement:** I thank Ravit Mandel for an informative conversation on practical aspects of swaptions.

[10] **Oscillating Convergence:** As δr is increased from 1bp, an oscillating convergence can sometimes be seen for Δ as a function of δr. Again, as with vega, the phenomenon observed here is typical of numerical noise. The amount of oscillation also depends on the type of curve shifted.

[11] **Spurious Numerical Instabilities and Management Meetings:** The presence of these occasional numerical instabilities can become a concern to nervous management. One expedient is just to smooth out the fluctuation by hand in the report and wait for the problem to go away tomorrow. Alternatively, the fluctuations can be smoothed out in the code. Of course the grid in the algorithm can be further refined, but this may take considerable time and detract from other activities. Such instabilities naturally provide a great opportunity to go to meetings to discuss the whole thing with everybody. At the end of the day, if the management is too wrapped up in focusing on spurious noise issues with little real risk, considerable time can be wasted. Some numerical noise will be present no matter how much the code is refined. Still, numerical noise should be constantly monitored by the quant group.

the cash rates, in particular if the short-end cash curve becomes inverted, e.g. the 1M cash rate is above the 6M cash rate, a situation that does sometimes occur.

The noise is also increased if the curve generation itself leads to discontinuities in the forward rates, that is, if the forward rates are not smoothed out in the curve-generation algorithm itself.

Delta and Vega Risk: Move Inputs or Forwards?

Delta Risk

First, consider delta risk. The type of rate curve that is moved is important. For example, if we move the inputs to the curve generation one at a time (notably swap rates) we can get wild forward rate fluctuations and big changes in swaption exercise probabilities. Such anomalies do not occur if we move individual forward rates. The situation is exhibited in the picture below:

Forward rates change wildly if 5-yr swap rate is increased, while holding other inputs fixed.

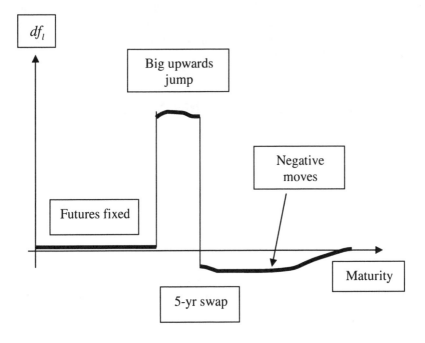

Say the forward curve is generated by 4 years of futures along with the 5, 7, 10, etc. year swap rates. Now consider as an example moving the 5-year swap rate R_{5-yr}, up say 10 bp, holding all other input rates (futures, other swap rates) fixed. Because the futures are held fixed, the forward rates out to 4 years are fixed. The only way to get $\delta R_{5-yr} = 10bp$ is if the forward rates between 4 and 5 years increase much more than 10 bp, roughly 50 bp. A pay-fixed swaption with a first exercise time at 4 years and forward swap between 4-5 years can become be highly effected. If this swaption starts out 50 bp OTM, it will become ATM under this scenario. Other swaptions can also be affected.

On the other hand, if individual forward rates f_l are increase by the nominal 10 bp, the swaption 50 bp OTM will still be 40 bp OTM after the shift and exercise will not be affected much.

An important point is that even though the 5-year swap is a hedge vehicle, in real markets swap rates and futures are highly correlated. Therefore while moving the 5-year swap rate seems like a good idea, it does not correspond to realistic market moves and it can lead to anomalous swaption behavior[12].

Vega Risk

A related conundrum involves the calculation of vega risk. Suppose that we ask for the response of a Bermuda swaption to moving one input volatility, say a 3-year volatility, keeping other input volatilities constant. The calculated local forward volatilities must change in a wild manner (similar to the picture above) in order to achieve these input vol changes. Again, apparently simple input changes can lead to anomalous behavior.

Swaptions and Corporate Liability Risk Management

We recall from the discussion of swaps that corporation *ABC* had issued a fixed-coupon bond and then entered into a swap with a broker-dealer *BD* to receive fixed and pay floating. If the bond is callable and market rates drop enough, *ABC* may decide to call the bond and then reissue debt at the more favorable current low market rates[13]. In this way, *ABC* is acting to minimize its liability risk. Consider the following diagram:

---

[12] **Discussions of the Delta:** Many discussions can occur regarding these considerations. This is not discussion of noise but substantive discussion of procedure in risk probing which is important in principle. It is sometimes difficult for people to grasp the issue of why something that seems so simple ("just move a swap rate") is in reality quite subtle.

[13] **Decision of ABC to Call the Bond and Refinance:** A decision of ABC to call the bond and refinance is not made lightly. There are a number of practical considerations, including costs of issuance, the market for new bonds in the particular sector

Interest-Rate Swaptions 145

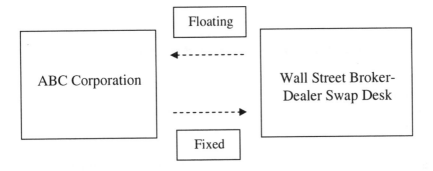

In tandem, *ABC* will want to cancel the swap at the same time. However, this is only possible if the swap contract has cancellation provisions. So to anticipate this, *ABC* has already placed into the swap contract the possibility to cancel the swap at the same times as in the bond call schedule. In this way, the bond and the rest of the swap disappear at the same time. This cancellation mechanism in the swap can be viewed as the exercise of an option (swaption) to enter into a forward swap to pay fixed and receive floating, canceling all cash flows of the original swap after the time of exercise.

From the point of view of the broker dealer, the swap has a receive-fixed swaption, also called a receiver's swaption or "call swaption" (the latter because *ABC* has in tandem called the bond).

In the section on swaps, we mentioned that a non-amortizing plain-vanilla swap can be written as the difference of a non-callable bullet bond and an FRN, and that at a swap reset date the FRN is at par, $FRN = 100$. We now show that the same relation holds for cancelable swaps and callable bonds, with the fixed swap rate being equal to the bond coupon.

If the swap is cancelable, an option to stop the swap exists. If the bond is callable, an option to stop the bond exists. Assume that the dates for exercise are the same for the swap and bond, and that these dates are at swap resets. At a given reset/exercise date, there is no difference between exercising the two (bond and swap) options, because at exercise the forward swap and forward bond just differ by a par FRN. In summary, at reset/exercise times the two options (swap option, bond option) are equal, so the cancelable swap and the callable bond

---

corresponding to ABC, changes in the ABC credit rating that might have occurred since the last issuance, etc. In general, there is a measure related to: (1) the discounted savings realized by refinancing, and (2) the loss of the option of the called bond.

differ by the FRN. For the receive-fixed swap in the example, this condition at resets/exercises is

$$S_{\text{Cancellable Receive-Fixed Swap}} = B_{\text{Callable Bond}} - 100 \quad (11.7)$$

Practical Example: A Deal Involving a Swaption

An entity[14] ABC announces that it will take bids on a plan to get an immediate amount of cash[ii]. ABC asks various broker-dealers (BD) to bid[15] to pay $\$C$ now to ABC for a Bermuda receivers swaption giving one BD the right to begin a swap with ABC at one of a set of designated times in the future. This forward swap is for BD to receive a fixed rate E and pay a floating rate. ABC will pay an above-market rate E in the swap, implying a cheap swaption price $\$C$ paid by BD. Here is the diagram of the deal:

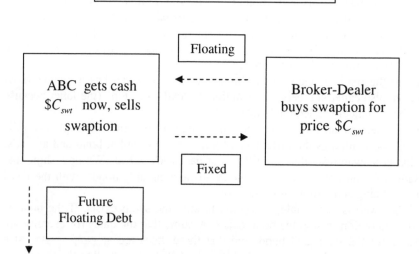

In strategic terms, ABC thinks rates are low now and thus E is relatively low (even given the extra spread), making it desirable to act now. BD gets the extra spread. Thus, the deal can appear attractive from both sides.

---

[14] **ABC:** In fact ABC was the Metropolitan Transportation Authority in New York. The announcement was in 1999. See Bloomberg News (ref.)

[15] **Beauty Contest:** The activity described in the text is called a "beauty contest". The winner is whichever broker-dealer comes in with the best deal for ABC.

ABC may regard the swap as preferable to issuing new debt now. This could be for a variety of reasons. First, outstanding ABC debt may not be callable for some time, and ABC may not want additional debt (for example ABC may not have the cash for the additional payments). Second, if ABC is a municipal authority, there are federal limits on bond refundings.

ABC can arrange the exercise dates on the swaption to match the call schedule of existing bonds, which form new potential bond refunding dates in the future. Actions are synchronized. If BD does exercise the swaption, then ABC can issue floating rate debt canceling the floating side of the forward swap. If BD does not exercise the swaption, ABC keeps the $\$C$ cash and just continues the current bonds.

"Net-net" as they say, after the dust has cleared, ABC winds up paying a reasonable fixed rate and gets the desired up-front cash $\$C$ regardless of the scenario, exactly as it wants.

BD buys the Bermuda swaption at the low price $\$C_{swt}$, which translates (through the BD pricing model) into an equivalent low volatility. Theoretically, even though the Bermuda swaption involves an extra premium relative to European options, the Bermuda volatility can be lower than the European volatility in practice, because ABC wants to get the deal done.

On the other hand, there is a liquidity issue. Once purchased from ABC, BD cannot sell the Bermuda swaption because there is no inter-dealer market for it. Thus, BD simply holds it, hedging it approximately with European swaptions that can be sold at market volatility, thus "locking in" a profit. The profit for BD is in return for providing the service to ABC as described above.

The initial analysis by BD goes like this. If rates do not go down, BD will not exercise its long receivers Bermuda swaption. In this case, from the point of view of BD, the premium $\$C_{swt}$ is lost. On the other hand, through judicious hedging with short positions in European receivers swaptions that are relatively liquid, the $\$C_{swt}$ premium along with a profit can be made back because in this case the short European receiver swaptions will gain value. If rates do go down, BD will exercise the Bermuda swaption, remove the hedges, and pick up the above-market fixed coupon E from ABC.

In addition to collecting the above market rate including the profit on the spread, BD might think that rates will decrease further, making the floating payments to ABC even lower than predicted by the forward rate curve on which the current pricing is based. Such "what-if" analyses would be based on trader market judgments involving macroeconomic or other considerations.

The relative risk-reward considerations leading to a bid by BD of a specific amount $\$C_{swt}$ for a given rate E involve other aspects. Standard considerations of BD include (high) costs for the traders, overhead costs for systems and reports for keeping track of the risk of the swaptions, transaction costs involved in the hedging, etc. Moreover, the swaps desk may have confining risk limits on the amount of volatility exposure to Bermuda swaptions that the management allows.

One low-probability but high-impact risk is that somehow BD will be forced to sell the Bermuda swaption in a hostile environment (e.g. the BD company decides to get out of the interest rate swaptions business) and thus loses the profit and more. Such examples have happened in the past.

Miscellaneous Swaption Topics

Liquidity and Basis Risk for Swaptions

As mentioned already, Bermudas and Americans are highly illiquid in the secondary market. Thus, considerable basis risk exists with European volatility. Typically, limits will be set depending on the risk tolerance for this basis risk.

Fixed Maturity vs. Fixed Length Forward Swap

Generally, the forward swap arising from swaption exercise has a fixed maturity date. Sometimes the forward swap lasts for a fixed time period after the date of exercise. For European swaptions, this is the same thing. For Bermuda or American swaptions, it is different because there are several possible exercise dates.

Caplets and One-Period European Swaptions

A one-period payers swaption (e.g. exercising into 6M Libor at a given date) is theoretically the same as a caplet, since the definitions coincide. However, for CMT the equivalence is not exact. With some assumptions regarding the absence of volatility in the discount factors, the equivalence in the CMT case can however be shown to be approximately true.

Advance Notice

There is an advance-notice feature, where the swaption holder announces, before the actual exercise date, that he is going to exercise the swaption. This period is usually 30 days. The uncertainty of exercise is thus eliminated before the exercise date, and the advance notice stops the diffusion process.

Skew for Swaption Volatility

In previous chapters, we have discussed skew for equity and FX options extensively. Swaptions also have skew complications. The price of a swaption has an extra skew dependence on the strike or exercise rate E. Specifically, if a

Interest-Rate Swaptions

single volatility σ is used to price two swaptions identical except for having different strike rates E_1 and E_2, the model prices C_1 and C_2 are not exactly market prices. To deal with this, the volatilities are made a function of the exercise rate to get an equivalent effective volatility, $\sigma(E)$. Skew here is the same idea as for equity or FX options. Really all that is going on is that a simple model is adjusted to reproduce market prices for the securities by modifying the model parameters. Essentially, the model really just serves as an interpolation scheme using $\sigma(E)$.

Because volatility as quoted in the broker market is LN (lognormal) as described above, we can use $\sigma_{LN}(E)$ with the Black formula. This $\sigma_{LN}(E)$ can be either a parameterized function or a numerical look-up table with interpolations between the (sometimes sparse) market-implied data. In order to reproduce the market values reasonably, several parameters may have to be used.

Another procedure is to change the process from LN to "not-LN". The skew effect as viewed from the LN model is replaced by the different dynamics from the not-LN model. For example, the "Lognorm Model" mixes LN and normal (Gaussian) processes. Because the rate changes and rate paths are different than in the LN model the probabilities of exercise become modified. Therefore, with a constant volatility σ_{Not_LN} for this "not-LN" model, we might hope to get market prices.

The behavior of skew can be inferred qualitatively. Say that the process that reproduces skew is some mix of Gaussian and lognormal. Because Gaussian rate probabilities are not suppressed at low rates, more Gaussian paths will go near $r = 0$ than in the LN model. For this reason, the equivalent LN volatility increases near $r = 0$. Therefore, for low strikes that probe the low-rate region, the effective LN volatility increases. In summary, the LN equivalent volatility with skew, $\sigma_{LN}(E)$, increases if strikes E decrease.

Since far-from-the-money swaptions do not trade regularly[16], model prices are used. Sometimes quants believe that their models give the "correct" prices when there are no market quotes—or sometimes even when the market quotes differ from the model[17].

---

[16] **Far From the Money Options:** Such options do not trade regularly, and so market quotes are often not available. In a sense, this is not terribly critical, since if an option is far ITM it will probably be exercised with a payoff independent of the volatility, while if it is far OTM it is not worth much regardless of the volatility. For transactions of far OTM options, a "nuisance" charge is sometimes applied because it costs money to monitor the option in addition to the small option value. A nuisance charge cannot be modeled by skew.

[17] **Aristotle's Horse and Correct Models:** While it may be difficult to obtain market quotes for illiquid instruments, it also puts the egoistic modeler in the same boat as the possibly apocryphal story about Aristotle. The story is that Aristotle theoretically (and

Model Dependence of Vega and Risk Aggregation Problems

Because volatility parameters in non-LN models are not standard LN volatilities, the volatility dependence (vega) for risk management is also model dependent[18]. This difference in vega between models for the same Bermuda swaption can be on the order of 10%. This can lead to aggregation problems for volatility risk between desks that use different models.

One way of avoiding such aggregation problems is to demand that desks uniformly carry out the same risk procedure. This procedure could be to move the input LN volatilities for European swaptions by 1%, revalue the whole swaption book using whatever non-LN model the desk has, and then take the difference of the revalued book with the value of the book using the original input LN European swaption volatilities.

Theta

The measurement of theta θ (time decay) for swaptions can be monitored. Revaluations can be done one day apart. In performing the revals, it must be specified whether rates are held constant or whether the rates are allowed to slide down the yield curve, as mentioned in previous chapters.

References

[i] **Black Swaption Formula**
Jamshidian, F., *Sorting Out Swaptions*. Risk, 3/96 and *LIBOR and swap market models and measures*. Finance and Stochastics Vol 1, 1997.

[ii] **Swaption Example**
Bloomberg News, *N. Y. Transit Agency Planning $360 Million Swaption*. 1/29/99.

---

incorrectly) deduced the number of teeth in a horse, and then refused to look at a real horse to count the teeth because he said his idea had to be correct. The decisive test for a model is whether the model price is near the price received if the option is sold.

[18] **Misunderstanding Vega:** The fact that vega is model dependent can lead to misunderstandings. Comprehension of the technical fine points of volatility in complicated models outside a quant group is generally shaky. Therefore, the problem can exist but go unnoticed.

12. Portfolios and Scenarios (Tech. Index 3/10)

Introduction to Portfolio Risk Using Scenario Analysis

In this chapter, we introduce the analysis of portfolio risk using scenarios. We first consider the definitions of portfolios and of scenarios. We discuss scenario types. These include instantaneous scenarios and scenarios in the future. We then discuss a scenario simulator. We end with comments on presentations of risk.

Definitions of Portfolios

The definitions of the portfolios depend on the goals. Among the variables are:

- Product
- Level of Granularity
- Level of Aggregation
- Type of Risk Analysis

These variables are all inter-related. An individual *product* can mean, for example, equity barrier options on the equity OTC derivatives desk, Libor swaps on the interest-rate swap desk, non-investment-grade bonds on the high-yield desk, CMOs on the mortgage desk, etc. A more refined level of *granularity* can mean, for example, Libor swaps with maturity only between 5 to 10 years. A high level of *aggregation* can mean, for example, the total vega in the Fixed Income Department across all desks. Many *types* of analysis can be done: at different aggregation or granularity levels, with different types of scenarios, with various reports, etc.

For a single product level on a desk, some refined analyses can be run that are prohibitive at an aggregate level. At an aggregate level, approximations may have to be made due to time and programming resource constraints. At the firm-wide level, the collection of risk measures can be a major, lengthy undertaking. For any aggregation, care should be taken systematically to collect enough information so that subsequent analyses using various data-view cuts of interest for different reports can be accomplished[1].

A special case is a single deal under consideration that needs to be analyzed for risk. Deal risk is also considered from the vantage of possible diversification of risk in larger portfolios[2].

Here is a simple portfolio of three hypothetical interest-rate derivative deals used in my CIFEr tutorial. The portfolio is called "Port 1". The definitions of the Greeks are the same as in the earlier chapters. The input curves, volatilities, etc. used for valuation and risks are from the date given. Each deal has its own file in which the specifics are listed, including the definition of the deal, the curves used for pricing, the hedging results, etc.

Portfolio Definition					
Name	Port 1				
Date	4/1/00				

ID	Type	Value $000	Delta	Gamma	Vega
S_0001	Swap	$3,163	1,656	-83	0
C_0001	Cap	$4,301	1,303	243	55
W_0001	Swaption	$1,362	684	68	3
TOTALS:		$8,826	3,643	228	58

All such information from portfolios should be stored in an appropriate database[3].

In addition, we can look at the risk ladders for each deal, as explained in previous chapters. The total ladders summed over deals would be examined to determine hedging requirements for the portfolio.

Realistic portfolios can have dozens, hundreds, or thousands of deals depending on the situation.

---

[1] **One-Off Analyses:** The alternative to systematic data collection can be the syndrome of frantically gathering information to answer a specific "one-off" management question.

[2] **Economic Capital:** See the chapter on Economic Capital for a discussion of the standalone and diversification risk issues.

[3] **Message to Systems: Please Store Both Prices AND Hedges in the Database, OK?** Historical series of both prices and hedges of deals in a portfolio are desirable in order to answer certain risk questions. The historical input curves should also be saved or archived. Early defective design decisions in the data model can have unforeseen restrictive consequences later. Bad decisions are sometimes made.

Definitions of Scenarios

By a *scenario* is meant a potential or hypothetical change in underlying variables. Some major considerations for scenarios are:

- The Scenario Date
- Variables in the Scenario
- Type of Scenario (Postulated, Historical, Statistical)

First, the scenario *date* needs to be specified. Often, this is just taken as today's date for an "instantaneous" change in the underlying variables. However, more information can be obtained by looking at scenario dates in the future.

The *variables* in the scenarios can be interest rates, volatilities, equity indices, specific stock prices, commodities, etc. The variables used depend on the situation and the resources available.

Here is an example of a scenario that called Scenario A (or Scen A for short), expressed in terms of the forward 3-month Libor rate changes.

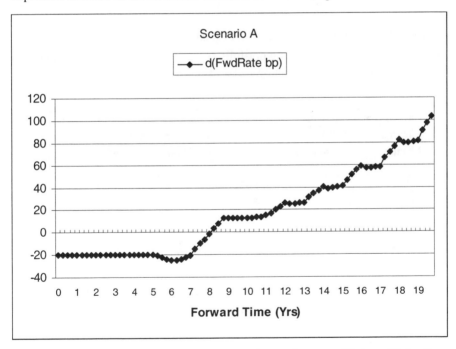

The scenario in the graph assumes that forward rates will drop 20 bp for a certain time and eventually increase up to 100 bp in 20 years[4]. Therefore, this is a

---

[4] **Long-Term Scenarios?** Naturally, no one has any idea of what the world will look like in 20 years. Nonetheless, a 20-year swap on 3-month Libor has forward rates going out to

yield-curve steepening scenario (the short-maturity end drops and the long-maturity end rises). The way the scenario is generated is that the underlying curve input parameters (from futures, treasuries, swaps spreads) are changed by certain amounts, the new curve is generated, and the difference is taken with respect to the old curve. This particular scenario assumed no changes in volatility.

Various *types* of scenarios exist. A scenario can be postulated (e.g. interest rates up 100bp). A historical scenario uses information from historical moves (e.g. spreads blowing out in the roiled markets in 1998). A theoretical statistical scenario uses a model (e.g. a path from a discretized multifactor model with random correlated fluctuations about forward curves).

Naturally, various opinions exist regarding the utility of these possibilities for scenarios. Information overload is a definite consideration with many scenarios. People just cannot absorb too much information. On the other hand, a myopic risk tunnel vision can result from a limited sample of scenarios (or no scenario analysis at all).

Resource and budget constraints often wind up severely limiting the number of scenarios run in practice.

Changes In Portfolios Under Scenario Assumptions

Here are the changes in the same portfolio (Port 1), assuming that the changes in the forward rates occur, as given by scenario A:

Change in Portfolio					
Name	Port 1		Scenario:	Scen A	
Date	4/1/00		Scen. Date	4/1/00	
ID	Type	Value $000	Delta	Gamma	Vega
S_0001	Swap	-$518	9	-1	0
C_0001	Cap	-$388	-45	50	2
W_0001	Swaption	-$222	-20	52	2
TOTALS:		**-$1,128**	**-56**	**101**	**4**

We see that under the steepening assumption of this scenario, all deals in Port 1 lost money[5]. Note that the hedges changed too. This scenario was instantaneous (the scenario date was the same as the date the analysis was done).

---

20 years. Note that not making a scenario at 20 years effectively means you assume that the Libor rate will be unchanged 20 years from now, which is itself a scenario.

[5] **My Favorite Job Interview Question:** Although I don't like to give pressure interviews to job candidates by asking them technical riddles and watching them freeze, I do usually ask one question. If the candidate doesn't know the answer, I just watch the reasoning process. Here's the question: What happens to the value of a cap when the curve of forward rates steepens? Naturally you know that the value of the cap goes up if

Many Scenarios with a Fixed Portfolio

The same sort of analysis can be done for the given portfolio with a collection of scenarios Scen A, Scen B, ..., Scen Z. The results for the individual changes in the portfolio can simply be tabulated in a report like this:

Portfolio	Port 1
Scenario	**Port 1 Change**
Scen A	$dP_1(A)$
Scen B	$dP_1(B)$
...	...
Scen Z	$dP_1(Z)$

Scenarios with Weights

Aggregation across scenarios for a fixed portfolio can be done by giving the scenarios probability weights. So Scenario A can have weight $w(A)$ etc with the weights summing to one. The weights might be fixed by hand using judgment. Alternatively, if a mathematical model is used for generating the scenarios, the model will specify the weights. Given the weights, we can also sum over the changes to get the average change. A report might look like this:

Portfolio	Port 1	
Scenario	**Weight**	**Weighted Changes**
Scen A	$w(A)$	$dP_1(A)$
Scen B	$w(B)$	$dP_1(B)$
...	...	...
Scen Z	$w(Z)$	$dP_1(Z)$
Total		Sum dP_1(All)

Many Portfolios and Scenarios

We can consider a variety of portfolios and scenarios. Here is a possible set up that might occur in a system. We have put in a little more detail.

---

rates go up because the cap, being insurance against rising rates, then has more probability to pay off. So why did the cap in the example Scenario A, a steepening scenario, lose money?

Run File for Several Portfolios and Several Scenarios

Today's date	4/1/00				
Scenario date	4/1/00				
Filename	Instantaneous.doc				

Portfolio	Want Port?	Book	Scenario	Want it?	Weight %
Port 1	Y	Book 1	Scen A	Y	25%
Port 2	N	Book 1	Scen B	Y	25%
Port 3	Y	Book 7	Scen C	Y	50%
...	...	...	...	...	...
...	...	...	Scen Z	N	0%
Port N_p	N	Book b(N_p)			

We have N_p portfolios. Each portfolio is in a "book" for aggregation. We can either include the portfolio or not in the run using the flag Y or N in the second column. For each scenario, specify whether we want to run that scenario, and put in a weight. All portfolios specified are included for each specified scenario.

A Scenario Date in the Future

The next set up is for a scenario in 10 days. Notice that the options allow changing the portfolios and/or scenarios as time progresses. We can also specify if we want to produce reports.

Second Run File for Scenario Date in Ten Days

Today's date	4/1/00				
Scenario date	4/10/00				
Filename	Ten_days.doc				

Portfolio	Want Port?	Book	Scenario	Want it?	Weight %
Port 1	N	Book 1	Scen A	N	0%
Port 2	N	Book 1	Scen B	Y	33%
Port 3	Y	Book 7	Scen C	Y	33%
...	...	...	...	...	...
...	...	...	Scen Z	Y	33%
Port N_p	Y	Book b(N_p)			

Instructions, Options		
Slide down fwd curve?		Y
Produce reports?		Y

Sliding or Not Sliding Down the Forward Curve

We can either "slide down the forward curve" or not (see the box near the bottom). First, consider just moving the time forward but leaving the rates "fixed". We actually have two possibilities for keeping rates "fixed" as time changes from "today" to the scenario date. Either the forward rate at a given date remains fixed, or the forward rate at a given time *interval* from the value date remains fixed. The latter possibility is called "sliding down the yield curve" because a rate at a given date after the time shift will be closer to the origin than the rate at the same given date before the time shift.

We can apply the same idea for a scenario. Having specified the scenario forward curve, we can slide down the curve to define the final scenario, or not.

A Scenario Simulator

In the above analysis, we kept the date fixed for each scenario run. We can now construct a scenario simulator in which we successively change the dates. Here is a setup where we run the instantaneous, ten-day, and one-month scenarios.

First Simulator Run File		
Filename	Sim_1.doc	
Run Files	**Date**	**Want it?**
Instantaneous	4/1/00	Y
Ten_days	4/10/00	Y
One_month	5/1/00	Y
Two_months	6/1/00	N
Three_months	7/1/00	N

Risk Analyses and Presentations

Given the results, we perform various statistical analyses from the reports generated by the system. We might look at averages, worst cases, confidence levels, plot histograms, etc. We can compare the results obtained from this analysis and compare what we got in the last analysis (e.g. last quarter). This is useful for trend analysis and strategic planning. Finally, we prepare presentations and communicate the results to the management[6].

---

[6] **Presentations and Communication:** For upper management, we need to summarize at a high level. We probably want to use PowerPoint with large font and a color printer. Details of the great analyses we did will probably just go into an appendix of a white-paper handout for reference. Please go back and reread the Amusing but Dead Serious Practical Exercise at the beginning of this book, and maybe even do the exercise!

PART III

EXOTICS, DEALS, AND CASE STUDIES

13. A Complex CVR Option (Tech. Index 5/10)

This chapter contains a case study of a complicated equity option called a CVR that was an important part of an M&A deal. The CVR will be considered in some depth in order to give an idea of the complexity that sometimes occurs [i, 1]. Many of the topics in this chapter are quite general. A variety of interesting theoretical points arose while pricing the CVR. These included conditions under which an option will or not be extended in time.

We use the present grammatical tense in order to dramatize the situation as it unfolded at the time. Letters ABC, DEF, XYZ are used generically to describe the players. Various specific topics are described in the footnotes.

The M&A Scenario

The ABC Corporation wants to acquire DEF Corporation. ABC is willing to pay the DEF stockholders cash along with ABC stock. From the point of view of DEF, the risk is that the ABC stock could decrease in value. This might happen for any of a number of reasons. ABC may have to issue extra debt, leaving it in a weakened condition. Moreover, ABC could be in competition with another firm XYZ to purchase DEF and thus might be forced to pay a high price. This in fact happens, with a fierce bidding war. To counter the XYZ bid, ABC after hesitation decides to offer DEF stockholders a sort of put option on ABC stock with certain side conditions[2]. Thus if ABC stock does go down, the DEF stockholders are compensated within certain limits, making DEF more willing to go through with the ABC proposal[3, 4].

---

[1] **Which Deal?** This high-profile M&A deal was front-page news around 1994. In fact, ABC was Viacom, DEF was Paramount Communications, and XYZ was QVC Network. See Refs. 1.

[2] **Put Option Structures:** These are generically called "price-protection collars".

[3] **The CVR Option and the Deal:** A complex extendable put option on the Viacom stock, called a CVR, was offered as a last-minute inducement to Paramount stockholders. It clinched the deal. CVR is short for "Contingent Value Right". The conditions defining the CVR were public record, and the CVR later traded.

[4] **The VCR Option.** Another option, present in a related deal between Viacom and Blockbuster was called a VCR, short for "Variable Common Right". Unofficially the

CVR Starting Point: A Put Spread

From ABC's point of view, a put option is a risky object with potentially quite expensive consequences. Therefore, side conditions are imposed on the put option restricting the payout. These side conditions boil down to requiring the DEF stockholders to give back some optionality to ABC. In the first complication, DEF gives back to ABC a less valuable put option, which has a strike price lower than the strike price of the put option given to DEF from ABC. Thus, including both put options (one long and one short); ABC gives DEF a put spread. The put spread starts to pay off to DEF linearly if the ABC stock descends below the upper strike. If the ABC stock descends below the lower strike, the payoff is held constant. In summary, in this first approximation, ABC gives DEF stockholders a put spread which exercises at time t_1^*.

CVR Extension Options and Other Complications

Adding more complexity, ABC also reserves the right to extend the put spread payoff to a second exercise date t_2^* past the first exercise date t_1^*. Thus, even though at t_1^* the put spread could be in the money for DEF, ABC could refuse to pay and simply extend the original put spread[5]. In the actual deal, the put spread if extended past t_1^* has modified strikes. In fact, the situation is even more complicated because ABC reserves the right to extend the option payoff a second time to a third date t_3^*.

CVR: Averaging Specifications

In fact, the stock options are not written on the stock price itself but on certain functions of the stock price. For t_1^*, a set of time windows (labeled by an index k) is defined starting at $t_k^{Start} = t_1^* - k\delta t$ [$\delta t = 1$ day[6], $k = 1...K_{max}$] with length Δt_{Window}. The average of the stock price is then taken over each of these

---

name VCR was said to have been made up by the investment bankers because Blockbuster's business includes rental and sale of prerecorded videocassettes played on VCRs. We will not look at the VCR in this book.

[5] **Conditions for Extension of the CVR:** The conditions for extension were hotly debated at the time, and will be considered below in some detail.

[6] **Window Days Definition:** These days are business days. The number of windows K_{max} was specified in the contract.

A Complex CVR Option

windows, namely $S_{ABC}^{Avg}(k) = \underset{t \in \left(t_k^{Start}, t_k^{Start} + \Delta t_{Window}\right)}{Average} \left[S_{ABC}(t)\right]$. Then the median of these averages over all windows is found, viz $S_{ABC}^{Median} = \underset{k=1...K_{max}}{Median}\left[S_{ABC}^{Avg}(k)\right]$.

This S_{ABC}^{Median} is then compared with the strike prices in the put spread. The same condition holds at the second strike date.

At the third exercise date, the maximum of the averages is compared with the strike prices instead of the median of the averages. This point, in the fine print, is of extra value to ABC since it inflates the effective ABC stock price and thus diminishes the payout to DEF.

The structure of the deal so far is summarized in the diagram below. The two boxes marked "Put Spreads" and "Extension Options" comprise the CVR[7].

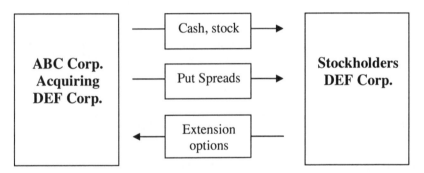

Basic Structure of the M&A Deal

Important additional complexity was present, including a "Chameleon Bond" and a warrant depending on it. These topics are in the footnotes below [8,9,10].

---

[7] **Specification of the CVR:** The CVR structure and parameters were specified by the investment bankers, who were provided with pricing information under various assumptions as described below.

[8] **The Viacom "Chameleon Bond" and its Behavior:** To help fund the Paramount deal, Viacom needed to issue merger debt. The debt was itself a contingent object, and was dubbed the "chameleon bond". This was because the bond could change its character depending on future events - notably the possibility of acquisition by Viacom of Blockbuster Entertainment Corp. The specifications included an exchangeable feature of the debt into preferred stock at Viacom's option if the acquisition of Blockbuster was not completed by a certain date. The reason was that Blockbuster had strong cash flow, and if acquired Viacom would have the additional cash to pay the debt coupons. Otherwise, Viacom needed to change the bond into stock. Various other bells and whistles were also present. The logical flow diagram for the chameleon bond took up an entire page with

The Arbs and the Mispricing of the CVR Option

Significant arbitrageur activity for a potential M&A deal is often present, where "arbs" bet on the deal going or not going through with appropriate positions in the ABC stock[11]. This can have an undesirable effect from the point of view of ABC. If the arbs own the CVR, which is some form of put, they want the ABC stock to go down near the exercise or payoff time. Therefore, near payoff, they may start to sell ABC stock or go short. This puts downward pressure on the stock, increasing the CVR payoff to the arbs. In this case, ABC tried to thwart the arbs, partially by making the CVR option so complicated that the arbs would have trouble understanding and pricing it. The result was that at first the CVR was *overpriced* by being considered by the arbs as a single one-period put spread, ignoring the negative effect on the CVR value through the complex extension conditions.

After the deal took place, the CVR was listed (on the ASE) and traded in the secondary market. At this point, the market wound up under-pricing the CVR relative to models. Hence, at both the start and afterwards, the CVR option was chronically "mispriced" with respect to "theoretical fair value". For discussion, see the footnotes[12,13].

---

small print, and was difficult to memorize. Nominally the bond was 8% debt with maturity 7/7/06. The bond was B1 credit, and so was a high-yield "junk" bond.

[9] **Junk Bonds and Stock:** The behavior of the price moves of the chameleon bond for an initial period after it was issued was highly correlated with Viacom's stock price moves. This is often the case with high-yield bonds, because the elevated credit risk for a high-yield bond is often regarded as making the bond risk similar to the risk for stocks in the eyes of investors.

[10] **The Bond-Dependent Warrant was Another Complication:** Warrants are call options that allow the warrant holder to purchase ABC stock when exercised. Warrants lead to some dilution, but they are less risky than CVR instruments because they are exercised when the stock is doing well, which lowers the dilution pain. The exercise can be performed, using cash or sometimes with other securities provided by the warrant holder, to ABC. Another complication in the Viacom-Paramount structure was a warrant that could be exercised either with cash or by using the chameleon bond under certain conditions. A precise valuation of this warrant including the convertibility dependence on the bond with its myriad uncertainties (including the probability of the Blockbuster merger) was not possible. In practice, this complex warrant was valued using the cash exercise assumption.

[11] **Acknowledgements:** I thank Doug Hiscano for informative discussions regarding arbs and on many other topics.

[12] **Important Philosophical Issue: To Be or Not To Be Mispriced:** So if the market price of a security disagrees with the theoretical valuation, who is right? An aggressive point of view announces that the market in such a case is "wrong", or "out of equilibrium", and given enough time the market will revert to the theoretical value. This sort of analysis is called "rich-cheap analysis". The time scales for the market to thus wise up and revert are often a bit hazy. If there is considerable empirical evidence in a

A Simplified CVR: Two Put Spreads with Extension Logic

Many of the basic issues arising for the CVR can be explained using two put spreads expiring at different times t_1^* and t_2^* with no averaging side conditions, since everything can be done explicitly. These two put spreads will be taken with the same parameters as occurred in the CVR[14]. We need back-chaining logic.

Back-Chaining Logic with Two Times

The back-chaining logic is needed to discover the "critical decision indifference points" in the stock price at each exercise date, determining academically if ABC pays out or extends. The procedure for the simplified example takes place at t_1^*. At this time, ABC has to make a decision on what to do. ABC first calculates the intrinsic value of the cash flow that it must pay at t_1^* due to the first put spread, if ABC does not extend. ABC also calculates the second put spread evaluated at t_1^*. This is the discounted expected value of future cash payments, which ABC may have to make at time t_2^*. The academic critical value or indifference point $\$S_{Ind}$ at t_1^* occurs when these two numbers are equal. At this point, ABC should be indifferent whether to pay out or extend. We solve for $\$S_{Ind}$ using this formula:

$$\$E_{1Upper} - \$S_{Ind}\left(t_1^*\right) = \$C_{\text{2nd Put Spread}}\left[\$S_{Ind}\left(t_1^*\right); \$E_{2Upper}, \$E_{2Lower}\right] \quad (13.1)$$

Non-Uniqueness of the Indifference Point for Two Put Spreads

Normally, there is one price region of extension and another price region where extension does not occur. A further complication occurs in this case, because it turns out that the indifference point is not uniquely determined mathematically

---

given case that the market oscillates around the theoretical price, then there are good grounds for such a view. However unusual non-standard securities may not have any equilibrium behavior, or more to the point there may be many equilibria.

[13] **Liquidity and Models:** With regard to the previous footnote, one can always say that the market is, in its infinite collective wisdom, making a "liquidity adjustment", or "low-demand adjustment", or "perceived credit adjustment", or whatever, to the theoretical price. These cannot be calculated (else they would be taken into account in the theoretical price to begin with). Only direct experience with the market allows evaluation.

[14] **Parameters:** The first two put spreads in the CVR had upper and lower strikes of ($48, $36) and ($51, $37). These are the put spreads in the simplified example. The third put spread in the CVR was ($55, $38). In the CVR, these numbers were compared against the functions of the stock price (median of averages, maximum of averages) discussed in the section on averaging.

from Eq. (13.1). This is because a put spread's intrinsic value at exercise can intersect another unexercised put spread (that has time value), at more than one point of the underlying stock.

The graph below shows an example. The intrinsic value of the first put spread with $\$E_{1Upper} = \$48$ and $\$E_{1Lower} = \$36$ is plotted against the second put spread with one year to exercise, and with $\$E_{2Upper} = \$51$ and $\$E_{2Lower} = \$37$. These are the relevant parameters for the CVR. The example shows the existence of two indifference points near $28 and $39. Hence, the "rational market" logic is more complex than usual.

The same phenonomenon occurs with the full CVR extension logic, although because of the complexity it is harder to visualize. The next graph shows the non-uniqueness of the indifference point in the example with two put spreads:

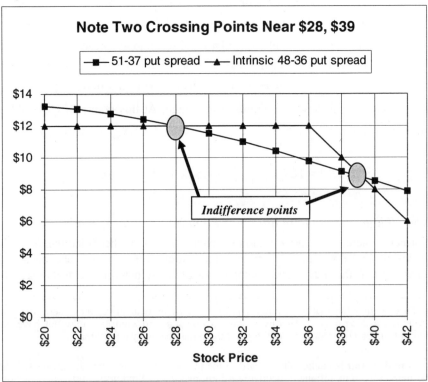

Non-Academic Corporate Decision for Option Extension

An additional complication in pricing the option in a realistic fashion involves a non-academic argument that may be important in practice[15]. The usual rational market academic dictum is not to extend if probabilistically more would have to be paid later. In that case, the standard academic logic would insist on a payout now, with no extension. However, ABC may want or need to ignore this academic criterion in favor of a more practical consideration.

Specifically, even if it would theoretically "save money" not to extend, ABC may find it preferable at a corporate level not to pay out substantial sums[16] in a weakened condition. For example, ABC may not have enough cash to pay out and may not be able to borrow the money easily, etc[17]. So ABC may take a "corporate decision" to extend.

Such a corporate decision is not a default of ABC, because no rational market logic is written into the contract with DEF; i.e. it is entirely at ABC's discretion to extend. The corporate decision is also not a scenario "what if" analysis[18], because all stock paths are included in the analysis. Stock paths are not "cherry-picked", i.e. we do not look only at low stock-price paths but rather at all paths. Instead, the rational-market decision process itself is modified.

Normally such corporate decision considerations are not included in option pricing[19]. However, the true worth of the option to the holder and to the issuer do depend on these considerations. The usual academic rational-market procedure

---

[15] **Disclaimer:** All material in this section relative to possible corporate behavior is merely surmised in the particular case involving Viacom, for illustrative purposes.

[16] **Payoff Amounts and Corporate Decision to Extend:** For Viacom, there were around 57MM CVRs that would need the maximum payoff of $12 at the end of the first year if the stock dropped enough, and without a corporate decision to extend, over $680 MM.

[17] **Loans to ABC to Pay off the Price Protection?** It may be problematic for a bank to extend credit to ABC in this sort of situation. First, ABC may already be weakened by purchasing DEF, and second ABC may need to pay out a large amount on a put option just when ABC's stock is dropping. A loan could be problematic for ABC also; the interest rate charged under such circumstances could be high, and the additional leverage might wind up putting some pressure to lower ABC's credit rating.

[18] **Scenario What-If Analysis:** A scenario what-if analysis corresponds to choosing a particular stock price path or a collection of paths. These analyses are extremely valuable, since economic considerations can be incorporated. However, for model option pricing we include all paths, with path densities in the stock price at a given time given by the model. Sometimes people may think of the consequences of some scenario and then generalize inappropriately to the option price.

[19] **Options, Cushions, and Corporate Decisions to Extend or Exercise:** Sometimes extra parameters are built into option pricing to include corporate decisions to exercise or extend, including cases involving embedded options in callable corporate and convertible bonds. These parameters are sometimes called "cushions".

determines the theoretical price. However if the rational-market procedure is not followed in reality, then an additional component of risk or value exists that cannot be determined by option theory. Without detailed knowledge of the individual corporate situation it is naturally a-priori problematic to assess the parameters. First, there is the probability $\wp_{CorpDecExt}$ that the corporate decision will be made. If so, there is the stock price $\$S_{CorpDecExt}$ at which the extension takes place. We do not try to estimate $\wp_{CorpDecExt}$, and simply price the CVR both ways—with and without the corporate decision to extend. If ABC does decide to extend we need $\$S_{CorpDecExt}$ and we need to assess the effect on the option price. It would be preferable to adopt a method that does not require detailed corporate information. It turns out that there is a feature in the case of extendable put options that provides a natural choice for the quantity $\$S_{CorpDecExt}$.

Compromise by Using the Two Indifference Points
Consider the next figure:

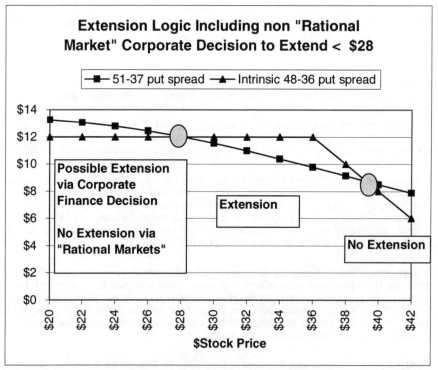

If two "indifference" points occur, a compromise solution can be found obeying both the rational market procedure and the corporate finance extension considerations in different regions. This graph shows that we can use the upper indifference point as corresponding to the rational market procedure, and choose the lower indifference point as corresponding to the low stock price $\$S_{CorpDecExt}$ at which the corporate decision to extend might be made. The possible corporate decision to extend is made when the intrinsic 48-36 put spread is at its maximum value, \$12. Possible extension when payout is maximal persists to the full CVR.

The CVR Option Pricing

We can now discuss the full complexities of the CVR pricing. The side conditions render the analysis complicated. Different paths in the stock price lead to completely different states and the result at times past an extension depend on what happened at previous possible exercise times. Twelve classes of paths exist.

First Year Paths
The following picture illustrates the first-year paths:

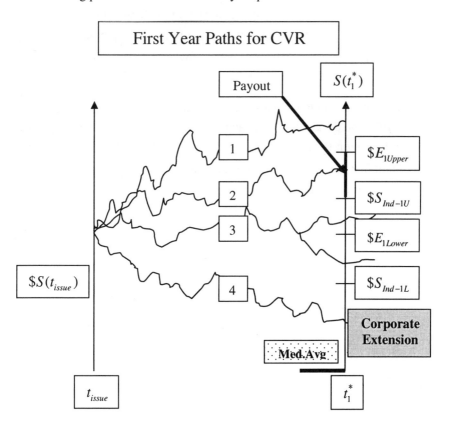

In the first year for the CVR, there are four classes of paths, starting at the spot stock price at the issue date t_{Issue} up to the first exercise time t_1^*, as shown above. We need to calculate the median of averages, as described above, during the period marked Med.Avg before t_1^* on the diagram.

The first-year paths are:

- *Class 1 paths* are such that the median of averages winds up above the upper strike price $\$E_{1Upper}$ for the first year put spread. In this case, the CVR is worth nothing.
- *Class 2 paths* have the median of averages below $\$E_{1Upper}$ but above the point $\$S_{Ind-1U}$. This is the "upper indifference point" that we get by rational market back-chaining logic. In this region, ABC pays DEF the value of the first put spread, $\$E_{1Upper} - \$S(t_1^*)$.
- *Class 3 paths* extend past t_1^* and follow the rational market logic. For these paths, ABC calculates a smaller economic penalty on the average for extending the option than for paying out at t_1^*, so extension occurs.
- *Class 4 paths* arrive below the "lower indifference point" $\$S_{Ind-1L}$ at t_1^*. Normally there is one indifference point. However, for two put spreads there are two indifference points, as we saw above. There are two possibilities.

According to the rational market logic, paths below $\$S_{Ind-1L}$ should stop, because ABC would calculate a greater economic penalty on the average for extending the option than for paying out at time t_1^*. If this happens, ABC pays the maximum first put spread intrinsic value, $\$E_{1Upper} - \$E_{1Lower}$.

However, just as in the simple example of two put spreads discussed above, a corporate decision to extend may occur. Again, this is not a feature of academic options pricing but because a weakened condition of ABC may occur.

Second Year Paths

The paths for the second year for the CVR are in the following diagram:

A Complex CVR Option

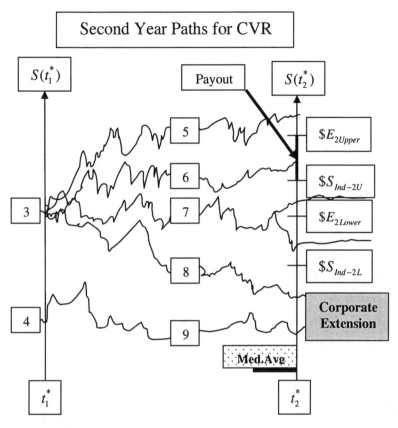

The second-year paths are:

- *Class-5 paths* finish out of the money and the CVR terminates at no cost.
- *Class-6 paths* are terminated at t_2^* with a payout equal to the intrinsic value of the second-year put spread down to the point $\$S_{Ind-2U}$ where the payout stops, similar to the class-2 paths at t_1^*.
- *Class-7 paths* extend past t_2^* through the standard back-chaining logic.
- *Class-8 paths* would stop at t_2^* by the rational market logic and pay the maximum amount $\$E_{2Upper} - \$E_{2Lower}$. However, these paths may extend past t_2^* by a corporate decision to extend, as explained above.
- *Class-9 paths* are a continuation of class-4 paths, which resulted from a previous corporate decision to extend at time t_1^*.

Third Year Paths

The paths for the third year for the CVR are in the diagram below.

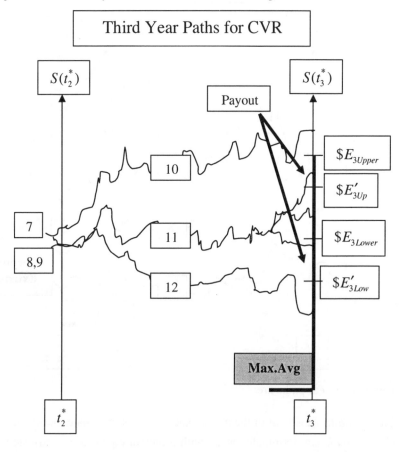

The third-year paths are:

- *Class-10 paths* are out of the money at t_3^* and are terminated without cost.
- *Class-11 paths* are paid out at t_3^* as a put spread. As explained below, the maximum condition may be approximately taken into account by lowering the put strikes $\$E_{3Upper}, \$E_{3Lower}$ for purposes of calculation to $\$E'_{3Up}, \$E'_{3Low}$.
- *Class-12 paths* are the continuation of the corporate-extension paths but which end at t_3^* with payouts as above.

Analytic CVR Pricing Methodology

An analytic method can be employed in practice. This method involves bivariate and trivariate integrals, with back-chaining logic. The bivariate integral occurs because the paths that terminate at t_2^* cannot range over all values at t_1^*. The trivariate integral occurs because the paths that terminate at t_3^* cannot range over all values at t_1^* and t_2^*. The method is not exact, mainly because of the complexities of the averaging. However, corrections can be made to the volatility to account for the media of averages and maximum of averages in a reasonable fashion, as discussed below.

Payout Details for the CVR
The payouts are as follows:

- For t_1^*, the payoff from the class-2 paths is a standard put-spread intrinsic value with upper strike $\$E_{1Upper}$. However, the put spread is cutoff to zero at price $\$S_{Ind-1U}$. This cutoff can be modeled as a short position in a digital option that pays $-\left(\$E_{1Upper} - \$S_{Ind-1U}\right)$ for $\$S\left(t_1^*\right) \leq \$S_{Ind-1U}$, or actually the median of the averages [20].

- For t_2^*, the payoff from class-6 paths again looks like that arising from a put spread minus a digital option. However there is an important difference, because the paths that extend past t_1^* are filtered. In some sense, we must multiply the payoff by the probability that the paths get past t_1^*. This translates into the bivariate integral[21].

- For t_3^* the buck stops, and payouts need to be made. The filters at the preceding times leave two extra integrals, and a trivariate integral appears[22]. This time the maximum of averages must be taken into account.

---

[20] **Digital Option:** A digital option just pays off a constant value. Hence, this digital option is just an appropriate integral over the Green function for the first year, with appropriate discounting. We discuss digital options in Ch. 15.

[21] **Bivariate Integral:** Technically, in the path integral framework the successive use of the convolution semi-group property over all intermediate values converts short-time Gaussian propagators into long-time Gaussian propagators. However, this breaks down when intermediate values are restricted, as they are at t_1^* so one additional integral remains. See the chapters on path integrals, Ch. 41-45.

[22] **Trivariate Integral:** The same remarks as made for the bivariate integral apply, except that two extra integrals remain due to some paths terminating at the two previous times.

The summary payout diagram for the whole CVR below should clarify the ideas:

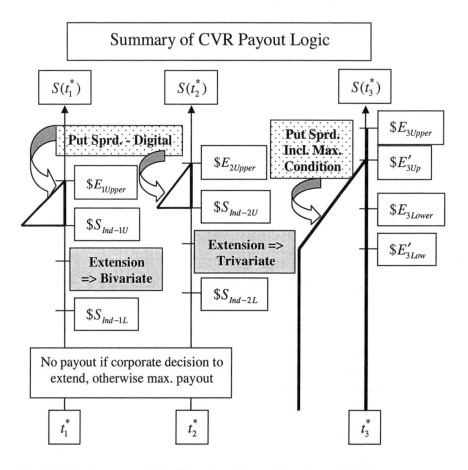

Possible Irrelevance of the First Two Lower Put-Spread Strikes

If ABC decides to extend the option at t_1^* and t_2^* below the lower indifference points $\$S_{Ind-1L}, \$S_{Ind-2L}$ then the lower strikes $\$E_{1Lower}, \$E_{2Lower}$ do not enter the payoff conditions. Hence, in that case these variables become irrelevant for the value of the CVR. Hence lowering the values of the lower strikes would not have increased Viacom's risk under the corporate extension decision assumption.

Back-chaining Logic for the CVR

The back-chaining logic for the CVR is a straightforward extension of the logic presented in the example for two put spreads. Again, back chaining is needed to discover the academic decision indifference points in the stock price at each exercise date, thus determining academically if ABC pays out or extends. It is easier to implement this logic with the analytic approach rather than the Monte-Carlo approach.

The procedure begins at the next to last exercise date t_2^*. ABC compares the intrinsic value of the cash flow that it must pay at t_2^* (if ABC does not extend) with the discounted expected value of all future cash payments. At t_2^*, the future cash payments are those at time t_3^*. The academic critical or indifference value of the stock price $\$S_{Ind}\left(t_2^*\right)$ at t_2^* occurs when these two numbers are equal.

A similar calculation is then performed at the first exercise date t_1^*. The expected discounted value of future cash flows at t_2^*, t_3^* must be evaluated at time t_1^*. This includes a bivariate integral to evaluate the expected discounted t_3^* cash flows at time t_1^*. The calculation performed at t_2^* clearly affects the future cash flows at t_1^*. In the back-chaining procedure, the results provide the standard academic criteria telling ABC when to pay out and when to extend.

The Averaging Conditions

For each of the first two years, the median of averages over a window is specified. Correspondingly, the volatility used in the analytic method can be reduced according to a standard approximation[23]. To be specific, the volatility in the presence of averaging in the time interval $T_{AvgPeriod}$ roughly leads to a reduction in the square of the volatility σ^2 by a factor of 1/3. Because variances add in the analytic model, we get the average volatility σ_{avg} given by

$$\sigma_{avg}^2 T_{Total} = \sigma^2 T_{NonAvgPeriod} + \sigma^2 T_{AvgPeriod} / 3 \qquad (13.2)$$

For the median of averages, the approximation of substituting the average of averages can be used, and the averaging method is then applied a second time to

---

[23] **Volatility Modifications for Average Options:** See derivation of volatility for average options using path-integral techniques (see Ch. 41-45). The factor 1/3 occurs in the limit of large numbers of observations in the averaging period.

the averaged volatility σ_{avg}. With the window interval Δt_{Window}, we get the median of averaged volatility σ_{MedAvg} as[24]

$$\sigma^2_{MedAvg} T_{Total} = \sigma^2_{avg}\left(T_{Total} - \Delta t_{Window}\right) + \sigma^2_{avg} \Delta t_{Window}\Big/3 \qquad (13.3)$$

For the third year specifying the maximum of averages, a 4-step binomial model was used in the window before t_3^* to get approximately the maximum of averages, and an effective lowering of the third year put-spread strikes was calibrated to agree with this model. Thus, an equivalent option can be defined by lowering $\$E_{3Upper}, \$E_{3Lower}$ to $\$E'_{3Up}, \$E'_{3Low}$ in order to take account of the maximum condition.

Some Practical Aspects of CVR Pricing and Hedging

In this section, we give some illustrative indications for the pricing and hedging methodology for the CVR[25].

Standard Rational, Market Logic

We first assume the canonical standard rational-market logic—i.e. no corporate decision to extend. On 6/7/94 (one day after being listed), the market value for the CVR was $7.625. The theoretical price of $8.96 and the market value $7.625 differed by about 15%, the market CVR price being "cheap" with respect to the model. This "cheapness" continued to be true as the CVR was traded in the market[26]. Evidently, the market discounted the value of the CVR below the model for some reason[12].

---

[24] **Numerical Volatility Modifications:** The Viacom-Paramount CVR contract specified $T_{Avg\ Period}$ = 20 days and Δt_{Window} = 60 days (business or trading days). We took T_{Total} = 250 days for one year, a common assumption (sometimes 252 or 260 days is used). The reduction in the input volatility σ was thus a factor 0.892 so σ = 28% became 25% in years 1 and 2.

[25] **Indicative Prices:** The theoretical numbers quoted in this section for 6/7/94 are intended only as an illustrative pedagogical example. Juan Castresana numerically implemented the full CVR analytic methodology.

[26] **Arbs Again:** It seems that at this time the arbs were no longer regarding the CVRs as 1-year put spreads, knew that the theoretical CVR price including the extension options was above the market price, and regarded the CVRs as cheap puts.

CVR Components

The CVR component parts that need to be priced correspond to the diagram of the payout logic above. Explicitly, these are:

- First year put spread minus digital
- First year low indifference point digital
- Second year filtered put spread minus digital (bivariate)
- Second year filtered low indifference point digital (bivariate)
- Third year double-filtered put spread (trivariate)

A simple one-year put spread was worth $10.17, so the (negative) extension value ($8.96 − $10.17) on 6/7/94 was worth −$1.21.

Indifference Points

It is important to calculate the values of the indifference points that we discussed above. These are:

- Upper indifference point at t_1^* : $$S_{Ind-1U}$
- Lower indifference point at t_1^* : $$S_{Ind-1L}$
- Upper indifference point at t_2^* : $$S_{Ind-2U}$
- Lower indifference point at t_2^* : $$S_{Ind-2L}$

On 6/7/94 these were $$S_{Ind-1U}$ = $43.25, $$S_{Ind-1L}$ = $24.55, $$S_{Ind-2U}$ = $44.98, $$S_{Ind-2L}$ = $28.18. Note that the lower indifference points were below the lower put-spread strikes. Therefore, e.g., a stock price at t_1^* below $$S_{Ind-1L}$ would correspond to the maximum put-spread payoff $$E_{1Upper} - E_{1Lower}. If a corporate decision to extend is made at t_1^*, the condition $$S_{CorpDecExt} = S_{Ind-1L} corresponds to avoiding this maximum payoff. This is similar to the two put-spreads example discussed above. We turn next to this possibility.

Non-Standard Corporate Decision to Extend

If, hypothetically, the corporate decision to extend is made, the value of the CVR corresponding to the above parameters increased by $0.44 to $9.40. Recall that the expected cash flow is increased because ABC decides not to pay, even if it probabilistically could cost more later. The digital options below the lower indifference points are eliminated for the first two years in this case.

The CVR component parts in this case are:

- First year put spread minus digital
- Second year filtered put spread minus digital (bivariate)
- Third year double-filtered put spread (trivariate)

Evidently, since the market price was lower than theoretical, either the market did not include this non-standard corporate decision to extend or else the discounting mechanism by the market from the model price already noted for the standard rational-market logic was increased.

Maximum Condition and Averaging Contributions.

We can look at the results with and without the third-year maximum condition. Recall that imposition of the maximum condition increases the effective ABC stock price at the third year, and so lowers the value of the price protection. The maximum condition lowered the CVR value by about -$0.28.

Given all the press about the complicated averaging conditions, it is of interest to value the CVR with and without averaging. The averaging contribution on this day was worth about $0.24.

CVR Valuation with Other Interest Rates

The interest rate r mainly affects the option through the forward stock price. In the absence of dividends, the forward price at time interval T from now is

$$\$S_{fwd}(T) = \$S_{spot} \exp(rT) \qquad (13.4)$$

The results quoted above were for the 3-year rate at the appropriate ABC credit. Although academically we are supposed to use the "risk-free rate", using Libor in fact gave an CVR model result on this day of around $9.70. This was farther from the market. In order to drive down the theoretical price to the market price, a rate over 13% had to be used. We might regard this high 13% rate as a measure of the inherent credit risk placed by the market on the CVR. On the other hand, we might regard it as the penalty for reduced demand[12].

CVR Hedging and Scenario Results

In order to trade, we need to know how to hedge, and so we need the CVR sensitivities. We also want to run scenario analyses, revaluing the CVR at different stock prices and vols. While it might seem redundant to do both, for a highly non-linear object like the CVR the usual sensitivities are not so exact. The hedges used were the usual quantities[27]:

---

[27] **Units and Risk Reports:** To emphasize it again, because risk quantities can be and are defined in different ways, it is always a good idea to get people to specify the definitions

A Complex CVR Option

- Δ = Delta (per \$1 stock change)
- γ = Gamma (per \$1 stock change)
- Vega (per % volatility change)

The hedging sensitivities calculated for 6/7/94 were $\Delta = -0.33$, $\gamma = -0.02/\$$, vega $= -\$0.08$ [28].

Physical Intuition

It is useful to think of stock price moves mentally to get intuition and a physical feel for the risk. At \$28 with a one-standard deviation move of around $\sigma\$S = \$7$, and including the forward stock price change from spot of \$1.30, the stock would likely be in the range \$22 to \$36 at the end of one year. This range covers the lower indifference point $\$S_{Ind-1L}$ to the lower strike $\$E_{1Lower}$.

Indicative Scenario Results, Comparison with the "Math Calculation"

Here are some scenario results. In practice, tables of scenario results may be provided to the trading desk. For fixed vols, interest rates, and time periods and stock price changes of \$4 and \$8, a "reval" calculation produces:

- If \$S = \$32, the CVR price = \$7.50, a change $\$\delta C_{Reval} = -\$1.46$
- If \$S = \$36, the CVR price = \$5.93, a change $\$\delta C_{Reval} = -\$3.03$

We can compare the "reval" results with the usual "math calculation" result

$$\$\delta C_{MathCalc} = \$\Delta \delta S + \tfrac{1}{2}\$\gamma(\delta S)^2 + \$vega\delta\sigma \qquad (13.5)$$

We get $-\$1.48$ and $-\$3.28$ for $\$\delta C_{MathCalc}$ in the above two cases. For small moves, we naturally expect $\$\delta C_{MathCalc} \approx \$\delta C_{Reval}$. The difference between the Math Calc and Reval is \$0.02 at $\$\delta S = \$4$, and \$0.25 at $\$\delta S = \$8$.

We can also look at the change in delta Δ. The conclusions are similar, $\delta\Delta_{MathCalc} = \gamma\delta S$.

---

including units. It is an even better idea to ask the programmers in the appropriate department to program the computer to print out this information in unambiguous, complete, recognizable English, free from local acronyms, directly on the risk reports.

[28] **Vega for CVR:** The sign of vega depended on the stock price. At other stock prices, vega was positive.

The CVR Buyback

It is clear that ABC would like to terminate the price protection. We have spent a lot of time in this chapter discussing under which conditions ABC would or would not be likely to extend, mainly concentrating on the dangerous (for ABC) low stock price paths. In fact, it turned out that the ABC stock price rose after the acquisition, and the ABC management then bought back the CVR option from the stockholders at a nominal fee[29]. The corporate risk management using the CVR option was successful from the point of view of ABC, and DEF stockholders enjoyed the partial price protection required to make the deal go through at the start. In sum, the CVR was a success.

A Legal Event Related to the CVR

Legal risk is an increasingly important topic. Quants might become involved in exceptional cases when complex securities are involved[30]. A legal event arose in connection with the CVR option. The information below is public record [ii].

An intellectual-property case was decided by New York Stock Exchange arbitration on 6/11/97 on a claim by T. Inman against Smith Barney regarding the Viacom/Paramount bid including the CVR. The total damages asked were over $15,000,000. Smith Barney was the advisor to Viacom, and the Smith Barney investment bankers designed the bid including the CVR.

The decision, in favor of Smith Barney, can be accessed on the Internet (NYSE, ref. ii).

References

---

[i] **Background Information on the Viacom-Paramount Deal**
Fabrikant, G., *Viacom Revises Its Offer – The Gap Narrows on Paramount Bids*. New York Times, Sect. D, 1/19/94.
Greenwald, J., *The Deal that Forced Diller to Fold – The inside story of how Viacom's Sumner Redstone placed a $10 billion bet against QVC's Barry Diller and finally won the long battle for Paramount*. Time, Vol 143, 2/28/94.

---

[29] **Viacom's CVR Buyback:** Viacom paid $83MM for around 57MM CVRs at a value of $1.44 @ on 7/7/95, corresponding to a median average stock price of $46.56. The buyback corresponded to the class-2 paths that paid off at t_1^* in the analysis. The initial valuation of the CVR was substantially larger than the actual CVR payout, so Viacom essentially realized a profit on the CVR.

[30] **History:** I was the quantitative advisor and the designated expert for Smith Barney during this arbitration. The experience was absorbing.

OBrien, R., *Market tumble slows resurgent M&A sector.* Investment Dealers' Digest, p. 14, 4/18/94.
Bowe, S. and Doherty, J. *Price of Viacom Bonds Rises in Advance Dealings Despite Confusion Over How to Value Them.* New York Times, Sect. D, 6/8/94.
Fabrikant, G., *Viacom to Pay Paramount Investors $83 Million.* NY Times, D, 6/28/95.
Dow Jones & Co., *Viacom Inc. to Retire a Security it Issued in Paramount Deal.* Wall Street Journal, B3, 6/28/95.
Falloon, W. *Equity Derivatives: The $48 and $52 Questions.* Risk, Vol 8 No 4, 1995.

[ii] Legal Case Related to the CVR Option

Siconolfi, M., *The Outsider: Did Smith Barney Cheat a Stockbroker to Clinch a Merger? Mr. Inman Claims Firm Stole His Finance Ideas to Save Viacom-Paramount Deal.* Wall Street Journal, page 1, 8/8/96.
New York Stock Exchange Arbitration, *Case: Thomas S. Inman v. Smith Barney Inc.*, www.nyse.com/pdfs/1995-005167.pdf, 6/11/97.

14. Two More Case Studies (Tech. Index 5/10)

In this chapter we consider two more case studies of structured exotic products. We first consider DECs and synthetic convertibles, including simple and complex varieties. We then go on to consider an exotic equity call option with a variable strike and expiration.

Case Study: DECS and Synthetic Convertibles

Some simple forms of convertibles are synthetic convertibles called DECs. A convertible bond is a security has both bond and embedded equity options with possibility of conversion to stock[i]. A synthetic convertible is generally simpler than a true convertible bond. A synthetic convertible example is the DECS ("Debt Exchangeable for Common Stock" or "Dividend Enhanced Convertible Stock"). It is common to use DEC for short pronounced "deck"; the plural is then (somewhat inconsistently) DECs. DECs are structures popular in corporate finance[1]. A DEC can be written explicitly as a sum of the underlying stock, coupon payments, and equity options. A true convertible bond is on the other hand a composite structure that requires complex pricing models.

We start with simple DECs and then proceed to a more complicated DEC (called DEC_{123}) that will be used to study several ideas in the same example. The discussion will also give an idea of what is sometimes asked of quants in practice for coming up with quick "indications" for ideas generated by Corporate Finance. We will walk through this calculation from soup to nuts. The DEC_{123} has these attributes:

- It is a "Synthetic Convertible" involving DECs.
- It has a "Best Of Option" [ii, 2] among three different DECs.

---

[1] **An Art of Corporate Finance**: There are many similar products: ELKS, ACES, EAGLES, MEDICS, SUNS, SPIDERS, FLIPS, etc. It seems that making up structured-security acronyms as words is a skill honed to a fine edge in Corporate Finance.

[2] **"Best-of Option"**: These clauses in the contract allow the choice of the "best of" a set of results by the investor X at a given time. In the present situation, X gets to choose the best result from three DECS. This provides additional value to X through diversification.

We shall look at these features in turn [3]. Those just interested in standard DECs can read the next section and skip the best-of option section.

Standard DECs (These are simple)

The idea is that the investor X gets an extra coupon above ABC dividends due to an exchange of equity options favorable to ABC. The stock on which these options are written may or may not be ABC stock[4]. At the same time the DECs allow ABC to issues debt in an advantageous manner. A prototype DEC has the following composition:

- A coupon is paid to X (calculated as a function of the structure). This is often re-expressed as a DEC-dividend yield.
- A European call option C_{ABC} is owned by ABC with a strike E_L generally set at current stock price S_0 at issuance.
- A European call option C_X is owned by X for a fraction $\eta < 1$ of a share, with a strike E_H set at a higher level[5] $E_H = \xi E_L$ with $\xi = 1/\eta$.
- The DEC behaves like the underlying below the current stock price S_0. Therefore the investor X suffers losses if the stock price drops.

The payout diagram of a DEC at maturity T^* from the point of view of the investor X is provided in the following diagram:

---

[3] **Remark:** Although DECs are common, this DEC_{123} invented by Corporate Finance was not marketed. Here it just serves as a good laboratory example.

[4] **Exchangeables:** If the stock on which the options are written is not ABC stock, the option is called an "exchangeable".

[5] **DEC Jargon**: *In general it is important to learn the jargon, because that is the language of the deal description. Moreover, the description may occur at rapid fire pace.* If $E_L = 100$ and $\xi = 1.25$, the DEC is said to be "100 up 25". The "conversion premium" is $\xi - 1$ or 25% and the number of shares η received by X upon exercise of the C_X option $\eta = 1/\xi = 0.8$ is the "participation" or "conversion ratio". The "conversion price" is E_H.

Two More Case Studies

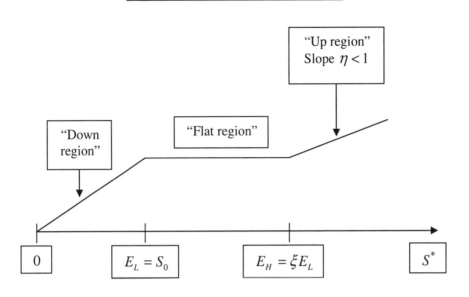

At the start of the deal (issuance) the stock price is called S_0. At T^* the stock price is denoted as $S^* = S(T^*)$, so if $S^* < S_0$, the investor X has a loss (the "down region"). For $E_H > S^* > S_0$ the payoff is flat due to X's short C_{ABC} option position. The payoff increases with slope η above E_H due to X's long C_X option at participation η.

Various products exist with different numbers of options, resettable strikes, callable structures[6], American options, etc. We will not consider them here.

Analysis of a Standard DEC

The analysis of the DEC described above is relatively straightforward. A diagram of a typical DEC structure is:

---

[6] **Callable DEC:** The issuer ABC can call the structure if call provisions are provided. These call provisions are similar to those of ordinary debt. They include a "call schedule" containing each call price or amount ABC will pay X if ABC exercises the call, when the call is allowed. There may be NC (No Call) provisions or "protection" during a time period when no call is allowed. ABC may call in order to "force conversion". In this case, X will have an American option allowing exercise at any time, and it will be better for X to exercise the option to convert to stock rather than accept the call price.

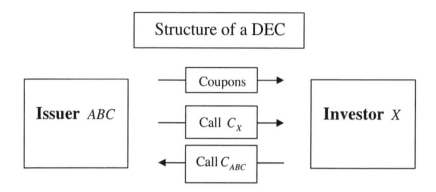

The present value (i.e. discounted) for the total sum of coupons paid to X up to maturity T^* (over and above any standard dividends) must, for fairness, be equal to the difference of the values of two options. These two options are the call C_{ABC} that X has given to ABC, and the call C_X that ABC has given to X. Now $C_{ABC} > C_X$ so X does get a positive coupon. The option values are calculated at the start of the deal when the coupon is specified. The options are generally simple European options that can be priced with the Black-Scholes formula. We have

$$PV(\sum \$Coupons) = \$C_{ABC} - \$C_X \qquad (14.1)$$

The DEC is quoted using equivalent dividend yield y_{DEC} (constant spread) that produces the present value of the sum of coupons. Actually, there is a "bogey" dividend yield $y_{DEC}^{"Bogey"}$, attractive enough to investors to make the deal sell. Denote r_{ABC} as the ABC corporate rate that is used to perform the discounting of coupons, and call f the number of coupons paid per year up to N years. Then the yearly $Coupon is the result of performing a standard geometric sum,

$$\$Coupon = y_{DEC}^{"Bogey"} \$S_0 = (\$C_{ABC} - \$C_X) r_{ABC} \frac{(1 + r_{ABC}/f)^{fN}}{(1 + r_{ABC}/f)^{fN} - 1} \qquad (14.2)$$

Credit Spreads, Discounting, Convertibles, and DECs
One thorny discussion topic is which interest rate to use for pricing the options in a convertible-type security[iii]. In particular we want to ask the following question:

- Which Interest Rate Should Be Used for Equity Options – the ABC Credit Rate r_{ABC}, or Libor, or Both?

Get ready for a confusing situation. This issue comes up in pricing both synthetic and standard convertibles. We start by breaking out the options from the DEC coupon—which could be done here. The options could be traded separately in a secondary market.

Textbook Argument: Use Libor
A textbook equity-option model specifies a "risk-free rate", but the question is more complicated in real markets. Note that increasing the interest rate r generally increases the price of the equity option. The forward stock price S_{fwd} at time T^* is proportional to $\exp(rT^*)$ and occurs inside the expectation integration over S^*. The increase of this term with increasing r generally is larger than the decrease due to discounting with discount factor $DF = \exp(-rT^*)$, an overall factor outside the S^* integration. A broker-dealer BD hedges equity options using BD funding (near Libor); this might argue $r = r_{Libor}$ both for the forward stock price and for discounting.

Other Arguments: Use Corporate Rate with the Credit Spread
If ABC were to hedge the options, the amount a bank or the markets would charge for ABC's hedging procedure would, indeed, include the ABC credit spread. This argues that from ABC's point of view, $r = r_{ABC}$. If the investor X could enter the argument he might say that since he is forced to take ABC credit risk, he would also use r_{ABC} everywhere. The options would then be priced higher, and justifiably from X's point of view would give X an increased DEC coupon, compensating X for the ABC credit risk. The BD could reach the same conclusion because these options are embedded in an ABC credit-sensitive structure and so carry ABC credit risk.

Compromise Point of View: Use Both Rates
An intermediate compromise point of view uses both r_{ABC} and r_{Libor} rates[7]. The argument here is that since S_{fwd} controls stock (an asset) appreciation and by no-

---

[7] **Two Rates in an Option Formula?** It is possible to have two interest rates in an option formula. For example, treasury-bond short-term options use both repo and Libor rates.

arbitrage should use a risk-free rate (Libor), so $S_{fwd} = \exp(r_{Libor}T^*)$. However the discounting of cash flows inside an ABC-issued credit sensitive security should be at r_{ABC}, so the discount factor should be $DF = \exp(-r_{ABC}T^*)$.

The discounting rates can be further refined depending on the final stock price S^*. If $S^* < S_0$ there is arguably more ABC credit risk at T^* than there is today, so the discounting rate should be above today's ABC rate, i.e. $r > r_{ABC}$. If $S^* > S_0$ there is less credit risk in the future than there is now, so $r < r_{ABC}$.

Derman et al. (Ref.) take the view that $r = r_{ABC}$ at low S^* while $r = r_{Libor}$ at high S^*, since at high enough S^*, conversion is certain at T^*, and the stock can then be hedged using a risk-free rate. Ref. iii suggests a blended discounting rate at intermediate S^*.

Finally, there are models that incorporate spreads for S_{fwd} but discount at Libor[8].

What's the Bottom Line?
Who is *"Right"*? As elaborated more fully later (Ch. 32, 33), there is no real Theory of Finance, just phenomenology. Different people support and use different procedures and different models[9].

D_{123} : The Complex DEC Synthetic Convertible

This section describes a more complicated DEC with a complicated option called a "best-of option". A best-of option payoff involves the choice of a maximum of several payoffs. To get the option value, this payoff is multiplied by the appropriate probability function (Green function), and then integrated and discounted. In fact the DEC_{123} structure involves three separate DEC_α (α = 1,2,3). The options are written on three separate stocks whose prices are called S_α. It is possible that none of them will be the ABC stock. The maturity date

---

[8] **Coupons and Discounting with the ABC Credit Rate:** The DEC coupon will be paid at regular intervals and act like coupons in a standard bond. The discounting of these coupons therefore should *include* the ABC credit spread over Libor, as any bond issued by ABC must. It is easy to see this. A par bond issued by ABC must give an extra coupon to X to compensate for ABC credit risk, and the bond is at par when the discount rate equals the coupon, thus including the credit spread. This extra credit-related amount built into the coupon should also be present in a synthetic structure issued by ABC.

[9] **Sociology:** People can sometimes act a little like religious zealots regarding models. Discussions can degenerate to "This model is right" and "That model is wrong". It can require diplomacy to avoid getting into turf model battles.

Two More Case Studies

$T*$ and the coupon paid are the same for each of the DECs. The best-of option in this case allows X to choose the best payoff from any of the three DEC_α. Recall that for the standard DEC, the investor X gets a coupon because X gives up a net optionality to ABC. Because the best-of feature helps X, the options become less of a liability to X. Therefore the coupon X gets paid by ABC will also be decreased[10].

Although the coupon has decreased relative to a standard DEC, the best-of options give X the chance for optimized up-side results. Although the overall value calculated at issue t_0 will not change due to this feature, X may have a scenario whereby he believes that the ultimate payoff will be more because of this best-of feature. If the stocks are chosen from different sectors, for example, better diversification will be achieved, giving potentially better results.

The best-of option in the structure is a composite object so the individual options can no longer be broken out. The coupon will be different than in the simple DEC, as we shall calculate. Here is a diagram of the DEC_{123} structure:

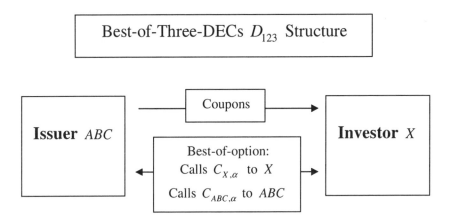

Pricing D_{123}: A Preview of Green-Function Techniques

In the rest of this chapter, we need to increase the mathematical firepower a bit. We will use some results from path integrals and Green-function techniques for pricing options. The interested reader can get a full exposition in Ch. 41-45.

Assuming lognormal stock dynamics the result for the Green function of n stocks is naturally an n-variate normal distribution [ii, 10]. We present it in general

---

[10] **Alternative:** Alternatively to lowering the coupon, ABC could increase the conversion price, i.e. move to higher value the strike of the $C_{X\alpha}$ options. This would make the $C_{X\alpha}$ options individually less valuable but get to the same total option value through the best-of feature.

form for use later. At the initial time t_a set $x_{a;\alpha} = \ln\left[S_\alpha(t_a)\right]$, set $x_{b;\alpha} = \ln\left[S_\alpha(t_b)\right]$ at the final time t_b, and set $T_{ab} = t_b - t_a$. The volatility of the time difference $d_t x_\alpha(t) = x_\alpha(t+dt) - x_\alpha(t)$ is $\sigma_{ab;\alpha}$ over the time window T_{ab}. The correlation[11] between $d_t x_\alpha$ and $d_t x_\beta$ is $\rho_{ab;\alpha\beta}$ over the time window T_{ab}. The correlation matrix is (ρ_{ab}) with matrix inverse is $(\rho_{ab})^{-1}$. Again a,b label the times and α, β label the matrix indices. The Green function is [12]:

$$G^{(n)}\left[\{x_{a;\alpha}\},\{x_{b;\alpha}\};t_a,t_b\right] = \frac{\Theta(T_{ab})\exp(-r_{ab}T_{ab})}{\left[\det(\rho_{ab})\right]^{1/2}\prod_{\alpha=1}^{n}\sqrt{2\pi\sigma_{ab;\alpha}^2 T_{ab}}}\exp\left[-\Phi_{ab}^{(n)}\right]$$

(14.3)

Here

$$\Phi_{ab}^{(n)} = \sum_{\alpha,\beta=1}^{n} D_t x_{ab;\alpha}\left(\rho_{ab}\right)^{-1}_{\alpha\beta} D_t x_{ab;\beta} \ /\ \left[2\sigma_{ab;\alpha}\sigma_{ab;\beta}T_{ab}\right] \quad (14.4)$$

Also[12]

$$D_t x_{ab;\alpha} = \left[x_{b;\alpha} - x_{a;\alpha} - \mu_{ab;\alpha}T_{ab}\right] \quad (14.5)$$

---

[11] **Volatility and Correlation Input Choices:** The choice of correlations and volatilities is critical. If historical volatilities are used they are often calculated over historical time windows with lengths equal to the forward time window of the deal $T_{ab} = T^* - t_0$ on the grounds that history over time periods of similar length to the future time period may be the most relevant. Implied stock volatilities can also be used, but it can easily happen that the time window for the deal may be longer than the maturities of existing available options. Finally, as emphasized later in the book, correlations are unstable over windows of interest and probably stochastic. Therefore stressed values of the correlations, consistent with a positive definite correlation matrix, need to be used in order to determine the correlation risk.

[12] **Notation: dx, $d_t x$ and $D_t x$:** We need to be careful to distinguish various differentials. First, $dx_\alpha(t)$ is the differential of values of $x_\alpha(t)$ at one single time t. This appears in the measure of the Green function written as a function of the $\{x_\alpha\}$. Next $d_t x_\alpha(t) = x_\alpha(t+dt) - x_\alpha(t)$ is the time difference of x_α at neighboring times t+dt and t. This is used to get volatilities and correlations. In practice dt will be a time small compared to T_{ab}; popular choices include dt from 1 day to 3 months. Finally $D_t x$ is the finite-time difference generalization of $d_t x$ over time T_{ab}. It arises when successive Gaussian propagations are convoluted (using the semi-group property) to get a Gaussian over finite T_{ab}. Which differential we are using will in any case be clear from the context.

Two More Case Studies

As usual the drift $\mu_{ab;\alpha}$ is given in terms of the interest rate r_{ab} over T_{ab}, the stock dividend yield $y_{ab;\alpha}$ and the Ito correction term due to the lognormal dynamics assumed as

$$\mu_{ab;\alpha} = r_{ab} - y_{ab;\alpha} - \sigma^2_{ab;\alpha}/2 \qquad (14.6)$$

The measure for given $\{x_{a;\alpha}\}$ integrating over $\{x_{b;\alpha}\}$ is the simple product of ordinary integrals, namely $\prod_{\alpha=1}^{n} \int_{-\infty}^{\infty} dx_{b;\alpha}$.

The Payoffs and Calculation of the Best-of Option

The calculation of the Best-of-Two $DEC_{\alpha\beta}$ value is the integral over $\left(S_\alpha^*, S_\beta^*\right)$ of the (α,β) bivariate $(n=2)$ Green function with integrand multiplied by the T^* payoff, $C_{DEC_{\alpha\beta}}(T^*) = \max\left(DEC_\alpha, DEC_\beta\right)(T^*)$. There are three of these $DEC_{\alpha\beta}$ with $(\alpha,\beta) = (1,2), (1,3)$ and $(2,3)$. The calculation of the Best-of-Three DEC_{123} option value is the integral over $\left(S_1^*, S_2^*, S_3^*\right)$ of the trivariate $(n=3)$ Green function integrand that is multiplied by the DEC_{123} payoff at T^*, $C_{DEC_{123}}(T^*) = \max\left(DEC_1, DEC_2, DEC_3\right)(T^*)$. All results are then discounted.

The payoff for the three-variable case is somewhat messy to administrate since there are 5 regions due to the three separate regions for each DEC_α, namely

1. All S_α^* in the "down" region. In this region, X has lost money due to downside risk, but at least he has minimized the loss.
2. At least one S_α^* in the "flat" region but no S_α^* in the "up" region. Here X gets back the notional even if some stocks have dropped.
3. One S_α^* in the "up" region. One of the $C_{X,\alpha}$ options is in the money.
4. Two S_α^*, S_β^* in the "up" region. Two $C_{X,\alpha}$ options are in the money.
5. All S_1^*, S_2^*, S_3^* in the "up" region. All $C_{X,\alpha}$ options are in the money.

The corresponding maximum payouts to X in the five regions are then[13]:

---

[13] **Normalization:** $N is the overall normalization or notional.

1. $NMax_\alpha \left(S^*_\alpha \big/ S_{0\alpha} \right)$

2. N

3. $N \left\{ 1 + \eta \left(S^*_\alpha \big/ S_{0\alpha} \right) \right\}$

4. $N \left\{ 1 + \eta Max_{\alpha,\beta} \left[\left(S^*_\alpha \big/ S_{0\alpha} \right), \left(S^*_\beta \big/ S_{0\beta} \right) \right] \right\}$

5. $N \left\{ 1 + \eta Max_{1,2,3} \left[\left(S^*_1 \big/ S_{01} \right), \left(S^*_2 \big/ S_{02} \right), \left(S^*_3 \big/ S_{03} \right) \right] \right\}$

Using $S_0 = \eta E_{H\alpha}$, we can simplify the payoff in the "up" regions using

$$\eta \left[\frac{S^*_\alpha}{S_0} \right] = 1 + \eta \left[\frac{S^*_\alpha - E_{H\alpha}}{S_0} \right] \qquad (14.7)$$

Indicative Pricing

An indicative price is a non-binding price transmitted by a marketer to a client or other counterparty that often has a critical time element, leaving little time for detailed quantitative modeling analysis. Indicative results therefore sometimes need to be obtained using simplified input parameters and/or an approximate numerical method. Perhaps a similar analysis has essentially already been done and can be quickly modified from this "off-the-shelf" model, but this is by no means always the case.

Regardless of the indication, if a product is utilized in a deal with a later secondary market for trading, a thorough numerical analysis has to be programmed along with hedging, and put into a risk-management system.

The origin of this chapter was in fact an indicative price for a deal that never got done, but the price had to be obtained quickly. The numerical approximation used was a coarse bucketing approximation for the payoffs, although calculations of the bivariate and trivariate normal probabilities were performed accurately.

Denote y^{avg}_{1-DEC} as the average of the DEC_α dividend yields, y^{avg}_{2-DEC} the average two-stock $DEC_{\alpha,\beta}$ dividend yield and y_{3-DEC} the DEC_{123} dividend yield. The results satisfied the inequalities expected physically:

$$y^{avg}_{1-DEC} > y^{avg}_{2-DEC} > y_{3-DEC}$$

Again, these DEC dividend yields are in addition to any standard stock dividends and are the equivalent yields to the coupons paid to X due to the option structures of the product in the various cases. The best-of options did have a significant effect. This effect increases with the number of choices, as was expected intuitively by Corporate Finance[14].

Case Study: Equity Call with Variable Strike and Expiration

This option has some complicated logic. It is a European call option with attached side conditions. The basic option has strike E and expiration date t^*. At a given decision date $t_{decision}$ the stock price $S(t_{decision})$ is compared with three given levels S_1, S_2, and E/λ with $S_1 < S_2 < E/\lambda$. Here $\lambda > 1$ is a parameter. The current price S_0 at time t_0 is in the band $(S_2, E/\lambda)$.

The logic is pictured below:

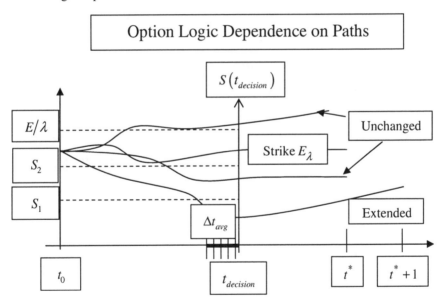

---

[14] **Corporate Finance and Intuition:** Quants should be aware that before being approached, a proposed transaction has been discussed enough in Corporate Finance so that these people have good intuition of what is going on. You should also try hard to get an intuitive feel. If your result does not agree with intuition, you may have misunderstood something, always possible in quick conversations. The ideal situation is to get a term sheet describing the proposed deal unambiguously, even for an indication, in order to help avoid such misunderstandings.

- $S(t_{decision}) < S_1$, the option expiration is extended to $t^* + 1$
- If $S_2 < S(t_{decision}) < E/\lambda$ the strike is lowered to $E_\lambda = \lambda S(t_{decision})$
- Else the option remains unchanged

The idea is to provide more value to the investor X. If the stock price drops substantially as specified by the first condition, additional time value is added with the possibility that the stock price will go back up. If the stock price is in the band specified by the second condition, the strike is lowered giving additional value to the call.

Note that there is no constraint on the paths before or after $t_{decision}$. Therefore, although the picture is not drawn this way, paths in any of the categories can cross back and forth over the horizontal boundaries. The classification of the path-dependence aspects of the option occurs only at $t_{decision}$.

Additional complexities also exist. The decision is made not on the value of the stock exactly at $t_{decision}$ but on the price averaged for a certain number of days Δt_{avg} before and up to $t_{decision}$. Such an averaging specification is quite common.

Theoretical Valuation of the Option

The European option can be valued formally using the path integral techniques presented later in the book [v,15]. The formalism pictorially follows the above diagram. Define the Green function [vi, 16]

$$G_{a,b} = G(x_a, x_b; t_a, t_b) = \frac{\Theta(T_{ab})\exp(-r_{ab}T_{ab})}{\sqrt{2\pi\sigma_{ab}^2 T_{ab}}} \exp[-\Phi_{ab}] \quad (14.8)$$

Here[17]

---

[15] **Path Integral Remarks:** You do not have to understand path integrals to follow this section. Just look at the diagrams that correspond directly to the formulae. For simplicity, we neglect the averaging before the decision time, although this can also be included. This section is based on standard lognormal dynamics. If you already know path integrals this is just free field theory with some boundary conditions. See Ch. 41-45 for details.

[16] **Semantics:** As was once explained to me, it makes sense to say *Green* function, not *Green's* function. You don't talk about a *Bessel's* function, right?

[17] **Variables:** Here t_a is an arbitrary initial time, $x_a = \ln S(t_a)$, t_b is an arbitrary final time, $x_b = \ln S(t_b)$, σ_{ab} is the volatility, r_{ab} the "risk-less" interest rate, $\mu_{ab} = r_{ab} - y_{div} - \sigma_{ab}^2/2$, y_{div} the dividend yield, $T_{ab} = t_a - t_b$ and Θ the Heaviside step function forcing $T_{ab} > 0$. Further explanation is given later in the book.

Two More Case Studies

$$\Phi_{ab} = [x_b - x_a - \mu_{ab}T_{ab}]^2 / [2\sigma_{ab}^2 T_{ab}] \quad (14.9)$$

The logic can be thought of as two exotic options given to X that expire at $t_{decision}$ and result in two knock-in options[18] under suitable conditions, namely:

- An extension option $C_{\text{Extd Time}}$ that knocks in if $S(t_{decision}) < S_1$
- A strike-lowering option $C_{\text{Lower Strike}}$, knocks in if $S_2 < S(t_{decision}) < E/\lambda$

The diagram below introduces some notation for the calculation:

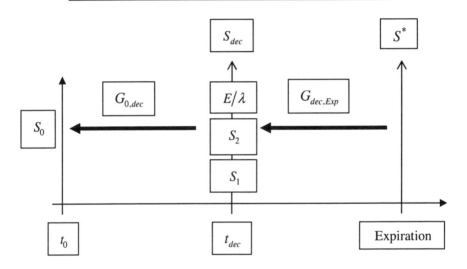

In the figure the arrows point backwards because we propagate cash-flow information backwards from expiration. The option can be broken into 3 parts.

$$C(S_0, t_0) = C_{\text{Unchanged}} + C_{\text{Extd Time}} + C_{\text{Lower Strike}} \quad (14.10)$$

Here the three component integrals are: [19]

---

[18] **Knock-in Options:** These are barrier options that spring into existence if the knock-in condition occurs. The formalism of barrier options is described in detail in Ch. 17-19.

[19] **Notation:** To save space, we write "S_{dec}" instead of $S(t_{decision})$ and use "Exp1" to denote the normal expiration t* where the stock price is called S_1^*. Similarly "Exp2" is the extended expiration t*+1 where the stock price is called S_2^*. All x variables are lnS with the appropriate labels.

196 Quantitative Finance and Risk Management

$$C_{\text{Unchanged}} = \int_{[S_{dec} \in (S_1, S_2)] \cup [S_{dec} > E/\lambda]} dx_{dec} \int_{\ln E}^{\infty} dx_1^* \cdot G_{0,dec} G_{dec,Exp1} \cdot (S_1^* - E)$$

$$C_{\text{Extd Time}} = \int_{S_{dec} < S_1} dx_{dec} \int_{\ln E}^{\infty} dx_2^* \cdot G_{0,dec} G_{dec,Exp2} \cdot (S_2^* - E)$$

$$C_{\text{Lower Strike}} = \int_{S_2 < S_{dec} < E/\lambda} dx_{dec} \int_{\ln(\lambda S_{dec})}^{\infty} dx_1^* \cdot G_{0,dec} G_{dec,Exp1} \cdot (S_1^* - \lambda S_{dec})$$

(14.11)

These integrals can be performed as bivariate normal integrals. Some approximations have to be implemented to include the averaging before $t_{decision}$. We shall not pursue it further, but go next to a Monte Carlo valuation.

Monte-Carlo Valuation

In order to price a complicated option like this, in general a Monte Carlo (MC) simulator can be used[20]. Then the logic tree can be implemented in a straightforward way[21, vii]. Given some parameters[22], the option value can be

---

[20] **Forward and Backward Propagation:** The Monte Carlo simulator propagates paths going forward starting at S_0. In practice this is more convenient than propagating paths backwards because only a small fraction of paths propagated backward will arrive at S_0 within any dS_0. The difference between forward and backward propagation only matters when there is a stock-dependent volatility $\sigma[S(t),t]$ such as occurs in "volatility skew". For a constant volatility such as assumed here there is no difference.

[21] **Implementation with Monte Carlo Simulation:** The numbers in this section were obtained with the software $Derivatool^R$ from Financial Engineering Associates, Inc. My understanding, based on a conversation with Mark Garman is that Derivatool works in a manner similar to the theoretical presentation above, jumping over all times where paths are unrestricted with no logic constraints with a single Green function. This is applicable from t_0 up to the $t_{decision} - \Delta t_{avg}$ and then from $t_{decision}$ to expiration (either $t^*_{Exp1} = t^*$ or $t^*_{Exp2} = t^* + 1$). The MC results are of course averages with statistical errors for a certain number of trials (here 10K). In practice the daily 20 bd averaging was simplified.

[22] **Input Parameters:** Starting stock price $\$S_0 = \$19.875$, lower boundary $\$S_1 = \$16.50$, middle boundary $\$S_2 = \$18.00$, parameter $\lambda = 1.2$, strike $\$E = \$25$, time from start to expiration $t^* - t_0 = 3$ yrs, time to decision $t_{decision} - t_0 = 0.79$ yrs, lognormal volatility (chosen constant) $\sigma = 0.31/\text{yr}^{1/2}$, interest rate $r = .0625/\text{yr}$ (continuous, 365), dividend yield $y_{div} = 0.012/\text{yr}$ (continuous, 365), averaging period $\Delta t_{avg} = 20\text{bd}$ (business days). These parameters can move around when working on a transaction for a variety of reasons (changes in the deal definition, changes in the market, refinements in the analysis etc). One refinement of the analysis would be to take a term structure of volatility $\sigma(t)$ or

found. In this case we get $\$C = \$3.71$ with 1sd statistical MC error $\pm \$0.08$. It is interesting to ask how much of the value was due to all this complexity. A plain vanilla 3-yr call with the same parameters is $\$C_{\text{Plain Vanilla}} = \$3.49$. The difference is therefore[23] $\$\delta C = \$C - \$C_{\text{PlainVanilla}} = \$0.22 \pm \$0.08$.

"Back of the Envelope" Calculation

Now we will do a back-of-the-envelope calculation for $\$\delta C$. This means we will make simplifications, and see if we can get the right "ball-park" answer[24], say within a factor 2 for $\$\delta C$. There are several good reasons for being able to do this.

- Estimates performed after-the-fact of doing the MC calculation helps us understand the answer[25].

---

even include some volatility skew (although this would be unlikely for a transaction involving a single stock).

[23] **"Errors and Uncertainties"**: Although the statistical MC error in $C was $0.08 (only 2% of $C), the resulting statistical error in δC is large (36% of δC). Some non-technical people will believe that if you report an "error" you have made a mistake. It may be better semantically to discuss "uncertainties" rather than "errors". On the other hand, there are also uncertainties in the input parameters to the calculation, simplifications in the lognormal model assumption (because real stock movements differ from lognormal), etc. Therefore it is in fact a mistake (a real mistake) to report a small MC error and imply that this is the only uncertainty in the calculation.

[24] **Back of the Envelope and Management:** Scientists and engineers will know what is meant by "back of the envelope". However a Wall Street manager may make up some accuracy target that is acceptable to him and then assume (or request) that your quick estimate will satisfy that accuracy. For this reason you need to be extremely careful about how you discuss approximate calculations, and make it clear that you only believe the results are accurate to whatever level you think they are accurate.

[25] **Why You Should "Understand" the Answer the Model Produces:** You should feel comfortable intuitively with the output of the model. Always perform sanity checks. Here are four good reasons:

First reason: Because your management or other interested parties may well want to know why the answer that you got in the model is so small or why it is so big. Here is the way this will happen. Person In Authority: "I don't believe that result". You: "...".

Second reason: The next time this sort of problem comes up, you will be better able to deal with it.

Third reason: A prototype rough calculation is always useful for sanity checks before the Real Production Model is developed.

Fourth reason: Maybe the model calculation actually has a real mistake (math, parameter input, programming, reporting, misapplication, ...) that you will discover by performing rough calculations. It is bad practice to believe an answer just because it is on a printout from a computer.

- If you are on the desk and the salesman wants to have some sort of "indication" (as he will call it) by 2pm, you *need* to be able to do this sort of calculation without waiting for 2 months while it is being programmed and getting Put In The System.

What is really going on? First, with definite probabilities there are transitions from S_0 to various regions of the S_{dec} axis where the logic occurs (again "*dec*" is short for decision). We can get these probabilities straightforwardly from the normal distribution. Then we can figure out what happens after t_{dec}.

Extension Option Contribution to $\$\delta C$

Consider first at the extension option. If $S_{dec} < S_1$, the option is extended. So the contribution of the extension is really the difference of a call from t_{dec} to $t^* + 1$ and a call from t_{dec} to t^*. In order to price this difference we need to assume a reasonable starting point for S_{dec} at t_{dec}. We collapse all $S_{dec} < S_1$ paths[26] to $S_{dec} = S_1$. We know this is an overestimate since paths starting at lower S_{dec} contribute less than paths starting at S_1. However the strike $\$E = \$25$ is above $\$S_1 = \$16.50$, so this part of the option is out of the money, and the contribution will not be very large anyway. The contribution to $\$\delta C$ from the extension option can then be approximated as the probability $P[S_{dec} < S_1]$ that $S_{dec} < S_1$ at t_{dec} for paths starting at S_0 times the difference of two ordinary call options:

$$\delta C_{\text{Extd Time}} \approx P[S_{dec} < S_1] \cdot \left\{ C\left[S_{dec} = S_1; t_{dec}, t^* + 1\right] - C\left[S_{dec} = S_1; t_{dec}, t^*\right] \right\}$$

(14.12)

Numerically this approximation turns out to give $\$\delta C_{\text{Extd Time}} \approx \$0.19$.

Lowered Strike Option Contribution to $\$\delta C$ and a Consistency Check

The strike gets lowered if S_{dec} is between S_2 and E/λ; numerically this means in the region $\$18$ to $\$20.83$. This is a band narrow enough so we can try using the

---

[26] **Approximation:** It is a good idea to keep track of the assumptions that you are making so you can refine them later if you have time and if it fits in the priority list.

average of these two numbers, $S_{dec}^{avg} = \$19.42$, to start this class of paths from t_{dec}. The lowered strike in this approximation is $\$E_\lambda = \lambda \$ S_{dec}^{avg} = \$23.30$. Hence the lowered strike option contributes to $\$\delta C$ in this approximation is

$$\delta C_{\text{Lower Strike}} \approx P[S_2 < S_{dec} < E/\lambda] \cdot$$
$$\cdot \left\{ C\left[S_{dec} = S_{dec}^{avg}; t_{dec}, t^*; E_\lambda \right] - C\left[S_{dec} = S_{dec}^{avg}; t_{dec}, t^*; E \right] \right\} \quad (14.13)$$

Here, $P[S_2 < S_{dec} < E/\lambda]$ is the probability that $S_2 < S_{dec} < E/\lambda$ at t_{dec} for paths that start at S_0. This approximation produces $\$\delta C_{\text{Lower Strike}} \approx \$0.10$. Hence the total for our BOE (back of the envelope) calculation is $\$\delta C_{\text{BOE Calc}} = \$\delta C_{\text{Extd Time}} + \$\delta C_{\text{Lower Strike}} \approx \$0.29$.

Now the result from the MC calculation was $\$\delta C_{MC} = \$0.22 \pm \$0.08$. We see that we are indeed in the same ballpark, and are even (mirabile dictu[viii]) within the MC errors. Because the biggest contribution was the extension option for which we know we made an overestimate, and because the calculation is bigger than the MC result, there seems to be a consistency[27].

References

---

[i] **Convertible Bonds**
Zubulake, L., *Convertible Securities Worldwide*. John Wiley & Sons, Inc. 1991.

[ii] **Best-of Options**
Johnson, H., *Options on the Maximum or the Minimum of Several Assets*. Journal of Financial and Quantitative Analysis, Vol 22 No. 3, 1987. P. 277.

[iii] **Mixed Prescription for Interest Rates in Pricing Convertibles**
Derman, E. et al, *Valuing Convertible Bonds as Derivatives*. Goldman Sachs, Quantitative Strategies Research Notes, unpublished 1994.

[iv] **Multivariate Green Functions and Related Topics**
Dash, J., *Path Integrals and Options*. Invited talk, SIAM Conference, 1993.

---

[27] **Why Did This Approximation Work?** I think it is mostly because $\$\delta C$ was small. In other cases, we may not be so lucky. Other cruder approximations did not work as well. All I would really want to claim is that the approximation to $\$\delta C$ is perhaps within a factor of 2.

[v] **Path Integrals and Options**
Dash, J. W., *Path Integrals and Options – I*. Centre de Physique Théorique, CNRS Marseille preprint CPT-88/PE. 2206, 1988.

[vi] **Green Functions**
Morse, P. M. and Feshbach, H., *Methods of Theoretical Physics I*. McGraw-Hill, 1953. Sect 7.4.
Jackson, J. D., *Classical Electrodynamics, 2nd Ed.* John Wiley & Sons, Inc. 1975.

[vii] **Monte Carlo, Derivatool**
Garman, M. Principal, FEA, Inc. Private communication.

[viii] **Latin**
Stone, J., *Latin for the Illiterati*. Routledge, 1996. See p. 59.

15. More Exotics and Risk (Tech. Index 5/10)

This chapter examines several exotic options and some related risk measures. The topics are:

- Contingent Caps (Complex and Simple)
- Digital Options: Pricing and Hedging
- Historical Simulations and Hedging
- Yield-Curve Shape and Principal-Component Options
- Principal-Component Risk Measures
- Hybrid Two-Dimensional Barrier Options
- Reload Options

Contingent Caps

Clients and Contingent Caps
In this section, we describe an exotic product called the contingent cap[1,2]. As a client, a corporation ABC wants to buy a cap from a broker dealer BD. As for any cap, ABC's motivation is to get insurance to protect against rising rates. The cost of a standard cap $\$C_{cap}$ is paid up-front in cash. This outlay can be substantial and is lost if ABC does not use the insurance (i.e. the cap does not pay off). The *Contingent Cap* is a product that keeps the same cap/insurance payoff, but makes the payment for the cap by ABC contingent on certain conditions. These conditions can be dependent on the underlying rate itself in such a way that ABC essentially only "pays for the insurance if the insurance pays out", or if the insurance has a good chance of paying out. In fact, ABC may wind up paying nothing at all. Of course the amount $\$C_{ContCap}$ that ABC pays if a payment is required will be more than $\$C_{cap}$.

---

[1] **History:** This was an instructive foray in new product development. The interest-rate contingent cap in the text was invented and developed by me in 1991. After an unpropitious start, the product actually sold reasonably well after 1993.

[2] **Acknowledgement:** I thank Bruce Fox for his enthusiasm for marketing contingent caps, and for helpful discussions on many topics.

Essentially with a contingent cap, ABC buys a cap from BD and sells a contingent payment option to BD of equal value, so that no up-front cash is exchanged. The contingent option, if exercised, produces payment $\$C_{ContCap}$ from ABC to BD.

The picture for the structure (cap, contingent option) of the contingent cap is:

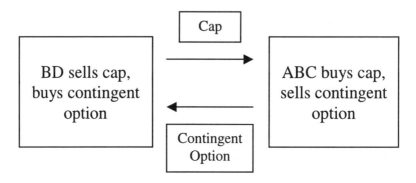

Structure of the Contingent Cap

The contingent cap is complicated. For a cap, there are a number of resets. If Libor on all resets is below a trigger rate K, then ABC pays nothing. Otherwise, the *first time* that $L > K$ on a reset date, ABC pays $\$C_{ContCap}$, and nothing after that[3]. We write, with $\chi > 1$, the payment equation $\$C_{ContCap} = \chi \cdot \$C_{cap}$. We generally have $\chi < 2$ with this product because the composite probability that ABC has to pay something is above ½. This follows since many reset trials (all requiring "success" from ABC's viewpoint) are needed if ABC "escapes" and winds up paying nothing[4].

The next picture gives the idea:

---

[3] **Connection of Contingent Caps With Barrier Options:** The contingent cap is really a knock-in discrete barrier option, with the discrete barrier being given by the trigger rate at the discrete set of reset times of the cap. See Ch. 17-19.

[4] **The Birthday Problem and Contingent Caps:** The birthday problem is: "How many people are needed in order that the probability is around 50% that two of the people have their birthday on the same day". The answer is only 22 or 23 people. The same situation exists for the contingent cap. Only a few resets near or at the money are needed in order that payment for the contingent cap has a reasonably high probability.

More Exotics and Risk 203

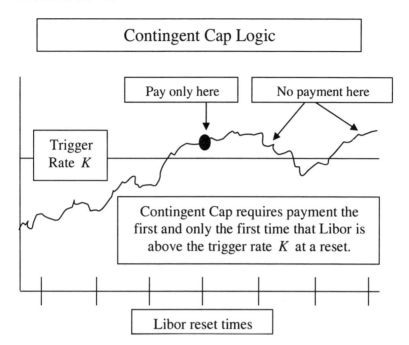

The value of the trigger rate K can for simplicity be chosen at the break-even rate R_{BE} for the swap whose resets are at the same times as those of the cap[5]. The strike E of the cap can be equal to K or different. The contingent payment naturally builds in a spread for the BD. There is downside to BD because dealing with an exotic requires extra hedging, operational, and systems costs.

A Simple "Contingent Cap" with Independent Caplets (Not This Product)
The contingent option first originated as a simple single-decision FX option[i]. There, payment is due if the exchange rate η is above a certain level η_{level} and not due if $\eta < \eta_{level}$. Roughly (ignoring interest-rate effects), the cost of such an option is double the price of a standard option. This is because the probabilities

---

[5] **Limitations on the Trigger:** If the trigger is far above the swap rate, the potential payments get large because the probability of escape is high. This limits the attractiveness of the single-level contingent cap. However, the 50-50 contingent cap with two triggers can have one trigger at a high level.

that η goes up or down are roughly ½. Hence 50% of the time ABC doesn't pay anything, so when ABC does pay, ABC has to pay double.

A direct extension of this single-decision option is just a basket of independent single-decision options. A simpler product that is just a set of independent contingent caplets (see Schap, Ref [i]) has also been called a contingent cap. Multiple payments for caplets in this product are due if Libor crosses the trigger at various resets[6].

This basket of independent contingent caplets is *not* the contingent cap described here. The contingent cap in this chapter is a composite structure that cannot be broken up into a simple sum, and requires the payment at most once (the first time that Libor crosses the trigger).

The 50-50 Contingent Cap

A variation of the contingent cap is a "50-50" contingent cap. There, the single level is split into two levels K_1, K_2. The upper trigger K_2 is set above the strike ($K_2 > E$) and the lower trigger K_1 below the BE rate ($K_1 < R_{BE}$). One payment is due the first time $L > K_1$ if that happens, and the second equal payment is due the first time $L > K_2$ if that happens. No other payments are due. The first payment, if it occurs, is less than the original cap price $\$C_{cap}$ would have been. The cap insurance must start to pay something before the second payment, if it occurs, becomes due.

Scenario Analyses and Total Return Comparisons for Clients

What else drives the sales of contingent caps? A desk dealing with interest-rate exotics at the broker-dealer BD will be speaking to a counterparty X, a financial expert in corporation ABC. If the yield-curve has positive slope, X may not believe that rates will increase as much as the theoretical statement made by the curve[7]. Still, X wants insurance protection against rate increases. However, X does not want to pay anything now. Also X does not want to have to go before the ABC management to tell them after the fact that he bought an expensive cap that didn't pay off. With the contingent cap, ABC may wind up paying nothing.

---

[6] **Simpler Independent Set of Caplets "Contingent Cap":** In this simpler product, for a rising forward rate curve, the forward rates that are below R_{BE} will have large contingent premium factors χ>>1 because the probability of escape is high. Clients tend to be reticent to gamble on large factors, whatever the theoretically small probabilities that they might have to pay. In the contingent cap described in the text, there is no large payment factor because only the correlated probabilities of escape are used.

[7] **Betting Against the Forward Curve:** Customer scenarios, which are sometimes different from the theoretical forward curve, drive many derivative transactions.

More Exotics and Risk 205

Even if ABC has to pay more later, there will be the satisfaction that the cap will start to pay off before ABC has to pay.

Scenario analysis can be performed to show under which conditions the contingent cap or its variations (such as the 50-50 contingent cap) will outperform or under perform a standard cap. The scenarios can be either theoretically based or historical. The historical scenarios involve running through time using data.

In general, the contingent cap outperforms the standard cap if rates do not rise above some level $R_{BigRateRise}$. Otherwise, the standard cap outperforms the contingent cap. The decision of ABC to purchase a contingent cap will rely on his judgment of future rate rises relative to this level. Scenario analyses are useful to help the client decide[8].

Contingent Option as a Correlated Set of Digital Options

We now consider the contingent payment option in more detail. This option as we have described it is a complex correlated set of digital options, with lumpy cash payments are made contingent on the underlying rate. The contingent cap needs to be priced using a MC simulator. Approximate analytic forms can be derived which can be used to obtain a reasonably good idea of the final result.

The risk management of the contingent cap is complicated because of the complex correlated digital nature of the payouts. To get an idea, we first consider the formalism for one digital option.

Digital Options: Pricing and Hedging

To give some background, we first spend a little time discussing single digital options. Consider a single digital call option[9] with unit payoff at time t^* if $x^* > E$, namely $C(x^*) = \Theta(x^* - E)$. The payoff can be regarded as the limit of a spread of two ordinary call options, one long and one short.

The idea is shown in the next picture:

---

[8] **Total Return Scenario Analyses:** A fair amount of time on the desk can be taken up running such scenarios. Sometimes clients specify scenarios that they would like to see.

[9] **Digitals Here, There, Everywhere:** The discussion in this section for single digital options holds for digitals of any sort – interest rate, FX, equity, commodity, etc. The discussion also holds for barrier options that in general can have digital components. In this section, we label x as the variable, not its logarithm.

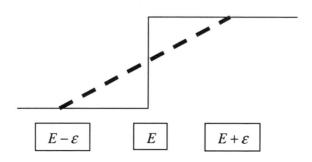

Digital option has strike E, and is approximated as a spread of two call options: Long at strike $E - \varepsilon$, Short at strike $E + \varepsilon$

This is the way the digital step option would be viewed for hedging purposes. We would short options at strike $E - \varepsilon$ and buy options at strike $E + \varepsilon$. The value of ε has to be chosen such that available hedges exist (e.g. 25 bp or 50 bp). The idea is rather like electrostatics where we view a dipole with dimension 2ε.

Here is a picture with of Δ and γ with $E = 0.2$,

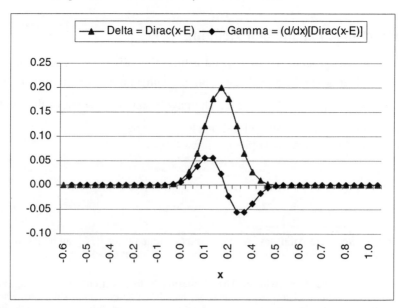

Mathematically for zero width we have

$$\Delta = \delta(x^* - E)$$
$$\gamma = \delta'(x^* - E)$$

Therefore, Δ for a digital call looks like γ for a standard call. Also γ for a digital call is the wildly-varying derivative of the Dirac delta function, or the smoothed out version of it from the spread option approximation.

Whipsaw Theoretical Scenario Analysis for Digitals

A time-dependent scenario to exhibit and drive home the dangerous nature of digital options is called "whipsaw" analysis. Here the underlying variable $x(t)$ is assumed to be alternately above and then below the strike of the digital option E as time passes. In this way, we alternately are closer to long and then short options, so the hedging oscillates dramatically. In the electrostatic analogy, we see alternately a net positive charge and a net negative charge from the dipole.

Hedging A Correlated Set of Digital Options

A correlated set of digital options such as seen in the contingent cap needs to be treated numerically. Still, the forward curve relative to the trigger level can be used to get a physical picture of the hedging properties.

The idea is to look at the time when the forward curve intersects the trigger level; this gives an average idea of when we can expect reasonable probability of the contingent option going into the money.

Historical Simulations and Hedging

An historical simulation can be run to test the efficacy of hedging strategies. We briefly outline the formalism. Take a portfolio $V(x,t)$ of the security $C(x,t)$ and hedge $H^{(k)}(x_k, t_k)$ at a number of successive points in the past (x_k, t_k) with historical data used for x_k. The hedge has a superscript because of a hedge rebalancing strategy (for example delta hedging), so the hedge function itself changes. At (x_{k+1}, t_{k+1}) the new hedge $H^{(k+1)}$ is the old hedge $H^{(k)}$ plus the change $\delta H^{(k,strategy)}$, viz

$$H^{(k+1)}(x_{k+1}, t_{k+1}) = H^{(k)}(x_{k+1}, t_{k+1}) + \delta H^{(k,strategy)}(x_{k+1}, t_{k+1}) \quad (15.1)$$

Revalue the portfolio $V_k = C(x_k, t_k) - H^{(k)}(x_k, t_k)$ successively moving forward in time using the time difference $d_t x_k = x_{k+1} - x_k$. Write the portfolio "reval" change between times t_k and t_{k+1} using the model as

$$\delta V_k^{(reval)} = V_{k+1} - V_k \quad (15.2)$$

Using the using the sensitivities from the model for the portfolio, calculate the "Math-Calc" change as

$$\delta V_k^{(MathCalc)} = \Delta_k \cdot d_t x_k + \frac{1}{2} \gamma_k \cdot (d_t x_k)^2 + vega_k \cdot d_t \sigma + \theta_k dt \quad (15.3)$$

Note that we do *not* replace $(d_t x_k)^2$ by dt. This replacement is only theoretically possible for the time-averaged or ensemble-averaged quantity, provided Gaussian statistics hold. Instead, we just record the results from the data for each time step.

The unexplained P/L or slippage is then the difference between the reval difference and the Math-Calc difference,

$$\delta V_k^{(Unexplained\_P/L)} = \delta V_k^{(reval)} - \delta V_k^{(MathCalc)} \quad (15.4)$$

The statistics over time can then be examined for the portfolio values $\{V_k\}$, the reval changes $\{\delta V_k^{(reval)}\}$, and the unexplained P/L, $\{\delta V_k^{(Unexplained\_P/L)}\}$. An informed judgment can then be made on the relative risk of the hedging strategy

The Process of Actually Running the Historical Simulation

Having said all this, it is a major undertaking to run a historical simulation[10]. Data must be collected and a front-end link from the model to the data established. Then the simulator must be run, results collected and analyzed, etc. The whole process easily may take weeks, depending on the set-up work required.

---

[10] **New Products and Simulations:** If the management wants to roll out a new product, it may consider risk simulations "nice to have but not really needed". A more conservative approach relies on positive simulation results before acting. In the limit, management that doesn't really want new products may use simulation requirements as a negative tactic.

Yield-Curve Shape and Principle-Component Options

One can write options on the yield-curve shape. Yield-curve options on the difference of two rates are standard. Options on butterflies have also been written. We now discuss a more sophisticated example of yield curve options based on principal components.

Principal-Component Options

Principal-component analysis leads to an equivalent description of yield curve movements, with the advantage that generally (although not always), a few principal components suffice to describe most of the yield curve shape changes. In Ch. 48, App. B we discuss the details of principal components or EOF analysis of the yield curve to which we refer the reader. Options can be written on principal components[11].

For example, we can look at three rates on the yield curve (maturity 2, 5, 10 years). The third-order "flex" component $\langle D_t Y^{(\beta=Flex)} \rangle$ of the yield-curve movements $\langle D_t r \rangle_T$ for maturity T, averaged over time, as observed in 1992 was

$$\langle D_t Y^{(\beta=Flex)} \rangle = \left[0.16 \langle D_t r \rangle_{2-yr} - 0.81 \langle D_t r \rangle_{5-yr} + 0.57 \langle D_t r \rangle_{10-yr} \right] \quad (15.5)$$

We use this as an example. In order to write an option on the flex component, consider the explicit time dependence of the flex component. For generality, we can put in arbitrary weights and maturities:

$$Y^{(\beta=Flex)}(t) = \left[w_{Short\_T} r_{Short\_T}(t) - w_{Middle\_T} r_{Middle\_T}(t) + w_{Long\_T} r_{Long\_T}(t) \right]$$
$$(15.6)$$

The sum of the squares of the weights is normalized to one, $\sum_T w_T^2 = 1$. There are various possible option payouts. For example, the payout for a flex call option at exercise date t^* would read (with $\mathcal{N}$ a normalization factor),

$$C_{call\_payout}^{(\beta=Flex)}(t^*) = \mathcal{N} \cdot \left[Y^{(\beta=Flex)}(t^*) - E \right]_+ \quad (15.7)$$

---

[11] **History:** The idea of writing principal-component yield-curve options occurred to me in 1992, or more generally on any linear combination of rates. Averaging was included as a feature. These options were called STAR options, short for "Swap Twist Average Rate" options. The combination was supposed to be a combination of three swap rates, time averaged. The complex yield-curve option idea was followed up and marketed in 1993.

To price this option, we can assume that the differential time change of the flex principal component $d_t Y^{(\beta=Flex)}(t)$ satisfies Gaussian or mean-reverting Gaussian statistics. This assumption is probably better than lognormal, because the flex can be either positive or negative. Standard option theory can then be used to construct the option value. The volatility would be an implied volatility taken from the market. The historical volatility for comparison would be constructed from the volatilities of the rates along with the correlations, as usual for a basket quantity like the flex principal component.

A summary of the approach was eventually published [12,ii].

Example of Potential Clients for Principal-Component Options

The appeal of the flex option can be for clients having portfolios with exposure to flexing of the yield curve. For example, insurance companies have SPDA products funded at the medium-term part of the curve (5-7 years), so they lose money if these rates go up. Their assets are however tied to other parts of the curve. For example, they may have longer-term mortgage assets and short-term floating-rate assets. In this case, there is some flex risk, which could be ameliorated by such options.

Principal-Component Risk Measures (Tilt Delta etc.)

We have been describing principal-component yield-curve options designed to hedge against various movements of the yield curve. In order to systematize the discussion of yield-curve movement risk, a class of risk measures can be constructed with principal components[13]. For example, we can define first-order quantities "tilt delta" or "flex delta" arising from constructing a tilt or flex movement of the yield curve and observing the change in the portfolio value under such a movement. Similarly, second-order gamma quantities can be constructed. For example, a "tilt gamma" can be defined as the difference of a "steepening tilt delta" and a "flattening tilt delta".

Because the principal components of the yield curve change as a function of time, it can be preferable to use stylized fixed definitions in order to separate out the risk of the portfolio cleanly. For example, the steepening tilt delta can be defined by tilting the yield curve counter-clockwise in a linearly-interpolated fashion by a certain amount, while holding one point on the curve (e.g. the 5-year point) fixed. These principal-component risk measures are specific scenario risks.

Various rate types can be used to define these risk measures. For example, we can use the forward-rate curve, the zero-coupon curve, etc.

---

[12] **Acknowledgement:** I thank Vineer Bhansali for discussions.

[13] **History:** The idea of principal-component risk measures in my notes dates from 1989.

More Exotics and Risk

Hybrid 2-Dimensional Barrier Options—Examples

The two-dimensional barrier options are single barrier options that are dependent on two different variables. They are part of a class of options also called "hybrid options"[14]. Hybrid options can be formulated in a variety of markets. Typically, what happens is that an option depending on one variable is knocked in or knocked out at a barrier depending on a second variable. The two variables are correlated together[15]. In this section, we give an example. The mathematical apparatus [16] is given in Ch. 19.

The idea is shown in the picture below:

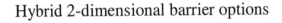

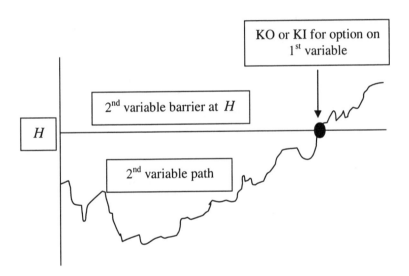

---

[14] **History:** This work was performed in 1993 while sitting on an exotic interest-rate desk. This included producing indicative prices and marketing these 2D hybrid options for a number of different pairs of underlying variables, including interest rates, equities, FX, and commodities for many quotes. Hybrids were then a new class of options. It was a fast-paced experience.

[15] **Correlation Risk:** One main risk of 2D hybrid options is correlation risk. It is necessary to price in some reserve corresponding to stressed values of the correlation that could cause difficulty with hedging.

[16] **Math Preview:** Under simple enough conditions including a constant barrier level, hybrid 2D options can be solved in closed form involving bivariate integrals. Monte-Carlo simulation is used if complicated side conditions are included. See Ch. 19.

The Cross-Market Knock-Out (XMKO) Cap

The cross-market knock-out cap, or XMKO cap, is a hybrid product that allows ABC to buy interest rate insurance at a cheaper value. The idea is to make the cap payout contingent on some event occurring in another market. Because ABC will not receive the cap payout under all conditions, the XMKO cap has a lower value than a standard cap. ABC for example might have assets that increase in value if (say) the price of oil increases. If the price of oil goes up enough, ABC can cover the increased cost associated with increasing interest rates and therefore does not need the cap insurance. Therefore, ABC may want to buy a XMKO cap that knocks out if the price of oil goes above a certain level.

Roughly speaking the XMKO cap can be thought of as ABC buying a standard cap and selling the knockout, thus lowering the cash ABC needs to pay up front. The idea is shown below:

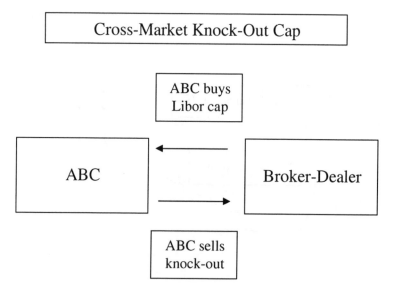

Clearly, the value of the XMKO cap is highly dependent on the correlation between the changes of oil prices (in this example) and the changes of Libor. If the correlation is higher, the knock out will be more probable and the XMKO cap value will be lower. Because correlations are unstable, some measure of correlation uncertainty will be included in the price ABC pays.

The XMKO cap is evaluated using the caplets in the cap. A separate XMKO option for each caplet is independently determined and the results added. Thus, independently, the payment for a given Libor caplet will knock out if the second variable crosses the barrier, but otherwise the Libor caplet will pay off if Libor goes above the strike rate of the cap.

Marketing XMKO Caps

One problem associated with the XMKO cap is that the asset and liability sides of corporations sometimes do not talk much to each other. Because this product is dependent on both the asset side (determining the knock out) and the liability side (involving debt issuance insurance), more people have to be involved.

Cross Market Knock-In Put Option and Correlation Risk

Here is an example of a cross-market knock-in option[17]. The example is from the commodities market [18], a 2-yr UI Al put 20% OTM with a barrier on NG up 20%. Naturally, the up-in put is cheaper than the standard put because of the restrictive condition on its existence. The details would be tailored to the customer whose business depends on both the variables.

We need to discuss correlation risk. The hybrid nature of the knock-in (KI) is strongly dependent on the correlation $\rho_{Al,NG}$ between Al and NG movements. The value of $\rho_{Al,NG}$, depending on the window, was between -0.3 and $+0.4$. This large variation in correlations happens in other variables, as we examine in other parts of the book (cf. Ch. 37).

If the put is to exist, NG has to increase enough so that the option knocks in. If $\rho_{Al,NG} > 0$, then Al will tend to increase also, making the KI put less valuable than if $\rho_{Al,NG} < 0$ where Al tends to decrease.

Results for the Illustrative XMKI Put Option

The results for the Cross-Market knock-in put price (expressed as a ratio to the standard put) show that the correlation dependence is strong.

The following is a table of the results:

---

[17] **Remark:** This was indicative pricing for a deal that didn't happen. You get used to it.

[18] **Translation:** The jargon means that the strike of the aluminum Al put option was 20% out-of-the-money OTM above its spot value. The put knocked in (up-in or UI) if natural gas NG went above the barrier 20% higher than its spot value. The date for the put exercise (if it got knocked in) was 2 years from the trade date. The results given correspond to the market in 1993.

Ratio of 2D knock-in put to standard put as a function of Al, NG correlation

Correlation	Ratio
-30%	80%
0%	66%
40%	46%

Standard Commodity Options

We have been looking at exotic commodity options. The last example had aluminum and natural gas. The hedging of these commodity exotics could involve standard commodity options. Forward commodity prices are obtained from exchanges (e.g. the LME for metals), and traditional option models are used for these standard options.

Reload Options

In this section, we consider qualitatively some aspects of certain corporate stock options called "reloads", which sometimes have been given to executives and other employees[iii,19]. The reload feature is part of the initial option specification, which gives back to the reload options holder a number of new reload options, equivalent to the number of shares needed for the costs of exercise and taxes. The reload options that result from exercise are generally granted at-the-money at exercise[20]. Typically, contractual or other restrictions apply[21, iv].

The idea is shown in the following picture:

---

[19] **History and Story:** I was the in-house corporate quantitative resource for a fascinating episode involving reload options that once played a role in an important proxy vote. The issue was the reload option value as obtained by a buy-side consulting service that advised institutional clients on how to vote their shares. The service made an error using a model written by another consultant. Once the mistake was ferreted out, the information had a positive impact on the vote. The lesson is that model risk can be significant at the corporate level. For other aspects of model risk in pricing executive options, see ref. iii.

[20] **Reload Strikes:** Because the reload options are granted at the money, these options have only time value at the time they are granted.

[21] **Restrictions of Exercise and Sale:** These can include contractual restrictions on the time period for vesting of the various options after they are obtained before exercise is allowed, a minimum value of the stock increase before exercise is allowed, restrictions on the time before stock obtained from exercise can be sold, etc. Sometimes executives make commitments not to exercise options even when allowed. See Ref. iv.

More Exotics and Risk

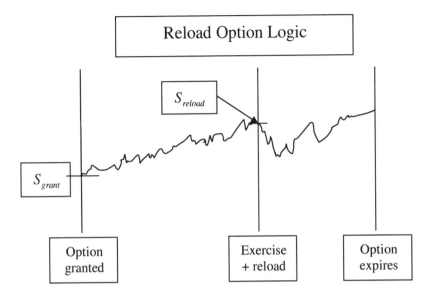

In valuing reload options, a successive approximation technique can be used. For simplicity, we begin with the approximation that there are no reloads or restrictions, and that early exercise is allowed as a standard American option[22]. Then we follow a perturbative approach by adding to the standard American option value an extra differential amount corresponding to the extra value for the possibility of reloading[23]. This extra amount is calculated using a Monte Carlo simulator, following the contractual logic of the reload options[24]. Contractual restrictions subtract from the value. We generate a number of stock paths into the

---

[22] **American Options basically give a Lower Bound for Reload Options:** American options allow for early exercise before the expiration date. The reload option (in the absence of restrictions) is at least as valuable as a standard American option. The reload option holder can exercise at any time, as with an American option, plus extra value exists from the reload provisions. Restrictions lower the value to some extent.

[23] **Reload Option FAS 123 Reporting:** Reload options have been valued for reporting purposes as short-term European options using a Black-Scholes model. The options may be valued at-the-money, using the stock price S_{grant} at the time that the options were granted or obtained. Reloads that arise from previous option exercises are included. This procedure satisfies FASB reporting requirements (ref. iv). The model in the text gives considerably larger values. Say at time t_0 that a given reload option is approximated as a European option up to an effective exercise time t^*. Then the value of future unrealized reloads for $t > t^*$ is ignored. This European approximation was not used in the story above. Marking the options to market with the current stock price $S(t_0)$ is a separate issue.

[24] **Technical Point Regarding Optimality of Exercise Including Reloads:** The approach requires some assumption regarding the timing of early exercise with reloads. Typically, it is assumed that exercise with reloads is performed as soon as allowed. It is plausible, but as far as I know unproved, that this is an optimal strategy. The strategy allows the option holder to "cash out" on the profit as soon as possible.

future. On each path, the characteristics of the reload contract are put in just as they would be if the stock price were realized in the real world. We then discount the extra value of reloads on each path, and average the results over all paths[25].

Although dependent on the details, the reload option without stock sale restrictions may be worth around 20% – 30% more than a standard American option. For a hypothetical example, if the total term of the reload option is 10 years, and if a 10-year American ATM option is worth $30, the total value of the reload option without stock sale restrictions could be worth $36 – $39.

Discounts due to Restrictions on Sale of Stock

The stock obtained by exercising company stock options may be restricted for sale by company regulations for a certain amount of time. The idea behind the modeling of the restriction discount value is shown in the following picture.

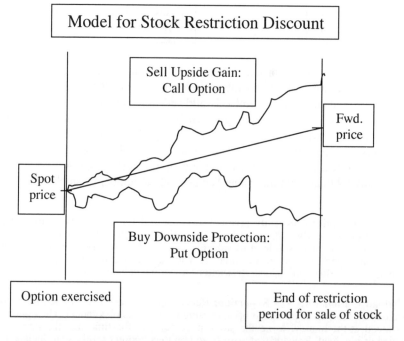

An option including this restriction is worth less because the option holder gives back some value, since he/she is not able to sell the stock during the restriction period. We next describe the model.

---

[25] **Effective Paths for Monte-Carlo Simulation:** An equivalent procedure is to generate effective paths with weights corresponding to the density of paths given by the probability distribution function generating the paths. These weights are specific from one local bin region in x(t) at time t to another local bin region in x(t + dt) at time t + dt. See Ch. 44 on numerical methods.

Option Model for Stock Restriction Discount

A model for this restriction discount is based on option theory. The idea is to estimate the hedging cost during the restriction period, such that the risk of stock price changes is eliminated, using options. To do this, recall that the no-arbitrage option theory tells us that the forward stock price is the market-expected stock price. We want to consider the forward stock price at the end of the restriction period.

If at the time the option is exercised and the restricted stock obtained, we were to sell a call option and buy a put option, both struck at the forward stock price, we would hedge out the stock price risk. This is because we buy protection from the put option if the stock price drops. However, we give up the gain from the short call option if the stock price increases. If the volatilities are the same, these two options have theoretically equal value. However, we must sell the call at the bid volatility and buy the put at the ask volatility. The resulting hedging cost gives the amount of the restriction discount.

Sometimes only a put option is used for the estimate of the restriction value. Downside protection by itself (i.e. just the put option) is very costly. However, including only downside protection ignores the upside gains.

As a rule of thumb, for a 2-year restriction, assuming 2% volatility bid-asked spread, the discount is around 10% of the stock value. The discounts increase with the amount of volatility spread and with the length of the restriction period.

References

[i] **Simple Set of Independent Contingent Option, Not the Contingent Cap in the Text:**
Editors, Derivatives Week, *Learning Curves: The Guide to Understanding Derivatives.* Institutional Investor, 1994, Vol I. See p. 90.
Schap, K., *Exotic Options Cap Interest Rate Fears.* Futures, 4/93, p. 32.

[ii] **Complex Yield-Curve Options**
Bhansali, V. and Seeman, B., *Interest Rate Basket Options or Yield Curve Options.* Learning Curve, Derivatives Week, 9/20/93.

[iii] **Model Risk in Pricing Options**
Leonhardt, D., *Options Calculus: Who Gets It Right?* The New York Times, 3/30/03, Section 3.

[iv] **Reload Options**
O'Brien, B., *Travelers Rule Prohibits Sales by Executives.* The Wall Street Journal, Aug. 7, 1996, pp. C1, C15.
Citigroup, *2003 Annual Report,* Pp. 137-139.
Financial Accounting Standards Board (FASB), *Statement of Financial Accounting Standards No. 123.* Financial Accounting Series, October 1995.

16. A Pot Pourri of Deals (Tech. Index 5/10)

Various securities, options, and risks are discussed in this chapter[1,i]. Some are exotic and some are not. We consider:

- TIPS, and an Important Statement for the Macro-Micro Model
- Muni Derivatives and Forward Muni Issuance with Derivatives Hedging
- Difference Options (Equity Index vs. Basket)
- Resettable Options: Cliquets
- Power Options
- Periodic Caps
- ARM Caps
- Index-Amortizing Swaps
- A Hypothetical Repo + Options Deal
- Convertible Issuer Risk

TIPS (Treasury Inflation Protected Securities)

Description of TIPS

TIPS, or Treasury Inflation Protected Securities, are Consumer Price Index (CPI) - indexed government securities[ii]. TIPS were introduced in 1997. Both the principal and the coupon increase to offset inflation. The main theoretical intuition is that a "real" yield r_{real} plus an "expected inflation" component g is supposed to be the observed nominal treasury market yield for the same maturity, $y = r_{real} + g = y_{Tsy}^{Market}$. The TIPS yield is supposed to be r_{real}, insulated from inflation. However the components r_{real}, g are not directly observable. The CPI year-over-year change is taken to give the inflation, namely $g(t) \cdot 1\text{yr} = \left[CPI(t) - CPI(t - 1\text{yr}) \right] / CPI(t - 1\text{yr})$.

---

[1] **History:** This work was mostly performed during 1988-97 (the last section in 2002). By the way, pot-pourri does not mean "rotten pot". According to the Larousse, pot-pourri is a "production littéraire formée de divers morceaux."

The TIPS bond price is $\$B_{TIPS} = \$\mathcal{P} \cdot \sum_{i=1}^{N} \eta^i \left[(c/2) + \delta_{iN} \right]$. Here, c is the initial coupon, $N/2$ years is the bond maturity, $\eta = (1 + g\Delta t)/(1 + y\Delta t)$, $\Delta t = \frac{1}{2} yr$, the initial notional is $\$\mathcal{P}$, and accrual interest is ignored. Hence

$$\$B_{TIPS} = \$\mathcal{P} \cdot \left\{ (c/2) \left[(\eta - \eta^{N+1})/(1 - \eta) \right] + \eta^N \right\} \tag{16.1}$$

The TIPS delta price risk is $\$\Delta B_{TIPS} = \delta g \cdot \left. \frac{\partial \$B_{TIPS}}{\partial g} \right|_y + \delta y \cdot \left. \frac{\partial \$B_{TIPS}}{\partial y} \right|_g$ which depends on inflation and yield changes δg and δy, along with the correlation[2] $\rho = \rho(\delta g/g, \delta y/y)$. Increased inflation $\delta g > 0$ increases the TIPS value, while increased yields $\delta y > 0$ decreases the TIPS value.

Now if a yield change δy occurs, we expect on the average a correlated inflation change as $\delta g = \beta \delta y$ where $\beta = \frac{\rho g \sigma_g}{y \sigma_y}$, and where ($\sigma_g, \sigma_y$) are the lognormal (inflation, yield) volatilities[3]. This particular idea is rather similar to that used in the CAPM model in another context[iii]. The above expression for δg is the "parallel" component of the inflation change along the δy direction. This can be seen by writing $\rho = \cos\theta$, as explained in Ch. 22. An idiosyncratic perpendicular component exists but we do not model it.

Noting that $\partial \eta / \partial g = -1/\eta \cdot \partial \eta / \partial y$, the TIPS delta price risk is therefore approximately

$$\$\Delta B_{TIPS} = \left[1 - \frac{\rho g \sigma_g}{\eta y \sigma_y} \right] \delta y \cdot \left. \frac{\partial \$B_{TIPS}}{\partial y} \right|_g \tag{16.2}$$

---

[2] **Inflation vs. Treasury Yield Correlation:** When TIPS were first issued in 1997 this correlation ρ was surprisingly low, under 20%. The simple theoretical idea is that the correlation should be high if inflation changes produce most of the market yield changes.

[3] **Lognormal Vols and Underlying Changes:** Recall that a lognormal volatility is defined as σ_{LN} = sqrt{< (dy/y)^2 >$_c$} so a change δy is on the order of y*σ_{LN} for some representative y.

The decreased risk due to inflation is manifest in the minus sign of the second term. Nonetheless, the risk is not zero. This means that the duration of the TIPS is not zero either. Hence a TIPS, while having more stable behavior than a treasury of the same maturity, is *not* the same as a floater at par[4].

The TIPS yield while relatively stable is not constant. For example, price data show that typical TIPS prices ranged around 94 to 103 over the period 4/15/99 to 7/31/01 (See Jarrow and Yildirim, Ref).

TIPS and an Important Statement for the Macro-Micro Model

One important point in this section is the analysis of Bertonazzi and Maloney[iv] based on data from 1997-1999 showing that TIPS yields are relatively stable compared to ordinary yields. They stated: "The implication is that inflation is the biggest component of variance in the yield on government bonds".

The Bertonazzi-Maloney statement is in philosophical agreement with the Macro-Micro Model[5]. The Macro-Micro Model, originating from a study of government bond data, states that most of the variance in bond yields arises from macroeconomic (Macro) causes, which naturally would include inflation. Macroeconomic effects other than inflation could still be present as trends in TIPS on the time scale of months.

A smaller highly mean-reverting and rapidly fluctuation component also exists, arising from trading (Micro) effects.

Municipal Derivatives, Muni Issuance, Derivative Hedging

Muni Derivatives

Muni derivatives are treated with formalism similar to that of Libor-based derivatives. The main assumption is that the Muni forward rates $f_{Muni}^{(\tau)}$ at different maturities τ are fractions $\xi^{(\tau)} < 1$ of Libor forward rates $f_{Libor}^{(\tau)}$, viz $f_{Muni}^{(\tau)} = \xi^{(\tau)} \cdot f_{Libor}^{(\tau)}$. The fractions can be written as $\xi^{(\tau)} = 1 - r_{Tax}^{(\tau)}$, with a "forward tax rate" $r_{Tax}^{(\tau)}$. The idea of course is that municipal cash bond interest is

---

[4] **Trader's Intuition?** Before the TIPS were issued, one influential trader forcefully promoted the idea that the TIPS would remain at par, and therefore have zero duration and no risk. The first week of trading sobered things up a bit. However, the TIPS are indeed relatively stable.

[5] **Macro-Micro Model:** This model is treated in detail later in this book, Ch. 47-51.

generally exempt from federal income tax, and possibly exempt from state and local taxes.

Muni swaps and options are defined similarly to Libor swaps and options. For example, the floating rates in a Muni swap might be specified as obtained from the J. J. Kenny municipal short-rate index.

The Muni rate risk can be found by writing the change in the muni forward rate composed of the two terms $\delta f_{Muni}^{(\tau)} = \xi^{(\tau)} \cdot \delta f_{Libor}^{(\tau)} + \delta \xi^{(\tau)} \cdot f_{Libor}^{(\tau)}$. The term with $\delta \xi^{(\tau)}$ gives the risk due to changes in the Muni/Libor ratio.

Muni Bond Futures and Options, and Comparison to the Treasury Market

Listed muni bond futures and some options on muni futures exist[6]. These are relatively illiquid compared to the treasury market. Relative strategies between the muni and treasury markets are often used, under the rubric MOB ("Municipal Over Bond") strategies. The pitfalls include the breakdown of the correlation between the two markets, which are driven by different dynamics.

Treasury-Bond Option Models

Muni-bond-future options models are similar to treasury-bond option models based on the Black-Scholes model (see the footnote[7]). The use of this simple model here is reasonable. This is because the bonds are insignificantly affected by the "pull to par" at maturity. To a good approximation, an American feature gives a small correction to the European result. This correction can be expressed as an equivalent change in the implied volatility of the option, and is generally only a fraction of one percent.

---

[6] **Muni Futures and Muni Options:** The muni futures are settled in cash as $1,000 times the Bond Buyer municipal bond index. The contracts are Mar, Jun, Sep, Dec and are traded on the CBOT (Chicago Board of Trade), but only the nearest future is liquid. Muni futures options are American style short-dated options. The Bond Buyer 40 index includes 40 high-quality actively traded tax-exempt revenue and GO bonds; the composition is changed from time to time.

[7] **Treasury-bond Option Models:** A modified Black-Scholes model gives a good approximation to short-dated options on long-maturity bonds. Some details include the use of an appropriate repo rate to get the forward bond price. A "risk-free-rate" is used for discounting (generally spread at or above the repo rate). A "dividend yield" is set as the annual coupon divided by the bond price. The clean bond price (without accrued interest) is used as the underlying variable. Other models are yield-based, but are very close in value to the price-based model. It should be noted that bond prices may be quoted as 100.nmr. Here nm = 00 to 31 indicate fractions of 1/32, and r = 0 to 7 indicates fractions of 1/8/32.

Short-dated muni-future options can similarly be reasonably priced using a modified Black model, using the underlying as the price of the muni future[8]. Again, the bonds on which the future is based for delivery have much longer maturities than the short-dated option, so the pull to par is negligible.

Example of Forward Muni Issuance With Derivative Hedging

Here is a hypothetical deal involving municipal bond issuance including hedging. Muni rates have dropped. The ABC municipality has an outstanding bond that it would like to refund given the savings implied by current low rates, but for various reasons ABC cannot call the bond. Therefore, ABC enters into a deal with a broker-dealer BD. The BD agrees to take the risk to underwrite an ABC bond issue at time t_{issue} in the future, with the constraint that ABC's yearly debt service $\$DS_\ell$ in each year t_ℓ will be specified now. Here[9], $\$DS_\ell$ is the principal re-payment $\$\delta P_\ell$ plus the interest payment $\$I_\ell$ on the unpaid principal for coupon c_ℓ. The amount $\$DS_\ell$ is advantageous from ABC's point of view, giving some compensation for the current low rates. The coupons c_ℓ will be set at issuance t_{issue}, and will be fixed (not depending on a floating rate).

The analysis of the risk of the deal from BD's point of view depends on the changes of muni rates between now t_0 and t_{issue}. If muni rates do not change, say that BD makes $\$C_{profit}^{(No\ Change)}$. If rates rise, in order to sell the deal to investors, the required coupons will rise, increasing interest payments. Since the ABC total debt service is fixed, the notional principal $\$P_0$ for the bond will have to decrease. Since BD has underwritten the deal at a fixed total amount $\$P_{deal}$, BD will lose money if rates rise enough, i.e. such that the profit or loss "P&L", is negative, viz $\$P_{P\&L} = \$P_0 - \$P_{deal} + \$C_{profit}^{(No\ Change)} < 0$. On the other hand, if rates decrease, then the principal $\$P_0$ increases, and then BD makes more profit. However, BD does not want to take a view on rates and therefore wants hedges the exposure to any rate change.

Hedging the Issuance

Possible hedges for the issuance include:

---

[8] **Black Option Model:** This formula, appropriate for an option on a future, can be obtained from the Black-Scholes model by including a fictitious "dividend" set equal to the risk-free rate, since the future is not an asset.

[9] **Time-Dependent Coupons and Principal Repayments:** We keep the notation general.

- Go Short Muni Bond Futures
- Go Long (Buy) Libor Payer's Swaption (and sell other swaptions)

The first possibility is to go short muni-bond futures. The short muni-bond futures hedge will increase in value as Muni rates increase, and the number of contracts can be arranged to stabilize $\$P_{P\&L}$, as determined by running various scenario analyses. The futures need to be rolled over if t_{issue} is past the next futures expiration date. We still have the basis risk $Basis = Cash - Future$ after putting on a futures hedge. Here the cash is not the Bond Buyer index on which the Muni bond future is written, but the ABC issuer bond $Cash_{ABC}$. The total basis risk is thus the usual (Index vs. Future) basis risk $Basis_{Usual} = Index - Future$ plus an idiosyncratic (Cash vs. Index) basis risk $Basis_{Cash} = Cash_{ABC} - Index$.

The second possibility is based on swaption hedging. The payer's swaption hedges the increase in rates by paying off more when rates increase. Now buying a swaption has an initial cost. This cost can be reduced or made zero by selling other swaptions, for example selling a payer's swaption at a higher strike and/or selling a receiver's swaption. The swaption parameters would be obtained by achieving a measure of stability of $\$P_{P\&L}$ using scenario rate analyses, along with basis risk assumptions. In this case, the basis risks are the Muni/Libor risk and the idiosyncratic cash risk.

There is also credit risk. If the credit of ABC drops before the time at forward issuance, the coupons that are needed to make the deal sell to investors will need to rise, also reducing $\$P_{P\&L}$. An estimate of this risk would be built into the nominal profit $\$C_{profit}$, and a credit reserve might be taken against this risk.

Difference Option on an Equity Index and a Basket of Stocks

This section discusses a classic European call option with payoff related to the difference of an equity index, e.g. S_{500} (S&P 500), and a basket of stocks S_B. So we need to consider the difference variable $S_{500} - S_B$. We can think of this option as a call option on the index, where the strike of the option is not a constant but equals the basket price. Going long this option hedges against underperformance of the basket relative to the S&P benchmark. Thus, if the basket outperforms the S&P, the call is worthless but the basket beats the benchmark. On the other hand if the basket under performs, the call is in the money and the payoff reduces the underperformance of the basket.

A Pot Pourri of Deals

The formalism[v] amounts to a straightforward application of the two-dimensional path-integral projected onto one dimension[10]. Here, we skip the formalism and only discuss the results and some intuition.

Assuming correlated lognormal dynamics for the index and basket, the difference call option is found to be[11]

$$C = \hat{S}_{500} e^{-y_{500}T} N\left[\Phi_{500}\right] - \hat{S}_B e^{-y_B T} N\left[\Phi_B\right] \qquad (16.3)$$

Here $\hat{S}_{500}$ (spot S&P 500 index), $\hat{S}_B$ (spot basket value), $T = t^* - t_0$ (time to expiration), y_{500} (S&P dividend yield), y_B (basket dividend yield), σ_{500} (S&P volatility), σ_B (basket volatility), ρ (correlation of S&P, basket returns), $\Phi_{500,B} = \left[\ln(\xi) \pm \tfrac{1}{2} \Lambda^2 T\right]/\sqrt{\Lambda^2 T}$, where $\xi = \left(\hat{S}_{500} e^{-y_{500}T}\right)/\left(\hat{S}_B e^{-y_B T}\right)$. Finally, the composite variance (the square of the total volatility Λ) is

$$\Lambda^2 = \sigma_{500}^2 + \sigma_B^2 - 2\rho\, \sigma_{500}\, \sigma_B \qquad (16.4)$$

Intuition for the Difference Option Formula

It is not hard to see why the option formula makes sense. The payoff at expiration t^* is the difference of the index and the basket, if that is positive, viz $\left[S_{500}^* - S_B^*\right]_+$. Hence the variable S_B^* acts like the strike at expiration and the basket spot price $\hat{S}_B$ acts like the strike at the current time. The formula for the difference option C has a Black-Scholes form with standard normal-function arguments. The discount factors $\exp(-rT)$ are cancelled by forward price factors $\exp(rT)$, occurring for both the index and the basket, and leaving the dividend-yield factors $\exp(-yT)$. The term $-2\rho\, \sigma_{500}\, \sigma_B$ in the composite volatility Λ has a minus sign because $S_{500} - S_B$ also has a minus sign. So you see the formula makes intuitive sense.

The hedge ratios N_{500} and N_B for the S&P and basket can be found by considering the portfolio $V = C + N_{500} S_{500} + N_B S_B$ and following standard logic as explained in Ch. 42.

In the previous chapter we discussed yield-curve options, simple forms of which resemble this option.

---

[10] **Path Integral Formalism:** See Ch. 42 and 45.

[11] **Homework:** Derive this formula.

Also, it is worth noting that the connection with options on the maximum of two assets is found using the formula $[A - B]_+ = \max(A, B) - B$, a case originally treated by Stulz. The generalization to many variables was done by Johnson [v].

Resettable Options: Cliquets

Resettable options are another type of option whose strike is not fixed. They are generally baskets of component options. Note that a "basket of options" is different from an "option on a basket", so we are talking about sets of options, not options on a composite variable. The component options have the feature that their strikes are unknown at the start, and are dependent on the behavior of the underlying variable at fixed times in the future. That is, the strikes are "reset", so these options generally could be called "resettable options". We have already looked at one example, the reload option, in a previous chapter.

"Momentum" caps are a set of interest-rate options that have resetting at-the-money (ATM) strikes at particular times. The holder of a momentum cap receives the rise in the interest rate from one reset to the next.

A "cliquet" is similarly a set of component options "cliquetlettes" expiring at successive times, formulated for equities or FX, and again generally reset ATM. Various bells and whistles are sometimes added to cliquet deals, including guaranteed returns, participation factors, strike percentages, and also cheapening the structure by replacing the component call options with call spreads. Cliquets can act as hedging vehicles for multi-year equity annuity programs that are based on equity returns, e.g. programs offered by insurance companies.

Pricing Resettable Options

We can get an exact solution for sufficiently simple deals. We use the language of stocks with stock price $S(t)$. If the strike E of a component option expiring at t^* is set ATM at the intermediate time $t_E < t^*$ then $E = S_E$. Hence the payoff of a call option at expiration is proportional to $\left(S^* - S_E\right)_+$. In order to value the option today t_0 we merely need to write the discounted expected value as the succession of two propagations in the two intervals (t_0, t_E) and (t_E, t^*), multiply by the payoff, and do the integrals. With the usual lognormal dynamics, these integrals are simple and closed-form solutions result.

The picture for a component of a generic resettable option gives the idea:

A Pot Pourri of Deals

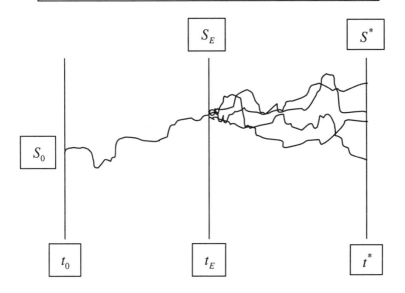

Resettable option expiring at t^* has strike E set at time t_E as the value S_E, at-the-money.

The form of the integration we need, with $x = \ln S$, is

$$C(x_0, t_0) = \int_{-\infty}^{\infty} dx_E \left[\exp(x^*) - \exp(x_E) \right]_+ G(x_0, x_E; t_0, t_E) G(x_E, x^*; t_E, t^*)$$

(16.5)

Here the discount factors and normalization are included in the Green functions. Since by assumption the Green functions are Gaussians in the x variables, we can do the integrals by completing the square using ordinary calculus. After a little algebra we get (with $\tau_E = t_E - t_0$) the result[12]

$$C(x_0, t_0) = S_0 \cdot \left[e^{-y \tau_E} N(d_{E+}) - e^{-r \tau_E} N(d_{E-}) \right] \qquad (16.6)$$

Here $d_{E\pm} = (r - y \pm \sigma^2 \tau_E / 2) / (\sigma \sqrt{\tau_E})$. Also (r, y, σ) are the discount interest rate, dividend yield, and volatility respectively. Note that the bracket

---

[12] **Homework again:** You can do the integrals, just try.

multiplying S_0 is independent of S_0, and is in fact the expression for an option over time interval t_E, with price and strike equal to 1.

Realistic Cliquet Deals

Cliquet deals, as mentioned above, are actually more complicated. We can have a guaranteed minimum return $\mathcal{R}_{Min}$, participation factors $\{\lambda_\ell\}$, persistence factors $\{\pi_\ell\}$, strike percentages $\{\xi_\ell\}$, and a maximum return $\mathcal{R}_{max}$. The ℓth return is $\mathcal{R}_\ell = \left[S_\ell^* - \xi_\ell S_{\ell-1}^* \right] / S_{\ell-1}^*$ and the ℓth notional is $\$\mathcal{N}_\ell = \pi_\ell \$P$. The ℓth component cliquetlette payoff equation at t_ℓ^* is defined as[13]

$$\$C_\ell^* = \$\mathcal{N}_\ell \max\left[\mathcal{R}_{Min}, \lambda_\ell \min\left(\mathcal{R}_\ell, \mathcal{R}_{max}\right)\right] \tag{16.7}$$

Defining the effective strikes $E_{eff}^{(A)} = S_0\left[1 + \mathcal{R}_{Min}/\lambda_\ell\right]$ and $E_{eff}^{(B)} = S_0\left[1 + \mathcal{R}_{max}\right]$ along with the time periods $\tau_\ell = t_\ell^* - t_0$ and $t_E = t_\ell^* - t_{\ell-1}^*$, the result for the expected discounted payoff is[14]

$$C_\ell(x_0, t_0) = e^{-r\tau_\ell} \mathcal{R}_{Min} + \frac{e^{-r\tau_\ell}\lambda_\ell}{S_0}\left[C_\ell^{(A)} - C_\ell^{(B)}\right] \tag{16.8}$$

Note that the difference of $C_\ell^{(A)}$ and $C_\ell^{(B)}$ defines the call spread $C_\ell^{(A)} - C_\ell^{(B)}$. The options have standard Black-Scholes forms,

$$C_\ell^{(A)} = S_0 e^{(r-y)t_E} N(d_{A+}) - E_{eff}^{(A)} N(d_{A-})$$
$$C_\ell^{(B)} = S_0 e^{(r-y)t_E} N(d_{B+}) - E_{eff}^{(B)} N(d_{B-})$$
$$d_{A\pm} = \left[\ln\left(S_0/E_{eff}^{(A)}\right) + \left(r - y \pm \sigma^2 \tau_E/2\right)\right] / \left(\sigma\sqrt{\tau_E}\right)$$
$$d_{B\pm} = \left[\ln\left(S_0/E_{eff}^{(B)}\right) + \left(r - y \pm \sigma^2 \tau_E/2\right)\right] / \left(\sigma\sqrt{\tau_E}\right)$$

$$\tag{16.9}$$

---

[13] **MaxMin:** You need to go through the payoff equation one step at a time starting from the inside expression to see the logic.

[14] **First Cliquetlette:** The first component option in a cliquet deal already done needs a separate calculation since its strike has already been set.

A Pot Pourri of Deals   229

If the rate r is taken from a term structure, the rate multiplying t_E in the forward stock price is the forward rate between $t^*_{\ell-1}$ and t^*_ℓ. Similar remarks hold for the dividend yield and the volatility. The rate multiplying τ_ℓ in the discount factor is the t^*_ℓ spot rate.

Example of a Cliquet Deal

In this deal, there are seven components or "cliquetlettes", with annual resets. The minimum and maximum returns are $\mathcal{R}_{Min} = 2\%$, $\mathcal{R}_{max} = 10\%$. The participating factors and persistence factors are $\lambda_\ell = \pi_\ell = 1$ and the strike percentages are $\xi_\ell = 100\%$ for simplicity. The rates are $r = 5\%$/yr, dividend yields $y = 1.3\%$/yr both continuous/365, and the volatility for the $C^{(A)}_\ell$ options is $\sigma = 40\%/\sqrt{\text{yr}}$. We put in some vol skew for the $C^{(B)}_\ell$ options, $\delta\sigma_{\text{Skew}}/\sigma = -10\%$. The notional is $\$P = \$100MM$. We pay at the end and reinvest the payments for each cliquetlette at a reinvestment rate $r = 5\%$/yr. The effective strikes are $E^{(A)}_{\text{eff}}/S_0 = 1.02$, or 2% out-of-the-money (OTM) for the long A options, and $E^{(B)}_{\text{eff}}/S_0 = 1.10$ or 10% OTM for the short B options.

We get the total 7-year cliquet as the sum of the cliquetlettes $\$C_{\text{Cliquet}}(x_0, t_0) = \sum_{\ell=1}^{7} \$C_\ell(x_0, t_0)$. We get $\$C_{\text{Cliquet}} = \$39.5 \text{ MM}$. The components are $\$C_{\text{Guarantee}} = \$11.5 \text{ MM}$ for the P.V. of the guarantee from $\mathcal{R}_{Min}$, and $\$C_{\text{Calls}} = \$28 \text{ MM}$ for all call options.

We can do a quick sanity check[15]. At average 3.5 years, the discount factor with continuous compounding is 0.84, so the guarantee should be worth around 0.84*7 yrs*2%/yr*$100MM, which checks $\$C_{\text{Guarantee}}$. The average per year for the call spread is $28MM/7 = $4 MM, or about 4% on the average for the 1-year call spreads. We find $16.36 for a 1-year option 2% OTM and $11.74 at 10% OTM, or $4.62 for the first call spread. Since this already includes one year of discounting, we multiply by 0.88 to get the average call spread, which checks $\$C_{\text{Calls}}$.

---

[15] **Sanity Checks:** Yes, you have to do sanity checks. For one thing, you want to have a one-liner to explain the results to the management. Also how do you know that your C++ programmer with only two months of experience on the Street hasn't screwed up?

Power Options

Power options have payoffs proportional to a power of the difference between the underlying variable and the strike. For example, we can define an interest-rate power floorlet (PF) with power $\lambda \geq 1$. Call f^* the forward Libor rate at option expiration t^*. If the floorlet has strike E, we write the power-floorlet payoff as $\$C^*_{PF} = \$K \cdot (E - f^*)^\lambda \theta[E - f^*]$ with $\$K = \$\mathcal{N} \cdot dt_{Libor}/E^{\lambda-1}$. Note that the payoff has units of $money only.

We can find the power floorlet value $\$C_{PF}(f_0, t_0)$ today t_0 with the spot rate f_0, assuming for example standard lognormal dynamics for the forward rate changes $d_t f(t)$. We first expand $\$C^*_{PF}$ in a power series in f^* (which truncates if λ is an integer), viz

$$(E - f^*)^\lambda = \sum_{k=0}^{\infty} (-f^*)^k (E)^{\lambda-k} \binom{\lambda}{k} \tag{16.10}$$

Now we use the standard discounted expected formalism corresponding to the lognormal Green function. Define $x^* = \ln f^*$, $x_0 = \ln f_0$, volatility σ, time to expiration $\tau^* = t^* - t_0$, and write $\xi_{max} = \left[\ln(E/f_0) + \sigma^2 \tau^*/2\right]/(\sigma\sqrt{\tau^*})$. After a little algebra, we find that we need the integral I_k where

$$I_k = \int_{-\infty}^{\xi_{max}} \frac{d\xi}{\sqrt{2\pi}} \exp\left[-\frac{\xi^2}{2} + k\xi\sigma\sqrt{\tau^*}\right] = \exp(k^2 \sigma^2 \tau^*/2) N\left[\xi_{max} - k\sigma\sqrt{\tau^*}\right] \tag{16.11}$$

This is enough to get the result,

$$\$C_{PF}(f_0, t_0) = \$K \cdot P^{(\tau^*)} \cdot \sum_{k=0}^{\infty} (-1)^k (E)^{\lambda-k} \binom{\lambda}{k} \exp\left[k\left(\ln f_0 - \sigma^2 \tau^*/2\right)\right] \cdot I_k \tag{16.12}$$

Here $P^{(\tau^*)}$ is the appropriate discount factor.

A Pot Pourri of Deals 231

Power-Barrier Options
It is possible to calculate barrier options with power law payoffs[16] by using the barrier Green functions, as described in Ch. 17. We leave this as an exercise.

Path-Dependent Options and Monte Carlo Simulation

This book has emphasized calculations that can be done at least partially analytically. The workhorse calculations that cannot be handled analytically generally are handled using Monte Carlo (MC) simulation. We do not spend much time on MC methods in this book, but we will mention a few applications here. They are:

- Periodic Caps
- ARM Caps
- Index Amortizing Swaps

Periodic Caps

A periodic cap is a collection of resettable caplets, but now including spreads. The caplet with expiration date t_ℓ^* has its strike set at the rate $r_{\ell-1}^*$ at previous reset date $t_{\ell-1}^*$ plus a spread $s_{\ell-1}$. Hence the payoff is $C_\ell^* = \left[r_\ell^* - r_{\ell-1}^* - s_{\ell-1} \right]_+$ where r_ℓ^* is the final rate at t_ℓ^*. Other types of resets have been proposed, including look-back minima[vi]. The pricing simply involves evaluating $C_\ell^{*,Path(\alpha)}$ on each $path(\alpha)$, applying the discount factor appropriate to that path, and then averaging the results over all paths[17].

ARM Caps

ARM Caps reflect the dynamics of the adjustable rate mortgage from which these options get their acronym. The payoff of one of the component caplets C_ℓ^* at

---

[16] **Inverse Power Options:** We can also have negative powers if we keep the underlying from approaching zero with a barrier.

[17] **General Statement for Monte Carlo – Boring But Brawny:** Get the cash flows on each path and discount them. Then average the results over paths. This brute-force method is really boring and inelegant, but it's the only way to fly for many situations.

time t_ℓ^* involves the difference between the Monte-Carlo (MC) rate r_ℓ^* and a calculated rate $r_\ell^{\text{Calculated}}$. There is a notional $\$\mathcal{N}_\ell$ and a time interval Δt (e.g. 6 months). The result is $\$C_\ell^* = \$\mathcal{N}_\ell \left[r_\ell^* - r_\ell^{\text{Calculated}} \right]_+ \Delta t$. Again we discount the cash flow $C_\ell^{*,Path(\alpha)}$ for each path and average the results over the paths.

The calculated rate functions like an ARM rate paid by a homeowner. It depends on a variety of complications including initial teaser rates, lifetime ceilings and floors for rates, and maximum allowed jumps up or down for rates. The calculated rate therefore depends on the path, including the various rules depending on the deal. Most of the work involves implementing the logic for the rules. The simulated MC rate plays the part of the market rate on which the ARM is based, and does not have anything to do with the rules.

Because the homeowner is long an embedded option like an ARM cap in an adjustable rate mortgage[vii], combinations involving ARMs and ARM caps packaged together can be made free of options[18]. The notionals $\{\$\mathcal{N}_\ell\}$ are then chosen as to reflect the anticipated amortization for the principal of the ARMs in the deal. This combined structure may appeal to investors who do not want to take option risk, but who may get extra pickup on the yield.

Index-Amortizing Swaps

Introduction to Index-Amortizing Swaps (IAS)

In this section we consider an "exotic" swap called an indexed-amortizing swap[19,viii] (IAS), also called an indexed-principal swap (IPS). In the section on swaps, we said that a different notional $\$N_l$ for each swaplet could occur. If these notionals decrease, the swap is called an amortizing swap. For the IAS, the swap amortizes according to a schedule fixed in advance. The schedule consists of an association between possible levels of an index floating rate F_{Index} and the amortization of the swap. The index rate F_{Index} is generally not the floating rate f_{Swap} in the swap. The notionals $\{\$N_l\}$ of the swap depend on the path that the index rate takes as time progresses. There is an initial "lock-out" period when no amortization occurs. The picture gives the idea:

---

[18] **ARM Floor:** There is also an embedded floor that the homeowner is short, reflecting the possibility that rates drop below the amount paid by the homeowner.

[19] **Acknowledgement:** I thank Alex Perelberg for helpful discussions on IAS, and on many other topics.

A Pot Pourri of Deals

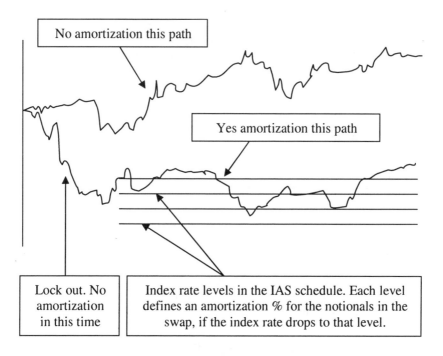

We write the notionals of the swap for a given path_α of the index rate F_{Index} as $\$N_l = \$N_l\left[F_{Index}(\text{path}_\alpha)\right]$ that amortizes according to the schedule. The swap along that path can be written as

$$\$S\left[F_{Index}(\text{path}_\alpha)\right] = \sum_{l=0}^{N_{swap}} \$N_l\left[F_{Index}(\text{path}_\alpha)\right] \cdot \left(\hat{f}^{(T_l)} - E\right) \cdot dt_{swap} \cdot P_{Arrears}^{(T_l)}$$

(16.13)

The entity ABC enters into the swap with the broker-dealer BD, receiving fixed and paying floating according to, say, Libor.

A typical IAS deal diagram looks like this:

Index-Amortizing Swap (IAS)

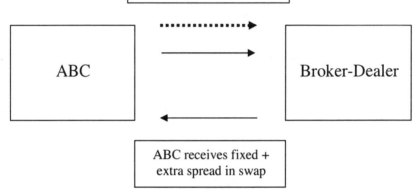

The extra spread and option are discussed below.

Here is an example of an amortization schedule, with the change in index rate given in bp from its starting value at the beginning of the deal:

Index Rate	Amort/yr
300	1%
200	2%
100	5%
0	15%
-100	20%
-200	30%
-300	35%

Connection of IAS with Barrier Options

Here is another schedule, just equivalent to a simple barrier knockout option if the index rate goes down 100 bp, at which point the notional would disappear:

Index Rate	Amort/yr
0	0%
-100	100%

Analysis of the IAS and its Embedded Options

If the index decreases enough to cross one or more levels of the schedule, the swap notionals increasingly amortize. ABC is receiving fixed including an extra spread and paying floating. If the index rate is decreasing, because of the generally positive correlations between rates, probably the floating rate that ABC is paying to BD is also decreasing. Hence the swap becomes in-the-money to ABC. Clearly ABC likes this arrangement. The amortization however is making the swap go away, to the benefit of BD. If the index rises, the BD does not want to lose the extra floating payment from ABC.

So, effectively, there is an embedded option in the IAS because ABC has given an option to BD to partially stop the swap. This option is, from BD's point of view, something like having a receiver's swaption, which gives BD the right to enter into a swap to receive fixed and pay floating. However this swaption is path-dependent, since it depends on the way the index is behaving with respect to the amortization schedule in the IAS.

In order to compensate for this embedded option that ABC gives to BD, ABC gets an extra spread in the swap.

Example of Application of the IAS : Mortgage Servicing Hedging

One application of index-amortizing swaps is to provide a hedge against prepayment risk for mortgage servicing[ix]. As rates drop, homeowners increasingly prepay their mortgages, and a mortgage servicer ABC loses some business on which it is making the servicing spread. ABC therefore wants a hedge that increases in value as rates drop, a floor-like product. Now the servicing spread is calculated including anticipated prepayments. The servicer pays for a hedge with a schedule consistent with these anticipated prepayments. The amount of the hedge decreases as rates drop since the amount of mortgage collateral decreases. The IAS provides these features.

The swap notional amortization depends on the index. For example, the index can be the 5-year Libor swap rate, a CMT rate, the prepayment on some GNMA mortgages, etc. If the index is tied to the mortgage rate driving refinancings and prepayments, then the IAS can work as a mortgage servicing hedge.

Pricing the IAS, a Volatility-Dependent Swap

It is clear from the description that a Monte-Carlo simulator is needed to price the IAS. The model in principle needs to simulate the index rate paths, and have a mechanism for correlating the swap floating rate with the index in order to price the swap.

No Simplification for IAS

Naturally, the IAS depends on the rate volatilities used in the MC simulator because these determine the probabilities that the paths will wind up crossing the schedule levels and changing the amortization.

Note that even if the notionals are independent of the swaplet index l, the IAS cannot be written as a bond and FRN. This is because the floating payment notionals depend on a stochastic index and are thus not constant, as needed to produce the FRN = par equivalence at reset.

Sociology

Typically, the quants will write the code and then deliver it to the systems people who put it in the system. Deals will be brought by the sales group to the desk where the trader or desk quant uses the system to price the deals[20].

A Hypothetical Repo + Options Deal

This example exhibits a hypothetical repo transaction with side conditions. For the deal, we do not have to know much about repo[x]. The risk involves a forward determination of repo specified in the transaction that is hedged in the ED futures market. There are also max/min constraints on the repo rate that amount to embedded repo options, and that are hedged in the ED futures options market. The main risk of the deal is the risk due to the uncertainty in the repo rates.

Definition of the Deal

The broker-dealer BD and investor X make the following arrangement.

1. BD pays repo to X at an agreed-on repo rate $r_{Repo}^{(Start)}$ for a period of time up to the next ED future-contract IMM date.

2. BD pays a "tail" repo rate $r_{Repo}^{(Tail)}$ from the next IMM date to deal maturity. The tail rate is defined as the $r_{Repo}^{(Start)}$ minus the change in the nearest ED future price (i.e. plus the change in the ED future rate).

3. Maximum and minimum repo rates are specified by the contract. These specifications amount to embedded options in the contract; one can think of them as contingent cash flows if r_{Repo} goes outside the boundaries.

---

[20] **History:** Index amortizing deals are now standard products that started around 1991. My activity (in two different jobs) was pricing them on an exotic interest-rate desk, and later supervising a quant group producing the numerical pricing code.

4. In return, X pays a fixed rate E_{Fixed} to BD.

Motivation and Risks for the Investor X
The motivation for the investor X can be that X thinks rates are likely to go up in the near term, or perhaps that X has some internal risk of rates going up and wants to hedge against that risk.

Motivation and Risks for the Broker-Dealer BD
The diagram summarizes the cash flows and options in the deal[21].

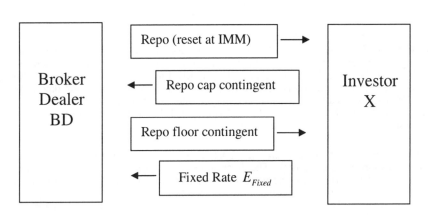

The deal can be attractive to BD if the $r_{Repo}^{(Start)}$ is low enough relative to E_{Fixed}. The BD pays repo r_{Repo} and receives the fixed rate E_{Fixed}, so the BD can make a profit through the "Carry" = $E_{Fixed} - r_{Repo}$. The main risk of the deal is the repo risk $r_{Repo}^{(Tail)}$ in the tail period from the next ED future IMM date to the deal

---

[21] **Repo Transaction Cash-flow Diagram:** The arrows indicate the cash flows. The BD pays the repo rate to X. This rate changes at the next IMM date. If the ED future rate on which the contract is based goes up (down), BD will pay a new repo rate up (down) by the same amount. This is called the "tail" rate because it is at the tail end of the deal in time. There are two options embedded in the contract, a repo cap and floor. The BD is long the cap and is short the floor. The repo rate paid by BD is limited by the repo cap (below the maximum in the contract). The repo rate received by X has the contract minimum provided by the repo floor. The combination of a long cap and short floor is called a collar. Finally, the BD receives the fixed rate E_{fixed} from X.

maturity. Scenarios would have to be run for variations in the ED future movement to assess the risk[22].

It should be clear that this looks like a swap[23] (fixed E_{Fixed} against floating r_{Repo}) with some extra options. In fact, the repo market mirrors the interest-rate derivatives market in many ways, the main difference being in the short-dated maturity of the repo transactions.

Hedging the Deal

The hedging of the deal by the BD consists of two parts:

1. BD shorts ED futures in order to hedge the repo reset risk[24].
2. BD purchases and sells appropriate ED futures ("pit") options to offset the repo cap and floor[25].

---

[22] **Risk Scenarios:** The risk in the deal can be assessed by calculating the results assuming various changes in the ED futures that in this contract determine the tail repo risk. This can be done by looking at historical ED future data movements at some confidence level. The analysis is done in time windows from the transaction date to the next IMM date. Alternatively, a scenario based on "judgment" can be applied. Given the change in the ED future, the risk of the deal is assessed. We will have a lot to say about risk measurement of this sort later in the book.

[23] **This Repo Deal is a Basically a Short-Dated Swap:** Sometimes a repo desk resembles a short-dated swap desk. The diagram in the text has been drawn to emphasize the repo dependence. By redrawing the picture to make the fixed rate "leg" (arrow) at the top of the column of arrows, the top two arrows look like a swap with BD receiving fixed and paying floating. The floating repo rate here is, however, not determined by the repo rates in the market at the reset date, but by the deal contract specifying the repo reset via the ED futures market.

[24] **Hedging the Tail Repo Risk with ED Futures.** BD is receiving fixed and paying floating, and therefore BD loses if rates go up. BD is "long the market". To hedge, BD has to go "short the market". BD shorts an appropriate number of the nearest ED future. This hedge money if the price P_f of the ED future decreases (i.e. if the ED rate r_{ED} goes up). This offsets the risk to BD of the potential increase in the repo rate.

[25] **Hedging the Repo Options with ED Futures Options or Pit Options.** The BD also performs some hedging transactions involving ED futures options ("pit options"). These options involve the same ED future as in the deal. The BD does this in order to flatten out the vega of the embedded repo cap and repo floor. In this way the overall desk portfolio will not have different vega after the deal is done. First BD *sells ED puts* with strike E_{put} which pay off if $P_f < E_{put}$ or if $r_{ED} > 100 - E_{put}$ which offsets the long BD repo cap. Then BD *buys ED calls* with strike E_{call} which pay off if $P_f > E_{call}$ or if $r_{ED} < 100 - E_{call}$ which offsets the short BD repo floor. However these hedges are not complete. The strikes ideally should be chosen so that one of the options pays off when a repo boundary defined in the deal is hit. However the ED pit option strikes are only available at certain discrete levels, so there is some residual risk of mismatch.

The following diagram summarizes the hedges put on by the BD.

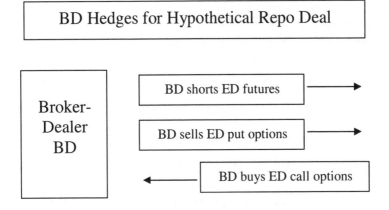

Convertible Issuance Risk

In this last section, we deal with risk involving the issuance of convertible bonds. The risk discussed here is a stock gap-risk. The issuer ABC of the convertible[xi] is often not investment grade, and the stock may not be traded in large volume. In particular the stock price can drop before the deal can be sold and/or hedged, leading to losses by the broker-dealer underwriter BD who has agreed to buy the bonds from ABC to sell to investors X. The problem is acerbated by the fact that if the stock price drops, the deal will not sell well at the original price.

A Risk Model

We first calculate the number N_{Iss} of shares of stock needed to short to hedge the issuance of N_{bonds} (actually, only the part of the issue that has not yet been sold). Generally, one convertible bond can be converted into a number of shares called the conversion ratio C_{ratio}. We need a convertible-bond model that gives the convertible bond price C as a function of (among other things) the stock price S. For a given move in the stock price δS, the number of shares needed N_{Iss} is determined by the convertible-bond delta $\Delta = \partial C / \partial S$ and to some extent by gamma $\gamma = \partial^2 C / \partial S^2$. Equating the change in the hedge with the change in the convertible, N_{Iss} is determined by

$$N_{Iss}\delta S = N_{bonds}C_{ratio} \cdot (\Delta \delta S + \gamma \cdot \delta S^2 / 2) \qquad (16.14)$$

We need the historical data for the daily volume N_{daily} for ABC stock trading as a function of time. If that information is not available, we can use a proxy stock or equity index closely related to ABC. For simplicity we can first assume Gaussian statistics for the relative stock price changes dS/S, requiring the average (Avg) and the standard deviation (σ).

We choose a confidence level CL^{Vol} for the daily volume N_{daily}. There is an associated parameter of the number of standard deviations k_{CL}^{Vol} (e.g. $k_{CL}^{Vol} = 1.65$ at $CL^{Vol} = 90\%$). Then we write

$$N_{daily}(k_{CL}^{Vol}) = Avg[N_{daily}] - k_{CL}^{Vol}\sigma[N_{daily}] \qquad (16.15)$$

The reason for the minus sign is to be pessimistic regarding the number of shares $N_{daily}(k_{CL}^{Vol})$ available for hedging in the market. It is bad if there are fewer shares because then it takes more time to hedge. The amount of time Δt_H needed to hedge the deal with this scenario is:

$$\Delta t_H = N_{Iss} / N_{daily}(k_{CL}^{Vol}) \qquad (16.16)$$

We assume some confidence level CL^{Stock} for stock price moves with associated parameter k_{CL}^{Stock} in the time Δt_H (e.g. 90% CL again). We need to obtain $\$\delta S(k_{CL}^{Stock}; \Delta t_H)$ from downward moves $\delta S < 0$ observed historically in time interval Δt_H at the confidence level for down moves chosen, CL^{Stock}.

The loss \$Loss in this pessimistic scenario is the number of shares needed to hedge times the drop in the stock price during the time Δt_H needed to put on the hedge. This is divided by two, because on the average during the total hedging time Δt_H, half the deal will be hedged, so the net loss is halved. We get

$$\$\text{Loss}(k_{CL}^{Vol}; k_{CL}^{Stock}; \Delta t_H) = N_{Iss} \cdot \$\delta S(k_{CL}^{Stock}; \Delta t_H)/2 \qquad (16.17)$$

Any sale of more bonds to customers will ameliorate this potential loss.

Preview of Coming Attractions: Math Coming Up Next

We have now gone over a number of deals and case studies in the book. The next few chapters go into the theory of exotic options in some depth.

References

[i] **Pot Pourri Definition**
Editors, *Petit Larousse illustré*. Librairie Larousse, 1976.

[ii] **TIPS**
Roll, R., *U.S. Treasury Inflation-Index Bonds: The Design of a New Security*. Journal of Fixed Income, Dec. 1996. Pp. 9-28.
Bloomberg Financial Markets News, *U.S. Treasury Inflation-Index Bonds*, 1997.
Wessel, D., *Treasury Plans to Sell 5-Year 'Inflation' Notes*. Wall Street Journal, 6/9/97.
Bertonazzi, E. and Maloney, M. T., *Sorting the Pieces to the Equity Risk Premium Puzzle*. Clemson U. working paper, 2002. Fig. 1 (TIPS yield data, 1997 – 1999).
Cox, J. C., Ingersoll, J. E. Jr., Ross, S. A., *A Theory of the Term Structure of Interest Rates*, Econometrica. Vol. 53, 1985, Pp. 385-407.
Jarrow, R. and Yildrim, Y., *Pricing Treasury Inflation Protected Securities and Related Derivatives using an HJM Model*. Cornell U. working paper, 2002. See Figure 1 (TII2 Market Prices, 1999 – 2001).

[iii] **CAPM**
Rudd, A. and Clasing, H. K. Jr., *Modern Portfolio Theory; The Principles of Investment Management*, 2nd Ed. Andrew Rudd, 1988. See Pp. 28-32, Ch. 1.

[iv] **TIPS and the Macro-Micro Model**
Bertonazzi, E. and Maloney, M. T., *Sorting the Pieces to the Equity Risk Premium Puzzle*. Clemson U. working paper, 2002. See P. 3.

[v] **Diff Options and Related Topics**
Margrabe, W., *The Value of an Option to Exchange One Asset for Another*, J. Finance, Vol. 33, 1978. Pp. 177-183.
Stulz, R., *Options on the Minimum or the Maximum of Two Risky Assets*. J. Financial Economics, Vol. 10, 1982. Pp. 161-185.
Johnson, H., *Options on the Maximum or Minimum of Several Assets*. J. Financial and Quantitative Analysis, Vol. 22,1987. Pp. 277-283.

[vi] **Resettable Options**
Editors, Derivatives Week, *Learning Curves*, 1994, Pp. 158-160.
Dehnad, K., *Lookback and Ladder Periodic Caps*. Derivatives Week, 10/25/93.

[vii] **ARM Caps**
Bartlett, W., *Mortgage-Backed Securities*. NY Institute of Finance, 1989. See Ch. 10.

[viii] **Index Amortizing Swaps**
Arterian, S., *Does Anybody Understand Index-Amortizing Swaps?* Derivatives Strategy & Tactics, Vol. 2, No 8., 1993. Pp. 1-9.

[ix] **Mortgage Servicing**
Kasaba, E., Mortgage Banking, February 1993. Pp.14-23 and Pp. 49-56.

[x] **Repo and Derivatives**
O'Keeffe, P., *Repo and Derivatives*. Learning Curve, Derivatives Week, 4/25/94.

[xi] **Convertible Bonds**
Zubulake, L., *Convertible Securities Worldwide*. John Wiley & Sons, Inc. 1991.

17. Single Barrier Options (Tech. Index 6/10)

In this chapter, we begin the discussion of barrier options using the path integral Green function formalism[1,i]. Barrier options are options that have the underlying process constrained by one or several boundaries called barriers. Usually these options are European options—that is, there is only one exercise date at which the option may pay off. If the underlying crosses a barrier, a "barrier event" occurs, and the option changes its character. For example, the option may disappear[2], be replaced by another option[3], be replaced by cash[4], etc. The barrier is usually continuous[5] although sometimes it is discrete[6] and it usually exists over the whole option period until exercise[7]. The barrier event usually is defined to

---

[1] **History:** Options for one constant continuous barrier seem to have been first catalogued by Rubenstein in 1990 (Ref). The connection with path integrals and other calculations in this chapter were performed by me in 1991.

[2] **Knock-out option:** This is an option that disappears if the barrier is crossed. If the barrier is below the starting value of the underlying, the option is called a "down-and-out" option. If the barrier is above the starting value of the underlying, the option is called an "up-and-out" option. If the option does not knock out, it will pay off like an ordinary option.

[3] **Knock-in options:** These are options that have no payoff but that make another option appear if the barrier is crossed. Generally, the option that appears is a standard European option. If the barrier is below the starting value of the underlying, the option is called a "down-and-in" option. If the barrier is above the starting value of the underlying, the option is called an "up-and-in" option.

[4] **Rebates:** Sometimes cash is given if the option knocks out, called a "rebate". The rebate may either be paid at the time when the knockout occurs, or the payment of the rebate may be deferred until some later date. Essentially the rebate is a knock-in step option. A knockout option with rebate therefore is really a portfolio of these two options.

[5] **Confusions:** Do not confuse the continuous nature of the boundary – at which barrier events may occur – with the single date of the European option exercise.

[6] **Discrete barriers:** The barrier may also exist at a certain number of points. For example, the contingent cap described in Ch. 16 has a payment that is essentially a discrete barrier option.

[7] **Partial barriers:** Barriers that exist only over part of the time period to exercise are imaginatively called partial barriers. Partial barriers are more complicated since they involve different propagations over the time periods when the barrier exists and when the barrier does not exist.

occur the first time the underlying crosses the barrier[8]. Usually the barrier is a constant value[9].

The barrier is given in advance, unlike the case in American options where the exercise boundary needs to be determined. For example, we may have a barrier at a given price of gold. For this reason, European barrier options are simpler to deal with than American options[10].

Naturally, there is an exotic zoo of barrier options. If more than one variable is involved, the option is called a "hybrid" barrier option. For example, we may have an interest-rate option that changes character as the price of gold passes a barrier value. There may be an upper barrier and a lower barrier. These are called double barrier options. Double-barrier options can have events that are different for each boundary. There can also be a set of barriers where progressive changes in the options occur. Such barriers are called soft barriers[11]. We can also have barrier options mixed with average options[12]. We will consider some of these exotics in more detail in Ch. 18-20.

Barrier options exist in all product types. There are stock barrier options, FX barrier options, interest-rate barrier options and commodity barrier options. Except for the presence of the barrier, the underlying dynamics are unchanged. Therefore, the same set-up as for standard options is appropriate, including the no-arbitrage constraints, and the various complexities of the different markets. We can specify different types of payouts. The risk measures are substantially more complex than for options without barriers and the risk itself is much harder to hedge. This is because the barrier options contain various discontinuities. We will discuss risk with barrier options.

We shall use path integrals and Green functions to describe the dynamics of barrier options[13]. Exactly as for ordinary options, if the parameters are simple

---

[8] **More complicated barrier events:** There are additional exotic types of barrier events. For example, it might be required that the underlying stay above a barrier a certain number of days.

[9] **Non-constant barriers:** These are generally harder to deal with and are much less common.

[10] **American barrier options:** If the option may be exercised at any time and if in addition there is a barrier, then all the complexities of American options including backward induction logic occur, but this time in the presence of a constraint because of the barrier.

[11] **Soft barriers:** One example is index-amortizing swaps for interest rates, described in Ch. 16. There, a schedule is provided with progressive events occurring as the index rate passes through the different barriers in the schedule.

[12] **Barrier-average options:** For example, we might require the 5-day average price to be above the barrier.

[13] **Path Integrals, Green Functions, and Options:** See Ch. 41-45 for details.

Single Barrier Options

enough, analytic results can be obtained. From the point of view of path integrals, this happens if the semigroup property allows the re-expression of successive propagation of the Green function over small-time increments in terms of a similar Green function over the total time interval.

Say G_{ab} is the Green function[14] over $(x_a, t_a; x_b, t_b)$, G_{bc} similarly is the Green function over $(x_b, t_b; x_c, t_c)$, and G_{ac} is the Green function over the total $(x_a, t_a; x_c, t_c)$ with $t_a < t_b < t_c$. The semi-group property in the absence of barriers is given by an ordinary integral over x_b:

$$G_{ac} = \int_{-\infty}^{\infty} G_{ab} G_{bc} dx_b \qquad (17.1)$$

The path integral exactly satisfies the semigroup property, as discussed in detail in Chapters 42-45. We get simple results if the Green function has a simple form, notably Gaussian. We discuss the semigroup property including a constant barrier below.

Knock-Out Options

Consider first a single barrier for a stock S at a value H and assume that $S < H$. Although we can formulate the general path integral for arbitrary functions $H(t)$, in order to get closed form results we need constant H to carry out the integrations in the path integral explicitly. Later on we will give the results for calculations assuming piecewise constant barriers of different types.

The quickest way to get at the problem is to use the classical method of images to solve the problem of the Green function for the diffusion equation that vanishes on a boundary[15, ii]. Transferring as usual to logarithmic variables, we write $x = \ln S$ and $x^{(bdy)} = \ln H$. The diffusion equation operator for lognormal

---

[14] **Notation:** For G_{ab}, $x_a = x(t_a)$ and $x_b = x(t_b)$ are the initial and final points, with time interval $t_{ab} = t_b - t_a > 0$. Also, we will call σ the volatility (assumed constant) and μ the drift constrained by standard no-arbitrage considerations.

[15] **Images:** Images result from the solution of the differential equation with the appropriate boundary conditions. The idea is that you replace the original problem with another problem for which images are included and the boundary is absent. The solution is valid in the "physical region", i.e. in the region for which the boundary is a barrier. The use of images in diffusion problems goes back at least to 1921 (see Carslaw, ref). Many image references do not have drift, which requires the extra factor $K(x_a)$ in the text.

dynamics is $\hat{O}_{diff} = \dfrac{\partial}{\partial t_a} + \dfrac{\sigma^2}{2}\dfrac{\partial^2}{\partial x_a^2} + \mu\dfrac{\partial}{\partial x_a}$. The Green function that solves the equation over the time interval (t_a, t_b) and vanishes on the boundary will be called the KO (knock-out) Green function; it has the form $\mathcal{G}_{ab}^{(KO)} = G_{ab} - G_{ab}^{(image)}$. In particular, we have $\mathcal{G}_{ab}^{(KO)}(x_a = x^{(bdy)}) = 0$. All paths that stay on the physical side of the boundary (here assumed as $S < H$) are contained in $\mathcal{G}_{ab}^{(KO)}$. If $S < H$, $\mathcal{G}_{ab}^{(KO)}$ is relevant to an up and out option. The first term G_{ab} in $\mathcal{G}_{ab}^{(KO)}$ is the Green function for the unconstrained problem that we have used repeatedly,

$$G_{ab} = \dfrac{1}{\left(2\pi\sigma^2 t_{ab}\right)^{1/2}} \exp\left\{-\dfrac{1}{2\sigma^2 t_{ab}}[x_a - x_b + \mu t_{ab}]^2\right\} \qquad (17.2)$$

The $G_{ab}^{(image)}$ term in $\mathcal{G}_{ab}^{(KO)} = G_{ab} - G_{ab}^{(image)}$ is the image Green function needed for the enforcement of the boundary condition. It is given by

$$G_{ab}^{(image)} = K(x_a) G_{ab}\left[x_a \to x_a^{(image)}\right] \qquad (17.3)$$

Here $x^{(image)}(t) = 2x^{(bdy)} - x(t)$ is the called the image path of $x(t)$. At the boundary, the path and the image path coincide[16]. Note that with $x < x^{(bdy)}$ in the physical region, we have $x^{(image)} > x^{(bdy)}$ so the image path is in the unphysical region on the other side of the boundary.

The factor $K(x_a)$ is needed to solve the equation $\hat{O}_{diff} G_{ab}^{(image)} = 0$ with drift present. It is given by

$$K(x_a) = \exp\left\{\dfrac{\mu}{\sigma^2}\left(x_a^{(image)} - x_a\right)\right\} \qquad (17.4)$$

---

[16] **What about the paths that cross over the boundary?** The reader may complain that, after all, the paths contributing to G_{ab} are unrestricted and some can cross over the boundary at $S = H$. The response is that we are not allowed to look at the solution in the unphysical region for the up and out option. Therefore, only those paths that stay below the barrier are relevant for the up and out option. The paths that cross the barrier are relevant to a different problem, the up and in option as we discuss next.

Note that we can also write $G_{ab}^{(image)} = \tilde{K}(x_b) G_{ab}\left[x_b \to x_b^{(image)}\right]$, with

$$\tilde{K}(x_b) = \exp\left\{-\frac{\mu}{\sigma^2}\left(x_b^{(image)} - x_b\right)\right\}, \; K(x_a)\tilde{K}(x_b) = \exp\left[\frac{2\mu}{\sigma^2}(x_a - x_b)\right].$$

The Semi-Group Property including a Barrier

In describing Green functions without barriers, the semigroup property is the critical ingredient in proceeding from infinitesimal time steps to finite time steps. The path integral in general satisfies the semigroup property. Here the knock-out (KO) Green function has been derived assuming a constant barrier. The KO Green function satisfies a semigroup property, but now we must be careful to restrict the region of integration in the semigroup equation to be the physical region. Physically this is just because we are following along paths that never cross the barrier. The up and out Green function satisfies the following semigroup property:

$$\mathcal{G}_{ac}^{(KO)} = \int_{-\infty}^{x^{(bdy)}} \mathcal{G}_{ab}^{(KO)} \mathcal{G}_{bc}^{(KO)} dx_b \qquad (17.5)$$

Here, the integral is over the up and out physical region below the barrier, namely $S < H$ or $x < x^{(bdy)}$, where again $x = \ln S$ and $x^{(bdy)} = \ln H$. This result can be proved explicitly by direct substitution and some algebra. In performing this, it is handy to use both expressions for the image Green function given above.

In the down and out case, for which the physical region is $S > H$ or $x > x^{(bdy)}$, we get the semigroup property satisfied by the KO Green function as

$$\mathcal{G}_{ac}^{(KO)} = \int_{x^{(bdy)}}^{\infty} \mathcal{G}_{ab}^{(KO)} \mathcal{G}_{bc}^{(KO)} dx_b \qquad (17.6)$$

Schematically we write the semigroup property for the KO Green function as

$$\mathcal{G}_{ac}^{(KO)} = \mathcal{G}_{ab}^{(KO)} \otimes \mathcal{G}_{bc}^{(KO)} \qquad (17.7)$$

The reader should stop and reflect about the semigroup property[17].

---

[17] **Suggested homework:** Do these calculations. They're not so hard and you will get a feeling for the important semigroup property. Try not to look at the useful integrals below at least for the first half-hour.

Calculating Barrier Options

We are now in a position to calculate knock out European constant-barrier options. These are given by taking the expectation of the option payoff C_{payoff} using the KO Green function, multiplying by the discount factor that we separate out explicitly. For the up-and-out option (with constant parameters) we have

$$C_{ab}^{(UpOut)} = e^{-rt_{ab}} \int_{-\infty}^{x^{(bdy)}} \mathcal{G}_{ab}^{(KO)} C_{payoff}\left(x_b, t_b\right) dx_b \qquad (17.8)$$

Similarly, a down and out option has the limits of integration from $x^{(bdy)}$ to ∞. Various payoffs can be considered. We may have a standard option payoff, a step payoff, a squared payoff, etc. The standard option can be a call or a put. We may have stock options, FX options, commodity options, etc. for the choice of the underlying variable. Using the integrals below we can evaluate the various possibilities. For example, with constant parameters (risk free rate r, volatility σ, no-arbitrage drift $\mu = r - \sigma^2/2$ in the absence of dividends), and initial price $S_0 = S(t_a)$, a stock down and out call option is[18]

$$C_{ab}^{(DownOut)} = C_{ab}^{(BlackScholes)} - C_{ab}^{(image)} \qquad (17.9)$$

Here, $C_{ab}^{(BlackScholes)}$ is the standard Black-Scholes call option formula[19]

$$C_{ab}^{(BlackScholes)} = S_0 N(d_+) - Ee^{-rt_{ab}} N(d_-) \qquad (17.10)$$

Here, $d_\pm = \left[\ln(S_0/E) + \left(r \pm \tfrac{1}{2}\sigma^2\right)t_{ab}\right]/\left(\sigma\sqrt{t_{ab}}\right)$. Also, $C_{ab}^{(image)}$ is the image term, which reduces the standard call value:

$$C_{ab}^{(image)} = \left(\frac{S_0^2}{H^2}\right)^{-\frac{r}{\sigma^2}+\frac{1}{2}} \left\{ S_0\left(\frac{H^2}{S_0^2}\right) N\left(d_+^{(image)}\right) - Ee^{-rt_{ab}} N\left(d_-^{(image)}\right)\right\} \qquad (17.11)$$

---

[18] **Notation:** The exercise time t* is t_b here, and. $t_{ab} = t_a - t_b$ is the time to expiration. Also, often the notation $d_1 = d_+$, $d_2 = d_-$ is used.

[19] **Black Scholes Homework:** You now have enough information to derive the Black-Scholes formula and the image formula yourself. More information is given in Ch. 42.

Here, $d_\pm^{(image)} = d_\pm \left[\ln S_0 \to \ln\left(H^2/S_0\right) \right]$. This replacement is the same as the replacement $x_a \to x_a^{(image)}$.

Knock-In Options

We now consider the KI knock-in Green function $\mathcal{G}_{ab}^{(KI)}$. This contains all paths that do cross the barrier. This set equals the set of all paths minus the set of all paths not crossing the barrier. The KO Green function contains the set of all paths not crossing the barrier. Therefore, $\mathcal{G}_{ab}^{(KI)}$ in the KO physical region must be the usual Green function G_{ab} minus the KO Green function $\mathcal{G}_{ab}^{(KO)}$. This is just the image Green function $G_{ab}^{(image)}$. Physically this is because we are seeing the image paths that have crossed over the barrier into the physical region for KO. In addition, $\mathcal{G}_{ab}^{(KI)}$ in the unphysical region for KO must just be G_{ab}. Physically this is because we see all the usual paths that have crossed over the barrier into the unphysical region for KO.

The idea is shown below for the paths contributing to an up and in option:

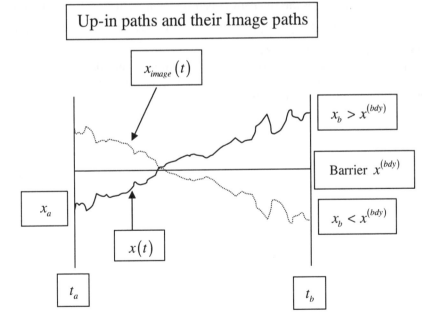

If a path crosses the barrier going up, its image path crosses the barrier going down. The final point after the transition across the barrier can be either above or below the barrier. Hence the up and in Green function consists of two parts, the standard Green function above the barrier and the image Green function below the barrier. Explicitly, the KI up-in Green function is

$$\mathcal{G}_{ab}^{(KI)} = \begin{pmatrix} G_{ab} & if\left(x_b > x^{(bdy)}\right) \\ G_{ab}^{(image)} & if\left(x_b < x^{(bdy)}\right) \end{pmatrix} \qquad (17.12)$$

The astute reader may complain again that the paths after the barrier crossing takes place have no restriction and may cross back over again. To satisfy this point, we consider the following consistency condition. We use the semigroup property to get the up and in Green function by following the paths. Consider the set of paths that start at time t_a, do not cross before time $t_{j-1} = t_j - dt$, then cross in time dt, and are then unrestricted. The paths that do not cross before time t_{j-1} are contained in the KO Green function from t_a to t_{j-1}. The Green function for crossing in time dt but otherwise unrestricted is the standard Green function propagating from $t_j - dt$ to t_j. The paths that are unrestricted after crossing are contained in the standard Green function from time t_j to the final time t_b. Plugging all this in and integrating over all possible crossing times does produce the up and in Green function.

Explicitly, for an upward crossing the convolution of the KO Green function and the Green function for crossing is found after some algebra including expanding in the infinitesimal dt to be

$$G_{aj}^{(crossing)} = \int_{-\infty}^{x^{(bdy)}} dx_{j-1} \int_{x^{(bdy)}}^{\infty} dx_j \left[G_{a,j-1} - G_{a,j-1}^{(image)} \right] G_{j-1,j}$$

$$= \left[x^{(bdy)} - x_a \right] \frac{dt}{\tau_{aj}} \cdot Gaussian\left[x_a \to x^{(bdy)} \text{ in time } \tau_{aj} \right] \qquad (17.13)$$

Multiplying by the standard Green function starting at $x^{(bdy)}$ at time t_j to t_b and integrating over the crossing time, with a little help from the useful integrals section below, then gives the up-in Green function.

European option values are obtained by multiplying the option payoff C_{payoff} by the KI Green function, along with the discount factor. There will be two

Single Barrier Options

contributions corresponding to the two possibilities for the paths after the barrier crossing being on one side or the other of the barrier at the exercise time.

Useful Integrals for Barrier Options

Here are some useful integrals for calculating constant single barrier options. Define

$$I_{AB}(x_a,\lambda) \equiv \int_{\ln A}^{\ln B} \frac{dx^*}{(2\pi\sigma^2\tau^*)^{1/2}} \exp(\lambda x^*) \exp\left\{-\frac{1}{2\sigma^2\tau^*}\left[x_a - x^* + \mu\tau^*\right]^2\right\}$$

(17.14)

Here λ is a parameter. Generally, $\lambda = 0,1$. We get after completing the square,

$$I_{AB}(x_a,\lambda) = \exp\left[\lambda(x_a + \mu\tau^*) + \lambda^2\sigma^2\tau^*/2\right]\left\{N(v_{\max}) - N(v_{\min})\right\}$$ (17.15)

Here, $v_{\max} = \frac{1}{\sigma\sqrt{\tau^*}}\left[x_a - \ln A + \mu\tau^* + \lambda\sigma^2\tau^*\right]$ and $v_{\min} = v_{\max}(A \to B)$.

We also have

$$I_{3/2}(a,b) = \int_0^{\tau^*} \frac{dt}{t^{3/2}} \exp\left(-\frac{a}{t} - bt\right) = \sqrt{\frac{\pi}{a}}\left\{e^{-2\sqrt{ab}}N(d_+) + e^{2\sqrt{ab}}N(d_-)\right\}$$ (17.16)

Here, $d_\pm = \frac{1}{\sqrt{\tau^*}}\left[-\sqrt{2a} \pm \sqrt{2b}\tau^*\right]$. Also,

$$I_{1/2}(a,b) = \int_0^{\tau^*} \frac{dt}{t^{1/2}} \exp\left(-\frac{a}{t} - bt\right) = -\frac{\partial}{\partial b} I_{3/2}(a,b)$$

$$= \sqrt{\frac{\pi}{b}}\left\{e^{-2\sqrt{ab}}N(d_+) - e^{2\sqrt{ab}}N(d_-)\right\}$$

(17.17)

Next,

$$J(a,\alpha) = \int_0^{\tau^*} \frac{dt}{t^{3/2}(\tau^*-t)^{1/2}} \exp\left[-\frac{a}{t} - \frac{\alpha}{(\tau^*-t)}\right]$$

$$= \sqrt{\frac{\pi}{a\tau^*}} \exp\left[-\frac{1}{\tau^*}\left(\sqrt{a}-\sqrt{\alpha}\right)^2\right]$$

(17.18)

Finally the key in unlocking the semigroup for barrier Green functions lies in the two following integrals

$$\int_{-\infty}^{x^{(bdy)}} G_{ab}^{(image)} G_{bc}^{(image)} dx_b = \int_{x^{(bdy)}}^{\infty} G_{ab} G_{bc} dx_b \qquad (17.19)$$

$$G_{ac}^{(image)} = \int_{-\infty}^{x^{(bdy)}} \left[G_{ab} G_{bc}^{(image)} + G_{ab}^{(image)} G_{bc}\right] dx_b \qquad (17.20)$$

Complicated Barrier Options and Numerical Techniques

In general, for complicated barriers, non-constant parameters, skewed volatility etc. the analytic expressions are no longer valid and numerical techniques have to be employed. These are usually Monte-Carlo or else discretization of the diffusion equation with the boundary conditions imposed. The discretization for a flat boundary or piecewise flat boundary is carried out most easily with a rectangular lattice in S – space, since this can be chosen to match the boundary.

A Useful Discrete Barrier Approximation

We have been focusing on a continuous barrier. For discrete barriers, aside from brute-force calculations, an analytic trick often gives reasonable approximations. The idea is that the image Green function is present for continuous barriers and absent for no barrier. Therefore, we look for an interpolating factor to multiply the image Green function that disappears as the continuous barrier becomes discrete with fewer and fewer points[20].

---

[20] **History:** This discrete barrier approximation was developed by me in 1997. The motivation was to come up with some sort of analytic approximation to perform sanity checks on Monte Carlo simulators for discrete barrier calculations. The approximation can be used for reasonable and quick approximate calculations.

Single Barrier Options

Let $\Delta t_{sampling}$ be the time between the discrete points of the barrier (assume barrier points are regularly spaced in time). Define the factor

$$\mathcal{X}^{(\gamma)}\left(\Delta t_{sampling}\right) \equiv \left[1-\left(\frac{\Delta t_{sampling}}{T_{ab}}\right)^{\gamma}\right] \qquad (17.21)$$

Here, T_{ab} is the total time over which the discrete barrier extends and $\gamma > 0$ is a parameter to be determined. Now write the discrete KO Green function approximation $\mathcal{G}_{ab}^{(KO,Discrete)}$ as

$$\mathcal{G}_{ab}^{(KO,Discrete)} \approx G_{ab} - \mathcal{X}^{(\gamma)}\left(\Delta t_{sampling}\right) G_{ab}^{(image)} \qquad (17.22)$$

For continuous barriers, $\Delta t_{sampling} \to 0$ and $\mathcal{X}^{(\gamma)} \to 1$. If the discrete barrier disappears, $\Delta t_{sampling} \to T_{ab}$ and $\mathcal{X}^{(\gamma)} \to 0$. Hence $\mathcal{G}_{ab}^{(KO,Discrete)}$ reduces to the appropriate limits. Numerically it turned out that $\gamma \approx 0.6$ provides a good approximation for a number of examples.

"Potential Theory" for General Sets of Single Barriers

We now use the semigroup property to obtain results for complex successions of single barriers[21]. In principle, there is no reason why the barriers cannot be at different levels. Repeatedly applying the semigroup property produces a sequence of convolution integrals that resembles a sort of "potential theory". The results can be expressed in terms of multivariate integrals.

An example is in the figure below for two barriers called $x_{bc}^{(bdy)}$ for times in $\left(t_b, t_c\right)$, and $x_{de}^{(bdy)}$ for times in $\left(t_d, t_e\right)$. We have the result:

$$\mathcal{G}_{ae}^{(Total)} = G_{ab} \otimes \mathcal{G}_{bc}^{(KO)} \otimes G_{cd} \otimes G_{de}^{(KI)} \qquad (17.23)$$

Here we use the notation $\otimes$ to signify convolution integration. The integrals have to be chosen to be over the appropriate limits for the multiple barrier conditions. For the example, the regions of integration are $x_b, x_c > x_{bc}^{(bdy)}$ and

---

[21] **History:** These results are in my SIAM talk in 1993; the work started in 1991.

$x_d < x_{de}^{(bdy)}$, while the final point satisfies $x_e > x_{de}^{(bdy)}$. So the integrals are of the form

$$\int_{x_{bc}^{(bdy)}}^{\infty} dx_b \int_{x_{bc}^{(bdy)}}^{\infty} dx_c \int_{-\infty}^{x_{de}^{(bdy)}} dx_d .$$

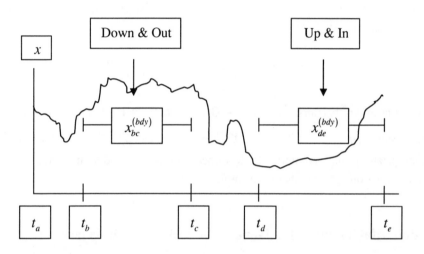

Step Barriers

A special case involves successive "steps", i.e. successive single barriers with different levels[22, iii]. The influence of a given step on paths depends on whether the path sees a step "up" or a step "down" as it moves along in time. For a step "up", some paths will hit the step and knock out. For a step "down", some paths will migrate downward depending on the drift and the volatility.

---

[22] **Barrier Step Calculations:** Some barrier step calculations were performed by me in 1996 that were inconclusive. These involved trying to calculate the effects of a step (i.e. two successive barriers at different levels but no intervening gap) explicitly. This was done by turning the step into a ramp over time dt. Before and after the step, the appropriate image solutions for constant barriers are used. The problem involved enforcing the zero boundary condition on the ramp. These considerations probably do not affect the validity of the potential theory given above. Vladimir Linetsky (Ref) has investigated step barrier options. He obtained results consistent with the potential theory.

Barrier Options with Time-Dependent Drifts and Volatilities

With time-dependent drift $\mu(t)$ and volatility $\sigma(t)$, the semigroup property can still be used to construct the path integral, but in general, the path integral cannot be evaluated analytically. One special case that yields tractable results occurs if $\mu^2(t)/\sigma(t)$ is constant if the underlying process is not an interest rate, or if the slope of $\mu^2(t)/\sigma(t)$ is -1 if the underlying process is the short rate (the change is due to discounting). However, these conditions are not realized in practice.

Barrier Gaussian Rate Model with No Negative Rates?

A long discussion of Gaussian interest rate models is in Ch. 43. While tractable, Gaussian models have the wart of negative interest rates. A practical problem for which it is tempting to use barrier technology involves modifying the Gaussian rate model by using an explicit rate barrier at zero, $r^{(bdy)} = 0$, along with the knock-out Green function including only paths for positive rates. The problem is that the shape of the forward curve produces time-dependent drifts that spoil the simple results[23].

One idea was to use a discrete step approximation to the forward curve. Then KO Green functions would be defined for each of the steps, and the potential theory convolutions would be used to get the KO Green function over large times, with only positive rates.

Explicitly, we can use the $r^{(bdy)} = 0$ barrier along with local drifts for KO Green-functions determined through consistency with the zero-coupon bond market data. Each local (t_a, t_b) drift μ_{ab} is calculated to be consistent with the corresponding forward bond price $P^{(t_b)}(t_a; t_0)$ obtained from the forward rate curve today t_0.

Consider expectations over $\mathcal{G}_{ab}^{(KO)}$ times a local discount factor $e^{-r_{ab}t_{ab}}$, integrating only over positive rates. Then μ_{ab} is determined such that the expectation of the maturity t_b terminal value 1 is $P^{(t_b)}(t_a; t_0)$ at t_a.

Once the local drifts are determined, the potential theory convolutions are used to generate the Green function over arbitrary times. Once the Green function is determined, contingent claims can be calculated. Alternatively, lattice numerical codes could be used.

---

[23] **Gaussian Non-Negative Rate Model using Barriers:** This idea was considered by me in 1991-93 and described at Merrill at a seminar in 1/93, but there was never time for numerical testing. I since heard that a similar idea was implemented somewhere.

References

[i] **Single Barrier Options**
Rubenstein, M., *Exotic Options*. working paper 11/90.
Rubenstein, M. and Reiner, E., *Breaking down the barriers*. Risk Magazine, 9/91, p. 28.
Derman, E. and Kani, I., *The Ins and Outs of Barrier Options*. Goldman Sachs, Quantitative Strategies Research Notes, 1993.
Taleb, N., *Dynamic Hedging – Managing Vanilla and Exotic Options*. John Wiley & Sons, Inc. 1997.

[ii] **Images for Diffusion with a Single Barrier**
Carslaw, H. S., *Introduction to the Mathematical Theory of the Conduction of Heat in Solids*. Second edition 1921; Dover Publications 1945. See p. 33.
Morse, P. M. and Feshbach, H., *Methods of Theoretical Physics, Vol 1*. McGraw-Hill Book Company, 1953. See p. 863.

[iii] **Ladder Barriers**
Linetsky, V., *The Path Integral Approach to Financial Modeling and Options Pricing*. Univ. Michigan report 96-7, 1996.

18. Double Barrier Options (Tech. Index 7/10)

In this chapter, we treat double barrier options[1]. These are options that depend on one underlying variable, but which have two barriers (upper and lower). The underlying process starts between the two barriers. The idea is in the figure:

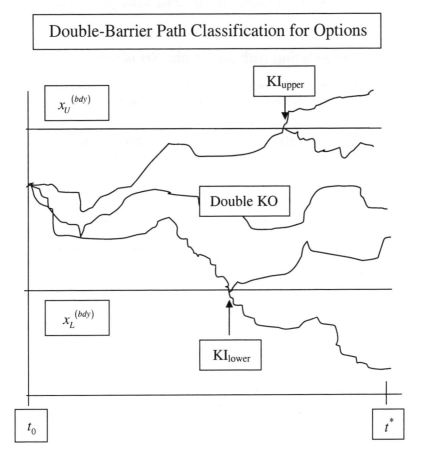

---

[1] **History:** The double-barrier Green function is in my notes from 1991. Examples were worked out in 1996.

The picture shows a double KO path and one mixed KO/KI path for each of the barriers. A double KI path (not shown) would cross both barriers.

The conditions on these two barriers $x_L^{(bdy)}$ and $x_U^{(bdy)}$ can be different. We may be interested in whether the paths always stay between the two barriers as a requirement[2] (a double KO), or whether the paths cross one boundary but not the other (a mixed KO/KI) or whether the paths can cross either barrier (a double KI). The image Green function has to satisfy the diffusion equation and in addition be zero on both boundaries

As usual, we assume constant continuous barriers and constant parameters so that we can get analytic results. Cases that are more general have a path integral solution, but require the usual Monte Carlo or lattice techniques for evaluation.

Double Barrier Solution with an Infinite Set of Images

The analytic solution to the single barrier problem has one image. The double barrier problem has an infinite set of images[1]. The first quartet is shown below:

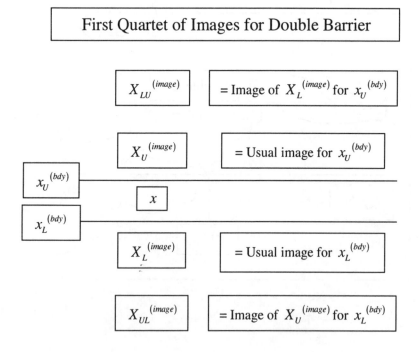

First Quartet of Images for Double Barrier

$X_{LU}^{(image)}$ = Image of $X_L^{(image)}$ for $x_U^{(bdy)}$

$X_U^{(image)}$ = Usual image for $x_U^{(bdy)}$

$x_U^{(bdy)}$

x

$x_L^{(bdy)}$

$X_L^{(image)}$ = Usual image for $x_L^{(bdy)}$

$X_{UL}^{(image)}$ = Image of $X_U^{(image)}$ for $x_L^{(bdy)}$

---

[2] **Notation:** Knock out (KO), knock in (KI), down and out (DO), up and in (UI), double knock out (DKO). Subscripts on a Green function G_{0*} mean that the initial time is t_0 and the exercise time is t* as shown in the figure. In this chapter, the Green functions do not include the discount factor.

Double Barrier Options

The construction of the images is a straightforward generalization by induction of the situation for one barrier. Consider[3] the upper barrier at $S = H_U$ first, ignoring the lower barrier $S = H_L$. At time t with $x = \ln S(t)$ along a given path, we construct the usual image $X_U^{(image)}$ for the upper barrier as $X_U^{(image)} = 2x_U^{(bdy)} - x$ where $x_U^{(bdy)} = \ln H_U$. Now consider the lower barrier H_L. We construct the usual image for the lower barrier as $X_L^{(image)} = 2x_L^{(bdy)} - x$ where $x_L^{(bdy)} = \ln H_L$. These are just like the single-barrier images.

Consider the upper barrier again. We now need an extra upper-barrier image $X_{LU}^{(image)}$ to cancel the effect of the usual lower-barrier image $X_L^{(image)}$ on the upper barrier, viz $X_{LU}^{(image)} = 2x_U^{(bdy)} - X_L^{(image)}$. Similarly, we need a new image $X_{UL}^{(image)}$ to cancel the effect of $X_U^{(image)}$ on the lower barrier, viz $X_{UL}^{(image)} = 2x_L^{(bdy)} - X_U^{(image)}$. Note that the new images $X_{LU}^{(image)}$, $X_{UL}^{(image)}$ are further away than are the usual images $X_U^{(image)}, X_L^{(image)}$. These new images will contribute with a plus sign to cancel out the negative sign of the usual images. This process clearly continues indefinitely. Naturally, we need to keep track of the signs of the terms in order to ensure cancellations on each boundary.

Each image is associated with its own image Green function. The image Green function is exactly of the form of the single barrier image Green function with the appropriate image parameters. Hence, the solution to the double barrier problem is given by an infinite set of single-barrier image Green functions with different arguments corresponding to the positions of the images[4].

---

[3] **Notation:** The barrier H is always quoted in the underlying, say stock price, but the images are constructed in x, the logarithm of the stock price. In addition, we mostly leave the time index off in this section to avoid confusion with the barrier labels.

[4] **Convergence of the Image Sums:** The astute reader no doubt is worried sick about how well the sums of the images converge. Because successive images move further and further away, the quadratic exponential damping of the Gaussians in the Green functions provides damping. In practical terms, the rate of convergence numerically depends on how close the boundaries are to each other, how close the starting point is to one of the boundaries, the magnitude of the volatility, whether the drift carries paths near a boundary, and so forth. The numerical convergence can be enhanced by grouping together quartets of images (two successive above and two successive below). This quartet grouping tends to cancel out oscillation instabilities that occur when images are added in one at a time. Although it is not of much help, infinite image sums are related to theta functions that occur in elliptic function theory (see Morse and Feshbach, Ref).

Double Barrier Option Pricing

The simplest double-barrier option is perhaps the double knockout DKO option (up-out OU across the upper barrier, down-out DO across the lower barrier), $C[UO(H_U), DO(H_L)]$. The DKO option is given by the usual Black-Scholes result corrected by image terms conveniently expressed via infinite sums of images arranged by quartets (we have explicitly exhibited the first quartet above). The sums contain Black-Scholes option expressions (multiplied by the extra $K(x_a)$ factors), with appropriate image parameters. There are also digital options that occur when a barrier stops a payoff from being able to go to zero when the underlying equals the strike.

For example, a double knockout (DKO) European call option is the integral between the barriers of the payoff at exercise multiplied by the DKO Green function $\mathcal{G}_{0*}^{(DKO)} = G_{0*} - G_{0*}^{(DKO\_image)}$, and discounted back from the exercise date,

$$C_{call}^{[DKO]}(x_0, t_0) = \int_{\ln H_L}^{\ln H_U} dx^* e^{-r\tau^*} \mathcal{G}_{0*}^{(DKO)} \left[e^{x^*} - E \right]_+ \qquad (18.1)$$

If the strike is between the barriers, $H_L < E < H_U$ (the usual case) we get the expression for the DKO call as

$$C_{call}^{[DKO]}(x_0, t_0) = e^{-r\tau^*} \int_{\ln E}^{\infty} dx^* \mathcal{G}_{0*}^{(DKO)} \left[e^{x^*} - E \right]$$

$$-e^{-r\tau^*} \int_{\ln H_U}^{\infty} dx^* \mathcal{G}_{0*}^{(DKO)} \left[e^{x^*} - H_U \right] - [H_U - E] e^{-r\tau^*} \int_{\ln H_U}^{\infty} dx^* \mathcal{G}_{0*}^{(DKO)}$$

$$\qquad (18.2)$$

$$= \sum_{Index=0, U, L, LU, UL...} \eta_{Index} K_{Index} \left\{ Call_{Index}(E) - Call_{Index}(H_U) - Digital_{Index} \right\}$$

$$\qquad (18.3)$$

Here the index runs over the initial t_0 point x_0 ($Index = 0$) and over all images ($Index = U, L, LU, UL,...$) at the initial time t_0. The signs are given by $\eta_{Index} = \pm 1$ and the extra image factors by K_{Index} (with $K_{Index=0} = 1$). We have

Double Barrier Options

rearranged the terms in order to exhibit the digital pieces[5] $Digital_{Index}$. The other terms marked $Call_{Index}$ are just Black-Scholes expressions with the strike indicated, and appropriate arguments corresponding to the index for the images[6].

It is important to calculate the probability $\mathcal{P}(DKO)$ that knock out does not occur, i.e. the probability that a path always stays between the two barriers. This is clearly given by

$$\mathcal{P}(DKO) = \int_{\ln H_L}^{\ln H_U} dx^* \, \mathcal{G}_{0*}^{(DKO)} \qquad (18.4)$$

This expression is evaluated as above using the infinite series of images. We get

$$\mathcal{P}(DKO) = \sum_{Index} \eta_{Index} K_{Index} \left\{ N\left[-d_{2,Index}(H_U)\right] + N\left[d_{2,Index}(H_L)\right] - 1 \right\} \quad (18.5)$$

Here, $d_{2,Index}$ is the usual Black-Scholes quantity, with arguments appropriate for the given index.

We also need the DKO digital call option $Call_{Digital}^{(DKO)}$ with strike E. This is an option that pays an amount $\mathcal{R}$, provided that the paths stay between the two boundaries and that at exercise $x^* > \ln E$. To obtain $Call_{Digital}^{(DKO)}$, the lower limit in $\mathcal{P}(DKO)$ is replaced by $\ln E$. The integral is multiplied by $\mathcal{R}$ and by the discount factor.

Simple path counting logic gives sum rules that provide the results for mixed KI/KO options. For example, a mixed KI/KO option consisting of up-in (UI) across the upper barrier and down-out (DO) across the lower barrier that we call $C[UI(H_U), DO(H_L)]$ is given by the KI/KO "sum rule",

---

[5] **Dangerous Digital Options:** Most of the complexities of barrier options arise from the digital components. If, as time passes, the underlying gets close to one of the barriers, the option will disappear (or get replaced by a rebate amount). This discontinuous change is naturally difficult to hedge. If the deal is large enough, a barrier that is has a probability of being crossed becomes a red flag and a focus for risk management. Some details about digital options and how they are hedged are given in Ch. 15.

[6] **DKO Option Expression:** The astute reader may be nervously wondering why call options with strikes at the upper barrier are present in the DKO expression when the DKO paths are restricted to remain between the barriers. As is clear from the derivation, these unphysical terms actually cancel out. The expression is useful for numerical computation.

$$C\left[UI(H_U), DO(H_L)\right] = C\left[DO(H_L)\right] - C\left[UO(H_U), DO(H_L)\right] \quad (18.6)$$

Here $C\left[DO(H_L)\right]$ is the single-barrier down-out option across the lower barrier and $C\left[UO(H_U), DO(H_L)\right]$ is the double barrier knockout option. What this means is that we count all paths that do not cross the lower barrier regardless of what happens at the upper barrier; this gives $C\left[DO(H_L)\right]$. Those paths that also stay below the upper barrier contribute to $C\left[UO(H_U), DO(H_L)\right]$ and those that cross the upper barrier contribute to $C\left[UI(H_U), DO(H_L)\right]$. The sum rule simply follows from tracking the paths.

Similar considerations give the double KI (DKI) options that get contributions from paths; each relevant path must cross both barriers. The DKI sum rule is

$$C\left[UI(H_U), DI(H_L)\right] = C^{[\text{No Barrier}]} - C\left[DO(H_L)\right] \\ - C\left[UO(H_U)\right] + C\left[UO(H_U), DO(H_L)\right] \quad (18.7)$$

Here, $C^{[\text{No Barrier}]}$ is just the European option with no barrier. If we ask for the option where either one or two knock-ins occur we need to add the two single barrier terms back in and watch out for double counting; we get another DKI sum rule,

$$C\left[UI(H_U) \text{ and/or } DI(H_L)\right] = C^{[\text{No Barrier}]} - C\left[UO(H_U), DO(H_L)\right] \quad (18.8)$$

There are special anomalous cases worth mentioning. These include a call with the strike below both barriers, or a put with strike above both barriers. Extra digital step options appear in those anomalous cases. There are some extra numerical problems involved with evaluating these digital options.

Rebates for Double Barrier Options

Rebates are sometimes given when a knockout occurs as a sort of consolation prize. Depending on the deal, a rebate $\mathcal{R}$ can be paid at the exercise time t^* of the original option, or the rebate can be paid "at touch" at the time a boundary is hit. Rebates for single barrier options are easily found. Rebates for double barrier options are more involved, as we now discuss.

Rebates for paths that hit a given boundary and that are paid at maturity are digital versions of KI options and are obtained with the KI/KO sum rule applied to digitals. If the rebate is paid if either boundary is hit, a simple result is obtained

$$C_{\text{Rebate}} = e^{-r\tau^*} \mathcal{R}\left[1 - \mathcal{P}(DKO)\right] \quad (18.9)$$

In principle one can also have two-sided rebates where both boundaries have to be crossed before the rebate is paid. If this rebate is paid at maturity, the DKI sum rule, applied to digitals, gives the result.

Rebates at touch require that time integrals be performed including the discount factor. A rebate at touch is worth more than a rebate paid at maturity because there is less discounting. Rebates at touch are difficult to evaluate analytically for double barrier options. A reasonable approximation can be obtained as follows. We replace the usual discount factor from expiration $\exp(-r\tau^*)$ with an "effective" discount factor $\exp(-r\tau_{\text{eff}}^*)$. We get the "effective time interval" τ_{eff}^* by multiplying the time to maturity τ^* by the probability that knockout does not occur. If the probability of knock-out decreases, τ_{eff}^* increases, reflecting the fact that then it would generally take longer for a knock-out to occur.

References

[i] **Infinite Set of Images**
Morse, P. M. and Feshbach, H., *Methods of Theoretical Physics, Part II*. McGraw-Hill Book Company, 1953. P. 1587.

19. Hybrid 2-D Barrier Options (Tech. Index 7/10)

Our purpose in this chapter is to set up the mathematics used to describe the two-dimensional hybrid option formalism.[1,2] We discussed these options in Ch. 15. The barrier options are assumed European with a single continuous barrier. There are two underlying variables.

The two underlying variables can be called $S(t)$ and $P(t)$. We assume that these two variables are described by correlated lognormal processes. So it is convenient to write $x(t) = \ln S(t)$ and $y(t) = \ln P(t)$. The first variable $x(t)$ contains all the information about the cash flows. The second "barrier" variable $y(t)$ contains the information about the barrier, defined for simplicity as a constant $y^{(bdy)} = \ln H$. We write[3] the volatilities as σ_x, σ_y and the drifts as $\mu_x = r_x - q_x - \sigma_x^2/2$, $\mu_y = r_y - q_y - \sigma_y^2/2$. The initial point and time are x_0, t_0 and the final (exercise) point and time are x^*, t^*. We write $\tau^* = t^* - t_0$ and call $(Dx)_{0*} = x_0 - x^* + \mu_x \tau^*$ and $(Dy)_{0*} = y_0 - y^* + \mu_y \tau^*$. It is also useful to define a quantity $(Dy_\perp)_{0*}$ that is the component of $(Dy)_{0*}$ geometrically perpendicular to $(Dx)_{0*}$ as

$$(Dy_\perp)_{0*} = (Dy)_{0*} - \rho \frac{\sigma_y}{\sigma_x}(Dx)_{0*} \qquad (19.1)$$

The 2D Green function in the absence of barriers that we call $G_{0*}^{(2D)}$ is given by

$$G_{0*}^{(2D)} = N_{0*} \exp\left[\Phi_{0*}^{(2D)}\right] \qquad (19.2)$$

---

[1] **History:** The hybrid option calculations in the text were done by me in 1993.

[2] **Acknowledgement:** I thank James Turetsky for helpful conversations.

[3] **Notation:** The parameters r_x and q_x are the risk-free rate and dividend yield for S, while r_y and q_y are the risk-free rate and dividend yield for P. The forms of the drifts are standard and follow from academic no-arbitrage considerations.

Here, $N_{0*} = \left[2\pi\sigma_x\sigma_y\tau^*\left(1-\rho^2\right)^{1/2}\right]^{-1}$ is the normalization, and $\Phi_{0*}^{(2D)}$ is nicely separated using the above variables as

$$\Phi_{0*}^{(2D)} = -\frac{\left[(Dx)_{0*}\right]^2}{2\sigma_x^2\tau^*} - \frac{\left[(Dy_\perp)_{0*}\right]^2}{2(1-\rho^2)\sigma_y^2\tau^*} \qquad (19.3)$$

We need the two-dimensional image Green function $G_{0*}^{(2D\_image)}$ generalizing the one-dimensional case. The idea is the same. We need to solve the diffusion equation appropriate to the assumptions made in setting up the problem, consistent with the boundary conditions implied by the existence of the barrier. The 2D image Green function $G_{0*}^{(2D\_image)}$ can be obtained by noting that, as far as the variable y is concerned, the x variable enters $\Phi_{0*}^{(2D)}$ through $(Dy_\perp)_{0*}$. So for fixed x^* we can think of the term $-\rho\frac{\sigma_y}{\sigma_x}(Dx)_{0*}$ as providing a modified drift in $(Dy)_{0*}$. We can then use the one-dimensional formalism for images. Define the image path for $y(t)$ as $y^{(image)}(t) = 2y^{(bdy)} - y(t) = 2\ln H - y(t)$. We replace y_0 in $\Phi_{0*}^{(2D)}$ by the image $y_0^{(image)} = 2y^{(bdy)} - y_0$ to get $\Phi_{0*}^{(2D\_image)}$. Explicitly, we write

$$\Phi_{0*}^{(2D\_image)} = -\frac{\left[(Dx)_{0*}\right]^2}{2\sigma_x^2\tau^*} - \frac{\left[\left(Dy_\perp^{(image)}\right)_{0*}\right]^2}{2(1-\rho^2)\sigma_y^2\tau^*} \qquad (19.4)$$

Here,

$$Dy_\perp^{(image)} = \left(y_0^{(image)} - y^* + \mu_y\tau^*\right) - \rho\frac{\sigma_y}{\sigma_x}(Dx)_{0*} \qquad (19.5)$$

We now write the 2D image Green function in the same form as for one dimension,

$$G_{0*}^{(2D\_image)} = N_{0*}K_{0*}^{(2D)}\exp\left[\Phi_{0*}^{(2D\_image)}\right] \qquad (19.6)$$

The extra factor $K_{0*}^{(2D)}$ is tricky. The one-dimensional analogy does not give the whole story. An x^*- independent term is required. This is obtained directly from the requirement that $G_{0*}^{(2D\_image)}$ solve the diffusion equation. We find

$$K_{0*}^{(2D)} = \exp\left\{\frac{2\left(y^{(bdy)} - y_0\right)}{\left(1-\rho^2\right)\sigma_y^2 \tau^*}\left[\rho^2\left(y^{(bdy)} - y^*\right) + \mu_y \tau^* - \rho\frac{\sigma_y}{\sigma_x}(Dx)_{0*}\right]\right\}$$

(19.7)

The first term in the square bracket proportional to ρ^2 is the extra term. With the 2D image Green function now determined, we write the total 2D Green function $\mathcal{G}_{0*}^{(2D\_KO)}$ for knock-out in the presence of the barrier as

$$\mathcal{G}_{0*}^{(2D\_KO)} = G_{0*}^{(2D)} - G_{0*}^{(2D\_image)}$$

(19.8)

The reader is no doubt anxiously awaiting the news that the 2D knockout Green function satisfies the semigroup property. It does.

The 2D knock-in Green function $\mathcal{G}_{0*}^{(2D\_KI)}$ is given by the same sort of expressions as in one dimension, involving both $G_{0*}^{(2D)}$ and $G_{0*}^{(2D\_image)}$ depending on the values of x^*. The sum of the integrated probabilities of the KO and KI Green functions has to be one, e.g.

$$\text{Prob}(Up + Out) + \text{Prob}(Up + In) = 1$$

(19.9)

Pricing the Barrier 2-Dimension Hybrid Options

We are now prepared to price barrier 2D hybrid options. Again, we are assuming constant continuous barrier and European-style options. Call the payoff cash flow at the exercise date $C(x^*, t^*)$. For example for a call option,

$$\$C(x^*, t^*) = \$P \cdot \varsigma_c \cdot \left[\exp(x^*) - E\right]_+$$

(19.10)

Here $P is a principal or notional amount and ς_c is an extra factor depending on the details. Examples of ς_c are one for a single stock option and $N_{days}/360 \approx 1/4$ for a 3 month Libor caplet. Call the discount factor (zero coupon bond) as $\hat{P}(t^*)$. The expected discounted value of $C(x^*, t^*)$ using the full 2D Green function $\mathcal{G}^{(2D)}_{0*}$ is the 2D option value $C(x_0, t_0)$ at today's time t_0. We get, with appropriate limits on the integrals,

$$\$C(x_0, t_0) = \hat{P}(t^*) \int_{x^*_{min}}^{x^*_{max}} dx^* \int_{y^*_{max}}^{y^*_{max}} dy^* \, \mathcal{G}^{(2D)}_{0*} \, \$C(x^*, t^*) \quad (19.11)$$

We may also get a combination of such integrals, depending on the circumstances. We avoid the temptation to list all the cases, but instead give some useful integrals, with which the interested reader can derive all cases [i].

Useful Integrals for 2D Barrier Options

Here are some useful integrals. The first one is

$$\int_{x^*_{min}}^{x^*_{max}} dx^* \int_{y^*_{max}}^{y^*_{max}} dy^* \exp\left\{-\left[A_1(x^*)^2 + A_2(y^*)^2 - Bx^*y^* + C_1 x^* + C_2 y^*\right]\right\}$$

$$= I_2(x^*_{max}, y^*_{max}) + I_2(x^*_{min}, y^*_{min}) - I_2(x^*_{max}, y^*_{min}) - I_2(x^*_{min}, y^*_{max})$$

(19.12)

Here, $I_2(\hat{x}, \hat{y})$ is given by

$$I_2(\hat{x}, \hat{y}) = \int_{-\infty}^{\hat{x}} dx^* \int_{-\infty}^{\hat{y}} dy^* \exp\left\{-\left[A_1(x^*)^2 + A_2(y^*)^2 - Bx^*y^* + C_1 x^* + C_2 y^*\right]\right\}$$

$$= \frac{2\pi e^{-D}}{\left[4A_1 A_2 (1-\rho^2)\right]^{1/2}} N_2\left[(\hat{x}+X)\sqrt{2A_1(1-\rho^2)}, (\hat{y}+Y)\sqrt{2A_2(1-\rho^2)}; \rho\right]$$

(19.13)

Here the correlation is $\rho = \dfrac{B}{2\sqrt{A_1 A_2}}$ and the other arguments are

$$X = \frac{C_2 B + 2C_1 A_2}{4A_1 A_2 - B^2}, \Upsilon = \frac{C_1 B + 2C_2 A_1}{4A_1 A_2 - B^2}, D = -\frac{\left(C_1^2 A_2 + C_2^2 A_1 + C_1 C_2 B\right)}{4A_1 A_2 - B^2}$$

(19.14)

Also, $N_2(a,b;\rho)$ is the usual bivariate integral,

$$N_2(a,b;\rho) = \frac{1}{2\pi\sqrt{(1-\rho^2)}} \int_{-\infty}^{a} dx \int_{-\infty}^{b} dy \exp\left[-\frac{1}{2(1-\rho^2)}\left(x^2 - 2\rho xy + y^2\right)\right]$$

(19.15)

The references list a number of methods for evaluating the bivariate integral numerically[ii]. A mix of these techniques is useful, depending on the parameters.

References

---

[i] **Hybrid Reference**
Heynen, R. and Kat, H., *Crossing Barriers*. Risk, Vol 7, June 1994. Pp. 46-51.

[ii] **Bivariate Integrals: Numerical Computation**
Abramowitz, M., Stegun, I., *Handbook of Mathematical Functions*. National Bureau of Standards, Applied Mathematics Series #55, 1964. See Ch. 26.
Borth, D., *A Modification of Owen's Method for Computing the Bivariate Normal Integral*. Applied Statistics Vol 22, No 1, 1973. Pp. 82-85.
Sowden, R. and Ashford, J., *Computation of the Bivariate Normal Integral*. Appl. Statistics, Vol 18, 1969, Pp. 169-180.
Curnow, R. and Dunnett, C., *The Numerical Evaluation of Certain Multivariate Normal Integrals*. Ann. Math. Statist., 1962. Pp. 571-579.
Owen, D.B., *Tables for Computing Bivariate Normal Probabilities*. Ann. Math. Statist., Vol 27, 1956. Pp 1075-1090.
Drezner, Z., *Computation of the Bivariate Normal Integral*. Mathematics of Computation, Vol 32, 1978, Pp 277-279.

20. Average-Rate Options (Tech. Index 8/10)

Options often contain arithmetic averaging features. In this chapter, we use path-integral techniques[i] to obtain some general results[1]. The reader is referred to the chapters on path integrals (Ch. 41-45) for more details.

To get an idea of the problem, consider the two figures below. The first figure has the valuation date t_0 during or inside the averaging period[2].

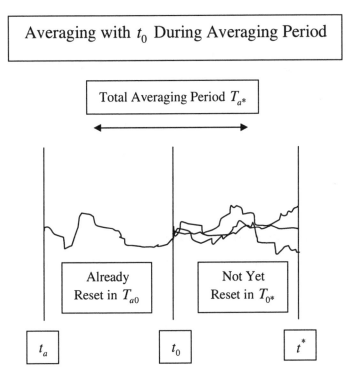

---

[1] **History:** These calculations were reported in my SIAM talk in 1993 and performed in 1989-93. Similar results can be obtained for equities and FX. The difference is that for interest-rate options the discount factor gets involved in the averaging calculations.

[2] **Notation:** Time intervals are denoted by $T_{ab} = t_b - t_a$.

The second figure has the valuation date t_0 before the averaging period starts.

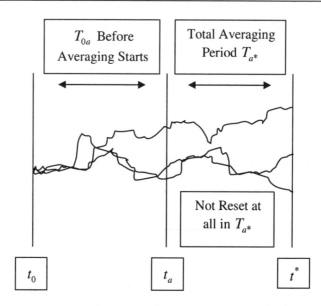

Arithmetic Average Rate Options in General Gaussian Models

Arithmetic-average rate options can be done exactly in Gaussian rate models. We exhibit the first part of the evaluation of average rate options with any arbitrary Gaussian model. Then we will get an explicit form using the Mean-Reverting Gaussian (MRG) model.

Again, we want to evaluate the option today at time t_0. Either t_0 is before the averaging period (t_a, t^*) starts at t_a or else t_0 is inside the averaging period. If t_0 is after the averaging period is over, there is nothing left to average.

Consider a time-partitioned discretization of the interest rate we want to average. If we are inside the averaging period, some rates have already been determined or "reset"[3]; call them $\{s_l\}$, $l = 1...L$. Call $\{r_j\}$, $j = 0...N - L$ the undetermined rates. Now define the general linear average as

---

[3] **Reset Rates, Systems, and the Back Office:** The back office is where the administration of the deals takes place. A well-oiled back office is critical to the

Average-Rate Options

$$F(r) = \frac{1}{(N+1)} \left[\sum_{l=1}^{L} \xi_l s_l + \sum_{j=0}^{N-L} \xi_j r_j \right] \tag{20.1}$$

We have included the possibility of some arbitrary linear coefficients $\{\xi_j\}$. The most common situation is just simple averaging, $\xi_j = 1$.

Next, we introduce a dummy variable $\hat{F}$ through a Dirac delta function, using the identity

$$1 = \int_{-\infty}^{\infty} \delta\left[\hat{F} - F(r)\right] \cdot d\hat{F} \tag{20.2}$$

The Fourier decomposition of the δ-function introduces another variable p,

$$\delta\left[\hat{F} - F(r)\right] = \int_{-\infty}^{\infty} \frac{dp}{2\pi} \exp\left\{-ip\left[\hat{F} - F(r)\right]\right\} \tag{20.3}$$

The oscillations kill the integral if $\hat{F} \neq F(r)$ and gives ∞ if $\hat{F} = F(r)$.

Now the value of the payoff C^* for the option at time t^* depends on the path that the underlying rate takes from t_0 to t^* and on the averaging function. We call the option payoff $C^*[F(r)]$. Now $C^*[F(r)]$ can be used as a "test function" for the δ-function, meaning simply that we can write the identity

$$\begin{aligned} C^*[F(r)] &= \int_{-\infty}^{\infty} C^*(\hat{F}) \cdot \delta\left[\hat{F} - F(r)\right] \cdot d\hat{F} \\ &= \int_{-\infty}^{\infty} C^*(\hat{F}) \cdot \int_{-\infty}^{\infty} \frac{dp}{2\pi} \exp\left\{-ip\left[\hat{F} - F(r)\right]\right\} \cdot d\hat{F} \end{aligned} \tag{20.4}$$

We call C the value of the option at the valuation date (today), t_0. This is obtained as the expectation value using the path integral for the Green function

---

functioning of a desk. One of the tasks is to maintain a system containing a database of the reset values that need to be input to the model in the system. By the way, this already shows one way that a real system is much more than just a model calculator containing the algorithm being described. System risk is discussed in Ch. 34-35.

$G(x_0, x^*; t_0, t^*)$, which in this chapter includes a discount factor. The path integral depends on all values of the paths from t_0 to t^*. We have

$$C = \int G(x_0, x^*; t_0, t^*) \cdot C^*[F(r)] dx^* \tag{20.5}$$

Using an expectation notation and using the above identity, we get

$$C = \int_{-\infty}^{\infty} d\hat{F} \cdot C^*(\hat{F}) \cdot \int_{-\infty}^{\infty} \frac{dp}{2\pi} \exp\{-ip\hat{F}\} \cdot \int_{-\infty}^{\infty} dr^* \mathcal{G}_{0^*}(p) \tag{20.6}$$

Here,

$$\int_{-\infty}^{\infty} dr^* \mathcal{G}_{0^*}(p) = \left\langle \exp\left(-\int_{t_0}^{t^*} r(t) dt\right) \exp\left[ipF(r)\right] \right\rangle \tag{20.7}$$

The expectation notation is just shorthand for the path integration and the discount factor is really discretized. Now we need the general integral that appears in the path integration for Gaussian models,

$$I_m(p) \equiv \left[\prod_{j=1}^{m} \int_{-\infty}^{\infty} dr_j\right] \left\{ \exp\left[\frac{ip}{N+1} \sum_{j=1}^{m} \xi_j r_j - \sum_{j=1}^{m} \beta_j r_j - \sum_{j,j'=1}^{m} r_j A_{jj'} r_{j'}\right]\right\}$$
(20.8)

This is evaluated by completing the square as

$$I_m(p) = \frac{\pi^{m/2}}{(\det A)^{1/2}} \exp\left[iap - \frac{1}{2}b^2 p^2 + \gamma\right] \tag{20.9}$$

Here,

$$a = \frac{-1}{2(N+1)} \sum_{j,j'} \xi_j (A^{-1})_{jj'} \beta_{j'}$$

$$b^2 = \frac{1}{2(N+1)^2} \sum_{j,j'} \xi_j (A^{-1})_{jj'} \xi_{j'} > 0 \tag{20.10}$$

$$\gamma = \frac{1}{4} \sum_{j,j'} \beta_j (A^{-1})_{jj'} \beta_{j'}$$

Finally, we need

Average-Rate Options

$$\int_{-\infty}^{\infty} \frac{dp}{2\pi} \exp\left\{-ip\hat{F} + iap - \frac{1}{2}b^2 p^2\right\} = \frac{1}{\sqrt{2\pi b^2}} \exp\left[-\frac{(\hat{F}-a)^2}{2b^2}\right] \quad (20.11)$$

With this, we obtain the exact result

$$C = \hat{P}^{(t^*)} \int_{-\infty}^{\infty} d\hat{F} \cdot C^*(\hat{F}) \cdot \frac{1}{\sqrt{2\pi b^2}} \exp\left[-\frac{(\hat{F}-a)^2}{2b^2}\right] \quad (20.12)$$

Here, $\hat{P}^{(t^*)}$ is today's zero-coupon bond of maturity t^* producing the discounting.

Examples: Average-Rate Caplet, Digital/Step Option
For a few special cases, consider an average-rate caplet payoff,

$$C^*[F(r)] = [F(r) - E]\Theta[F(r) - E] \quad (20.13)$$

We get the average-rate caplet value in this model as

$$C = \hat{P}^{(t^*)}\left\{(a-E)N\left(\frac{a-E}{b}\right) + \frac{b}{\sqrt{2\pi}}\exp\left[-\frac{(a-E)^2}{2b^2}\right]\right\} \quad (20.14)$$

Another example is the average all-or-nothing step or digital option. The payoff is

$$C^*[F(r)] = \Theta[F(r) - E] \quad (20.15)$$

The average-rate all-or-nothing step or digital option is

$$C = \hat{P}^{(t^*)} N\left(\frac{a-E}{b}\right) \quad (20.16)$$

Results for Average-Rate Options in the MRG Model

In order to obtain explicit results for the quantities a and b^2, we use the mean-reverting Gaussian model with constant volatility σ and mean reversion ω. We also take uniform averaging, $\xi_j = 1$, for simplicity. We obtain two separate results depending on whether t_0 is before the averaging starts or is in the averaging period. The calculations are messy but straightforward.

The main trick is to combine all terms in the discretized interest rate r_j,

$$\exp\left[-r_j dt + ipr_j/(N+1)\right] = \exp\left[-r_j h(p) dt\right] \qquad (20.17)$$

Here, $h(p) = 1 - ip/T_{a*}$ where $T_{a*} = t^* - t_a$ is the complete averaging period. Now we change variables, scaling rates by $h(p)$, and we define p-dependent "volatilities" by $\sigma^2(p) = h^2(p)\sigma^2$. The integrations can then be carried out.

Valuation date Before Averaging Period Starts

If t_0 is before the averaging period starts, we get

$$b^2 = \sigma^2 \left[T_{0a}\varsigma(\omega T_{0a}, \omega T_{a*}) + \frac{1}{3}T_{a*}\mathscr{Y}(\omega T_{a*}) \right]$$

$$a = \frac{1}{(N+1)}\sum_{j=0}^{N} f_j - \frac{1}{2}b^2 T_{a*} \qquad (20.18)$$

Here, f_j is the forward rate at time t_j (as determined today), $T_{0a} = t_a - t_0$ is the time interval up to the start of the averaging and again $T_{a*} = t^* - t_a$ is the complete averaging period. The generalized volatility b gets a full contribution from the period T_{0a}. It is suppressed by a factor 1/3 in the averaging period T_{a*} up to some extra factors[4]. The extra factors are

---

[4] **Estimates of volatility when part of the period involves averaging:** These formulas can be used for quick estimates of the total volatility in deals.

Average-Rate Options

$$\varsigma(\omega T_{0a}, \omega T_{a*}) = \frac{1}{2(\omega T_{0a})(\omega T_{a*})^2}\left(1-e^{-2\omega T_{0a}}\right)\left(1-e^{-\omega T_{a*}}\right)^2$$

$$\mathcal{Z}(\omega T_{a*}) = \frac{3}{(\omega T_{a*})^3}\left[\omega T_{a*} + 2\left(e^{-\omega T_{a*}} - 1\right) - \frac{1}{2}\left(e^{-2\omega T_{a*}} - 1\right)\right] \quad (20.19)$$

Valuation date During the Averaging Period

If t_0 is inside the averaging period we get the results

$$b^2 = \frac{\sigma^2}{3} \frac{T_{0*}^3}{T_{a*}^2} \mathcal{Z}(\omega T_{0*})$$

$$a = \frac{1}{(N+1)}\left\{\sum_{l=1}^{L} s_l + \sum_{j=0}^{N-L} f_j\right\} - \frac{1}{2}b^2 T_{a*} \quad (20.20)$$

Now the volatility is suppressed by 1/3 times an extra factor.

Simple Harmonic Oscillator Derivation for Average Options

We remark that the problem of average options can be done at least two other ways. The second method is using the quantum mechanic simple harmonic oscillator SHO analogy[ii]. We need to replace the SHO in quantum mechanics using $\omega \to i\omega$, where $i = \sqrt{-1}$. This turns the Schrödinger equation into the diffusion equation. It also gets rid of the oscillations that occur in quantum mechanics, but not in finance.

As described in Ch. 43, App. A, we use the MRG stochastic equation at each time t_j, namely $\eta_j = \left[x_{j+1} - x_j(1+\omega dt)\right]/(\sigma dt)$ in $\exp\left[-\frac{1}{2}\int_{-\infty}^{\infty}\eta_j^2 dt\right]$, which appears in the Gaussian measure. Here, $x_j = r_j - r_j^{(cl)}$ measures the random difference of the short term rate from the classical path $r_j^{(cl)}$ at time t_j. The classical path is given by the forward rate f_j up to a convexity correction.

In carrying out the calculation, we need to make sure we include the surface terms, namely the cross terms. These are generally unimportant, but here the surface terms cannot be ignored. After some algebra, we get the same result as obtained above using explicit multivariate integration.

Thermodynamic Identity Derivation for Average Options

The third method is to use thermodynamic equalities, valid for Gaussian models. We have

$$\langle \exp(X) \rangle = \exp\left[\langle X \rangle + \frac{1}{2} \langle X^2 \rangle_c \right] \quad (20.21)$$

Here, the "connected part" is $\langle X^2 \rangle_c = \langle X^2 \rangle - \langle X \rangle^2$. We need to choose

$$X = -\int_{t_0}^{t^*} r(t)\,dt + ipF(r) \quad (20.22)$$

Contact with the first derivation is established by recalling that

$$\langle \exp(X) \rangle = \int_{-\infty}^{\infty} dr^* \mathcal{G}_{0^*}(p) \quad (20.23)$$

We can evaluate $\langle X^k \rangle$, $(k = 1, 2)$ explicitly and obtain the same answer.

Average Options with Log-Normal Rate Dynamics

Most applications assume lognormal behavior. However, the sum of lognormal variables is not lognormal[iii]. For this reason, analytic techniques cannot be used for exact results for arithmetic average options in a lognormal world.

There is a closed form for a *geometric-average* options using lognormal dynamics, derived by Turnbull and Wakeman[iv].

Turnbull and Wakeman showed that a moment expansion could be used for reasonable numerical approximations to an arithmetic-average option with underlying lognormal dynamics; see also Levy [iv].

This moment method was criticized by Milevsky and Posner [iv] as being inaccurate for long dated trades or cases with high volatility. These authors suggest using a reciprocal gamma probability distribution, in which case exact results are obtained for arithmetic-average options.

Berger et. al. document a number of Asian options and the techniques used to solve them.

Gaussian into Lognormal Using a Simple Trick

Here we show one simple way to get approximate results for lognormal dynamics from Gaussian model results. Consider the results for the average caplet and average digital step option above. It is easy to see that

$$C_{caplet} = \left[a - E + b^2 \frac{\partial}{\partial a} \right] C_{step} \qquad (20.24)$$

The forms for the parameters a,b are given above. Now set $y = (a-E)/b$, $\delta y = b/(a+E)$. Assume $a \approx E$ and $b \triangleleft a + E |$, so the options are near at-the-money and the volatility is small. Then we have the approximations

$$\left[a - E + b \frac{d}{dy} \right] N(y) \approx aN(y + \delta y) - EN(y - \delta y) \qquad (20.25)$$

$$\frac{(a-E)}{b} \approx \frac{1}{(b/E)} \ln\left(\frac{a}{E} \right) \qquad (20.26)$$

With these approximations, we get the result

$$C_{caplet} \approx \hat{P}^{(t^*)} \left[aN(d_+) - EN(d_-) \right] \qquad (20.27)$$

Here, $d_\pm = (E/b) \left[\ln(a/E) \pm b^2/(2E^2) \right]$. Using $T_{0*} = t^* - t_0$, we define the average lognormal volatility σ_{AvgLN} as

$$\sigma_{AvgLN} \sqrt{T_{0*}} = b/E \qquad (20.28)$$

This redefinition of the volatility then puts the above expression for C_{caplet} into the canonical lognormal form[5].

---

[5] **Magic? Gaussian into Lognormal:** This sort of trick can often be used. Physically it is based on the observation that for small volatility and near the money options it does not really matter whether you use Gaussian or lognormal dynamics because you are far from the zero level where the difference is appreciable.

References

[i] General Arithmetic Average Options in Gaussian Models
Dash, J. W., *Path Integrals and Options*. Invited talk, Mathematics of Finance section, SIAM Conference, Philadelphia, July 1993.

[ii] Path Integrals
Feynman, R. P. and Hibbs, A. R., *Quantum Mechanics and Path Integrals*. Mc-Graw-Hill Book Company, 1965. See Ch. 8.

[iii] Sums of Lognormal Variables
Barakat, R., *Sums of independent lognormally distributed random variables*. J. Opt. Soc. Am., Vol. 66, No. 3, 1976. PP. 211-216.

[iv] Moment Expansion, Geometric-Average Options
Turnbull, S. M. and Wakeman, L. M., *A Quick Algorithm for Pricing European Average Options*. Journal of Financial and Quantitative Analysis, Vol. 26, No. 3, 1991. Pp. 377-389.

Levy, E., *Pricing European average rate currency options*. Journal of International Money and Finance, 11, 1992. Pp. 474-491.

Milevsky, M. and Posner, S., *Closed-Form Arithmetic Asian Option Valuation*. Derivatives Week Learning Curve, 7/14/97.

Berger, E., Goldshmidt, O., Goldwirth-Piran, D., *Modeling Asian Options*. Bloomberg Financial Markets working paper, 1998.

PART IV

QUANTITATIVE RISK MANAGEMENT

21. Fat Tail Volatility (Tech. Index 5/10)

In this chapter, we look at fat tails in distributions of underlying variable moves from a practical perspective. We are especially concerned with obtaining some sort of volatility for fat tails. We introduce the idea using some examples, and deal with practical questions at the end.

Gaussian Behavior and Deviations from Gaussian

It has been known for many years that the probability distribution of underlying changes of financial variables $d_t x(t) = x(t+dt) - x(t)$ over time interval dt is at best only approximately Gaussian. Practically all production models in finance use Gaussian assumptions, either associating $x(t)$ directly with a financial variable or using some simple transformation like $x(t) = \ln r(t)$ which leads to the lognormal model[1]. The description of deviations from Gaussian behavior forms a large part of this book. In this chapter, we will focus on jump outliers, giving rise to "fat tails".

Gaussian behavior means we assume $d_t x(t) = [\mu(t) + \sigma(x,t)\eta(t)]dt$. The stochastic variable $\eta(t)$ satisfies $\langle \eta^2(t) \rangle = 1/dt$, and has probability distribution $\mathcal{P}[\eta(t)] = \sqrt{dt/2\pi}\, \exp[-\eta^2(t)dt/2]$. In addition, for different times we have the delta-function normalization $\langle \eta(t)\eta(t') \rangle = \delta(t-t')$. An equivalent notation is $\eta(t) = dz(t)/dt$. The drift function $\mu(t)$ is assumed deterministic and fixed by no-arbitrage arguments. The substitution of the stochastic equation

---

[1] **Is there a Transformation of the Data to Gaussian?** A possible exercise would be to try to find a transformation of variables so that the data exhibit a Gaussian probability distribution. Because there are so many ways that the Gaussian behavior is broken, it is possible that such an exercise would be highly unstable and idiosyncratically dependent on the individual variables and the time intervals. The existence of macroscopic large time scale parameters and the presence of jumps or gapping behavior at short time scales are essential complications.

in the probability distribution for $\eta(t)$ leads to the Gaussian probability distribution for $d_t x(t)$, namely[2]

$$\mathcal{P}^{Gaussian}\left[d_t x(t)\right] = \sqrt{\frac{1}{2\pi\sigma^2(t)dt}} \cdot \exp\left\{-\frac{\left[d_t x(t) - \mu(t)dt\right]^2}{2\sigma^2(t)dt}\right\} \qquad (21.1)$$

Outliers and Fat Tails

Example of a Fat-Tail Event in the Real World

The reader might need to be convinced that there is a fat-tail problem. To this end, here is the example used in my CIFEr tutorials. On June 2, 1995 at 8:30 am, the 10-year US treasury yield dropped 35 bp in half an hour. Assume Gaussian behavior of the logarithm of yield changes, scale this move by $\sqrt{time}$ for one year[3] and write the result as $d_t r^{(dt=1yr)} \equiv 35bp \cdot \sqrt{1yr/30\min} = k_\sigma r\sigma\sqrt{1yr}$. This produces $k_\sigma \approx 20$ standard deviations, occurring with probability 10^{-40}.

To drive home what this means, 10^{-40} happens to be on the order of the probability of specifying at random a tiny volume containing one proton in the whole observable universe. Since all this is absurd, the conclusion is that these fat tails manifestly violate the Gaussian assumption.

Notable fat-tail events include the 1987 crash that was the winner for percent changes in equity-land. For FX, huge fat-tail moves occur when a pegged currency (with an artificially small volatility with respect to USD) is unpegged. Other examples exist in emerging markets, commodities, etc[4].

---

[2] **Path Integrals:** There is an additional step involving integrations over the x(t+dt) variable and the η(t) variable that we have not shown. For the details, see Ch. 42, 43 on Path Integrals. All you have to do to get the general path integral is to repeat the above procedure at each time t, take the product, and then integrate over the physical region of the underlying variable at each time.

[3] **Fat-Tail Event Parameters:** Recall, the mathematical definition is that Gaussian behavior of changes which is equivalent to Brownian motion is supposed to hold at all time scales. We used the lognormal yield volatility σ = 15%, 252 business days/yr, 6.5 trading hrs /day along with a rate of 6.2%/yr, producing 21.5 standard deviations.

[4] **Story: Many Fat-Tail Examples:** I used to collect printouts of great fat-tail events pinned up on a bulletin board. Eventually, there were so many pieces of paper that the pin couldn't hold them. They all fell down on the floor.

Gaussian Random Numbers for MC Simulations in Excel

We will need to know how to run simple Gaussian Monte Carlo simulations, and we will learn a little about Excel in the process. Gaussian random numbers are obtained via $R_G = (-2\ln U_1)^{1/2} \cos(2\pi U_2)$, $R_G = (-2\ln U_1)^{1/2} \sin(2\pi U_2)$ with $U_{1,2} = U(0,1)$ uniform random numbers [5,6,i].

Simple Illustrative Fat-Tail Monte-Carlo Model in Excel

Here is a simple pedagogical example illustrating fat tails using a volatility that depends on the changes, $\sigma = \sigma[d_t x]$. Gaussian random numbers $\{G_\lambda\}$, $\lambda = 1,...,N_{MC}$, are sorted from smallest to largest[7] so $G_\lambda < G_{\lambda+1}$. This orders the $\{d_t x\}$ numerically. A volatility σ_λ is assigned to G_λ as deterministically increasing, viz $\sigma_\lambda < \sigma_{\lambda+1}$. The graph of the volatilities will be shown below. The illustrative model is $d_t x^{(FatTails)}(G_\lambda; \sigma_\lambda) = \sigma_\lambda G_\lambda$. The histogram below[8] gives the results, along with a standard model using a constant volatility σ_0 and the same random numbers, viz $d_t x^{(Std)}(G_\lambda; \sigma_0) = \sigma_0 G_\lambda$.

---

[5] **Random Numbers in Excel:** To get a uniform random number in Excel, type rand() and hit Enter. You should check the mean and standard deviation of a set of random numbers to see if they meet your accuracy criteria. An annoying feature of Excel is that the random numbers change every time you enter something in any cell unless you specify "manual calculation" under Tools\Options. Apparently, there is no control over the seed. Hence, you may want to save a set of random numbers. To do this, put the curser on the first random number in a column, hold down Shift-Ctrl and press the down arrow key to highlight all the random numbers. Copy the numbers using Ctrl-C, click the mouse in an empty column, and then use Edit\Paste-Special with the "Values" option.

[6] **Question to the Reader:** You are using Excel (or some equally facile program) to do your prototyping, right?

[7] **Sorting in Excel:** Highlight the column to be sorted, hit Data\Sort\Ascending. You probably already knew that.

[8] **Parameters:** In the example, N_{MC} = 100 and σ_0 = 1.5 and a deterministic fat-tail vol prescription starting at σ =1.0 at the smallest $d_t x$ to σ =3.0 at the largest $d_t x$.

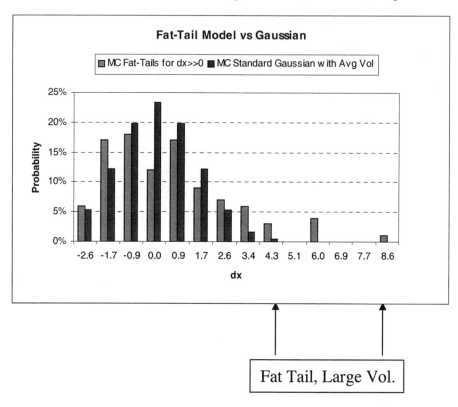

Fat Tail, Large Vol.

The fat tails are seen as more events or higher probability than the standard model would predict at large values of $d_t x$, due to the assumed larger volatilities at larger $d_t x$.

Inversion for the Fat-Tail Volatility at Definite CL in the Model

This simple model also serves to illustrate another point. We know, by construction for the fat tail model, which volatility corresponded to which point. Now suppose we announce we would like to look at the 99th percent confidence level (CL). For a large number of points, we know that the Gaussian with unit width has $k_{99\%CL} \approx 2.33$. Therefore we can calculate the volatility at the 99% CL for any distribution as $\sigma^{99\%CL} = d_t x^{99\%CL} / 2.33$. Similar remarks hold at other CL[9].

---

[9] **Confidence Levels:** We obviously cannot choose CL around $d_t x = 0$ with zero standard deviations. There is also an edge problem for the CL. To make the distribution with 100 points symmetric, the first point has CL 0.5% and the last point CL 99.5%. The results will fluctuate; these fluctuations decrease as the number of points in the MC simulation

The graph below shows the results for the above model for all CL above 68%. In the example with definite input volatilities, the calculated volatilities roughly reproduce the input volatilities, as they should.

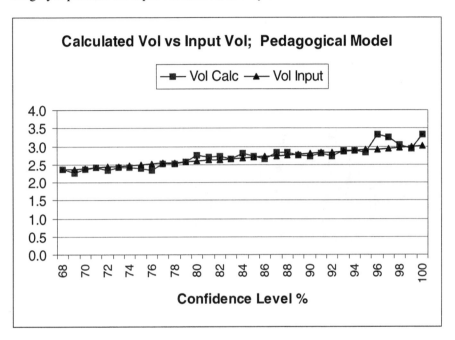

Use of the Equivalent Gaussian Fat-Tail Volatility

In the real world, $d_t x$ is not generated by a Gaussian with any volatility. Still, the procedure we use [10,ii] is similar. It consists of the following steps:

1. Choose a tail CL, e.g. 99% CL. Get $d_t x^{99\%CL}$ from the data.
2. Get the 99% CL "Fat-Tail" vol as $\sigma^{99\%CL} = d_t x^{99\%CL} / 2.33$.
3. Generate the "Fat-Tail" Gaussian using this "Fat-Tail" vol.
4. As a refinement, define separate FT vols for $d_t x > 0$, $d_t x < 0$.

---

increase. Since only generated 100 points were generated, the graph is just meant for illustration.

[10] **History:** Fat-tail Gaussians were used in 1988-89 ago to describe interest-rate outliers; c.f. Dash and Beilis (Ref). A physicist colleague Naren Bali earlier had the idea for fat-tail Gaussians. While we were fitting high-energy data in 1974, he said "You can fit your grandmother with Gaussians". Later the wavelet mathematicians got the same idea.

By definition, we get the Fat Tail Gaussian empirical distribution as correct at the chosen tail CL. We use the Fat-Tail Gaussian to model the rest of the tail. The idea is in the figure below:

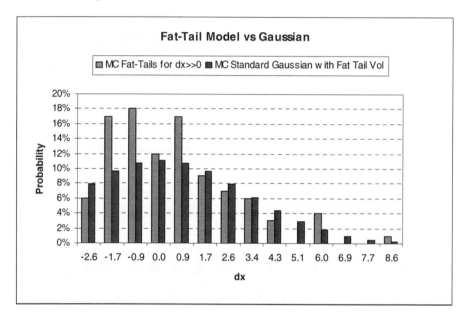

Practical Considerations for the Fat-Tail Parameters

The discerning reader at this point is no doubt thinking of many objections to the above procedure. Why choose the 99% CL? What about asymmetric distributions? Why use Gaussians at all? We have no easy answers, and we do not think any easy answers exist. There is no theory of fat tails that is convincing. We offer the following observations.

Why Use Fat-Tail Gaussians and not Other Distributions?

One justification of using fat-tail Gaussians follows a long history of using simple functions to describe variations phenomenologically in science. In the end, the fat-tail Gaussians are simple to use and explain to management. There is no good reason to use functions that are more complicated in the absence of convincing phenomenology[11]. Moreover, with only a few points in the tail to fit,

---

[11] **Convincing Theory or Phenomenology of Fat Tails?** A convincing theory is more than just a lot of high-level mathematics. We also need to see a lot of phenomenology in many markets. We also would want stability of the parameters over time, and also stability with respect to different time windows. Having said that, we believe that the Reggeon Field Theory is a promising candidate to consider, as described in Ch. 46.

we have not found the robustness of the description using other functions any more convincing than FT Gaussians.

How do you know that Fat-Tail Gaussians are Really Fat?

Good question. A possible (though rare) occurrence is that the $d_t x^{99\%CL}$ in the data is actually smaller than $2.33\sigma_{std}$ with σ_{std} the usual standard deviation for all $d_t x$ data. Therefore, a conservative approach is to define the FT vol as the maximum of σ_{FT} as defined above and σ_{std}.

How do you justify the use of the 99% CL for the Fat Tail Vol?

There is no unique choice. Practically speaking, the 99% CL is a reasonable compromise. Three years of data (750 data points) represents a large effort of data collection when all markets are considered. The 99% CL for 750 points interpolates between the 7th and 8th biggest move and leaves 7 points for the tail, which "sounds" reasonable. A good procedure is to try different tail CLs and look at the variation in the final risk results. To be really conservative, you can use the maximum of the Fat-Tail vols over, say, points at and beyond the 99% CL. The sensitivity depends on the question being asked (see the next section).

Bis: How do you justify the use of 99% CL for Fat Tail Vol?

When MC simulations are run over diversified portfolios at a very high CL (e.g. 99.97% CL [12]) past the tail CL of 99%, numerically large contributors as a rule of thumb tend to have no more than around 2 fat-tail standard deviations. Therefore, some internal consistency is achieved by defining the fat-tail volatility at a tail CL of around 99% CL. Moreover, in practice, not much change in the results at a 99.97% CL is observed with different choices of the tail CL defining the fat-tail vol. However, substantial changes *can* be observed in the maximum loss.

What About the Central Part of the Distribution?

The fact that the central part of the $d_t x(t)$ distribution is not described by a Gaussian with the FT vol is a correct observation. However, the central part of the distribution contributes little to the outlier risk. If the FT vol is used in simulation there will be some overstatement of the risk, but in practice, this overstatement is not very significant.

---

[12] **The 99.97% CL:** This "3 in 10,000" level is a popular confidence level for Economic Capital for an AA (Aa) rated company, as will be described in Ch. 39.

A refinement of the fat tail approach would be to explicitly model the central distribution with another Gaussian and use the composite distribution. However, this would probably not add much in practice to the description of outlier risk.

Note that we are *not* talking about adding up two Gaussian random noises; that would just be Gaussian. What we are talking about is adding together two Gaussian probability density functions with different volatilites; the result for this sum is not just a Gaussian.

Serial Correlation Problem and a Lemma to the Data Theorem

Serial correlations occur with overlapping data windows. The reader should note that, for daily differences using daily data, the windows used to define $d_t x$ do not overlap, and there is no serial correlation. We recommend the daily difference procedure anyway (see discussion on No Credit for Mean Reversion below).

However, for the choice of differences over more than one day, overlapping data windows pose a thorny practical problem. The reader should note that, independent of mathematical purity, in the real world there usually is no choice. The use of overlapping windows is unavoidable if there are not enough data to use independent non-overlapping windows. In fact, we have a Lemma:

Lemma to the Fundamental Black-Hole Theorem of Data

A strong Lemma to the Black-Hole Data Theorem states: "There are not enough data points for independent non-overlapping windows, almost everywhere". This is *not* a joke.

The Red Herring

The serial correlation problem is in one sense a red herring. If differences over time window intervals $Dt > 1$ day are used, numerical experiments show that overlapping windows do not substantially degrade the determination of $d_t x^{CL}$ at a given CL[13]. So the good news is that numerically the use of overlapping windows does not have much effect, *provided* that the size of the windows is small compared to the total length of the data.

To test this statement in a simple example, consider a series of 500 Gaussian R# (equivalent to 2 years of data). We first look at $Dt = 10$-day overlapping moving windows, add up the 10 random numbers in each window to define 491 values of $d_t x$, sort-order these values, and thereby extract the $d_t x^{CL}$ at each

---

[13] **Window Caveats:** Naturally, the window size has to be small with respect to the total data sample for this to work. In addition, the ratio of risks from CL_1 to CL_2 for real data (not a Gaussian model) has been examined. A similar statement holds: overlapping windows do not degrade the analysis as long as the window size is small compared to the total data time series length.

possible CL. We then look at independent non-overlapping 10-day windows. Naturally, there are fewer CLs that can be examined using independent windows (CL $=2\%,...,98\%$). There are ten sets of 49 independent windows corresponding to the ten possible starting days for each set, and therefore there are ten results for $d_t x^{CL}$ at each CL. For each CL we plot $d_t x^{CL}$ from the overlapping windows and we plot the average $d_t x^{CL}$ from the ten independent windows along with the standard deviations.

The results are in the graph below:

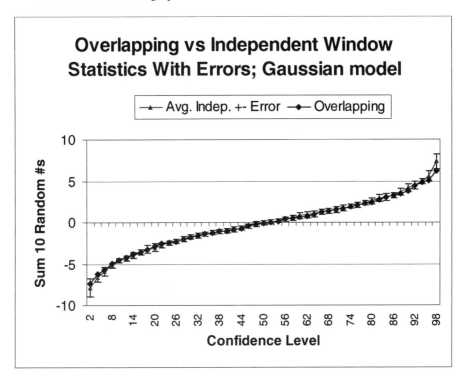

It is seen that the overlapping window results $d_t x^{CL}$ across the CL spectrum are within the errors of the ten independent window estimates for $d_t x^{CL}$.

Let's look at the highest CL (98%) available here, $d_t x^{98\%CL}[Dt=10]$. The plot below shows that the results for ten sets of independent windows oscillate around the result for the overlapping windows. This means that the use of overlapping windows does not give misleading results.

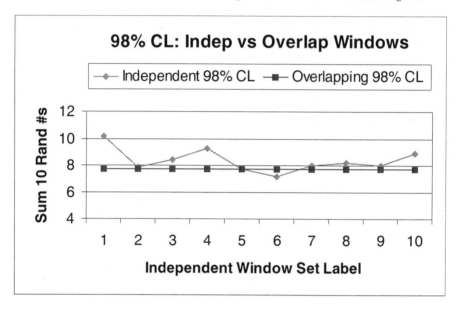

We conclude with a remark about worst-case moves. Naturally, if we are looking at worst case moves within any 10-day window we *want* to look at overlapping windows to get the largest number of possible states. The worst case will lie somewhere in the independent sets of non-overlapping independent windows.

What about the Difference Time and Mean Reversion?

The difference time dt for $d_t x$ is not unique. If the series $d_t x(t)$ exhibits mean reversion, a common assumption, then at larger $Dt > dt$ we get some damping $d_t x[Dt] < \sqrt{Dt/dt} \cdot d_t x[dt]$. Conservatively however we may not want to allow credit for mean reversion in performing tail-risk analysis[14]. This argues for

---

[14] **No Mean Reversion Credit for Traders in a Turbulent Market:** Assumptions about the existence of mean reversion are nefarious for risk. Attempts by traders to use mean reversion to minimize risk should be treated with suspicion. When mean reversion comes up in the discussion (as it will), it will go like this: "Markets are mean reverting. That's how we run our business. Therefore your assumption of no mean reversion is way too conservative". You might ask the traders the sarcastic question of how well the assumption of mean reversion worked for the convergence plays of the fabulous traders at some world-renown hedge funds, before those strategies crashed and burned in 1998. The serious issue is that in a turbulent environment corresponding to severe outlier risk - at which you *do* want to look - assumptions of mean reversion and convergence break down. There's no free lunch.

Fat Tail Volatility

square-root-time scaling up from daily differences, i.e. $dt = 1$ Day and $d_t x[Dt] = \sqrt{Dt/1 \text{ Day}} \cdot d_t x[1 \text{ Day}]$.

What about Up/Down Data Move Asymmetry?

The data for $d_t x$ are usually asymmetric for positive vs. negative moves. Two FT signed vols $\sigma_{FT}^{(\pm)}$ can be chosen independently at the tail CL for positive and for negative moves. The signed FT vols can and should be used in MC simulations. If a positive (negative) R# is generated, use $\sigma_{FT}^{(+)}$ ($\sigma_{FT}^{(-)}$).

What is the Relation of the Fat Tail Vol with Huge Jumps?

Naturally, jumps make the empirical volatility change as a function of time. If there is a big jump, it will increase the volatility for the time that the jump remains in the time window. This is one way to look at the dynamics behind the increase in volatility as $d_t x$ increases. To illustrate the idea, consider the following 500 Monte-Carlo Gaussian changes with unit daily vol, including a big 10 standard deviation 1-day jump added in the middle of the time series. The results for the 65-day windowed vols are shown below[15]:

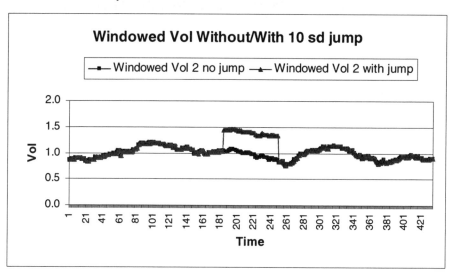

The windowed vol fluctuates around the input vol of one, except for the windows in which the big jump appears. The big jump shows up in the histogram

---

[15] **Label:** The label Vol 2 is there for reference in Ch. 25.

of $d_t x$ at the maximum point, past the 99% CL. Therefore it does not effect the determination of the fat tail vol $\sigma^{99\% CL}$.

The MaxVol

We could define a "MaxVol" using the jump in the above paragraph as $\sigma_{MaxVol} = d_t x^{Jump} / k_{Max}$, where k_{Max} is the number of standard deviations corresponding to the maximum CL obtainable from the total time series. For the most conservative available historical estimate, this MaxVol can be used. The MaxVol corresponds to using a Gaussian that fits the worst-case move.

References

[i] **Random Number Generation**
Abramowitz, M. and Stegun, I., *Handbook of Mathematical Functions, Ch. 26.* National Bureau of Standards Applied Mathematics Series 55, 1967.

[ii] **Fat-Tail Gaussian Component of Interest Rates**
Dash, J. W. and Beilis, A., *A Strong Mean-Reverting Yield-Curve Model.* Centre de Physique Théorique CPT-89/PE.2337 preprint, CNRS Marseille, France, 1989.

22. Correlation Matrix Formalism; the $\mathcal{N}$-Sphere (Tech. Index 8/10)

The Importance and Difficulty of Correlation Risk

Correlation risk is one of the most dangerous and least-analyzed risks. In order to come to grips with correlation risk with many variables, we need to be able to deal with the problem of stressed correlation matrices. Stressed correlations are needed for robust risk management because correlations are notoriously unstable[1]. In particular, during stressed markets, correlations often increase dramatically[i]. Therefore, for risk assessment, we need to stress the correlations.

Unfortunately, the procedure of stressing the correlations is problematic. Often, stressed correlation matrices are non-positive definite (*NPD*) matrices because they arise from inconsistent stressing of the various correlation matrix elements that are interdependent and constrained[2].

In this section, we discuss the representation of correlations in N dimensions, i.e. correlation matrices with elements as correlations between pairs of N variables. One of the main problems with correlations is that they are dependent. By reformulating the problem geometrically, we recast the correlations into functions of independent angular variables[3]. These angles are the natural spherical co-ordinates for an $\mathcal{N}$ - sphere where $\mathcal{N} = N - 1$. In this way, the

---

[1] **Correlation Risk, Timescales, Stressed Correlations and Stochastic Correlations:** Correlation risk is insidious and needs to take a more prominent place in risk management. Scenario stress analysis or a stochastic model for correlations is needed to determine correlation risk.

[2] **Non-Positive Definite Matrices in Ordinary Data Collection:** NPD correlation matrices may also arise from somewhat inconsistent data, without any stress procedure at all.

[3] **History:** My formulation of this geometrical approach was developed in 1993. The idea was mentioned in my CIFEr tutorial starting in 1996. I noticed the connection with the Cholesky decomposition in 1999. The basic idea came from running Monte Carlo simulations of n-dimensional phase space in high-energy particle physics in the 1960's.

problem of describing positive definite (PD) correlation matrices becomes tractable. With this, we are then able to deal with stressed correlation matrices.

In general, an arbitrarily stressed correlation matrix will be non-positive-definite (NPD). The idea that we will present in Ch. 24 is to get a best PD matrix fit to the NPD matrix, using least-squares fit in the angular variables. The problem can be cast into the geometrical problem of determining the point on an $\mathcal{N}$-sphere (corresponding to the PD stressed correlation matrix) that is the closest to a point off the $\mathcal{N}$-sphere (the target NPD correlation matrix).

One Correlation in Two Dimensions

We will proceed slowly, starting with low dimensional cases and gradually building up, making sure that the physical intuitive picture is always present. The basic starting point is the recognition that a correlation corresponds to the cosine of an angle $\rho_{\alpha\beta} = \cos(\theta_{\alpha\beta})$. These angles are, however, dependent because they are constrained by laws of cosines. Our goal is to find a set of independent angles to recast the correlations in a consistent manner.

We start with a set of orthogonal unit vectors $\{\hat{e}_\alpha\}, \alpha = 1...N$ forming a basis in Euclidean N-space[4]. We imagine that the time changes $d_t x_1(t)$ of the first variable[5] x_1 are measured along the first axis with unit vector $\hat{e}_1$. The time changes $d_t x_2(t)$ of the second variable are measured at an angle θ_{12} with respect to the first axis in the plane formed by the first two unit vectors $(\hat{e}_1, \hat{e}_2)$. The time runs over the time window producing N_w time differences for both variables $d_t x_1(t_\ell), d_t x_2(t_\ell)$ (e.g. over 3 months). The correlation and angle θ_{12} are given by the usual expression

$$\rho_{12} = \cos(\theta_{12}) = \frac{1}{\sigma_1 \sigma_2 (N_w - 1)} \sum_{\ell=0}^{N_w - 1} \left[d_t x_1(t_\ell) - \langle d_t x_1 \rangle \right] \left[d_t x_2(t_\ell) - \langle d_t x_2 \rangle \right]$$

(22.1)

---

[4] **Orthogonality in 3D:** Position your right hand with your thumb pointing up, your index finger pointing straight ahead, and your middle finger pointing to the left.

[5] **Change of Variables:** The variable can be $x_1 = \ln(r_1)$ for returns, where r_1 is the physical first variable, etc.

Here the $d_t x_1(t)$ time average is $\langle d_t x_1 \rangle$ and its volatility is σ_1. Notice that this looks exactly like the dot product of two vectors in an N_w-dimensional "time" space to form the cosine of the angle θ_{12}. This "time" space is associated with, but is not the same as, the space that we are use for the independent angles[6].

Two Correlations in Three Dimensions; the Azimuthal Angle

Adding another variable x_3 leads to fluctuations of the time differences $d_t x_3$ partly in the $(\hat{e}_1, \hat{e}_2)$ plane and partly out of that plane, in the direction $\hat{e}_3$. We introduce the azimuthal angle φ_{23} is to fix the projection of the 3rd variable $d_t x_3$ on the axis $\hat{e}_2$ by a rotation in the $(\hat{e}_2, \hat{e}_3)$ plane by φ_{23}, leaving the projection on $\hat{e}_1$ fixed. We have three correlations $\rho_{12} = \cos(\theta_{12})$, $\rho_{13} = \cos(\theta_{13})$, and $\rho_{23} = \cos(\theta_{23})$. Each of these angles is measured between the time-averaged changes in the corresponding pair of variables. These angles are dependent because of the law of cosines, viz

$$\cos(\theta_{23}) = \cos(\theta_{12})\cos(\theta_{13}) + \sin(\theta_{12})\sin(\theta_{13})\cos(\varphi_{23}) \quad (22.2)$$

Directly in terms of the correlations,

$$\rho_{23} = \rho_{12}\rho_{13} + \sqrt{1-\rho_{12}^2}\sqrt{1-\rho_{13}^2}\cos(\varphi_{23}) \quad (22.3)$$

This constraint is the essential problem. Imagine what happens if we try to stress the correlations in some way. We cannot arbitrarily choose $\{\rho_{12}, \rho_{13}, \rho_{23}\}$ because if we do, the law of cosines will in general not be satisfied for any choice of the azimuthal angle $\varphi_{23} \in [0, 2\pi)$. If the law of cosines is not satisfied, the fluctuations of the three variables cannot physically exist[7].

---

[6] **Fibers:** The N_w values of the time changes of each variable exist along a geometrical fiber above the space that we are going to be using. The space of fibers is N-dimensional. That's it for differential geometry, guys.

[7] **The Forbidden Hyperbolic Geometry, Relativity, and Imaginary Azimuthal Angles:** If the law of cosines is satisfied, fluctuations for the x variables are associated with points on the 2-sphere in 3-space. Transformations on the 2-sphere producing different correlations are described by the rotation group O(3). If the law of cosines is not satisfied with a real azimuthal angle, the azimuthal variable will be imaginary. Then fluctuations of the x variables are associated with points on an unphysical non-compact

The clue to the puzzle is already here. The trick is to swap the dependent variable ρ_{23} for the azimuthal angle φ_{23}. That's all we have to do.

Now if we were in fact to describe the correlations using the variables $\{\theta_{12}, \theta_{13}, \varphi_{23}\}$ then we would have an independent set of angles. We could vary these angles arbitrarily. Each set of angles corresponds to a definite set of correlations $\{\rho_{12}, \rho_{13}, \rho_{23}\}$ using the above relations, and each set of correlations corresponds to an independent set of angles.

For convenience, we define unit vectors $\{d_t \hat{y}_\alpha\}$ by dividing out the volatility from the fluctuations about the average for each variable, namely

$$d_t \hat{y}_\alpha = \left[d_t x_\alpha - \langle d_t x_\alpha \rangle \right] / \sigma_\alpha \tag{22.4}$$

We can represent each $d_t \hat{y}_\alpha$ (over the collection of times in the time window) by a point on the 2-sphere using spherical co-ordinates,

$$\begin{aligned} d_t \hat{y}_1 &= \hat{e}_1 \\ d_t \hat{y}_2 &= \cos(\theta_{12}) \hat{e}_1 + \sin(\theta_{12}) \hat{e}_2 \\ d_t \hat{y}_3 &= \cos(\theta_{13}) \hat{e}_1 + \sin(\theta_{13}) \left[\cos(\varphi_{23}) \hat{e}_2 + \sin(\varphi_{23}) \hat{e}_3 \right] \end{aligned} \tag{22.5}$$

This gives the connection of the formalism with spherical geometry in 3D.

The Degenerate World of FX Triangles

Before going ahead with higher dimensions, we pause for a brief description of the degenerate world of FX triangles. The degeneracy is because three variables are linearly dependent. Consider the trio of FX rates:

$$\eta_{XYZ} = \frac{\#Units(XYZ)}{OneUSD}, \quad \eta_{UVW} = \frac{\#Units(UVW)}{OneUSD}, \quad \eta_{UVW/XYZ} = \frac{\#Units(UVW)}{OneXYZ} \tag{22.6}$$

---

hyperbolic space with an imaginary axis similar to Einstein's special relativity theory. Mathematically, transformations on this hyperbolic space need generalized orthogonal groups. In three dimensions, instead of the compact group $O(3)$ we get the non-compact group $O(2,1)$. Naturally, all this is forbidden in finance. There, did that wake up all you high-energy theorists?

Let $x_1 = \ln \eta_{XYZ}, x_2 = \ln \eta_{UVW}, x_3 = \ln \eta_{UVW/XYZ} = x_2 - x_1$. Then the $(\hat{e}_1, \hat{e}_2)$ plane contains $d_t x_3 = d_t x_2 - d_t x_1$, and this degeneracy produces $\varphi_{23} = 0$. After a little arithmetic we find $\rho_{23} = (\sigma_2 - \sigma_1 \rho_{12})/\sigma_3$, $\rho_{13} = -(\sigma_1 - \sigma_2 \rho_{12})/\sigma_3$ and $\sigma_3^2 = \sigma_1^2 + \sigma_2^2 - 2\sigma_1 \sigma_2 \rho_{12}$, producing $\rho_{23} = \rho_{12}\rho_{13} + \sqrt{1-\rho_{12}^2}\sqrt{1-\rho_{13}^2}$.

Here is a picture of an FX triangle for the three $d_t x_\alpha$ (with the averages assumed subtracted). The angles can be determined by the historical correlations. Closing the triangle then gives the lengths of the legs, i.e. the volatilities expected from the correlations. Conversely, given recent moves of the variables, we can look at the angles needed to make the triangle close.

Triangle Relation for FX Rates

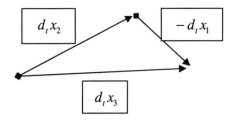

FX Option-Implied Correlations

Another version[ii] of the above FX triangle is to consider the lengths of the lines as given by implied volatilities $\sigma_\alpha^{(implied)}$, $\alpha = 1, 2, 3$. These would be the vols of the FX options corresponding to each of the x_α variables (naturally at consistent strikes and maturities). Then, option-implied correlations are obtained from the angles such that the three legs close.

This Triangle Doesn't Close

Of course if think you have a handle on specifying both the volatilities (line lengths) and correlations (angles), in general the triangle won't close. It is possible that some FX traders have a lot of fun trying to find arbitrage opportunities using this sort of approach.

Correlations in Four Dimensions

Returning to the task, we need only extend our formalism from three to N dimensions to get the result for an arbitrary correlation matrix using standard spherical co-ordinates. As we add another variable, we just increase by one the dimensionality of the space and write down the extra spherical co-ordinates necessary to describe this new variable. Here is a picture of the 4-D case.

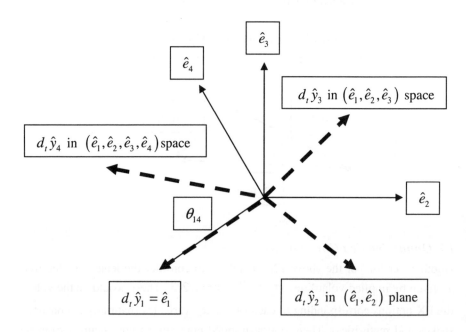

We need to introduce a compact notation to keep the equations from running off the page. We set $C_{\theta\alpha\beta} = \cos(\theta_{\alpha\beta}), S_{\theta\alpha\beta} = \sin(\theta_{\alpha\beta})$, etc. We also put subscripts on the brackets like $\left[\left[\ \right]_2\right]_1$ to keep them straight.

Introducing a fourth variable with similar notation $d_t\hat{y}_4$ for the unit vector of the changes, we also have to introduce a 4th orthogonal unit vector $\hat{e}_4$, an extra polar angle θ_{14} and two extra azimuthal angles $\varphi_{24}, \varphi_{34}$. The polar angle θ_{14} gives the projection of $d_t\hat{y}_4$ on $\hat{e}_1$. Then we rotate around $\hat{e}_1$ by angle φ_{24} in the $(\hat{e}_2, \hat{e}_3)$ plane to fix the projection of $d_t\hat{y}_4$ on $\hat{e}_2$. Finally, we rotate by φ_{34}

about $\hat{e}_2$ in the $(\hat{e}_3, \hat{e}_4)$ plane to fix the projection of $d_t \hat{y}_4$ on $\hat{e}_3$. The remainder is projected on the new dimension $\hat{e}_4$.

Therefore, we get the decomposition of the unit vector $d_t \hat{y}_4$ for the fourth variable as

$$d_t \hat{y}_4 = C_{\theta 14} \hat{e}_1 + S_{\theta 14} \left[C_{\varphi 24} \hat{e}_2 + S_{\varphi 24} \left[C_{\varphi 34} \hat{e}_3 + S_{\varphi 34} \hat{e}_4 \right]_2 \right]_1 \quad (22.7)$$

We now can reconstruct all the correlation matrix elements in four dimensions. They are all given by $\rho_{\alpha\beta} = d_t \hat{y}_\alpha \cdot d_t \hat{y}_\beta$. Therefore, we merely need to plug in the spherical decompositions. There are three new correlations. They are

$$\begin{aligned}
\rho_{14} &= C_{\theta 14} \\
\rho_{24} &= C_{\theta 24} = C_{\theta 12} C_{\theta 14} + S_{\theta 12} S_{\theta 14} C_{\varphi 24} \\
\rho_{34} &= C_{\theta 34} = C_{\theta 13} C_{\theta 14} + S_{\theta 13} S_{\theta 14} \left[C_{\varphi 23} C_{\varphi 24} + S_{\varphi 23} S_{\varphi 24} C_{\varphi 34} \right]
\end{aligned} \quad (22.8)$$

The correlations ρ_{14}, ρ_{24} have the same form as what we had in the 3D case for ρ_{13}, ρ_{23}. This is not surprising because we could have introduced the third and fourth variables in the opposite order. The correlation ρ_{34} has a new and more complicated form.

We see by construction that all correlations are specified once we specify the angles $\{\theta_{1\beta}\}, \{\varphi_{\alpha\beta}\}$. Moreover, we can now see a simple sequential algorithm to get the angles from the correlations. For example, we get φ_{24} from the correlations $\{\rho_{12}, \rho_{14}, \rho_{24}\}$. Given φ_{24} (and φ_{23} which we already know), we get φ_{34} from the set $\{\rho_{13}, \rho_{14}, \rho_{34}, \varphi_{23}, \varphi_{24}\}$.

Correlations in Five and Higher Dimensions

We now can see the general pattern. We continue to introduce new dimensions along with more angles. For five dimensions, we need one more polar angle θ_{15} and three more azimuthal angles to specify the position of $d_t \hat{y}_5$. The equations for $\{\rho_{25}, \rho_{35}, \rho_{45}\}$ successively determine these new azimuthal angles $\{\varphi_{25}, \varphi_{35}, \varphi_{45}\}$. The equations for the new correlations in five dimensions are

$$\rho_{15} = C_{\theta 15}$$
$$\rho_{25} = C_{\theta 25} = C_{\theta 12}C_{\theta 15} + S_{\theta 12}S_{\theta 15}C_{\varphi 25}$$
$$\rho_{35} = C_{\theta 35} = C_{\theta 13}C_{\theta 15} + S_{\theta 13}S_{\theta 15}\left[C_{\varphi 23}C_{\varphi 25} + S_{\varphi 23}S_{\varphi 25}C_{\varphi 35}\right] \quad (22.9)$$
$$\rho_{45} = C_{\theta 45} = C_{\theta 14}C_{\theta 15} + S_{\theta 14}S_{\theta 15} \cdot$$
$$\cdot \left[C_{\varphi 24}C_{\varphi 25} + S_{\varphi 24}S_{\varphi 25}\left[C_{\varphi 34}C_{\varphi 35} + S_{\varphi 34}S_{\varphi 35}C_{\varphi 45}\right]_2\right]_1$$

For example, consider the 5x5 correlation matrix:

Corr matrix (ρ) =	1	2	3	4	5
1	1	0.80	0.50	0.20	0.20
2	0.80	1	0.80	0.50	0.20
3	0.50	0.80	1	0.80	0.50
4	0.20	0.50	0.80	1	0.80
5	0.20	0.20	0.50	0.80	1

The various angles for this matrix are given by:

cos(θ₁₂) =	0.80						
cos(θ₁₃) =	0.50	cos(φ₂₃) =	0.7698				
cos(θ₁₄) =	0.20	cos(φ₂₄) =	0.5784	cos(φ₃₄) =	0.7293		
cos(θ₁₅) =	0.20	cos(φ₂₅) =	0.0680	cos(φ₃₅) =	0.6580	cos(φ₄₅) =	0.8627

In general, for the αth variable $d_t\hat{y}_\alpha$ we need one more polar angle $\theta_{1\alpha}$ to specify the projection on $\hat{e}_1$, along with $\alpha - 2$ more azimuthal angles $\{\varphi_{2\alpha}, \varphi_{3\alpha}, ..., \varphi_{\alpha-1,\alpha}\}$ to specify the projections on the remaining prior axes $(\hat{e}_2, \hat{e}_3, ..., \hat{e}_{\alpha-1})$. Given $d_t\hat{y}_\alpha$ and (in exactly the same way) $d_t\hat{y}_\beta$, we get the correlation as the dot product $\rho_{\alpha\beta} = d_t\hat{y}_\alpha \cdot d_t\hat{y}_\beta$.

The counting is that there are $N-1$ polar angles and $(N-1)(N-2)/2$ azimuthal angles adding up to the total number of correlation matrix elements $N(N-1)/2$.

The angles $\{\theta_{1\beta}\}, \{\varphi_{\alpha\beta}\}$ completely determine the correlations. Conversely, the same argument as given above sequentially determines the angles given the correlations. The mapping is one to one.

The $\mathcal{B}$ Variables and a Recursion Relation

We next give a useful recursion relation for $\gamma = 1,...,\alpha-2$; $2 < \alpha < \beta$. The relation can be used to get the general expressions by giving a name for the innermost bracket and proceeding outward by recursion. We write

$$\mathcal{B}^{(\alpha,\beta)}_{\alpha-\gamma} = \cos(\varphi_{\alpha-\gamma,\alpha})\cos(\varphi_{\alpha-\gamma,\beta}) + \sin(\varphi_{\alpha-\gamma,\alpha})\sin(\varphi_{\alpha-\gamma,\beta})\mathcal{B}^{(\alpha,\beta)}_{\alpha-\gamma+1} \quad (22.10)$$

The angles used in the spherical geometry are given by

$$\begin{aligned}\cos(\theta_{1\beta}) &= \mathcal{B}^{(1,\beta)}_1 & (2 \le \beta) \\ \cos(\varphi_{\alpha\beta}) &= \mathcal{B}^{(\alpha,\beta)}_\alpha & (2 \le \alpha < \beta)\end{aligned} \quad (22.11)$$

All the original correlations are given by

$$\rho_{\alpha\beta} = \cos(\theta_{\alpha\beta}) = \mathcal{B}^{(\alpha,\beta)}_1 \quad (22.12)$$

Spherical Representation of the Cholesky Decomposition

The above result is neatly summarized in the following observation. Recall[iii] that the Cholesky decomposition of a positive definite correlation matrix is the square root of the matrix. But we explicitly construct the correlation matrix by taking the square: $\rho_{\alpha\beta} = d_t\hat{y}_\alpha \cdot d_t\hat{y}_\beta$. Therefore, we have generated the Cholesky decomposition by construction. Write the two decompositions $d_t\hat{y}_\alpha = \sum_{\gamma'}(\hat{e}_{\gamma'} \cdot d_t\hat{y}_\alpha)\hat{e}_{\gamma'}$, $d_t\hat{y}_\beta = \sum_{\gamma}(\hat{e}_\gamma \cdot d_t\hat{y}_\beta)\hat{e}_\gamma$, note the orthogonality relation $\hat{e}_\gamma \cdot \hat{e}_{\gamma'} = \delta_{\gamma\gamma'}$, and identify terms in $\rho_{\alpha\beta} = \sum_\gamma (\rho^{1/2}\Omega^T)_{\alpha\gamma}(\Omega\rho^{1/2})_{\gamma\beta}$.

Here, Ω is the appropriate rotation to the co-ordinate system that we have constructed by hand. After a little index manipulation we get the Cholesky result in spherical co-ordinates, namely

$$(\Omega\rho^{1/2})_{\alpha\beta} = \hat{e}_\alpha \cdot d_t\hat{y}_\beta \quad (22.13)$$

The entries of the αth row are the $\hat{e}_\alpha$ coefficients of all $\{d_t\hat{y}_\beta\}$ and the entries of the βth column are the coefficients of $d_t\hat{y}_\beta$ for all $\{\hat{e}_\alpha\}$. Explicitly in terms of the angles, the matrix is

$$\left(\left(\Omega\rho^{1/2}\right)_{\alpha\beta}\right) = \begin{pmatrix} 1 & C_{\theta 12} & C_{\theta 13} & \cdots & C_{\theta 1\beta} & \cdots \\ 0 & S_{\theta 12} & S_{\theta 13}C_{\varphi 23} & \cdots & S_{\theta 1\beta}C_{\varphi 2\beta} & \cdots \\ 0 & 0 & S_{\theta 13}S_{\varphi 23} & \cdots & S_{\theta 1\beta}S_{\theta 2\beta}C_{\varphi 3\beta} & \cdots \\ 0 & 0 & 0 & \cdots & \cdots & \cdots \\ 0 & 0 & 0 & 0 & S_{\theta 1\beta}\prod_{\alpha=2}^{\beta-1} S_{\varphi\alpha\beta} & \cdots \\ 0 & 0 & 0 & 0 & 0 & \cdots \end{pmatrix} \qquad (22.14)$$

The determinant of this triangular matrix is just the product of the diagonals, so the determinant of the full correlation matrix is its square,

$$\det(\rho) = \left[\prod_{\beta=2}^{N} S_{\theta 1\beta} \prod_{\alpha=2}^{\beta-1} S_{\varphi\alpha\beta}\right]^2 \qquad (22.15)$$

This determinant is of course positive. All sub determinants also have positive determinants. This completes the description of the geometry of the correlation matrix.

Numerical Considerations for the $\mathcal{N}$- Sphere

The complexity of the geometrical approach of the correlations is not as severe as it might appear. There are a few technical subtleties [8], but using the recursion relations, the programming is actually straightforward. Practically speaking, a correlation matrix with hundreds of variables can be decomposed on a single machine.

We also note that we can find explicit representations for the derivatives of the correlations $\rho_{\alpha\beta}$ with respect to the angles $\{\theta_{1\beta}\}, \{\varphi_{\alpha\beta}\}$. This will be useful in the sequel when we construct a least-squares measure to obtain an optimal correlation matrix $\{\rho_{\alpha\beta}\}$ relative to a "target" correlation matrix.

---

[8] **Sign of the Sine:** There is a tricky point regarding the sign of the sine of an azimuthal angle. We have to take the positive branch of the square root + $(1-\cos^2\phi)^{1/2}$ and we cannot write the expression $\sin(\cos^{-1}(\cos\phi))$. This requirement may be due to similar conventions in canned numerical routines.

Again, since the angles are all independent, they can be varied independently allowing a consistent representation of modifying positive-definite correlation matrices in general.

References

[i] **Unstable Correlations in Stressed Markets**
Bookstaber, R., *Global Risk Management: Are We Missing the Point?* The Journal of Portfolio Management, Spring 1997. Cf. *Correlations between markets during market events increase dramatically*, p. 104.

[ii] **FX Triangular Relations**
Taleb, N., *Dynamic Hedging – Managing Vanilla and Exotic Options*. John Wiley & Sons, 1997. Pp. 438-444.

[iii] **Cholesky Decomposition**
Press, W. H., Teukolsky, S. A., Vetterling, W. T., Flannery, B. P., *Numerical Recipes in Fortran: The Art of Scientific Computing*, 2nd Edition, Cambridge University Press, 1992. Chapter 2.9.

23. Stressed Correlations and Random Matrices (Tech. Index 5/10)

In this chapter, we first consider various methods for dealing with a matrix of stressed correlations. We start with scenario analysis to define target stressed correlations, motivated by data (see also Ch. 37). We introduce the concept of the average correlation stress. Naturally, these are target correlation stresses for which the correlation matrix will not be positive definite. The technique of finding an optimal positive-definite approximation to a non-positive-definite target correlation matrix is treated in the next chapter.

We then show how to generate random correlation matrices using two techniques. The first method is a direct application of historical data. Historical correlation matrices are in principle positive definite, and in practice are close to positive definite. The second method uses historical data to construct a model for stochastic random correlation matrices. The model contains Gaussian multivariate techniques on the space of angles of the $\mathcal{N}$-sphere that is equivalent to the correlation matrices. This second method always yields a positive definite correlation matrix.

Correlation Stress Scenarios Using Data

In this section, we look at data for correlations to get an idea of the variability of the correlations $\left(\rho_{\alpha\beta}\right)$ in practice over many variables. We need this in order to perform the stressed correlation matrix analysis. We need to get a target stressed matrix $\left(\rho_{\alpha\beta}\right)^{(Target)}$ that has the property that individual matrix elements are stressed using information from historical data. We will then find the best-fit positive-definite correlation matrix $\left(\rho_{\alpha\beta}\right)^{(BestFit, PosDef)}$ to be used in the risk analysis, such as the Stressed VAR, as described in Ch. 27. The amount of the stress in the correlations for the result $\left(\rho_{\alpha\beta}\right)^{(BestFit, PosDef)}$ will in general be less than the amount of stress for $\left(\rho_{\alpha\beta}\right)^{(Target)}$. This is because the various correlations are dependently constrained.

Therefore, the correlation matrix elements cannot consistently be stressed independently.

As one example, we look at the maximum and minimum $\rho_{\alpha\beta}^{(Max)}$, $\rho_{\alpha\beta}^{(Min)}$ correlation of each pair of variables α, β over all windows of a given length running over the time series[1]. This is easy to accomplish. We now can get a scale for the maximum fluctuation of a correlation using $\Delta\rho_{\alpha\beta}^{(Max,Min)}$ where $\Delta\rho_{\alpha\beta}^{(Max,Min)} = \rho_{\alpha\beta}^{(Max)} - \rho_{\alpha\beta}^{(Min)}$.

We could stress the current correlations[2] to get the target stressed correlations by assuming that the historically biggest change for each correlation is the stress target change, i.e. $\rho_{\alpha\beta}^{(Target)} = \rho_{\alpha\beta}^{(Current)} \pm \Delta\rho_{\alpha\beta}^{(Max,Min)}$. The sign $\pm$ in this equation is the same as the sign of $\rho_{\alpha\beta}^{(Current)}$. That is, we stress positive current correlations upward towards one and negative current correlations downwards toward minus one. This makes sense, since in a stressed environment correlations between risk factors tend to become more marked in a signed sense[3].

We naturally need $\rho_{\alpha\beta}^{(Target)}$ constrained in the range [-1, 1]. If $\Delta\rho_{\alpha\beta}^{(Max,Min)}$ is too big we cut off the stress and impose the [-1, 1] constraint by hand. This should be understood in all the correlation stress equations.

We can also define the average of the max and min correlations $\rho_{\alpha\beta}^{(Avg)}$ as $\rho_{\alpha\beta}^{(Avg)} = \frac{1}{2}\left[\rho_{\alpha\beta}^{(Max)} + \rho_{\alpha\beta}^{(Min)}\right]$. A less conservative choice for the target stress is to use $\Delta\rho_{\alpha\beta}^{(Max,Avg)} = \rho_{\alpha\beta}^{(Max)} - \rho_{\alpha\beta}^{(Avg)} = \frac{1}{2}\Delta\rho_{\alpha\beta}^{(Max,Min)}$. From this viewpoint, we stress the correlations by the biggest fluctuations about the historical average. These are half as big as the maximum fluctuations. Using this assumption, we define

$$\rho_{\alpha\beta}^{(Target)} = \rho_{\alpha\beta}^{(Current)} \pm \frac{1}{2}\Delta\rho_{\alpha\beta}^{(Max,Min)} \qquad (23.1)$$

---

[1] **Time Windows:** The time length of the windows is determined by financial relevance. See the discussion of correlations and data in Ch. 37.

[2] **Current Correlations:** These are the correlations measured over a standard window ordinarily used for risk management. Our goal, again, is to take into account correlation variability risk. In order to do this, we need to stress the correlations.

[3] **Correlation Stresses for Hedges, Pairs Trading:** For hedged (long – short) positions on underlyings x_α and x_β with $\rho_{\alpha\beta} > 0$, the most loss would occur if $\rho_{\alpha\beta}$ decreased. For pairs trading or convergence trades with $\rho_{\alpha\beta} < 0$, the most loss would occur if $\rho_{\alpha\beta}$ increased. These correlation moves are opposite to what is assumed in the text for most variables. A more refined risk correlations would take the major correlation risks physically into account.

We remind the reader that the stressed-correlation target matrix is a scenario, and any measure provides a possible target. Since the best-fit positive-definite stress will be less than the target stress, it is perhaps better to adopt a conservative stress for the target stress.

The Average-Target-Correlation Stress

It is simpler to adopt a procedure where the correlations are stressed by a fixed average amount $\Delta\rho_{Avg}^{(Target)}$ independent of α, β. This has the disadvantage that the average stress is less accurate than detailed stresses. However, the procedure has the distinct advantage that it is easier to implement and easier to explain to management. At the end of this section, we will refine this average in different ways. We write, again with the $\pm$ being the same as the sign of $\rho_{\alpha\beta}^{(Current)}$, namely $\rho_{\alpha\beta}^{(Target)} = \rho_{\alpha\beta}^{(Current)} \pm \Delta\rho_{Avg}^{(Target)}$.

In order to get $\Delta\rho_{Avg}^{(Target)}$, we can plot the distribution in α, β of $\Delta\rho_{\alpha\beta}^{(Max,Min)}$ and take half the average value over $\{\alpha, \beta\}$, viz $\Delta\rho_{Avg}^{(Target)} = \frac{1}{2}\left\langle \Delta\rho_{\alpha\beta}^{(Max,Min)} \right\rangle$, up to some error that could taken as the width of the $\Delta\rho_{\alpha\beta}^{(Max,Min)}$ distribution.

Exposure-Weighted Definition

It is more logical and relevant for the averaged stressed correlation to receive contributions from "more important" variables. The importance can be quantified by the exposures[4] $\left\{ {}^{\$}\mathcal{E}_\alpha \right\}$. Therefore, a better definition for $\Delta\rho_{Avg}^{(Target)}$ would be the exposure-weighted sum, namely

$$\Delta\rho_{Avg}^{(Target)} = \frac{1}{{}^{\$\$}\Gamma} \sum_{\alpha\neq\beta} {}^{\$}\mathcal{E}_\alpha {}^{\$}\mathcal{E}_\beta \Delta\rho_{\alpha\beta} \qquad (23.2)$$

Here $\Delta\rho_{\alpha\beta}$ is the measure chosen to represent the correlation uncertainty and the normalization is ${}^{\$\$}\Gamma = \sum_{\alpha\neq\beta} {}^{\$}\mathcal{E}_\alpha {}^{\$}\mathcal{E}_\beta$.

Subset Definitions, Arbitrary Weights

More refined definitions involve summing over a subset of the possible correlation pairs. This can be done to emphasize the correlation instabilities of a

---

[4] **Exposures:** Exposures are measures related to the positions, e.g. DV01. They are described in Ch. 26.

particular subset S of variables, with sums running over $\{\alpha, \beta \in S\}$. For example, S could be chosen as all those correlations with correlation values above a certain threshold, or as correlations inside a certain sector (e.g. fixed income, equities, commodities, emerging markets, etc.). We can also include a weight $w_{\alpha\beta}$, for example using the exposures. The result for $\Delta\rho_{Avg}^{(Target)}$ is dependent on the subset of variables S and the set of weights w. Setting $\Gamma_S = \sum_{\alpha,\beta \in S} w_{\alpha\beta}$, we have

$$\Delta\rho_{Avg}^{(Target)}(w; S) = \frac{1}{\Gamma_S} \sum_{\alpha,\beta \in S} w_{\alpha\beta} \Delta\rho_{\alpha\beta} \qquad (23.3)$$

Applying the Subset Average-Target Correlation Stress
An obvious choice for the application of the subset average target stress can be for only those $\{\alpha, \beta \in S\}$, leaving the other correlations fixed. However, the stress can in fact be applied to *all* correlations regardless of whether they are in the subset S or not. This is consistent and it may be preferable. There is a good reason for this point of view.

Consider the result obtained using a subset S of relatively volatile correlations. This means that $\Delta\rho_{Avg}^{(Target)}(S)$ will be larger than the target stress $\Delta\rho_{Avg}^{(Target)}(All)$ for *all* variables. Now consider the final average stress $\Delta\rho_{Avg}^{(Final)}$ for the final matrix after the least-squares optimization, subtracting the current correlation matrix $\rho_{\alpha\beta}^{(Current)}$,

$$\Delta\rho_{Avg}^{(Final)} = \frac{1}{\Gamma} \sum_{\alpha \neq \beta} \left| \rho_{\alpha\beta}^{(Final)} - \rho_{\alpha\beta}^{(Current)} \right| \qquad (23.4)$$

Here, $\Gamma = N(N-1)/2$. The final stress is generally less than the target stress, so $\Delta\rho_{Avg}^{(Final)} < \Delta\rho_{Avg}^{(Target)}(S)$. Since also $\Delta\rho_{Avg}^{(Target)}(All) < \Delta\rho_{Avg}^{(Target)}(S)$, it can happen that $\Delta\rho_{Avg}^{(Final)}$ may be approximately equal to $\Delta\rho_{Avg}^{(Target)}(All)$. That is, aiming at a high target correlation stress can lead to consistency with the average correlation stress once the positive-definite constraints are imposed [5].

---

[5] **Remark:** This situation has occurred in practice.

Pedagogical Example for the Average-Target Correlation Stress

The first drawing is an illustration of a possible distribution among the different (α, β) for the maximum correlation $\rho_{\alpha\beta}^{(Max)}$, defined with respect to the various choices of the starting times of the fixed-length windows in the data series[7]. The maximum correlation $\rho_{\alpha\beta}^{(Max)} = 0.1$ to 0.2 is the most probable range for this illustration. Significant high-correlation tail effects are present all the way out to $\rho_{\alpha\beta}^{(Max)} = 1$. A few events with $\rho_{\alpha\beta}^{(Max)} < 0$ can exist[8].

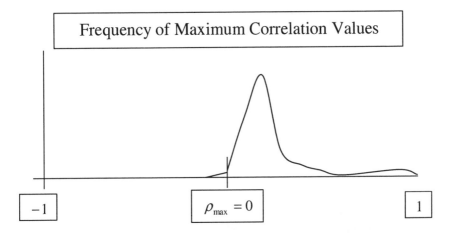

An illustrative distribution of minimum correlations $\rho_{\alpha\beta}^{(Min)}$ defined with respect to the various choices of the starting times of the fixed-length windows, again for the different choices (α, β), is in the next drawing. For the minimum, $\rho_{\alpha\beta}^{(Min)} = -0.1$ to -0.2 is the most probable range. Tail effects are present, though not to the extent of $\rho_{\alpha\beta}^{(Max)}$.

---

[7] **Acknowledgements:** The figures are intended only to provide a pedagogical example. They are based on data for $N_f = 100$ risk factors in 1999-2000. We thank Citigroup for the use of the statistics, run by M. Rodriguez.

[8] **Correlation Sign Warning**: The signs of some correlations are arbitrary. A correlation involving an FX rate will change sign depending on the quoting convention, for example $\eta = 100$ Yen/USD or $\xi = 0.01$ USD/Yen, because $d\eta/\eta = -d\xi/\xi$. This makes sense because these definitions correspond to mirror statements about the currencies. See the chapter on FX options (Ch. 5).

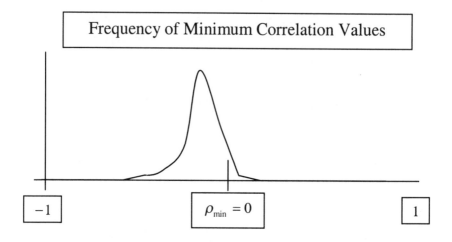

The target stress for the correlations can be taken from the max, min correlation difference $\Delta\rho_{\alpha\beta}^{(Max,Min)}$. An illustration of this distribution is in the drawing below.

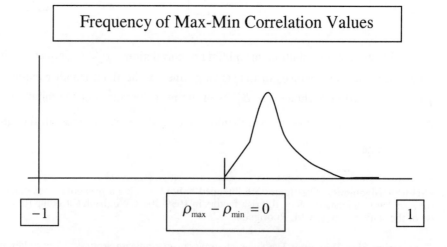

The most probable range for Max-Min correlation difference $\Delta\rho_{\alpha\beta}^{(Max,Min)}$ is 0.2 to 0.3, and again tail effects are present.

In practice, it turns out that if the target stress of 0.3 is used for all correlations, then the final correlation stress, after obtaining a positive-definite matrix, is about half this number, $\Delta\rho_{Avg}^{(Final)} \approx 0.15$. It also happens that if we were to use a subset S containing rather volatile correlation fluctuations, say $\langle \Delta\rho_{\alpha\beta}^{(Max,Min)}(S) \rangle \approx 0.6$, along with the fluctuations around the average, then we would get $\Delta\rho_{Avg}^{(Target)}(S) = 0.30$. Using the volatile correlation subset S as a target, the final average correlation stress is close to the target average stress, around the average for all variables.

Stressed Random Correlation Matrices

We have seen that a variety of correlation stress scenarios can be adopted. A statistical treatment of stochastic correlations generalizing the scenario assumptions can be carried out in principle. There are at least two ways that stochastic correlation matrices can be generated. The first is a direct historical approach, and the second uses the geometrical approach, as presented in the previous chapter[9].

Random Correlation Matrices Using Historical Data

For a direct historical determination of random correlation matrices, we first need to adopt a time window of length N_w, say 3 months or 65 business days. As usual, we often want variables, e.g. $x_\alpha = \ln r_\alpha$ to deal with returns. Then for each window starting date t_j and for variables x_α, x_β the windowed correlation is

$$\rho_{\alpha\beta}^{(Hist)}(t_j) = \frac{1}{\sigma_{\alpha j}\sigma_{\beta j}(N_w-1)} \cdot \sum_{\ell=j-N_w-1}^{j-2} \left[d_t x_\alpha(t_\ell) - \langle d_t x_\alpha \rangle_j \right]\left[d_t x_\beta(t_\ell) - \langle d_t x_\beta \rangle_j \right]$$

Here $d_t x_\alpha(t_\ell) = x_\alpha(t_{\ell+1}) - x_\alpha(t_\ell)$, $\langle d_t x_\alpha \rangle_j = \frac{1}{N_w} \cdot \sum_{\ell=j-N_w-1}^{j-2} d_t x_\alpha(t_\ell)$, and the

variance at t_j is $\sigma_{\alpha j}^2 = \frac{1}{(N_w-1)} \cdot \sum_{\ell=j-N_w-1}^{j-2} \left[d_t x_\alpha(t_\ell) - \langle d_t x_\alpha \rangle_j \right]^2$.

---

[9] **Third Method for Stressed Correlations Using Principal-Components:** There is, theoretically, a third method for getting random correlations using principal components. The idea is to generate changes in the eigenvalues, along with rotations of the eigenfunctions. However, this method is far removed from the changes in the physical correlations, because many variables contribute to each principal component. For this reason, we prefer the methods in the text dealing directly with correlations.

The historical correlation matrix $\left(\rho_{\alpha\beta}^{(Hist)}(t_j)\right)$ for all $\{\alpha,\beta\}$ using the above prescription is not guaranteed to be positive definite, but it should be close. The usual scheme to be described in Ch. 24 (SVD coupled with throwing out the negative or zero eigenvalues coupled with renormalizing the eigenfunctions), but without least-squares optimization, is good enough for our purposes. Note that we need to go through this procedure for each time t_j.

The resulting positive-definite historical correlation matrices are then used to get random correlation matrices simply by random selection. Take a set of random numbers $\{R\#\}$ uniformly distributed over the unit interval $(0,1)$. Also say we have N_{data} historical correlation matrices at t_j, $j=1...N_{data}$. Then we bin $(0,1)$ into N_{data} bins. Throwing the dice, we get the matrix from a particular time t_j with probability $1/N_{data}$. For a particular random number $R\#$ and its associated time t_j we get

$$\left(\rho_{\alpha\beta}^{(Hist)}(R\#)\right) = \left(\rho_{\alpha\beta}^{(Hist)}(t_j)\right) \quad (23.5)$$

These random matrices $\left(\rho_{\alpha\beta}^{(Hist)}(R\#)\right)$ can then be used as the stochastic correlation matrices in a simulation, for example a stressed VAR simulation including stochastic correlations.

Advantages and Disadvantages of Random Historical Matrices

The advantage of this method is simplicity. There are some drawbacks. First, the number of obtainable correlation matrices is limited by the length of the time series of the data. Further, if the windows overlap in time, the correlation matrices obtained, while random, are dependent. Finally, the extremes in correlations that could potentially exist and for which the risk could be very large, may not be present (and probably are not present) in the data available or used for the analysis.

We turn next to a possibly more powerful method of obtaining random correlation matrices that avoids these difficulties.

Stochastic Correlation Matrices Using the $\mathcal{N}$-sphere

The idea here is that, starting with some base correlation matrix $\left(\rho_{\alpha\beta}^{Base}\right)$, we generate randomly fluctuating new matrices $\left(\rho_{\alpha\beta}^{New}\right)$ from this base matrix. The stochastic model for random correlation matrices uses the $\mathcal{N}$-sphere representation of correlation matrices described in the previous chapter. This technique will guarantee that positive definiteness is respected.

Review of the $\mathcal{N}$ - Sphere and the Variables

Given N variables with correlation matrix $(\rho_{\alpha\beta})$, there is an associated $\mathcal{N}$-sphere with $\mathcal{N} = N-1$, embedded in N-dimensional Euclidean space. We call the polar and azimuthal angles describing points on this sphere $\{\theta_{1\beta}\}, \{\varphi_{\alpha\beta}\}$. For this discussion, the reader does not need the details.

A one-to-one mapping can be explicitly constructed between a positive-definite correlation matrix and these angles. This mapping is just the Cholesky decomposition in spherical co-ordinates. Moreover, the mapping can be inverted.

Thus, we have for each matrix element the representation in terms of the angles, namely $\rho_{\alpha\beta} = \rho_{\alpha\beta}\left[\{\theta_{1\beta'}\}, \{\varphi_{\alpha'\beta'}\}\right]$. We also have the inverse expression for each angle $\theta_{1\beta} = \cos^{-1}(\rho_{1\beta})$ and $\varphi_{\alpha\beta} = \varphi_{\alpha\beta}\left[\{\rho_{\alpha'\beta'}\}\right]$ in terms of $\{\rho_{\alpha\beta}\}$.

The Angle Volatilities and the Angle-Angle Correlations

In order to proceed, we need the "angle volatilities" for the polar angles $\sigma(\theta_{1\beta})$ and the azimuthal angles $\sigma(\varphi_{\alpha\beta})$. We also need the "angle-angle correlations" between polar angles $\rho(\theta_{1\beta}, \theta_{1\beta'})$, between azimuthal angles $\rho(\varphi_{\alpha\beta}, \varphi_{\alpha'\beta'})$, and between polar-azimuthal angles $\rho(\theta_{1\beta'}, \varphi_{\alpha\beta})$.

The angle volatilities and angle-angle correlations can be obtained either using a scenario assumption or using historical data. The scenario assumption just means that numerical values for these quantities are assigned directly. The historical data approach would use input from the historical correlation matrices generated using the procedure outlined above, but for non-overlapping time windows[10].

---

[10] **Non-Overlapping Time Windows, Angle Volatilities, and Correlations:** If we are picking the correlation matrix with the biggest effect we want overlapping time windows to increase the phase space of available matrices. Here the windows should be non-overlapping. Otherwise, the angle volatilities will be too small. Correlations arising from overlapping windows are close to each other only because they use some of the same information. The multivariate model can generate arbitrarily many random matrices.

The Random Matrices

Now the idea of generating the stochastic matrix elements is quite straightforward. We use a multivariate Gaussian model to generate the random angles for points on the $\mathcal{N}$-sphere. We then construct the random correlation matrix elements.

Specifically, given the angle volatilities and angle-angle correlations we use a Gaussian multivariate model to generate random points $\mathcal{P}^{(R\#)}$ on the $\mathcal{N}$-sphere. Each point is defined with a given set $\{R\#\}$ of random numbers, viz $\mathcal{P}^{(R\#)} = \mathcal{P}\left[\left\{\theta_{1\beta}^{(R\#)}\right\}, \left\{\varphi_{\alpha\beta}^{(R\#)}\right\}\right]$. We then reconstruct the random correlation matrix for a given set $\{R\#\}$, viz $\rho_{\alpha\beta}^{(\text{Stochastic Angles})}(R\#)$.

Note that we are *not* constructing a Gaussian multivariate model for the stochastic correlations matrix elements, but rather for the angles. This guarantees that every matrix is positive definite, because every point on the $\mathcal{N}$-sphere given by the angles corresponds to a positive definite matrix. Had we tried to model the correlation matrix elements directly, we would get matrices far from positive definite that would require considerable computer processing.

Picture of Random Correlation Matrices Using the $\mathcal{N}$-sphere

The following picture gives the idea:

---

[11] **Cyclic Conditions:** We use cyclic boundary conditions to associate points in the angles differing by multiples of π for polar angles and 2π for azimuthal angles.

Random Correlation Matrices Using the $\mathcal{N}$-sphere

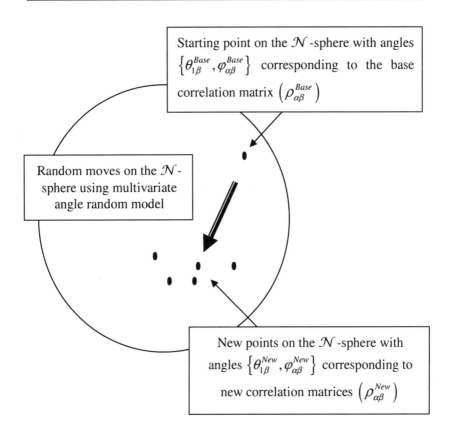

24. Optimally Stressed PD Correlation Matrices (Tech. Index 7/10)

In this chapter, we deal with the problem of finding an optimal positive-definite (PD) approximation for a given correlation matrix. Consider a non-positive-definite (NPD) correlation matrix $(\rho)^{NPD}$. We call NPD matrices "illegal" and positive-definite matrices "legal". Correlation matrices that are NPD can arise from various sources. As discussed before, stressed correlation matrices are desirable to probe correlation risk. Such stressed matrices are produced by moving the individual correlation matrix elements $\rho_{\alpha\beta}$ from their current values $\rho_{\alpha\beta}^{Current}$ by amounts $\Delta\rho_{\alpha\beta}$.

We did present a way in the last chapter for the $\Delta\rho_{\alpha\beta}$ to be chosen while preserving PD constraints by using the $\mathcal{N}$ - sphere geometrical formalism.

However, as also described in the last chapter, we might want to move $\Delta\rho_{\alpha\beta}$ by hand using (e.g.) historical volatilities, maximum moves, etc. However, the constraints among the various correlations make it difficult (read impossible) to preserve the PD constraint. The stressed matrix is therefore most probably NPD.

One good reason to stress the correlations is in order to use them in Monte-Carlo simulations generating stressed-correlated movements of underlying variables[1]. However, if a correlation matrix is NPD, it is useless for simulation. This is because a NPD matrix has negative eigenvalues and so the real square root matrix needed for simulations does not exist[2]. Hence, a method is needed to render NPD matrices positive definite in such a way that the stressed character of the correlations is preserved as much as possible.

Using the $\mathcal{N}$ - sphere geometrical formalism, an optimal technique is presented here for finding a legal PD matrix that is the "best approximation" to an illegal NPD target matrix. The NPD matrix is called the "target" because in

---

[1] **Stressed VAR Simulation with Stressed Correlations:** We will examine the use of stressed correlations in simulations when we consider the Stressed VAR in Ch. 27.

[2] **NPD Matrix Misuse:** Of course, a non positive definite matrix could be used, e.g., in a quadratic-form VAR (c.f. Ch. 26). However, such a procedure would be logically inconsistent. If ad-hoc correlation matrices in quadratic-form expressions are used to aggregate risk, they should be checked for being positive definite.

fact we may want to consider correlations stressed in a certain way for physical reasons (the target), and we want to get as close to this target condition as possible, maintaining the condition of PD legality.

The idea is to use a two-step procedure[3]:

1. Get a starting point for the matrix, $(\rho)^{PD;Start}$ that is positive definite using an algorithm involving the singular value decomposition SVD, plus an eigenvalue and eigenfunction renormalization, as described below.

2. Perform a least-squares fit moving the matrix elements, *always maintaining positive definiteness*, until we get close to the NPD target matrix $(\rho)^{NPD;Target}$.

The procedure can be illustrated by the following flow chart:

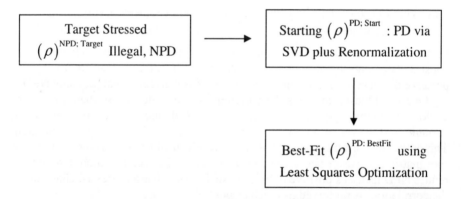

The technique used to obtain the best-fit approximation to the target correlation function proceeds using the geometrical spherical representation described in Ch. 22. For the present purposes, it suffices to know that any PD correlation matrix between N variables corresponds to a set of $N(N-1)/2$ angles $\{\theta_{1\beta}\}, \{\varphi_{\alpha\beta}\}$ on an $\mathcal{N} = N - 1$ sphere in N dimensions.

---

[3] **History:** The theory for this two-step procedure for optimally stressed positive-definite correlation matrices was developed by me in 1999-2000. It was implemented numerically by Mark Rodriguez and Juan Castresana.

Least-Squares Fitting for the Optimal PD Stressed Matrix

The least squares approach is used to provide a measure for minimizing the difference between the target and best-fit matrix as the angles are varied around the sphere. This procedure always produces a PD matrix, because any correlation matrix parameterized by the $\mathcal{N}$ - sphere angles is positive definite, as we discussed in Ch. 22. The picture gives the idea.

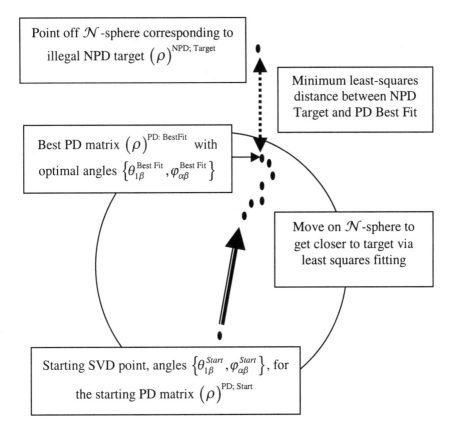

We write the chi-square as a function of the various angles to be moved around on the sphere, as follows:

$$\chi^2\left[\{\theta_{1\beta}\},\{\varphi_{\alpha\beta}\}\right] = \sum_{\alpha'<\beta'=2}^{N} w_{\alpha'\beta'}\left[\rho_{\alpha'\beta'}^{(\text{NPD;Target})} - \rho_{\alpha'\beta'}\left[\{\theta_{1\beta}\},\{\varphi_{\alpha\beta}\}\right]\right]^2$$

(24.1)

We move the angles $\{\theta_{1\beta}\},\{\varphi_{\alpha\beta}\}$, beginning at the starting point $\{\theta_{1\beta}^{Start},\varphi_{\alpha\beta}^{Start}\}$, using a least-squares routine until χ^2 is as small as is practical, winding up with the best-fit angles $\{\theta_{1\beta}^{Best\,Fit},\varphi_{\alpha\beta}^{Best\,Fit}\}$.

Weights in the Least-Squares Fitting

The weights $\{w_{\alpha\beta}\}$ can be chosen to emphasize those correlations that the fit will try to best reproduce. It is desirable to choose these correlations where the most risk is perceived. An example is to take the weights proportional to the absolute value of the product of the exposures of the underlying variables, or the product of the fat-tail volatilities[4] and the exposures. The variables that do not really contribute much to the risk will have small weights, and the fit will not bother to get these corresponding irrelevant correlations close to the target. Therefore, we can choose either one of the two following expressions:

$$w_{\alpha\beta} = \left|{}^{\$}\mathcal{E}_{\alpha}{}^{\$}\mathcal{E}_{\beta}\right| \bigg/ \sum_{\alpha',\beta'}\left|{}^{\$}\mathcal{E}_{\alpha'}{}^{\$}\mathcal{E}_{\beta'}\right|$$

(24.2)

$$w_{\alpha\beta} = \sigma_{\alpha}\sigma_{\beta}\left|{}^{\$}\mathcal{E}_{\alpha}{}^{\$}\mathcal{E}_{\beta}\right| \bigg/ \sum_{\alpha',\beta'}\sigma_{\alpha'}\sigma_{\beta'}\left|{}^{\$}\mathcal{E}_{\alpha'}{}^{\$}\mathcal{E}_{\beta'}\right|$$

(24.3)

Numerical Considerations for Optimal PD Stressed Matrix

Practically speaking, the number of variables N that can be dealt with are a few hundred up to around a thousand at the maximum. The time required to perform the least squares search for a few hundred variables can be run overnight. Typically, the convergence is rapid for a beginning period and then bogs down[5].

---

[4] **Fat-Tail Vols:** These result from Gaussian fits to outlier fat tails. Fat-tail volatility is described in detail in Ch. 21.

[5] **Parallelization:** The computer code can be parallelized, which would improve convergence.

Example of Optimal PD Fit to a NPD Stressed Matrix

Here is a pedagogical example. The details are unimportant, but the results are representative. The first figure[7] gives an illustration of the histogram frequency of the values of the best-fit stressed correlations (dotted line) vs. unstressed correlations (solid line). Consistent with the requirement of positive definiteness, correlations can undergo substantial stress.

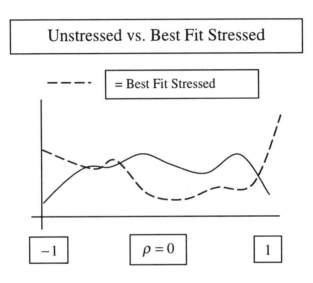

The stressed target matrix can be obtained by a scenario. As discussed in Ch. 23, one scenario has all positive (negative) correlations increased (decreased) by some $\pm \Delta \rho_{Avg}^{(Target)}$, up to the limits $[-1,1]$. This represents a breakdown in the markets envisioned where correlation increases due to flight-to-quality effects, etc. There is a "hole" in this NPD target by construction, where no correlations exist for $\left(-\Delta \rho_{Avg}^{(Target)}, \Delta \rho_{Avg}^{(Target)}\right)$. The target stressed matrix is generally not positive definite, containing negative eigenvalues.

The next drawing illustrates the matrix elements of the NPD target matrix $(\rho)^{NPD; Target}$ (solid line) and of the best-fit matrix $(\rho)^{PD: BestFit}$ (dotted line) for the case $\Delta \rho_{Avg}^{(Target)} = 0.3$. This amount of correlation stress is representative of

---

[7] **Data, Acknowledgements:** The illustrative figures are based on data from 20 time series during 1999-2000, for which I thank Citigroup. The least-squares optimization was implemented by M. Rodriguez.

changes in volatile correlations. The stressed correlations from the NPD matrix are reasonably reproduced by the best fit correlations. The hole at small correlations in the NPD target is filled in by the best fit.

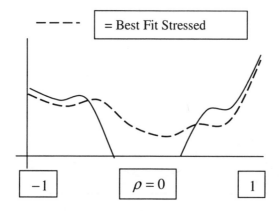

The last figure illustrates the histogram of values of the correlations in the PD starting point matrix $(\rho)^{\text{PD; Start}}$ (solid line) compared with the best-fit stressed result (dotted line). The starting matrix is obtained using the SVD procedure described in the next section. The least-squares optimal procedure results in more stress, closer to the target than the starting matrix.

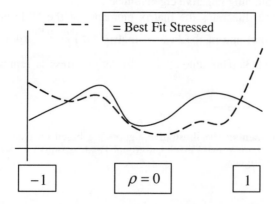

Stressed Correlations for Equity, Currency, or Rate Baskets

The above example illustrates the idea that stressed correlation analysis can be done with baskets. This includes many practical deals. The baskets that can be profitably analyzed for correlation risk include baskets of equities, baskets of currencies, baskets of rates, etc.

If a multifactor forward-rate model is used, the effect of stressed correlations between forward rates on interest-rate derivatives can be examined.

In practice, the analysis can be done with up to hundreds of variables.

SVD Algorithm for the Starting PD Correlation Matrix

Summary of the Starting Algorithm (SVD + Renormalization)

A useful algorithm that produces a positive definite correlation matrix $(\rho)^{PD}$ from a non-positive-definite stressed target matrix $(\rho)^{NPD}$ uses the Singular Value Decomposition (SVD). We can use this to provide the PD starting point in the least-squares procedure outlined above. We can also use it to transform NPD historical correlation matrices that have some data inconsistencies into legal matrices.

The procedure is as follows[8,9]:

- Run the Singular Value Decomposition[10,i] (SVD) to obtain the eigenvalue spectrum of the NPD matrix.
- Set all non-positive eigenvalues to a small positive value. The resulting matrix is PD.
- Renormalize the eigenfunctions to get unit diagonal correlation matrix elements.
- Rerun the SVD to transform the renormalized eigenfunctions from the previous step into an orthonormal set.

---

[8] **History:** This SVD + Renormalization algorithm was formulated independently by me in 1998. It turns out to be a common procedure.

[9] **Acknowledgements:** I thank Eduardo Epperlein and Kevin Jian for informative discussions on correlations.

[10] **The SVD and the Cholesky Decomposition:** The Cholesky decomposition cannot handle NPD matrices; the SVD can. For PD matrices the results of the two algorithms are the same.

The Singular Value Decomposition (SVD)

The SVD is a time-honored method for dealing with NPD matrices. Here, we have a symmetric matrix, simplifying the SVD. We write the matrix equation $(\rho)^{NPD} = U \Lambda V^T$ where Λ is the diagonal matrix of eigenvalues $\lambda^{(a)} = \Lambda_{aa}$, and U is the matrix of eigenfunctions $\psi^{(a)}$, the eigenfunctions having components $\psi_\alpha^{(a)} = U_{\alpha a}$.

Two Important Technical SVD Points

There are two important technical details[11]. First, in our case with a symmetric $(\rho)^{NPD}$, the matrix V is the same as U except for one important trap arising from a convention. The convention is, for example, in the SVD routine in the Numerical Recipes book[10]. If a negative eigenvalue $\lambda^{(a)} \leq 0$ exists, the routine reassigns the absolute positive value to that eigenvalue and hides the minus sign in the matrix V.

Second, degeneracies may occur in the correlation matrix because identical degenerate time series may be present[12]. Some SVD routines do not handle degeneracies well. Therefore "Magic Dust" can be added to the degenerate time series in order to break the degeneracies[13].

The Renormalization of the Eigenvalues and Eigenfunctions

The whole problem with a NPD stressed matrix $(\rho)^{NPD}$ is that it contains negative or zero eigenvalues. Therefore, a straightforward trick to restore positive definiteness is to renormalize (i.e. change by hand) all non-positive eigenvalues of Λ to a small positive number $\varepsilon > 0$. Although this may seem arbitrary, our only goal here is to get *some* PD matrix to use as a starting point in the least-squares fit[14]. Therefore, we replace the diagonal matrix of eigenvalues Λ by this

---

[11] **Psychology:** If you don't know the keys to these details, you can go bonkers.

[12] **Degenerate Time Series:** Degeneracies can occur if not enough data exist, and one time series is being used as a proxy for two variables.

[13] **"Magic Dust":** This is an added random amount small enough to be past all significant digits of the numbers in each copy of the time series producing the degeneracy. This addition causes no change in the time series to its existing accuracy and it removes the degeneracies.

[14] **VAR:** This first-stage SVD + Renormalization technique, without the least-squares fitting, is often used for standard VAR applications, where some NPD problems arise from somewhat dirty or inconsistent data.

strictly positive diagonal matrix $\Lambda^{(PD)}$, i.e. $\Lambda \to \Lambda^{(PD)}$. If the number of bad non-positive eigenvalues is N_{bad}, then $\Lambda^{(PD)}$ contains the original positive eigenvalues of Λ, along with N_{bad} copies of ε.

Having performed this eigenvalue change, $V = U$ holds in the SVD. However, the reconstruction of the correlation matrix through the prescription $U\Lambda^{(PD)}U^T$ with the new $\Lambda^{(PD)}$ matrix is not satisfactory. This is because the resulting putative diagonal correlation elements will not be one.

In order to guarantee the correlation matrix has ones on the diagonal, we need to renormalize the eigenfunctions. We therefore renormalize the matrix of eigenfunctions U into a new matrix $U^{(UnitDiag)}$, i.e. $U \to U^{(UnitDiag)}$. Here $U^{(UnitDiag)}_{\alpha a} \equiv \Gamma_\alpha U_{\alpha a}$, where the renormalization factor Γ_α is defined as

$$\Gamma_\alpha = \frac{1}{\left[\sum_{b=1}^{N} U_{\alpha b}^2 \Lambda_{bb}^{(PD)}\right]^{1/2}} \tag{24.4}$$

Note the quantity under the square root is strictly positive; we naturally take the positive square root. In addition, the sum effectively stops at $N^{eff} = N - N_{bad}$. The correlation matrix (for this algorithm so far) is then

$$(\rho)^{PD} = U^{(UnitDiag)} \Lambda^{(PD)} U^{(UnitDiag)T} \tag{24.5}$$

As can be seen by a one-line calculation, the diagonal elements are now one, i.e. $(\rho)^{PD}_{\alpha\alpha} = 1$, as required.

The Square-Root Correlation Matrix Using SVD or Cholesky

In order to run simulations, we need the square root of a PD correlation matrix. Now the matrix $U^{(UnitDiag)}$ we just constructed is *not* a matrix of eigenfunctions of the correlation matrix due to the renormalization factors $\{\Gamma_\alpha\}$. Hence, we need to perform a rotation by running the SVD again in order to reorganize the expression for $(\rho)^{PD}$ as $(\rho)^{PD} = U^{(Final)} \Lambda^{(PD)} U^{(Final)T}$. The final matrix $U^{(Final)}$ is the (orthogonal) matrix of eigenfunctions of $(\rho)^{PD}$.

Then we get the desired square root matrix in the usual way, viz $\sqrt{(\rho)^{PD}} = U^{(Final)} \left[\Lambda^{(PD)}\right]^{1/2} U^{(Final)T}$. Because $(\rho)^{PD}$ is indeed a positive definite matrix, this square root can be found using either SVD or the Cholesky decomposition[15].

PD Stressed Correlations by Walking through the Matrix

This section describes an alternative method[16] for obtaining stressed PD correlation matrices. The idea is to walk through the correlation matrix by steps along a path, successively stressing correlations. We ensure that each step is taken such that the resulting stressed matrix at that step is PD. Hence, at the first step we stress matrix element $\rho_{\alpha 1, \beta 1}^{\text{Step 1}}$ in some row $\alpha 1$ and column $\beta 1$. Moving along a path to step $\#m$, we stress $\rho_{\alpha m, \beta m}^{\text{Step m}}$. The allowed stress amount for a matrix element at a given step is naturally constrained by the stresses along the chosen path up to that step. In practice, as we walk deeper into the matrix, it turns out that the amount of allowed stress for matrix elements becomes more reduced.

References

[i] **SVD and Cholesky Algorithms**
Press, W. H., Teukolsky, S. A., Vetterling, W. T., Flannery, B. P., *Numerical Recipes in Fortran: The Art of Scientific Computing*. 2nd Edition, Chapter 2.6, 2.9. Cambridge University Press, 1992.

---

[15] **Dimensional Considerations:** However, it appears in practice that the SVD is more stable for large dimensions than is the Cholesky decomposition. For this reason, it may be preferable to forget about the Cholesky decomposition entirely.

[16] **Path Walking vs. Least Squares:** The path-walking method was devised by me earlier than the least squares approach. Walking the matrix can be useful for specific stress scenarios. In particular, if we are most concerned with postulating definite stressed values for only a few correlations, this method is useful. Recall that in the least-squares approach, we can influence the fit by choosing weights. However, we cannot guarantee any specific stressed value for any particular correlation matrix element. On the other hand, path walking is clumsy to implement for large matrices and was therefore replaced by the more tractable LS approach for corporate VAR applications.

25. Models for Correlation Dynamics, Uncertainties (Tech. Index 6/10)

In this chapter, we look at some models for the underlying dynamics of correlations, and for correlation uncertainties[1]. This includes the time dependence of correlations. Our point of view is that correlations have an intrinsic meaning, independent of historical time series. As an analogy, volatility is often treated as a dynamical variable in stochastic volatility models, independent of historical time series. In a similar way, correlations can be treated as dynamical variables. In the last chapter, we were concerned with a different issue, namely given some correlations, to find ways to consistently stress the given correlations. Our purpose here is to investigate some possibilities for the underlying dynamics of the correlations themselves.

"Just Make the Correlations Zero" Model; Three Versions

We start with the simple zero-correlation heuristic in three versions A, B, C [2].

Zero Correlations, Version A: "No Reason, No Correlation"

Some people think that correlations should be set equal to zero if you cannot think of a rationale or reason that nonzero correlations should be present. What these people effectively have in mind is that scenarios should be put together for correlations. That is, if there is no motivation for a particular correlation to be nonzero, their scenario would be a default of zero correlation.

A good response to this attitude is that while correlation scenarios are indeed useful, real-world non-zero correlations exist whether or not we have an underlying theory to describe them. It is perhaps presumptuous and fallacious to think that we are smart enough to give a reason for each non-zero correlation. We ignore potentially dangerous correlations at our peril, whether we understand why they are non-zero or not.

---

[1] **Numerical Problems Extracting Correlations:** In Ch. 37, we deal further with the numerically induced instabilities of correlations, from an historical data point of view.

[2] **Zero-Correlation Adherents:** All these reasons (sic) for setting correlations equal to zero have been proposed, utilized, and vociferously defended.

An example of correlation danger is provided by failures of diversification. Diversification assumes that by holding products in many markets, portfolio uncertainties can be reduced to small values through cancellations. However, in a stressed market environment, high correlations appear when investors panic, and a flight to quality ensues. Various spread products become correlated regardless of markets, just because they are in fact spread products that are all being sold, as investors collectively rush to safe havens like US Treasuries. The false assumption of small correlations led to the demise of famous hedge funds and Arb units in 1998.

Zero Correlations, Version B: "We Need Long Time Scales"

A variation on the theme of "just make the correlations zero" is to look at very long time scales (e.g. years). At such long time scales, historical correlation fluctuations can average out to small values. Using long time scales has the added apparent theoretical benefit that averaging over long times suppresses windowing errors, revealing the stable correlation value if there is one. One defense of this approach is that "there is more information in a longer time series, so we ought to use long time series to determine the correlation". This idea is appropriate for investors with long time horizons, on the order of years.

The proponents of this argument make an important point. There is indeed more information in longer time series. This information can and should be used to generate information about the uncertainties in the correlations.

The basic problem however is that the underlying assumption of the existence of a stable correlation is incorrect. Correlation breakdowns and instabilities involve intrinsic uncertainties independent of the numerical windowing uncertainties. The associated problem is that the assumption of long time scales is manifestly inappropriate for investors or strategies that have shorter time scales (e.g. months or weeks). Therefore, long time series should not be used to estimate the correlations themselves.

The bottom line is that market disasters involving correlation changes do not wait for mathematical theorem conditions of long times to be satisfied. In spite of windowing noise, there is no choice but to look at correlations over limited time intervals. Moreover, there is good evidence that no well-defined underlying stable correlations exist. During times of market stress, a "strategy" based on the assumption that there is a long-term stable correlation cannot be carried out without considerable financial damage[3].

---

[3] **What? Trader Angst?** Traders, with not much measurable angst over mathematical niceties for long time scales, react quickly to big or sudden changes in correlations in the markets. In this way, they retain their jobs.

Zero Correlations, Version C: "Its Too Hard Otherwise"

Another example of the zero-correlation model is the forced assumption when the number of market variables becomes so large that dense correlation matrices cannot be treated by the computer software and/or hardware[4]. In that case, some block-diagonal assumptions have to be made, setting the inter-block correlations to zero. For example, all intra FX correlations can be put into one block and all intra interest-rate correlations in another block, with inter FX-interest rate correlations set to zero.

Such zero-correlation inter-market assumptions will be unable to pick up the risk of, for example, an inflationary environment affecting both FX and interest rates. The trade-off is that the intra-market correlations are more accurately described with large numbers of variables.

The Macro-Micro Model for Quasi-Random Correlations

In this section, we present a new model for underlying correlation dynamics, motivated by the Macro-Micro (MM) model. In Ch. 47-51, we describe the MM model in detail. Macroeconomics provides strong long-term influences on correlations. The Macro component of the MM model attempts to model these macroeconomic influences in a parsimonious fashion, leaving the task of constructing the real macroeconomic underpinnings to the future. We will focus on the Macro component here. The Micro component just provides for an additional very-short time scale noise that would average out to zero in any windowed measure of correlation. The Macro component does not however average out to zero except possibly at very long times.

To illustrate the idea[5], we deal with a single correlation $\rho_{\alpha\beta}$. The Macro component of the MM model for correlations proceeds according to the prescription of quasi-random correlation slopes or correlation drifts:

- Use a Gaussian probability distribution for the correlation slope or drift.
- Specify the probability distribution of time intervals between slope changes, with minimum time interval defined by a minimum time cutoff.
- Bound the correlations by ±1.

---

[4] **Large Correlation Matrix PD Problems:** In particular, the imposition of positive definiteness for large correlation matrices becomes problematic when the number of variables is above a thousand, as an order of magnitude. In that case, block-diagonal assumptions can be used.

[5] **Extension of the MM Correlation Model to Correlation Matrices:** The application to a whole correlation matrix would involve rendering the matrix positive definite, as described in previous chapters. This would have to be done at each point in time.

The Macro Quasi-Random Slope for Correlation

The model for the time change of the correlation involves a Macro Correlation Slope $\mu_{\rho;\alpha\beta}^{CorrSlope}(t)$. The defining equation reads:

$$d_t \rho_{\alpha\beta}(t) = \rho_{\alpha\beta}(t+dt) - \rho_{\alpha\beta}(t) = \mu_{\rho;\alpha\beta}^{CorrSlope}(t) dt \qquad (25.1)$$

The quasi-random correlation slope $\mu_{\rho;\alpha\beta}^{CorrSlope}(t)$ changes the correlation $\rho_{\alpha\beta}$ as time changes. The idea is shown in the picture below:

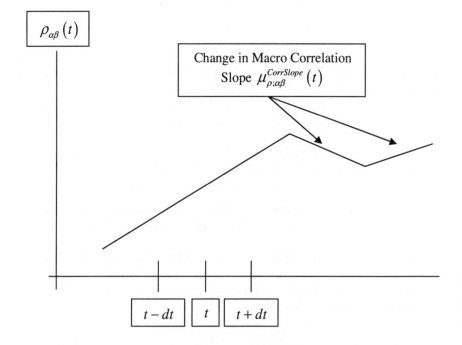

The Macro correlation slope $\mu_{\rho;\alpha\beta}^{CorrSlope}(t)$ is not a function. It changes in a quasi-random fashion at discrete times $\left\{ t_{ChangeSlope}^{(k)} \right\}$, as defined next:

Models for Correlation Dynamics, Uncertainties   333

$$\mu_{\rho;\alpha\beta}^{CorrSlope}(t) = \begin{cases} \mu_{\rho;\alpha\beta}^{CorrSlope}(t-dt)\Theta[1-\chi] & \text{if } t \neq t_{ChangeSlope}^{(k)} \\ \Im[\rho_{\alpha\beta}(t)] \cdot N\left(\mu_{Avg,\rho;\alpha\beta}^{CorrSlope}, \sigma_{\rho;\alpha\beta}^{CorrSlope}\right) & \text{if } t = t_{ChangeSlope}^{(k)} \end{cases} \quad (25.2)$$

When $\mu_{\rho;\alpha\beta}^{CorrSlope}$ changes at some $t_{ChangeSlope}^{(k)}$, the possible new values are drawn randomly from a normal distribution $N\left(\mu_{Avg,\rho;\alpha\beta}^{CorrSlope}, \sigma_{\rho;\alpha\beta}^{CorrSlope}\right)$. The normal or Gaussian pdf for $\mu_{\rho;\alpha\beta}^{CorrSlope}$ with correlation-slope volatility $\sigma_{\rho;\alpha\beta}^{CorrSlope}$ is

$$\wp\left(\mu_{\rho;\alpha\beta}^{CorrSlope}\right) = \frac{1}{\sigma_{\rho;\alpha\beta}^{CorrSlope}\sqrt{2\pi}} \exp\left\{-\frac{\left(\mu_{\rho;\alpha\beta}^{CorrSlope} - \mu_{Avg,\rho;\alpha\beta}^{CorrSlope}\right)^2}{2\left(\sigma_{\rho;\alpha\beta}^{CorrSlope}\right)^2}\right\} \quad (25.3)$$

The function[6] $\Theta[1-\chi]$ with $\chi = \text{sgn}\left(\mu_{\rho;\alpha\beta}^{CorrSlope}(t)\right) \cdot \rho_{\alpha\beta}(t)$ prevents the correlation from exceeding ± 1 if there is no correlation slope change. The function $\Im[\rho_{\alpha\beta}(t)]$ prevents the correlation from exceeding ± 1 if there is a correlation slope change. We can, for example, specify $\Im$ using the ansatz[7] $\Im[\rho_{\alpha\beta}(t)] = \sqrt{1-\rho_{\alpha\beta}^2(t)}$, which vanishes at ± 1.

We are not interested in small time-scale noise, and we therefore set a possible Micro stochastic noise component $\sigma_{\rho;\alpha\beta}^{Micro\_Corr\_Slope\_Noise} \eta_{\alpha\beta}(t)dt = 0$ in Eqn. (25.1).

The Time Intervals for Macro Correlation Slope Changes
We need to specify the (discrete) times at which the correlation slope changes. In the MM model, the distribution for time-change intervals is taken as (cf. Ch. 50):

$$\wp(\Delta t_{Macro}) = \mathcal{N} \cdot \exp\left\{-\frac{1}{2\left(\sigma_{\rho;\alpha\beta}^{MacroTime}\right)^2} \cdot \ln^2\left[\frac{\Delta t_{Macro} - \tau_{Cutoff}}{\tau_{MacroAvg} - \tau_{Cutoff}}\right]\right\} \quad (25.4)$$

---

[6] **Notation:** This function is one if the argument is positive and zero otherwise. Also, sgn means the sign +- . Finally, note that the argument of $\rho_{\alpha\beta}$ is at time t, not t − dt. Thus we look to see if the result for $\rho_{\alpha\beta}(t)$ would be illegal, and if so we impose the constraint.

[7] **Motivation:** This square-root form of the factor is just phenomenology. Any other function vanishing at the endpoints ρ = +-1 would also be possible.

Here, τ_{Cutoff} is the minimum Macro time interval cutoff [8]. The normalization is $\mathcal{N} = 1/\left(\sigma_{\rho;\alpha\beta}^{MacroTime}\sqrt{2\pi}\right)$.

Example of Quasi-Random Macro Correlation Slope Changes
Here is a picture of the idea. We will specify the details below. The quasi-random correlation slope moves the correlation with a starting value $\rho_{\alpha\beta}(t_0)$ and long-term average $\langle \rho_{\alpha\beta} \rangle_{Time}$ of 0.1.

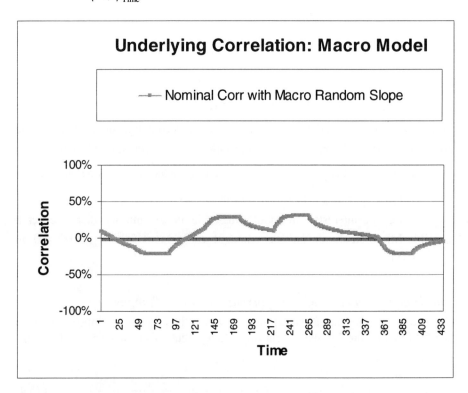

Details of the Numerical Results
Here are the details for the example above. For illustration, we simplify eq. (25.4) using a specific time interval $\tau_{MacroAvg}$. That is, we take

---

[8] **Brownian Limit of the MM Model:** As the cutoff time goes to zero, the Macro slope model reduces to ordinary Brownian motion with the additional factor limiting the correlations inside +- 1.

$\wp(\Delta t_{Macro}) = \delta(\Delta t_{Macro} - \tau_{MacroAvg})$ where δ is the Dirac delta function. Hence, the correlation slope changes in this simplified model take place at regular time intervals, $t^{(k)}_{ChangeSlope} = t^{(k-1)}_{ChangeSlope} + \tau_{MacroAvg}$. The changes of the underlying correlation over time were evaluated in a Monte-Carlo simulation using the simplified quasi-random Macro correlation slope model. The time $\tau_{MacroAvg}$ was taken as two months (44 days) and the average slope $\mu^{CorrSlope}_{Avg,\rho;\alpha\beta}$ was taken as zero. As before, the average correlation over time was taken as $\langle \rho_{\alpha\beta} \rangle_{Time} = 0.1$. The preceeding graph gave the results for one MC realization.

Correlation Dependence on Volatility

A complication that exists in modeling correlations is that correlations can depend on volatility. Such a dependency is natural in stressed markets. When markets become stressed, volatilities increase, jumps occur, and correlations become more pronounced away from zero as investors undergo various forms of panic behavior and a flight to quality ensues.

The behavior of correlations under this circumstance is difficult to extract. Consider a jump in a variable x_α increasing its volatility. The magnitude and sign of the correlation change $\delta \rho_{\alpha\beta} \propto d_t x_\alpha^{Jump}(t) \cdot d_t x_\beta(t)$ depends on the amount of change in the other variable, i.e. $d_t x_\beta(t)$.

There is also a feedback on the total volatility of a variable depending on the correlation from the correlation instabilities.

Numerical Example of Correlation Dependence on Volatility
We take two variables, use the MM model for the correlation ρ described above, and write a bivariate model for the time changes $d_t x_1$, $d_t x_2$ excluding jumps:

$$d_t x_1 = \sigma_1 \eta_1 dt$$
$$d_t x_2 = \sigma_2 \left[\rho \eta_1 + \sqrt{1-\rho^2} \cdot \eta_2 \right] dt \qquad (25.5)$$

The notation is $\eta(t) = dz(t)/dt$. We also include the 10 SD jump for $d_t x_2$ over one day in the middle of the time series.

The results are pictured as follows:

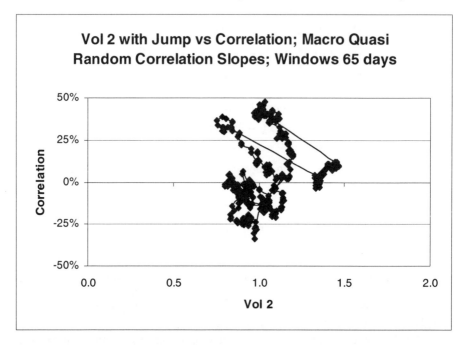

This is a scatter plot of the windowed correlation vs. the total volatility "Vol 2" of $d_t x_2$ (including the jump and volatility effects on the changes in correlation slope).

The effect of the jump is clearly visible in this scatter plot. Non-trivial dependences between the correlation and the total volatility are seen, with a number of string-like continuous regions.

To take this sort of effect into account in risk management, correlation stress analysis is desirable, since no single correlation value is likely to represent a stable parameter.

Windowed Correlations with a Macro Quasi-Random Slope

We can ask the question of how well windowed correlation measurements reproduce the given average correlation $\langle \rho_{\alpha\beta} \rangle_{\text{Time}} = 0$ in the presence of the Macro quasi-random correlation slopes. The comparison in the figure below is between the series labeled "Windowed Corr, No jump" and "Nominal Corr with Macro Random Slope" (the latter the same as above). The windowed correlations qualitatively follow the underlying correlation quasi-random behavior, with random fluctuations.

We also exhibit the windowed correlation with the big 10 sd jump as above in $d_t x_2$ to give an idea of windowing correlation uncertainties vs. jump-volatility-induced correlation uncertainties. The jump in the underlying produces a clear and large jump in the correlation. The MC run exhibited was chosen to illustrate this point. Other Monte Carlo runs did not exhibit such a large jump effect.

Here are the illustrative results from the Monte Carlo run:

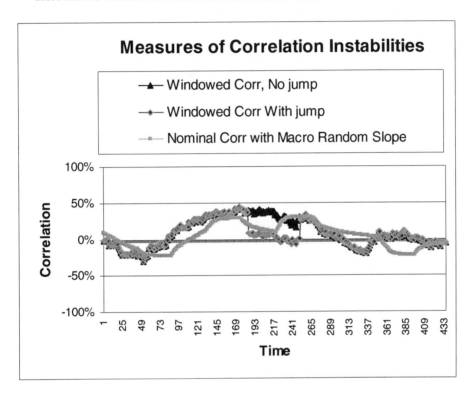

Should We Forget About Correlations?

Some people think that because there are complications in extracting correlations, we should not think about correlations at all.

The problem really is that correlations are the natural language. If we want to think about relations between movements of gold and Libor, we need to consider the gold-Libor correlation. Therefore, we cannot forget about correlations.

Concurrently, we need to worry about correlation uncertainties. Because we use models for pricing and risk that contain volatilities and correlations, the uncertainties in the correlations must be considered.

Can Stochastic Volatility Explain Correlation Instabilities?

Some people think that stochastic volatility can explain the instabilities in correlations. Indeed, we have been exhibiting correlation changes induced by jumps.

However, we do not believe that stochastic volatility can explain, e.g., a flight to quality, when many correlations tend to increase in magnitude. If the correlation uncertainties were simply due to volatility uncertainties, there would be no reason for correlation changes to occur in a correlated fashion.

In any case, because most models do not use stochastic volatility, we are again back to the position of using deterministic volatilities and correlations. Whatever the cause of the correlation uncertainties, we need to assess the correlation risk in the framework of the models we use. Again, we need to stress the correlations to assess this risk.

Implied, Current, and Historical Correlations for Baskets

Although this chapter is primarily concerned with direct models for correlations between market variables, we consider briefly the connection with implied correlations. Implied correlations are found by the inverse procedure of utilizing market prices of correlation-dependent securities to back out the correlations. The idea is the same as for implied volatilities. While consistency with the market is desirable, the difficulty is that correlation-dependent securities are often illiquid. Hence, market prices may difficult to obtain, leading to large uncertainties in the implied correlations[9].

Market price information, even if available, may be insufficient to determine the whole matrix of implied correlations. For example, consider a basket defined as $B(t) = \sum_\alpha w_\alpha S_\alpha(t)$ of securities (stocks, interest rates, etc) all assumed lognormal. The basket volatility σ_B given a reference time t_0 is approximately

$$\sigma_B^2 dt \equiv \left\langle [dB(t)/B(t)]^2 \right\rangle_c \approx \sum_{\alpha,\beta=1}^{N} (\sigma_\alpha w_\alpha S_{\alpha 0}) \rho_{\alpha\beta} (\sigma_\beta w_\beta S_{\beta 0}) dt / B_0^2 \qquad (25.6)$$

Here, $B_0 = B(t_0)$, $S_{\alpha 0} = S_\alpha(t_0)$, $S_{\beta 0} = S_\beta(t_0)$. Now suppose we approximate this using an average correlation, $\rho_{\alpha\beta} = \rho_{avg}$, $\alpha \neq \beta$. For large N and assuming that $\rho_{avg} > 1/N$, the off-diagonal terms dominate and we approximately get the basket vol as proportional to $\sqrt{\rho_{avg}}$, viz

---

[9] **One-sided Markets:** For example, the Street may only be selling the product to customers without any buying through secondary trading. In that case, there is no market-determined "mid point". Estimates of the other side of the market can break down under stressed market conditions or "fire sales".

$$\sigma_B \approx \sqrt{\rho_{avg}} \cdot \sum_{\alpha=1}^{N} \sigma_\alpha w_\alpha S_{\alpha 0} \bigg/ B_0 \qquad (25.7)$$

Given the basket volatility and the independent component volatilities, we can approximately extract the average correlation.

As time progresses, the correlation risk depends on the current correlation. Historical correlation uncertainties can be used to estimate the correlation hedging risk. Conversely, this estimated hedging risk (or part of it) could be used in the pricing of a new deal, thus specifying the initial implied correlation.

26. Plain-Vanilla VAR (Tech. Index 4/10)

In this chapter, we discuss begin a discussion of VAR, an acronym for Value at Risk[1]. The "Plain-Vanilla VAR" *(PV-VAR)* along with its incarnation as a quadratic form *(QPV-VAR)* is a standard risk measure that we discuss first. PV-VAR is rather blunt and unrefined[1]. In the next chapter, we will discuss refinements. The first refinement stage defines the "Improved Plain-Vanilla VAR" *(IPV-VAR)*. Further refinements produce the "Enhanced/Stressed VAR" *(ES-VAR)* [2]. We give a few previews in the footnotes, which also contain other important points.

The various components of the VAR, called CVARs, will also be discussed. The CVARs are useful because they give a consistent picture of the composition of the risk. We show that the CVARs have uncertainties (i.e. there is a CVAR volatility) and we show how to calculate these uncertainties[3]. The CVAR volatility is useful because it shows the uncertainty in different possible compositions for a given total risk.

Plain-Vanilla VAR (PV-VAR)

We first describe "Plain-Vanilla VAR" *(PV-VAR)*. Basically, PV-VAR is a one-step simulator in time that measures risks at a given confidence level of a portfolio[4] C using a variety of simplifying assumptions. The portfolio C is a function of underlying variables $\{x_\alpha\}$, $\alpha = 1...n$. The variable x_α can be a physical variable (e.g. interest rate, stock price, etc.), or x_α can be a function,

---

[1] **Dignity for Plain-Vanilla VAR:** We by no means imply to denigrate the enormous and justifiable effort that may be needed to implement this risk measure on a firm-wide basis.

[2] **Notation:** The names "Plain Vanilla VAR", "Improved Plain Vanilla VAR", and "Enhanced/Stressed VAR" are intended to convey a sense of relative sophistication, and are my own jargon.

[3] **History:** The theory of CVAR volatility described here and in Ch. 27-29 was discovered and developed by me in 1999-2002.

[4] **Which Portfolio?** The portfolio can be at the level of a product, desk, business-unit, or central firm-wide Risk Management.

commonly $x_\alpha = \ln y_\alpha$, where y_α is the physical variable[5]. The time difference is written $d_t x_\alpha(t) = x_\alpha(t+dt) - x_\alpha(t)$, where dt is a convenient time unit[6]. If $x_\alpha = \ln y_\alpha$, the time difference $d_t x_\alpha$ defines the return $d_t y_\alpha / y_\alpha$ over time dt. The probability distribution function of the time differences is modeled; the usual assumption is multivariate Gaussian. The volatilities $\{\sigma_\alpha\}$ and correlations $\{\rho_{\alpha\beta}\}$ occurring therein are obtained from (hopefully clean) data[7].

A risk "exposure"[8] denoted as $^\$\mathcal{E}_\alpha$ describes a change in value[9] of the portfolio $^\$C$ for a given change in the underlying x_α. A few examples of exposures include bond DV01 and spread risks, the Greeks for options, mortgage prepayment risk, etc[10]. The risk exposures must be obtained from models or other

---

[5] **Principal Component VAR:** Another possibility is to take the variables x_α, or some of them, as principal components (PC). One example is to take the leading PCs for the yield curve, or to add a few PCs orthogonal to the parallel yield-curve shift that is already present as DV01. This PC-VAR idea is in my notes from 1999. Unfortunately, there is not enough space in the book to include it, but the reader can fill in the blanks.

[6] **Time Interval dt:** Popular choices are 1 day, 10 (business) days, or 1 quarter. For the Gaussian assumption of the $d_t x$ distribution, the sqrt(dt) scaling translates between these choices.

[7] **Getting the Underlying Data:** Obtaining appropriate and clean data can be a huge issue. One of the two irreverent but big theorems in this book is "Data = Black Hole".

[8] **Disclaimer:** No information regarding numerical values of exposures of any firm is in this book.

[9] **USD:** By convention here, the US dollar is used here as the reporting currency. The VAR calculated in USD would include FX risk for assets held in non-USD currencies.

[10] **Idiosyncratic Risks:** Some risks that are called idiosyncratic in this book cannot or are not obtained using models, exposures, and underlying data. These risks can be large. Idiosyncratic risks must be estimated using judgment if they are to be included. They are by definition left out of the plain-vanilla VAR. Examples are discussed in the next chapter, along with a way of including idiosyncratic risks in a more refined VAR.

sources[11]. The first approximation is the "linear" assumption, $^\$\mathscr{E}_\alpha = \partial^\$ C/\partial x_\alpha$, which ignores the second derivative, or convexity[12].

Because the portfolio C changes all the time as securities are bought and sold, a time t_{Frozen} is chosen when a "snapshot" of the portfolio is taken and the calculations are performed on this "frozen" portfolio[13].

The VAR calculation is performed using a Monte-Carlo (MC) simulation, or in the simplest case, a quadratic approximation[14]. The calculation is done to a given confidence level (CL) of loss. We call k_{CL} the number of standard deviations (sd) corresponding to the given CL. Different assumptions are made corresponding to the application. For standard reporting, a 99% CL is often used (so k_{CL} = 2.33). For Economic Capital, which we discuss in Ch. 39, a more stringent CL is used (e.g. 99.97% for an AA firm, with k_{CL} = 3.43).

Calculations of VAR[15] can be done frequently or infrequently, depending on the need and other considerations [16].

---

[11] **Getting the Exposures:** Getting the exposures for a large institution can be a huge issue. If models from the trading desks are used, feeds for the exposures must be provided to firm-wide Risk Management. If independent models are desired by Risk Management, these models must first be constructed, which is an even bigger issue.

[12] **Convexity:** The contribution of convexity depends on the situation. Bonds do not have much convexity if they do not contain strong call features. Therefore, for a bond desk, ignoring convexity may not matter much. The same statement holds for plain-vanilla swaps. On the other hand, ignoring convexity is a terrible approximation for an options desk. We discuss ways of including convexity in the next chapter.

[13] **What Time for the Portfolio?** Usually the frozen time is specified by management; e.g. end of quarter. If the portfolio at the frozen time is not representative, the risk calculated will not be representative either. We can also envision portfolio changes through exposure scenarios in the future. We describe this idea in the next chapter.

[14] **Historical VAR Simulation:** An historical VAR simulation can also be run that merely chooses the changes in the underlying variables randomly from a historical set of such changes. These changes are applied to the frozen portfolio. The historical simulation has the advantage that no assumptions about correlations have to be made. It has the disadvantage that the number of states and their composition are limited to the available data. We will not discuss the historical VAR simulation further, although it is widely used. The historical VAR simulation can be used to obtain additional information. Since no risk measure is complete, this is actually a good idea.

[15] **Acknowledgement:** I thank D. Humphreys for many discussions and calculations of VAR during the period that my Quantitative Analysis group was responsible for producing the VAR at Smith Barney.

[16] **Frequency and Timeliness of VAR Calculations:** Variables include the availability of data and exposures, the calculational engine efficiency, the amount of budget and manpower dedicated, etc. The timeliness depends for what purpose the calculation is

Sometimes scenarios are used instead of or in addition to performing such calculations[17].

Jamshidian and Zhu have formulated a simulation methodology involving a discretization of multivariate distributions, and applied it to calculating VAR[ii]. Further discussion of this sort of approach is in Ch. 44.

VAR, as commonly defined, is supposed to reflect risk measured as the difference of the total loss and the usual or expected loss. VAR thus is supposed to represent an unusual, rather rapid, fluctuation in the markets leading to a bad loss. The expected risk is assumed to be constant or slowly varying in time[18].

Quadratic Plain-Vanilla VAR and CVARs

This simplest of all versions of VAR uses a quadratic form $^\$ VAR^{Quad}$ made up of the exposures, volatilities and correlations to get risk at an overall confidence level (CL) corresponding to k_{CL} standard deviations (SD) for time interval dt. This ignores convexity and uses the other assumptions mentioned above. For a quick review, we write $d_t x_\alpha = \sigma_\alpha dz_\alpha$ with volatility[19] σ_α. This ignores the drift term that does not (by definition) contribute to the VAR. Here the correlated

---

used. If the VAR is used as an indicative risk benchmark for reporting purposes, timeliness is not as important as if it is used as a qualitative benchmark for active risk management. For active risk management, stale positions and data are about as useful as an old newspaper.

[17] **Scenario-based Risk:** Sometimes scenario analysis to obtain risk is used for a variable $d_t x_\alpha$. This means that one value (or a small set of values) is assumed for $d_t x_\alpha$. The loss of the portfolio is then calculated using the scenario. This is a degenerate form of Monte-Carlo simulation where the randomness is replaced by the certainty of the scenario. The scenario for $d_t x_\alpha$ is often taken as the worst-case move over time dt, or the move at some high confidence level CL. In some cases, no assumption is made for $d_t x_\alpha$, and a scenario is assumed directly for the portfolio loss. Scenario-based risk cannot be consistently evaluated at a given overall CL. However, scenarios are necessary if data do not exist for $d_t x_\alpha$ or for exposures, as we discuss later for idiosyncratic risk.

[18] **Expected Losses, Pricing, Reserves and Economics:** Expected-loss risk is not included in VAR. Expected losses are supposed to be included in pricing, expected-loss reserves, etc. If appropriate reserves do not exist or do not reasonably reflect expected losses, a mistake in the firm's risk assessment will be made. One problem is that changing economic conditions can and probably will change the expected losses. In addition, the average loss in the MC simulation used to get the VAR may or may not reasonably reflect the expected loss in reality.

[19] **Volatility Specification for VAR:** This volatility can be taken as the ordinary volatility or as a fat-tail vol, as described in Ch. 21, and as elaborated in the next chapter.

Plain-Vanilla VAR

Gaussian measures satisfy[20] $\langle dz_\alpha dz_\beta \rangle = \rho_{\alpha\beta} dt$. The expectation of the product of the time changes is $\langle d_t x_\alpha \cdot d_t x_\beta \rangle = \sigma_\alpha \rho_{\alpha\beta} \sigma_\beta dt$.

Now we set the change in the total portfolio value to its linear approximation $d^\$ C \approx \sum_{\alpha=1}^{n} (\partial^\$ C / \partial x_\alpha) d_t x_\alpha$. This has variance $\langle [d^\$ C]^2 \rangle \approx [^\$ VAR^{Quad}]^2 dt$, where $^\$ VAR^{Quad}$ is the total volatility in the familiar quadratic-form expression[21]

$$^\$ VAR^{Quad} = \left[\sum_{\alpha,\beta=1}^{n} {}^\$\mathcal{E}_\alpha \sigma_\alpha \rho_{\alpha\beta} {}^\$\mathcal{E}_\beta \sigma_\beta \right]^{1/2} \quad (26.1)$$

The QPV-VAR result, defined as k_{CL} standard deviations of this total portfolio volatility over time dt, is

$$^\$ VAR^{QPV} = k_{CL} \sqrt{dt} \cdot {}^\$ VAR^{Quad} \quad (26.2)$$

Squaring the $^\$ VAR^{Quad}$ equation and isolating the sum $\sum_{\alpha=1}^{n}$ produces

$$^\$ VAR^{Quad} = \sum_{\alpha=1}^{n} {}^\$ CVAR_\alpha^{Quad} \quad (26.3)$$

Here $^\$ CVAR_\alpha^{Quad}$, the quadratic-plain-vanilla "component VAR", QPV-CVAR, is defined as

$$^\$ CVAR_\alpha^{Quad} = \frac{{}^\$\mathcal{E}_\alpha \sigma_\alpha}{{}^\$ VAR^{Quad}} \sum_{\beta=1}^{n} \rho_{\alpha\beta} {}^\$\mathcal{E}_\beta \sigma_\beta \quad (26.4)$$

Unlike the positive $^\$ VAR^{Quad}$, a CVAR can be either positive or negative, depending on the signs of the exposures and correlations. This point will come up again when we discuss the allocation of risk.

---

[20] **Expectation Values of Products of Gaussian Measures:** The textbook result holds only in the limit as we generate an infinite number of Gaussian random numbers. Finite Monte-Carlo simulations produce fluctuations about this theoretical limit.

[21] **Homework:** It is not hard to get the result for quadratic-form VAR.

Monte-Carlo VAR

We can relate the above QPV-VAR to a Monte Carlo simulation at the overall portfolio risk CL with k_{CL} standard deviations [22]. We use independent Gaussian measures by taking the square root of the correlation matrix[23] as $dz_\alpha = \sum_{\gamma=1}^{n} \sqrt{\rho}_{\alpha\gamma} dz_\gamma^{Indep}$. Here $\{dz_\gamma^{Indep}\}$ are independent Gaussian measures satisfying $\langle dz_\gamma^{Indep} dz_{\gamma'}^{Indep} \rangle = \delta_{\gamma\gamma'} dt$.

The States and Portfolio Changes of the MC Simulation

From a MC simulation, we obtain N_{MC} different states $\{\aleph\}$ for the movements of the underlying variables that are generated by different values of the random numbers. For example, a given state $\aleph$ might have a move up of +12 bp for 5-year Libor swaps, a move down of -24 points for the S&P 500, etc.

We look at the histogram of portfolio changes $\{d^\$ C^{(\aleph)}\}$ for the different states, viz $d^\$ C^{(\aleph)} = \sum_{\alpha=1}^{n} {}^\$ \mathcal{E}_\alpha (d_t x_\alpha)^{(\aleph)} = \sum_{\alpha,\gamma=1}^{n} {}^\$ \mathcal{E}_\alpha \sigma_\alpha \sqrt{\rho}_{\alpha\gamma} dz_\gamma^{Indep(\aleph)}$. Next, we pick out the value of $d^\$ C_{CL}$ at the specified CL corresponding to the particular state $\aleph(k_{CL})$ producing the total loss at that CL. For example, for the 99% CL and N_{MC} = 10,000 states, we pick out the 9,900th worst loss. This reproduces the above QPV-VAR result under the linear assumption. Up to standard MC noise of $O(N_{MC})^{-1/2}$, we have

---

[22] **Confidence Levels for the Total Portfolio VAR and For Individual Variables:** Note that k_{CL} is the number of SD for the overall CL of the total portfolio risk. This overall CL can be attained if each volatility σ_α is replaced by $k_{CL} \sigma_\alpha$. However, in a Monte Carlo calculation, the number of sd $k_{CL;\alpha}$ for each variable $d_t x_\alpha$ is different. Moreover, this $k_{CL;\alpha}$ varies with the MC run, as explained in the text below.

[23] **Correlation Matrix Square Root, the Cholesky Decomposition, the SVD, and Block Diagonalization:** The story of taking the square root of the correlation matrix is long and grimy. Because the correlation matrix taken from the data contains data noise, it is rarely (read never) positive definite. Therefore, a real square-root matrix does not exist and the Cholesky decomposition breaks down. Instead, a procedure using the Singular Value Decomposition and optimal least-squares fitting can be used, as described in Ch. 24. For large matrices involving many variables (e.g. thousands), optimization can only be performed crudely. Further, because of machine memory and other limitations, assuming a block-diagonal form may be required.

Plain-Vanilla VAR 347

$$d^\$ C_{CL} \approx k_{CL}\sqrt{dt} \cdot {}^\$ VAR^{Quad} \qquad (26.5)$$

Hence, in the limit of an infinite MC simulation, in the linear approximation, we recover the QPV-VAR result, $\lim_{N_{MC} \to \infty} d^\$ C_{CL} = {}^\$ VAR^{QPV}$. While it may seem that this is like using a hammer to swat a fly, MC simulation is the only method available when we start refining the VAR assumptions.

Backtesting

A *backtesting* procedure is often used to provide a test of the daily VAR calculations, i.e. $dt = 1$ day. The historical simulation[14] on the frozen portfolio[24] is used for comparison. The changes $\{d^\$ C(t_\ell)\}$ due to historical changes $\{d_t x_\alpha(t_\ell)\}$ over the backtesting period, e.g. 1 year (250 days) are rank ordered along with the calculated $d^\$ C_{CL}$. If the CL is 99%, then the backtesting "works" and is consistent provided that there are no more than two or three "exceptions" such that $|d^\$ C(t_\ell)| > |d^\$ C_{CL}|$.

Monte-Carlo CVARs and the CVAR Volatility

We now discuss the CVARs generated by MC simulation. The MC state $\aleph(k_{CL})$ that corresponds to the overall CL of k_{CL} standard deviations in the total VAR contains values of all the variables $\{(d_t x_\alpha)^{\aleph(k_{CL})}\}$. In the state $\aleph(k_{CL})$, we have by definition the result for the change in the portfolio at this CL, given the exposures $\{{}^\$\mathcal{E}_\alpha\}$ and the changes $\{(d_t x_\alpha)^{\aleph(k_{CL})}\}$, viz

$$d^\$ C_{CL} = \sum_{\alpha=1}^{n} {}^\$\mathcal{E}_\alpha (d_t x_\alpha)^{\aleph(k_{CL})} \qquad (26.6)$$

---

[24] **Backtesting Using Daily Portfolios?** The use of daily P&L from the daily portfolios as they occurred during the backtesting period is easier than calculating the historical simulation of the frozen portfolio. All that has to be done is to compare the VAR to the P&L directly. This procedure is only valid if the transactional portfolio-changing effects due to buying and selling, changes in reserves, etc. are small.

Now the VAR is, by construction, the sum of the CVARs. Hence, by identifying terms, the CVARs in the state $\aleph(k_{CL})$ are just the terms in the sum eq. (26.6) for $d^\$ C_{CL}$. That is,

$$\left({}^\$ CVAR_\alpha\right)^{\aleph(k_{CL})} = {}^\$\mathcal{E}_\alpha\left(d_t x_\alpha\right)^{\aleph(k_{CL})} \tag{26.7}$$

Non-Uniqueness of the CVARs
However, *a given CVAR generated by MC simulation is not unique, even with an infinite number of states* $\{\aleph\}$. This is because a given fixed value of a sum can have different amounts of the components making up the sum no matter how accurately the sum is generated [25].

To see why the CVARs are not unique, we define Monte-Carlo runs labeled with an index $\mathcal{R}$. To emphasize that the CVAR uncertainty has nothing to do with MC noise, we specify that each run $\mathcal{R}$ is comprised of an *infinite* number of states $\{\aleph_\mathcal{R}\}$. Exactly the same value $d^\$ C_{CL}$ at the given k_{CL} can be generated, but with *different* states $\aleph_\mathcal{R}(k_{CL})$, for different runs. A *given* variable $d_t x_\alpha$ will have *different* values $(d_t x_\alpha)^{\aleph_\mathcal{R}(k_{CL})}$ in *different* runs $\mathcal{R}$, all for a given total $d^\$ C_{CL}$. A given term, i.e. a given CVAR, in the sum $\sum_{\alpha=1}^{n} {}^\$\mathcal{E}_\alpha\left(d_t x_\alpha\right)^{\aleph_\mathcal{R}(k_{CL})}$ is different for different runs.

Summary for MC Simulation so far
Summarizing, take a MC run $\mathcal{R}$ and the state $\aleph_\mathcal{R}(k_{CL})$ at which $d^\$ C_{CL}$ has the specified CL corresponding to k_{CL} SD. Then the CVAR for variable $d_t x_\alpha$ as generated in that particular MC run, called $\left({}^\$ CVAR_\alpha\right)^{\aleph_\mathcal{R}(k_{CL})}$, is given by $\left({}^\$ CVAR_\alpha\right)^{\aleph_\mathcal{R}(k_{CL})} = {}^\$\mathcal{E}_\alpha\left(d_t x_\alpha\right)^{\aleph_\mathcal{R}(k_{CL})}$. For the MC run $\mathcal{R}$, the change in portfolio value $d^\$ C_{CL}$ at the specified CL is given by

---

[25] **Grocery Bag Components Analogy for CVAR Uncertainty:** A useful and visual analogy for the CVAR variability is that a given and exact total amount of money like $80.00 (VAR) can a-priori be spent buying different amounts (variability) of different things (CVARs). A drawback of this analogy is that negative CVARs are hard to picture.

Plain-Vanilla VAR

$$d^{\$}C_{CL} = \sum_{\alpha=1}^{n} {}^{\$}\mathcal{E}_{\alpha}(d_t x_\alpha)^{\aleph_{\mathcal{R}}(k_{CL})} = \sum_{\alpha=1}^{n} \left({}^{\$}CVAR_\alpha\right)^{\aleph_{\mathcal{R}}(k_{CL})} \tag{26.8}$$

Connection of the Monte-Carlo and Quadratic CVARs

The connection of the Monte-Carlo CVAR, $\left({}^{\$}CVAR_\alpha\right)^{\aleph_{\mathcal{R}}(k_{CL})}$, with the quadratic CVAR defined above is accomplished by averaging over runs and inserting the factor $k_{CL}\sqrt{dt}$, viz

$$k_{CL}\sqrt{dt} \cdot {}^{\$}CVAR_\alpha^{Quad} = \underset{\mathcal{R}}{Avg}\left[\left({}^{\$}CVAR_\alpha\right)^{\aleph_{\mathcal{R}}(k_{CL})}\right] \tag{26.9}$$

The CVAR Volatility

Because different runs produce different amounts of a given component risk, we want to define the volatility for a given ${}^{\$}CVAR_\alpha$. The CVAR-volatility $\sigma\left({}^{\$}CVAR_\alpha\right)$ can be defined by the standard deviation of $\left({}^{\$}CVAR_\alpha\right)^{\aleph_{\mathcal{R}}(k_{CL})}$ over different MC runs, viz

$$\left[\sigma\left({}^{\$}CVAR_\alpha\right)\right]^2 = \sum_{\mathcal{R}}\left[\left({}^{\$}CVAR_\alpha\right)^{\aleph_{\mathcal{R}}(k_{CL})}\right]^2 - \left\{\sum_{\mathcal{R}}\left[\left({}^{\$}CVAR_\alpha\right)^{\aleph_{\mathcal{R}}(k_{CL})}\right]\right\}^2 \tag{26.10}$$

We stress again that the CVAR volatility implying the CVAR uncertainty has *nothing whatever* to do with uncertainty due to finite-statistics Monte-Carlo noise. MC noise exists independently of (and in addition to) the uncertainty we have been discussing.

In Ch. 28, we give a closed form expression for the CVAR volatility $\sigma\left({}^{\$}CVAR_\alpha\right)$ using continuous multivariate statistics, consistent with an infinite number of states.

Obtaining CVAR Volatilities in Practice; the "Ergodic Trick"

In order to get the CVAR volatility numerically, we use an "ergodic trick". Rather than producing many MC runs (which would be prohibitive numerically), we note that for a Gaussian model, we can connect the total portfolio losses at two different confidence levels CL and CL' by

$$d^\$C_{CL} = \frac{k_{CL}}{k_{CL'}} d^\$C_{CL'} \tag{26.11}$$

Thus, for a given CL generated at state $\aleph(k_{CL})$ we can take some neighboring states $\{\aleph(k_{CL'})\}$ at different CL' and scale up each $(d_t x_\alpha)^{\aleph(k_{CL'})}$ by $k_{CL}/k_{CL'}$. For example to scale up $(d_t x_\alpha)^{\aleph(k_{CL'})}$ at $CL' = 98\%$ to $CL = 99\%$ we would just use $k_{CL}/k_{CL'} = 2.33/2.05$. We then calculate the CVARs for these scaled neighboring states[26]. The CVAR volatility is defined as the standard deviation of these CVARs. See the Ch. 38 for more mathematical details.

Confidence Levels for Individual Variables in VAR

Consider the confidence level $CL_\alpha(k_{CL})$ for an individual variable, i.e. for the change $(d_t x_\alpha)^{\aleph(k_{CL})}$ in the state $\aleph(k_{CL})$. This $CL_\alpha(k_{CL})$ is not the same as the confidence level k_{CL} for the total VAR. For example, in the 99% CL state for the total VAR, it can happen that a particular spread moves up by 1 SD, representing the 84% CL for that spread.

Generally, given each variable change $(d_t x_\alpha)^{\aleph(k_{CL})}$ in the state $\aleph(k_{CL})$ producing the VAR at k_{CL}, we can get the number $k_\alpha^{(k_{CL})}$ of SD using $(d_t x_\alpha)^{\aleph(k_{CL})} = k_\alpha^{(k_{CL})} \sigma_\alpha$. Then the confidence level $CL_\alpha(k_{CL})$ for $(d_t x_\alpha)^{\aleph(k_{CL})}$ can be found by inverting the normal distribution at the value $k_\alpha^{(k_{CL})}$.

Individual Variables generally do NOT contribute at High CL
It turns out in practice that even for very high CL for VAR like 99.97% where $k_{CL} = 3.43$, individual $CL_\alpha(k_{CL})$ values for diversified portfolios are usually at or below the 99% CL, corresponding to individual $k_\alpha^{(k_{CL})} \leq 2.33$. This is an

---

[26] **Neighboring States and Convexity:** Convexity corrections will be roughly equal for neighboring states, provided the convexity is not too large. In practice the ergodic trick works reasonably well in the presence of some convexity.

important point, and it means that the "worst-case" moves for individual variables do *not* contribute to the VAR even at a high overall CL[27].

References

[i] **VAR Textbook**
Jorion, P., *Value at Risk*. McGraw-Hill, 2001.

[ii] **Scenario Simulation and VAR**
Jamshidian, F. and Zhu, Y., *Scenario Simulation: Theory and methodology*. Finance and Stochastics, 1, 1997. Pp. 43-67.

---

[27] **Exception – Lack of Diversification:** The exception is when a single exposure dominates the calculation. In that case, the CL for the variable corresponding to the dominant single exposure will tend to be around the CL for the total VAR. This is because the total risk just degenerates into the single exposure risk. We will see the same phenomenon when we discuss issuer credit risk if there is a dominant exposure to a single issuer in the portfolio.

27. Improved/Enhanced/Stressed VAR (Tech. Index 5/10)

In this chapter, we discuss various stages of refinements of the plain-vanilla VAR discussed in Ch. 26. Increasingly realistic aspects will be included, with the final aim to obtain a risk measure that is more useful in active risk management. The first set of improvements give what is termed in this book "Improved Plain Vanilla VAR" *(IPV-VAR)*. We then define a series of further improvements to produce "Stressed VAR" and finally "Enhanced/Stressed VAR" *(ES-VAR)*[1]. We close with some miscellaneous topics including subadditivity issues, and also an integrated form of VAR.

Improved Plain-Vanilla VAR *(IPV-VAR)*

The following table summarizes the next stage, including refinements past the PV-VAR to obtain the *IPV-VAR*, or Improved Plain-Vanilla VAR. These refinements are often included in current implementations of VAR.

Quantity Compared	Plain Vanilla VAR	Improved PV VAR
Convexity	Not Included	*Included via Grid*
Time Scale dt	Uniform (10 days)	*Variable (liquidity)*
Cutoffs for $d_t x_\alpha$	Not included	*Included (Judgment)*
Time Period: x_α Data	Recent (1 to 3 yrs)	*Recent or Variable*

We describe these improvements in the IPV-VAR one at a time.

Convexity and the Grid

Convexity exists in all option products, and even to some extent in discount factors. Convexity effects can be included in a VAR calculation if a grid of exposures is available. A given variable x_α is changed by discrete amounts to

---

[1] **History:** I developed some of these improvements to VAR between 1999-2002. Most of Improved Plain-Vanilla VAR is now routinely included in VAR analyses. The various Enhanced and Stressed refinements are less common.

values on a grid, $\left\{x_\alpha^{(\text{Grid})}\right\}$, for example $x_\alpha^{(\text{Grid})} = x_\alpha \pm k_\alpha^{(\text{Grid})} \sigma_\alpha$, with various values for $k_\alpha^{(\text{Grid})}$. Before starting the VAR calculation, the portfolio is revalued at each grid point.

The idea is that, for a given $d_t x_\alpha$ arising from a given throw of the dice, we pick out the appropriate exposure on the grid.

Explicitly, consider a MC simulator run producing the state $\aleph(k_{CL})$ at the CL with k_{CL} SD for the total VAR. Now consider, for a given variable x_α, the number $k_\alpha^{(k_{CL})}$ of SD for the changes $d_t x_\alpha$ in the state $\aleph(k_{CL})$. Then we can use the corresponding exposure from the grid at $k_\alpha^{(\text{Grid})} \approx k_\alpha^{(k_{CL})}$. Procedurally, a lookup interpolation table for the grid can be established.

Another and simpler possibility is just to choose the exposure at a conservative level of, say, two SD.

Example for the Grid

For example, suppose that we have the $DV01$ of a portfolio at five grid points: the current interest rate r_0, $r_0 \pm 50bp$, and $r_0 \pm 100bp$, obtained by direct revaluation. Now suppose that we happen to get a change $d_t r_0^{\aleph(k_{CL})} \approx 50bp$ in the state $\aleph(k_{CL})$ corresponding to the 99% CL for the total VAR. Then we would use the $DV01$ at $r_0 + 50bp$.

This procedure clearly includes convexity effects, since the $DV01$ chosen depends on the rate level, here $r_0 + 50bp$.

Caveats

The procedure involving one-dimensional grids naturally does not include cross-convexity terms between different variables.

Considerable effort generally has to be expended in order to generate the grid in the first place. The lucky situation will be if the front-office risk systems already generate the grid. Otherwise, things become murky[2].

---

[2] **Murky:** This is a technical term, possibly implying a long interaction to get (or not get) the grid. In addition, note that a real grid is not the same as a useless grid with the change in portfolio value just scaled up arithmetically, thus ignoring the convexity and missing the whole point. You might want to make that clear in the discussions.

Time Scale dt, Liquidity, and Product Types

Besides the daily $dt = 1$ day, common assumptions are $dt = 10$ (business) days or $dt = 1$ quarter. For the Gaussian assumptions behind the VAR, the translation between the various assumptions is just made using $\sqrt{dt}$ scaling. However, real losses in real firms with real traders and real strategies may have nothing to do with an assumed $\sqrt{dt}$ scaling of a frozen portfolio.

It is a good idea to step back to see what the parameter dt is supposed to represent. We can think of dt in two distinct ways:

- *Assumption #1*. The time dt is a *market perturbation* time over which an "unusual" large disruption occurs, after which the market returns to a "normal" state.
- *Assumption #2*. The time dt a *liquidity* time, i.e. the time it takes traders to sell or hedge the risky exposures in a generally turbulent market.

Assumption #1: dt = market perturbation time. The problem with this assumption is that bad perturbations across markets can occur for times greater than, e.g. 10 days. Markets that become roiled can stay turbulent and volatile for a long time. The relaxation time is rather long to arrive at a calm state from a market that has made a phase transition into a panic-driven state, where clever trading strategies collapse, mean reversion becomes a fantasy, and investors jump en masse onto the flight-to-quality bandwagon.

Assumption #2: dt = liquidity time. If dt represents the time it takes to sell or hedge the risk, there is no single number that corresponds to such action. Government bonds, short-dated plain-vanilla swaps, FX forwards and other such liquid instruments have a short liquidity dt. On the other hand, illiquid securities with limited transaction volume have a long liquidity time dt even in a calm market, and arguably an even longer dt in a turbulent market.

Therefore, denoting product type by the label $\Im$, we have to include the $\Im$ dependence as $dt^{(\Im)}$. Note that the same underlying $d_t x_\alpha$ may contribute to several different product types with different liquidities. For example, Libor can contribute two-year IMM swaps or to index-amortizing swaps, with different liquidities. Hence, to be accurate we should define an exposure corresponding to a given product type, $^\$\mathcal{E}_\alpha^{(\Im)}$, with $^\$\mathcal{E}_\alpha = \sum_\Im {}^\$\mathcal{E}_\alpha^{(\Im)}$ being the total exposure for the underlying $d_t x_\alpha$.

We can then include liquidity square-root-time effects by defining effective exposures by product and underlying[3] as $^\$\mathcal{E}_\alpha^{\mathit{Eff}(\mathfrak{I})} = {}^\$\mathcal{E}_\alpha^{(\mathfrak{I})} \sqrt{dt^{(\mathfrak{I})}}$, while dropping the now irrelevant overall factor $\sqrt{dt}$.

The main problem with this assumption is that there is no guarantee that the risk will actually be hedged or eliminated in the time $dt^{(\mathfrak{I})}$. As we discuss below, there are decision times to act during which losses can accumulate, and there are sound business reasons why the risky inventory may not be eliminated or hedged. Moreover, there is the underlying problem of defining the value of $dt^{(\mathfrak{I})}$ in a turbulent environment[4].

Cutoffs for Underlying Variable Moves

The Monte-Carlo simulation is rather stupid in at least two ways. First, a volatility input derived from data may have an "unreasonable" value, and the simulator will use any input vol. Second, the normal constraints for spreads, for example, may be "unreasonably" violated by the MC state $\aleph(k_{CL})$ generating the total VAR.

The volatility can be "unreasonable" for several reasons. First, there can be bad data points in the time series. Second, the time period of the data series may be over a particularly violent period or a particularly calm period (see below for further discussion). In such cases, modification of the volatility may be desirable[5].

Constraints occur among spreads between different markets, involving credit, liquidity, and other factors. While the details are complicated, we naturally want obvious constraints maintained, for example high-yield or emerging markets spreads being larger than high-grade corporate spreads. However, the MC simulator only knows about the underlying variable statistics, the multivariate Gaussian pdf, the exposures, the liquidity time(s), etc. Therefore, the MC state $\aleph(k_{CL})$ may have violations of these obvious constraints. For example, if there is a large short exposure to a low-credit rate, the simulator will perversely find the state of loss where this rate rallies, and for a high volatility typical of low-

---

[3] **Where to Put the Time Square Roots?** We could also multiply the sqrt[dt$^{(\mathfrak{I})}$] factors into volatilities, but then the underlying simulation itself would depend on the product and become unwieldy.

[4] **What Liquidity Time?** Traders may want to associate dt$^{(\mathfrak{I})}$ with the normal liquidity time for normal business operations in normal markets. This may have nothing to do with the abnormal liquidity time for abnormal business operations in abnormal markets.

[5] **What Volatility Modification?** There is naturally no unique answer. One procedure would involve a collective discussion with Risk Managers and Traders.

credit rates, this rate may rally lower than the high-grade corporate rate in the MC state $\aleph(k_{CL})$.

For this reason, cutoff logic on the moves $d_t x_\alpha$, while messy to implement, can be desirable.

Data Time Periods and a Measure of VAR Uncertainty

One practical issue is that the time periods available will be different for different data. Correlations with short time series require that longer time series be truncated. However, the longer time series can be used to find the corresponding volatilities.

There is a more profound issue. As mentioned above, a particular time period of the data may be over a particularly violent period or a particularly calm period. This can be turned to advantage to define a VAR uncertainty. The idea is simple. First, different time periods $\{\Delta T_\lambda^{DATA}\}$ with different market environments are defined. Then, the VAR for each environment is calculated to get VAR_{Calm}, $VAR_{Turbulent}$ etc. In this way, the VAR uncertainty from these different VAR results is exhibited.

We stated at the beginning of this book that it would be highly desirable to have a handle on the uncertainty in the risk measures themselves. The uncertainty in assumptions itself poses a risk, and this assumption risk is not included in the risk calculated under a given set of assumptions. The calculation of the uncertainty of the VAR would go some distance in this desirable direction.

Enhanced/Stressed VAR *(ES-VAR)*

The ES-VAR, or Enhanced/Stressed VAR, is the most refined version of VAR that we shall consider. The "Stressed" attribute means that risks further out in the tails will be considered. The "Enhanced" attribute means that other attributes lending a more realistic aspect to the VAR will be included.

The ES-VAR includes the refinements of the IPV-VAR given above, plus the items in the following table:

Quantity Compared	**Improved PV VAR**	*Enhanced/Stressed VAR*
VAR Confidence Level	99%	*High, e.g. 99.97%*
Volatility Input	Standard Deviations	*Fat-tail Vols*
Correlation Input	Usual Definition	*Stressed Correlations*
Idiosyncratic (no data)	Not Included	*Estimated (Judgment)*
Liquidity Penalty	Zero	*Nonzero for Hostile Market*
$T(Start)$ Expos Reduct.	Start of Risk Period	*Any Time in Risk Period*

Starting Expos. Level	At Frozen Time	Adjust to Expected Level
Ending Exposure Level	Zero	Nonzero (Judgment)
Number of Time Steps	One Step	Several Steps, Composite Vol

We discuss these enhancements one at a time.

Higher VAR CLs for Stressed VAR and Economic Capital

The first improvement to get the stressed VAR is straightforward, and just involves raising the CL from the canonical 99% level. We will discuss Economic Capital (EC) in Ch. 39. For the moment, we merely note that EC is generally defined at a very high CL, for example 99.97%. The VAR can be run at this high CL, naturally if enough MC events are generated. For example, if 30,000 events are generated, we take the ninth worst loss[6].

The CVAR uncertainties can still obtained using the ergodic trick for the states $\{\aleph(k_{CL'})\}$ surrounding the 99.97% CL state $\aleph(k_{CL})$ with $k_{CL} = 3.43$. For 30,000 events, for example, we can use 15 states, including seven just above and seven just below the 99.97% CL state.

Fat-Tail Vols for Stressed VAR

We have discussed fat-tail (FT) vols in Ch. 21. Here we note an important practical consistency of the MC simulation of the Stressed VAR with the definition of the FT vols. The FT vol assumption involves fitting the tail of the $\{d_t x_\alpha(t_\ell)\}$ e.g. histogram at the 99% CL to obtain $\sigma_\alpha^{FatTail} = d_t x_\alpha^{99\%CL}/2.33$.

The consistency with the Stressed VAR for diversified portfolios is that *the values of the underlying moves* $(d_t x_\alpha)^{\aleph(k_{CL})}$ *producing the most risk are observed in practice to be around and rarely above the 99% CL*. Therefore use of the FT vols in the MC simulator using a 99% CL is a consistent assumption[7].

---

[6] **Economic-Capital and VAR scaling-factors:** Often Economic Capital is defined for market risk using a scaling of standard 99% CL-VAR with a numerical "scaling-factor", usually taken as 3 or 4. In our opinion there is little justification for any such specific a-priori assumption. A better approach is to deal head-on with the issues, which is what we do in this book. After the dust settles, a "scaling-factor" could be defined by the ratio of the calculated Economic Capital to the standard VAR.

[7] **Fat-Tail Vols for All Variables vs. Stochastic Volatility**: Using the FT vols for all moves will overestimate the risks of the less risky underlyings – but since by definition these are less risky, not much error is produced in the high CL Stressed VAR. A better assumption could be to use a stochastic volatility fitting the fat tails, although this is difficult to implement.

Stressed Correlations for Stressed VAR

In previous chapters, we discussed stressed correlations at great length. Here we first note that the use of stressed correlations $\rho_{\alpha\beta}^{(Stressed)}$ in practice generally results in the increase in the risk, relative to the use of unstressed correlations. Hence, we at least want to use some scenario for stressed correlations.

Ultimately, the use of stochastic correlation matrices would provide an even richer set of states. The idea would be to construct a set $\left\{ \rho_{\text{Matrix }\mathcal{M}}^{(Stressed)} \right\}$ of stressed positive-definite matrices. Then a randomly selected matrix $\rho_{\text{Matrix }\mathcal{M}}^{(Stressed)}$ would be used for each state of the set $\{d_t, x_\alpha\}$, and the 99.97% CL state from all such states would then be chosen for the total Stressed VAR[8].

Idiosyncratic Risk Inclusion for Enhanced VAR

By "idiosyncratic risk" is meant a risk that is not included in the VAR as defined above. A short and incomplete list of examples includes:

- Illiquidity risks of low-credit bonds
- Risks of one-off options with a one-way market
- Various types of basis risks for spreads or volatilities
- Volatility skew effects
- Exposures lasting only a short time and not captured at the frozen time t_{Frozen}
- Specific risks in the zoo of mortgage products
- Unusual political uncertainty effects in emerging markets
- Anomalous yield-curve shape-change effects

The estimation and approximate quantification of such risks involves advanced risk management. Analysis requires microscopic and deep knowledge of individual markets. Inclusion of idiosyncratic risk into the VAR first involves a judgment call for the stand-alone magnitude of each such risk.

The correlations between different idiosyncratic risks $\rho_{Idio,Idio}$, and between idiosyncratic risks and normal risks $\rho_{Idio,Normal}$ need to be specified. One simple assumption is to use one value $\rho_{Idio,Any}$ for all such correlations, and then take the maximum value $\rho_{Idio,Any}^{Max}$ consistent with positive definiteness for the total correlation matrix. One way to do this is to get the stressed correlation matrix for the normal variables, add in $\rho_{Idio,Any}$ for all idiosyncratic-risk correlations, and

---

[8] **Stressed Stochastic Correlation Matrices:** Such a model that maintains positive definiteness was discussed in Ch. 23.

then increase $\rho_{Idio,Any}$ to get the biggest possible result $\rho_{Idio,Any}^{Max}$ still keeping the total correlation matrix positive definite[9]. As a rule of thumb, an idiosyncratic correlation of around 0.3 with all other variables is a reasonable approximation in practice. Compared to the histogram of normal correlations, even stressed, this is a rather large and therefore conservative correlation.

The VAR is then run with the idiosyncratic risks added in deterministically as above, and with the MC simulation generating the normal risks as usual. The state with the desired CL for the total VAR (including the idiosyncratic risk) is then picked out.

The CVARs for the idiosyncratic risks can be defined using quadratic sum approximations, as we show below.

Illiquidity Penalty for Enhanced VAR

The reduction of exposure over the period $dt^{(\Im)}$ for product types $\Im$ in a turbulent market environment usually involves additional losses for spread products and any other product with illiquid aspects. There are two reasons for these additional losses. The first is the inevitable flight to quality reducing demand for illiquid $\Im$. There is also a magnification effect since *many* firms will be trying to reduce the same $\Im$ exposures at the same time under the circumstances, thus reducing secondary trading possibilities[10].

For this reason, a *liquidity penalty* $\lambda_\alpha^{(\Im)}$ depending on product type $\Im$ and/or underlying $d_t x_\alpha$ can be introduced. Thus, we increase the risk due to the exposure by the appropriate extra penalty. Depending on preference, the liquidity penalty can be expressed as a $ loss, additional spread bp, percentage change, etc. If $\lambda_\alpha^{(\Im)}$ is defined as a percentage change, then logically this quantity can be added onto the exposure by replacing

$$^{\$}\mathcal{E}_\alpha^{(\Im)} \rightarrow {}^{\$}\mathcal{E}_\alpha^{(\Im)}\left(1+\lambda_\alpha^{(\Im)}\right) \qquad (27.1)$$

---

[9] **Positive-Definite Total Correlation Matrix:** If the positive-definite condition is violated, you can get the total risk calculated as larger than the sum of the stand-alone normal risk and the stand-alone idiosyncratic risk. This makes no sense, violating the necessary conditions of real azimuthal angles in the geometric construction of Ch. 22. It also violates the condition of "subadditivity", discussed below.

[10] **Death of a Strategy:** Here is a rerun of a story. Once, a swaps desk had the idea of using mortgage derivatives in a certain strategy. A sudden adverse change of interest rates led to decisions to sell these derivatives by every broker-dealer on the Street, with no buyers at model prices. The sale price was so low that the effective prepayments implied by the models were astronomically higher than historical prepayments. Thus, there was illiquidity at usual price levels, and a huge liquidation penalty.

We will shortly distinguish exposures that are sold in the liquidity period and those that are not sold. The liquidity penalty then will apply to the exposures that are sold, since these correspond to realized losses.

Starting Time for Exposure Reduction, Asteroids, Decisions

We interpreted the liquidity time $dt^{(3)}$ as the time needed for exposure reduction of product type $\Im$. However, we have been vague about the details regarding the actual sequence of events leading to exposure reduction. To illustrate, suppose that we are calculating the VAR for one quarter (3 months). Call this one-quarter time period the "Risk Period" $\Delta T_{RiskPeriod}$.

Now a specific liquidity time interval $dt^{(3)}$ may be only, say, 5 days. We have effectively assumed that the start of this 5-day period (and every other liquidity time interval) occurs at the beginning of the risk period. Effectively we have assumed that the turbulent bad market condition—call it an "asteroid"—starts at the beginning of the risk period.

Essentially this assumes that a "red flag" goes up for all desks simultaneously and that all desks start reducing exposures simultaneously at the beginning of the risk period.

In reality, there is no reason that simultaneous risk reduction should occur. There are at least two good reasons why the liquidity time intervals can start at various non-simultaneous times inside the risk period rather than at the beginning:

1. Bad market events happen at different times in different markets. A quick look at data will confirm this assertion.

2. A non-zero "decision" or reaction time $\tau^{(3)}_{Decision}$ to a bad market will occur in order to decide to reverse the current strategy, which led a desk to take on the exposure for the product $\Im$ in the first place. The desk may want to "wait for awhile to see what happens", for example.

The non-simultaneity of stressed conditions in different markets may or may not increase the risk, depending on the assumptions. If it turns out that some markets will not get stressed within $\Delta T_{RiskPeriod}$, the risk is lowered. On the other hand, if risk builds up and then an asteroid hits within $\Delta T_{RiskPeriod}$, then the risk is increased.

Because risk can accumulate from the beginning of the risk period to the starting time of exposure reduction, the liquidity interval $dt^{(3)}$ does *not* give an accurate description of the total risk. Consideration of the decision time in general increases the risk.

Effectively, the liquidity time interval that should be used is the *longer* time $dt_{Total}^{(3)}$, including the decision time, viz

$$dt_{Total}^{(3)} = dt^{(3)} + \tau_{Decision}^{(3)} \qquad (27.2)$$

The value of $\tau_{Decision}^{(3)}$ depends on the strategy and desk. The default value (assuming that the decision to defease does occur sometime in the risk period) would be $\tau_{Decision}^{(3)} = \Delta T_{RiskPeriod}/2$, so $dt_{Total}^{(3)} = dt^{(3)} + \Delta T_{RiskPeriod}/2$.

A picture of the idea is shown below with an hypothesized decision time to sell of 1 month and a liquidity time of 5 days, all within a 3-month risk period:

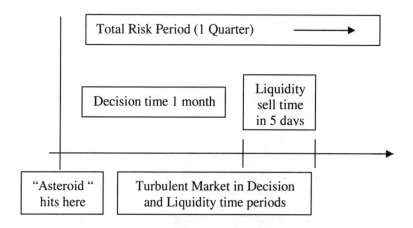

Starting Exposure Level; Corrections to the Expected Level

The starting exposure levels $\{{}^{\$}\mathcal{E}_\alpha\}$ before exposure reduction are taken as the exposures at the frozen time t_{Frozen}. However, these exposures may not be representative of the exposure levels $\{{}^{\$}\mathcal{E}_\alpha^{(Expected)}\}$ expected during the next risk period $\Delta T_{RiskPeriod}$. For this reason, it may be reasonable to correct the exposures input into the MC simulation to these expected levels. The expected levels would

Improved/Enhanced/Stressed VAR 363

generally require knowledge of desk strategy, or data of historical exposures that could be used for the expected exposure, etc.

In Ch. 40, a procedure is described to correct risk for unused exposures relative to their limits. The first step in this procedure is to perform exactly the above correction to expected exposure levels. However, it would be more accurate to perform the correction in the VAR. This is because the state $\aleph(k_{CL})$ for the VAR is itself dependent on the exposures assumed.

The Ending Exposure Level is Not Zero

We have assumed that, following the decision time, the starting exposures for product $\Im$ are reduced to zero in liquidity time interval $dt^{(\Im)}$. This is by no means a realistic assumption. For example, a high-yield bond desk would not want to sell off its entire inventory even in a turbulent market, nor could the desk hedge the entire high-yield spread risk. The desk may have a profitable customer-flow business, and also may need to make markets. That is, for business reasons, the ending exposure may not be zero after exposure reduction. The fraction of the exposure sold $f_{Sold}^{(\$\mathcal{E}_\alpha)}$ is therefore a parameter. However, the remainder of the exposure (unsold) has risk that accumulates throughout the entire risk period $\Delta T_{RiskPeriod}$.

A Component Model for Risk Exposures

A "Noise + View + Hold" risk model for exposures is a reasonable framework to begin to model a more realistic description of the different types of exposure.

The noise component of a given exposure, $\left[{}^\$\mathcal{E}_\alpha^{(\Im)}\right]_{Noise}$, would represent the part of the exposure due to day-to-day operations, customer transactions, normal hedging activities, etc. The noise component would fluctuate as a function of time around a mean (e.g. zero) with some characteristic time τ_{Noise}. There would be no decision time for this exposure. For purposes of risk calculations, the noise exposure component would be taken from its frozen level and dropped to zero in time τ_{Noise}. This is because τ_{Noise} is the representative time that the exposure noise would fluctuate to zero anyway. Properties of the noise component could be determined through historical statistics of the exposure.

The view component $\left[{}^\$\mathcal{E}_\alpha^{(\Im)}\right]_{View}$ would represent an exposure held because of a certain strategy of the desk (e.g. yield curve steepening, currency weakening, commodity forecasting, etc.). The view exposure would be expendable in time τ_{View} (including a decision time) if the market turned sufficiently against the strategy.

The hold component $\left[{}^{\$}\mathcal{E}_\alpha^{(3)}\right]_{Hold}$ would represent the core component that would be held regardless of the market just in order to stay in business or for some other reason. It would be held for the entire risk period $\Delta T_{RiskPeriod}$. Each of these exposure components produces risk over a different period τ_ζ. Therefore, each exposure would be associated with its own $\sqrt{\tau_\zeta}$ scaling factor.

A picture of the idea follows:

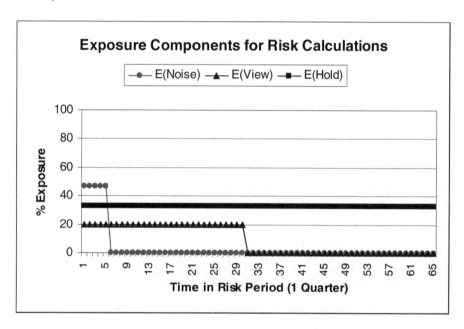

Exposure: Reduction vs. Double-Up in Turbulent Markets

We have been assuming that at the end of the liquidity period, the exposure has been reduced from its current value. In real life, even in a turbulent market environment, traders may see opportunity and want to increase the exposure, or "double up". The trade offs have to do with buying cheap versus watching the market decline even further, and the tolerance of management for losses.

We discuss the problem of additional risk for unused limits in Ch. 40. There, a model for exposure time dependence is postulated. This time dependence is supposed to occur from the current time up to the time that the turbulent market starts. At that point, the reduction or double-up behaviors as a reaction to the turbulent market begins. Since up to now we have only used a one time-step simulator, all these features become overlaid. We now turn to some comments regarding the extension of VAR to a multi-step simulator.

Increasing the Number of Time Steps for VAR-Type Simulation

We have introduced some explicit time dependence in the Enhanced VAR through the parameters defined above: $dt^{(3)}$, $\tau_{Decision}^{(3)}$, $\Delta T_{RiskPeriod}$, etc. Still, the MC simulator used so far is effectively a one-step simulation. In principle, we can use a multi-step MC simulation with intermediate time assumptions related to all the above points. Such a simulation would require a high level of risk management sophistication. It would also be extremely costly, numerically intensive, and require more assumptions. Still, such a tool would provide the most realistic assessment of risk possible[11].

In the next chapter, we present a summary of the extension to multiple time steps using path integral techniques.

As a partial improvement along these lines without involving the complexity of the explicit treatment of multiple steps, composite volatilities can be defined reflecting different events in at intermediate times. For example, we can picture two time periods during which the diffusion occurs with two different fat-tail volatilities $\sigma_\alpha^{(1)}$ and $\sigma_\alpha^{(2)}$. Over the first period of time ΔT_1, the noise exposure component fluctuates to zero and the sale of the view exposure component is achieved. In the second period of time ΔT_2, the hold exposure component continues to the end of the risk period. An effective volatility $\sigma_\alpha^{[Eff]}$ taking all this into account would be defined through the variance equation

$$^\$\mathcal{E}_\alpha^{[Total]} \sigma_\alpha^{[Eff]} \sqrt{\Delta T_{RiskPeriod}} = \left\{ \left[^\$\mathcal{E}_\alpha^{[Total]} \sigma_\alpha^{(1)} \sqrt{\Delta T_1} \right]^2 + \left[^\$\mathcal{E}_\alpha^{[Hold]} \sigma_\alpha^{(2)} \sqrt{\Delta T_2} \right]^2 \right\}^{1/2}$$

(27.3)

Other VAR Topics

VAR and Subadditivity

Heath and colleagues[i] have pointed out that under some conditions a subadditivity property can be violated. Basically, subadditivity asserts the condition that the sum of the risks of stand-alone portfolios should be greater than the risk of the composite portfolio with all securities in all portfolios. This is physically reasonable. Physically, diversification and risk cancellation can occur in the composite portfolio, but risk enhancement theoretically should not occur.

---

[11] **Color Movie for a Multi-Time Step Risk Simulator?** As mentioned at the beginning of the book, I have been waiting for over 15 years for such a sophisticated risk simulation, in color, as a movie. Maybe it will come soon. Then again, ...

Equivalently, the effective correlations between the risks of the individual portfolios must have magnitudes $\rho^{(a,b)} \leq 1$.

As a corollary, if the potential loss is doubled by doubling the amount of a security, the risk measure of this loss should linearly double[12].

Heath's Example of Subadditivity Problem

Convexity effects can cause subadditivity violations. Heath's example involves a short digital call $C_{\text{Short Call}}$ and a short digital put $C_{\text{Short Put}}$ just before expiration. Individually, at a given CL, the risks may be zero. Yet, the composite risk of the total portfolio $C_{\text{Short Call, Put}}$ at the same CL may be nonzero, violating subadditivity.

Consider a MC simulation. A fraction f_{up} of paths arriving above the call strike produces a loss, but if $CL < 1 - f_{up}$, the risk of $C_{\text{Short Call}}$ at that CL is zero. Similarly, the risk of $C_{\text{Short Put}}$ is zero if $CL < 1 - f_{down}$ for paths arriving below the put strike. However, in the composite portfolio, the total fraction of paths causing a loss is $f_{tot} = f_{up} + f_{down}$. If $CL > 1 - f_{tot}$, the loss of $C_{\text{Short Call, Put}}$ is not zero. The presence of both sets of paths and the discrete nature of the risk causes the problem in this example.

One resolution of this problem suggested by Heath is to calculate risk from a limited set of paths using scenarios.

Practical Resolution of VAR and Subadditivity

Subadditivity is satisfied for VAR including fat-tail vols and stressed correlations, if convexity is approximated by using a fixed "scenario" value for delta from a grid. We have discussed several such possibilities above.

The proof of subadditivity here rests on the existence of the positive-definite correlation matrix between underlyings. If idiosyncratic risk is included, the expanded correlation matrix involving these risks must also be positive definite.

Confidence-Level Integrated VAR Measures

We have been discussing risk at a given confidence level. Even the ergodic trick required the approximate transformation of states with different CLs into a given

---

[12] **Risk Enhancement – Other Aspects:** Actually this academic statement neglects the issue of volume effects that can non-linearly increase risk. If a desk owns a substantial fraction of the total issue of some security, there can be severe liquidity problems if the desk decides (or is forced) to sell. A portfolio with twice as much of a security can have a real risk of far greater than twice as much. The collapse of LTCM and various Arb desks in 1998 was in large measure due to exactly this problem.

Improved/Enhanced/Stressed VAR

CL. Here we consider a different idea, an integrated $^\$VAR_{avg}$, calculated as an average between two confidence levels CL_{min} and CL_{max}. Such an integrated risk measure has smoother properties than risk measured at an isolated CL. Basically, integrals always smooth out functions. As in the ergodic trick, information from more paths is sampled in the averaging process.

One of the difficulties of the integrated average VAR is that backtesting would become somewhat problematic and confusing.

We shall meet the CL-integrated risk idea again when we consider issuer risk.

Example of CL-Integrated VAR

To illustrate the idea, consider the simple quadratic VAR. The integrated average $^\$VAR_{avg}$ between CL_{min} (at $k_{CL\,min}$ SD) and CL_{max} (at $k_{CL\,max}$ SD) is obtained by calculating $^\$VAR_{avg} = \left\langle \sum_{\alpha=1}^{n} {}^\$\mathcal{E}_\alpha \cdot d_t x_\alpha \right\rangle_{(k_{CL\,min},\,k_{CL\,max})}$. The result is the same as $^\$VAR$ if k_{CL} is replaced by $k_{CL}^{eff} = \varsigma(k_{CL\,min}, k_{CL\,max})$. This k_{CL}^{eff} is the effective number of SD, and is given by[13]

$$\varsigma(k_{CL\,min}, k_{CL\,max}) = \frac{1}{\sqrt{2\pi}} \frac{\left[\exp(-k_{CL\,min}^2/2) - \exp(-k_{CL\,max}^2/2)\right]}{\left[N(k_{CL\,max}) - N(k_{CL\,min})\right]} \quad (27.4)$$

Here $N(\cdot)$ is the usual normal integral. We get the result

$$^\$VAR_{avg} = \varsigma(k_{CL\,min}, k_{CL\,max})\sqrt{dt} \cdot {}^\$VAR^{Quad} \quad (27.5)$$

We can consider, e.g. the $^\$VAR_{avg}$ at $CL = 99\%$, as an average between appropriately chosen CL_{min}, CL_{max}.

Another possibility is to take $CL_{max} = 100\%$. In that case, the $^\$VAR_{avg}$ risk is considered as averaged over all paths in the tail past CL_{min}. Of course in that case we get all the outliers and $k_{CL}^{eff} > k_{CL}$.

In the limit $CL_{min} = CL_{max} = CL$, we get $\varsigma(k_{CL\,min}, k_{CL\,max}) \to k_{CL}$, restoring the usual formalism at a single confidence level.

---

[13] **Homework:** Read the next chapter, then come back and do this as an exercise. The appropriate expected value averaged between the two confidence levels is normalized by the denominator, which is the integrated measure between the two confidence levels.

References

[i] **VAR and Subadditivity**
Heath, D., Talk 5/22/96, ISDA, New York.
Artzner, P., Delbaen, F., Eber, J-M., Heath, D., *A Characterization of Measures of Risk*, Cornell U. working paper, 1996.

28. VAR, CVAR, CVAR Volatility Formalism (Tech. Index 7/10)

In this chapter, we present a formal functional derivation of the VAR and CVAR equations for the linear case. We pay particular attention to the CVAR volatility. The derivation is done for in the continuous multivariate framework. This shows that CVAR uncertainties are present in the limit of an infinite-length Monte-Carlo (MC) simulation run. We indicate extensions for non-linear exposures (convexity) to VAR, as discussed in the last chapter. We end with a summary of the extension to multiple time steps[1].

Set-up and Overview of the Formal VAR Results

To perform the calculations, we need the multivariate Gaussian probability distribution for one time step. The time difference of an underlying variable x_α is $d_t x_\alpha(t) = x_\alpha(t+dt) - x_\alpha(t)$ at fixed time t. This is the return if $x_\alpha = \ln r_\alpha$. To apply the formalism to the Stressed VAR, we would specify the vol σ_α of $d_t x_\alpha$ as a fat-tail vol $\sigma_{\alpha;FT}$, as discussed in Ch. 21. We would also specify the $d_t x_\alpha$, $d_t x_\beta$ correlation $\rho_{\alpha\beta}$ as the stressed correlation $\rho_{\alpha\beta}^{(Stressed)}$, as described Ch. 23, 24.

The probability is then an integral over all possible values of each $d_t x_\alpha$, involving the measure $d(d_t x_\alpha)$. This horrible notation just means that we actually integrate over values of $x_\alpha(t+dt)$ at fixed $x_\alpha(t)$, using the measure $dx_\alpha(t+dt)$.

For simplicity of notation in this chapter, we do not indicate factors of $\sqrt{dt}$ in the intermediate formulae explicitly. This can be corrected simply by inserting a factor $\sqrt{dt}$ for σ_α, $^\$ CVAR_\gamma^{Quad}$, and $^\$ VAR^{Quad}$.

---

[1] **History:** These calculations were performed by me between 1998 and 2002.

The probability integral for calculating expectation values under the above assumptions is the usual multivariate Gaussian in the underlying internal indices for one time step [2]

$$\wp = \frac{1}{\sqrt{\det \rho}} \left\{ \prod_{\alpha=1}^{n} \int_{-\infty}^{\infty} \frac{d(d_t x_\alpha)}{\sigma_\alpha \sqrt{2\pi}} \right\} \exp\left(-\Phi\left[\{d_t x_\alpha\}\right]\right) \quad (28.1)$$

Here,

$$\Phi\left[\{d_t x_\alpha\}\right] = \frac{1}{2} \sum_{\alpha,\beta=1}^{n} \left[\frac{d_t x_\alpha}{\sigma_\alpha} \rho^{-1}_{\alpha\beta} \frac{d_t x_\beta}{\sigma_\beta} \right] \quad (28.2)$$

With the exposure ${}^\$\mathcal{E}_\alpha$, the change in value of the portfolio due to the change $d_t x_\alpha$ is just ${}^\$\mathcal{E}_\alpha \cdot d_t x_\alpha$, by definition. For different values of $d_t x_\alpha$ we will get different values of ${}^\$\mathcal{E}_\alpha \cdot d_t x_\alpha$. The total VAR will be the value of the sum of all such contributions, at the specified CL corresponding to k_{CL} standard deviations:

$${}^\$\dot{VAR} = \left[\sum_{\alpha=1}^{n} {}^\$\mathcal{E}_\alpha \cdot d_t x_\alpha \right]_{k_{CL}} \quad (28.3)$$

We next consider the CVARs. By definition, the contribution to the total VAR from ${}^\$\mathcal{E}_\alpha \cdot d_t x_\alpha$ is just ${}^\$ CVAR_\alpha$. On the average, ${}^\$ CVAR_\alpha$ is the expectation value $\left\langle {}^\$\mathcal{E}_\alpha \cdot d_t x_\alpha \right\rangle$ over the $\wp$ distribution. We will show in fact that up to a factor, $\left\langle {}^\$\mathcal{E}_\alpha \cdot d_t x_\alpha \right\rangle$ is just the quadratic form ${}^\$ CVAR_\alpha^{Quad}$ described in a previous section.

We will further show that $\left\langle \left[{}^\$\mathcal{E}_\alpha \cdot d_t x_\alpha \right]^2 \right\rangle$ is nonzero. Therefore ${}^\$ CVAR_\alpha$ has an uncertainty or volatility $\sigma\left({}^\$ CVAR_\alpha\right)$, which we will calculate. We will also show that $\sigma\left({}^\$ CVAR_\alpha\right)$ has a nice geometrical interpretation.

---

[2] **Path Integral for multi-time-step VAR:** The VAR we discuss here involves just one time step. Inclusion of risk for many time steps can be directly handled using path integrals, which begin by discretizing the time axis into many time steps. Hence, VAR can be directly generalized to many time steps in a direct fashion theoretically. At the end of this chapter, we discuss the matter a little further. The reader is also invited to read the detailed discussion of path integrals and finance in this book, especially Ch. 45.

Calculation of the Generating Function

We now present the details. We define the generating function, depending on conjugate variables $\{J_\alpha\}$ as

$$\wp[\{J_\alpha\}] = \frac{1}{\sqrt{\det\rho}} \left\{ \prod_{\alpha=1}^{n} \int_{-\infty}^{\infty} \frac{d(d_t x_\alpha)}{\sigma_\alpha \sqrt{2\pi}} \right\} \exp\left\{ -\Phi[\{d_t x_\alpha\}] + \sum_{\alpha=1}^{n} \left[{}^\$\mathcal{E}_\alpha d_t x_\alpha \cdot J_\alpha \right] \right\}$$

(28.4)

We note that $\wp = \wp[\{J_\alpha = 0\}]$.

We will need to get a variable that looks like the VAR. To this end, we introduce a "P+L" variable F and write the Dirac delta function constraint equation[3]

$$1 = \int_{-\infty}^{\infty} dF \cdot \delta\left(F - \sum_{\alpha=1}^{n} {}^\$\mathcal{E}_\alpha \cdot d_t x_\alpha \right) \qquad (28.5)$$

We insert the factor 1 into the integrand of $\wp[\{J_\alpha\}]$ and rewrite 1 as $\int_{-\infty}^{\infty} dF \cdot \delta\left(F - \sum_{\alpha=1}^{n} {}^\$\mathcal{E}_\alpha \cdot d_t x_\alpha \right)$. We then use the Fourier representation of the Dirac delta function[4],

$$\delta\left(F - \sum_{\alpha=1}^{n} {}^\$\mathcal{E}_\alpha \cdot d_t x_\alpha \right) = \int_{-\infty}^{\infty} \frac{d\omega}{2\pi} \exp\left(-i\omega F + i\omega \sum_{\alpha=1}^{n} {}^\$\mathcal{E}_\alpha \cdot d_t x_\alpha \right) \qquad (28.6)$$

We can now do all the $\{d_t x_\alpha\}$ integrals. We have

---

[3] **$Units:** The VAR-variable F has units $. The conjugate variables J_α and Fourier variable ω have units 1/$. This notation is not in the equations to avoid clutter.

[4] **What, Again?** The reader who has followed this book will notice a repetition of the same theme regarding the generating function or functional, introducing the constraints through Dirac delta functions, using the Fourier transform, and then doing the integrals.

$$\frac{1}{\sqrt{\det \rho}} \left\{ \prod_{\alpha=1}^{n} \int_{-\infty}^{\infty} \frac{d(d_t x_\alpha)}{\sigma_\alpha \sqrt{2\pi}} \right\} \exp\left\{ -\Phi\left[\{d_t x_\alpha\}\right] + \sum_{\alpha=1}^{n} \left[{}^{\$}\mathcal{E}_\alpha \cdot d_t x_\alpha \cdot (i\omega + J_\alpha) \right] \right\} =$$

$$= \exp\left\{ \frac{1}{2} \sum_{\alpha,\beta=1}^{n} {}^{\$}\mathcal{E}_\alpha \sigma_\alpha \rho_{\alpha\beta} {}^{\$}\mathcal{E}_\beta \sigma_\beta \cdot \left[-\omega^2 + i\omega(J_\alpha + J_\beta) + J_\alpha J_\beta \right] \right\}$$

(28.7)

Now from Ch. 26, we recall that ${}^{\$}CVAR_\alpha^{Quad} = \frac{{}^{\$}\mathcal{E}_\alpha \sigma_\alpha}{{}^{\$}VAR^{Quad}} \sum_{\beta=1}^{n} \rho_{\alpha\beta} {}^{\$}\mathcal{E}_\beta \sigma_\beta$ and

$${}^{\$}VAR^{Quad} = \sum_{\alpha=1}^{n} {}^{\$}CVAR_\alpha^{Quad}.$$

With these substitutions, we find that the Fourier transform integral is

$$\int_{-\infty}^{\infty} \frac{d\omega}{2\pi} \exp\left[-i\omega \left(F - {}^{\$}VAR^{Quad} \sum_{\alpha=1}^{n} J_\alpha {}^{\$}CVAR_\alpha^{Quad} \right) - \frac{\omega^2}{2} \left({}^{\$}VAR^{Quad} \right)^2 \right] =$$

$$= \frac{1}{{}^{\$}VAR^{Quad} \sqrt{2\pi}} \exp\left[-\Psi(F;\{J_\alpha\}) \right]$$

(28.8)

Here

$$\Psi(F;\{J_\alpha\}) = \frac{1}{2\left({}^{\$}VAR^{Quad} \right)^2} \left[F - {}^{\$}VAR^{Quad} \sum_{\alpha=1}^{n} J_\alpha {}^{\$}CVAR_\alpha^{Quad} \right]^2 \quad (28.9)$$

We also write ${}^{\$\$}\sigma_{\alpha\beta}^2 = \cdot {}^{\$}\mathcal{E}_\alpha \sigma_\alpha \rho_{\alpha\beta} {}^{\$}\mathcal{E}_\beta \sigma_\beta$. The result for the generating function is then

$$\wp[\{J_\alpha\}] = \int_{F=-\infty}^{F=\infty} d\mu(F) \exp\left[\Omega(F;\{J_\alpha\}) \right] \quad (28.10)$$

Here the measure $d\mu(F)$ in the VAR-variable F reads

$$d\mu(F) = \frac{dF}{^\$VAR^{Quad}\sqrt{2\pi}} \exp\left[-\frac{F^2}{2\left(^\$VAR^{Quad}\right)^2}\right] \quad (28.11)$$

Note that this Gaussian measure for F has volatility $^\$VAR^{Quad}$. Therefore, at k_{CL} standard deviations in F, we have $F = k_{CL} \cdot ^\$VAR^{Quad}$.

The exponent $\Omega(F;\{J_\alpha\})$ is

$$\Omega(F;\{J_\alpha\}) = \frac{F}{^\$VAR^{Quad}}\sum_{\alpha=1}^n J_\alpha\,^\$CVAR_\alpha^{Quad}$$
$$-\frac{1}{2}\left[\sum_{\alpha=1}^n J_\alpha\,^\$CVAR_\alpha^{Quad}\right]^2 + \frac{1}{2}\sum_{\alpha,\beta=1}^n J_\alpha\,^{\$\$}\sigma_{\alpha\beta}^2 J_\beta \quad (28.12)$$

This completes the calculation of the generating function $\wp[\{J_\alpha\}]$.

Calculation of the Exposure*Underlying Change Moments

We take derivatives with respect to $\{J_\alpha\}$ and then set $\{J_\alpha = 0\}$ to get the moments.

Recall we set $\langle d_t x_\gamma \rangle = 0$ because the VAR is supposed to represent fluctuations away from the average. Therefore, up to the first two moments, we have

$$\left\langle ^\$\mathcal{E}_\gamma \cdot d_t x_\gamma \right\rangle = \left[\partial\wp[\{J_\alpha\}]/\partial J_\gamma\right]_{\{J_\alpha=0\}} \quad (28.13)$$

$$\left\langle ^\$\mathcal{E}_\gamma \cdot d_t x_\gamma \,^\$\mathcal{E}_{\gamma'} \cdot d_t x_{\gamma'} \right\rangle = \left[\partial^2\wp[\{J_\alpha\}]/\partial J_\gamma \partial J_{\gamma'}\right]_{\{J_\alpha=0\}} \quad (28.14)$$

After a little algebra we find

$$\left[\partial\Omega[\{J_\alpha\}]/\partial J_\gamma\right]_{\{J_\alpha=0\}} = \frac{F}{^\$VAR^{Quad}} \cdot ^\$CVAR_\gamma^{Quad} \quad (28.15)$$

$$\left[\partial^2\Omega[\{J_\alpha\}]/\partial J_\gamma \partial J_{\gamma'}\right]_{\{J_\alpha=0\}} = ^{\$\$}\sigma_{\gamma\gamma'}^2 - ^\$CVAR_\gamma^{Quad} \cdot ^\$CVAR_{\gamma'}^{Quad} \quad (28.16)$$

Therefore

$$\left\langle {}^\$\mathcal{E}_\gamma \cdot d_t x_\gamma \right\rangle = \int_{F=-\infty}^{F=\infty} d\mu(F) \cdot \frac{F}{{}^\$VAR^{Quad}} \cdot {}^\$CVAR_\gamma^{Quad} \qquad (28.17)$$

Also,

$$\left\langle {}^\$\mathcal{E}_\gamma \cdot d_t x_\gamma \Box {}^\$\mathcal{E}_{\gamma'} \cdot d_t x_{\gamma'} \right\rangle = \int_{F=-\infty}^{F=\infty} d\mu(F) \cdot$$
$$\cdot \left[\frac{F^2 \cdot {}^\$CVAR_\gamma^{Quad} \cdot {}^\$CVAR_{\gamma'}^{Quad}}{\left({}^\$VAR^{Quad}\right)^2} + \left({}^{\$\$}\sigma_{\gamma\gamma'}^2 - {}^\$CVAR_\gamma^{Quad} \cdot {}^\$CVAR_{\gamma'}^{Quad}\right) \right]$$
(28.18)

VAR, the CVARs, and the CVAR Volatilities

The total ${}^\$VAR(k_{CL}, dt)$ is equal to VAR-variable F at a prescribed CL (e.g. 99%) corresponding to k_{CL} standard deviations in F, and for a specific time interval dt. Hence, we drop the integration $\int_{F=-\infty}^{F=\infty} d\mu(F)$ and substitute $F = k_{CL}\sqrt{dt} \cdot {}^\$VAR^{Quad}$ in the integrand (with the factor of $\sqrt{dt}$ restored). So we get

$${}^\$VAR(k_{CL}, dt) = k_{CL}\sqrt{dt} \cdot {}^\$VAR^{Quad} \qquad (28.19)$$

The CVAR is just the average risk for a specific underlying at k_{CL} standard deviations for the total VAR, that is

$${}^\$CVAR_\gamma = \left\langle {}^\$\mathcal{E}_\gamma \cdot d_t x_\gamma \right\rangle_{k_{CL}} \qquad (28.20)$$

This is, restoring the $\sqrt{dt}$ factor, ${}^\$CVAR_\gamma = k_{CL}\sqrt{dt} \cdot {}^\$CVAR_\gamma^{Quad}$.

The second moment at k_{CL} standard deviations for the total VAR is

$$\left\langle {}^\$\mathcal{E}_\gamma \cdot d_t x_\gamma \cdot {}^\$\mathcal{E}_{\gamma'} \cdot d_t x_{\gamma'}\right\rangle_{c;k_{CL}} = {}^{\$\$}\sigma^2_{\gamma\gamma'} - {}^\$ CVAR^{Quad}_\gamma \cdot {}^\$ CVAR^{Quad}_{\gamma'} \quad (28.21)$$

Here, the connected part of the two-point function at k_{CL} is defined as

$$\left\langle {}^\$\mathcal{E}_\gamma \cdot d_t x_\gamma \cdot {}^\$\mathcal{E}_{\gamma'} \cdot d_t x_{\gamma'}\right\rangle_{c;k_{CL}} = \left\langle {}^\$\mathcal{E}_\gamma \cdot d_t x_\gamma \cdot {}^\$\mathcal{E}_{\gamma'} \cdot d_t x_{\gamma'}\right\rangle_{k_{CL}}$$
$$- \left\langle {}^\$\mathcal{E}_\gamma \cdot d_t x_\gamma \right\rangle_{k_{CL}} \left\langle {}^\$\mathcal{E}_{\gamma'} \cdot d_t x_{\gamma'}\right\rangle_{k_{CL}} \quad (28.22)$$

The CVAR Volatility

If we set $\gamma' = \gamma$ in this equation, we get the square of the CVAR volatility ${}^\$\sigma_{CVAR_\gamma}$, namely

$$^{\$\$}\sigma^2_{CVAR_\gamma} = \left\langle \left[{}^\$\mathcal{E}_\gamma \cdot d_t x_\gamma\right]^2 \right\rangle_{c;k_{CL}} = \left[{}^\$\mathcal{E}_\gamma \sigma_\gamma\right]^2 - \left[{}^\$ CVAR^{Quad}_\gamma\right]^2 \quad (28.23)$$

Notice that the CVAR volatility is not zero. Therefore, there is an uncertainty in each CVAR. For the linear case, Eqn. (28.23) is the exact formula for the CVAR volatility. Also, note that the CVAR volatility is independent of k_{CL}. This is because the second bracket in Eqn. (28.18) is independent of F.

Using the "ergodic trick" of Ch. 26, the validity of the above formula can be (and has been) checked using Monte Carlo simulation for linear portfolios.

The Total VAR has Zero Volatility

Even though there is a volatility for each CVAR, there is no volatility at all for the total VAR, i.e. ${}^\$\sigma_{VAR} = 0$. This is easily seen by summing over all (γ, γ') in the two-point function above to get

$$^{\$\$}\sigma^2_{VAR} = \left\langle \left[\sum_{\gamma=1}^n {}^\$\mathcal{E}_\gamma \cdot d_t x_\gamma\right]^2 \right\rangle_{c;k_{CL}}$$
$$= \left[{}^\$ VAR^{Quad}\right]^2 - \left[\sum_{\gamma=1}^n {}^\$ CVAR^{Quad}_\gamma\right]^2 = 0 \quad (28.24)$$

Since the VAR volatility is zero, the VAR is exactly determined. This shows clearly that the CVAR volatility has nothing to do with Monte-Carlo statistical noise for a finite number of events, since all the above calculations have been done in the continuous limit corresponding to an infinite Monte-Carlo simulation.

The CVAR Volatility Triangle

From Eqn. (28.23) we have a nice geometrical right triangle involving the CVAR volatility, as shown in the picture below.

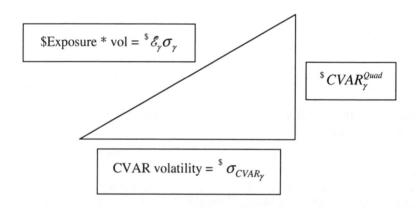

The CVAR Volatilities for the Nonlinear Risk Case

The existence of the CVAR volatility is not at all limited to the linear case. The uncertainty in the components of risk for a given total risk is a general concept.

Moreover, it turns out that even for somewhat non-linear portfolios, the CVAR volatility formula (28.23) is a reasonably good approximation. This has been explicitly checked by Monte-Carlo generated CVAR volatility with convexity approximated by using a grid. Only for highly convex option portfolios does Eqn. (28.23) break down.

Effective Number of SD for Underlying Variables

From the value of $^\$CVAR_\gamma = \left\langle ^\$\mathcal{E}_\gamma \cdot d_t x_\gamma \right\rangle_{k_{CL}}$, we can back out an effective number of standard deviations $k_\gamma^{(\it{eff})}$ for the variable $d_t x_\gamma$. This is done simply using the (ergodic average) value of $d_t x_\gamma$ in the state giving the total VAR at k_{CL} SD. We write

$$k_\gamma^{(eff)} = \langle d_t x_\gamma \rangle_{k_{CL}} / \sigma_\gamma \qquad (28.25)$$

Because of diversification due to correlations, the effective number of standard deviations $k_\gamma^{(eff)}$ is less than the VAR k_{CL}, i.e. $k_\gamma^{(eff)} < k_{CL}$. For example consider $k_{CL} = 3.43$ SD for a Stressed VAR calculation at the 99.97% CL. Suppose[5] $\langle d_t x_\gamma \rangle_{k_{CL}} = 1$ and $\sigma_\gamma = 0.5$. Then the number of standard deviations on average for this variable $d_t x_\gamma$ is $k_\gamma^{(eff)} = 2$. Values of $k_\gamma^{(eff)}$ around 2 are attained by a few of the riskiest variables at the overall VAR 99.97% CL for diversified portfolios, as mentioned before.

Fat-Tail Vol Consistency with the Effective Number of SD $k_\gamma^{(eff)}$

There is a consistency check of the definition of fat-tail (FT) vols with the effective number of standard deviations. As discussed in Ch. 21, we defined FT vols at the 99% CL or 2.33 SD. This is in fact roughly consistent with the most important variables having $k_\gamma^{(eff)} \approx 2$. For the less important variables, the use of the FT vol will increase the risk somewhat, but since less important variables have less risk, the total risk is not much affected.

Calculation of $k_\gamma^{(eff)}$ in the Linear Risk Case

Because we have the expression for the CVAR for the linear case, we can write down $k_\gamma^{(eff)}$. We get

$$k_\gamma^{(eff)} = \frac{k_{CL}}{\$VAR^{Quad}} \sum_{\beta=1}^{n} \rho_{\gamma\beta} \$\mathcal{E}_\beta \sigma_\beta \qquad (28.26)$$

To get an idea, we evaluate this in the case of equal exposures, equal vols, and constant correlation ρ_0. The correlation matrix is $\rho_{\gamma\beta} = \delta_{\gamma\beta} + \rho_0 (1 - \delta_{\gamma\beta})$.

---

[5] **Time Units:** If we are doing, say quarterly VAR calculations where dt = 65 days, then the volatility is a quarterly volatility, either scaled up by sqrt(65) from the daily vol or else defined using windows of 65 days.

It is not hard to see that $\det \rho = (1-\rho_0)^{n-1}[1+(n-1)\rho_0]$ where so n is the number of variables. Hence, we need $\rho_0 > -1/(n-1)$ for a positive definite correlation matrix. Substituting, we find

$$k_\gamma^{(\text{eff})} = k_{CL}\sqrt{[\rho_0(1-1/n)+1/n]} \qquad (28.27)$$

We see that the positive-definite constraint keeps $k_\gamma^{(\text{eff})}$ as a real number. For large n we get $k_\gamma^{(\text{eff})} \approx k_{CL}\sqrt{\rho_0}$. Even for correlations around 0.25, which is on average is a large correlation, we see that $k_\gamma^{(\text{eff})} \approx k_{CL}/2$, which is still substantially less than k_{CL}. If $\rho_0 = 0$ we get $k_\gamma^{(\text{eff})} = k_{CL}/\sqrt{n}$ which goes to zero.

Extension to Multiple Time Steps using Path Integrals

Although we have no illusions about the practicality of performing a multiple step MC simulation with hundreds (if not thousands or even tens of thousands) of variables as needed for corporate-wide VAR, we can nonetheless easily extend the formalism using standard path integral techniques. We outline the linear case. The interested reader will have no difficulty filling in the steps. All the improvements to the VAR discussed in previous chapters can be implemented in principle.

The time labels are $\ell, m = 1...N$, and we retain the internal index labels $\alpha, \beta, \gamma = 1...n$. The $^\$\mathcal{E}_\alpha$ exposure at t_ℓ is $^\$\mathcal{E}_{\ell;\alpha} =\, ^\$\mathcal{E}_\alpha(t_\ell)$ and the x_α time difference over dt_ℓ at t_ℓ is $d_t x_{\ell;\alpha} \equiv d_t x_\alpha(t_\ell) \equiv x_\alpha(t_\ell + dt_\ell) - x_\alpha(t_\ell)$. Step by step, we proceed exactly as above. The probability density exponent becomes

$$\Phi[\{d_t x_{\ell;\alpha}\}] = \frac{1}{2}\sum_{\ell=1}^{N}\sum_{\alpha,\beta=1}^{n}\left[\frac{d_t x_{\ell;\alpha}}{\sigma_{\ell;\alpha}}(\rho_\ell^{-1})_{\alpha\beta}\frac{d_t x_{\ell;\beta}}{\sigma_{\ell;\beta}}\right] \qquad (28.28)$$

Here the volatilities contain the factor $\sqrt{dt_\ell}$. The conjugate variables become $\{J_{\ell;\alpha}\}$. The time-local CVAR variables also pick up the time index, $^\$CVAR_{\ell;\gamma} = \langle^\$\mathcal{E}_{\ell;\gamma} \cdot d_t x_{\ell;\gamma}\rangle_{k_{CL}}$, still at the specified overall CL for the total

VAR. The total VAR over the complete time interval $T = \sum_{\ell=1}^{N} dt_\ell$ for k_{CL} standard deviations is $^\$ VAR = \left[\sum_{\ell=1}^{N} \sum_{\gamma=1}^{n} {}^\$ CVAR_{\ell;\gamma} \right]_{k_{CL}}$.

Repeating the algebra for the crossed second moment, we find that

$$\left\langle {}^\$\mathcal{E}_{\ell;\gamma} \cdot d_t x_{\ell;\gamma} \cdot {}^\$\mathcal{E}_{m;\gamma'} \cdot d_t x_{m;\gamma'} \right\rangle_{c;k_{CL}} = \delta_{\ell m} {}^{\$\$}\sigma^2_{\ell;\gamma'} - {}^\$ CVAR^{Quad}_{\ell;\gamma} \cdot {}^\$ CVAR^{Quad}_{m;\gamma'} \tag{28.29}$$

Notice that the subtracted term has two factors at different times t_ℓ, t_m.

The local CVAR volatility at fixed time obtained by setting $\ell = m$ and $\gamma = \gamma'$ obeys the same triangle relation as found above for the single time step case.

29. VAR and CVAR for Two Variables (Tech. Index 5/10)

The CVAR Volatility with Two Variables

Here, we restrict our attention to two variables. We begin with the CVAR volatility. Here is a picture of the geometry:

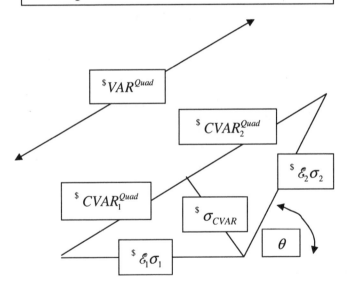

Geometry for the CVAR vol, CVARs, and Exposures*Vols for $n = 2$ variables

The CVAR volatility[1] turns out to be the same for both variables. Both triangles with $^{\$}CVAR_1^{Quad}$, $^{\$}CVAR_2^{Quad}$ have a common leg, the CVAR volatility $^{\$}\sigma_{CVAR}$. Writing the correlation $\rho_{12} = \cos\theta$, the CVAR volatility is

---

[1] **Synopsis:** For those of you who just tuned in, CVAR volatility measures the uncertainty in the contribution of risk of the corresponding variable to the total VAR. The

$$^{\$}\sigma_{CVAR} = \frac{\left|{}^{\$}\mathcal{E}_1 \sigma_1 {}^{\$}\mathcal{E}_2 \sigma_2 \sin\theta\right|}{{}^{\$}VAR^{Quad}} \quad (29.1)$$

Geometry for Risk Ellipse, VAR Line, CVAR, CVAR Vol

The following diagram gives the idea for the geometry. The details are below:

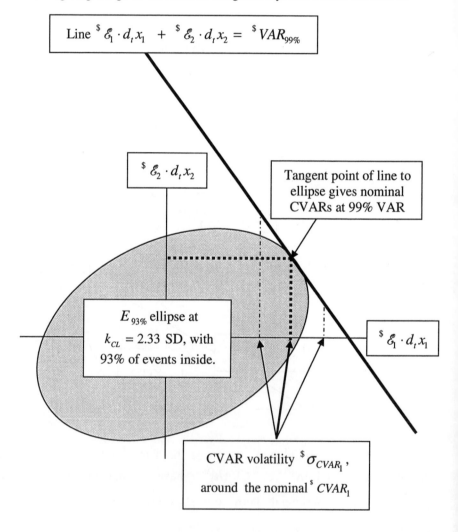

superscripts "Quad" indicate quadratic forms appropriate in the case of linear risk. For notational simplicity, dt = 1 here. See preceding VAR chapters for details.

VAR and CVAR for Two Variables

The axes are for the two individual risks $^\$\mathcal{E}_1 \cdot d_t x_1$ and $^\$\mathcal{E}_2 \cdot d_t x_2$. For illustration, we used a 99% CL.

The VAR Line for Two Variables

The line $^\$\mathcal{E}_1 \cdot d_t x_1 + {}^\$\mathcal{E}_2 \cdot d_t x_2 = {}^\$VAR_{99\%}$ is made up of points that, in a MC simulation, produce the value of the VAR at the 99% CL. This line defines a region to the left of the line containing 99% of probability, i.e. 99% of the MC events. To see this, recall that the integrated probability depending on the P&L variable F to be less than some given value F_{CL} is[2]

$$\wp[F \le F_{CL}] = \int_{-\infty}^{F_{CL}} \frac{dF}{{}^\$VAR^{Quad}\sqrt{2\pi}} \exp\left[-\frac{F^2}{2\left({}^\$VAR^{Quad}\right)^2}\right] \quad (29.2)$$

Setting $F_{CL} = k_{CL} \cdot {}^\$VAR^{Quad}$, we get $\wp[F \le F_{CL}] = N(k_{CL})$. For example, we get the usual 0.99 for $k_{CL} = 2.33$.

The quadratic form VAR is given by the usual 2-variable expression

$$\left({}^\$VAR^{Quad}\right)^2 = \left({}^\$\mathcal{E}_1 \sigma_1\right)^2 + \left({}^\$\mathcal{E}_2 \sigma_2\right)^2 + 2 {}^\$\mathcal{E}_1 \sigma_1 {}^\$\mathcal{E}_2 \sigma_2 \cos\theta \quad (29.3)$$

The Risk Ellipse E for Two Variables

The Risk Ellipse is an ellipse E whose boundary has a constant probability. At $n = 2$, the total probability distribution is the Gaussian (c.f. Ch. 28)

$$\wp^{(n=2)} = \frac{1}{\sqrt{1-\rho_{12}^2}} \left\{ \prod_{\alpha=1}^{2} \int_{-\infty}^{\infty} \frac{d(d_t x_\alpha)}{\sigma_\alpha \sqrt{2\pi}} \right\} \exp\left(-\Phi\left[\{d_t x_\alpha\}\right]\right) \quad (29.4)$$

We will wind up naturally not integrating over the full range of the variables. Writing the exponent as $\Phi\left[\{d_t x_\alpha\}\right] = R^2/2$, and setting $\rho_{12} = \cos\theta$ we have

---

[2] **The VAR pdf:** Again for those who skipped the formalism, under the linear risk assumption the possible values of the VAR are distributed as a Gaussian. The width of this Gaussian is just the quadratic-form VAR. To put in the dt dependence, the quadratic-form VAR is multiplied by sqrt(dt).

$$R^2 \sin^2 \theta = \left(\frac{d_t x_1}{\sigma_1}\right)^2 + \left(\frac{d_t x_2}{\sigma_2}\right)^2 - 2\frac{d_t x_1}{\sigma_1}\frac{d_t x_2}{\sigma_2}\cos\theta \qquad (29.5)$$

The surface traced out by the equation $R = $ constant is an ellipse in the $(d_t x_1, d_t x_2)$ plane, and a different ellipse E in the $({}^\$\mathcal{E}_1 \cdot d_t x_1, {}^\$\mathcal{E}_2 \cdot d_t x_2)$ plane.

Relation of VAR and the CVARs to the Risk Ellipse

The ellipse E has the property that if $R = k_{CL}$, then E is tangent to the VAR line, and this point is an extremal point for the risk on E. Moreover, the CVARs at this point are exactly the nominal values. We show this next.

We start by changing variables to elliptic co-ordinates, $d_t x_1 = \sigma_1 R \cos\beta$, $d_t x_2 = \sigma_2 R \cos(\beta - \theta)$. The probability measure is just $R\exp(R^2/2)dRd\beta$, independent of β. The total risk on the ellipse at fixed R is therefore ${}^\$C_E(R,\beta) = {}^\$\mathcal{E}_1 \sigma_1 R\cos\beta + {}^\$\mathcal{E}_2 \sigma_2 R\cos(\beta - \theta)$. We want the risk maximum on the ellipse by moving the angle β. Setting the derivative of ${}^\$C_E$ with respect to β equal to zero at β_{Max}, we get ${}^\$\mathcal{E}_1 \sigma_1 \cos\beta_{Max} = {}^\$ CVAR_1^{Quad}$ and ${}^\$\mathcal{E}_2 \sigma_2 \cos(\beta_{Max} - \theta) = {}^\$ CVAR_2^{Quad}$. Here the quadratic CVARs are the usual expressions (again, $\rho_{12} = \cos\theta$):

$$\begin{aligned}{}^\$ CVAR_1^{Quad} &= {}^\$\mathcal{E}_1\sigma_1\left({}^\$\mathcal{E}_1\sigma_1 + \cos\theta {}^\$\mathcal{E}_2\sigma_2\right)/{}^\$ VAR^{Quad} \\ {}^\$ CVAR_2^{Quad} &= {}^\$\mathcal{E}_2\sigma_2\left({}^\$\mathcal{E}_2\sigma_2 + \cos\theta {}^\$\mathcal{E}_1\sigma_1\right)/{}^\$ VAR^{Quad}\end{aligned} \qquad (29.6)$$

We have ${}^\$ VAR^{Quad} = {}^\$ CVAR_1^{Quad} + {}^\$ CVAR_2^{Quad}$. If further we set $R = k_{CL}$ we get ${}^\$\mathcal{E}_1 \cdot d_t x_1\big|_{\beta_{Max}} = k_{CL}{}^\$ CVAR_1^{Quad}$ and ${}^\$\mathcal{E}_2 \cdot d_t x_2\big|_{\beta_{Max}} = k_{CL}{}^\$ CVAR_2^{Quad}$. But these are just the nominal CVARs at k_{CL} standard deviations for the total risk. The maximum total risk on the ellipse E is therefore ${}^\$ C_E(k_{CL}, \beta_{Max}) = k_{CL}{}^\$ VAR^{Quad}$, and this is just the total VAR.

Therefore, we have shown that the maximum risk on the ellipse is just the total VAR. On the other hand, the risk along the VAR line is also the total VAR. Therefore, the VAR line is tangent to the ellipse with $R = k_{CL}$ at the point β_{Max}. This is shown by the figure above. As we go around the ellipse, the risk goes up and down, reaching a maximum at β_{Max}. Note that when $\beta = \beta_{Max} + \pi$ (on the opposite point of the ellipse), the sign of the risk changes and we make (instead of lose) the maximum amount of money on the ellipse[3].

It is worth noting that the percentage of events inside E generated by a MC simulation is less than 99%, because 99% of the events lie to the left of the VAR line and some of these events lie outside E. The integrated probability $\wp[R \le R_{Max}]$ inside E for $R \le R_{Max}$ is

$$\wp[R \le R_{Max}] = \int_0^{R_{Max}} \exp(-R^2/2) R \, dR \cdot \int_0^{2\pi} \frac{d\beta}{2\pi} \qquad (29.7)$$

Taking $R_{Max} = k_{CL} = 2.33$, we get $\wp[R \le k_{CL}] = 1 - \exp(-k_{CL}^2/2) \approx 93\%$ as the percentage of events lying inside E. This is why we notated the Risk Ellipse by $E_{93\%}$ in the figure.

The CVAR Volatility for Two Variables (bis)

We can now get some more insight into the CVAR volatility. Walking along the 99% CL VAR line away from the tangent point with $E_{93\%}$, the total risk stays the same (equal to VAR). However, the projections on the axes change. Since these projections are just the CVARs, we see that the CVARs can change with the total risk being unchanged.

We already know the form of the CVAR volatility $^\$\sigma_{CVAR}$ because we just calculated it. It turns out that if we look at a bigger risk ellipse $E_{99\%}$ with 99% of the events, then the projections of the intersections of $E_{99\%}$ with the 99% VAR line are related to the CVAR volatility. In particular, if we set $\cos\gamma = k_{CL}/R$ where R now corresponds to $E_{99\%}$, then the length $\mathcal{L}$ of the VAR line segment between its intersection points with $E_{99\%}$ is given by $\mathcal{L} = 2\sqrt{2} R \sin\gamma \cdot ^\$\sigma_{CVAR}$.

This ends the discussion of VAR and CVAR in the two-variable case.

---

[3] **Notation and Signs:** Note that by convention here, positive VAR means a loss.

30. Corporate-Level VAR (Tech. Index 3/10)

In this chapter, we consider additional topics related to applications of VAR and CVAR for corporate-level risk management. We first discuss aggregation issues. We then discuss implied correlations between business unit P&Ls. We end with a consideration of aged inventory.

Aggregation, Desks, and Business Units

Corporate Structure and Practical Aggregation Difficulties

Large banks and broker-dealers have a complex internal structure involving hundreds of products dealt with on many desks. The desks are arranged in a hierarchy into business units and/or divisions[1]. For corporate risk management over this entire structure, it is necessary to aggregate the risks of the individual components. A plethora of problems or difficulties can arise, both technical and non-technical.

Some technical difficulties, not necessarily in order of importance and certainly not complete, include:

- Data for time series: Availability, consistency, completeness etc.
- Systems: Hundreds of feeds, thousands of variables, legacy issues etc.
- Risk measures: Availability, timeliness, consistency, completeness etc.
- Calculation: Level of sophistication, huge correlation matrices etc.

We have spent a fair amount of time in this book discussing these technical issues in some detail. Other formidable difficulties are non-technical, including budgets, priorities, time limitations, personnel, communication, sociology, etc. Moreover

---

[1] **Example:** A business unit or division can be fixed income, equities, etc. The fixed income business unit has different desks, e.g. U.S. mortgages, corporate bonds, etc. The mortgage desk has a substructure of desks trading different mortgage products. The nomenclature and hierarchical details depend on the institution. Businesses of a completely different character within the corporation can include insurance, commercial credit cards, etc.

in this age of acquisitions and mergers, the corporate structure can change[2], requiring flexibility. Regulator requirements also exist that exert pressure. The bottom line is that these real-life issues can make corporate-level risk aggregation a gigantic, long, painful effort.

VAR Aggregation

Conceptually there is no difficulty in writing down VAR aggregation. In Ch. 27, we indicated the need to distinguish different products labeled by the symbol $\Im$. We wrote the exposure for a given underlying variable change[3] $d_t x_\alpha$ as the sum $^\$\mathcal{E}_\alpha = \sum_\Im {}^\$\mathcal{E}_\alpha^{(\Im)}$. We can use the same idea to designate different business units, desks, etc. in the corporate hierarchy. For simplicity, we will drop the product label in the following discussion, although it can easily be put back.

The index $a = 1...A$ will indicate corporate structural components, which for simplicity we just call "desks", or sometimes "business units". Inclusion of various levels of hierarchy does not change the logic, only requiring more indices. A total exposure $^\$\mathcal{E}_\alpha$ for $d_t x_\alpha$ (e.g. Libor DV01) is decomposed between desks as[4]

$$^\$\mathcal{E}_\alpha = \sum_{a=1}^{A} {}^\$\mathcal{E}_\alpha^{(a)} \tag{30.1}$$

The result for the total VAR is then obtained exactly as before, for example by Monte-Carlo simulation. The increasing levels of sophistication including fat-tail vols, stressed correlations, idiosyncratic risks, convexity, liquidity, etc. can in principle be included in the same way as we have described earlier.

---

[2] **Mergers and Acquisitions:** Each M&A can produce big changes in personnel, systems, etc. I personally lived through two large M&A events and a number of smaller ones.

[3] **Returns:** Again, the variable x_a can be the logarithm of the physical variable to describe lognormal dynamics, with the time difference $d_t x_\alpha$ producing returns.

[4] **Zero Exposures:** Naturally if a desk does not have a particular exposure there is no contribution to the sum. However, it is convenient to keep the formalism general.

Desk CVARs and Correlations between Desk Risks

CVARs and Stand-Alone Risks at the Desk Level

The risk for a given desk a can be defined by summing over internal variables and products for that desk. When the total VAR for the firm is evaluated at the specified CL at k_{CL} SD, we will get the CVAR for the desk, which we call $^\$CVAR^{(a)}$. So we write

$$^\$CVAR^{(a)} = \sum_{\alpha=1}^{n} \left\langle ^\$\mathcal{E}_\alpha^{(a)} \cdot d_t x_\alpha \right\rangle_{k_{CL}} \tag{30.2}$$

Now the state for the total VAR at the specified k_{CL} depends on the exposures from all desks. Hence, the CVAR for desk a also depends on exposures $\left\{ ^\$\mathcal{E}_\alpha^{(b)} \right\}$ for all desks $\{b\}$. We can see this clearly if we write down the quadratic form for the linear-risk case, now summed over desks:

$$\left(^\$VAR^{Quad} \right)^2 = \sum_{a,b=1}^{A} \sum_{\alpha,\beta=1}^{n} {}^\$\mathcal{E}_\alpha^{(a)} \sigma_\alpha \rho_{\alpha\beta} {}^\$\mathcal{E}_\beta^{(b)} \sigma_\beta \tag{30.3}$$

Then the total VAR at the specified CL at k_{CL} SD, over time dt, is given as usual by

$$^\$VAR(k_{CL}, dt) = k_{CL} \sqrt{dt} \cdot {}^\$VAR^{Quad} \tag{30.4}$$

To get $^\$CVAR^{(a)Quad}$ for desk a, we can use the same trick we used before to get the CVAR for a specific underlying risk, pulling off the appropriate sum $\sum_{a=1}^{A}$ so that $^\$VAR^{Quad} = \sum_{a=1}^{A} {}^\$CVAR^{(a)Quad}$. We get

$$^\$CVAR^{(a)Quad} = \sum_{\alpha,\beta=1}^{n} {}^\$\mathcal{E}_\alpha^{(a)} \sigma_\alpha \rho_{\alpha\beta} \sum_{b=1}^{A} {}^\$\mathcal{E}_\beta^{(b)} \sigma_\beta \Big/ {}^\$VAR^{Quad} \tag{30.5}$$

We also have the "Stand-Alone" risk, which is just the expression for the risk from a desk by itself. The quadratic stand-alone risk $^\$ SA^{(a)Quad}$ is also the desk volatility $^\$\sigma^{(a)}$, viz

$$\left(^\$\sigma^{(a)}\right)^2 = \left(^\$SA^{(a)Quad}\right)^2 = \sum_{\alpha,\beta=1}^{n} {}^\$\mathcal{E}_\alpha^{(a)} \sigma_\alpha \rho_{\alpha\beta} {}^\$\mathcal{E}_\beta^{(a)} \sigma_\beta \qquad (30.6)$$

Correlations between Desk Risks and P&L Correlations

Because the CVAR for a given desk depends on the other desks, it is clear that correlations exist between desk risks. Note that unless the idiosyncratic risks are included as described in Ch. 27 for Enhanced VAR, only "normal" risks will be included. Further, various business-related activities are not captured by either design or omission. These include commissions, changes in reserves, new transactions, etc. Hence, it may or may not be the case that the correlations calculated here approximate the actual P&L correlations between desks. That is, the actual P&L correlations between desks may have little to do with the correlations appropriate for VAR calculations. Further problems occur if the customer-related P&L is mixed in with the trading P&L if it is desired to treat the risks of these different activities separately.

We can get an approximation to correlations between desk risks using the linear formalism. With the above caveats, we denote the P&L for desk a (at time t over time dt for given moves $\{d_t x_\alpha\}$ in the underlying variables) as $^\$ PL^{(a)}$, where

$$^\$ PL^{(a)} = \sum_{\alpha=1}^{n} {}^\$\mathcal{E}_\alpha^{(a)} \cdot d_t x_\alpha \qquad (30.7)$$

Over time, assuming Gaussian statistics, we have $\langle d_t x_\alpha \cdot d_t x_\beta \rangle_c = \sigma_\alpha \sigma_\beta \rho_{\alpha\beta} dt$. Hence, we obtain the second cross moment for the P&Ls as $\langle ^\$ PL^{(a)} \cdot {}^\$ PL^{(b)} \rangle_c = \sum_{\alpha,\beta=1}^{n} {}^\$\mathcal{E}_\alpha^{(a)} \sigma_\alpha \rho_{\alpha\beta} {}^\$\mathcal{E}_\beta^{(b)} \sigma_\beta \cdot dt$, with the usual definition $\langle uv \rangle_c = \langle uv \rangle - \langle u \rangle \langle v \rangle$. We define the a,b desk-desk P&L correlation $\rho^{(a,b)}$ by

$$\rho^{(a,b)} = \frac{\langle ^\$ PL^{(a)} \cdot {}^\$ PL^{(b)} \rangle_c}{^\$\sigma^{(a)} \cdot {}^\$\sigma^{(b)}} \qquad (30.8)$$

where $^{\$}\sigma^{(a)} = \left\langle \left(^{\$}PL^{(a)} \right)^2 \right\rangle_c^{1/2}$. We then get the result[5]

$$\rho^{(a,b)} = \frac{\sum_{\alpha,\beta=1}^{n} {}^{\$}\mathcal{E}_\alpha^{(a)} \sigma_\alpha \rho_{\alpha\beta} {}^{\$}\mathcal{E}_\beta^{(b)} \sigma_\beta \cdot dt}{{}^{\$}\sigma^{(a)} \cdot {}^{\$}\sigma^{(b)}} \qquad (30.9)$$

Aged Inventory and Illiquidity

The aged inventory is a set of transactions, mostly with definite and pronounced illiquid attributes. We focus here on those positions that have been on the books for a long time and are tagged as being part of the aged inventory, and which cannot be transacted without potential substantial losses[6]. The aged inventory is naturally of concern to management and is monitored regularly.

Not all transactions that have been on the books for a long time are illiquid, since long-term strategies can exist for liquid markets. Liquid securities held for strategic purposes can be sold quickly without loss (except for some perceived opportunity cost of abandoning the strategy). Interesting as these strategies may be, we are only concerned here with illiquidity. Of course, it is possible that the strategy can fail and the result can be that the supposed liquid securities involved in the strategy suddenly become somewhat illiquid, and can be sold only with some losses.

The calculation of VAR for aged inventory is uncertain. It is difficult to deal with the risk of the aged inventory by its very nature. First, accurate current prices are hard to obtain, since by definition illiquid securities are not selling well in the market. Sometimes pricing sources will disagree significantly on pricing[7]. In exceptional cases, there may be no price at all. More importantly, there may not be direct risk exposure information, e.g. DV01, requiring extra assumptions.

---

[5] **Alternate Method:** There is also a method involving least squares fitting that produces desk-desk correlations to fit a given set of CVARs produced by MC Simulation. This method also works in the nonlinear case involving a convexity grid.

[6] **What is this Toxic Waste and Why is It There?** There are many types of illiquid securities. They can be bonds concentrated in lower credits or subordinated bonds in corporate or emerging market sectors, illiquid mortgage products, tranches of structured deals, etc. Illiquid securities can exist for many reasons, including. (1): Securities may be left over from underwriting deals that didn't sell out to investors, (2): There may be odd lot amounts from deals with customers, (3): There can be a drop in market demand which lowers liquidity, (4): There may be a decrease in credit quality that lowers liquidity, etc.

[7] **Disagreements of Pricing Sources:** Occasionally prices can differ widely, e.g. 30% for illiquid securities, including some mortgage derivatives, etc. Such pricing uncertainty reflects uncertainty in models, softness in the market, etc.

Variables for Aged Inventory

These include:

- *The horizon time* $t_{Horizon}$. This is the calendar date before which the desk plans to sell the security.
- *The liquidation time interval* Δt_{Liq}. This is the time that it would take to sell a security at a given price after the desk decides to sell it.
- *The liquidation penalty* λ_{Liq}. This is the penalty incurred selling into a stressed environment.

These variables are clearly related. Generally (though not always) there will be a buyer at a low enough price. Taking longer to sell (larger Δt_{Liq}) may give more opportunities for sale with less penalty λ_{Liq}. A later $t_{Horizon}$ can lead to a reversal of market factors leading to better liquidity. Of course, pessimistically, the reverse can happen.

Aged Inventory Reports

An aged inventory report might list a description of the illiquid deals, some remarks related to the volume of similar deals trading in the market, pricing, etc.

Quantitative Analysis of Aged Inventory

The difficulty of performing detailed quantitative analysis should be clear from the previous remarks. However, some analysis is possible in some circumstances. This can include the following:

- Statistical analysis can be performed for downward price moves at a very high confidence level on the tails of historical data, if relevant or analogous data exist. It is not enough just to look at ordinary standard deviations. The "fat tail" volatility, which we discuss in Ch. 21, should be used.
- In addition there can be slow but significant downward-moving price movements that are missed in any standard deviation calculation. If these are judged important, or for conservative estimates, they should be added.
- Judgmental scenarios based on proxy examples or analogous situations or events can be used. Sometimes such a scenario is all that is available. This method requires intimate knowledge and expertise of the local situation by the risk manager.

31. Issuer Credit Risk (Tech. Index 5/10)

In this chapter we discuss issuer credit risk[1] for bonds or other securities[2]. We will also consider the relation of issuer credit risk and market risk, and discuss why and how sometimes these are calculated separately. We also present a straightforward method of defining a unified credit + market risk measure.

An issuer is typically a corporation or a government that issues debt. Issuer credit risk for a bond[3] is the risk that the issuer of the bond suffers a credit downgrade or default. The quantitative determination of issuer credit risk for a particular bond relies on a variety of parameters:

- The starting credit α of the bond, determined by a rating agency[4, i].
- The probability for credit change $p_{\alpha \to \beta}$ from credit α to credit β.
- The probability for default $p_{\alpha \to Default}$.
- The recovery rate $\mathcal{R}$ for the bond in case of default.

Issuer risk for a portfolio of bonds relies in addition on

- The confidence level CL for the calculation.
- The portfolio of bonds: its composition, size, etc.

We will discuss various aspects of all these parameters. Given them, the issuer credit risk for a portfolio is the loss[5] determined, for example, by a Monte-

---

[1] **Acknowledgement**: I thank Jack Fuller and Jim Marker for many informative discussions on issuer credit risk. I thank Rick Stuckey for helpful conversations. I also thank Citigroup for providing Moody's transition/default matrices.

[2] **History**: The formalism of stressed transition/default matrices and the unified market + credit risk simulation described here was done by me in 2000-01.

[3] **Bonds and Loans:** We use the word "bond" in the text, but similar considerations with somewhat different parameters apply to loans.

[4] **Credit Ratings:** For the purposes of this chapter, we take the ratings of the rating agencies as defining the credit and assume that rating changes are timely within the time scale of the calculation, nominally one year. Some complications arise. For example, the credit rating of a bond in a foreign distressed environment will be negatively affected, independent of ratings of bonds by the same issuer in non-distressed environments.

Carlo simulation at the given confidence level CL. If sufficient reserves or other considerations exist to compensate for expected losses[6], the unexpected issuer credit risk is measured from the expected level. In this chapter we will mostly be concerned with unexpected issuer credit risk.

For the expected issuer credit risk, the same calculation is done and the average loss is picked out. This expectation can either be the numerical average or the median, although the interpretation of the unexpected loss is easier if the expectation is the numerical average.

A lot of confusion can be avoided at the start of a discussion if the participants would (please) specify what parameters and definition they mean to imply by the words "issuer risk".

The reader should also note that issuer risk is not counterparty risk[7].

Transition/Default Probability Matrices

The transition and default probability credit matrix contains the corresponding probabilities, starting at some time t over a given time period τ (e.g. 1 year). Using the Moody credit notation Aaa etc., the matrix for general credits α, β is

$$
\left(p_{\alpha \to \beta} \right) = \begin{pmatrix} p_{Aaa \to Aaa} & \cdots & p_{Aaa \to \alpha} & \cdots & p_{Aaa \to \beta} & \cdots & p_{Aaa \to C} & p_{Aaa \to Default} \\ \cdots & \cdots & \cdots & \cdots & \cdots & \cdots & \cdots & \cdots \\ p_{\alpha \to Aaa} & \cdots & p_{\alpha \to \alpha} & \cdots & p_{\alpha \to \beta} & \cdots & p_{\alpha \to C} & p_{\alpha \to Default} \\ \cdots & \cdots & \cdots & \cdots & \cdots & \cdots & \cdots & \cdots \\ p_{C \to Aaa} & \cdots & p_{C \to \alpha} & \cdots & p_{C \to \beta} & \cdots & p_{C \to C} & p_{C \to Default} \end{pmatrix}
$$

(31.1)

---

[5] **Currency Units:** Although the results for all bonds ultimately have to be expressed in the reporting currency, e.g. USD, we do not indicate currency units here.

[6] **Reserves and Expected Losses Consistency:** This consistency needs to be checked. The determination of the amount of the reserves and the calculations of the expected losses are not always performed by the same people.

[7] **Counterparty Risk is not Issuer Risk:** A counterparty X of a deal between X and Y is just the other party to Y. Counterparty risk is the risk that the counterparty defaults on some condition of the deal. This is not issuer risk. The counterparty can default but the issuer of the bond can be solvent, or vice-versa. We mentioned counterparty risk in Ch. 8.

Examples (Bad Year, Average, Worst Cases)

Probabilities are tabulated by rating agencies. For example, for 1990 (a particularly bad year), the 1-yr transition/default matrix for U.S. corporates was [8]:

1990 Historical 1-year Credit Transition and Default Matrix, Moody's							
	Aaa	Aa	A	Baa	Ba	B	Caa-C	Default
Aaa	**88.8%**	11.2%	0.0%	0.0%	0.0%	0.0%	0.0%	0.0%
Aa	0.0%	**90.0%**	9.8%	0.3%	0.0%	0.0%	0.0%	0.0%
A	0.0%	2.6%	**89.0%**	6.9%	1.2%	0.4%	0.0%	0.0%
Baa	0.0%	0.3%	3.4%	**90.4%**	4.9%	0.6%	0.3%	0.0%
Ba	0.0%	0.0%	0.0%	2.4%	**78.4%**	14.8%	0.7%	3.8%
B	0.0%	0.6%	0.3%	0.6%	3.0%	**74.7%**	3.3%	17.5%
Caa-C	0.0%	0.0%	0.0%	0.0%	0.0%	0.0%	**42.9%**	57.1%

Moody has tabulated these matrices since 1970. The average transition/default matrix (over 1970-99) naturally shows less risk than for the bad year 1990:

Transition & Default Matrix Average 1970 - 1999							
	Aaa	Aa	A	Baa	Ba	B	Caa-C	Default
Aaa	**92.1%**	7.3%	0.6%	0.0%	0.0%	0.0%	0.0%	0.0%
Aa	1.4%	**90.6%**	7.6%	0.3%	0.1%	0.0%	0.0%	0.0%
A	0.1%	2.2%	**92.2%**	4.9%	0.5%	0.1%	0.0%	0.0%
Baa	0.1%	0.2%	5.4%	**88.4%**	4.9%	0.7%	0.1%	0.1%
Ba	0.0%	0.0%	0.4%	5.4%	**86.2%**	6.1%	0.4%	1.3%
B	0.0%	0.0%	0.1%	0.4%	6.9%	**83.7%**	2.0%	6.8%
Caa-C	0.0%	0.0%	0.0%	0.7%	2.0%	4.2%	**67.3%**	25.8%

Here is the number of cases in each cell of the 1990 matrix:

Number of Cases for Each Credit Transition or Default, 1990							
	Aaa	Aa	A	Baa	Ba	B	Caa-C	Default
Aaa	**159**	20	0	0	0	0	0	0
Aa	0	**332**	36	1	0	0	0	0
A	0	15	**515**	40	7	2	0	0
Baa	0	1	11	**293**	16	2	1	0
Ba	0	0	0	10	**333**	63	3	16
B	0	2	1	2	10	**248**	11	58
Caa-C	0	0	0	0	0	0	6	8

---

[8] **Matrices with Sub Grades or lumped Whole Grades:** There are also matrices for subgrades with more credit-level refinement (and naturally fewer cases per cell). Matrices can also be defined with whole grades lumped together. The matrices in the text have all C-grade bonds lumped together.

For a given portfolio in an historical simulation, the worst-case loss will result from one of the historical matrices. We could call this matrix the "Historical Worst Case" matrix.

The year for the worst case for a given $p_{\alpha \to \beta}$ matrix element differs depending on the matrix element. Using this fact, a "Theoretical Historical Worst Case" matrix can be constructed that is even worse than the "Historical Worst Case". We first define the quantity $p_{\alpha \to \beta}^{MaxMin}$. This is the maximum probability in the case of downgrades or default, or the minimum probability in the case of unchanged credit or upgrades, measured over time, viz

$$p_{\alpha \to \beta}^{MaxMin} \equiv \begin{pmatrix} \max_t \{p_{\alpha \to \beta}(t)\} & \text{if Downgrade, Default} \\ \min_t \{p_{\alpha \to \beta}(t)\} & \text{if Unchanged, Upgrade} \end{pmatrix} \quad (31.2)$$

Then the theoretical worst-case matrix element $p_{\alpha \to \beta}^{TheorWorstCase}$ can be defined by normalizing the total probability to one, namely

$$p_{\alpha \to \beta}^{TheorWorstCase} = p_{\alpha \to \beta}^{MaxMin} \Big/ \sum_{\gamma=Aaa}^{Default} \left[p_{\alpha \to \gamma}^{MaxMin} \right] \quad (31.3)$$

The result of performing these operations is given in the table below:

Theor. Worst Case (Max downgrade, Min unchanged or upgrade; Renormalized)								
	Aaa	Aa	A	Baa	Ba	B	Caa-C	Default
Aaa	**66.3%**	23.2%	9.9%	0.0%	0.5%	0.0%	0.0%	0.0%
Aa	0.0%	**78.5%**	18.1%	1.8%	0.7%	0.3%	0.0%	0.6%
A	0.0%	0.0%	**81.1%**	14.7%	2.4%	1.2%	0.2%	0.3%
Baa	0.0%	0.0%	1.2%	**86.2%**	2.4%	6.2%	1.6%	2.5%
Ba	0.0%	0.0%	0.0%	2.3%	**58.4%**	28.1%	5.0%	6.2%
B	0.0%	0.0%	0.0%	0.0%	0.0%	**68.4%**	7.0%	24.6%
Caa-C	0.0%	0.0%	0.0%	0.0%	0.0%	0.0%	**0.0%**	100.0%

Models for Stressed Transition/Default Probability Matrices

In this section we will be interested in constructing a stressed matrix for a bad credit environment away from the average, using a model approach. Since we naturally are going to be looking at losses, we want to increase the probability of downgrades or default, and decrease the probability of upgrades or unchanged credit. To this end, we define the sign flag $\eta_{\alpha \to \beta}$ as

$$\eta_{\alpha \to \beta} = \begin{pmatrix} +1 & \text{if Downgrade, Default} \\ -1 & \text{if Unchanged, Upgrade} \end{pmatrix} \quad (31.4)$$

We denote as $\sigma_{\alpha \to \beta}$ the historical volatility, defined for each $p_{\alpha \to \beta}$, as given by the standard deviation of $p_{\alpha \to \beta}(t)$ over time from the available data.

Examples of Model Transition/Default Matrices

As a simple approach to a stressed matrix, we can specify a certain number $k_{\alpha \to \beta}$ of standard deviations $\sigma_{\alpha \to \beta}$ for each transition (and similarly for defaults). We then define $\delta p_{\alpha \to \beta}$ by the following model ansatz:

$$\delta p_{\alpha \to \beta}(\sigma_{\alpha \to \beta}, k_{\alpha \to \beta}) = \max(\eta_{\alpha \to \beta} k_{\alpha \to \beta} \sigma_{\alpha \to \beta}, 0) \quad (31.5)$$

We use $\delta p_{\alpha \to \beta}$ to perturb the time-averaged transition/default matrix $\langle p_{\alpha \to \beta} \rangle$. After renormalization to unit probability, we get the model stressed matrix element $p_{\alpha \to \beta}^{Stressed}$ as

$$p_{\alpha \to \beta}^{Stressed} = \left[\langle p_{\alpha \to \beta} \rangle + \delta p_{\alpha \to \beta} \right] \Big/ \sum_{\gamma = Aaa}^{Default} \left[\langle p_{\alpha \to \gamma} \rangle + \delta p_{\alpha \to \gamma} \right] \quad (31.6)$$

For example, suppose $k_{\alpha \to \beta} = k_{\alpha \to Default} = 1.0$. We get in this case the model stressed matrix shown below:

Model Stressed Matrix, # stdev are	$k_{\alpha\beta}$ =			1.0	and $k_{\alpha,Default}$ =			1.0
	Aaa	Aa	A	Baa	Ba	B	Caa-C	Default
Aaa	83.8%	13.4%	2.7%	0.0%	0.1%	0.0%	0.0%	0.0%
Aa	0.2%	86.3%	12.2%	0.8%	0.3%	0.1%	0.0%	0.1%
A	0.0%	0.0%	90.5%	7.8%	1.2%	0.4%	0.0%	0.1%
Baa	0.0%	0.0%	0.6%	87.7%	9.3%	1.7%	0.3%	0.5%
Ba	0.0%	0.0%	0.0%	2.5%	81.7%	11.6%	1.3%	2.8%
B	0.0%	0.0%	0.0%	0.0%	3.3%	80.3%	4.1%	12.3%
Caa-C	0.0%	0.0%	0.0%	0.0%	0.0%	0.0%	46.6%	53.4%

Referring back, we see that this stressed probability matrix bears a qualitative resemblance to the historical bad 1990 matrix.

We can generate other model matrices by varying the parameters $\{k_{\alpha\to\beta}\}$. For example, if we set $k_{\alpha\to\beta} = k_{\alpha\to Default} = 2.3$, so each matrix element is taken at a 99% CL with the cutoff in the model ansatz, we get (after renormalization) the following stressed matrix:

Model Stressed Matrix, # stdev are $k_{\alpha\beta}$ =			2.3	and $k_{\alpha,Default}$ =			2.3
	Aaa	Aa	A	Baa	Ba	B	Caa-C	Default
Aaa	**73.3%**	21.2%	5.2%	0.0%	0.3%	0.0%	0.0%	0.0%
Aa	0.0%	**79.3%**	17.9%	1.5%	0.5%	0.1%	0.0%	0.6%
A	0.0%	0.0%	**85.2%**	11.6%	2.1%	0.7%	0.1%	0.3%
Baa	0.0%	0.0%	0.0%	**79.3%**	15.1%	3.1%	0.6%	1.9%
Ba	0.0%	0.0%	0.0%	0.0%	**70.3%**	18.2%	2.4%	9.1%
B	0.0%	0.0%	0.0%	0.0%	0.0%	**63.3%**	6.1%	30.6%
Caa-C	0.0%	0.0%	0.0%	0.0%	0.0%	0.0%	**6.2%**	93.8%

This stressed matrix is qualitatively similar to the theoretical worst-case historical matrix.

Actually we have not been trying to fit anything. Better fits could be obtained by refining the choice of the parameters $\{k_{\alpha\to\beta}\}$. We know roughly how to choose the $\{k_{\alpha\to\beta}\}$ to get the historically bad and worst-case matrices. Hence we can move these parameters around in a sensible to generate many stressed model matrices. We can also generate matrices further out on the tail even than the theoretical historical worst-case matrix. Note that there can be smoothing issues. In this matrix, the stressed default probability of Aa is larger than that of A, which is not reasonable.

Stochastic Transition/Default Matrices

A model to generate stochastic matrices can be envisioned as an extension of the scenario stressed matrix approach just described. We would need a multivariate formalism, including correlations $\rho_{\alpha,\beta;\alpha',\beta'} = \rho(\delta p_{\alpha\to\beta}, \delta p_{\alpha'\to\beta'})$ between changes $\delta p_{\alpha\to\beta}$, $\delta p_{\alpha'\to\beta'}$ from the average of different matrix elements. We also need to replace $\{k_{\alpha\to\beta}\}$ with random numbers. Finally we need constraints or a parameterization to ensure a rough monotonic decrease of transitions away from the diagonal. This just means that at least most of the time, two-level transitions should be less probable than one-level transitions, etc.

In this way we can establish a method to generate stochastic transition/default credit matrices.

Issuer Credit Risk

Calculation of Issuer Risk—Generic Case

We can determine the issuer credit risk due to downgrade and default by straightforward simulation. We merely run through the portfolio[9] one bond at a time[10], get the distribution of losses for the portfolio, pick out the portfolio loss at some given confidence level, subtract the average loss, and write up a nice report for the management with some pretty color plots[11].

We will explicitly consider a pedagogical simple case of a portfolio of plain-vanilla bonds, a single transition/default matrix, a single recovery rate in case of default, and well defined distinct spread levels. We generalize to more complicated cases below.

Bond Price Changes and Spreads; the Spread DV01

We quickly set up some generic notation. A bond's price $\mathcal{B}(s)$ depends on its spread $s = y_\mathcal{B} - y_{Base}$, where $y_\mathcal{B}$ is the bond's yield and y_{Base} is the base yield (e.g. treasury) at the same maturity. The bond price for credit rating α is $\mathcal{B}_\alpha = \mathcal{B}(s_\alpha)$ where s_α is the spread for bonds with credit α.

A bond's spread s_α, its credit rating α, and the values of the probabilities $\{p_{\alpha \to \beta}\}$ are naturally closely related. The change in the bond value $\delta \mathcal{B}_\alpha$ is determined by changes in spreads, as we next describe.

The "spread DV01" is $\Delta_\alpha = 10^{-4} \cdot \partial \mathcal{B}_\alpha / \partial s_\alpha$ with spreads measured in bp (or more exactly, bp/yr). The change in value of the bond for credit change $\alpha \to \beta$, with corresponding spread change $\delta s_{\alpha\beta} = s_\beta - s_\alpha$, is $\delta \mathcal{B}_\alpha^{CreditChange} = \Delta_\alpha \cdot \delta s_{\alpha\beta}$. If there is no credit change ($\alpha \to \alpha$), then $\delta \mathcal{B}_\alpha^{NoCreditChange} \approx 0$, i.e. to first approximation there is no price change. We will amplify this statement when we discuss market risk.

---

[9] **Portfolio Data:** Hopefully you will receive pristine data, all the credit ratings will be present, current and consistent, no bonds will be missing, and the files will not have any formatting errors. In the contrary case, please refer to the Black Hole Data Theorem.

[10] **Netting of Short and Long Bond Positions for Given Issuer:** All bond positions (long plus short) of a *given* issuer at a *given* credit rating in the *same* portfolio are netted before performing the calculation. However, a net short position of one issuer behaves very differently than a net long position of another issuer, as we shall see.

[11] **Color Plots in Reports:** Don't knock them. You will be happier if the management likes your presentations.

Defaulting Bond Price Changes

For default ($\alpha \to Default$), the bond loses its value[12] $\mathcal{B}_\alpha$, and it gains the recovery rate $\mathcal{R}$ fraction of the notional $\mathcal{N}$, so $\delta\mathcal{B}_\alpha^{Default} = -\mathcal{B}_\alpha + \mathcal{R}\cdot\mathcal{N}$.

Monte-Carlo Simulations of Credit Issuer Risk at a given CL

The unit interval is partitioned into segments of lengths $\{p_{\alpha\to\beta}\}$ and $p_{\alpha\to Default}$ (the lengths adding to one). Perform Run_1, the first portfolio "run". Given a bond $\mathcal{B}_\alpha$ with credit α, get a random number $r(\mathcal{B}_\alpha; Run_1)$ from the uniform distribution $U(0,1)$. This picks out a final state (credit β or Default), with corresponding matrix element $p_{\alpha\to\beta}$ or $p_{\alpha\to Default}$. The change in value of the bond, $\delta\mathcal{B}_\alpha(Run_1)$, is given by the appropriate case listed above.

Continuing successively with each bond and adding up the changes to get the total portfolio change for the first run, $\delta\mathcal{P}(Run_1) = \sum_{AllBonds} \delta\mathcal{B}_\alpha(Run_1)$, which we save. We repeat the whole procedure for many runs, obtaining the set of portfolio changes $\{\delta\mathcal{P}(Run_\lambda)\}$. We then pick out the change $\delta\mathcal{P}^{CL}$ at the specified confidence level CL for the calculation.

It needs to be emphasized that this calculation is *not* the expected credit risk[13] $\langle\delta\mathcal{P}\rangle$. In fact, the reported risk is the difference between $\delta\mathcal{P}^{CL}$ and $\langle\delta\mathcal{P}\rangle$,

$$\delta\mathcal{P}^{Reported} = \delta\mathcal{P}^{CL} - \langle\delta\mathcal{P}\rangle \tag{31.7}$$

Transition/Default Probability Uncertainties

If we have many possible transition/default matrices $\{(p_{\alpha\to\beta}^{(X)})\}$ labeled by an index X, we simply precede each run by throwing the dice to choose a particular matrix. For example, the historical simulation is obtained by choosing a matrix

---

[12] **Bond Value Change Under Default:** Sometimes the bond is taken to lose value equal to its *notional* and get back the recovery fraction of its notional. This is not appropriate for a bond that is marked to market, as in a trading portfolio. For a bond in a holding portfolio that is carried at notional value, it is appropriate. More details are given below.

[13] **Expected Credit Loss vs. High-CL Credit Loss:** Some people are used to calculating expected credit losses <δP> but not high-CL unexpected credit losses δP^{CL}. Intuition is quite different in these two cases. Statements like "That bond will never default. I don't believe your calculation" may be appropriate for expected loss but not for high-CL loss.

Issuer Credit Risk

from some year t at random, $\left(p_{\alpha\to\beta}(t)\right)$ to start a run, repeating such a draw to get a matrix for each run through the portfolio. If the transition/default matrices are generated through a model, the same procedure is adopted.

Recovery Rate Uncertainty

The recovery rate in case of default is not constant in time. If economic conditions are bad, less will be recovered because less is available to be recovered. The recovery rate also depends on the type of security (e.g. bonds or loans), on the seniority of the bond (senior vs. junior subordinated debt), and on complex legal issues. The uncertainty in the recovery rate can be modeled by a distribution $\wp(\mathcal{R})$ with a draw from the distribution every time a default is signaled in the simulation.

Spread Uncertainty for a Given Credit

So far we have assumed that a spread s_α for a given credit α is unique. This is not true. There is first some uncertainty δs_α in the spread at a fixed credit α at a given time. There is also a time dependence of a spread over time interval Dt. These features can be modeled. In the historical simulation, we choose a spread $s_\alpha(t)$ at time t including an extra random uncertainty δs_α. Then, for a credit transition $\alpha \to \beta$ over time interval Dt, we take the spread change $\delta_t s_{\alpha\beta}(t) = s_\beta(t+Dt) - s_\alpha(t)$, also including some extra random uncertainty for the final spread $s_\beta(t+Dt)$. We consider this further when we discuss market risk below.

Short Positions: Why They Don't Alleviate the Risk at High CL

We have already mentioned[10] that short and long positions of a *given* issuer for a *given* credit in the *same* portfolio are netted out to define what we have been calling a "bond" in this chapter, before performing the issuer-risk calculation. A net short position in a bond contributes to the average or expected risk. *However, generally a net short position in a bond hardly contributes at all to the unexpected loss at a high CL*. In a high-CL loss state for unexpected issuer risk, short positions do not default. This is because the gains due to the default of a short position show up in states that are less risky than the state specified at the high-risk CL. For this reason, it is misleading to try to get a "ballpark" unexpected risk number for a portfolio by subtracting the short notional from the long notional and using some average credit value.

Concentration (Large Position) Risk

The calculations for a high CL can be lumpy in an important sense. If we have one bond $\mathcal{B}^{Huge}$ in a portfolio with an exceptionally large notional, the result at a high CL can be that this bond defaults. The concentration risk that this implies can be very large, much larger than the loss at a high CL without that bond. Moreover, as we move the CL down, at some point the $\mathcal{B}^{Huge}$ bond will not default, and the issuer risk can suddenly jump down to a much lower level. Conversely, if we move the CL up, at some point the $\mathcal{B}^{Huge}$ bond will default in the calculation, and the traders may start screaming[13].

Dependence of the Results on Portfolio Definitions

Naturally if the portfolio with the $\mathcal{B}^{Huge}$ bond is put into another portfolio with other "huge" bonds, the relative importance of the $\mathcal{B}^{Huge}$ bond is less, and at the given high CL this bond may no longer default.

Therefore the portfolio structure, which may depend on arbitrary definitions, can be very important for the calculation of credit risk. We will exhibit this sort of effect in the example of issuer risk below.

Credit Derivatives and Issuer Risk

Credit derivatives are instruments that pay off under some sort of credit event. The event can be a default, downgrade, spread change, etc. We will not treat credit derivatives in this book. However, we do need to mention that risk offsets occur due to the netting of credit derivatives with bonds. The degree of risk netting has to be determined on a case-by-case basis. Offsets should be included appropriately in credit issuer-risk calculations.

Credit Correlations

If we try to model transitions and default from an ab-initio perspective, the transitions $\alpha \to \beta$ and default $\alpha \to Default$ from an initial state α are in general dependent on the transitions $\alpha' \to \beta'$ and default $\alpha' \to Default$ from another initial state α'. That is, there are correlations between different transitions and between different defaults. Such correlations are critical for some credit derivatives. The modeling of credit correlations is a topic of intense research, which however we will not discuss in this book.

If we base the issuer-risk credit calculations on historical credit transition/default matrices, this is not a problem. The correlations that existed historically are already built into the historical transition and default probability numbers. The stressed matrices retain a measure of these correlations. Hence, the bonds can be treated as independent in this approach.

Other Refinements (Exposure Changes, Liquidity, Etc.)

In the same way that we discussed refinements to Plain-Vanilla VAR for market risk in another part of this book, we can also discuss liquidity times, exposure scenarios, multi-step MC simulations and other refinements for credit issuer risk.

Example of Issuer Credit Risk Calculation

Here is an illustrative simple example. We take two business units and eight "bonds" representing desk portfolios that are investment grade or junk. The rows in the table below give the "bond" names, their loss in case of default (in $000), their average credit, their probabilities of default, and their business unit names. This simple calculation ignores non-default transitions.

Bonds, Loss if default ($000), Credit, Prob. Default, Business Unit							
B1	B2	B3	B4	B5	B6	B7	B8
$1,800	$15,000	$1,200	$12,600	$17,400	$150,000	$2,700	$4,020
Ba	Aa	Ba	Aa	Ba	Aa	Ba	Ba
0.0301	0.0014	0.0301	0.0014	0.0301	0.0014	0.0301	0.0301
a	a	a	b	b	b	b	b

There are 256 possible states, where each bond can default or not. Each state has a composite probability obtained by multiplication (e.g. 0.0301 if $B1$ defaults, times 0.9986 if $B2$ doesn't default, etc). We can generate a pseudo Monte-Carlo simulator by listing states and probabilities. We assign to each state a fraction of the total number of MC "paths" equal to that state's probability. We list the losses in decreasing order to get the loss for each CL by counting the appropriate number of "paths" for that CL (or as close to it as possible).

Now lets add one bond at a time and calculate the losses at the 99.6% CL, as an example. We get the results below:

Losses for different portfolios successively adding bonds at CL = 99.6%							
B1	B2	B3	B4	B5	B6	B7	B8
$1,800							
$1,800	$ -						
$1,800	$ -	$ -					
$1,800	$ -	$ -	$ -				
$ -	$ -	$ -	$ -	$17,400			
$ -	$ -	$ -	$ -	$17,400	$ -		
$ -	$ -	$1,200	$ -	$17,400	$ -	$ -	
$ -	$ -	$ -	$12,600	$ -	$ -	$2,700	$4,020

The first bond $B1$ alone by itself defaults at the 99.6% CL. This is because the probability of default is 3.01% for this junk bond, well above this CL threshold. Now a curious thing happens. As we successively add one bond at a time, the issuer risk at this CL remains the same up to $B4$. Physically, it is clear that a portfolio with $(B1...B4)$ has more risk than a portfolio with just $B1$. Nonetheless, at this CL, the risk does not change. When we add $B5$, only $B5$ defaults. This says that the issuer risk has shifted from business unit a to business unit b. Adding $B6$ does not change the issuer risk. When $B7$ is added, $B3$ and $B5$ both default. Finally with the full portfolio $(B1...B8)$, three bonds $B4$, $B7$, and $B8$ default.

How is this possible? The lumpy nature of the portfolio with a small number of bonds has exaggerated the problems, but all MC simulations do exhibit some of the same characteristics to some extent.

Now lets look at losses of the full portfolio $(B1...B8)$ as a function of the confidence level. Here are some results:

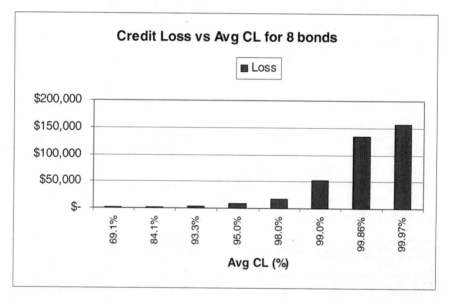

Actually what we have done is to calculate the average integrated risk between confidence levels with the averages shown on the graph. See the end of Chapter 27 on VAR for an introduction.

We can also look at the individual risks of the two business units, the sum of the standalone risks, and the diversified risk as a function of the CL. Here is a graph of the results:

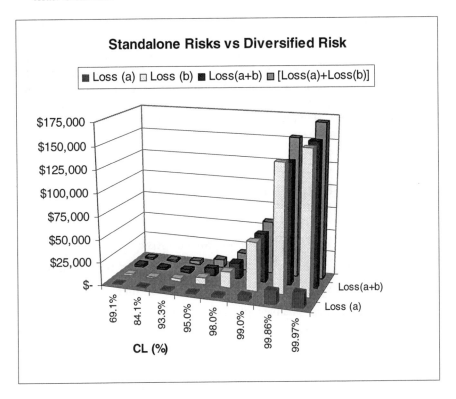

The sum of the standalone losses has to be bigger than the loss of the total portfolio. This is satisfied at the upper six CL in the graph, but violated in the lower two CL in the graph. While this violation is theoretically unsatisfactory, the risks happen to be small here. We consider these considerations below.

Credit Factors

In order to sidestep the lumpy nature of MC issuer-risk simulations, a credit factor approach can be taken. In this approach, a credit factor f_α is used for each bond $\mathcal{B}_\alpha$ with credit α. We then write $\$\delta C(\mathcal{B}_\alpha) = f_\alpha \cdot \$N(\mathcal{B}_\alpha)$ for the issuer risk $\$\delta C(\mathcal{B}_\alpha)$ of the bond $\mathcal{B}_\alpha$, with notional $\$N(\mathcal{B}_\alpha)$. The credit factors can be chosen such that the total issuer risk is the same as the issuer risk obtained at the desired CL in a MC simulator at a given time with a given total portfolio. As a new deal is added, the additional credit issuer risk is simply added using the credit factor.

The advantage of this approach is that it is smooth, avoiding the lumpy character of the MC simulations. It also gives equal risk to the same bond that might administratively find itself in one portfolio or a different portfolio, avoiding internal arbitrage situations.

In practice a number of issues arise. These are related to the changing nature of the portfolio over time (the credit factors start to lose their association with a MC simulation at a given CL), concentration issues (bigger deals should have bigger factors), recovery rate uncertainties, etc. Periodic updates have to be made to the credit factors. Finally, the normalization (sum of standalones, total diversified, etc.) has to be specified.

Issuer Credit Risk and Market Risk: Separation via Spreads

Spread Uncertainty at Fixed Credit and Market Risk

The uncertainty δs_α in the spread at fixed credit and fixed time along with its time dependence $d_t s_\alpha(t) = s_\alpha(t + Dt) - s_\alpha(t)$ over time Dt introduces a nonzero value in the change of bond value for no credit change, $\delta \mathcal{B}_\alpha^{NoCreditChange} \neq 0$. We can, *purely by convention*, call "market spread risk" the change in the portfolio with no credit changes and no defaults. This risk is then *not* to be included in the credit issuer risk.

Spread Gaps Between Different Credits and Credit Risk

The separation of the total risk into market risk with unchanged credit $\alpha \to \alpha$ and credit risk with credit changes $\alpha \to \beta$ and defaults $\alpha \to Default$ makes sense. First, spreads are affected by many technical market factors having incalculable (if any) relations or correlations with credit[ii]. Second, spread changes for credit transitions are generally much larger than spread changes for a given credit either for changing time or for fixed time. That is,

$$\delta_t s_{\alpha\beta} > d_t s_\alpha \qquad (31.8)$$

$$\delta s_{\alpha\beta} > \delta s_\alpha \qquad (31.9)$$

Separation of Market and Credit Risk

Because of the inequalities above, the market spread risk can be rather cleanly separated from the credit spread risk. The market risk calculation is done with statistics for the spread changes that are not big enough to involve credit changes.

The risk separation also may be driven from "sociology", as market risk managers and credit risk managers may be in different departments[14].

Separating Market and Credit Risk without Double Counting

We need to avoid double counting. This can be done since the small spread uncertainty at fixed credit associated with market risk is much less than the large spread gap between different credits giving credit issuer risk, as shown below:

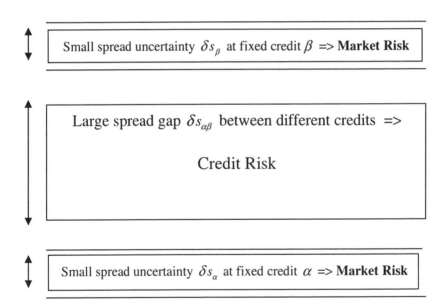

To account for all the spread risk, we have to include all values of spread changes. The relatively small spread changes associated with fixed credit $\alpha \to \alpha$ over one year do *not,* arguably, have any credit component. This is because rating

---

[14] **Myopia:** Separation of market and credit risk departments can lead to a situation where some people may not see the advantage in a consistent unified credit + market risk calculation, or fully understand what it means.

agencies do not change their definitions for rating criteria over one year for a fixed credit.

Although there is a credit component in the spread s_α, this credit component arguably *cancels out* in the difference $s_\alpha(t+1yr) - s_\alpha(t)$. For much larger spread changes, credit changes $\alpha \rightarrow \beta$ do occur. Such very large spread changes are associated with issuer credit risk. Since all spread changes (small and large) are included, and since any spread change is assigned uniquely, there is no double counting. In this way, double counting is avoided.

Example for High CL Credit and Market Risk

In this section we give a numerical example. For high credit risk, we need to choose a bad year and junk credit. At the same time, the market risk is also high since spreads for a given credit then increase a lot. For illustration, we choose the bad year 1990, and we consider the transition from the junk credit $\alpha = Ba$ in 1989 to the even lower credit $\beta = B$ in 1990. The picture below gives the idea[15]:

Large Spread Change in 1989-1990 Credit Transition from Ba to B

$\beta = B$ 850 bp

$\alpha = Ba$ 270 bp

1989 $\longrightarrow$ 1990

The credit risk for the change in spread for this drop in junk credit from Ba to B was thus $850 - 270 = 580$ bp/yr, or in \$ using the spread DV01,

$$\$\delta C_{Ba \rightarrow B}^{\text{Credit Risk}} = 580 \text{ bp/yr} \cdot \$\Delta_{Ba}^{\text{Spread DV01}} \qquad (31.10)$$

---

[15] **Acknowledgement:** We thank Citigroup for the use of these spread data. Numbers rounded off.

We need to compare this credit spread change with the change in spread at fixed credit due to market risk. For the market risk, we look at the volatility $\sigma(s_{Ba})$ of the fixed credit Ba spread over one year. From the data between 1986-1999 we get $\sigma(s_{Ba}) \approx 100$ bp/yr.

From MC simulations, we know[16] that even at a very high overall CL for the total risk, the CL for an individual variable is generally at or less than 99%. Hence, for a bad level of risk for spreads at a given credit, consistent with the Economic Capital CL of 99.97%, we take the 99% CL, or $k_{99\%CL} \approx 2.3$ SD. The stressed market risk for the Ba spread is thus $k_{99\%CL}\sigma(s_{Ba}) \approx 230$ bp/yr, or in $ using the spread DV01,

$$\$\delta C_{Ba}^{\text{Market Risk}} = 230 \text{ bp/yr} \cdot \$\Delta_{Ba}^{\text{Spread DV01}} \qquad (31.11)$$

Hence, at stressed levels for both market risk and credit risk, the market risk is 40% of the credit risk in this example. There is no double counting since the regions of spread change are separated.

Advantages of Separating Market and Credit Issuer Risk

An advantage of the separation is that market risk is generally calculated with many variables besides spreads (e.g. volatilities etc.) as we discussed in the chapters on VAR (Ch. 26-30). Credit issuer risk calculations do not involve these variables. Because we want to include spread risk consistently with risk from these other variables, the market/credit risk separation again makes sense.

There are also idiosyncratic risks that the market risk managers may include that are not present in the spread data used by the credit risk calculation.

Disadvantages of Separating Credit and Market Risks

The first disadvantage of this separation is that at the corporate level, the market and credit risks have to be reassembled. Inconsistent assumptions and procedures between the market risk and the credit risk calculations can hinder this reassembly. In the next section, one way to achieve the combination is shown.

Another downside is a different kind of double-counting error. This is due to administration, rather than anything fundamental. Market risk managers, while correctly assessing spread risk for unchanged credit ratings, may not include the fact that the probability is not one that the credit does not change. Market risk managers may be consistent in their assumption that credit does not change by

---

[16] **Individual Variable CL:** An individual variable's CL is generally much less than the overall CL for the total risk. In practice, even the CL for an important variable is generally no larger than 99%, even with the overall CL being at the Aa-credit Economic Capital value 99.97%. See the chapters on VAR earlier in this book.

choosing spread changes that are not big enough to be associated with credit changes. However, in the end the market spread risk for credit α needs to be multiplied by the diagonal probability $p_{\alpha \to \alpha}$.

There are reporting issues. Should the reduction of the market spread risk by factors $(1 - p_{\alpha \to \alpha})$ be put into market risk? If so, the market risk managers have to be concerned with the credit risk calculation of $(1 - p_{\alpha \to \alpha})$. Alternatively, since the factor $(1 - p_{\alpha \to \alpha})$ is a credit factor, should the reduction in market risk be subtracted from the credit risk? If so, how do we include correlations? These conundrums are absent in the unified credit + market approach, described next.

A Unified Credit + Market Risk Model

In this section we show one way of how to resolve the difficulties of the separation of market and credit risk while still preserving the advantages of that separation. For general descriptions and specific aspects, see e.g. the talks by N. Marinovich, S. Turnbull, D. Rosen, T. Tracy, B. Selvaggio, C. Monet, F. Iacono, and N. Sparks at the RISK 2000 Conference[iii]. A plethora of other references exist.

The real problem is that large MC simulators may exist in the market-risk world, and separately in the credit-risk world. A composite market/credit dual simulator requires a large effort. The calculation we suggest here is done *without* the need to construct such a huge dual simulation.

We therefore imagine that two MC simulations exist, one for market risk (including all variables, not just spreads) and one for issuer credit risk as described above. We tabulate all the states of the credit simulation $\{\aleph^{Credit}\}$ and the states of the market simulation $\{\aleph^{Market}\}$.

We use spread changes $d_t s_\alpha(t) = s_\alpha(t + Dt) - s_\alpha(t)$ that are assumed present and known in both simulations in order to label the states consistently.

We then form composite states $\{\aleph^{Composite}\}$, each composite state corresponding to a credit loss and to a market loss. We tabulate the total loss histogram for these composite states, and find the total credit + market loss at a given CL.

The idea is shown in the following figure.

Issuer Credit Risk

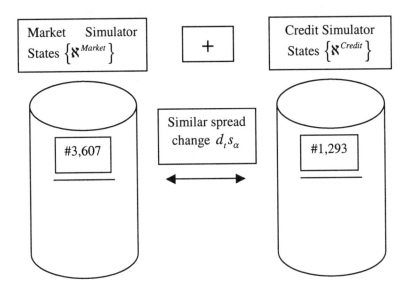

In the illustration, the state #3,607 in the market simulation turns out to have a spread increase $d_t s_\alpha$ close to that of the state #1,293 in the credit simulation. The composite state containing all the information in both states is defined and given the notation $\left(Mkt_{\#3,607}, Credit_{\#1,293}\right)$.

Random selections of market and credit states are used to define the composite states. The losses in all the composite states are tabulated and the loss at the desired CL is picked out.

In practice, bins of spread increases have to be defined. The spread used for the connection can be a weighted average of spreads, with weights corresponding to the losses in the individual simulations.

Geometry for Credit + Market Risk

As in the chapters on VAR, there is a geometrical interpretation associated with the two types of risk (credit, market).

The following picture gives the idea:

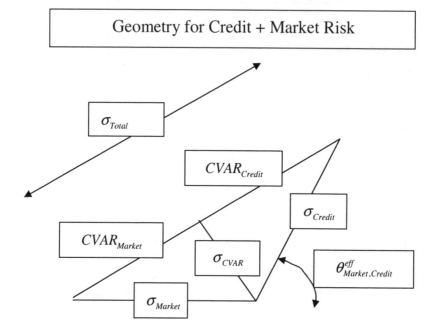

For a review, see Ch. 29. The quantity σ_{CVAR} is the CVAR volatility, or the uncertainty in the composition of the fixed total risk from the individual (credit and market) CVARs. Also, the correlation is $\rho^{eff}_{Market,Credit} = \cos\left(\theta^{eff}_{Market,Credit}\right)$, as we next describe.

Effective Correlation between Market and Credit Issuer Risks

As a final topic, we note that an effective correlation $\rho^{eff}_{Market,Credit}$ between credit issuer risk and market risk can be calculated using the composite formula

$$\sigma^2_{Total} = \sigma^2_{Market} + \sigma^2_{Credit} + 2\rho^{eff}_{Market,Credit}\sigma_{Market}\sigma_{Credit} \qquad (31.12)$$

Here, σ_{Market} is the VAR market risk from all market variables, σ_{Credit} the credit issuer risk, and σ_{Total} the total market + credit risk from the unified calculation. It should be noted that usually the logic runs the other way. That is, an ad-hoc assumption has to be made for $\rho^{eff}_{Market,Credit}$ in order to get σ_{Total} from the independent calculations of market and credit issuer risk.

References

[i] **Agency Credit Ratings**
Global Credit Research, *Moody's Rating Methodology Handbook.* Moody's Investors Service, 2000.

[ii] **Spreads and Their Default Risk Component**
Delianedis, G. and Geske, R., *The Components of corporate credit spreads: Default, Recovery, Tax, Jumps, Liquidity and Market Factors.* Anderson Graduate School of Management, UCLA Working paper 11025, Rev. 12/2001.

[iii] **Market and Credit Risk (Talks at RISK Conference 11/00)**
Marinovich, N., *Integrating market and credit risk within a VAR framework.*
Turnbull, S. and Turetsky, J., *Integrating EDF's into loan pricing,*
Rosen, D., *An integrated market and credit risk portfolio model.*
Tracy, T., *Sourcing and processing data within an integrated market and credit risk framework.*
Selvaggio, B., *Integrating market and credit risk within a RAROC framework.*
Monet, C., *Wrong way credit exposure: clculating counterparty credit exposure when credit quality is correlated with market prices.*
Iacono, F., *Case study: Addressing market risk in a credit risk hedge.*
Sparks, N., *Consistent pricing of banking and capital market structures.*

32. Model Risk Overview (Tech. Index 3/10)

This short and non-technical chapter contains some observations on models with an emphasis on risk. Model Quality Assurance will be treated in the next chapter. You can read this chapter without having to read the rest of the book.

Summary of Model Risk

We start with the obvious comment that models are now an indispensable part of modern finance. Securities and derivatives require model pricing. Therefore, models are indeed indispensable. Finance could not live without them.

Nonetheless, in spite of the best efforts of many talented and smart people, model risk is constantly present at some level and is due to many causes. One risk is the variability in model assumptions, none of which can be proved in any rigorous way[1]. Some important effects on prices can be modeled only imperfectly, if at all. No financial model has the status of a "law of physics", even if physics-based diffusion models and other concepts are used. Further, there is no "best" model, regardless of whose ego is involved. Different firms can and do have different models for the same instrument.

Model risk is hidden unless model-to-model or model-to-market comparisons are made. The risk is highest for illiquid, long-dated options. For highly liquid instruments, models are standardized with slight variations. Substantial losses due to model risk have occurred even for plain-vanilla products, however [2].

Model risk includes the risk of using approximate or inappropriate parameter types, or using the model in inappropriate parameter regimes. Models are used in practice to parameterize securities in some approximate way, and are usually only to be trusted for some short extrapolation from the region where market data are available. These parameters include maturity length, strike values for options, etc.

The intentional use of inaccurate parameter values is a separate problem[3].

---

[1] **Rigor Mortis?** The model risk problem is not alleviated by mathematical rigor applied to models, since models contain non-rigorous assumptions that cannot be proved.

[2] **My Favorite Options Model:** My favorite options model is the theory as described to me by a trader as "Picking up dimes in front of a steam roller".

[3] **Operational Risk:** We do not count fraud due to intentional mismarking by deliberately choosing off-market parameters as model risk. This would be placed in operational risk.

Numerical approximations are unavoidable and are a function of available time and resources, but they can lead to difficulties.

Part of model risk lies in the pitfalls of software development. See Ch. 34. A host of mundane but important issues exists: coding errors, computer malfunctions, misinterpretations, communication snafus, inconsistencies, etc. Anyone who wants to get an idea of the difficulty is invited to sit down at the computer and give it a try[4].

Model-generated hedging predictions are more problematic than pricing, since hedging involves taking differences of prices under changing market conditions. Models differ more in the hedges than in the prices.

Model risk issues have arisen in a number of contexts. The interested reader is invited to consult the literature and the references[i].

Model Risk and Risk Management

Models stand at the cornerstone of risk management. Models characterize the behavior of financial instruments under different possible environments. This information is used to determine the risk of these instruments, and thus of departments, and ultimately of the corporation with respect to the markets. We have exhibited a variety of model calculations. We need to understand the model limitations. These limitations translate into a risk associated with the very models used for assessing risk. Model risk results from model limitations.

Time Scales and Models

Financial markets exhibit different behaviors at different time scales. No model used today accurately describes the dynamics in all the various time-scale regimes. The incorporation of these time scales is the biggest challenge in financial modeling. Models often assume that functions of underlying variables (interest rates, stock prices, FX rates) follow some sort of Brownian or random walk diffusion with constrained drift. In the real world, the parameters are often hard to determine or identify. Moreover, many effects break the Brownian assumption. Manifestly different dynamics in financial markets occur at different

---

If model risk is counted as operational risk (it has to go somewhere), then fraudulent use of the models should be listed separately from the quantitative risk discussed above.

[4] **Homework: Experience is the Best Teacher for Constructing Models and Prototype Systems:** Try putting together some model code into a prototype system, including input/output, a GUI etc., that you design and program yourself. You will learn a great deal about models, systems, data, perseverance, and probably life itself.

time scales: short, medium, and long. In principle, this means that risk management on different time scales should be different.

Mean Reversion

Time scales that are often used are connected with mean reversion in a Brownian framework. Mean reversion can depend on a number of market variables, which can change from one period to another. This includes the disappearance of mean reversion altogether under adverse conditions. This happened for example in the fall on 1998, to the consternation of some institutions (e.g. the famous hedge fund LTCM) that had placed large leveraged bets on mean reversion. Of course the big question with mean reversion is: "Reversion to What, When?"

Jumps, Gaps, and Nonlinear Diffusion

Clear violations of Brownian motion with or without mean reversion occur on short time scales with "gap or jump" behavior. Jumps are often parameterized by a Poisson distribution. Jumps have to do with the response of traders to some sort of important unexpected news. We have no reliable way to model jumps. Actually, jumps may arise from collective non-linear phase-transition effects over short times. Possibly non-linear diffusion including a modified version of the Reggeon Field Theory (RFT) used in elementary particle physics may be helpful; this approach generalizes Brownian motion. The RFT is described in Ch. 46. An alternative approach involves the non-linear dynamics of chaos.

Long-Term Macro Component with Quasi-Random Behavior

Risk management for long-term securities, or over a long term, need to deal with realistic dynamics over a long term. Violations of Brownian motion occur on long time scales with "quasi-equilibrium random macro" behavior. Here, macroeconomics affects financial variables in a smooth but changing fashion, which explicitly involves time scales on the order of months or more. This "macro" behavior dependent on macroeconomics cannot be described with Brownian motion. We describe the "Macro-Micro" model in Ch. 47-51, which is targeted at coping with these issues.

Liquidity Model Limitations

There is a variety of other problems not included in models. Effects related to supply and demand, the trading volume, and the time needed to sell a security are lumped together into "liquidity", and there are no good way to model these effects. Often they are just left out of the model price.

Bid-offer spreads are a related issue. If there is only a one-sided market so that (for example) the selling price is not known, then these spreads must be estimated independent of the model algorithm.

Which Model Should We Use?

Because there is no real theory of finance in the sense of physics, financial models are not unique. Different institutions, especially for illiquid financial products, often use different models. If one model is used in place of another, the differences in the values of the securities and the differences in the sensitivities with respect to movements in the underlying variables become an issue in risk management. For example, if one model reports the interest-rate dependence of a partially hedged position is near zero while another model just as sophisticated and defended with at least as much exuberance reports that the interest-rate dependence of the same position is large, which statement do you believe?

Sometimes a proprietary desk model is used for trading, and another model with simpler assumptions but widely used on the Street is used for corporate risk management reporting. Which model should be used to measure risk? The real answer is that there is model risk. Different models give different results. Therefore, there is an uncertainty in risk reporting due to the very existence of different models. A corporate goal should be the quantification of this model risk, creating an uncertainty of risk management itself.

Psychological Attitudes towards Models

The psychological attitudes toward models are not to be ignored. People who do not understand the limitations of models ask for the "best" model, and some people who should know better may believe they have the "best" model. Some people trust the models to such an extent that if their model disagrees with the market they assume the market is "wrong" and will eventually agree with the model. Sometimes this attitude pays off and sometimes it results in disaster. Sophisticated players understand and even fear the limitations of the models, sometimes using them only as a guide in difficult markets for illiquid products.

Model Risk, Model Reserves, and Bid-Offer Spreads

Models used for risk management are themselves risky to some extent. A corporate reserve could be taken to account for this model risk. This is difficult to convey to accountants, who want to know exactly how much the model risk is and exactly when or under which conditions they should apply the reserve. Because model risk will really show up when relatively illiquid positions are sold in difficult market conditions under pressure at someone else's model price, the risk is hard to quantify. Still, model risk is not zero and may be very large.

Alternatively, if known, model risk can be used to estimate part of the bid-offer spread for illiquid products.

Model Quality Assurance

The best way to quantify model risk is through a model "quality assurance" (QA) program[5] (c.f. Ch. 33). Model QA now exists in most large financial institutions, although model risk is generally only determined in an incomplete fashion. Model assumptions and procedures are documented. Sensitivities to different parametric assumptions can be examined. Models can be assessed and compared.

Models and Parameters

Because there is no financial model that proceeds from first principles that are unambiguously correct, models are largely driven by parameters. These parameters are chosen through a combination of somewhat conflicting goals. The parameters are chosen such that the model into which the parameters are placed produces prices that, at least approximately, fit selected market data. The difficult problem is for cases where there is little or no market information. Models differ partially because the market constraints placed on the models can be chosen in different ways. The number of parameters is a compromise between fitting the known market prices and unwieldy complexity.

Having chosen the parameters to fit the market approximately, the models can be viewed essentially as providing an extrapolation or interpolation methodology. Thus, if a deal comes up that has parameters not currently quoted in the market, which is often the case for over-the-counter deals, the model is used to derive a price for that deal. The models, through extrapolation or interpolation, price illiquid instruments in a portfolio.

It should be emphasized the types and numbers of parameters in reality form an integral part of a model. In a profound sense, the parameters cannot be separated or isolated from the assumption of the underlying dynamics and the implementation of the mathematics through some computer algorithm[6]. For

---

[5] **Model Quality Assurance or Model Validation?** Because no financial model is "valid" in a real scientific sense, we believe that the appellation "Model Validation" is inaccurate and conveys a false sense of security. There can be different models giving different estimates of the risk, which however are all "validated". However, what is important is that some form of the activity gets done. See the next chapter.

[6] **Types vs. numerical values of parameters:** The types of parameters form a part of the model. The *numerical values* of the parameters, as distinct from their *types*, are dictated

example, if one volatility is used to describe a diffusion process rather than several volatilities, this is a model assumption. In fact, some models are just shells into which complex parametric functions are inserted. The Black-Scholes equity option formula is currently used in practice with a breathtaking richness of the parameterization of the volatility "surfaces" describing different options including "skew" effects[7]. The volatilities cannot be separated from the assumption of simple diffusion and the algorithms used.

References

[i] **Model Risk Articles: A Sampling**
Derman, E., *Model Risk.* Risk Magazine, 5/96, Pages 34-37.
Weiss, G., *When Computer Models Slip on the Runway.* Business Week, 9/21/98, p. 120.
Coy, P., Wooley, S., *Failed Wizards of Wall Street.* Business Week 9/21/98, pp. 114-120.
Cheyette, O. and Kahn, R., *Derivatives Trading Again: Finance Pros Take on Physicists.* Letter to the Editor, Physics Today, July 1996, p. 91.
Hull, J. and Suo, W., *A Methodology for the Assessment of Model Risk and Its Application to the Implied Volatility Function Model.* Talk, Courant Institute, 11/00.
Bloomberg, L. P.*Bank One Tax Court Fight over Derivatives Value Nears End.* 11/28/01.
Kremer, V., *Is the Other Shoe About To Drop at UBS?* Derivatives Week, 11/15/97, p. 1.
Bloomberg, L. P., *UBS 'Very Confident' Derivatives Losses Won't Be Repeated.* 11/21/97.
Falloon, W., *Rogue Models and Model Cops.* Risk Magazine, 9/98. Pp. 24-31.
Dunbar, N., *Commerzbank's Model Problem.* Risk Magazine, 2/00. P. 9.
Margrabe, W., *Comparing Add-In Accuracy.* Derivatives Strategy, 8/98. P. 40.
Gapper, J., *Deep-Rooted Reasons for Nat West Loss.* Financial Times, 3/4/97.
Shirreff, D., *Lessons from Nat West.* Euromoney, 5/97. P. 42.
McGee, S., *Bank of Tokyo Blames Loss on Bad Model.* Wall Street Journal 3/28/97.
The Economist, *Banking's Bad Jokes.* 4/12/97.
Clark, D., *Smile and Dial.* Euromoney, 4/97.
Staff Reporter, *First Boston's Loss Of $7Million Blamed on Mismarked Options.* Wall Street Journal, 10/11/96.
Raghavan, A., Pulliam, S., *Bond Hit at Morgan: $25 Million.* WS Journal 10/18/96, C1.

---

by the external market conditions and do *not* form part of the model. The distinction between the types of parameters and their numerical values is sometimes not well understood and can have significant consequences. For example, the numerical values of parameters chosen can be audited by a specialized "rate reasonability" group while the computer algorithms can be checked by a model quality assurance group, but the types and numbers of parameters chosen in the first place is a separate issue, which if not examined can lead to a gap in the control process.

[7] **Skew, Volatility Surfaces, and Ptolemy:** The complexity of the description of skew and volatility surfaces approaches Ptolemaic epicycles. We discuss skew in Ch. 5, 6.

33. Model Quality Assurance (Tech. Index 4/10)

We have discussed models for various markets and purposes, and we have spent some time talking about model risk. In this chapter, we deal with procedures and activities designed to cope with some aspects of model risk[1], variously denoted as *"Model Quality Assurance"*, *"Model Review"*, or *"Model Validation"*. We use Model Quality Assurance or Model QA for short. This is partly because the term "Quality Assurance" is used across the software industry[i].

Whatever it is called, the idea is to reduce model risk, increase the understanding of models for Risk Management and possibly for the desks themselves, reduce personnel risk, produce documentation, etc. Although the beginnings were sometimes rocky, regulators and official bodies now recommend or even require such activities[ii]. Groups of quants performing Model QA now exist across the industry[2].

Model Quality Assurance Goals, Activities, and Procedures

A summary of Model QA goals, activities, and procedures follows. This is a representative list, but different policies exist.

Desk-Level Model QA

Some (even extensive) model QA is done by desk quant model developers. Model testing in various situations takes place before the model is put on the desk. The traders then use the model in the real world, comparing prices to the

---

[1] **History and a Story: The Trader's Flying Putter:** My involvement with Model QA started around 1995 at Smith Barney, and later in 1997-99 at Salomon Smith Barney, where I instituted and ran firm-wide Model QA for models for Fixed Income Derivatives, Mortgages, Equity Derivatives, and FX Derivatives. Part of these early pioneering efforts involved selling the idea, sometimes to indifferent or frankly hostile audiences.

A good war story involved an influential trader, who exhibited his objections in a meeting to some aspects of Model QA by throwing a putter. It missed, but not by much.

[2] **Acknowledgements:** I thank members of my Quantitative Analysis Group: A. Beilis, J. Castresana, T. Gladd, and A. Lapidus, for their diligent work while this group was in charge of performing firmwide Model QA at Salomon Smith Barney. I thank the Salomon business-unit desk quants, systems people, and risk managers for their co-operation in making the SSB Model QA effort a success. Finally, I thank E. Picoult and L. P. Chan, who now lead Model Validation at Citigroup, for discussions.

market, noting whether the predicted hedges are or are not realistic, etc. Battle-hardened models are developed over years of use. Desk quants may have the attitude that, since they understand the models (perhaps better than anybody else) and have already tested them, model QA performed by an independent group is redundant. In a sense, the desk quants can perform the activities of a model QA group. They have a point. When I was developing desk models or supervising their development, I had a bit of the same attitude. However, independent Model QA—which is the real topic of this chapter—is indeed useful.

If vendor models are used, desk quants will generally perform some model QA activities to test the vendor models.

Independent Model QA Group

Model QA at a corporate level is often performed by an independent group of quants, usually PhDs in engineering, science or math with some finance background. This group is not under the control of the business units. The personnel have to have enough background to understand the models. Depending on the situation, they may reproduce the desk models or produce alternative independent models. Programming skills are essential. Systems support is desirable, although not always present at the desired level.

A hybrid version is to use desk-level QA to do the work, with the independent quant group performing a control or supervisory role.

Model Risk Testing Using Various Independent Models

Our philosophy is that perhaps the most important issue is to quantify model risk through comparison of the results of different models. Then, when firm A (using its best model) has to sell its inventory to the street, and firm B buys this inventory (using its best model, probably different from A's model), part of the bid-ask spread is the difference in the model valuations.

In order to get a handle on this issue, the independent Model QA group can either construct or otherwise obtain models that are different from the desk models. One way of proceeding is to use relatively simple Street-standard "Ford" models as the independent models, to compare with the whiz-bang "Mercedes" proprietary models on the desk. This is actually reasonable, since if the desk is forced to sell its inventory (acting as firm A) then the buyer (firm B) may indeed be using a Street-standard model and so the desk may be forced to sell its inventory using its Mercedes model at the level given by the Ford model[3].

It is a good idea to quantify the results from different models using "Diff reports". These are simply catalogs of results comparing the output from the

---

[3] **Question to the Professors:** OK, so in this case, which model is Right?

models, day-by-day, for a representative set of deals[4]. The differences should be noted for the hedges, not just the prices.

Test Suites

Test suites of examples are useful to run periodically for control purposes. This activity will indicate if any model changes have occurred.

Documentation, Personnel Risk, and All That

With some notable exceptions, typical desk quants are "too busy" to document models. From their point of view, since they understand exactly what they did, the value of documentation is low. From a corporate point of view, the value of documentation is very high. One reason is personnel risk. When the quant leaves, who understands the model? [5]

The documentation should include full disclosure of the models. This includes the theory, equations, parameters, numerical techniques, and practical utilization examples. Model limitations should be noted. Model inaccuracies, when known or found, should be listed. Software design criteria, including code comments[6], should be noted.

The documentation should be made available to corporate risk management. This is so that the risk managers can better understand the models producing the risks with which they are dealing.

Proprietary concerns that the desk might have are fully justified[7]. Therefore, appropriate security measures should be taken.

Documentation by the independent group by definition means that the independent Model QA quants need to talk to the desk quants. This should be done with mutual respect. One way to establish a good rapport is to have regular time-limited discussions, where the agenda is understood in advance. The time allocated to the meeting should also be specified. The meetings should take place in a quiet office, off the trading floor.

---

[4] **Systems Help?** If you're in a QA group and are lucky, maybe you can get some help constructing such reports in an automated fashion from the friendly systems group. Otherwise, you may be reading computer files in obscure formats, cutting/pasting, etc.

[5] **New Head Desk Quant:** A variation occurs when a new head quant arrives. Documentation of the existing models can be very useful in that case.

[6] **Code Comments:** Of course, *you* always document *your* code with explicit, easy to read, complete sentences in every code section, don't you?

[7] **Disclaimer for Proprietary Models:** Although I obviously know a lot about Salomon's proprietary models from having run firm-wide Model QA at Salomon Smith Barney, no information is in this book regarding these models.

Independent Reproduction of the Same Model

One way to understand exactly what is in the desk model is to try to reproduce the desk model in the independent Model QA group. The QA quants reproduce all equations independently. Reproducing (i.e. recoding) the model computer code is an important part of this philosophy. This is useful in that it can reveal model limitations or assumptions perhaps not otherwise evident (or understood). It also controls for software coding errors. Coding models is generally a very time-consuming task. One issue that arises is how model QA should be performed on the recoded models by the independent QA group.

Judgment on Assumptions, Methodology, Algorithms

The independent Model QA quants should have enough experience and sophistication so they can form independent judgment on relevance and reliability of various model assumptions, methodology, numerical algorithms etc.

Which Models need Model QA?

Generally, the models that produce hedges that affect corporate risk management should require independent Model QA, as well as the models that produce prices for the official books and records of the firm.

Prototype models not yet in production, trading models used for decisions on the desk, quick calculators in spreadsheets, etc. may or may not fall under Model QA, depending on policy, resources, negotiations, etc.

Model Testing Environment

If possible, it is good to quantify model risk present for pricing and hedging through comparative testing in a "real world" context. Portfolios of representative securities should be used with realistic input parameters. Testing should occur over a period of time[8].

Feedback to the Business Units

Maybe the independent quants will find out something that the desk didn't know about its models. Feedback to the desks is therefore a positive idea.

Model QA: Sample Documentation

Here is a sample suggestion of what might be considered as ideal Model QA documentation. The objective here is to illustrate the issues raised in model

---

[8] **Story:** My quant group performing model QA once spotted an unannounced model change on a desk through a big change in valuation differences that was reported one day.

development and systems implementation. This documentation is much more thorough than what is usually done. However, such a document could be relevant for complex models. Most models only need a fraction of this information.

The format below is an outline to be filled out. Other formats naturally can accomplish the same ends. There are three sections:

- User Section

To be filled in by the users of the model (trading, sales, etc.)

- Quantitative Section

To be filled out by the Quantitative Group designing the model, and (when applicable) coding the model.

- Systems Section

To be filled out by systems personnel involved with model coding, integration, maintenance, etc.

User Section of Model QA Documentation

U1. Model Usage Specification
U1A. Trading and Sales
- List the trading applications of the model, if any
- State the context in which the model is used.

U1B. Risk Management Reporting
- Specify if the model is used for risk management reporting, especially if different from the model used for desk hedging.

U1C. Risk Management and Desk hedging
- Specify if the model is used for desk hedging. State which parameters are different than for corporate risk management reporting, if any.

Quantitative Section of Model QA Documentation

Q1. Model: Fundamental Aspects
Q1A. Theoretical Description Summary
- Give a short (<100 word) abstract. A sensibly and reasonably complete description, in good English and with formulas should be

attached. Include a categorization of model variables, parameters, stochastic assumptions, analytic model valuation, approximations.

Q1B. Domains of Applicability of Model
- Include products or features describable by the model in its present version. Also, describe enhancement plans, if any.

Q1C. Limitations of Model
- No financial model has the status of a physical law in physics. Describe the approximations and limitations of the model. Describe products, for which this model does not apply. Especially relevant are products that may be appropriate for a future version.

Q1D. Hedging Aspects for Model
- Specify hedging aspects of the model, if used for risk management.

Q1E. Definitions of sensitivities from the model
- Give summary here. The attached model document should give the definitions of model sensitivities to each risk parameter, precisely in terms of measurable quantities or in terms of internal model parameters with reference to appropriate equations.
- This should be supplemented with at least one clear numerical example, with all terms defined.

Q1F. Comparative Model Analysis and Model Consistency across the Firm.
- When known, state how this model is consistent or inconsistent with models used to price the same securities in other areas of the firm.

Q1G. Model Peer Review Description

Q1H. Model Noteworthy Aspects (e.g. discontinuous payouts, exotic features)

Q2. Model Parameterization

Q2A. Market Input Parameters (rates, prices, vols, dividends, correlations,...)
- For market parameters, specify the method used to extract the parameter as well as uncertainties. Specify the source for the market parameters, if known.

Q2B Model Input Parameters (Strike, barrier level, expiration, etc.)
- Specify the model input parameters. Give the acceptable numerical regions of the model parameters for the model.

Q3. Model Numerical Algorithms

Q3A. Analytic or Semi-analytic Numerical Methodology
- Describe in detail the relevant algorithms. Include references, formulas and parameters.

Q3B. Non-Analytic Numerical Methods—Overall Description
- Specify the numerical method (e.g. Binomial, Monte Carlo, PDE, Path Integral, ...). Also, give the parameter specifications relevant for the numerical analysis.

Q3C. Convergence Criteria
- Convergence criteria should preferentially be expressed in terms of pricing and when relevant, hedging with respect to appropriate variables. Give the parameters describing the accuracy and convergence. These can include the time-step discretization, including parametric dependences (e.g. monthly or SA, specific dates, interpolation, day count conventions).
- Specify any flexibility to obtain convergence (e.g. changing time step amounts, grid enhancement). If Monte Carlo is used, give the number of paths if specified in running the model. For binomial convergence, specify odd vs. even number of time steps. Specify other parameters: maximum rate, minimum time step, maximum # paths allowed, etc.

Q4. Model Implementation ("Black Box")

Q4A. Requirements Documentation Description
- Describe the requirements documentation that you have assembled.

Q4B. Prototype Code
- Include reference for prototyping the model, if appropriate. Prototypes that evolve into production code are no longer prototype.

Q4C. General Specification (Compiled, Spreadsheet, etc.)

Q4D. Intent (for other code, for sales, for trading, etc.)

Q4E. Design, Architecture of Model Implementation
- Include modules or object description at a high level. Specify the data structures at a high level. Specify the language for the model and language for the wrapper for inclusion in the system. Specify any other tools for model development used, if any.

Q4F. Model Coding
- Describe steps that were taken to ensure that "good practice coding" has taken place to enhance maintenance and extensibility.
- Well-documented code means that what is going on is clear without having to go through the code line by line. This includes code readability by a person OTHER than the actual coder. Comments (that are up to date and accurate) should exist in *addition* to self-documenting variable names. A metric is the percentage of comments/total lines in the code. Each module should at least have a header explaining its function and results returned. Describe the code documentation here.
- List any peer code review that has taken place.

Q4G. Spreadsheets for Models
- If model is used as an add-in, specify here. State how changes to spreadsheet models are controlled (passwords, locked cells, etc.)

Q5. Model Changes or Enhancements
Q5A. Requirements Specifications
Q5B. Implementation of Model Changes
Q5C. Model-Change Documentation

Q6. Model Quality Assurance "Alpha" Unit Testing
- "Alpha QA" testing means testing by the quantitative group responsible for the model and by the personnel coding the model. Include independent pricing and hedging QA measures, checks with other systems, hand calculations, sanity checks, case checking, etc.

Q7. Personnel Aspects
- Specify the quantitative personnel responsible for the theoretical specification of the model, parameters, hedging properties, etc
- Specify the personnel responsible for prototyping the model, if any
- Specify the personnel responsible for coding the black box model
- List the model maintenance personnel.

Q8. Vendor Model Software
Q8A. Vendor Model Software
- Specify if the model is part of a vendor system, (which system?)

Q8B. Vendor Model Quality Assurance / Evaluation
- Specify procedures for vendor model evaluation, acceptance criteria for vendor models, and details for vendor model testing.

Q9. Model Library and Software Reuse
- Specify any documentation for the library into which the model will be put. Specify, when appropriate, library routines used in this model.

Systems Section of Model QA Documentation

S1. Model/System Integration
S1A. Plan or Details of Integrating Model into System
- The model, if it is not a standalone pricer, will be incorporated into a system. Name the system and the version.

S1B. Environmental Specification of System

- Operating System (including version), live feeds specification, network, database including version, GUI builder, compilers, Case tools, version control software (e.g. SCCS), hardware compatibility, any parallel processing capability, etc.

S1C. System Language
- The system language need not be the same as the model language, since one language can call another.

S1D. System/Model Integration Specifications—Other Details
- Include model diagnostics from system when running the model. Indicate diagnostics stored in the database or as files that can allow for interpretation of the model results, quality assurance, etc.

S1E. "Good Practice Coding"
- Describe steps that were taken to ensure "good practice coding".

S1F. Time Scale Estimation
- Describe how time estimates are established for model incorporation into the system, and whether in the past these time estimates have been accurate.

S1G. Maintenance

S1H. Changes to Model: System Integration Perspective

S1I. Contacts with Quantitative Personnel for System Implementation
- Describe the contacts with the quantitative modeling personnel in order to ensure accuracy of the model as implemented in the system.

S1J. Model Distribution
- Include procedures for model distribution and version control, according to geographical location.

S1K. System Upgrades and Model Re-integration, if appropriate
- Describe plans for re-incorporation of models into new or upgraded system versions.

S1L. System Disaster Recovery Plan

S1M. Systems Integration for Spreadsheets
- If the model is implemented in a spreadsheet, or a spreadsheet add-in, give the details of the systems integration, if any:

S2. Model Changes or Enhancements—Systems Aspects

S2A. Requirements—Systems Aspects
S2B. Implementation—Systems Aspects

S3. Model Quality Assurance Testing—Systems Aspects

S3A. Independent Checks with other Systems
S3B. System Integration Testing—Procedure
- Include description of the test suite. Describe the consistency of the system testing with the unit testing.

S3C. Regression Testing Procedure when changes are made to model/system
S3D. Feedback Procedures, Forms for Testing
S3E. Quality Assurance Documentation—Systems integration of model
- Include legible English commentary, screen printouts, etc.

S4. Personnel Descriptions—Systems Aspects
- Specify the personnel responsible for the general design, the integration of the model into the system, and the coding.
- Specify personnel responsible for QA of system integration, including testing if model or system is changed.
- Specify personnel responsible for maintenance of the model.
- Describe steps taken to avoid personnel risk.

S5. Model Library and Software Reuse
- Describe library, if any, including resources.

References

[i] **Quality Assurance, Software Development, and Related Topics**
Perry, W. H., *Quality Assurance for Information Systems*. John Wiley & Sons, 1991.
McConnell, S., Code Complete – *A Practical Handbook of Software Construction*. Microsoft Press, 1993.
Carnegie Mellon University Software Engineering Institute, *The Capability Maturity Model – Guidelines for Improving the Software Process*. Addison-Wesley, 1995.
Harrington, H. J. and Mathers, D. D., *ISO 9000 and Beyond – From Compliance to Performance Improvement*. McGraw-Hill, 1997.
Sodhi, J., *Software Requirements Analysis and Specification*. McGraw-Hill, Inc., 1992.
Schulmeyer, G. G. and McManus, J. I., Editors, *Total Quality Management for Software*, Van Nostrand Reinhold, 1993.
Webster, B. F., *Pitfalls of Object-Oriented Development*. M&T Books, MIS:Press, 1995.
Thielen, D., *No Bugs*. Addison-Wesley Publishing Co., 1992.
Wiener, L. R., *Digital Woes – Why We Should Not Depend on Software*. Addison-Wesley Publishing Co., 1993.
Jenkins, G., *Information Systems – Policies and Procedures Manual*. Prentice Hall, 1997.
Byte Magazine: "How Software Doesn't Work and What You Can Do About It", 12/95.
Dowd, K,, *High Performance Computing*, O'Reilly & Associates, 1993.
IEEE Software Magazine *Quality*. Vol. 13, No. 1, 1/96.
Kiczales, G., *Why Black Boxes are So Hard to Reuse* .... OOPSLA '94 Conference, 1994. Univ. Video Communications; ACM; http://www.xerox.com/PARC/spl/eca/oi.html.

[ii] **Model QA and the Fed Guidelines**
Federal Reserve System, *Trading and Capital-Markets Activities Manual*. 2/98. Section 2100.1, Pp. 4,5.

34. Systems Issues Overview (Tech. Index 2/10)

This chapter contains a qualitative and non-technical overview of problems with Systems. Before starting, it should be emphasized that each organization tries to produce (or buys and customizes) the best systems it can, consistent with time pressures and resource constraints. There are very successful systems that work well, and are used every day. However, the development of systems can be problematic. This essay deals with some reasons for these problems.

Advice and a Message to Non-Technical Managers

You are an expert in your field. Still, you may have to approve or disapprove expensive software projects and/or expensive computer hardware purchases, but you have little computer background. Sometimes high-level managers approve what turn out to be large, badly designed, and quite expensive homegrown systems that lead to all kinds of friction in the organization. You will feel more comfortable and make better decisions if you take the time to learn something about computer hardware and about software development so you have a gauge to use when confronted with such a decision. This chapter is a quick summary of some of the issues.

What are the "Three-Fives Systems Criteria"?

Many computer programs with entirely different scopes and capabilities are called "systems". One definition of a system uses what I call the *"Three-Fives Systems Criteria"*. Namely, the system development uses up $50MM, employs 50 people, and takes 5 years to become useful and functional. Some people disagree with these criteria, arguing that the numbers should be larger. Other people use "system" to represent a smaller effort, and so come up with smaller numbers.

Calculators are not Systems

By a "calculator" is meant computer code that calculates the results of models along with some risk measures. Calculators are much cheaper than systems. Sometimes people misidentify a calculator with a system. A system can have calculators as modules, but generally a system will also have a graphical user front end for input and output, connection to a database, network communications, be employed by many users, etc.

What is the Fundamental Theorem of Systems?

Systems are a Black Hole of Wall Street[1].

What are Some Systems Traps and Risks?

There are many success stories of software development. Systems development on Wall Street struggles against many pitfalls of systems development that are in fact chronic to the entire software development industry. There are four conflicting goals:

- Fast development
- Cheap development
- Complete development
- Reliable development

Refinements include over-optimism, under budgeting[2], redefinitions of goals in midstream by the end-users or by the systems group, misunderstanding, miscommunication, upper management incomprehension, egoism, programming errors, absence of software quality assurance, inappropriate design architecture, disparate hardware, incompatible databases, programmers getting reassigned, etc.

Unclear language is rampant: "All right, we really need to get it done by next month". The phrase "get it done", which uses the dangerous pronoun "it", is always to some extent undefined and can change meaning.

An often overlooked problem is personnel risk, where only a few people actually know how individual parts of the system are built, and who are "too busy" to document anything. This particular problem manifests itself when these people leave, at which time part of the systems organization may go into some version of crisis mode, mixed with temporary stagnation in system development.

---

[1] **Exceptions:** Of course, maybe your system is different.

[2] **The Fudge Factor for Systems Budgeting:** This applies to proposed projects that look big but are underestimated in cost, manpower, and time. This chronic problem can be partially avoided by taking the best rational estimates and multiplying them by a Fudge Factor FF to give the Real Answer. Based on historical experience FF = 3 is a reasonable estimate for each of these variables. What do you think?

Specific Communication Issues

A generic issue is the lack of effective communication between the computer personnel and the end-users, including quants. This is made worse by the fact that these different groups of people—by their training—speak different languages and only roughly glimpse the problems faced by the others. A good tactic is to train the quants in systems issues and to train the systems programmers in business and quantitative issues so that they start to speak the same language. In reality, the most valuable personnel do have the ability to speak technical and business languages fluently. However, this takes time and is often only partially successful.

For PhD quants involved with models, the most common situation is that the quant needs to understand the mathematics, the numerical algorithms, "enough" of the finance, and the parameters; he/she then actually does the programming for the model. This model is then inserted as a black box into a system programmed by system programmers. This paradigm presents its own set of communication difficulties between the quants and the system programmers. These difficulties are sometimes nefarious, and must be worked out patiently to be successful.

One potential problem is programmer turnover with the consequence of inefficiencies due to training and ramp-up time[3].

Another issue is communication among the various systems people themselves, including their hierarchy. Possibilities of miscommunication grow exponentially with the number of people involved. An entire layer of mid-level systems management is needed to cope with this problem and to interact with the end users. The efficiency of system development using this paradigm depends critically on the systems expertise of the manager along with his/her knowledge of the business. Sometimes this layer helps, but unfortunately, sometimes it just gets in the way of effective communication between the end user and the programmer.

The Birth and Development of a System

It is instructive to see the different perspectives regarding how systems begin and develop. Systems developers often (and justifiably from their perspective) want "specs" written by the end users that completely and accurately describe the requirements for the system soup to nuts. Once the systems group establishes the

---

[3] **System Personnel Risk:** Here is one trap. The systems management might have the naïve attitude that programmers are functionally identical, and therefore programmers can be put into a pool and switched around between projects without degrading anything.

specs, they establish "milestones" along a well-defined path to the final delivered "solution" product. The whole thing is monitored using software[4].

The end users generally have very little time for these "extra" activities. First, they usually have no idea what the computer people mean by the word "spec", and only reluctantly participate. In fact, end users often cannot describe the requirements accurately. This does not necessarily imply a deficiency or inattention on the part of the end user, because planning a system is extremely complex. Compounding the problem, different users, who appear at different times, can have different requirements. Further, the requirements of a given user (justifiably from his/her point of view) can change with time. Requirements may change for good reasons connected with optimizing the business, or because of an enhancement that for some reason only becomes envisioned after the development is underway.

Often, thinking through a complete system is impossible because the human brain simply cannot work through the logic. That means that writing a definitive once-and-for-all spec is sometimes literally impossible.

Development difficulties are ameliorated if the system is written in a modular or object-oriented fashion, but even in this case there can be real planning and execution problems for large systems.

Once a system gets to a certain stage, the system is often difficult to change. Systems programmers are usually reluctant to accept the changes because they compromise the system milestones, and they get annoyed at the end users for not putting in these items at the beginning.

The problems are further exacerbated because end users rarely have any accurate concept of the difficulties of the programming, and do not understand "Why It Is Taking So Long"[5].

Programming Difficulties

Programmers have to cope with a myriad of extremely difficult issues. Programming is both an art and a science. Extreme concentration, exceptional discipline, difficult training, and high mental ability are prerequisites for a good programmer. Moreover, programming generally requires communication and co-ordination with others working on the same system. The best programmers are worth their weight in gold.

Programmers, on the other hand, sometimes do not have enough financial background to work as efficiently as they might, or to know when a result coming out of the program makes sense or is manifest garbage.

---

[4] **Gantt and Pert Charts:** These utilities can be useful for tracking systems progress, but they may be difficult to formulate and update, and they can even become a distraction. My reaction to a systems manager who shows up with one is usually "Whoopee".

[5] **Programming Difficulties:** You will get a much better sense of the problems if you try programming something yourself. Come on, its not Beneath your Dignity.

A common programming problem exists in modifying or interacting with large systems that others have written, and which the current programmers only understand incompletely. This can be especially true for programmers writing custom code to insert into a large system, raising up consistency issues.

Prototyping

In practice, prototyping is often used. A quick prototype is constructed that is intended to be a "trial balloon" that then will be reprogrammed once the end users sign off. The end-user starts to use the prototype. By definition, a prototype does not have full functionality and the end-user wants to see more, quickly, without waiting for the rewrite. An initially unintended consequence is that the prototype steadily grows larger, and because of technical problems due to shortcuts, becomes incapable of the flexibility required to meet growing business needs. Many prototypes become ensnared in this global attractor to become the Final System, warts and all.

A common problem is that the prototype, a small endeavor, starts to grow and eat up resources and money on the long rocky road to becoming a True System. Often, management does not recognize the scope or uncertainty of the required resources, time, and money.

Who Controls the Systems?

Because a system is a complex product, it is often constructed in an independent Systems Technology Organization STO (or whatever the systems department is called). The STO can implement a rigorous systems environment that is desirable from many points of view, including regulatory aspects.

Alternatively, a decision by the business units to control the system process locally in the departments can lead to better flexibility and communication, but this can also lead to shortsighted technical decisions for a variety of reasons.

Not the least problem is related to the boundless ego of some traders who believe they can infallibly direct the construction of the World's Best System.

Structurally, isolated departmental efforts can become problematic. Centralized corporate requirements (for example centralized risk management reporting) can become a gigantic exercise in co-coordinating information from iconoclastic isolated departmental systems. The business units, while co-operating, can be less than enthusiastic about this expensive activity.

Systems in Mergers and Startups

In recent years, with mergers becoming a staple of life, a variety of difficulties related to retrofitting old systems or building communication layers between existing systems from the different companies in the merger becomes a large and unwieldy enterprise. The temptation to "scratch it all and build it over right this time" is seductive. Sometimes this is a viable approach, but can lead back to the Black Hole Systems Theorem.

Startups present an entirely different picture, since there is a blank slate, which does not constrain development. Unfortunately, if startups are controlled by management that is inexperienced regarding systems development, potential systems pitfalls can become reality.

Vendor Systems

One alternative, often adopted, is to use vendor systems in which development costs are divided among the vendor's customers. Vendor products have real advantages in some cases, especially with startups. A vendor system is "turnkey", so that long, expensive initial in-house system development is avoided. Some vendor systems for derivatives, for example, have hundreds of man-years of development[6]. While the cost for a vendor system can seem large, it can be much more economical than startup in-house development from scratch. Vendors develop their systems to appeal to market participants. Some vendors have clients who are broker dealers and banks and they get valuable feedback from these users. Sometimes alliances are formed for mutual development projects.

Some problems with vendor systems can surface. For example, once the honeymoon period is over after the contract is signed, the service can drop in quality. Communication difficulties with the vendor bureaucracy can lead to the vendor product not being delivered completely in line with what the client wants.

Still, all in, a vendor solution is one that is often successful in fulfilling the needs of the client in a reliable and perfectly reasonable manner.

Evaluating Vendor Systems

In order to evaluate vendor systems, a concerted effort involving quants, systems, trading, back office, etc. must be undertaken. A good deal of time and effort is required to examine and compare thoroughly the systems of the various vendors. A vendor review document should be systematically constructed for evaluation.

---

[6] **Man-Years and Risk:** For example, 100 man-years can mean 20 programmers working for 5 years. The major expense and long time, along with many difficulties that need to be overcome, are the risks that the vendor assumes. The clients share the development costs and are largely insulated from the risks.

In-House Developers and Vendor Systems

In-house developers usually oppose vendor systems. There is a logical basis for this attitude, since the in-house system code is more transparent, and the in-house system will generally contain the customization desired with good feedback.

Vendor systems can be simple or problematic for customization. Most systems allow proprietary models to be inserted, and customized risk reports can be generated. However, dealing with interfaces to systems, which are essentially black boxes to the customer, in some cases can be difficult and clumsy in practice. In other cases, the interfaces are well designed, and adding models is quite simple. It is a good idea to test-drive the capability of the system before buying, in order to see just how simple or difficult it is in practice to add a model to the system.

Do you REALLY want the Source Code? Maybe Think Again.

The possibility of buying the vendor source code is sometimes seductive to ambitious in-house programmers. The advantage of owning the source code is that the in-house programmers have complete control and transparency, and can modify the source code for customization.

Source code however is a double-edged sword. Upgrades of the vendor software can require a lengthy and painful porting exercise for the customized code. Moreover, the in-house programmers need to spend substantial time initially understanding the vendor system, which can be huge (the scale can be a million lines of code or more). Compatibility problems between the in-house code and the vendor code can surface.

A red herring that has been used to justify source code purchase is that the human mind can only process a few things simultaneously. The argument is that therefore hooking up many models to a vendor system without source code is problematic. However, models can be hooked up independently. Moreover, the argument can backfire since the complexity of the source code can be a trap.

New Paradigms in Systems and Parallel Processing

New paradigms in systems are important. While it is important to be on the "leading edge", it is desirable to avoid the "bleeding edge"[7]. The most significant potential development is the Internet. Except for transmitting and displaying results, e-mail etc., quantitative development today mostly remains in more traditional areas. The biggest potentially useful paradigm in systems is parallel processing. The perennial increase in hardware power has meant that simple networked sets of workstations can be used in many cases for co-coordinated

---

[7] **Acknowledgement:** I thank Gregg Rapaport, a super back-office guy, for this quote.

computing. Parallel processing is sometimes used, but could be exploited more than it is now. We discuss parallel processing in the next chapter.

Languages for Models: Fortran 90, C++, C, and Others

Computer languages are unfortunately often viewed emotionally a bit like religion[8]. The programmer skill and the style of programming are, however, paramount. Good or bad code can be (and has been) written in any language.

Model code is a separate issue from system development. Models tend to be structured as technically difficult mathematically-intense modules that connect to a larger system for input and output.

Do We Really Need to Write in C++ for Model Code?

Systems tend to be written in C++. Models may be, but do not have to be, written in C++. Because model coding is extremely difficult, it is most efficient for a quant to write model code in his/her most comfortable language (C, C++, or Fortran), instead of being forced into using C++ (which may be an unfamiliar language). Writing in the language of highest proficiency for the quant saves time and reduces bugs.

Note that there is no "Tower of Babel" since models written in one language can be readily interfaced with systems written in another language.

Fortran 90

Fortran was, and maybe still is, the most common language used for science/engineering codes, and is the first language of many quants. High-end supercomputing is often done with Fortran, which partially due to its superior compiler technology, remains the fastest and perhaps the most convenient language for numerical computing[9].

Fortran is often badly misunderstood by otherwise well-informed systems gurus. Many people do not know that Fortran was transformed into a modern language years ago (Fortran 90), and the language is being further improved [i]. The bottom line is Fortran 90 contains most of the features of C, C++, and APL [10].

---

[8] **Viewpoint:** This section is written from a contrarian, perhaps heretical, viewpoint.

[9] **A Fortran vs. C++ Speed Story:** The president of a large C++ derivatives vendor software company was once in my office for a demo of my derivatives model prototype system. He asked in what language the software was written. When he heard "Fortran", he said "Wow, *that's* why it's so fast".

[10] **Fortran 90 Is Modernized, Containing Most Features of C, C++, and APL:** Today, Fortran 90 has object-oriented features including modules, encapsulated data and procedures, private and public attributes, general structures, and operator overloading.

C

C is of course routinely used for numerical model code. In addition, entire systems are often written in C. However, extra care should be taken since the C language can be unpleasantly problematic[11].

Other Languages

Java has recently appeared, and is rapidly becoming popular. Java is useful for simple models on the Internet, but the language has not made inroads into industrial-strength model code. Visual Basic is a popular language, enthusiastically promoted and used for model prototypes by some, but again not for industrial-strength model code. Proprietary languages in commercial packages (Mathematica, MatLab and others) are also used profitably for modeling, though mostly for prototyping or occasional basis.

The Most Common Prototype Platform

The most common prototyping method remains spreadsheets, mostly Excel and (especially in the past) Lotus. Spreadsheets can function as front ends, with add-in functions to perform the actual model calculations.

Many APL-like parallel array or vector functions are provided that are convenient and useful for scientific programming. This means a one-line statement (e.g. A = 0.) can apply to an entire matrix. Procedures (subroutines, functions) now have call by value in addition to call by reference, optional arguments, recursive procedures, and arbitrary argument order. Pointers exist and have an explicit target attribute for efficient and controlled usage, dynamic memory allocation, and pointer operations. Character/string enhancements include new operators. Execution constructs include case, cycle, and do while. Other improvements include enhanced read/write I/O features, free-format code, 31 character names, bit manipulation, explicit obligatory typing, new random number generator features, and arbitrary array indexing. Compatibility with Fortran 77 exists, so code migration can take place deliberately, compatible with numerical libraries. Fortran can be interfaced with Excel. There are compatible screen-drawing programs. The Fortran language remains easy to read. Check it out.

[11] **Bad News C-Code Bugs, Heisenbugs, and a Story:** Destructive, time-wasting episodes with bugs in C-code written by skilled quants and programmers in my quant groups occurred over the years. These included compiler incompatibilities, memory leaks, and pointer problems. The "Heisenbug" is a mysterious C bug that shows up only when the program is unfortunately "disturbed" by the computer environment. Here's the story: The programmers at one derivatives shop refused to use C at all because they considered the language "too dangerous". While that is perhaps an extreme viewpoint, I heard it with my own ears.

What's the "Systems Solution"?

Systems-speak uses the word "solution" to characterize a system. Actually, there is no really good solution. Human beings are analog animals. No computer or robot can reproduce the capabilities of human vision or touch. On the other hand, human evolution has not required the precise logical thinking of the type required for digital computer programming. Programming often consists of a giant logical exercise with severe consistency and interdependence difficulties that are only partially resolved by modular or object-oriented design[12].

Are Software Development Problems Unique to Wall Street?

No. No. No.

It may be (but probably is not) a consolation to understand that the problems of system development faced by Wall Street are endemic across the entire software industry, where system projects are often incomplete, over budget, buggy, and late. Computer technical journals are replete with articles on the subject. Perhaps the reader has had some direct experience with the problem.

Books have been written about "best practices" in programming, and there are metrics (for example the Capability Maturity Model [ii]), which have been devised to measure systems quality. In practice, such good advice is followed only to some extent, mainly because "there is no time" and "it is too hard".

References

[i] **Fortran 90 and Further Developments**
Adams, J. C., Brainerd, W. S., Martin, J. T., Smith, B. T., Wagener, J. L., *Fortran 90 Handbook: Complete ANSI/ISO Reference*. McGraw-Hill Book Company, Intertext Publications/Multiscience Press, Inc., 1992.
Brainerd, W. S., Adams, J. C., Goldberg, C. H., *Programmer's Guide to Fortran 90*. Intextext Publications, Mc-Graw-Hill Book Company, 1990.
Computational Science Education Project, *Fortran 90 and Computational Science*, csep1.phy.ornl.gov/CSEP/PL/PL.html .
Chivers, I. and Sleightholme, J., *Fortran 90, 95 and 2003 Home Page*, www.kcl.ac.uk/kis/support/cit/fortran/f90home.html.

[ii] **The Capability Maturity Model**
Carnegie Mellon University Software Engineering Institute, *The Capability Maturity Model – Guidelines for Improving the Software Process*. Addison-Wesley, 1995.

---

[12] **Consistency Issues in Programming:** You will understand the seriousness of the consistency problems if you have tried programming a prototype system, or have witnessed expert programmers grappling with an industrial-strength system.

35. Strategic Computing (Tech. Index 3/10)

This chapter is concerned with an overview of strategic directions for numerical financial computing[1, i]. Parallel processing will be emphasized. The need is for rapid and cost-effective valuation of large portfolios of options, mortgage-backed securities, bonds, etc. in order to enhance competitiveness in trading, risk management, and sales. A description is given of the utility of parallel-processing computers, distributed workstation environments, and technological advances pertaining to new directions in financial numerical computing[2].

The material in this chapter is self-contained, and can be read separately from any other material in this book.

---

[1] **History + Stories.** This chapter is based on my 1989 preprint (ref i). The history started in 1987-89 at Merrill Lynch. At that time, as anyone over 40 will remember, mainframes dominated, Unix workstations were just beginning, and PC's were not yet competitive. Following the lead of Prof. Harold Shapiro visiting from the Courant Institute and who had set up a SUN workstation, I started a project called "cost-effective computing". The idea was that quants could contribute to the firm's bottom line through innovative thinking about computing. The process included a systematic evaluation of vendor workstations, holding meetings, giving presentations, and trying to convince the management that workstations would be an effective paradigm. This led to the first workstation network at the firm for quantitative work, among the first on the Street. The whole process was rather exhilarating.

I then started a serious investigation of parallel-processing machines. At the time, Intel had an initiative for high-end parallel financial computing. A parallel machine was eventually adopted at the firm for CMO mortgage calculations.

The experience was at times dangerous. One war story will suffice. A meeting in 1988 was commanded by the Heads of the firm's Systems Group (read Mainframes), all sitting at one end of a huge polished wood table. These clearly annoyed officials wanted to know why workstations were not a waste of money. My manager up two layers deflected the attack in a masterful fashion by characterizing workstations as "just big calculators". The dinosaurs slipped on this banana peel and the project continued.

Much later, I was having coffee in the World Financial Center. A manager in the firm's systems group (not one of the above) came over and said "You changed the way we thought about computing". I was proud of that.

[2] **Acknowledgements:** I thank Jon Hill for many informative discussions on systems, and also for carrying out many calculations in my groups over the years. I also thank Mike Driscoll for his enthusiastic help at Merrill with the workstation project. Finally, I thank all the systems programmers and managers over the years that contributed to the success of the quantitative efforts with which I was involved.

Introduction and Background

Numerical financial computing includes a number of important issues. Banks, broker-dealers, insurance companies etc. must deal with calculating the worth of large portfolios of financial instruments. These include mortgage-backed securities, corporate bonds, options, complex structured products, etc. Often the securities contained in these portfolios are not traded on the open market. For risk-management and portfolio total-return scenario analysis in the presence of underlying variable (e.g. interest-rate) movements, these securities need to be re-evaluated many times using models. Trading and sales activities also require securities pricing using models, preferably in real time.

Financial securities often contain embedded options that the holder or issuer of the security may exercise, and the options models are often sophisticated and numerically intensive. Typically, this involves diffusion with complex boundary conditions. The valuation is carried out using analytic approximations, Monte-Carlo simulations, PDE algorithms, etc.

For optimal financial strategies, these complex models need to run fast for large numbers of securities and, for a robust evaluation, under a selection of possible economic or risk stress scenarios. Moreover, all this must be done at a reasonable cost.

An extremely promising direction for some time has been parallel processing. Many numerically-intensive financial applications parallelize easily. These include portfolio calculations involving models. More generally, applications including many repetitive calculations are natural candidates for parallel processing. The most natural application is Monte Carlo simulation. There are other uses for parallel architecture, including large database applications. Feature recognition and comparisons of different time series with each other for arbitrage trading purposes can parallelize, including real-time feed applications. Optimization applications are more complicated.

Illustration of Parallel Processing for Finance

To illustrate parallel financial computing, here is an example. Imagine that an analyst or risk manager wants a portfolio of securities to be run each day with a number of different yield curves as scenarios for risk analysis. A parallel platform can be constructed to do these calculations. The basic idea is simply to choose a platform with several "nodes", each node capable of performing a calculation of a security. The same architecture can be used for a Monte-Carlo simulation of portfolio risk, mortgage models, etc. Each node is independently capable of performing calculations for one security. The analyst interacts with the nodes through a host machine. The host machine sends information to the nodes, receives results for display, produces reports, etc. The picture below gives the idea:

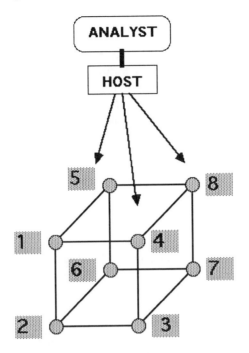

Some Aspects of Parallel Processing

Parallel Processing Computers
The explosion in the technology of microprocessor chips is the driving force behind the viability of distributed and parallel computing[ii] as feasible solutions to increasing computing requirements. Advances in solid-state physics and engineering, and in chip design, have contributed to this development[iii]. This revolution is by no means over, and developments will dramatically further increase the power of microprocessors. Since the chips are cheap, the

price/performance of such systems is extremely favorable. Further advances in chip technology and networking implies that the situation will continue to improve. Decreasing the transistor size on the chips is challenging, involving x-ray lithography and advanced heat-sink technology.

A parallel distributed computing environment composed of advanced chips and a network can be packaged as a separate board, set of boards, or parallel processing computer. The networking can be made efficient for such parallel machines.

Supercomputers now invariably have parallel architectures [3,iv].

Distributed Parallel Workstation Networks

A workstation network can itself be used to perform parallel computing because one workstation can call on other available workstations in the network to perform simultaneous computations. Issues to be addressed in comparing such a style of parallel computation include the availability (or unavailability) of enough workstations at appropriate times to perform the required calculations, the capacity of the network connecting the workstations together, and the relative cost/performance of adding another workstation to the network. Workstation networks are naturally cheap, because they use existing machines.

A drawback in practice can be that runs using many machines have interruptions due to system administrator actions that are not coordinated with the run schedule. Further, the time delay for calculations in an ordinary network is greater than for a customized back-plane network that couples processors together in a parallel machine.

Nodes

Central to the discussion of parallel processing is the concept of a "node". A medium-sized node is comparable to a workstation, complete with local memory, and possibly I/O [Input/Output] capability, but not necessarily a monitor. N of these nodes are then coupled together in a network. The number (e.g. N = 2, 4, ..., 128, ...) depends on the type of node and the overall cost. Especially important is the network-management software. The network is then attached to the host.

The nodes can be smaller or larger. A few mainframes can be networked together in a unified system; here the (huge) node is a mainframe. At the other end is a computer with thousands of very small nodes. The size of the node is referred to as the "granularity" of the machine. The choice of the parallel hardware, and in particular the granularity, depends critically on the nature of the

---

[3] **State of the Art Parallel Supercomputer:** The biggest supercomputers have been parallel machines for a long time. IBM recently announced plans for two supercomputers consisting of 130,000 microprocessors in one case and 12,000 in the other. The goal is to reach petaflop speeds.

applications. The basic points are that one node must be able to handle an independent calculation of interest, and that node power should be used optimally. This means that a large and powerful node should not be used to perform many repetitive small calculations, and a small node should not be called upon to perform large calculations. For many financial applications involving models, a medium-grained node machine is probably optimal.

Trivial or Linear Parallelization

The parallel processing architecture is most appropriate for those applications where individual calculations are independent of each other (such applications are said to "linearly" or "trivially" parallelize). Thus, these individual calculations can be carried out by all nodes at the same time. The results are transmitted to the host. In this linear-parallelizing case, the conversion of software to take advantage of parallel processing can be simple. Parallel processing can be applied to applications requiring inter-node communication, but the coding is naturally more complicated. A large class of financial applications linearly parallelizes.

Vectorization and Parallelization

It should be noted that "parallelization" and "vectorization" are not the same. Vectorization means that calculations are arranged if possible in a mode involving mathematical vectors; a typical example is a do-loop, if successive trips through the loop are independent of each other. Vector hardware processes these calculations in a pipeline or assembly-line fashion[v], with on the order of one operation performed every clock cycle. Parallel hardware, on the other hand, can perform many operations every clock cycle since there are many clocks, one for each processor (though these clocks may run slower). The relative efficiency of vector and parallel hardware depends on the application. Actually, vectorization and parallelization are complementary ideas. An advanced architecture combines vectorizing hardware in parallel, conceptually consisting of parallel copies of assembly lines.

Parallel Languages and High Performance Fortran

Software products can aid in the identification of portions of existing code that can parallelize. In particular, some languages have been extended with parallelization. The best example is High Performance Fortran or HPF.[vi] HPF is an extension of the new Fortran 90 standards, and is often used for high-end scientific computationally-intensive applications. The idea is that the compiler sets up scheduling of parallel operations that exist in the code.

Parallel Communication and the MPI

The MPI ("Message-Passing Interface") is a robust standard for communication between different processors to implement parallelized calculations[vii]. MPI was based on work at IBM, Intel and other companies that developed message-passing languages. MPI is designed to be an efficient and portable communication interface for Fortran and C code. MPI can be used with MIMD (Multiple Instruction, Multiple Data) programs. Communicating computers have their own local memory and send messages to each other to co-ordinate tasks. Software called PVM ("Parallel Virtual Machine")[viii] can be used in conjunction with MPI.

Parallel Computing and Data

The existence of large databases and the need to maintain security, integrity, and cohesiveness is a challenging issue in distributed or parallel computing. The distribution of the data (the "load") between workstations, or between chips in a parallel computer, has to be determined on a case-by-case basis.

PCs or workstations individually do not have the flexibility and power of mainframes or large servers for handling large amounts of data. A common solution is a parallel network attached to a large server that handles the large databases.

Technology, Strategy and Change

New computing paradigms are ignored by a financial institution at the peril of losing the leading edge. At the same time prudence and reliability, along with the existing investments in software and hardware must temper implementation of new technologies. Still, investment in an installed computing base should not be allowed to serve as an absolute inertial barrier to the examination and deployment of new powerful and cost-effective technology beneficial to long-term interests. Strategic decisions involve change; change is not without disruption, but the alternative may be stagnation.

Systems Groups, End Users, and New Technology

A proactive role of end-users with the corporate computer Technology or Systems Groups to implement new technologies in an optimal fashion is desirable. For maximum synergy in dealing with the difficult issues and decisions involved, the end-users have to understand systems issues and the systems groups have to understand end-user requirements. Unfortunately, the reality is generally that these two tribes understand each other's language at best imperfectly.

References

[i] **Strategic Computing**
Dash, J. W., *Strategic Directions in Numerical Financial Computing: Parallel-Processing Computers and Distributed Workstation Environments*. Centre de Physique Théorique, CNRS Marseille preprint CPT-89/PE. 2335, 1989.

[ii] **Parallel Processing**
Hwang, K., Briggs, F., *Computer Architecture and Parallel Processing*. McGraw-Hill Book Company, 1984.
Catanzaro, B., *Multiprocessor System Architectures – A Technical Survey of Multiprocessor/Multithreaded Systems using $SPARC^R$, Multilevel Bus Architectures and $Solaris^R$*. Sun Microsystems, 1994.

[iii] **Scientific References**
National Research Council, *Scientific Interfaces and Technological Applications – Physics through the 1990s*. National Academy Press, 1986.
Thompson, J. M. T., *Physics and Electronics – Visions of the Future*, Cambridge University Press; The Royal Society, 2001.
National Research Council, *Physics in a New Era*. National Academy Press, 2001.

[iv] **Parallel Supercomputers**
Merkoff, J., *I.B.M. plans a Computer that will set Power Record*. NY Times 11/19/2002.

[v] **Pipelining**
Kogge, P., *The Architecture of Pipelined Computers*. Hemisphere Publishing, 1981.

[vi] **High Performance Fortran with Parallelization**
Koelbel, C., Loveman, D., Schreiber, R., Steele Jr., G., Zosel, M., *The High Performance Fortran Handbook*. The MIT Press, 1994.

[vii] **MPI References**
Group, W., Lusk, E., Skjellum, A., *Using MPI – Portable Parallel Programming with the Message-Passing Interface*. The MIT Press 1994.
Snir, M., Otto, S., Huss-Lederman, S., Walker, D., Dongarra, J., *MPI – The Complete Reference*. The MIT Press, 1996.

[viii] **PVM Reference**
Geist, A., Beguelin, A., Dongarra, J., Jiang, W., Manchek, R., Sunderam, V., *PVM – Parallel Virtual Machine. A Users' Guide and Tutorial for Networked Parallel Computing*. The MIT Press, 1994.

36. Qualitative Overview of Data Issues (Tech. Index 2/10)

A complex and knotty problem faced by financial risk management at the corporate level pertains to obtaining consistent, reliable and complete financial data[1]. Data problems produce sources of uncertainty for risk management[2]. Specific issues with data are discussed in other chapters in this book. Here, we deal with some overall issues in a qualitative fashion.

Data Consistency

Because of resource limitations and historical development, data are typically fragmented across the organization in departmental databases. The firm-wide "back-office" or "books and records" databases contain the firm's official positions and their values, which are used for reporting purposes. These data are typically obtained from "feeds" from the "front-office" trading databases. The traders use their departmental databases to calculate their risks.

The departmental risks need to be aggregated at a corporate level. Corporate risk management likes to use official data. Unfortunately, a back-office database often does not contain critical information needed for risk management. This leads to a number of serious problems. Either a firm-wide database is constructed with the needed risk information, or feeds containing required risk information are brought in from the front office. Neither solution is very good. Consistency problems can arise between databases if a separate corporate database is

---

[1] **What is the Fundamental Theorem of Data?** As mentioned several times in the book, this apparently tongue-in-cheek but actually serious statement says: "Data constitute another Black Hole of Finance". Some people think that this Fundamental Data Theorem is more profound than the Fundamental Theorem of Systems (cf. Ch. 34).

[2] **Data Sentence Practice for the Connoisseur and A Point of Grammar**: Here are some handy data sentences to practice, all of which can and have been used: "I don't believe your data", "My data are better than your data", "You're using the wrong data series", "The data don't imply what you're claiming", and "My judgment is more important than your data". Others: "We don't have the data", "The data feed broke again", and "We'll have access to the data next month", which can be used periodically.

As an annoying point of grammar, the word "data" is actually a plural word; the singular from Latin is "datum". Thus, it is improper to say "this data stinks"; the correct version would be "these data stink".

constructed. On the other hand, front-office feeds must be ironed out to avoid misunderstanding. Front-office conventions for data representation are sometimes unclear, often undocumented in writing, and may change without warning. Moreover, the tolerance of the traders for data work done for corporate risk management is limited, because this activity does not make any money for them.

Data Reliability

Reliability of the data in any database is a serious potential issue, requiring dedicated personnel that understand the database structure and as well as the financial content of the data. This optimal deployment of personnel is often not present or inadequate. Data errors (sometimes from human input error, sometimes from feed error, sometimes from lack of input at all) exist. This makes it difficult sometimes to distinguish a true outlier event from a simple error.

Data Completeness

Completeness of data is a serious issue. If the front office does not trade in a particular sector, traders will generally see no reason to put data corresponding to that sector in the database. Then later, if trading does occur in that sector, data available for historical comparison of risk will be limited for that sector. This is an issue, because misleading results can be obtained by performing risk analysis using limited data. In particular, if data only go back during periods when the markets were relatively stable, the potential market stresses from turbulent periods will not be realistically evaluated. This is true no matter what statistical tests or which high confidence levels are used.

Data Vendors

Data vendors exist and are often used in absence of, in place of, or to complement, in-house information. However, inconsistencies sometimes exist between data from different vendors. Moreover, the algorithms used by the vendors to construct their data time series from their market sources are sometimes not available. This can lead to problems since one algorithm or set of sources can be used for one time series of data, and another algorithm or set of sources for another time series.

Still, it can be said that the data vendors live or die by their product, and therefore they do their best to cope with an inherently extremely difficult problem.

Historical Data Problems and Data Groups

It takes a long time to discuss problems with historical data. These include distinguishing outliers from bad data points, coping with missing data points, inconsistent measurement frequencies for different time series, effects of overlapping windows needed when not enough data are present, inconsistent data from different sources, strings of zeros entered by traders who do not happen to have positions in that variable at the moment, etc.

This highlights the need for a separate group that is dedicated to dealing with these perhaps boring but critically important and thorny data issues. A data quality group is not the same as the group of systems programmers that handle the representation of the data in the computer using a database. It is sometimes neither easy to get management to understand these issues nor to obtain the appropriate resources to deal with the problems adequately.

Preparation of the Data

Even given pristine data, we must still be concerned about preparation of the data as input to analysis. Preparation can include smoothing techniques (various choices of moving averages or centered moving averages; splines, Padé approximants), weighting techniques emphasizing recent data or data from a particular time period which might be considered more relevant, possibly putting caps restricting large moves which happened in the past but "can never happen again", etc. The definitions of the variables to use in the analysis need to be specified; these can be various functions of the underlying time series (e.g. returns, simple differences, or other).

Bad Data Points and Other Data Traps

Distinguishing bad data points from valid outliers can require an in-depth knowledge of the market. The presence of typos or zeros in the series is an obvious problem, but there are more subtle issues of various types of data traps.

For example, bad data can occur for a spread constructed by subtracting two time series that are individually reasonable but constructed with inconsistent methodologies (probably undocumented). It is moreover common that proxy spread index data are used for a particular spread, either because the data do not exist or because there are too many spreads for the risk model (e.g. a VAR simulator) to handle explicitly. The outliers for the proxy data may not be outliers for the spread – or conversely. Naturally there can be lively discussions regarding the appropriateness or inappropriateness of the proxy index used – or why the simulator is not using more indices.

Gee, we seem to be back to the Black-Hole Data Theorem again[1].

37. Correlations and Data (Tech. Index 5/10)

In this chapter, we deal qualitatively with some important aspects of correlations and data, giving some examples. We discuss windowing uncertainties, including overlapping vs. non-overlapping windows. We also discuss uncertainties due to the limited amount of data relative to the number of variables. Finally we discuss intrinsic dynamically generated correlation uncertainties. See also Ch. 23-25.

Fluctuations and Uncertainties in Measured Correlations

Correlations have several sources of instability, both statistical and dynamical[i]. The existence of uncertainties or fluctuations for correlations cannot be in doubt. An example below shows the correlation[1] between silver Ag and the German mark DM currency exchange rate to USD in 1992-93. The reader can see that the correlation varied widely.

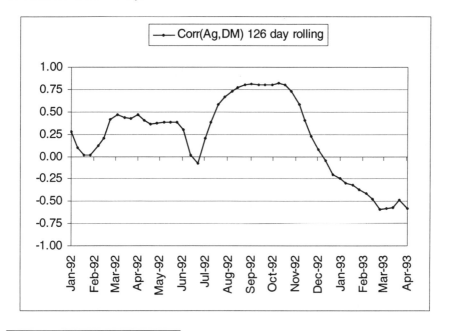

---

[1] **Acknowledgements:** The Ag, DM correlation is the same example as in my CIFEr tutorials 1996-2001. We thank Citigroup for the use of the data.

Time Windowing

We now discuss issues related to the time windows in which correlations are measured. One is the window size, and another is whether overlapping or non-overlapping windows are used. We treat these in turn.

Window Size

The time length of the windows used to get the correlations should be determined by financial relevance. We need to use finite-sized time windows in risk management because we cannot wait forever to analyze why a strategy might be losing money. Time scales for instabilities in correlations can be short. The collective panic and flight to quality in markets that are suddenly disturbed by bad news lead to sudden changes in correlations.

A convenient window size for corporate risk management is the reporting interval of three months. Analyses that use long-term averages over correlations are only valid for long-term buy-and-hold management.

The standard measure for the uncertainty in a correlation due to windowing noise is the Fisher result. If the measured correlation (using samples of size N) is denoted as r, and if a true correlation value ρ exists, then the expected value of r is ρ and the uncertainty of r is approximately (cf. Ch. 38)

$$\sigma_r^{Fisher} = \left(1 - r^2\right) / \sqrt{(N-3)} \qquad (37.1)$$

This statistical uncertainty is only part of the story since strong intrinsic uncertainties in correlations exist that have nothing to do with the above formula.

Overlapping and Non-Overlapping Windows

The second issue is whether or not to use overlapping windows. The windows generally used for risk management are overlapping. The problem is that we usually do not have enough data to be able to afford the luxury of non-overlapping windows. Measurements in a sequence of overlapping windows are serially correlated because the same measurement exists in different windows.

If we are interested in the maximum and minimum correlations, then we *want* to use overlapping windows. This is because each of the windows represents a realized physical state in history for a correlation, and we do not want to throw any states away, serially correlated or not.

The fluctuations for correlations that are measured with overlapping windows are smaller than correlations using non-overlapping windows for a given number of windows. This is because the same measurements occur many times in overlapping windows, suppressing the fluctuations, whereas different measurements occur in non-overlapping windows, enhancing fluctuations.

Example of the Effects of Windowing

In this example all the effects listed above are shown. The example uses 500 points of a bivariate Gaussian distribution with fixed correlation $\rho = 0.1$. We then attempt to measure the correlation using windows of different sizes, both overlapping and non-overlapping. The results for the measured correlation uncertainties along with the Fisher uncertainty are shown in the figure below:

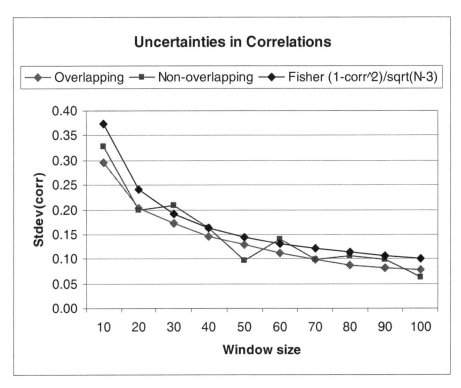

All measures of the uncertainty are similar. The Fisher uncertainty is largest The non-overlapping window uncertainties are somewhat larger than the overlapping window uncertainties as expected. For corporate risk reporting using 3-month (65 business day) windows, the overlapping uncertainty is about 80% of the Fisher uncertainty. For large window sizes, only a few non-overlapping windows are present. The overlapping-window uncertainties have the advantage of forming a smooth curve.

Example for Effect of Jumps on Correlations

We next show the results when we insert a 10 standard deviation jump up at one time t_{jump} in the second time series, otherwise generated by the above bivariate

Monte-Carlo. See also Ch. 25 where a similar exercise was performed. The effect of the jump on the correlation depends on what the first variable does at time t_{jump}. For example, if the first variable does not change at t_{jump} then there is no effect on the correlation. However there can be substantial effects if the first variable does move. The following plot shows the results for a particularly large effect on the correlation by the jump in this model (using overlapping windows).

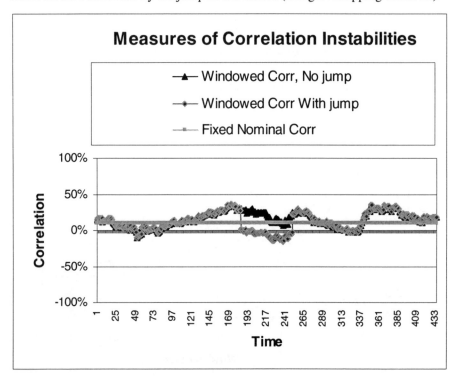

Correlations, the Number of Data Points, and Variables

It seems self-evident that the number N of data points used to determine correlations should be greater than the number of variables p in order to have a well-defined data procedure. Mathematically, $N > p$ is required to have non-degenerate correlations, and for the Wishart Theorem (c.f. Ch. 38). However, this condition may not be satisfied in practice in risk management. For example if we have two years of data (and even that is sometimes hard to get), then $N = 500$. However the number of variables p can easily run into the thousands (including various interest rates and spreads in various currencies, individual stocks, commodities, idiosyncratic risks for various products, and so on).

Correlations and Data

We have several remarks. First, if we actually measure correlations with $N < p$ and compare these correlations to the case when $N > p$ we find that there is not much difference in the values of the correlations. Mathematically what happens is that some correlations become degenerate, i.e. take on the same value, as other correlations when $N < p$. However, for large N and most data, many correlations with $N > p$ tend to lie close to one another. When $N < p$, the changes in correlations that occur due to the degeneracies are not large.

Example with Different Numbers of Observation and Variables

Here is an illustrative pedagogical example. We can run a multivariate Monte Carlo simulation with fixed input correlations. We can vary the number of observations N and the number of variables p, and look at what happens to an individual specific correlation $\rho_{specific}$. If there is no big effect, then regardless of the values of N and p, as we run the simulation we should get the input $\rho_{specific}$ on the average, with an standard deviation error roughly given by the Fisher windowing uncertainty.

The particular correlation chosen for comparison in the exercise was taken as $\rho_{specific} = 0.53$, with other values for the other correlations[2]. The value 0.53 is a high correlation. Therefore, if there is a problem with the correlation moving because of the degeneracies, we should notice it.

The output Monte-Carlo average correlations for different N, p are within errors of the input correlation 0.53. Nothing special or catastrophic happens as the $N = p$ point is crossed. The correlation errors are consistent with the Fisher uncertainties $\sigma_r^{Fisher} = [1 - (0.53)^2]/\sqrt{N-3}$. Here are illustrative results:

Average MC correlation (nb: within 0.53 ± error)	N = 126	N = 252	N = 504
p = 2	0.49	0.49	0.49
p = 181	0.58	0.56	0.56
p = 279	0.53	0.52	0.52

σ_ρ MC statistical error	N = 126	N = 252	N = 504
p = 2	0.07	0.05	0.03
p = 181	0.06	0.04	0.03
p = 279	0.06	0.04	0.03
σ_r^{Fisher} uncertainty	0.06	0.05	0.03

---

[2] **Acknowledgements:** The input correlations come from data ending in 1999. We thank Citigroup for the use of the data statistics, run by J. Hill.

Bottom Line: In Practice, N > p is a Red Herring

The results illustrate what is probably a general property. Although the condition $N > p$ is required mathematically, the degeneracies that result from the measurements with $N < p$ do not affect the numerical values of the correlations much.

The bottom line is that, practically speaking, the academic constraint $N > p$ appears to be of little significance, and can be disregarded. The apparent need for $N > p$ is a red herring. The procedures of corporate risk management are not endangered.

Intrinsic and Windowing Uncertainties: Example

We have buffeted the reader with the statement that correlations have more uncertainty than just that given by windowing statistical noise. The idea is below.

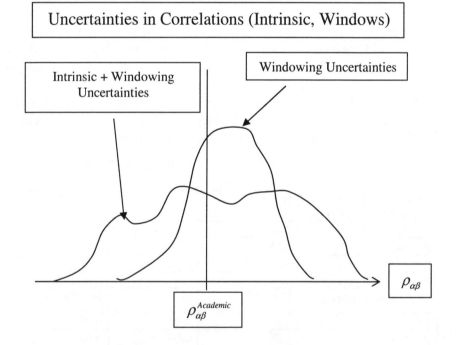

The picture implies that the deviations from a putative academic correlation $\rho_{\alpha\beta}^{Academic}$ are due to both intrinsic and windowing uncertainties. An illustrative example is the Ag, DM correlation given at the beginning of this section. This correlation is measured using overlapping windows. Hence the correlation uncertainty at one standard deviation should be less than or at most on the order of the Fisher uncertainty if there is no intrinsic correlation instability. On the other hand, if there is intrinsic instability, the observed uncertainty will be greater than the Fisher uncertainty.

The Fisher Transform for Correlations

The Fisher transform R^{Data}_{Fisher} of the correlation data r is defined as[3]

$$R^{Data}_{Fisher} = \frac{1}{2}\ln\left[(1+r)/(1-r)\right] \qquad (37.2)$$

Call $R^{AvgData}_{Fisher}$ the average of the Fisher-transformed data with standard deviation $\sigma(R^{Data}_{Fisher})$. Then set $R_{\pm} \equiv R^{AvgData}_{Fisher} \pm \sigma(R^{Data}_{Fisher})$, and transform back using $r_{\pm} = \tanh R_{\pm}$. Finally, define the observed uncertainty in the correlation data at the 68% CL using $\pm\sigma_r^{Data} = \pm(r_{+} - r_{-})/2$.

Data Uncertainties Compared with Fisher Uncertainty

Here are the results for the Ag, DM correlation:

Observed Uncertainty of Correlation	Fisher Uncertainty for Correlation
±47%	±9%

The observed 68% CL uncertainty in the correlation from the data is an average of ±47% (40% up and -53% down from the average correlation). The observed maximum (82%) and minimum (-59%) are about ±1.7 SD from the average (91% CL).

The observed uncertainty in this correlation from the data cannot be explained by the 68% CL Fisher windowing uncertainty of ±9%, obtained using $\sigma_r^{Fisher} = (1-r^2)/\sqrt{(N-3)}$, with the average correlation used for r. At the 91% CL the Fisher uncertainties predict that the correlation should lie in the restricted range (12%, 41%).

The bottom line is that the Fisher uncertainty cannot account for the observed variability in these correlation data.

---

[3] **Fisher's Transform:** We discuss the Fisher transform formalism in Ch. 38.

The Importance of Intrinsic Correlation Uncertainties

We believe that the observed uncertainty in this example clearly indicates the existence of intrinsic dynamically generated correlation instabilities. We have looked at other examples. Some instabilities seem large, as in the Ag, DM example. Other instabilities are smaller. A consequence in that case is that no "true" correlation exists, regardless of the size of the windows and the length of time we wait to measure the correlation.

We believe that intrinsic instability of correlations should become an important modeling issue in the future for financial risk. In Ch. 25 we looked at some models for dynamical variations in correlations. In Ch. 23, 24 we discussed the stressed correlation matrices, used in the Stressed VAR.

Two Miscellaneous Aspects of Data and Correlations

The Importance of Conventions

You should keep track of the conventions and be able to defend them. As we have emphasized, time windows of different length (e.g. 3 months or 1 year) can produce different results, and are relevant under different circumstances. Even minor conventions: 260 bd/yr or 250 bd/yr (2 weeks of holidays) are important.

There are potentially dangerous sociological situations that can be associated with conventions. Non-technical people can latch onto small differences in calculations and assign illogical importance to these differences. In particular, time can be wasted ferreting out the reasons for small differences in badly documented alternative calculations that depend on different conventions.

On the other hand, sometimes such differences are large and warrant scrutiny.

Data Problems and Positive-Definite Correlations

Any "garbage" in the data will produce a breakdown of the correlation matrix away from its theoretical positive definite attribute. Dealing with non-positive definite matrices requires the use of the singular value decomposition procedure, as described in Ch. 24.

On the other hand, cleaning up the data can be and generally is a big issue, and can involve internal political ramifications far beyond the scope of this book.

References

[i] **Correlation Instabilities**
Dash, J. W., *Evaluating Correlations and Models*. Talk, RISK Conference, 11/94

38. Wishart's Theorem and Fisher's Transform (Tech. Index 9/10)

In this chapter, we consider the mathematics and theory for three topics. They are (1): the Wishart theorem, (2): the Fisher Transform, and (3): Implications for correlation uncertainties due to sampling error[1]. John Wishart, in a brilliant exposition[i], generalized earlier work[ii] to obtain the distribution of standard deviations and correlations obtained from a sample of N measurements of p variables, assuming that all variables obey a multivariate Gaussian distribution.

For the case $p = 2$, we will discuss theoretically the correlation uncertainties due to sampling error using finite windows, and give a simple derivation of the results using the Fisher transformation. We will also discuss the Wishart theorem using Fourier transforms that we believe gives some insight[2].

Of course, statistical uncertainties exist. It is important to have a handle on the statistical sampling uncertainties. Still, other non-statistical uncertainties are probably bigger than statistical uncertainties, especially in stressed markets.

Although the Wishart distribution is well defined only if the number of measurements exceeds the number of variables, i.e. $N > p$, risk management at the corporate level often assumes multivariate Gaussian behavior for many thousands of variables but yet generally works with only a few years of data at most, so in practice $N < p$ is unavoidable. As was mentioned in the previous chapter, the use of $N < p$ merely implies numerical degeneracies of correlations reducing the dimension p down to some effective dimension $p_{\mathit{eff}} < p$ satisfying $N > p_{\mathit{eff}}$. We presented some numerical examples that indicate that these degeneracies do not significantly impact risk analysis[3]. The significant uncertainties (due both to intrinsic instabilities and windowing noise)

---

[1] **Acknowledgement:** I thank Ardavan Nozari for pointing out the Wishart theorem, and for informative conversations.

[2] **History:** The work described in this section was mostly done in 1999, with additional work on the Wishart distribution in 2002-03.

[3] **Corporate VAR:** The statement that N < p does not really cause problems is good news for corporate risk managers who need to quote a risk measure like VAR, and who live in a world where the number of data points available is less than the number of variables needed. Otherwise, the Wishart theorem would rule out their jobs.

overshadow the smaller uncertainties due to the imposition of partial correlation degeneracies due to $N < p$.

Warm Up: The Distribution for a Volatility Estimate

In this section, at the risk of boring the reader, we derive the well-known results for the distribution of a volatility estimate, mostly in order to introduce the Fourier transform formalism that used below to discuss the Wishart theorem.

We consider a variable x_ℓ (with $\ell = 1...N$ being a time index). By x_ℓ we really have in mind time differences or returns, but to simplify notation we do not indicate the time difference operator d_t. We simplify notation further by taking the probability distribution of each x_ℓ to be Gaussian with unit variance and zero mean, so the integrated probability distribution is

$$\mathcal{P}_N^{(Integrated)} = \int d\mathcal{P}_N = \prod_{\ell=1}^{N} \int_{-\infty}^{\infty} \exp(-x_\ell^2/2) \frac{dx_\ell}{\sqrt{2\pi}} \qquad (38.1)$$

Now $\mathcal{P}_N^{(Integrated)}$ is exactly 1, but we will soon restrict the integration region. The estimated variance is $s^2 = \sum_{\ell=1}^{N} x_\ell^2$. This formula contains a factor $\sqrt{N-1}$ in the definition to avoid cluttering up the page (at the end of this section we will redefine the variance in the more conventional notation). We want the distribution $\mathcal{P}_N(s^2)$ with $\mathcal{P}_N^{(Integrated)} = \int \mathcal{P}_N(s^2) ds^2$. To this end[4], we insert the factor 1 in the above integral and substitute for it the right-hand side of the Dirac delta function identity

$$1 = \int_0^\infty \delta\left(s^2 - \sum_{k=1}^{N} x_k^2\right) \cdot ds^2 \qquad (38.2)$$

We then use the Fourier transform representation of the delta function

$$\delta\left(s^2 - \sum_{k=1}^{N} x_k^2\right) = \int_{-\infty}^{\infty} \exp\left[i\omega\left(s^2 - \sum_{k=1}^{N} x_k^2\right)\right] \frac{d\omega}{2\pi} \qquad (38.3)$$

---

[4] **Dirac Delta Function Trick and G. Parisi:** By now the reader should be familiar with this trick, used many times in this book. The idea comes from a paper (long lost), written by the physicist Giorgio Parisi around 1975.

Wishart's Theorem and Fisher's Transform

We remove the integral $\int_0^\infty ds^2$ to get

$$\mathcal{P}_N\left(s^2\right) = \int_{-\infty}^{\infty} \frac{d\omega}{2\pi} e^{i\omega s^2} \prod_{\ell=1}^{N} \int_{-\infty}^{\infty} \exp\left[-(i\omega+1/2)x_\ell^2\right] \frac{dx_\ell}{\sqrt{2\pi}} \qquad (38.4)$$

We can do each Gaussian integral to get

$$\mathcal{P}_N\left(s^2\right) = \int_{-\infty}^{\infty} \frac{d\omega}{2\pi} e^{i\omega s^2} \frac{1}{2^{N/2}(i\omega+1/2)^{N/2}} \qquad (38.5)$$

Now we have $s^2 > 0$ so we can do the ω integral using complex variable integration. We close the contour in the upper half (UH) ω complex plane with $\mathrm{Im}\,\omega > 0$. The integral vanishes for the semi-circle at infinity due to the factor $\exp(-s^2 \,\mathrm{Im}\,\omega) \to 0$. If $N = 2M$ is even, there is a multiple pole at $i/2$ on the imaginary ω axis.

To proceed we introduce an auxiliary parameter ζ and write

$$\mathcal{P}_{2M}\left(s^2\right) = \frac{(-1)^{M-1}}{2^M(M-1)!} \cdot \frac{\partial^{M-1}}{\partial \zeta^{M-1}} \int_{-\infty}^{\infty} \frac{d\omega}{2\pi} e^{i\omega s^2} \frac{1}{(i\omega+\zeta+1/2)} \bigg|_{\zeta=0} \qquad (38.6)$$

Here, the parameter ζ is kept as $\zeta > -1/2$, and then is set to zero after the derivatives are performed.

This integral has a single pole at $\omega^{(0)} = i(\zeta+1/2)$ in the UH ω plane, producing $\exp\left[-(\zeta+1/2)s^2\right]$. Performing the derivatives, using the gamma function[iii] definition $\Gamma(M) = (M-1)!$ and replacing $M = N/2$, we get the usual result

$$\mathcal{P}_N\left(s^2\right) = \frac{1}{2^{N/2}\Gamma(N/2)} \cdot e^{-s^2/2}\left(s^2\right)^{\frac{N}{2}-1} \qquad (38.7)$$

Although derived for N even, we can analytically continue the result in Eqn. (38.7) to N odd. We note $\Gamma(z+1) = z\Gamma(z)$ and $\Gamma(1/2) = \sqrt{\pi}$.

Then $\mathcal{P}_N^{(Integrated)}\left[s^2 \leq \chi^2\right] = \int_0^{\chi^2} \mathcal{P}_N(s^2) ds^2$ is the probability that the variance estimate is less than or equal to χ^2.

We now eliminate the simplifications made above. We redefine the Gaussian probability distribution to have a mean μ and volatility σ. We need to include the average $\bar{x} = \frac{1}{N}\sum_{k=1}^{N} x_k$. We do this by inserting the additional constraint

$$1 = \int_{-\infty}^{\infty} d\bar{x} \int_{-\infty}^{\infty} \frac{d\bar{\omega}}{2\pi} \exp\left[i\bar{\omega}\left(\bar{x} - \frac{1}{N}\sum_{k=1}^{N} x_k\right)\right] d\bar{x} \quad (38.8)$$

We also set $s^2 = \frac{1}{N-1}\sum_{\ell=1}^{N}(x_\ell - \bar{x})^2$, restoring the usual definition. Performing the extra two integrals results in an extra factor replacing the order of the singularity $N/2 \to (N-1)/2$. Hence, with all factors, setting $\lambda = 1/(N-1)$, we now get

$$\mathcal{P}_N(s^2) ds^2 = \frac{ds^2}{2\sigma^2 \lambda \Gamma((N-1)/2)} \left(\frac{s^2}{2\sigma^2 \lambda}\right)^{(N-3)/2} \exp\left(-\frac{s^2}{2\sigma^2 \lambda}\right) \quad (38.9)$$

This produces $\langle s^2 \rangle = \sigma^2$. Note that we need $N \geq 2$ in order that the singularity in s^2 is integrable as $s^2 \to 0$.

The Wishart Distribution

Consider a p-dimensional world with variables $\{x_{\alpha\ell}\}$, where $\alpha = 1...p$ is the label for internal degree of freedom (e.g. interest rates, commodities, etc.) and $\ell = 1...N$ the time index. Again, we have in mind applying the formula to time differences or returns, but again for simplicity, we do not indicate the d_t notation. Define the quadratic form $V_{\alpha\beta}$ by[5]

---

[5] **Comment on the V matrix**: Note that there is no correlation matrix in the definition.

$$V_{\alpha\beta} = \sum_{\ell=1}^{N} x_{\alpha\ell} x_{\beta\ell} \tag{38.10}$$

As above, we first simplify the arithmetic by assuming that the $\{x_{\alpha\ell}\}$ variables are distributed according to a multivariate Gaussian distribution with unit volatilities, ignore the average constraints, and denote the given correlation matrix by $(\rho_{\alpha\beta})$. The Wishart distribution for the probability that $V_{\alpha\beta} = \rho_{\alpha\beta}$ in volume $\prod_{\alpha\leq\beta} dV_{\alpha\beta}$ is [6]

$$W_N(V) = \frac{(\det V)^{(N-p-1)/2} \exp\left[-Tr(V\rho^{-1})/2\right]}{(\det \rho)^{N/2} \cdot \mathcal{K}} \tag{38.11}$$

Here the "kinematic factor" $\mathcal{K} = 2^{pN/2} \pi^{p(p-1)/4} \prod_{\alpha=1}^{p} \Gamma\left[(N-\alpha+1)/2\right]$ is a constant depending on N, p and not on the dynamical matrices V, ρ.

Note that if $p = 1$ we have $\det V = s^2$ and $\rho = 1$ so we just reproduce the distribution we had for the estimated volatility above. To keep the distribution finite for general p as $\det V \to 0$ we need the restriction $N > p$.

If we impose the average constraints, we replace $N \to N-1$. The unit volatility assumptions are relaxed through scaling, and the normalization with $\lambda = 1/(N-1)$ can be included explicitly as above.

The expected value of $V_{\alpha\beta}$ is just $\rho_{\alpha\beta}$. Including volatilities $\{\sigma_\alpha\}$, the expected value of $V_{\alpha\beta}$ is $\sigma_\alpha \rho_{\alpha\beta} \sigma_\beta$. The uncertainty of $V_{\alpha\beta}$ around its expected value is the subject of the Wishart distribution.

We discuss the form of the Wishart distribution below.

The Probability Function for One Estimated Correlation

Before launching into a general discussion, we consider the case $p = 2$ with one off-diagonal correlation element ρ_{12} in the original 2-dimensional Gaussian probability distribution for each $x_{1\ell}, x_{2\ell}$. We (hopefully not too confusingly)

---

[6] **Notation:** Please do not confuse the Latin letter p used for the internal dimension with the Greek letter ρ used for the correlation matrix.

denote ρ_{12} by ρ, so the determinant of the correlation matrix $\det \rho$ in this notation is $\det \rho = 1 - \rho^2$. We want to integrate over the volatility degrees of freedom to find the distribution for the measured sample correlation r relative to the given correlation ρ in the probability distribution. Here r is given as

$$V_{12} = \sum_{\ell=1}^{N}(x_{1\ell} - \bar{x}_1)(x_{2\ell} - \bar{x}_2) = rs_1 s_2 (N-1) \qquad (38.12)$$

We replace $N \to N-1$ in the $p=2$ Wishart distribution to include the $x_{1\ell}, x_{2\ell}$ averages. The resulting pdf is originally a result of Fisher [ii]. We integrate over $ds_1^2 ds_2^2$ by changing to elliptic co-ordinates $s_1 = \sigma_1 \xi \cos \psi$, $s_2 = \sigma_2 \xi \sin \psi$ with $\psi \in [0, \pi/2]$. All dependence on the volatilities eventually cancels, so we can take unit vols. During the algebra, we encounter the Γ function

$$\int_0^\infty e^{-y} y^{N-2} dy = \Gamma(N-1) \quad \text{where} \quad y = N\xi^2(1 - r\rho \sin 2\psi)/(2 \det \rho).$$ We get

the integral $I = \int_0^{\pi/2} d\psi \dfrac{(\sin 2\psi)^{N-2}}{(1 - r\rho \sin 2\psi)^{N-1}}$. This integral can be expanded in a series in $r\rho$ and integrated term-by-term using the Euler beta function [iii]. We obtain the result for the probability distribution $\wp(r)$ of the correlation r with N measurements, given a "true" value for the correlation ρ, as

$$\wp(r) = \frac{(1-\rho^2)^{(N-1)/2}(1-r^2)^{(N-4)/2}}{\sqrt{\pi}\,\Gamma\big[(N-1)/2\big]\Gamma\big[(N-2)/2\big]} \sum_{k=0}^{\infty} \frac{\Gamma^2\big[(N+k-1)/2\big](2r\rho)^k}{k!}$$

(38.13)

Fisher's Transform and the Correlation Probability Function

The above sum is, to say the least, unenlightening. Progress can be made using the Fisher transformation

$$R_{Fisher} = \frac{1}{2}\ln\big[(1+r)/(1-r)\big] \qquad (38.14)$$

The inverse is $r = \tanh R_{Fisher}$. As $r \to \pm 1$ we have $R_{Fisher} \to \pm\infty$. We also write $\rho = \tanh R_0$.

As the reader may surmise, we are after a Gaussian approximation to $\wp(r)$ as a function of R_{Fisher}. To this end, we assume that N is large and use various asymptotic formulae. We also need the WKB approximation, which we discuss next.

WKB Approximation and Fisher's Gaussian Approximation

Notice that, apart from the squared Γ function, the sum Eqn. (38.13) for $\wp(r)$ looks like the exponential sum. As a warm-up, we note that the WKB method [iv] can be used to obtain the dominant region in the exponential sum, $\exp v = \sum_{k=0}^{\infty} v^k/k!$. We use Stirling's asymptotic formula [iii] $k! \approx k^k e^{-k} \sqrt{2\pi/k}$, along with an integral approximation to the sum. This produces $\sum_{k=0}^{\infty} v^k/k! \approx \int dk \cdot \sqrt{k/2\pi} \exp[\Phi(k)]$, where the function in the exponential is $\Phi(k) = -k \ln k + k + k \ln v$. The WKB method involves expanding $\Phi(k)$ to quadratic order around the stationary point k_0 where $\Phi'(k_0) = 0$. We find $k_0 = v$ as the stationary point. Carrying out the Gaussian integral, we obtain $\exp v = \sum_{k=0}^{\infty} v^k/k!$ consistently.

We return to the evaluation of $\wp(r)$. The asymptotic formula $\Gamma(z+a) \approx z^a \Gamma(z)[1 + a(a-1)/(2z)]$ is used, with $z = (N-3)/2$ and with appropriate values of a for the various gamma functions. We then have an approximation to the sum for $\wp(r)$ as an exponential sum. Here, $v = (N-3) \tanh R_{Fisher} \tanh R_0$. Using the WKB result, we replace $k \to v$ for non-dominant terms at large N. We use Taylor expansions to first order, $\ln(\cosh^2 R) \approx R^2$ and $\tanh R \approx R$ in order to get the leading quadratic terms in the exponent. We then obtain Fisher's approximate Gaussian result

$$\wp(r) \approx 1/\sqrt{2\pi\sigma_R^2} \cdot \exp\left\{-\frac{1}{2\sigma_R^2}(R_{Fisher} - R_0)^2\right\} \qquad (38.15)$$

Here, $\sigma_R^2 = 1/(N-3)$.

Fisher's Volatility: Statistical Noise of Estimated Correlation

Fisher's result is an approximate Gaussian behavior in the variable R_{Fisher} with an expected value of $R_0 = \frac{1}{2}\ln\left[(1+\rho)/(1-\rho)\right]$ and an approximate width of $\sigma_R = 1/\sqrt{N-3}$. This result can be used conveniently to set confidence levels for correlations in measuring the windowing noise. We can re-express the Fisher width for the estimated correlation r at one SD uncertainty using

$$\left(\sigma_r^{Fisher}\right)^2 = \left\langle (r-\rho)^2 \right\rangle = \left(\frac{dr}{dR_{Fisher}}\right)^2 \left\langle (R_{Fisher} - R_0)^2 \right\rangle = \left(1-r^2\right)^2 / (N-3)$$

(38.16)

The quantity σ_r^{Fisher} was used in the previous chapter in the numerical correlation uncertainty study. We notice that the $(1-r^2)$ factor makes σ_r^{Fisher} vanish at ± 1. Naturally as $N \to \infty$, the statistical uncertainty is zero and the value $r = \rho$ is recovered.

We re-emphasize that this entire framework assumes that there exists in fact one given value for the correlation. Perversely, in the real world, correlations have important non-stationary intrinsic instabilities that have nothing to do with the statistical uncertainties described by the above results.

The Wishart Distribution Using Fourier Transforms

We now give some details regarding the derivation of the Wishart distribution. Wishart used a geometrical proof. Here, we follow the Fourier transform method, generalized from the above discussion of the $p=1$ case for the chi-squared distribution for the estimated volatility. The Fourier generating function of the Wishart distribution is straightforward to obtain and agrees with the known result. Evaluating the integrals of the Fourier generating function to get the Wishart distribution is a difficult exercise in multi-complex-variable contour integration, and the proof is not complete. Nonetheless, the origin of all the component factors in the Wishart distribution can be seen in a rather straightforward fashion. This gives insight into the inner workings of the Wishart distribution.

Outline of the Derivation, and the Dynamic Components in Wishart's Distribution

We begin with the assumed multi-normal form for the $\{x_{\alpha\ell}\}$ variables (with unit volatilities and assumed zero average value to simplify the notation),

$$d\mathcal{P}_N = \prod_{\ell=1}^{N} d\mathcal{P}_\ell = \frac{1}{(\det \rho)^{N/2}} \exp\left[-\frac{1}{2}\sum_{\alpha,\beta,\ell} x_{\alpha\ell}\rho^{-1}_{\alpha\beta}x_{\beta\ell}\right]\prod_{\alpha,\ell}\frac{dx_{\alpha\ell}}{\sqrt{2\pi}} \qquad (38.17)$$

We immediately see one factor in the Wishart distribution, $(\det \rho)^{-N/2}$.

The symmetry in the internal variable labels α, β and the fact that only $\alpha \le \beta$ variables are independent plays an important, if annoying, role. We use the Dirac delta function trick for the independent $V_{\alpha\beta}$ with $\alpha \le \beta$,

$$1 = \prod_{\alpha \le \beta} \int_{-\infty}^{\infty} dV_{\alpha\beta}\, \delta\left(V_{\alpha\beta} - \sum_{\ell=1}^{N} x_{\alpha\ell}x_{\beta\ell}\right) \qquad (38.18)$$

Introducing the variables $\{\omega_{\alpha\beta}\}$ we write the Fourier transform as

$$\delta\left(V_{\alpha\beta} - \sum_{\ell=1}^{N} x_{\alpha\ell}x_{\beta\ell}\right) = \eta_{\alpha\beta}\int_{-\infty}^{\infty}\frac{d\omega_{\alpha\beta}}{2\pi}\exp\left[i\eta_{\alpha\beta}\omega_{\alpha\beta}\left(V_{\alpha\beta} - \sum_{\ell=1}^{N} x_{\alpha\ell}x_{\beta\ell}\right)\right]$$

$$(38.19)$$

Here, $\eta_{\alpha\alpha} = 1$, and $\eta_{\alpha\beta} = 2$ if $\alpha < \beta$. We now rewrite the integrand to get rid of $\eta_{\alpha\beta}$ in favor of summing over all α, β with the understanding that $\omega_{\beta\alpha} = \omega_{\alpha\beta}$,

$$\prod_{\alpha \le \beta}\exp\left[i\eta_{\alpha\beta}\omega_{\alpha\beta}\left(V_{\alpha\beta} - \sum_{\ell=1}^{N} x_{\alpha\ell}x_{\beta\ell}\right)\right] = \prod_{\text{All }\alpha,\beta}\exp\left[i\omega_{\alpha\beta}\left(V_{\alpha\beta} - \sum_{\ell=1}^{N} x_{\alpha\ell}x_{\beta\ell}\right)\right]$$

$$(38.20)$$

The reason for this step is that identities we need only hold for sums over all α, β. Still, the integrations are only over those $\{\omega_{\alpha\beta}\}$ with $\alpha \le \beta$.

We note that $\prod_{\alpha,\beta}\exp\left[i\omega_{\alpha\beta}V_{\alpha\beta}\right]=\exp\left[iTr(\omega V)\right]$, which appears on the right hand side of this equation. If we replace the matrix ω with the matrix $\omega^{(0)}=i\rho^{-1}/2$, we would get the quantity $\exp\left[-Tr(V\rho^{-1})/2\right]$. This is another factor in the Wishart expression. In fact, this replacement is exactly what happens during the integration over the $\{\omega_{\alpha\beta}\}$ variables in the multidimensional complex plane.

The last factor in the Wishart distribution, $(\det V)^{(N-p-1)/2}$, arises from derivatives of the quantity $\exp[-\zeta V]$ with respect to the determinant of derivatives of an auxiliary matrix ζ. The auxiliary matrix generalizes the $p=1$ auxiliary variable whose derivatives are needed to turn a simple pole into higher-order singularities.

Further Details; the Wishart Fourier Generating Function

As just mentioned, we need an auxiliary matrix ζ with matrix elements $\zeta_{\alpha\beta}$; these are eventually set to zero. We multiply the probability distribution by the factor $\exp\left[-\sum_{\alpha,\beta=1}^{p}\zeta_{\alpha\beta}\sum_{\ell=1}^{N}x_{\alpha\ell}x_{\beta\ell}\right]$. The matrix $\rho^{-1}+2i\omega+2\zeta$ then appears in the quadratic form; we rewrite this as $2i\left(\omega-\omega^{(0)}-i\zeta\right)$ where $\omega^{(0)}=i\rho^{-1}/2$. The integral over the $\{x_{\alpha\ell}\}$ can be done immediately and produces the factor $1/\left[\det\left(\omega-\omega^{(0)}-i\zeta\right)\right]^{N/2}$. The result for the Fourier transform, i.e. the integrand of the $\{\omega_{\alpha\beta}\}$ variables, up to a constant $2^{p(-N+p-1)/2}(i)^{-Np/2}$ is then

$$G_N = (\det\rho)^{-N/2}\prod_{\alpha\leq\beta}\int_{-\infty}^{\infty}dV_{\alpha\beta}\cdot\exp\left[iTr(\omega V)\right]\cdot\left[\det\left(\omega-\omega^{(0)}-i\zeta\right)\right]^{-N/2}$$

(38.21)

This is the Fourier generating function for the Wishart distribution and agrees with the literature (see Evans et. al.)[i].

Performing the Multiple Fourier Integrals

We now have to do the Fourier integrals. Counting variables, we have $p(p+1)/2$ integrals over the $\{\omega_{\alpha\beta}\}$ variables with $\alpha \leq \beta$. Just as in the $p=1$ case, the order of the singularity needs to be reduced to correspond to the order of the integration. We want $\left[\det\left(\omega-\omega^{(0)}-i\zeta\right)\right]^{-(p+1)/2}$ since the determinant itself is of order p. We need to differentiate with respect to $\det\left(\partial/\partial\zeta_{\alpha\beta}\right)$. A difficulty, discovered by hand calculations and Mathematica[7], is that $\det\left(\partial/\partial\zeta_{\alpha\beta}\right)\left[\det\left(\omega-\omega^{(0)}-i\zeta\right)\right]^{-1} = 0$. However, we have the more general result, a special case of Cayley's theorem [v],

$$\det\left(\partial/\partial\zeta_{\alpha\beta}\right)\left[\det\left(\omega-\omega^{(0)}-i\zeta\right)\right]^{-1+\lambda} = -\lambda(1+\lambda)\left[\det\left(\omega-\omega^{(0)}-i\zeta\right)\right]^{-2+\lambda} \quad (38.22)$$

We set $\lambda = (p-1)/2$. To get $\left[\det\left(\omega-\omega^{(0)}-i\zeta\right)\right]^{-N/2}$ we differentiate $\left[\det\left(\omega-\omega^{(0)}-i\zeta\right)\right]^{-(p+1)/2}$ by $\det\left(\partial/\partial\zeta_{\alpha\beta}\right)$ a total of $(N-p-1)/2$ times. We then pull the derivative $\left[\det\left(\partial/\partial\zeta_{\alpha\beta}\right)\right]^{(N-p-1)/2}$ outside the $\{\omega_{\alpha\beta}\}$ integrals.

Multiple Cauchy Theorem

We are left with the task of integrating $\left[\det\left(\omega-\omega^{(0)}-i\zeta\right)\right]^{-(p+1)/2}$. The determinant $\det\left(\omega-\omega^{(0)}-i\zeta\right)$ consists of a sum of terms, each term containing p factors. We need to look at the zeros of the determinant in order to apply the generalization of Cauchy's theorem[vi]. We can get the determinant to vanish if we set to zero each term in the sum comprising the determinant. Consider the change of variables:

$$\left(\omega-\omega^{(0)}-i\zeta\right)_{\alpha\beta} = \sqrt{\varepsilon_\alpha \varepsilon_\beta} \cdot \gamma_{\alpha\beta} \cdot \exp\left[i\left(\phi_\alpha+\phi_\beta\right)/2\right] \quad (38.23)$$

---

[7] **Acknowledgement:** I thank Tom Gladd for verifying this special case of Cayley's theorem using Mathematica.

Each term in $\det(\omega - \omega^{(0)} - i\zeta)$ will contain $\prod_{\alpha=1}^{p} \varepsilon_\alpha \exp(i\phi_\alpha)$ as a common factor. This factor can then be pulled out of the determinant. Taking ε_α as a small number defines a small circle in each $\omega_{\alpha\beta}$ complex plane about the point $(\omega^{(0)} + i\zeta)_{\alpha\beta}$. All factors $\{\varepsilon_\alpha\}$ cancel. Cauchy's theorem can then be used for each $\omega_{\alpha\beta}$ separately. We recall that $\omega^{(0)} = i\rho^{-1}/2$. For nonzero results we close in the $\omega_{\alpha\beta}$ upper half plane if $\rho^{-1}_{\alpha\beta} > 0$, which requires $V_{\alpha\beta} > 0$ for the vanishing of the contour at infinity. In the other case $\rho^{-1}_{\alpha\beta} < 0$ that can result if $\alpha < \beta$, we need $V_{\alpha\beta} < 0$ in order to close in the $\omega_{\alpha\beta}$ lower half plane. The $\gamma_{\alpha\beta}$ matrix prevents the vanishing of the sum of the Levi-Civita alternating signs $\varepsilon^{LC}_{\alpha_1\alpha_2...\alpha_p} = \pm 1$ in the determinant. Any $\gamma_{\alpha\beta}$ with $\det \gamma \neq 0$ can be used.

Applying the multiple Cauchy theorem on the $p(p+1)/2$ variables $\{\omega_{\alpha\beta}\}$ means that we make the replacement $\omega \to \omega^{(0)} + i\zeta$ in the factor $\exp[iTr(\omega V)] \to \exp[-Tr(V\rho^{-1})/2]\exp[-Tr(\zeta V)]$. Using the following identity $(N-p-1)/2$ times then completes the argument:

$$\det(\partial/\partial \zeta_{\alpha\beta}) \cdot \exp[-Tr(\zeta V)] = (-1)^p \det V \cdot \exp[-Tr(\zeta V)] \qquad (38.24)$$

This concludes the discussion of the origin of the dynamical terms in the Wishart distribution.

Limitations of the Proof

Although we have obtained the Fourier generating function and the dynamical terms in the Wishart distribution, our proof using the Fourier-transform method is not complete. The first problem is deriving the constant $\mathcal{K}$. It seems that $\mathcal{K}$ can only be obtained here by requiring that the total probability, the integral over $\prod_{\alpha \leq \beta} \int_{-\infty}^{\infty} dV_{\alpha\beta}$, is one[8]. The second problem involves the multiple Cauchy theorem. First, we placed conditions on the signs of $\{V_{\alpha\beta}\}$ for $\alpha < \beta$ depending on the

---

[8] **Remarks:** We cannot set $\lambda = (p-1)/2$ in the coefficients because some derivatives are then zero. Instead, we keep λ free in all constant factors. We also have a dependence on the γ matrix above.

signs of $\rho^{-1}_{\alpha\beta}$. Second, the determinant $\det\left(\omega - \omega^{(0)} - i\zeta\right)$ can vanish through cancellations *between* its various terms. Treating the resulting interdependent singularities in the $\{\omega_{\alpha\beta}\}$ integrals is extremely difficult [vi].

I would appreciate learning if a derivation resolving these difficulties in evaluating the Fourier transforms exists in the mathematical literature, and if so, where.

References

[i] **The Wishart Distribution**
Wishart, J., *The Generalised Product Moment Distribution in Samples from a Normal Multivariate Population*. Biometrika Vol. XXA, 1928. Pp. 32-52.
Johnson, R. A. and Wichern, D. W., *Applied Multivariate Statistical Analysis, 3rd Edition*. Prentice Hall. See p. 150.
Evans, M., Hastings, N. and Peacock, B., *Statistical Distributions, 3rd Edition*. John Wiley & Sons, 2000. See #43, p. 204.

[ii] **Earlier Related Work in Lower Dimensions**
Fisher, R. A., *Frequency Distribution of the Values of the Correlation Coefficient in Samples from an Indefinitely Large Population*. Biometrika, Vol. X, 1915. Pp. 507-521.
"Student", Biometrika, Vol. VI, 1908. Pp. 4-6.

[iii] **Gamma, Beta Functions**
Erdélyi, A. et. al., *Higher Transcendental Functions*, Vol. I, Ch. 1. McGraw-Hill, 1953.

[iv] **WKB Approximation**
Morse, P. M. and Feshbach, H., *Methods of Theoretical Physics*. McGraw-Hill Book Company, 1953. See Part II, Pp. 1092-1106.

[v] **Cayley's Theorem**
Turnbull, H. W., *The Theory of Determinants, Matrices, and Invariants, 3rd Ed*. Dover Publications, Inc, 1960. See p. 114 and problem 12, p. 123.

[vi] **Multi-Dimensional Complex Variable Integration**
Gunning, R. C. and Rossi, H., *Analytic Functions of Several Complex Variables*. Prentice-Hall, Inc., 1965. See Ch. 1.
Bertozzi, A. and McKenna, J., *Multidimensional Residues, Generating Functions, and Their Application to Queueing Networks*. SIAM Review, Vol. 35, No. 2, 1993. Pp. 239-268.

39. Economic Capital (Tech. Index 4/10)

In this chapter, we discuss Economic Capital, mostly in a qualitative fashion[1]. We describe standard procedures and assumptions as well as problems and issues. Many relevant quantitative issues that play important roles in the actual calculations are discussed in detail elsewhere in the book.

Basic Idea of Economic Capital

Economic Capital[2] (abbreviated EC in this chapter) is now becoming a common barometer for corporate risk. EC can be defined as the amount of liquid capital needed to enable a firm ABC to survive (i.e. not default) under extreme and unexpected adverse conditions. These conditions are taken to last for a given period of time τ_{EC}. This definition, turned around, implies that ABC should be able to lose an unexpected amount of money EC in time τ_{EC} without defaulting. Often $\tau_{EC} = 1$ yr is chosen[3]. Other definitions for EC are sometimes used, e.g. the estimated amount of capital needed to obtain a given credit rating by a rating agency[4]. Another is the hypothetical premium needed for an outside party to insure against default. These definitions are related but not equivalent.

Although no pot of money is literally set aside as economic capital, we will regard EC as real concrete assets needed to cover losses. Following Moody, we assume that these assets are *"permanent and immediately available to absorb losses before general creditors are affected in any way"*.[i]

---

[1] **History:** Most of the work for this chapter was done in the period 1999-2001.

[2] **Units:** Economic Capital is measured in USD, or whatever the reporting currency is.

[3] **Why one year?** Sometimes it is said that one year is a "reasonable" period to require for solvency before the company can access the capital markets to get more capital.

[4] **Rating Criteria:** The rating criteria in reality are much more complex than just capital. Stable revenue and sufficient cash flows are essential. See *Moody's Rating Methodology Handbook* (Ref). For this reason alone, Economic Capital can only be a rough measure for ratings. In this chapter, it is sometimes assumed for illustration that ABC is an Aa (AA) rated bank or broker-dealer.

There are Sharpe ratios, i.e. return/risk ratios, that utilize EC for the risk in the denominator, giving return on economic capital as a measure of business success. Thorny problems of whether or not to include diversification effects for a given business unit exist, and will be discussed below.

Adverse Conditions for EC, and an Insurance Analogy

A critical variable is the choice of the meaning of "adverse" conditions or events for EC. A-priori there is no "right" or "wrong" definition. We will consider some examples below. It is convenient to think of EC as considered to act as the capital backing a fictitious "insurance policy" for ABC, with the policy written by ABC to cover itself, ABC. Adverse events are supposed to be covered by this "insurance". Naturally, the extent to which these adverse events are far out on the statistical tail is a critical consideration. If such an event (call it an "asteroid"[5] or a bad "earthquake") does occur, losses depending on the exposure of ABC will occur.

Either the EC is enough to cover these losses or not (in which case ABC may indeed default). The EC also has to be in a form that will enable ABC to cover the losses and still stay in business. If illiquid assets need to be sold, probably into a hostile market, substantial additional losses may occur. On the other hand, keeping the EC permanently in liquid assets in case it might be needed, means that a low return will be suffered, and business opportunities may be missed.

Example of the Insurance Analogy

A mundane illustration may help. Assume that you do not have an earthquake insurance policy with an outside insurer. It is neither right nor wrong to have enough assets (your "economic capital") to insure yourself to cover potential severe earthquake damage to your house (your "exposure"). The probability for an earthquake may be far out on the probability distribution, and may never have occurred in your area.

If, nevertheless, an earthquake does occur, you need enough capital to rebuild the house to a livable state. Your capital may be liquid (e.g. cash in a money market account that returns little) or your capital may be tied up in illiquid assets (some of which you may well have to sell at a big discount). At the end, if you don't have enough money to rebuild the house to a livable state, then the rule says that you "default".

---

[5] **Asteroids:** The word "asteroid" is used in this book for dramatic effect just to indicate the sudden onset of a severe problem from stressed markets.

Stress Testing and EC

Stress tests can include what-if scenarios, historical scenarios, and statistical measures. We have discussed these earlier in the book, and provide some more insight below. A variety of these tests is performed at various institutions[ii]. Each of them can be used for a part of Economic Capital.

What-if Scenarios (WIS) as Indicators

What-if scenarios (WIS) can be envisioned for quantification of EC in some cases. WIS have been and will remain a staple of risk management. At the same time, WIS are clearly just indicators. WIS can be formulated in a number of ways. These include:

- Numerical (specified changes in stock indices, bonds, FX, commodities, etc.)
- Economic or political (deep recession, default of large banks, wars, etc.)

It is clear that there is no point in considering something like a "Rand Corporation nuclear-war" scenario, i.e. a scenario in which the adverse climate is so severe that ABC will not exist at all regardless of any EC consideration.

Therefore, the scenarios we want to consider are essentially serious "mid-level" disasters.

Numerical What-if Scenarios

Numerical WIS are very common. Based on history or an estimated projection, or some other idea, changes $\left\{\delta x_\alpha^{(\text{Scenario }\beta)}\right\}$ are postulated for variables $\left\{x_\alpha\right\}$ for scenario β. For example, scenario β might assume that (over a period $\tau_{EC} = 1$ yr), the S&P500 drops 40%, oil prices rise 30%, the USD drops 10% with respect to major currencies, gold rises 25%, etc. Note that correlations are built into such a what-if scenario (i.e. stocks down, oil up means $\rho_{Stock,Oil} < 0$ over that time period).

Underlying Economic or Political What-if Scenarios

Economic or political WIS involves questions like: "How much capital do we need for ABC to survive under a deep recession?" In order to quantify this, we clearly need a definition of a "deep recession" and we need to attempt to calculate the result of the existence of a deep recession. This still means specifying changes $\left\{\delta x_\alpha^{\text{(Recession Scenario)}}\right\}$ and the time interval over which these changes occur [6].

It should be noted that although a deep recession may not have occurred in the last 30 years since the invention of modern finance, this does not mean that a deep recession cannot occur. The same holds true for other possible disasters.

Probabilities, Entropy, and What-if Scenarios

It is useless to try to calculate the probability of some given definition of a WIS such as a given numerical scenario or the results of a deep recession. This is not so important. It is a red herring to argue that if you cannot calculate the probability of a scenario, you should not think about it. A given WIS specifying many details will have a very small probability. *However, the possible number of ways that some disaster can occur (the "entropy") is very large.* Therefore is perfectly reasonable to consider a representative disaster using a WIS.

Historical Scenarios (HS) as Indicators

Historical scenarios involve questions like: "How much capital do we need for *ABC* to survive under a series of events like the stock market crash of 1987?" If so, we calculate what the consequences would be if an historical 1987 scenario were to repeat. Of course, history never exactly repeats, so HS are also just indicators.

Statistical Measures (SM) as Indicators

Statistical measures involve questions like: ""How much capital do we need for *ABC* to survive at a probability level of 99.97%?" This latter number is the average one-year default probability for Aa-rated companies from Moody[iii] over 1970-1998[7]. SM have to be defined with respect to some statistical calculations, and there are a variety of uncertainties regarding such calculations. We have spent a lot of time in this book discussing various sorts of SM, e.g. Stressed VAR with fat tail jumps and stressed correlations. In the end, SM are also indicators.

---

[6] **Economists and Financial Variable Changes:** Economists may naturally be cautious about specifying too much detail about the changes in the variables needed for a risk calculation. This means some ad-hoc assumptions will still be needed.

[7] **What's so Special About the 99.97% Confidence Level?** Other time intervals in the data naturally produce different numbers, e.g. 99.93% over 1920-1996. See Moody (ref).

The Classification of Risk Components of Economic Capital

Economic Capital has many components. After all, EC is supposed to represent capital needed to survive against all risks. Traditionally three risk categories are used for classification. They are

- Market Risk
- Credit Risk
- Operational Risk

We have spent considerable time in this book on market risk. The best candidate for a realistic assessment of the market risk component of EC, in our judgment, is the Enhanced/Stressed VAR (cf. Ch. 27). Some attention has been paid to credit risk (cf. Ch. 31 on issuer risk). Almost nothing has been said about operational risk.

The enumeration of all risks is very large and by definition of Murphy's Law[8], incomplete. Moreover, the placement of a given risk in these three categories is sometimes unclear [9].

Consistent vs. Inconsistent Calculations of EC Components

The practical calculations of the various components of risk comprising EC are done in a variety of ways. This is because it is not possible to calculate all risks in a consistent framework. No single methodology is rich enough to cover the estimation of disasters and it is sometimes not possible to ensure that risk assumptions are internally consistent from one area to another. Judgment and policy therefore play a role in practice.

---

[8] **What's the Next Risk Type Surprise?** While market risk has large gaps from time to time, and while credit failures can be spectacular, the really dangerous unknowns lie in operational risk (which is where everything else goes).

[9] **Would Linnaeus Agree that This Classification is Complete Enough?** With the triumvirate market-credit-operational classification, any risk has to be shoehorned in somewhere. One issue regards the multitude of possible risks. A conference speaker once wrote a long catalog of risks on one slide for emphasis. The font was very small and the slide appeared black.

A pesky issue regards combination risks. An example is model risk for convertible bonds. The model risk shows up as part of the bid-ask spread (market risk), depends on corporate credit spreads (credit risk), and several different possible models could be chosen (operational risk). One procedure is to just put model risk into the catch-all operational risk.

Exposures for Economic Capital: What Should They Be?

EC is generally calculated using exposures as of a given date. However, a forward-looking measure is desirable. In the next chapter, we will discuss a framework for an estimate of EC for unused limits, that is, for businesses to change the exposures within their limits during the period τ_{EC} in the future.

Attacks on Economic Capital at High CL

Various critiques have been levied at Economic Capital, sometimes by smart traders pushing back, and sometimes by quants. The main issue is that Economic Capital is expensive, focuses on rare events that are hard to measure, and might be used in assessments of risk that influence compensation.

We present three arguments, which we call "attacks" on EC, because that is what they are. These arguments have varying degrees of relevance.

First Attack Misses: Lascaux Cave Paintings and an Ergodic Statement

The use of high confidence levels is often attacked using what amounts to an ergodic statement. For example, suppose that we have a 1-year time frame for EC and that we take the Moody's 1970-98 Aa default $CL = 99.97\%$. This translates into a default probability of $3/10,000$. The argument says that this is an absurd measure because (and this is the ergodic statement) we cannot look at the worst 3 out of the last 10,000 years, taking us back halfway to the time of the Paleolithic Lascaux cave paintings[iv].

This argument is a red herring. Naturally, we do not want to argue that anything that happened in prehistoric times has much to do with (e.g.) swaps traders. In fact, we do *not* want to use such an ergodic statement, and we are *not* forced into using it by false consistency.

The proper argument is that we have some information (albeit imperfect) about default statistics from what happened in the 20th century. We use this information for probabilities of default in 10,000 states starting now. These 10,000 states *can* be generated, by Monte Carlo simulation.

Second Attack is Closer: Not Enough Companies

The second argument is that using this high CL is absurd because we do not have enough Aa companies in the historical data for robust probability estimates. This is a better argument. Moody's (exhibit 34)[iii] shows that only one company defaulted from 1970-1998 that had an Aa rating at one-year prior, DFC Financial (Overseas) Ltd on 10/3/89. Different periods of time do produce different results for default probabilities.

Third Attack Hits the Target: Is EC Related to Default?

The third argument is that even if we use such a high CL for default, there is no apparent reason to use the same CL for movements of the underlying variables. In other words, the default of an AA company in the real world may not be correlated, e.g., with equivalently large moves of market variables. This is an excellent argument.

There has not been any real attempt to achieve consistency in the philosophy of EC between potential causes of default and the fact that the calculation of EC is based on default probabilities.

Indeed, defaults often seem be caused by liquidity cash-flow problems in practice, not capital. That is, the firm misses an interest payment on debt. For example in 1998, 123 public corporations defaulted with 66 defaults due to missed interest payments [iii]. A firm can have cash flow problems causing default, still with plenty of capital[10].

However, it can logically be assumed that a minimum amount of capital on the order of EC is needed to avoid default over an extended period if a loss on the order of EC occurs.

Therefore, while this third attack is disturbing, it does not kill the high CL approach to EC.

Given All That, What do we Actually do for Economic Capital?

At the end of the day, a conservative measure is adopted, policy is approved, difficult data collection is done, calculations are performed, presentations are given, and attention focuses elsewhere.

Allocation: Standalone, CVAR, or Other?

Suppose we are given a calculation of Economic Capital EC for the firm, and we accept the results. We still have to allocate the firm's EC between desks or business units (BU). Allocation is a difficult topic. Several possibilities exist that we discuss in turn[11]. See also the discussion in Ch. 30 for corporate-level VAR.

---

[10] **Acknowledgement:** I thank Tom Schwartz for illuminating conversations on this and many other topics.

[11] **Acknowledgement:** I thank Jim Marker and Jack Fuller for helpful conversations on this and other topics.

Standalone Risk for Allocation?

We might want to look at each BU as a separate entity. In that case, we would use the stand-alone result $EC_a^{(SA)}$ for BU_a, containing risk from BU_a positions and intra-BU diversification inside BU_a only, but not any inter-BU diversification (BU_a, BU_b) for other business units BU_b.

However, assume that we write the total EC as the sum of the stand-alones,

$$EC^{(\text{Sum of SA})} = \sum_a EC_a^{(SA)} \qquad (39.1)$$

Now we have a problem. The sum of stand-alones $EC^{(\text{Sum of SA})}$ does not correctly asses the firm's risk because it allows no diversification offsets, and therefore $EC^{(\text{Sum of SA})} > EC$. Indeed, the major impetus for much of modern corporate strategy is *precisely* to take advantage of inter-BU diversification.

There is also a serious consistency problem. Suppose there is an administrative reallocation of risks leaving the total risk unchanged. For example, take a hedged position with zero risk. Put the long position in BU_a and the short hedge in BU_b. The standalone risks change. Hence, without changing the risk of the firm, the total standalone $EC^{(\text{Sum of SA})}$ also changes.

The bottom line is that Sum of Standalones approach has the virtue of dealing with each business unit separately. However, it has the vice of being neither a consistent nor a realistic measure of economic capital for the firm.

CVARs for Allocation?

Given the firm's EC including inter-BU diversification, the CVARs provide a consistent methodology to allocate total risk[12]. Indeed, if we set $EC_a = CVAR_a$ then we are guaranteed by construction that $EC = \sum_a EC_a$ consistently.

The complication here is that also by construction, EC_a for BU_a is naturally dependent on correlations with the risks included in EC_b for BU_b. This means that a given BU_a can do nothing different (or even do absolutely nothing at all) and wind up with it's EC_a being changed due to activities of a different BU_b.

---

[12] **CVAR:** The CVAR methodology is described in detail in Ch. 26-30.

Economic Capital

An unusual but possible case is that $CVAR_a < 0$ is negative, implying that BU_a is hedging out other risk in the firm. Hence EC_a allocated for BU_a using the CVAR approach will also be negative.

Businesses and upper management want to view each individual business-unit risk as due to its own individual activities. While the CVAR approach is internally consistent, the sociological problems using CVARs can be non-negligible.

Compromise Recipe for Allocation?

A possible compromise procedure is to list the total firm's EC correctly calculated as $EC^{\text{With Diversification}}$, with diversification reductions, when dealing with firm-wide reporting. Allocation to each BU_a is performed using its own stand-alone economic capital $EC_a^{(SA)}$.

The disadvantage of this compromise is again that the origins of corporate strategy of diversification are not present in the allocations.

The Cost of Economic Capital

Assume that we regard EC as an amount of traditional capital to be kept in liquid assets in order to avoid cash-flow problems[13], in case of a loss equal to EC. Then we can define a cost of economic capital as being related to a "Lost Opportunity Spread" $s_{\text{Lost-Opportunity}}$, defined as

$$s_{\text{Lost-Opportunity}} = \mathcal{R}_{\text{Illiquid}} - \mathcal{R}_{\text{Liquid}} \qquad (39.2)$$

This spread is the difference in between the return $\mathcal{R}_{\text{Illiquid}}$ (that could have been obtained by investing EC in illiquid assets) and the smaller return $\mathcal{R}_{\text{Liquid}}$ (for holding EC in liquid assets). $\mathcal{R}_{\text{Illiquid}}$ is related to the marginal efficiency of capital [v], which is the yield earned by the last additional unit of capital (here associated with EC).

---

[13] **Avoiding Loss of Investor or Consumer Confidence:** The presence of enough traditional liquid capital presumably also serves to retain investor and consumer confidence.

Presumably the spread $s_\text{Lost-Opportunity}$ would be related to the "cost of insurance", if such insurance were hypothetically available from a reliable third party to cover losses equal to EC.

An Economic-Capital Utility Function

Consider[14] a firm-wide utility function Ψ related to return[15] $\mathcal{R}$ and economic capital with risk coefficient λ_{EC},

$$\Psi = \mathcal{R} - \lambda_{EC} \cdot EC \tag{39.3}$$

A firm might try to adopt a corporate strategy that maximizes Ψ for a given risk tolerance λ_{EC} for losses on the order of EC. An example of λ_{EC} could be the magnitude of the lost opportunity spread, $\lambda_{EC} = \left| s_\text{Lost-Opportunity} \right|$.

Constraints on important issues such as leverage limitation, minimal diversification, core business requirements, business costs, and sufficient liquidity could be imposed to prevent runaway unphysical solutions[16].

Sharpe Ratios

The Sharpe return/risk ratio S_Firm for the firm could be taken as the utility function (the risk-adjusted return) divided by the risk (measured by a form of Economic Capital). It is most convenient to use $EC^{(\text{Sum of SA})}$ in the denominator, viz

$$\begin{aligned} S_\text{Firm} &= \Psi / EC^{(\text{Sum of SA})} \\ &= \left(\mathcal{R} - \lambda_{EC} \cdot EC \right) / EC^{(\text{Sum of SA})} \end{aligned} \tag{39.4}$$

---

[14] **Acknowledgement:** Santa Federico has many sophisticated ideas for the utility function. I thank Santa for helpful discussions on this and many other topics.

[15] **Units:** The units of the return, the economic capital, and the utility function are USD/year. The Sharpe ratio has no units.

[16] **When Will We See Calculations Using the Firm's Utility Function?** Probably at about the same time as the appearance of a real-time movie of firm-wide risk in color.

Economic Capital

The business-unit BU_a Sharpe ratio S_a could be similarly defined. For the S_a numerator, an amount should be subtracted from the BU_a return $\mathcal{R}_a$ equal to the charge for EC_a based on the firm-wide risk coefficient λ_{EC}, or on the lost-opportunity spread $s_{\text{Lost-Opportunity}}$. The standalone $EC_a^{(SA)}$ should be used for the S_a denominator to avoid potential problems with negative (or zero) EC_a obtained with the CVARs. So S_a is

$$S_a = \left(\mathcal{R}_a - \left|s_{\text{Lost-Opportunity}}\right| \cdot EC_a\right)/EC_a^{(SA)} \tag{39.5}$$

In the limit that BU_a is the whole firm, $S_a \to S_{Firm}$ if $\lambda_{EC} = \left|s_{\text{Lost-Opportunity}}\right|$.

Revisiting Expected Losses; the Importance of Time Scales

Standard Assumption: No Uncertainty for Expected Losses

We have so far taken the conventional definition of EC as involving only unexpected losses[17]. Implicit in this definition are two statements:

- Expected losses $C_{\text{Expected Losses}}$ do not depend on the stressed environment.
- Sufficient reserves C_{Reserves} over pricing margins cover expected losses.

The idea behind these assumptions is that expected losses are not risky because, after all, they are known and can be dealt with deterministically.

More Realistic: Expected Losses Do Have Some Uncertainty

There is a problem with the standard assumption. Think of the time dependence of loss as being composed of two parts, a drift and a volatility. The expected loss acts as the drift. *The problem is that, due to the stressed environment, the expected loss can change, perhaps substantially, over a one-year period.*

For this reason, the first statement, that expected losses do not depend on the stressed environment, is dubious. For example, if we enter a recession environment, the average expected losses could increase due to reduced

---

[17] **Acknowledgement:** I thank Evan Picoult for helpful discussions on this and many other topics.

consumer demand. The second statement regarding reserves and pricing margins may or may not be true, depending on the details[18].

It may be better not to make these assumptions and to write the explicit expression for the difference $\delta C_{\text{Expected}}$, instead

$$\delta C_{\text{Expected}} = C_{\text{Reserves}} - C_{\text{Expected Losses}}^{(\text{Stressed Environment})} \qquad (39.6)$$

The amount $\delta C_{\text{Expected}}$ could then be included in a revised definition of EC. If indeed it turns out that $\delta C_{\text{Expected}} = 0$, then the EC will be unchanged. If reserves are greater than expected losses in a stressed environment, then the EC will be decreased because these extra reserves could be used. If however reserves are smaller than expected losses in a stressed environment or if pricing margins dropped, the real EC needed will logically be expected to increase[19].

Therefore, the point is essentially that uncertainty in the expected losses should be included in the uncertainty leading to EC.

Time Scales Are Again the Issue

The problem lies in the time scales. The EC calculations are usually envisioned as being due to short-term "asteroid-like" adverse conditions. Discussions on EC in this sense revolve around the length of time for hedging, the amount of risk hedged, etc. These topics were discussed at length in the chapter on Enhanced/Stressed VAR in Ch. 27.

On the other hand, the uncertainties in the expected losses are due to longer term "getting stuck in the mud" adverse conditions. These can be quite different but no less severe.

In Ch. 47-51, we discuss the Macro-Micro model that incorporates uncertainties in macro components of variations of underlying variables over long time scales. The difficulties discussed for economic capital here arise from exactly the same point.

---

[18] **Pricing Margins and Expected Losses:** The inclusion of expected losses in the pricing of goods and services may be problematic in a stressed environment where increased competition may exert pressure to lower prices exactly at the same time that expected losses are gradually increasing.

[19] **Consumer Business:** These considerations could be important for risk for consumer businesses (e.g. credit cards) where major shocks are unlikely and the main risk is slow but important degradation due to changing economic conditions.

Summary for Time Scales and EC

The high-level summary is that there is not enough attention paid to the time scales of risk. The dynamics are completely different for short and long time scales.

It would be more realistic if Economic Capital assumptions and calculations would take into account these time scales in an explicit fashion.

Cost Cutting and Economic Capital

If EC is regarded as a measure of default in a literal sense, another refinement enters. Namely, some fraction f of returns $\mathcal{R}$ could be made available to cover mandatory cash-flow payments and help solve liquidity problems. This involves a transfer from spending for variable costs, essentially through cost cutting, and is exactly the procedure followed by corporations with liquidity difficulties. Since this amount $f \cdot \mathcal{R}$ replaces forced sales of some assets to cover mandatory cash flows, it can be viewed as replacing part of the capital needed to avoid default, and therefore could be included to reduce the EC.

Normally, no influence of returns is present in the calculations for EC. Again, this presents a consistency problem if high CL calculations are used (e.g. 99.97% for Aa credit) that are motivated by default statistics, while on the other hand the dynamics of real-world default involving cash-flow liquidity problems are ignored for EC calculations.

It would be more realistic to change the procedure for calculations of EC to make the EC more relevant to real-world considerations.

Traditional Measures of Capital, Sharpe Ratios, Allocation

We have been discussing Economic Capital EC. To review, EC is calculated capital needed to offset unexpected loss for adverse events according to some conservative criteria for market, credit, and operational risks.

On the other hand, traditional capital measures exist, as explained in texts on corporate finance. For example, the return on common equity uses common equity capital. Common equity capital is defined in corporate finance as common stock at par + capital surplus + retained earnings. A closely related capital is book value, which is share capital + additional paid-in capital + retained earnings. [v]

Traditional Capital is Not Economic Capital

It is clear that these traditional capital measures are not equal to EC. In a sense, traditional capital is "capital you have", while EC is "capital you need" to survive stressed environments. The connection is supposed to be that enough traditional capital is needed to prevent default if unexpected losses on the order of EC occur.

If the management desires a traditional measure of capital to be allocated or to be used in Sharpe return/risk ratios, the calculation of EC would not seem to be of much relevance. We have already seen the difficulties of allocating EC itself. It is unclear how to perform allocation of other forms of capital that are not involved in the economic capital calculations.

For example, simple numerical scaling of the allocations by the ratio of book value to economic capital is simple to write down, but has uncertain meaning.

References

[i] **Economic Capital Definition**
Pinkes, K. et. al., *Moody's Rating Methodology Handbook*. Moody's Investors Service, 2000. P. 209.

[ii] **Stress Testing**
Committee on the Global Financial System, *A survey of stress tests and current practice at major financial institutions*. Bank for International Settlements, April 2001.
Fender, I. and Gibson, M., *The BIS census on stress tests*. Risk, May 2001, Pp. 50-52.

[iii] **Moody Default Probabilities**
Global Credit Research, *Historical Default Rates of Corporate Bond Issuers, 1920-1998*. Moody's Investors Service, 1/99. See Exhibits 31, 34, 35, and 8.

[iv] **Lascaux Cave Paintings**
Thoraval, J., Pellerin, C., Lambert, M., Le Solleuz, J., *Les Grands Étapes de la Civilisation Française*. Bordas, 1972. See Fig. 1, p. 9.

[v] **Fundamental Analysis**
Series 7, *General Securities NYSE/NASD Registered Representative Study Manual*, Securities Training Corp., 1999. Chap. 20.
Bloomberg, L. P., *Help Equity FA and Help Corp*, Bloomberg Professional, 2001.
Downes, J., Goodman, J. E., *Dictionary of Finance and Investment Terms*. Barron's Educational Series, Inc. , 1987.

40. Unused-Limit Risk (Tech. Index 6/10)

In this chapter, we deal with exposure-change risk as an extension to risk calculations and Economic Capital. Most risk assessments use existing portfolios and exposures. We do want to gauge the historical accuracy of our risk assessments through backtesting. Nonetheless, we are really interested in assessing future risk. *After all, the future is risky, not the past.* Therefore, we are (or should be) interested in the risk due to potential changes in risk exposures, consistent with limit constraints. In this book, we will use a forward + option approach in order to model this potential exposure-change risk[1,2].

General Aspects of Risk Limits

In order to discuss exposure-change risk, we need to discuss limits. Considerable effort needs to be expended in order to accomplish the various goals and activities involving the establishment and the monitoring of limits.

Types of Limits

Limits constraining risk exposures that can be assumed by desks are imposed in different ways. For example, detailed limits may be set for a given exposure $^\$\mathcal{E}_\alpha^{(a,\Im)}$ of a given product $\Im$ on a given desk a that depends on the underlying variable x_α or α for short. For example, we can have a limit on vega exposure ($^\$\mathcal{E}$) for Libor ($\alpha$) Bermuda swaptions ($\Im$) on the exotic options desk (a).

A limit can be imposed on some measure $^\$\mathcal{E}^{(\Im)}$ of a product $\Im$ depending on several underlying variables, for example the composite notional of Latin American bonds.

On the other hand, a specific exposure may not have a specific limit. An example might be the 10-year AA credit spread risk of industrials, although these bonds would be included in more general limits.

---

[1] **Acknowledgements:** I thank Dave Bushnell for insightful comments that greatly facilitated this work. I thank Andy Constan for a related discussion. I thank the Market Risk Managers at Citigroup for helpful conversations on this and many other topics.

[2] **History:** This unused-limit risk model was developed by me in 1999-2000.

An overall limit may exist on the total exposure $^\$\mathcal{E}_\alpha^{(a)}$ of the variable x_α on desk a, summed over all product types $\mathfrak{I}$. For example, we can consider delta for the S&P index on the equity options desk.

Limits can also be imposed on exposures summed across desks in a division $^\$\mathcal{E}_\alpha^{(Division)} = \sum_{a \in Division} {}^\$\mathcal{E}_\alpha^{(a)}$. An example could be the total spread DV01 across all fixed-income desks.

Setting and Monitoring Limits

Setting limits depends on choosing the most important and relevant risk exposures, performing risk scenarios or calculations at some level, specifying the amount of loss to be tolerated, specifying business requirements, and other aspects. Intensive discussions and negotiations between Risk Management and the business units may take place to define and to set parameters for specific limits.

Systems need to be constructed to monitor the limits efficiently. Otherwise, the monitoring has to be done by hand, which is time consuming.

The number of exposures used in VAR or other risk calculations can be very large. Setting limits on all such exposures would be tedious to monitor and counter-productive to impose. For this reason, the number of limits can be much less than the number of possible exposures. Generally, some aggregation is used in setting limits, e.g. spread DV01 for investment-grade corporates. Still, the collection of limit specifications can produce a large document.

In practice, as opportunities arise and as portfolios change, exceptions to limits may (or may not) be granted. Periodic review and possible resetting of limits can occur.

Example of a Conundrum with Detailed Limits

We need to be careful in order that the limits do measure real risk. Here is a simple example of detailed limits that backfire. Suppose we have limits on two buckets #1 and #2 in maturity. For example, bucket #1 could be 0-2 years and bucket #2 could be 2-5 years.

Imagine that initially we have a hedged position with a "calendar spread" in bucket #2. For example, we can have a long position at a slightly shorter maturity than a short position, both in bucket #2. Assume that either position individually would violate the limit, but that together the risks cancel out, giving zero risk in bucket #2. Assume nothing is in bucket #1.

As time progresses, the long position can move into the shorter maturity bucket #1, but with the short position still staying in the longer maturity bucket #2. At this point, the limits in both buckets are violated. However, the total risk has not changed (modulo possible risks explicitly associated with moving to shorter maturities). Therefore, in this case red lights and alarms go off in the system monitoring the limits, even though the real risk may be small.

The Unused Limit Risk Model: Overview

The model is formulated in terms of fractions of exposures with respect to limits. First, we present the model with only one exposure $^\$\mathcal{E}$, and generalize it below. Call $^\$\mathcal{L}_\mathcal{E}$ the limit for this exposure and write the fraction of the limit utilized by the exposure as [3]

$$f_\mathcal{E} = {^\$\mathcal{E}}/{^\$\mathcal{L}_\mathcal{E}} \qquad (40.1)$$

If the limits are respected, as we shall assume[4], we have $f_\mathcal{E} \leq 1$. Therefore, this means there is a barrier at $f_\mathcal{E} = 1$.

Exposure Fractions: Time Dependent Decomposition

The unused limit risk model relies on an estimate of an exposure $^\$\mathcal{E}(t)$ decomposed into a drift term and a volatility term at time t, $^\$\mathcal{E}(t) = {^\$\mathcal{E}^{\text{Drift}}}(t) + {^\$\mathcal{E}^{\text{Vol}}}(t)$. We divide this decomposition by the limit to get a model for the fraction of the used limit,

$$f_\mathcal{E}(t) = f_\mathcal{E}^{\text{Drift}}(t) + f_\mathcal{E}^{\text{Vol}}(t) \qquad (40.2)$$

---

[3] **Positive and Negative Limits; the case of Gamma:** There can be both positive and negative exposure limits, not necessarily equal. If the exposure is negative $\$E < 0$, then we choose the negative exposure limit $\$L_E$ to define the fraction f_E. The fraction f_E is always non-negative. For example, consider gamma. Only negative gamma is a risk. Positive gamma is an asset for which you have to pay. Therefore, the definition is to consider only negative gamma exposure with a negative gamma limit. The fraction for gamma is still between 0 and 1.

[4] **Limit Exceptions, Leaky Barriers, and the Three Strikes Rule:** Including limit exceptions would be a messy task and involve the barrier at f =1 being leaky or porous. This violates my Three Strikes Rule, namely being (1) difficult to have intuition, (2) difficult to get parameters and to calculate, and (3) difficult to explain to management. For these reasons, refinements of the model may not be as desirable as might appear academically.

The fraction starts at its current, or spot level $f_{\mathcal{E};spot}$. The drift term $f_{\mathcal{E}}^{Drift}(t)$ over the time period assumed for economic capital gives the forward expected or most likely exposure fraction, $f_{\mathcal{E};fwd}$. This expected value is to be specified by someone who understands the general business strategy and the likely behavior of the desk.

The fraction volatility term $f_{\mathcal{E}}^{Vol}(t)$ describes the uncertainty $df_{\mathcal{E}}$ about the forward expected level. We shall discuss the details of this term below.

The idea is shown in the picture below:

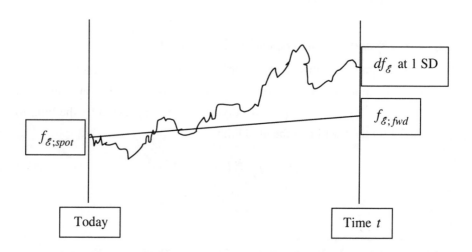

The Two Components of the Unused Limit Risk Model

With these two terms specified, the model consists of two related components:

- A "forward" denoted C^{Fwd}, depending on the expected exposure level from the drift. This would be present even if there were no volatility—i.e. certainty in the change in the exposure.
- An "option" denoted $C^{UpOutCallOption}$, depending on the volatility of the exposure level. The desk owns a call option. The option is to increase, if the desk likes, its exposure up to the limit. Because of the limit, the option is an up & out call barrier option.

Therefore, the model is

$$C^{UnusedLimits} = C^{Fwd} + C^{UpOutCallOption} \quad (40.3)$$

The forward exposure can be either above or below the current spot level. If the forward level is below the current level, the "forward" proportional to $\left(f_{\mathcal{E};fwd} - f_{\mathcal{E};spot}\right)$ will be negative. That means that the desk gets a contribution reducing its future risk. See below for more discussion on this point.

In order to proceed, we need a model for the volatility term. The reader will not be astonished to learn that we propose using a lognormal model for the exposure vol term, or because the limit is assumed constant, a lognormal model for the fraction of used limit[5]. With the forward value of the fraction being specified, the model can be cast into the familiar framework of an up-and-out European call option with a constant continuous barrier. The option is struck at the forward $E = f_{\mathcal{E};fwd}$, not the spot[6]. The option uses the economic capital at the limit, $^{\$}EC_{LimitMax}$, as the notional amount $^{\$}N$. There is a risk-free rate r for discounting over the option period τ. There is also an effective "dividend yield" y_d. This is not "real"; it is just used to reproduce the forward fraction, viz $y_d = r - \left[\ln\left(f_{\mathcal{E};fwd}/f_{\mathcal{E};spot}\right)\right]/\tau$. The barrier fraction level is the maximum fraction $H = f_{\mathcal{E};Max} = 1$.

The standard up-out call option model is then used [7],

---

[5] **Why Lognormal Dynamics for the Exposure Fractions?** There is some empirical evidence that a fraction is reasonably approximated as lognormal (e.g. scatter plots of $d_t f$ vs f exhibiting some linearity). Different behaviors are seen for other exposure fractions, including double peaks (at a low fraction during times of substantial hedging and a high fraction otherwise). The model uses a mean and width of the exposure distribution.

It is possible that even if the model were refined, the mean and width would not be substantially different, giving similar results. In any case, the model refined along the lines of including more realistic exposure distributions would suffer from the same three-strikes problem described above. The model in the text reaches a reasonable compromise.

[6] **Alternate Model for Unused Limit Risk:** An alternate model contains only an up-out call option, but struck at the spot or current fraction. The extra Economic Capital from this alternate model is always positive. However, this alternate model does not allow for the deterministic reduction in risk when, for example, a desk is deliberately pursuing an exposure reduction strategy or policy.

[7] **Standard Barrier Option Model:** See the discussion in Ch. 17. The fact that the model for unused limits can be cast in familiar form is a distinct advantage in explaining it to traders and management.

$$^\$C^{UpOutCallOption} = {}^\$C[\text{Standard UO Call Formula}] \qquad (40.4)$$

Basket Approach to Multiple Exposures and Limits

We have discussed a single limit so far. Although the number of limits is less than the number of exposures, considerable simplification still has to be made in order to get a tractable calculational scheme. To this end, it is convenient to use a basket option approach. The most important risky exposures $\{^\$\mathcal{E}_\alpha\}$ are specified, defining the most important risky fractions $\{f_{\mathcal{E}_\alpha}\}$. Positive weights $\{w_{\mathcal{E}_\alpha}\}$ are specified, with $\sum_\alpha w_{\mathcal{E}_\alpha} = 1$. The fraction used in the model is the basket fraction, i.e. the weighted sum of fractions,

$$f_\mathcal{E} = \sum_\alpha w_{\mathcal{E}_\alpha} f_{\mathcal{E}_\alpha} \qquad (40.5)$$

Since each $f_{\mathcal{E}_\alpha} \leq 1$, we still have $f_\mathcal{E} \leq 1$ constrained to be below the barrier.

Illustrative Example for Unused Limit Economic Capital

Here is an illustrative example. The exposures $\{^\$\mathcal{E}_\alpha\}$ of the exotics desk with the Backflip Options portfolio[8] are mostly DV01, spread, vega, and FX. The Market Risk Manager for that desk, who is intimately familiar with the risk, specifies relative importance risk weightings $\{w_{\mathcal{E}_\alpha}\}$ of 20%, 60%, 10%, and 10% respectively for these exposures. The current weighted fraction, which functions as the spot value, is $f_{\mathcal{E};spot} = 20\%$. The lognormal volatility of the fraction is $\sigma(df_\mathcal{E}/f_\mathcal{E}) = 45\%$ for the period of time of the calculation (say 1 year). This would be either estimated or determined from the historical utilization data. At one SD, the uncertainty in the fraction is $\pm 45\% * 20\% = \pm 9\%$.

The Economic Capital as determined from the spot exposures is $^\$EC_{spot} = {}^\$25MM$. With the limits saturated, we therefore would get a result

---

[8] **Backflip Options?** Recall the amusing but dead-serious practical exercise for the reader in Ch. 3, which of course you already did.

five times larger, $^{\$}EC_{LimitMax} =^{\$} 125MM$. This result is not reasonable unless it is highly likely that the desk will in fact have exposures saturating the limit[9].

However the desk exposures are far from the limit and are not likely to reach anywhere near the limit. The risk manager determines that the most likely value for the fraction at the end of the time period for the Economic Capital (e.g. one year) is $f_{\mathcal{E};fwd} = 28\%$. Because this is judged the forward most likely value for the exposure, the desk is charged an additional amount (after discounting with discount factor DF) of the first component forward[10],

$$^{\$}C^{Fwd} = \left(f_{\mathcal{E};fwd} - f_{\mathcal{E};spot}\right) \cdot DF \cdot ^{\$}EC_{LimitMax} \approx 7.6\% \cdot ^{\$}EC_{LimitMax}$$

The second component (the up & out call option) describes the uncertainty in the risk managers judgment due to the volatility in the exposures on the desk. The option has the strike at the forward, $E = 28\%$ and the notional $^{\$}N =^{\$} 125MM$. With the 45% lognormal volatility, the one SD range of the forward fraction is around $(19\%, 37\%)$. Note that with these parameters it is highly unlikely that the exposure will get near the maximum level at $f_{\mathcal{E};Max} = 1$. The up-out call option in this case is therefore close to the call option with no barrier at all. We get

$$^{\$}C^{UpOutCallOption} \approx 4.6\% \cdot ^{\$}EC_{LimitMax}$$

The Economic Capital from both components due to the unused limit risk is therefore $4.6\% + 7.6\% = 12.2\%$ of the maximum economic capital, viz[11]

$$^{\$}C^{UnusedLimits} \approx 12.2\% \cdot ^{\$}EC_{LimitMax} \approx ^{\$}15MM$$

The total Economic Capital for the desk is therefore not $25 MM based on spot exposures, but rather

---

[9] **High Limits do *NOT* Necessarily Imply a Bigger Economic Capital:** This possible problem is resolved by the model. If the limits of some desk were to increase but the desk's exposures were not projected to get anywhere near the limit, then the economic capital will not increase. This is because the limit barrier is essentially invisible.

[10] **Other Parameters:** Here, r = 5% ctn., 365. The "effective dividend yield" was − 28.65%. Again, there are no dividends here; this parameter is just present to reproduce the given forward fraction value.

[11] **Alternate Model:** The alternate model with only one component (as mentioned in a footnote above) produces around 8.8%, rather than 12.2%, for these parameters.

$$^\$ EC_{Tot} =\; ^\$ EC_{spot} +\; ^\$ C^{UnusedLimits} \approx\; ^\$ 40MM$$

Notice that although the Economic Capital has increased substantially due to the future exposure considerations, the result is still much lower than the maximum amount $^\$ EC_{LimitMax} =\; ^\$ 125MM$.

EC Can be Reduced if Exposures are Expected to Decrease

Say that the risk manager had decided that the most likely forward fraction was *lower* than spot $f_{\mathcal{E};fwd} < f_{\mathcal{E};spot}$. Then it is possible that the forward $^\$ C^{Fwd} < 0$ could have a larger magnitude than the positive $^\$ C^{UpOutCallOption}$, resulting in a *reduction* in Economic Capital, $^\$ EC_{Tot} <\; ^\$ EC_{spot}$.

A reduction in Economic Capital for deterministic risk reduction is eminently reasonable. For example, suppose that corporate management decides to wind down certain exposures. Then future risk will certainly decrease. This sanity feature is provided by this two-component model.

Exposure Scenarios: Comparison to VAR Exposure Reduction

In Ch. 27, we discussed enhancements to VAR involving scenarios for exposure reduction under assumed stressed market environments. The situation here is a bit different. The most likely forward exposure scenario, as intended here, is supposed to start from the current market environment. If the current market environment is not stressed, the current exposure level is not constrained by a stressed market, and the forward estimate would be made under normal business conditions.

In a time-dependent simulation, the example presented in the text would have the current exposure fraction of 20% under normal conditions estimated to increase to 28% under normal conditions. Then, if an "asteroid" hits the market, the desk would presumably start at some later time to reduce exposure in that stressed market environment. This could all be treated explicitly if we used time-dependent simulations. The present model just approximates the effects using a simple add-on procedure.

Consider the drawing below, which should illuminate these ideas:

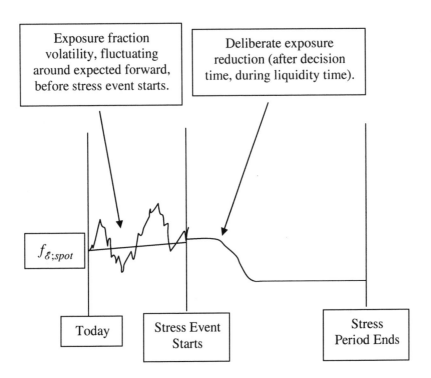

Unused Limit Economic Capital for Issuer Credit Risk

The above model focused on market risk. Exactly the same formalism in principle can be used for unused limits for issuer credit risk.

Credit limits can be formulated by geographical region (e.g. Latin America), industry (e.g. industrials), credit level (e.g. high yield), specific issuer (e.g. GM), or any other criteria used to classify bonds.

The exposure corresponding to a given credit limit would have risk due to the same sort of decomposition, $^\$\mathcal{E}(t) = {}^\$\mathcal{E}^{\text{Drift}}(t) + {}^\$\mathcal{E}^{\text{Vol}}(t)$. The drift component would be specified as the most likely forward credit exposure, and the volatility component would be specified as credit exposure fluctuations about the forward credit exposure. Again the credit model would consist of two components, the credit forward and the knock-out UO credit call option.

The parameter estimations for the credit unused limits could follow a similar procedure to that explained above for market risk.

PART V

PATH INTEGRALS, GREEN FUNCTIONS, AND OPTIONS

41. Path Integrals and Options: Overview (Tech. Index 4/10)

In previous chapters in this book, path-integral techniques were in fact used repeatedly for valuation. In this part of the book, we deal directly with the formalism of path integrals as applied to finance. Those who already know path integrals and who want to jump-start into finance might start with these chapters. The finance discussion is self-contained. For those who are unfamiliar with path integrals, the presentation will have appropriate background material.

At the same time, because path integrals are in fact Green functions, the discussion will be relevant to the Green function approach to options.

Path Integrals and Physics

Path integrals were developed by Feynman as a technique used in his Nobel-prize winning work related to relativistic quantum mechanics. Path integrals constitute a powerful and elegant framework for treating problems containing random or stochastic variables. This framework is very general. There is a long history of path integrals applied to practical problems in physics.

Path Integrals and Finance

Path integrals applied to finance provide a powerful, understandable approach. Path integrals are useful for options. This is because finance theory involves diffusion equations, based on assumptions using random variable models for interest rates, stock prices, exchange rates, etc. The diffusion equation is solved directly and exactly by a path integral [1].

The following points are basic:

- An explicit feature is the picturesque idea of future paths of the underlying variables as time progresses. This increases physical intuition.

---

[1] **Relation to Quantum Mechanics and Rigor:** Some finance professors have erroneously concluded that the path integral approach to finance is not rigorous, perhaps misunderstanding the difference here with quantum mechanics. The diffusion equation is mathematically simpler than the Schrödinger equation, having solutions with no oscillations in time. The Schrödinger equation can be turned into a diffusion equation through a so-called Wick rotation. Issues of rigor (uninteresting as they are), are basically nonexistent in finance relative to quantum mechanics.

- The fundamental idea of a contingent claim being equal to the expectation value of the discounted terminal value, consistent with boundary constraints, is manifest.
- Complicated obscure mathematics is avoided. The reader may be comforted to know that mathematics background for most analytic calculations with path integrals requires only an ability to perform Gaussian integrals by completing the square. Schwartz distributions (Dirac delta functions) connect the stochastic calculus with path integrals in a straightforward fashion.
- The no-arbitrage conditions are implemented simply by specifying drift parameters in the path integral through external constraints.
- Consistency is obtained with the standard no-arbitrage hedging recipes.
- Path integrals can be evaluated numerically, e.g. using all the usual techniques, including binomial (or multi-nomial) discretizations, grid discretizations, and Monte-Carlo simulation. General path-integral discretization, pioneered in finance by Castresana and Hogan, provides an efficient and flexible numerical approximation technique.
- Generalization to N dimensions for applications to multi-factor models is straightforward.

A Few Basic Details of Path Integrals

The path integral evaluates the consequences of fluctuations of random variables $x_\alpha(t)$ as a function of time t. The probability distribution function (pdf) of the fluctuations $d_t x_\alpha(t) \equiv x_\alpha(t+dt) - x_\alpha(t)$ has to be specified. Averaging or finding expectation values of some quantity C then involves merely integrating C times the pdf over the possible values of the random-variable fluctuations as time progresses. This is just the path integral.

The path integral gives the propagation of information in time by consecutive small (or in the limit, infinitesimal) time steps of size dt in such a way that the underlying diffusion equation is manifestly satisfied at each step. Each such small step is accomplished by including a "propagator". The path integral in fact is the Green function solution to the underlying diffusion equation.

It is important to understand that the path integral is essentially just a convolution of standard-calculus integrals.

A standard physics approximation consists of the WKB semi-classical approximation. In finance, (up to a convexity term) this starts with a deterministic forward path of stock prices, interest rates, etc. depending on the application. The size of the fluctuations around the forward path is measured by volatility.

In simple cases, the path integral can be evaluated explicitly. Whenever an analytic solution exists, it can be derived using path-integral techniques. Discretization provides a natural base for numerical approximations.

Summary of the Chapters on Path Integrals

The chapters on path integrals that follow are:

- *Path Integrals and Options I*. This chapter presents an introductory overview of the use of path integrals in options pricing including some pedagogical examples. The connection with stochastic equations is exhibited. We give a transparent proof of Girsanov's theorem. We deal with no arbitrage and hedging in the language of path integrals. Finally, we give some results for local volatility in perturbation theory.

- *Path Integrals and Options II*. This chapter contains the path-integral framework for one-factor term-structure interest rate models, including Gaussian and mean-reverting Gaussian. Results for general models including arbitrary rate-dependent volatilities are given. Models with memory effects are also presented. It is shown explicitly how the stochastic equations for rate dynamics are built directly into the path integral.

- *Path Integrals and Options III*. This chapter presents some aspects of numerical methods for options based on path-integral techniques. The fundamental path-integral discretization method originated by Castresana and Hogan is described. We emphasize that standard binomial and multinomial approximations, Monte-Carlo simulations, etc. are just techniques for evaluating the path integral.

- *Path Integrals and Options IV*. This chapter presents options in the presence of many random variables, including principal component path integrals.

- *Reggeon Field Theory (RFT)*. This chapter contains an introduction to this theory of nonlinear diffusion. The RFT is soluble under certain conditions, and can produce non-Brownian critical exponents and scaling laws, calculable in certain approximations. We conjecture that the RFT may be applicable for calculating aspects of fat-tailed distributions.

42. Path Integrals and Options I: Introduction (Tech. Index 7/10)

Summary of this Chapter

Path integrals are widely used in physics for treating problems with stochastic variables. In particular, diffusion equations have path integrals as exact solutions. This chapter presents an introductory overview[1] of the use of path integrals in options pricing.[2,3]. We begin with European options, the venerable Black-Scholes model. We exhibit how Bermuda and American options fit into the path-integral framework. A list of references is at the end of the chapter [i, ii].

Green functions and semigroup techniques are used throughout. The connection of path integrals with stochastic equations is exhibited explicitly. We also give a transparent proof of Girsanov's theorem. Finally, we deal with no-arbitrage, with which path integrals are fully consistent, and discuss hedging in the language of path integrals.

In other chapters of this book, we apply path integrals to term-structure interest-rate models, barrier options, options on several variables and other cases.

Mathematics background for most of the material in this chapter will not require much more than an ability to perform Gaussian integrals. Facts regarding differential equations, Fourier transforms, and Dirac delta-functions will be explained as needed.

---

[1] **History and Acknowledgements:** This chapter is largely taken from the first paper in Ref.i, and is based on work done in 1986-1987 as a consultant to Merrill Lynch, while on leave from the French CNRS. I thank Santa Federico for asking the question to establish the connection between stochastic equations and path integrals that landed me on Wall Street. I also thank Andy Davidson and Mike Herskovitz for support during this time.

[2] **Already Know About Path Integrals? Already Know the Models?** Those familiar with path integrals will find the discussion of path integrals trivial; they should focus on the finance. Those who already know the finance should focus on the path integral formalism. Very few people know both well.

[3] **To the Quants: Don't freak out. You already know something about path integrals:** Anybody who has done Monte Carlo simulations, constructed binomial models, solved diffusion equations using analytic methods etc. has essentially been using path integrals. Hopefully, the general framework and connection between these ideas will become clear.

The reader is assumed somewhat familiar with the financial models [ii], though for convenience in the presentation we shall include enough finance information to keep the discussion self-contained.

Introduction to Path Integrals

Path integrals were developed by Feynman as a technique used in his Nobel-prize winning work related to relativistic quantum mechanics [iii]. Further work by Kac [iv] and others, along with many applications to physics [v] soon followed. Path integrals provide a powerful, elegant framework for treating problems containing stochastic variables. Any diffusion equation has a path integral solution [vi].

The special case of the path integral that we will be using is also called the Wiener integral [vii].

Standard options pricing models involve (backward Kolmogoroff) diffusion equations, based on considerations using random variable models for interest rates, stock prices, exchange rates, etc. These can be Brownian, possibly with mean reversion [viii], or other.

In general, the path integral is useful because it affords a natural framework to visualize physical situations and to carry out calculations.

The path integral has proved useful as a realistic calculation tool in finance. In simple cases, e.g. the Black-Scholes model (a free-diffusion Gaussian model), the path integral can be evaluated explicitly. Somewhat more generally, path integrals can be analytically evaluated with Gaussian dynamics if the boundary conditions and parameters are simple enough. This is because in that case, simple consecutive convolutions of Gaussians occur, and the result is again a Gaussian.

When the path integral cannot be evaluated analytically, standard numerical methods are used. These include Monte Carlo simulations, binomial (or multinomial) approximations, etc. PDE solvers of diffusion equations fit in also, because as we just said the path integral is the solution of the diffusion equation.

The formalism of path integrals applied to options is known to a few members of the quantitative-finance community—see especially the early work of Geske and Johnson [ix, 4]. One aim among others of this introductory chapter on path integrals is pedagogical, in order to make path integral concepts comfortable to the reader by using explicit examples, and in order to emphasize the generality of the approach.

Basic Idea of Path Integrals

The basic idea of a path integral is the propagation of information in time by an infinite set of infinitesimal time steps in such a way that the underlying

---

[4] **Congnoscenti:** Steve Ross tells me that he knew path integrals were relevant. There may be others that also had this realization. Still, from my experience, the path-integral formalism is not well known in the general finance community.

differential equation is manifestly satisfied at each step. This idea is illustrated[5] in Fig. 1 at the end of the chapter. Information about the option exercise value at the expiration or strike date t^* is propagated backward to the present time t_0 by a series of consecutive time steps $\Delta t < 0$, where Δt is finite. We also use the notation $dt > 0$ that will always be infinitesimal. At the end of a calculation, dt and/or Δt may taken formally to zero to get the continuous limit[6].

The Propagator or Green Function

Each such small Δt step is accomplished by including a "propagator", which is the Green function solution to the underlying diffusion equation over time Δt. Paths are thus generated in the co-ordinate x-space between the present time t_0 and time t^*. The dependence of x on the financial variables is specified by the model. For example, in the Black-Scholes (BS) model [x], x is the logarithm of the stock price on which the option is written. Each path is associated with a probability measure or weight specified by the model[7], and all paths are summed over by the path integral. Probabilities for different paths may or may not exhibit a degeneracy, depending on details like whether the volatility and other parameters are or are not x-dependent.

The Path Integral is Just a Convolution of Ordinary Integrals

The path integral is actually a functional [iii]. That is the path integral depends on paths in co-ordinate x-space. The paths themselves are functions depending on the time. Thus, the path integral is not a standard integral, but rather a large multidimensional integral (formally infinite dimensional in the limit $\Delta t \to 0$), consisting of a convolution of ordinary integrals.

---

[5] **Figures:** The numbered figures, taken from Ref. 1, are at the end of the chapter.

[6] **Notation: Δt and dt:** Δt in this discussion is a time interval that may remain finite, while dt is always eventually to be taken infinitesimal. Formally, we can set $\Delta t = -k dt$ for some k, and when $dt \to 0$ we let $k \to \infty$ to keep Δt finite. Sometimes we will take $\Delta t \to 0$. The circumstances will always be made clear.

[7] **Weights and Probabilities for Paths:** In a Monte Carlo simulation of a path integral, each path generated has weight = 1. However, the probability of generating a given path (passing through a set of "bins") depends on the model probability distribution function. It can be useful for numerical approximation to group together paths in bunches or "effective paths", which are then associated with appropriately integrated probabilities. For more discussion, see Ch. 44.

Approximations to Path Integrals

A standard approximation consists of searching for a "best" Gaussian approximation. This is usually either the WKB semi classical approximation [iii] or a non-interacting free diffusion approximation. One can then perform an expansion around the Gaussian in a perturbation series. If the resulting fluctuations are small as measured by some small parameter, the perturbation series can prove to be quite useful numerically whether or not it converges formally, as exhibited by Quantum Electrodynamics [xi].

In finance, the classical approximation is replaced by fluctuations about deterministic forward quantities. The size of the fluctuations is measured by volatility. In general, either the formalism is simple enough so that analytic solutions are possible or else numerical techniques are adopted. Perturbation theory itself is usually not performed.

In a more general setting, the discretization of the path integral itself provides a natural base for reasonable numerical approximation to a theory when analytic results are not available[8]. Monte-Carlo methods constitute a popular tool for the numerical evaluation of difficult path integrals [xii].

In the best of cases, there is a preferred path (or set of paths), about which fluctuations are small.

Heretical Remarks on Rigor and All That

We use notation close to that of Feynman [iii]. For a straightforward presentation and in the spirit of Ref. [iii], we shall not follow an unprofitable mathematically rigorous development, which is not required for applications [9, xiii]. Appropriately sophisticated analysis is performed when needed [xiv].

---

[8] **Numerical Path Integral Applications – the Castresana-Hogan Approach:** Juan Castresana and Marge Hogan have shown how to discretize path integrals explicitly in practice. The numerical methods based on this discretization are reliable, flexible, and fast. These methods have been used in production on the desk for Bermuda swaptions, among other products. This approach is discussed in Ch. 44.

[9] **Too Much Mathematical Rigor in Finance?** The whole path-integral discussion can, if desired, be put on a much more mathematically rigorous basis. Path integrals in finance are simpler than for quantum mechanics because there are no oscillations, making the theory a-priori mathematically well defined. See Glimm and Jaffe, Ref, p. 44. No errors are made using the path integral applied to finance.

However, following Feynman (Feynman and Hibbs, ref. p. 94), it is difficult see the utility of a full-court press for rigor when financial models are only approximate, i.e. various assumptions behind the models are manifestly violated in the real world.

There is, moreover, a serious case against too much mathematical rigor in finance. Rigor can hide irrelevance. Rigor teaches us nothing new of practical importance. Rigor can be counterproductive because it makes the subject appear harder than it really is. The worst is that rigor gives a false sense of model validity.

The use of excessive rigor in finance parallels physics in the 1960's for the mathematically rigorous axiomatic field theory. One paper (Gell-Mann et. al, Ref.) put

The Rest of this Chapter

The organization of the rest of this chapter is as follows. In the next section, the Black-Scholes model is discussed in some detail for orientation, followed by the inclusion of dividends. For generality, we allow arbitrary dividends, even stochastic and time dependent. Then, we give the general form for options with a multiple (put/call) schedule, and for American options.

Appendix A presents two straightforward and related derivations of the Girsanov theorem [xv] using path integrals. One demonstration follows directly from the incorporation of the stochastic equations as delta function constraints in the path integral, and then carrying out the change of variables explicitly by hand to isolate the terms involving the drift[10].

Appendix B contains a discussion of no arbitrage, hedging and path integrals.

Appendix C contains calculations using a local volatility and perturbation theory. See Ch. 6 for an introduction to local volatility and skew.

In Ch. 43 and 45, we present a discussion of stochastic interest rates and options that depend on several stochastic variables. A picture of the 2-dimensional case is in Fig. 7 at the end of the chapter.

Path-Integral Warm-up: The Black Scholes Model

The reader may already be familiar with the Black-Scholes (BS) model. The goal here is partly to put old wine in new bottles and to exhibit the path integral formalism for those who are unfamiliar with it. We start with demonstrating the compatibility of path integrals with the standard "no-arbitrage" framework. At the end of the section, we re-derive the same results using a more compact and more straightforward approach in which "no-arbitrage" appears as a simple parameter specification. Appendix B contains more no-arbitrage details.

Therefore, we begin with stock options. Similar models are used for FX (foreign exchange) options, commodity options, and some other types of options.

---

the situation in perspective: "In particular, the contribution of axiomatic field theory to calculations has been less than any pre-assigned positive number, however small".

Nonetheless, I repeat that the application of path integrals applied to finance can be made as rigorous as you like.

[10] **The Stochastic Equations are in the Path Integrals**: For details, see the end of this chapter, Appendix B, and also the next chapter "Path Integrals and Options II"

Textbook Discussion in a Path Integral Framework

The usual discussion[11] starts with assuming that N_S shares of stock with stochastic price $S(t)$ per share at time t are contained in a portfolio along with N_C options with price per option $C(S,t)$. The portfolio value V is

$$V = N_S S(t) + N_C C(S,t) \qquad (42.1)$$

In order that there is "no arbitrage", the return of V has to be the same as holding risk-free securities[12], so V is assumed to satisfy

$$\frac{dV}{dt} = r_0 V \qquad (42.2)$$

Here r_0 is the risk-free interest rate (presumed constant for simplicity)[13]. We define the volatility σ_0, also held constant in time for the moment[14]. We assume

---

[11] **More No Arbitrage and Hedging:** See Appendix B for a general approach discussing no-arbitrage and hedging.

[12] **No Arbitrage Warm-up and a Joke:** The reader might argue that no arbitrage is nonsense. If one cannot do better than buying treasuries that produce a risk-free rate, why would people go to the trouble of buying options and dynamically hedging them with stock? Why should we assume that time-averaged stock returns are equal to the risk-free rate, when we all know that stock is riskier than debt, so the stockholder deserves a greater return than the bondholder (who because of corporate credit risk, already receives a coupon above the risk-free rate). Nonetheless, options are priced using no arbitrage. The answers to the questions are what you need to understand to become a quant or a trader.

Here is the no-arbitrage joke. The professor and the trader are walking along when they both spot a $10 bill on the sidewalk. The professor says, "This is impossible as demonstrated by no arbitrage; it must be a mirage". The trader picks up the $10.

[13] **The "Risk-Free Rate":** This rate is assumed constant in this section. It is actually specified over a time period relevant for the option. The type of rate is not unique. It can for example be taken as a treasury rate, Libor, a cost-of-funds rate based on Libor plus a spread, Fed Funds, a stock rebate rate, etc. Libor is the Street standard. The appropriate Libor rate for a given option is obtained by interpolation from the Eurodollar futures and swaps markets. For FX options there are two interest rates – the "domestic" and the "foreign" rate that must be considered. See earlier chapters for details.

[14] **A Little Essay on Volatility:** For those readers starting the book here, we give a practical synopsis of volatility. Relaxing the constant volatility assumption is one of the central complicating features of options pricing and hedging. The volatility is taken as different for different times ("volatility term structure"). It may also include stock-price effects to produce "skew", needed to match market prices of options with different strikes. The volatility is sometimes taken as obeying a stochastic equation, with a "volatility of volatility" describing fluctuations of the volatility itself. The volatility takes significance from the model in which it is defined, and models are not unique because no

the lognormal stochastic equation $dS/S = \mu_S dt + \sigma_0 dz(t)$ with the Wiener measure satisfying[15, 16] $(dz(t))^2 = dt$. Next, we use the expansion for the full time derivative of the option value C,

$$\frac{dC}{dt} \approx \frac{\partial C}{\partial t} + \frac{\partial C}{\partial S}\frac{dS}{dt} + \frac{1}{2}\frac{\partial^2 C}{\partial S^2}\frac{(dS)^2}{dt} \qquad (42.3)$$

The Option Diffusion Equation

Setting the hedge ratio $N_S/N_C = -\partial C/\partial S$ cancels out the stochastic quantity dS/dt. This produces the diffusion equation for $C(S,t)$ as[17]

---

model describes the statistical properties of the underlying variable except in approximation.

In practice, the option volatility is backed out from interpolating values of the volatility needed to obtain agreement with options trading in the market; this defines the "implied" volatility. Only a small fraction of possible options actually trade – and your option may not trade at all - so the implied volatility may be an interpolated or extrapolated quantity.

The implied volatility is usually compared with the volatility of the stock price observed in the past (the "historical" volatility). It is often said that the implied volatility is the market's estimate of future historical volatility, and this is the assumption made in the option pricing formalism. However there are all kinds of technical issues affecting option prices, and therefore affecting implied volatilities (option supply/demand being an example). Therefore, it is hard to know to what approximation this association is true. Another complication is that the value of the historical volatility depends on the size of the data window.

Traders naturally hedge options with stock, and therefore the relation of the implied volatility to the historical volatility forms an obsessive topic in determining whether trading makes or loses money. Sometimes the stock of the hedge is the same as the stock (or index) on which the option is written, but often for practical reasons it isn't.

---

[15] **Path Integrals and (dz(t))² = dt:** This is actually just a statement of the width of individual Wiener measures that begin the path integral approach. There is nothing mysterious about it at all. This equation is valid for the expectation value, not for some individual pick of a random number, of course.

[16] **Brownian Motion Limitations:** The infinitesimal limit $dt \to 0$ with the expectation $(dz(t))^2 = dt$ assumes that a Brownian-motion diffusion random-walk process of the underlying stochastic variable occurs down to the smallest time scales. In the real world this idealization of infinitesimal time scale price changes cannot occur (not even a computer can react in one picosecond, and people go to sleep sometimes). Following the standard literature on options models, we temporarily ignore this problem along with other issues of importance, such as discontinuous jumps, possible feedback non-linearities in the options price itself, effective phase transitions from disordered to coherent actions among investors, etc.

[17] **Extension of the no-arbitrage derivation:** See Appendix B.

$$\frac{\partial C}{\partial t} = r_0 C - \mu_0 \frac{\partial C}{\partial x} - \tfrac{1}{2}\sigma_0^2 \frac{\partial^2 C}{\partial x^2} \qquad (42.4)$$

Here $x = \ln(S)$ acts as a co-ordinate[18], while $\mu_0 = r_0 - \tfrac{1}{2}\sigma_0^2$ functions as a drift. The stock-specific drift μ_S does not enter since dS/dt terms cancelled. Also, Ito's rule (or alternatively the need to obtain the same diffusion equation under change of variable) means the stochastic variable satisfies $dx = dS/S - \tfrac{1}{2}\sigma_0^2 dt$, while $\dfrac{\partial}{\partial x} = S\dfrac{\partial}{\partial S}$ and $\dfrac{\partial^2}{\partial x^2} - \dfrac{\partial}{\partial x} = S^2 \dfrac{\partial^2}{\partial S^2}$ as ordinary variables.

Solution of the Option Diffusion Equation

The solution to this equation is classic. Let us take a moment to recall its derivation. Write the formal Taylor expansion for a time step Δt as

$$C(x, t+\Delta t) = e^{\Delta t \cdot \partial_t} C(x,t) \qquad (42.5)$$

Here, $\partial_t = r_0 - \mu_0 \partial_x - \tfrac{1}{2}\sigma_0^2 \partial_{xx}^2$, where $\partial_x = \partial/\partial x$, $\partial_{xx}^2 = \partial^2/\partial x^2$, $\partial_t = \partial/\partial t$.

We continue using Fourier Transform (FT) methods [xvi]. We set

$$C(x,t) = \int_{-\infty}^{\infty} \frac{dk}{2\pi} e^{ikx} \tilde{C}(k,t) \qquad (42.6)$$

Note that $\partial_x = ik$, while $\partial_{xx}^2 = -k^2$ when operating on $\exp(ikx)$. We get

---

[18] **The logarithmic change of variable and the Ito, Stratanovich prescriptions:** The Black-Scholes discussion could proceed using the stock price S instead of its logarithm x. If this is done, we need quadratic terms in the expansion of the return of the stock price. Setting $S_j = x(t_j)$ using a time discretization, we have

$$(S_{j+1} - S_j)/S_j = \exp(x_{j+1} - x_j) - 1 \approx (x_{j+1} - x_j) + \tfrac{1}{2}(x_{j+1} - x_j)^2$$

in order to reproduce the results using x. The quadratic term is replaced by its average, $\sigma_0^2 dt/2$, which is valid as $dt \to 0$. Keeping this quadratic term is equivalent to the Ito prescription (which we follow), while dropping it is equivalent to the Stratanovich prescription. To emphasize it again, we use the Ito prescription.

Path Integrals and Options I: Introduction

$$C(x, t+\Delta t) = e^{r_0 \Delta t} \int_{-\infty}^{\infty} \frac{dk}{2\pi} \exp\left\{\Delta t\left[-i\mu_0 k + \tfrac{1}{2}\sigma_0^2 k^2\right] + ikx\right\} \cdot \tilde{C}(k,t) \quad (42.7)$$

Note that the integral makes sense only if $\Delta t < 0$, since the integral must converge at $k \to \pm\infty$. Hence, we will be propagating information *backward* from some boundary condition given in the future at time t^* (for European options, this is the strike date or expiration date). Write the inverse FT formula at $t = t^*$,

$$\tilde{C}(k,t) = \int_{-\infty}^{\infty} dx^* \exp(-ikx^*) \cdot C(x^*, t) \quad (42.8)$$

Set $\Delta t = t_0 - t^*$, $\Delta t = t_0 - t^*$, $x_0 = \ln S(t_0)$, $x^* = \ln S(t^*)$. This produces

$$C(x_0, t_0) = \int_{-\infty}^{\infty} dx^* G_f\left(x_0 - x^*; \Delta t\right) \cdot C(x^*, t^*) \quad (42.9)$$

The Free Green Function or Propagator

Here, the "free propagator" Green function $G_f\left(x_0 - x^*; \Delta t\right)$ is[19]

$$G_f\left(x_0 - x^*; \Delta t\right) = \frac{e^{r_0 \Delta t}}{\left[-2\pi\sigma_0^2 \Delta t\right]^{1/2}} \exp\left\{-\frac{\left[x_0 - x^* - \mu_0 \Delta t\right]^2}{-2\sigma_0^2 \Delta t}\right\} \theta(-\Delta t) \quad (42.10)$$

Here $\theta(-\Delta t)$ is equal to one for $\Delta t < 0$, zero for $\Delta t > 0$, and 1/2 for $\Delta t = 0$. The definition $G_f = 0$ for $\Delta t > 0$ has been made for convenience while the result for $\Delta t < 0$ follows from standard Gaussian integration in k.

Note that $G_f\left(x_0 - x^*; \Delta t\right)$ is a function of $|x_0 - x^* - \mu_0 \Delta t| / \sqrt{|\Delta t|}$, which is the canonical Brownian motion, square-root scaling[20].

---

[19] **Notations for the free propagator:** G_0, G_f and sometimes just G are interchangeable notations in the book.

[20] **Non-Brownian Scaling Models:** Other possibilities for scaling involving powers other than ½ are possible in non-Brownian dynamics. However, these models are difficult to work with and difficult to understand. The practice on the Street is to use Brownian motion with various parameters fit to the market, warts and all. We discuss these considerations in Ch. 46 when we discuss the Reggeon Field Theory.

The terminal or exercise-date boundary condition for a call option with strike price E is

$$C(x^*, t^*) = \left(e^{x^*} - E\right)_+ \equiv \left(e^{x^*} - E\right)\theta\left(e^{x^*} - E\right) \qquad (42.11)$$

The Classic Black-Scholes Formula

The usual Black-Scholes (BS) model result for a call option is then obtained by straightforward integration[21]

$$C(x_0, t_0) = S(t_0) N(d_+) - E e^{-r_0 \tau} N(d_-) \qquad (42.12)$$

Here $\tau = -\Delta t$ is the positive time difference from valuation to expiration[22], and the "d functions" are [23]

$$d_- = \{\ln[S(t_0)/E] + \mu_0 \tau\}/(\sigma_0^2 \tau)^{1/2} = \{\ln[S(t_0)/E] + (r_0 - \sigma_0^2/2)\tau\}/(\sigma_0^2 \tau)^{1/2}$$
$$d_+ = d_- + (\sigma_0^2 \tau)^{1/2} = \{\ln[S(t_0)/E] + (r_0 + \sigma_0^2/2)\tau\}/(\sigma_0^2 \tau)^{1/2} \qquad (42.13)$$

The standard normal integral is

$$N(\xi) = \int_{-\infty}^{\xi} \frac{du}{\sqrt{2\pi}} \exp(-u^2/2) \qquad (42.14)$$

---

[21] **Normalization:** There is often an additional normalization factor to convert the equations to real prices. For example, S&P index options have a multiplier of $100/contract.

[22] **Complexity and times in the real world:** To give an idea of real-world complexities, in practice there are several different times used. The "diffusion time" τ_{diff} may be used for the time from valuation to the date the option decision for exercise is made, the "discounting time" τ_{disc} may be used for the time from valuation to the date that cash is actually paid out for the option exercise, etc. Sometimes even fractions of a day are included which the options quant will tout as being "more accurate", although given the uncertainties in the volatility this seems like splitting hairs. This sort of detail can be particularly annoying if you need to reproduce the results of a black-box model whose details are unknown (e.g. the model developers have disappeared).

[23] **Notation: the "d" functions:** These functions are ubiquitous in standard options theory because they result from the Gaussian integrations over the limits specified by the options constraints. Another common notation is $d_1 = d_+$, $d_2 = d_-$.

Handy Integrals

Some useful integrals to avoid the calisthenics of completing the square are:

$$I_0 = \int_{-\infty}^{\xi_{max}} \frac{d\xi}{\sqrt{2\pi}} \exp\left(-\frac{\xi^2}{\alpha} + \gamma\xi\right) = \sqrt{\frac{\alpha}{2}} \exp\left(\frac{\alpha\gamma^2}{4}\right) N(\psi_{max}) \quad (42.15)$$

$$I_1 = \int_{-\infty}^{\xi_{max}} \frac{d\xi}{\sqrt{2\pi}} \xi \exp\left(-\frac{\xi^2}{\alpha} + \gamma\xi\right) = \frac{\partial I_0}{\partial \gamma} = \frac{\alpha\gamma}{2} I_0 - \frac{\alpha}{\sqrt{8\pi}} \exp\left(-\frac{\psi_{max}^2}{2} + \frac{\alpha\gamma^2}{4}\right) \quad (42.16)$$

Here, $\psi_{max} = (\xi_{max} - \alpha\gamma/2)\sqrt{2/\alpha}$.

Sometimes the parameters are taken differently for physical reasons. A useful integral in that case (with all time intervals $\tau > 0$) is

$$C(x_0, t_0) = e^{-r_d \tau_d} \int_{\ln(E_a)}^{\infty} \frac{dx^*}{(2\pi\sigma_a^2 \tau_a)^{1/2}} (e^{x^*} - E_b) \exp\left\{-\frac{(x_0 - x^* + \mu_b \tau_b)^2}{2\sigma_c^2 \tau_c}\right\}$$

$$= \sqrt{\frac{\sigma_c^2 \tau_c}{\sigma_a^2 \tau_a}} \left[S(t_0) e^{r_b \tau_b - r_d \tau_d + \frac{1}{2}(\sigma_c^2 \tau_c - \sigma_b^2 \tau_b)} N(d_+) - E_b e^{-r_d \tau_d} N(d_-) \right] \quad (42.17)$$

Here $\mu_b = r_b - \frac{1}{2}\sigma_b^2 \tau_b$, while $d_- = \{\ln[S(t_0)/E_a] + \mu_b \tau_b\} / (\sigma_c^2 \tau_c)^{1/2}$ and $d_+ = d_- + (\sigma_c^2 \tau_c)^{1/2}$.

The reader should carefully note that the main point of emphasis is *not* this standard textbook option result, but rather the more general importance of the path integral formalism and the free propagator G_f.

The Semi-Group Property

The Green function or propagator G_f satisfies the "semi-group" or "reproducing-kernel" property as can easily be seen by direct integration or FT techniques

$$G_f(x_0 - x_2; t_0 - t_2) = \int_{-\infty}^{\infty} dx_1 G_f(x_0 - x_1; t_0 - t_1) G_f(x_1 - x_2; t_1 - t_2) \quad (42.18)$$

Note that the *same function* G_f appears on *both sides* of this equation. Physically this means that free propagation from (x_2, t_2) to (x_1, t_1) followed by free propagation from (x_1, t_1) to (x_0, t_0) integrated over all x_1, is equivalent to free propagation from (x_2, t_2) directly to (x_0, t_0) as illustrated in Fig. 2 at the end of the chapter.

Building up the Path Integral from the Semi-Group Equation

It is important to recall that the definition of simple European options solved by the BS model does involves integration over all x-values for all times between the present and the expiration date[24]. In general a European stock option allows any intermediate stock price with $t \in (t_0, t^*)$, with $S(t) = \exp(x(t))$ from 0 to ∞ in principle, since the investor cannot exercise a European option at times before t^* by definition regardless of what the stock price is. Now using the semi-group property for G_f we may iterate an arbitrary number of times, obtaining[25]

$$G_f(x_0 - x^*; t_0 - t^*) = \left[\prod_{j=1}^{n-1} \int_{-\infty}^{\infty} dx_j \right] \left[\prod_{j'=0}^{n-1} G_f(x_{j'} - x_{j'+1}; t_{j'} - t_{j'+1}) \right] \quad (42.19)$$

There are $n-1$ integrals and n propagators. Here, $x^* = x_n$ and $x_j = x(t_j)$ for t_j between t_0 and $t^* = t_n$, as illustrated in the Fig. 1 at the end of this chapter.

---

[24] **What happens if the integrations have constraints?** If constraints on the integration over intermediate states exist, the discussion becomes more complicated. If the constraints are simple enough, closed-form solutions can still be obtained. This is the case with simple "barrier" options as described in Ch. 17-19. Under fairly general conditions, the iterated path integral satisfies the semi-group formula. With constraints inserted at intermediate times, general options can be evaluated using numerical approximations (Monte Carlo simulations etc).

[25] **Notation for Labels and Indices in Path Integrals – README!** For this equation *only* there are big brackets and distinguished indices j and j'. In general, by convention the brackets and the different labels are to be understood and will not be exhibited. The extension of a dummy index labeled **j** in the first product is intended to extend only locally in the formula only up to the second product with a separate dummy index which can be labeled by the same letter **j**. This convention is common practice in physics papers and avoids cluttering up the page with bracket signs and a plethora of different labels. Properly understood, there should be no confusion.

Formal Continuous Limit for the Path Integral

The formal continuous path integral is defined by the $n \to \infty$ limit of the successive propagations with fixed $t^* - t_0$. Defining the "velocity" $dx(t)/dt$ as
$$\frac{dx(t)}{dt} = \left(x_{j+1} - x_j\right)/\left(t_{j+1} - t_j\right) \text{ at } \left(x_j, t_j\right) \text{ as } t_j - t_{j+1} \to 0,$$
and substituting the Gaussian form for local propagation for G_f, the path integral for the Green function for propagation over the whole interval is written as

$$G_f\left(x_0 - x^*; t_0 - t^*\right) = e^{-r_0 \tau} \int_{\substack{\text{Paths with} \\ x(t_0) = x_0, \\ x(t^*) = t^*}} [Dx(t)] \exp\left\{-\int_{t_0}^{t^*} \frac{dt}{2\sigma_0^2}\left[\frac{dx(t)}{dt} - \mu_0\right]^2\right\}$$

(42.20)

The option price $C(x_0, t_0)$ as a path integral is obtained by inserting the path integral expression for G_f into Eqn. (42.9). The above equation for G_f is in the standard Lagrangian form of the path integral.

So far, it might seem that we have merely succeeded in somehow rendering the simple BS model much more complicated. However, the path integral formalism that we have presented is fundamental. The simplicity of the BS model containing just free propagation allows the path integral to be evaluated in a trivial fashion, simply by undoing the steps leading to Eqn. (42.20).

More General Parameters and the Path Integral

Now consider replacing the constant drift μ_0 by a general price and time dependent drift function $\mu(x,t)$. This is, for example, produced by the general dividend model, which can produce jumps, and requires the path integral apparatus or an equivalent approach. Similarly, allowing the volatility σ_0 to become a function $\sigma(x,t)$ in order to include "skew" effects likewise requires the path integral. The American option restricts the class of paths in a non-trivial way involving some complicated optimization logic, and again necessitates the path integral.

In each of these cases, every propagation is forced to occur in infinitesimal Δt. At the end of the section, there is a diagram to illustrate this point.

Convolution of these propagations to get a *finite-time* propagator in the general case cannot be evaluated analytically. The semigroup property is true, but now holds only for the path integral itself if finite time intervals $t_0 - t_1$, $t_1 - t_2$

are considered. Since, for the BS case, the path integral itself is just the free propagator with modified parameters, the result is trivial to obtain.

The Path Integral Satisfies the Diffusion Equation

It is profitable to see how the path integral solution for $C(x,t)$ satisfies the diffusion differential equation (42.4). We first start with the free diffusion equation with constant parameters. By direct algebra it is not hard to see that the Green function G_f satisfies the following equation[26]

$$\left[\partial_t + \mu_0 \partial_x + \tfrac{1}{2}\sigma_0^2 \partial_{xx}^2 - r_0\right] G_f(x-x';t-t') = -\delta(x-x')\delta(t-t') \quad (42.21)$$

Dirac Delta Functions

Here $\delta(\varsigma)$ is the Dirac delta-function [xvii], mathematically a Schwartz distribution, which is defined by the formula[27]

$$\int_{-\Lambda}^{\Lambda} f(\varsigma)\delta(\varsigma)d\varsigma = f(0) \quad (42.22)$$

for any suitable "test" function $f(\varsigma)$, and any Λ. We will also need the formula[28]

$$\delta(t-t') = -\partial_t \theta(t'-t) \quad (42.23)$$

Details: How the Green function satisfies the Diffusion Equation

The fact that G_f satisfies the singular diffusion differential equation can be seen by expanding $G_f(x-x';t+dt-t')$ in a series in dt to perform the time partial derivative. Again, our notation is $dt = -\Delta t > 0$. We need to take the limit $t \to t'$ to see the singular behavior. We need the replacement of the square

---

[26] **Homework:** Show this. Try at first not to look at the comments below.

[27] **Dirac Delta Function and Schwartz Distributions:** This is put on a rigorous basis using Schwartz distribution theory. The interested reader who is unfamiliar with the theory of distributions is invited to consult the references. Only simple manipulations will be required here, and will be explained when needed.

[28] **Homework:** Show this. It will give you some insight into the Dirac delta function.

displacement by its average, $(x'-x)^2 \approx -2\sigma_0^2 \Delta t$, valid in the infinitesimal limit [xviii, 29]. We ignore higher order terms $(x-x')G_f$ or $(t-t')G_f$ relative to the leading $-\delta(x-x')\delta(t-t')$ term. Note the $\delta(x-x')$ behavior of a Gaussian as $(t-t') = \Delta t \to 0$. As we take the limit $t-t' \to 0$ through negative values, the Green function becomes the delta function, $G_f \to \delta(x-x')$. This produces the required boundary condition at the expiration date with $t' = t^*, x' = x^*$.

The No-Arbitrage World and the Fictitious Stock Prices

Next, we give a formal but simple argument leading to the path integral. First, the diffusion equation (42.4) implies that $\exp(r_0 t)C(x,t)$ is related[30] to the probability $\tilde{\wp}(\tilde{S}, t)$ for a "fictitious" stock price $\tilde{S}$ to have the value $S(t)$ at time t in a "no-arbitrage world". The stochastic equation for $\tilde{S}$ that yields the diffusion equation for $\tilde{\wp}(\tilde{S}, t)$, namely Eqn. (42.4) with the term $r_0 C$ removed and $\tilde{\wp}$ substituted for C, is just

$$\frac{d\tilde{x}(t)}{dt} = \mu_0 + \sigma_0 \eta(t) \tag{42.24}$$

Here, $\tilde{x} = \ln(\tilde{S})$ and $\eta = dz/dt$ is the formal derivative of $z(t)$. This is the stochastic equation for $x = \ln(S)$ but with the replacement of the stock return μ_S by the risk free rate r_0 as we found from no arbitrage. The return μ_S is an irrelevant variable and does not enter in the determination of the option C.

Notation: $dx(t)$, $d_t x(t)$, $\eta(t)$ and Brownian Motion

It is useful to focus on $\eta(t) = dz(t)/dt$, which is a random Gaussian slope variable. It is also necessary to avoid confusion regarding differentials. The symbol "$dx(t)$" unfortunately can appear with two different meanings, between which we need to distinguish.

---

[29] **Limiting Process:** As indicated below, we really need a two-step limiting process. First we set dt = -Δt = Lδt, and let δt → 0 with L → ∞, such that Δt is constant. This allows the replacement (x'-x)² by -2σ₀² Δt. Then we let Δt → 0.

[30] **Option, Probability Relation:** The relation involves the second derivative of the option with respect to its strike.

For fixed time t, we can integrate over all possible values of the variable x at time t. The measure in this integral contains the ordinary integration measure dx specified at time t, or $dx(t)$. On the other hand, the stochastic Langevin-Ito equation tells us how x at time t differs from x at time $t + dt$ for fixed but infinitesimal dt, i.e. $x(t+dt) - x(t)$, which is the time difference, unfortunately commonly also called $dx(t)$. Strictly speaking, we need to use another symbol, which we call $d_t x(t)$.

We can draw a straight line between $x(t)$ and $x(t+dt)$, for a given path realization. The slope of this path segment line, which we call $\eta(t)$, is a Gaussian random variable. This is because a random walk occurs even for the infinitesimal interval between t and $t + dt$. Brownian motion assumes that a random walk occurs inside any time scale, no matter how small.

The figure below gives the idea:

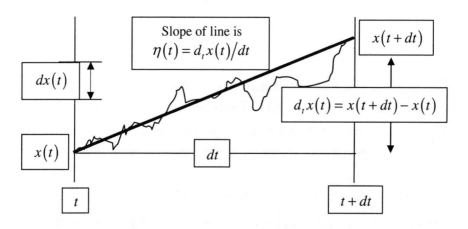

Inside are still an infinite number of steps. The slope $\eta(t) = d_t x(t)/dt$ is a Gaussian random variable if the time difference $d_t x(t)$ is Gaussian. Unfortunately, $d_t x(t)$ is often called $dx(t)$. This causes confusion when doing integrals over the $x(t)$ variable, since $dx(t)$ is just the ordinary measure on the $x(t)$ axis. Discretizing avoids confusion.

Now integration over $x(t)$ for all times t between two times t_a, t_b is equivalent to integration over all paths between $x(t_a)$ and $x(t_b)$. One can picture this in two ways. First, one can integrate over all values of each $x(t)$ at each value of t. Alternatively, one can integrate over all slopes of all intermediate path segments.

Any confusion as to the physical interpretation of what is going on can be resolved by taking finite discrete time partitions. It is sometimes helpful to think of x as the co-ordinate for a particle undergoing a random walk; the straight-line path segments can be viewed as free flight between successive scatterings with incremental scattering angles given by a Gaussian probability distribution (see Feynman, Williamson, Refs).

The figure below generated by computer may also help with the intuition:

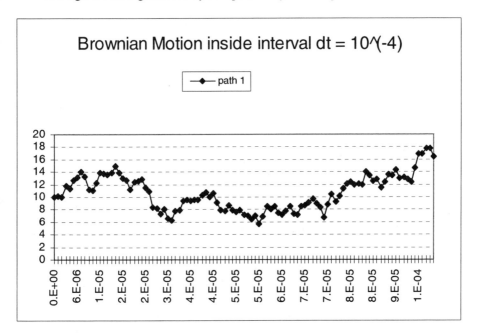

Connection of Path Integral with the Stochastic Equations

The path integral is not only fully consistent with the stochastic equations, the stochastic equations are directly used to get the path integral. Now in fact it is $\eta(t)$ that is the fundamental starting point. We can start with the statement that $\eta(t)$ is a Gaussian random variable and then use the stochastic equation to introduce $d\tilde{x}(t)$.

Consider the total conditional probability, which we call $\wp(x_0, x^*; t_0, t^*)$. This $\wp(x_0, x^*; t_0, t^*)$ by definition is the product of all the $\eta(t)$ Gaussians, integrated over each $\eta(t)$ with $t \in (t_0, t^*)$ and over each $\tilde{x}(t)$ variable[31] such that $\tilde{x}(t_0) = x_0$ and $\tilde{x}(t^*) = x^*$. In the discretization $\eta_j = \eta(t_j)$ with $j = 0, \ldots, n-1$ and with $t^* = t_n$, the final η_{n-1} is determined by the requirement $\tilde{x}(t^*) = x^*$, viz $\eta_{n-1} = \dfrac{1}{\sigma_0}\left[\dfrac{x^* - \tilde{x}(t_{n-1})}{dt} - \mu_0\right]$.

The next formula for the Dirac delta function is the key to what we need:

$$1 = \int_{-\infty}^{\infty} d\tilde{x}_{j+1} \delta\left(\tilde{x}_{j+1} - \tilde{x}_j - \mu_0 dt - \sigma_0 \eta_j dt\right)$$

$$= \int_{-\infty}^{\infty} d\tilde{x}_{j+1} \frac{1}{\sigma_0 dt} \delta\left\{\eta_j - \frac{1}{\sigma_0}\left[\frac{\tilde{x}_{j+1} - \tilde{x}_j}{dt} - \mu_0\right]\right\}$$

(42.25)

We get[25]

$$\wp(x_0, x^*; t_0, t^*) = \prod_{j=0}^{n-2} \int_{-\infty}^{\infty} d\eta_j \cdot \prod_{j=0}^{n-1} \frac{1}{(2\pi/dt)^{1/2}} \exp\left(-\tfrac{1}{2}\eta_j^2 dt\right) \quad (42.26)$$

$$= \prod_{j=0}^{n-2} \int_{-\infty}^{\infty} d\eta_j \cdot \int_{-\infty}^{\infty} d\tilde{x}_{j+1} \frac{1}{\sigma_0 dt} \delta\left\{\eta_j - \frac{1}{\sigma_0}\left[\frac{\tilde{x}_{j+1} - \tilde{x}_j}{dt} - \mu_0\right]\right\}$$

$$\cdot \prod_{j=0}^{n-1} \frac{1}{(2\pi/dt)^{1/2}} \exp\left(-\tfrac{1}{2}\eta_j^2 dt\right) \quad (42.27)$$

$$= \prod_{j=0}^{n-2} \int_{-\infty}^{\infty} d\tilde{x}_{j+1} \prod_{j=0}^{n-1} \frac{1}{(2\pi/dt)^{1/2}} \exp\left[-\frac{1}{2\sigma_0^2 dt}\left(\tilde{x}_{j+1} - \tilde{x}_j - \mu_0 dt\right)^2\right] \quad (42.28)$$

Here the product is over all $t_j \in (t_0, t^*)$ in the discretized version as the time step vanishes, $dt \to 0$. The Dirac δ functions eliminate the $\eta(t)$ integrals and

---

[31] **Notation:** In this section only, we keep the tilde ~ notation to remind the reader that we work in the fictitious no-arbitrage world with the stock return replaced by the risk free rate.

replace $\eta(t)$ by $\dfrac{1}{\sigma_0}\left[\dfrac{d\tilde{x}(t)}{dt} - \mu_0\right]$ in the Gaussians. To avoid confusion over labels it is best to use the discretized version.

This result for $\wp(x_0, x^*; t_0, t^*)$ is exactly the same as for the path integral for the Green function $G_f(x_0 - x^*; t_0 - t^*)$ up to the discount factor $e^{-r_0\tau}$, since the $\{\tilde{x}(t)\}$ are dummy variables.

No Arbitrage: An Equivalent Approach

The entire no-arbitrage discussion for the path integral is equivalent to specifying the drift parameter in the Green function. We could have started with an arbitrary drift and then insisted that the portfolio V (consisting of N_S shares of stock and N_C options) have a return equal to the risk-free rate. This constraint would then specify the drift. Viewed in this fashion, no arbitrage just consists of specifying parameters. This direct simplicity is characteristic of the path-integral approach. First the dynamics are specified and the Green function derived. Then appropriate parameters are specified by physical market constraints. These constraints constitute the no-arbitrage conditions. For further discussion, see App. B.

We will encounter exactly the same idea when we apply path integrals to interest rate options.

This concludes the discussion of the basic path integral framework and the Black-Scholes model.

Dividends and Jumps with Path Integrals

The incorporation of the dividends provides the first example where the path integral—or some approximation to it—is in general needed, even with European options. Dividends in the general case produce stock-dependent effects that cannot be treated analytically, except in a simple "dividend-yield" approximation or in deterministic cash dividends. Cash payments at specified dates produce jumps in the stock price across these dates, because the stock price includes a potential dividend before it is paid and the stock is less valuable after a dividend has been paid.

It is worth stating that jumps in the stock price over short times can happen for a variety of causes besides dividends—bad news, announcement or cancellation of a takeover, etc. The formalism presented in this section is applicable to these effects also in some approximation.

Let $D(x,t)$ be the dollar dividend per share of stock per unit time. The normalization is made for convenience. $D(x,t)$ will in general not be a continuous function, since dividends are paid at discrete times. The stock price changes by an extra amount $-D(x,t)$ per unit time. Hence, we place this extra term divided by $S(t)$ in the drift function for the return $d_t S/S$ of the stock price, where $d_t S(t) = S(t+dt) - S(t)$. Define the "effective drift function" $\mu(x,t)$ by

$$\mu(x,t) = r_0 - \tfrac{1}{2}\sigma_0^2 - D(x,t)e^{-x(t)} = \mu_0 - D(x,t)e^{-x(t)} \qquad (42.29)$$

Again $x(t) = \ln S(t)$. We first assume $D(x,t)/S(t) = D(x,t)e^{-x(t)} = D_0$ is a constant. D_0 is called the dividend yield[32]. This common approximation makes the problem analytically soluble. For this case, the drift is constant, viz $\mu(x,t) \equiv \mu_D = \mu_0 - D_0 = r_0 - D_0 - \tfrac{1}{2}\sigma_0^2$, and so we immediately get the option in the same way as in the last section[33]. Noting the identity

$$\exp\left\{ x^* - \frac{[x_0 - x^* - \mu_D \Delta t]^2}{-2\sigma_0^2 \Delta t} \right\} = \exp\left\{ x_0 - (r_0 - D_0)\Delta t - \frac{[x_0 - x^* - \hat{\mu}_D \Delta t]^2}{-2\sigma_0^2 \Delta t} \right\}$$

(42.30)

---

[32] **Notation:** The dividend yield denoted called D_0 here is also called y and sometimes q.

[33] **Dividend Yield vs. Cash Dividends, Dividend Models, Dividend Risk, and a Story:** The dividend yield might seem less accurate than the cash dividend. The cynic would say that future dividends are just assumptions anyway (do *you* know what cash dividends IBM will pay in the future?). There are various models of dividends (historical, growing at a constant rate, growing with the forward stock price, etc.). Various assumptions can lead to significantly different option prices. Related disasters occasionally hit the news. See e.g. the article describing a big loss at a broker-dealer related to dividend models: "Blind faith", *The Economist*, 1/31/98, page 76.

Complications do occur if trading occurs around the date of a dividend payment. I once heard an emerging markets trader recount with joy a trade with a competitor who was screwed (a technical term used by the trader), because the competitor used dividend yield and not cash dividends. For index options, there is less of an issue since dividends are paid by the various stocks in the index at different times. The risk, e.g. the change of option value with stock price S, depends on cash vs. yield dividends - since in one case D is constant and in the other D/S is constant.

Here is the story. I was once called in to settle a problem. I walked into the conference room chock full of people, including the department head, who were baffled in various degrees by a risk report produced by their risk system. The explanation was the point above regarding whether D or D/S is held constant as S is moved. In retrospect, the incident now seems somewhat amusing. At the time, it all had the air of a serious courtroom drama, with arrows being slung in various directions.

Here, $\hat{\mu}_D = \mu_D + \sigma_0^2$ produces the European call option[34] result with $d_\pm$ as before but with μ_0 replaced by μ_D

$$C(x_0, t_0) = S(t_0) e^{-D_0 \tau} N(d_+) - E e^{-r_0 \tau} N(d_-) \qquad (42.31)$$

Note that if the dividend yield is large enough, the option value $C(x_0, t_0)$ is less than the intrinsic value $C_{\text{intrinsic}}(x_0, t_0) = S(t_0) - E$. Hence, the option holder would, if he could, exercise the call and pick up a profit $C_{\text{intrinsic}} - C$ from getting the intrinsic value while losing the option. For a European option, such early exercise is by definition not possible, but for American or Bermuda options that allow early exercise, this plays an important role. Basically the idea is that it may be better to buy the stock now at cost E and sell it, rather than wait to buy the stock at cost E after large dividends have reduced its value, overcoming the possible higher stock prices from random diffusion. We will study American options in the next section.

The forward stock price, which is the average no-arbitrage stock price at some future time t, is always used in discussions of equity options. This is

$$S_{fwd}(t) = \langle S(t) \rangle = \int_{-\infty}^{\infty} dx G_f(x - x_0; t - t_0) \cdot S(t) \qquad (42.32)$$

The result follows immediately from the identity Eqn. (42.30), using the fact that the integral over a Gaussian over infinite limits is one. We get

$$S_{fwd}(t) = S_0 \exp\left[(r_0 - D_0)(t - t_0)\right] \qquad (42.33)$$

It is instructive to get this result in another way. We have for any Gaussian theory

$$\langle \exp(x) \rangle = \exp\left(\langle x \rangle + \tfrac{1}{2} \langle x^2 \rangle_c\right) \qquad (42.34)$$

Here, the second-order "connected part" with subscript $\langle \; \rangle_c$ is the usual quantity

$$\langle x^2 \rangle_c = \langle (x - \langle x \rangle)^2 \rangle = \langle x^2 \rangle - \langle x \rangle^2 \qquad (42.35)$$

---

[34] **Options on futures:** This case is reproduced by taking $D_0 = r_0$ so that the future F has zero average return <d_tF/F> = 0. Of course, futures do not have dividends but neither do futures have a deterministically appreciating component since they are not assets, so this formal trick gives the correct result.

Integrating the stochastic equation for $x = \ln(S)$ produces

$$x(t) = x_0 + \mu_D (t - t_0) + \sigma_0 \int_{t_0}^{t} \eta(t') dt' \qquad (42.36)$$

Here, $\eta(t) = dz(t)/dt$ as explained above, with the expectation of the product $\eta(t')\eta(t'')$ being a Dirac delta function[35]

$$\langle \eta(t')\eta(t'') \rangle = \delta(t' - t'') \qquad (42.37)$$

We get $\langle x \rangle = x_0 + \mu_D (t - t_0)$, $\langle x^2 \rangle_c = \sigma_0^2 \int_{t_0}^{t} dt' \int_{t_0}^{t} dt'' \langle \eta(t')\eta(t'') \rangle = \sigma_0^2 (t - t_0)$.

Hence the forward stock price (in the no-arbitrage world) with constant dividend yield is, as above,

$$S_{fwd}(t) = S_0 \exp\left[(r_0 - D_0)(t - t_0)\right] \qquad (42.38)$$

Note that the forward price is independent of the volatility. The option price can be rewritten in terms of the forward stock price if desired. This produces useful insight. In the absence of volatility, the single deterministic path along the forward stock price path gives the dynamics. The volatility can then be viewed as producing fluctuations about the forward stock price.

For deterministic and discrete cash dividends (i.e. specified dividends at specified times in the future), we can proceed as follows. If a first cash dividend D_1 is paid at time t_1 then the stock price will drop by D_1 at t_1. The forward stock price just after the dividend is paid is $S_{fwd}(t_1) = S_0 \exp\left[r_0(t_1 - t_0)\right] - D_1$. Then $S_{fwd}(t_1)$ is moved forward to the next dividend payment at t_2 by a factor $\exp\left[r_0(t_2 - t_1)\right]$. Continuing this logic, if a number of cash dividends $\{D_\beta\}$ are paid at times $\{t_\beta\}$ before time t we get the forward stock price at time t as

$$S_{fwd}(t) = S_0 \exp\left[r_0(t - t_0)\right] - \sum_{t_0 < t_\beta < t} D_\beta \exp\left[r_0(t - t_\beta)\right] \qquad (42.39)$$

---

[35] **Delta-function expectation of** $\eta(t')\eta(t'')$: This formula is equivalent to $(dz)^2 = dt$ as can be seen by straightforward integration.

Taking the present value of this equation back from time t, we get the value of the stock with the effect of dividends subtracted out; call it $S_{NoDividends}$. This is the spot stock price S_0 minus the sum of the present value of the dividends. Dividends are discounted back from payment dates to t_0.

$$S_{NoDividends} = S_0 - \sum_{t_0 < t_\beta < t} D_\beta \exp\left[-r_0\left(t_\beta - t_0\right)\right] \quad (42.40)$$

Consider the figure below.

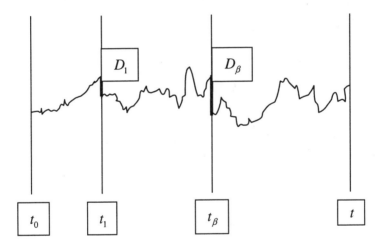

It should be clear from the preceding discussion that in general, jumps in the stock price can be handled in the same way as for cash dividends.

In order to get the option with deterministic discrete cash dividends we can simply replace the forward stock price using the dividend yield by this forward stock price using cash dividends[36].

---

[36] **Another model for options on stocks with Dividends:** Sometimes the assumption is made for Brownian motion of the logarithm of $S_{NoDividends}$. This is reasonable because, after all, the dividends are not stochastic. However this leads to different results for a given volatility than we have presented because the volatility part of $d_t S/S$ is not the same. In order to get the same result the volatility would have to be modified by the ratio $S_{NoDividends}/S$. Just another ambiguity.

If the payment times are uncertain but the distribution $\Phi_{time}\left(\{t_\beta\}\right)$ of payment times is known, we would need to perform the appropriate time averaging. Similarly, if the dividend amounts are only known probabilistically with distribution $\Phi_{Div}\left(\{D_\beta\}\right)$ we need to perform dividend averaging.

For general dividend functions, the Green function is now dependent on all $\{\mu(x,t)\}$. We call the Green function $G_D(x,x';t,t')$. The Green function is no longer a simple function of co-ordinate and time differences. In fact it depends on all intermediate co-ordinates and times[37]. Now, for an infinitesimal time step the propagation is free, i.e. $G_D(x,x';t,t') \approx G_f(x-x';t-t')|_{\mu_0 \to \mu(x,t)}$ for small enough $t-t'$. Physically, "small enough" is the time scale over which the dividend function varies appreciably. For dividends paid at discrete times, infinitesimal time steps are not needed since the semigroup property can be used to combine the free propagation between the discrete dividend payments. So, if and only if $\Delta t = t - t' \approx 0$, G_D is given by the free form $G_D^{(0)}$ where

$$G_D^{(0)}(x,x';t,t') = \frac{e^{r_0 \Delta t} \theta(-\Delta t)}{\left[-2\pi\sigma_0^2 \Delta t\right]^{1/2}} \exp\left\{-\frac{\left[x-x'-\mu(x,t)\Delta t\right]^2}{-2\sigma_0^2 \Delta t}\right\} \qquad (42.41)$$

The formal expression for the path integral is defined as the limit of the successive propagations with dividends paid over infinitesimal time steps, exactly as in the previous discussion without dividends. We get for finite number of steps n the result

$$G_D(x_0,x^*;t_0,t^*) = \prod_{j=0}^{n-2} \int_{-\infty}^{\infty} dx_{j+1} \prod_{j=0}^{n-1} G_D^{(0)}(x_j,x_{j+1};t_j,t_{j+1}) \qquad (42.42)$$

The result has the same form as for the free diffusion case but now with the general drift. In the limit $n \to \infty$ we get the formal functional result

$$G_D(x_0,x^*;t_0,t^*) = e^{-r_0 \tau} \int_{\substack{\text{Paths with}\\x(t_0)=x_0,\\x(t^*)=t^*}} [Dx(t)] \exp\left\{-\int_{t_0}^{t^*} \frac{dt}{2\sigma_0^2}\left[\frac{dx(t)}{dt} - \mu(x,t)\right]^2\right\}$$

---

[37] **Notation:** We have left off the tildes on the x variables since by now the reader should be used to the idea that we are using the no-arbitrage dictum.

(42.43)

The (exact) equation satisfied by $G_D(x,x';t,t')$ is [38,39]

$$\left[\partial_t + \mu(x,t)\partial_x + \tfrac{1}{2}\sigma_0^2 \partial_{xx}^2 - r_0\right] G_D(x,x';t,t') = -\delta(x-x')\delta(t-t')$$

(42.44)

Proof of the Path Integral Satisfying the Semi-Group Property

It is important to note that the semigroup property over finite times is satisfied by the full path-integral expression for G_D exactly. Thus, for *finite* time intervals (t_0,t_1) and (t_1,t_2) we have the *exact* result

$$G_D(x_0,x_2;t_0,t_2) = \int_{-\infty}^{\infty} dx_1 \, G_D(x_0,x_1;t_0,t_1) G_D(x_1,x_2;t_1,t_2) \quad (42.45)$$

The proof is immediate, as illustrated in Fig. 3 at the end of the chapter. $G_D(x_0,x_2;t_0,t_2)$ contains consecutive propagations with integrations over all $x(t)$ with $t_0 < t < t_2$. $G_D(x_0,x_1;t_0,t_1)$ contains consecutive propagations with integrations over all $x(t)$ with $t_0 < t < t_1$. Finally, $G_D(x_1,x_2;t_1,t_2)$ contains consecutive propagations with integrations over all $x(t)$ with $t_1 < t < t_2$. All the propagations are present, and all integrations but one over x_1 are accounted for. Therefore, the x_1 integration must be inserted over the product of the two

---

[38] **The Backward Diffusion Equation:** Because options involve backward-moving logic, we exhibit backward (Kolmogorov) diffusion equations, which do not involve derivatives of the drift or volatility functions. Forward (Fokker-Planck) equations do involve such derivatives. It would seem like x-derivatives of µ(x, t) could appear in the backward equation that contains differentiation with respect to initial variables (x, t) because the Green function does after all depend on µ(x, t). In fact, such derivatives do appear at intermediate stages but then these terms cancel out at the end.

[39] **No Unitarity here:** A point regarding backward and forward propagation involves "unitarity" which holds in quantum mechanics (QM). If we propagate forward and then backward over the same time interval in QM, the result is the Dirac delta function (zero unless we arrive back to the same point). For real diffusion, which is the case here, unitarity does not hold even for constant drift and volatility. Propagating forward and then backward leads to a nonzero result regardless of the initial and final spatial x values.

propagators $G_D(x_0,x_1;t_0,t_1)G_D(x_1,x_2;t_1,t_2)$ in order to get the overall propagator $G_D(x_0,x_2;t_0,t_2)$. This is just Eqn. (42.45).

The equation for the option price $C(x_0,t_0)$ is found simply by integrating the payoff value multiplied by the Green function exactly as in the Black Scholes model. The equation is the same backward equation as satisfied by the Green function, since the differential operators wind up acting only on the initial step in the multiple product defining the Green function. The result is the Black-Scholes equation replacing the constant drift μ_0 by $\mu(x_0,t_0)$, and is

$$\frac{\partial C}{\partial t_0} = r_0 C - \mu(x_0,t_0)\frac{\partial C}{\partial x_0} - \tfrac{1}{2}\sigma_0^2 \frac{\partial^2 C}{\partial x_0^2} \qquad (42.46)$$

More development of the general drift formalism will be given in the discussion of the Macro-Micro model in Ch. 47-51.

Discrete Bermuda Options

We now consider an option with multiple strike or exercise dates. These multiple dates are said to form a schedule. If the schedule is discrete, the option is called a Bermuda option. If exercise is possible at any time (a continuous schedule) the option is called an American option. Relative to a European option, an option with multiple exercises is clearly worth more, since the option holder has the right to exercise at more times than just once. Determining this extra "premium" uses rather messy logic and takes a lot of numerical effort. The logic is called *backward induction*. To gain insight, we present the general formalism using path integrals[40]. A special case was first treated by Geske and Johnson [ix].

The essential problem is in the boundary condition at each strike date t_ℓ^* where $\ell = 1,\cdots,L$. An *optimality* condition is used to determine whether the option is exercised[41]. The idea is that the option holder, as the time on his watch

---

[40] **Dividends:** We will include the dividend-yield case for simplicity in the development, but as we shall show the results can easily be generalized. We need dividends for call options because otherwise there is no premium of American over European options. For put options, there is a premium even if no dividends are present.

[41] **Do American option holders *really* employ the optimality condition to decide to exercise:** Unless the option is liquid and the market price is known, probably not, because most option holders do not have software to price American options and therefore could not figure out what the optimal condition would say. Sometimes in practice, exercise decisions are probably made on the basis of some scenario for the underlying believed (or feared) by the option holder, not by averaging over all paths using complex logic. Real-

moves forward, asks at each t_ℓ^* whether holding or exercising the option is more profitable. The answer again involves boundary-condition information given by convention in the future. This information is propagated backward in time by the model equations from the future times where the boundary conditions are applicable to the time on the watch.

Consider a call option. Suppose that the stock price at t_ℓ^* satisfies $S(t_\ell^*) > E_\ell$, with E_ℓ the strike price at time t_ℓ^*. Then the option is in the money and so can be exercised profitably. Theoretically, this exercise may not be the most profitable. However, it is possible that the anticipated return from the collective possibility of future exercise at the later strike dates $\{t_j^*\}$ with $j > \ell$ is greater than profit equal to the intrinsic value $S(t_\ell^*) - E_\ell$ from exercising at time t_ℓ^*. We need to find the points at which there is no difference in value between these two possible courses of action. We next turn to this problem.

The Atlantic or Critical Path for Bermudas

In order to discuss this problem it is convenient to divide each x_ℓ^* axis into two sections, the ℓth "European" section and "American" section, separated by a point $\hat{x}_\ell$. For lognormal dynamics, $x_\ell^* = \ln(S(t_\ell^*))$. We will soon discuss American options defined by an infinite call or put schedule (i.e. $L \to \infty$), and the points $\{\hat{x}_\ell\}$ will merge to form a path $\hat{x}(t)$, which we shall call the "Atlantic path"[42]. The Atlantic path separates $x(t)$ space into the "American region" and the "European region" formed by the continuum of the American sections and European sections. For the discrete case, $x_\ell^* > \hat{x}_\ell$ for calls and $x_\ell^* < \hat{x}_\ell$ for puts defines the ℓth American section, where the option is exercised at time t_ℓ^*. The ℓth European section, where the option is held at least to time $t_{\ell+1}^*$, is $x_\ell^* < \hat{x}_\ell$ for calls and $x_\ell^* > \hat{x}_\ell$ for puts. The Atlantic path then consists of the discrete set of points $\{\hat{x}_\ell\}$. The idea is illustrated in Figs. 4A, 4B at the end of the chapter.

---

world financial aspects may enter as well. See the discussion of the Viacom CVR described in Ch. 13 for an example.

[42] **Why the name Atlantic Path?** This name is intuitive because this path lies midway between the American or Bermudan region where early exercise takes place and the European region where exercise does not yet occur. Usually this is called the critical path or the free boundary.

Path Classes

Leave aside the problem of actually getting the discrete Atlantic path $\{\hat{x}_\ell\}$ for a moment. Then it is easy to see that paths break up into classes. If exercise occurs at time t_ℓ^*, not before, clearly the paths must cross all x_j^* axes with $1 \leq j < \ell$ such that $x_j^* < \hat{x}_j$ for calls, $x_j^* > \hat{x}_j$ for puts. Since exercise at $t = t_\ell^*$ means the procedure stops at t_ℓ^*, no path continuation for $t > t_\ell^*$ is relevant. Between successive times (t_j^*, t_{j+1}^*) and before exercise, the paths are unrestricted. For this reason, the propagation between such successive times is free and is therefore effected with the free propagator $G_f(x_j^* - x_{j+1}^*; t_j^* - t_{j+1}^*)$ with $\mu = \mu_0 - D_0$. This is also true for $j = 0$ if we interpret $x_0 = x_0^*, t_0 = t_0^*$.

The idea is illustrated in Figs. 5 and 6 at the end of the chapter. Fig. 5 shows the path classes for a call schedule and Fig. 6 corresponds to a put schedule.

Contibutions from Path Classes for Bermuda Options

Hence, we write the option value at present time t_0, with present stock price $S(t_0) = \exp(x_0)$, as a sum over contributions from path classes labeled by ℓ,

$$C(x_0, t_0) = \sum_{\ell=1}^{L} C^{(\ell)}(x_0, t_0) \tag{42.47}$$

For a call option specified by a schedule with possible exercise at t_ℓ^*, we have[43]

$$C_{call}^{(\ell)}(x_0, t_0) = \int_{\hat{x}_\ell^{call}}^{\infty} G_{call}(x_0, x_\ell^*; t_0, t_\ell^*)\left[S(t_\ell^*) - E_\ell\right] dx_\ell^* \tag{42.48}$$

Here, the Green function G_{call} is built up by a product of free propagators $G_f^{(\mu)} = G_f \big|_{\mu=\mu_0-D_0}$ with drift $\mu = \mu_0 - D_0$. Thus,

$$G_{call}(x_0, x_\ell^*; t_0, t_\ell^*) = \prod_{j=1}^{\ell-1} \int_{-\infty}^{\hat{x}_j^{call}} dx_j^* \prod_{j=0}^{\ell-1} G_f^{(\mu)}(x_j^* - x_{j+1}^*; t_j^* - t_{j+1}^*) \tag{42.49}$$

---

[43] **Intrinsic Value Positivity:** The intrinsic value in the region of integration is positive.

Path Integrals and Options I: Introduction

This holds if $\ell > 1$ (for $\ell = 1$ no integrals are present).

For a put option specified by a schedule with possible exercise at t_ℓ^*, we have

$$C_{put}^{(\ell)}(x_0, t_0) = \int_{-\infty}^{\hat{x}_\ell^{put}} G_{put}(x_0, x_\ell^*; t_0, t_\ell^*) \left[E_\ell - S(t_\ell^*) \right] dx_\ell^* \tag{42.50}$$

Here for $\ell > 1$,

$$G_{put}(x_0, x_\ell^*; t_0, t_\ell^*) = \prod_{j=1}^{\ell-1} \int_{\hat{x}_j^{put}}^{\infty} dx_j^* \prod_{j=0}^{\ell-1} G_f^{(\mu)}(x_j^* - x_{j+1}^*; t_j^* - t_{j+1}^*) \tag{42.51}$$

In order to evaluate the discrete Atlantic path $\{\hat{x}_j\}$, we will need the option values at times t_{L-m}^* and points x_{L-m}^*, with $m = 1...L$. We have

$$C(x_{L-m}^*, t_{L-m}^*) = \sum_{\ell=L-m+1}^{L} C^{(\ell)}(x_{L-m}^*, t_{L-m}^*) \tag{42.52}$$

For a call,

$$C_{call}^{(\ell)}(x_{L-m}^*, t_{L-m}^*) = \int_{\hat{x}_\ell^{call}}^{\infty} G_{call}(x_{L-m}^*, x_\ell^*; t_{L-m}^*, t_\ell^*) \left[S(t_\ell^*) - E_\ell \right] dx_\ell^* \tag{42.53}$$

Here, for $\ell = L - m + 1, ..., L$ the call Green function is

$$G_{call}(x_{L-m}^*, x_\ell^*; t_{L-m}^*, t_\ell^*) = \prod_{j=L-m+1}^{\ell-1} \int_{-\infty}^{\hat{x}_j^{call}} dx_j^* \prod_{j=L-m}^{\ell-1} G_f^{(\mu)}(x_j^* - x_{j+1}^*; t_j^* - t_{j+1}^*) \tag{42.54}$$

For puts, similar equations hold, viz

$$C_{put}^{(\ell)}(x_{L-m}^*, t_{L-m}^*) = \int_{-\infty}^{\hat{x}_\ell^{put}} G_{put}(x_{L-m}^*, x_\ell^*; t_{L-m}^*, t_\ell^*) \left[E_\ell - S(t_\ell^*) \right] dx_\ell^* \tag{42.55}$$

Here, the Green function for puts is

$$G_{put}\left(x_{L-m}^{*},x_{\ell}^{*};t_{L-m}^{*},t_{\ell}^{*}\right) = \prod_{j=L-m+1}^{\ell-1} \int_{\widehat{x}_{j}^{put}}^{\infty} dx_{j}^{*} \prod_{j=L-m}^{\ell-1} G_{f}^{(\mu)}\left(x_{j}^{*}-x_{j+1}^{*};t_{j}^{*}-t_{j+1}^{*}\right) \quad (42.56)$$

Semigroup Properties for the Call and Put Green Functions for Bermudas

We note in passing that G_{call} and G_{put} satisfy different semigroup properties, written again for finite time intervals $(t_0 - t_1), (t_1 - t_2)$:

$$G_{call}(x_0, x_2; t_0, t_2) = \int_{-\infty}^{\widehat{x}_1^{call}} dx_1 G_{call}(x_0, x_1; t_0, t_1) G_{call}(x_1, x_2; t_1, t_2) \quad (42.57)$$

$$G_{put}(x_0, x_2; t_0, t_2) = \int_{\widehat{x}_1^{put}}^{\infty} dx_1 G_{put}(x_0, x_1; t_0, t_1) G_{put}(x_1, x_2; t_1, t_2) \quad (42.58)$$

Two generalizations of these results are possible. First, with appropriate integration limit changes, a mixed schedule of calls and puts can be accommodated. The $\{\widehat{x}_{\ell}\}$, which are different for calls and puts, would also change. Second, the general dividend case with $\mu(x,t)$ discussed earlier can evidently be accommodated by replacing each free propagator $G_{f}^{(\mu)}\left(x_{j}^{*}-x_{j+1}^{*};t_{j}^{*}-t_{j+1}^{*}\right)$ with the path integral $G_{D}\left(x_{j}^{*}-x_{j+1}^{*};t_{j}^{*}-t_{j+1}^{*}\right)$ of Eqn. (42.42). In using this equation, the index j on the right-hand side should be replaced by $j' = 1,...,n$ for the partition between times $\left(t_{j}^{*}, t_{j+1}^{*}\right)$ into n intervals in an obvious fashion.

Atlantic Critical Path Algorithm

We turn to the determination of the discrete Atlantic or critical path formed by the set of points $\{\widehat{x}_{\ell}^{call}\}$ for a call schedule and $\{\widehat{x}_{\ell}^{put}\}$ for a put schedule; here $\ell = 1,...,L$. This procedure is well known, but we present it anyway for completeness using this language. At time $t = t_{L}^{*}$, the last possible exercise date, $\widehat{x}_{L} = \ln(E_{L})$ for both calls and puts since no ambiguity is left to the option holder who holds his option to t_{L}^{*} (either the option is in the money or it isn't, and no further possible decisions exist).

Now consider $t = t^*_{L-1}$. An option holder who holds the option until this time will get $\pm\left[S\left(t^*_{L-1}\right) - E_{L-1}\right]$ for a call (+) or put (−) if the option is in the money and if he exercises. If he doesn't exercise, he keeps his option with value equal to $C^{(L)}\left(x^*_{L-1}, t^*_{L-1}\right)$ defined by Eqn. (42.53) or Eqn. (42.55) with $\ell = L$, equal to $C\left(x^*_{L-1}, t^*_{L-1}\right)$, which is the option value diffused back to t^*_{L-1} from t^*_L. The value of $\widehat{x}_{L-1}$ is defined as that value such that no difference results from exercising or not exercising at t^*_{L-1}. For a call this is

$$\exp\left(\widehat{x}^{call}_{L-1}\right) - E_{L-1} = C^{(L)}_{call}\left(\widehat{x}^{call}_{L-1}, t^*_{L-1}\right) > 0 \tag{42.59}$$

For a put we have the corresponding condition:

$$E_{L-1} - \exp\left(\widehat{x}^{put}_{L-1}\right) = C^{(L)}_{put}\left(\widehat{x}^{put}_{L-1}, t^*_{L-1}\right) > 0 \tag{42.60}$$

This process is continued. At t^*_{L-2}, the option holder may expect to get $C\left(x^*_{L-2}, t^*_{L-2}\right)$, the sum of two terms from the anticipated possibility of exercising at times t^*_{L-1} ($\ell = L-1$) or t^*_L ($\ell = L$). When the positive quantity $C\left(x^*_{L-2}, t^*_{L-2}\right)$ is set equal to the option intrinsic value at t^*_{L-2} as in Eqn. (42.59) or (42.60), $\widehat{x}_{L-2}$ is determined. In general, $\widehat{x}_{L-m}$ is determined by setting the quantity $C\left(x^*_{L-m}, t^*_{L-m}\right)$ equal to the option intrinsic value at t^*_{L-m}. All paths start in the American sectors of the $x^*_{L-m+\ell}$ axes and cross intermediate x^* —axes back to x^*_{L-m} in their European sectors. In this way, all the parameters of the discrete Atlantic path $\left\{\widehat{x}^{call}_\ell\right\}$ for a call or $\left\{\widehat{x}^{put}_\ell\right\}$ for a put are determined. Again, the idea is illustrated in Figs. 5,6 at the end of this chapter.

This completes the determination of the discrete Atlantic path for a call schedule or a put schedule. Some options have mixed conditions—at some times the option decision is a call and at other times a put. For such a mixed call/put schedule, careful tracking of the various American and European sectors of the x^* -axes must be done to get the Atlantic path points $\left\{\widehat{x}_\ell\right\}$.

Expression for a Bermuda Put Option

To close this section, we exhibit the form of the put option price including L puts in a schedule, assuming lognormal diffusion and a dividend yield.

The case $L = 3$ for puts was written down by Geske and Johnson [ix], as we describe below. They also saw that the American put option would be the $L \to \infty$ limit with $E_\ell = E$, a constant American put option strike price.

We use constant effective drift $\mu = \mu_0 - D_0$ and volatility σ_0. As derived above, the put option price breaks up into a sum of terms. For simplicity, we set $\tau = t^*_{j+1} - t^*_j$ equal to a constant. Then the ℓth put term in Eqn. (42.50) has the explicit form[43]

$$C^{(\ell)}_{put}(x_0, t_0) = e^{-r_0 \ell \tau} \prod_{j=1}^{\ell-1} \int_{\hat{x}^{put}_j}^{\infty} dx^*_j \qquad (42.61)$$

$$\cdot \int_{-\infty}^{\hat{x}^{put}_\ell} dx^*_\ell \frac{\left[E_\ell - S(t^*_\ell)\right]}{(2\pi\sigma_0^2\tau)^{\ell/2}} \exp\left[-\sum_{j=0}^{\ell-1} \frac{(x^*_{j+1} - x^*_j - \mu\tau)^2}{2\sigma_0^2\tau}\right]$$

Now make the changes of variables to $y_j(x^*_j, \mu) = (x^*_j - x_0 - j\mu\tau)/\sqrt{j\sigma_0^2\tau}$ for $j = 1, 2, \ldots, \ell$. There is no interlocking of limits in the variables of integration since x_0 is fixed. We find

$$C^{(\ell)}_{put}(x_0, t_0) = e^{-r_0 \ell \tau} E_\ell I_\ell\left(\{\hat{y}^{put}_j\}\right) - e^{-D_0 \ell \tau} S(t_0) I_\ell\left(\{\hat{y}^{put}_{j+}\}\right) \qquad (42.62)$$

Here $\hat{y}^{put}_j = y_j(x^*_j = \hat{x}^{put}_\ell, \mu_D)$, and $\hat{y}^{put}_{j+} = y_j(x^*_j = \hat{x}^{put}_\ell, \hat{\mu}_D)$. As before for dividend yields, $\mu_D = \mu_0 - D_0 = r_0 - D_0 - \tfrac{1}{2}\sigma_0^2$ and $\hat{\mu}_D = \mu_D + \sigma_0^2$. We also have defined the integral

$$I_\ell\left(\{\hat{y}^{put}_j\}\right) = \prod_{j=1}^{\ell-1} \int_{\hat{y}^{put}_j}^{\infty} dy_j \sqrt{\frac{j}{2\pi}} \int_{-\infty}^{\hat{y}^{put}_\ell} dy_\ell \sqrt{\frac{\ell}{2\pi}}$$

$$\cdot \exp\left\{-\frac{\ell y_\ell^2}{2} - \sum_{j=1}^{\ell-1}\left[jy_j^2 + \sqrt{j(j+1)}\, y_j y_{j+1}\right]\right\} \qquad (42.63)$$

In addition, we have used the identity

$$\exp\left\{x_\ell^* - \sum_{j=0}^{\ell-1}\frac{\left[x_{j+1}^* - x_j^* - \mu_D \tau\right]^2}{2\sigma_0^2 \tau}\right\} =$$

$$\exp\left\{x_0 - (r_0 - D_0)\ell\tau - \sum_{j=0}^{\ell-1}\frac{\left[x_{j+1}^* - x_j^* - \hat{\mu}_D \tau\right]^2}{2\sigma_0^2 \tau}\right\}$$

(42.64)

The integrals of Eqn. (42.62) can be related to the multivariate integral notation by identification of variables. For $\ell = 3$, the trivariate integral appears as related to our quantity I_3. Denoting the trivariate integral as $N_3(h, k, j; \rho_{12}, \rho_{13}, \rho_{23})$ we have

$$I_3\left(\hat{y}_1^{put}, \hat{y}_2^{put}, \hat{y}_3^{put}\right) = N_3\left(-\hat{y}_1^{put}, -\hat{y}_2^{put}, -\hat{y}_3^{put}; \tfrac{1}{\sqrt{2}}, -\tfrac{1}{\sqrt{3}}, -\sqrt{\tfrac{2}{3}}\right) \quad (42.65)$$

The multivariate notation is potentially confusing, since only correlations between nearest neighbor co-ordinate variables actually exist in the propagators by construction. However, the notation for the trivariate integral seems to suggest the existence of a next-to-nearest-neighbor correlation ρ_{13} between variables 1 and 3. In fact, the y_1, y_3 terms cancel out. This completes the discussion of discrete-schedule Bermuda options. We turn next to American options.

American Options

The American call or put option, as mentioned above, is just the infinite L limit of a call-schedule option or a put-schedule option. All the work has been done. Integrals over all the $\{\hat{x}_\ell\}$ become integrals over $x^*(t)$ for all t, and since these are dummy integration variables, we may remove the star $*$ and write integrals over all $x(t)$, i.e. over all values of x at each t. The Atlantic path $\hat{x}(t)$ is formed by the continuous limit of the $\{\hat{x}_\ell\}$ as the difference between the decision times goes to zero, $t_\ell^* - t_{\ell+1}^* \to 0$. Now propagation in the European region means integration over $x(t)$ such that $x(t) < \hat{x}(t)^{call}$ for a call, $x(t) > \hat{x}(t)^{put}$ for a put.

The reader may wonder about the propagation between $\left(t_\ell^*, t_{\ell+1}^*\right)$ in the continuous limit. For the discrete case, $x(t)$ was *unrestricted* for $t_\ell^* < t < t_{\ell+1}^*$. Now a path $x(t)$ must stay *restricted* to the European region except at its starting point. The resolution is that the excursions of a path $x(t)$ away from European sections between $\left(t_\ell^*, t_{\ell+1}^*\right)$ are suppressed by the Gaussian propagator to a greater and greater extent as $t_\ell^* - t_{\ell+1}^* \to 0$. This is because there is no time for the path to perform a random walk away from the European section at one decision time and still get back to the European section at the next decision time. Finally, the discrete numbers E_ℓ are replaced by E, the American option strike if constant or by the appropriate time-dependent strike price $E(t)$.

Appendix A: Girsanov's Theorem and Path Integrals

In this Appendix, we give two straightforward and related derivations of Girsanov's theorem using path integrals [xv].

First Derivation of Girsanov's Theorem

The essential point is contained in the simple remark that a shift in the drift of a Gaussian produces the original Gaussian multiplied by some factors. Thus, consider the typical Green function used throughout this paper for propagation with drift $\mu = \mu[x(t),t]$ and volatility $\sigma = \sigma[x(t),t]$ in an infinitesimal negative time interval $\Delta t = t - t'$ with space displacement equal to $x - x' = x(t) - x(t - \Delta t)$,

$$G^{(\mu)}(x,x';t,t') = \frac{1}{\left[-2\pi\sigma^2 \Delta t\right]^{1/2}} \exp\left[-\frac{(x-x'-\mu\Delta t)^2}{-2\sigma^2 \Delta t}\right] \quad (42.66)$$

We have indicated the dependence of G on μ explicitly by the superscript. Aside from technical points of convergence of integrals, the exact dependence of μ and σ on $x(t)$ and t is general and irrelevant to the discussion.

Consider the path-integral expectation-value of some arbitrary quantity $Y\left[\{x(t),t\}\right]$ with respect to $G^{(\mu)}$ for times between t_a and t_b, defined as the $\Delta t \to 0$ limit of

Path Integrals and Options I: Introduction

$$\langle Y[\{x(t),t\}]\rangle_{G^\mu} = \prod_{t=t_a}^{t_b} \int_{-\infty}^{\infty} dx(t) G^{(\mu)}\left[x(t), x(t-\Delta t); t, t-\Delta t\right] Y[\{x(t),t\}] \quad (42.67)$$

E.g., Y could be a discount factor, or an option payout depending on $x(t)$.

Now suppose we rewrite μ as consisting of two pieces,

$$\mu = \mu_0 + \mu_1 \quad (42.68)$$

Here, μ_0 and μ_1 may depend on $x(t)$ and t. A little algebra produces

$$G^{(\mu)}(x,x';t,t') = \exp\left[-\frac{\mu_1}{\sigma^2}(x-x'-\mu_0\Delta t) + \frac{\mu_1^2 \Delta t}{2\sigma^2}\right] G^{(\mu_0)}(x,x';t,t') \quad (42.69)$$

Here, $G^{(\mu_0)}$ is the same Gaussian as $G^{(\mu)}$ but with μ replaced by μ_0.

Now the interpretation of $G^{(\mu)}$ as the infinitesimal-time diffusion propagator is consistent with the statement that the stochastic equation for $x(t)$ is

$$x(t) - x(t-\Delta t) - \mu[x(t),t]\Delta t = -\sigma[x(t),t]\Delta z(t) \quad (42.70)$$

with $\Delta z(t)$ a Gaussian or Wiener random variable.

On the other hand, the stochastic equation producing the diffusion equation for which $G^{(\mu_0)}$ is the infinitesimal propagator is

$$\tilde{x}(t) - \tilde{x}(t-\Delta t) - \mu_0[\tilde{x}(t),t]\Delta t = -\sigma[\tilde{x}(t),t]\Delta z(t) \quad (42.71)$$

Here we have put a tilde over the variable x as a label to indicate that the drift is μ_0 rather than μ. Actually, since both $x(t)$ and $\tilde{x}(t)$ serve only as dummy integration variables, this label is not really needed.

In the integral Eqn. (42.67) we may change variables from $\{x(t)\}$ to $\{\tilde{x}(t)\}$. In performing this change of variables, we need to keep the probability density weight for a given path unchanged. That is, for a given path specified at times $\{t_j\}$ by numbers $\{x_j^{(path)}\}$, the integrands of the original integral at

$x_j = x_j^{(path)}$ and the transformed integral at $\tilde{x}_j = x_j^{(path)}$ must have the same value. This is clearly satisfied when we insert Eqn. (42.69) for each infinitesimal propagator. We may then use the stochastic equation for $\tilde{x}(t)$ in Eqn. (42.71). We denote the positive infinitesimal time step $dt = -\Delta t$, take the limit as $dt \to 0$, and call $dz = \Delta z$. Denote the drift function $\tilde{\mu}_1 = \mu_1[\tilde{x}(t), t]$ and the volatility function $\tilde{\sigma} = \sigma[\tilde{x}(t), t]$. We obtain

$$\langle Y[\{x(t),t\}]\rangle_{G^{(\mu)}} = \prod_{t=t_a}^{t_b} \int_{-\infty}^{\infty} d\tilde{x}(t) \exp\left[\frac{\tilde{\mu}_1}{\tilde{\sigma}} dz(t) - \frac{\tilde{\mu}_1^2}{2\tilde{\sigma}^2} dt\right] \quad (42.72)$$
$$\cdot G^{(\mu_0)}[\tilde{x}(t), \tilde{x}(t+dt); t, t+dt] Y[\{x(t),t\}]$$

$$\equiv \left\langle \exp\left[\int_{t_a}^{t_b} \frac{\tilde{\mu}_1}{\tilde{\sigma}} dz(t) - \int_{t_a}^{t_b} \frac{\tilde{\mu}_1^2 dt}{2\tilde{\sigma}^2}\right] Y[\{\tilde{x}(t),t\}] \right\rangle_{G^{(\mu_0)}} \quad (42.73)$$

Eqn. (42.73) is the desired result, Girsanov's theorem.

A Second, Quicker Derivation of Girsanov's Theorem

A more succinct derivation uses the stochastic equations as delta-function constraints in the path integral, as described in the text. We just write down the path integral expectation including the stochastic equation as a constraint twice.

- First, use the full drift μ.
- Second, use only part of the drift μ_0, but include an extra factor $\exp\{\xi[\tilde{x}(t), t]\}$. This extra factor is chosen to force the identity of these two procedures.

Carrying out the two steps produces the identity

$$\langle Y[\{x(t),t\}]\rangle_{G^{(\mu)}} = \lim_{dt \to 0} \prod_{t=t_a}^{t_b} \int_{-\infty}^{\infty} dx(t) \int_{-\infty}^{\infty} d\eta(t) \frac{1}{[2\pi\sigma^2 dt]^{1/2}} \exp\left[-\tfrac{1}{2}\eta^2(t) dt\right]$$
$$\cdot \delta\left\{\eta(t) - \frac{1}{\sigma}\left[\frac{dx(t)}{dt} - \mu\right]\right\} Y[\{x(t),t\}] \quad (42.74)$$

Path Integrals and Options I: Introduction

$$\langle Y[\{x(t),t\}]\rangle_{G^{(\mu)}} = \lim_{dt\to 0}\prod_{t=t_a}^{t_b}\int_{-\infty}^{\infty}d\tilde{x}(t)\int_{-\infty}^{\infty}d\eta(t)\frac{1}{\left[2\pi\tilde{\sigma}^2 dt\right]^{1/2}}\exp\left[-\tfrac{1}{2}\eta^2(t)dt\right]$$

$$\cdot\delta\left\{\eta(t)-\frac{1}{\tilde{\sigma}}\left[\frac{d\tilde{x}(t)}{dt}-\mu_0\right]\right\}Y[\{\tilde{x}(t),t\}]\cdot\exp\{\xi[\tilde{x}(t),t]\} \quad (42.75)$$

Since the Dirac delta functions just eliminate the integrals over the $\{\eta(t)\}$ variables, it is easy to find this factor e^ξ just by comparing the exponents in the integrands. The answer is clearly just the same as found in Eqn. (42.73), again reproducing Girsanov's theorem.

Rewriting $\langle Y\rangle = E(Y)$ produces another standard notation. Identification of μ_0, μ_1, and Y in various circumstances provides the explicit connection between the notation used in this chapter and other notations.

For $Y = 1$ we just recover the path integral for the finite-time propagator.

Appendix B: No-Arbitrage, Hedging and Path Integrals

This appendix[44] deals with no arbitrage and hedging in the framework of path integrals. It is important to understand that the usual no-arbitrage conditions are directly implied by the path integral formalism when coupled with the standard hedging recipes. The basic reason is that the Green function or path integral is itself the general solution of the diffusion equation. The diffusion equation and its boundary and terminal conditions determine everything about an option. Therefore, any statement about no-arbitrage or market consistency of the option must be possible by adjustment of suitable parameters in the path integral.

Although equivalent, the procedure here is somewhat different from the textbook procedure. We first construct the Green function directly. The form of the Green function does *not* need any hedging argument, just the stochastic equations and their Gaussian nature. Then, through a suitable choice of the drift parameter in the Green function along with the usual assumption of standard Δ-hedging, we arrive at the no-arbitrage result. Hence, there is consistency, though perhaps this approach can lead to a somewhat modified philosophy.

---

[44] **History:** The next two appendices were added for the book; they were not in my 1988 paper (Ref i).

For illustration[45], consider the simple zero-dividend stock option with lognormal dynamics using the variable $x = \ln S$. The Green function probability density measure $G_{0*}dx^*$ for the transition from (x_0, t_0) to (x^*, t^*) in dx^* with $T_{0*} = t^* - t_0$ is the familiar quantity

$$G_{0*}dx^* \equiv G(x_0, x^*; t_0, t^*)dx^* = \frac{\Theta(T_{0*})e^{-r_{0*}T_{0*}}}{\sqrt{2\pi\sigma_{0*}^2 T_{0*}}} \exp[-\Phi_{0*}]dx^* \quad (42.76)$$

Here,

$$\Phi_{0*} = \left[x^* - x_0 - \mu_{0*}T_{0*}\right]^2 / \left[2\sigma_{0*}^2 T_{0*}\right] \quad (42.77)$$

A European option $C_0 \equiv C(S_0, t_0)$ with terminal value $C(S^*, t^*)$ is

$$C_0 \equiv C(S_0, t_0) = \int_{-\infty}^{\infty} G(x_0, x^*; t_0, t^*) C(S^*, t^*) dx^* \quad (42.78)$$

The Path Integral and the Green Function G_{0*}

It is important to realize that the Green function G_{0*} arises straightforwardly from Gaussian integration and the usual stochastic equations. For convenience, we again summarize the procedure. We partition the time $T_{0*} = t^* - t_0$ by $N+1$ points t_j, and we set $t_0 = t_{j=0}$, $t^* = t_{j=N}$, and $x_j = \ln(S_j)$. The Gaussian integrals are over N independent Gaussian variables η_j ($j = 0, ..., N-1$) with width $1/dt_j$, which simply defines the probability measure. The stochastic equation of motion is $d_t x_j = x_{j+1} - x_j = \left(\mu_j^{(S)} + \sigma_j \eta_j\right) dt_j$. Here, $\mu_j^{(S)}$, the physical drift at time t_j, is not the no-arbitrage drift and will be eliminated.

We insert a Dirac delta function constraint arising from the stochastic equation in each η_j integral[46], viz

---

[45] **Generality of the No-Arbitrage and Green Function/Path Integral Method:** The method is general. For no-arbitrage in fixed income curve construction using path integrals, see the next chapter.

[46] **Dirac Delta Function:** To remind the reader: the delta function δ(w) is a Schwartz distribution, equal to 0 except at its support (w = 0) where it equals infinity, and has an integral of 1 when integrated over an interval including its support.

Path Integrals and Options I: Introduction

$$1 = \int_{-\infty}^{\infty} dx_{j+1} \delta\left[x_{j+1} - x_j - \left(\mu_j^{(S)} + \sigma_j \eta_j\right) dt_j \right] \quad (42.79)$$

We rewrite the δ-function to eliminate each η_j in favor of dx_{j+1} and perform the Gaussian integrations over the x_{j+1} variables.

We also insert the product of the discount factors $\prod \exp(-r_j dt_j)$ needed to properly discount cash flows[47]. The formula above for G_{0*} then emerges in a straightforward fashion. It is a function of the total variance $\sigma_{0*}^2 T_{0*} = \sum_{j=0}^{n-1} \sigma_j^2 dt_j$, the total drift $\mu_{0*} T_{0*} = \sum_{j=0}^{n-1} \mu_j^{(S)} dt_j$ and also $r_{0*} T_{0*} = \sum_{j=0}^{n-1} r_j dt_j$. Ultimately, we take the continuum limit $dt_j \to 0$. We take this after physical quantities have been calculated[48]. This makes everything well defined.

The Elimination of the Drift $\mu^{(S)}$ for the No Arbitrage Drift μ

So far, we have been working with $\mu^{(S)}$. This is actual future drift of $\ln S$, and is not related to other parameters. However, this is not what we want for the final answer. For financial reasons, we want to impose a market no-arbitrage constraint. We can easily do this. We remove the stock drift $\mu^{(S)}$ from G_{0*} and in its place simply insert a drift parameter μ that we specify to satisfy the no arbitrage constraint[49]. We also need make the usual assumption of Δ–hedging.

---

[47] **Discounting:** To get the usual formalism, we use continuous discounting.

[48] **Continuum Limit Again: Is it Relevant?** The continuum limit is a fiction. All sorts of time scales exist in trading and in numerical calculations that require partitioning of the time axis. A good part of this book is related to dealing with time scales.

[49] **More about the "Fictitious World of No Arbitrage" and The Real World:** Sometimes the replacement of the parameter μ for the physical log-stock drift $\mu^{(S)}$ is said to place the calculation in a fictitious world. Our philosophy is more mundane. We merely state that, since no one would agree on the stock drift anyway, markets are facilitated by normalizing the calculation to produce a financial result that everyone does agree on – namely the no arbitrage condition. Still, the motivation of customers and traders for transactions with options live very much in the real world where people have their own favorite scenarios for the future average stock behavior $\mu^{(S)}$, which may differ markedly from the no-arbitrage drift. The same philosophy drives transactions in fixed-

The hedging assumption is separately imposed, as in the standard textbook argument. The result is the removal of the uncertainty in the return for a portfolio of stock and an option. We next give the details

No-Arbitrage, Hedging and Path Integrals: Result

The no-arbitrage condition and the standard hedging results at t_0 for a stock option are equivalent to the following statements[50]:

1. Replace the log-stock drift $\mu_0^{(S)}$ with the parameter $\mu_0 = r_0 - \sigma_0^2/2$
2. Hedge the option with the usual $N_S = -\Delta$ shares, where $\Delta = \dfrac{\partial C_0}{\partial S_0}$

Proof. The Green function G_{0*} satisfies a diffusion equation[51]. Because the option C_0 is a given by a suitably convergent integral over G_{0*}, differentiation can be interchanged with integration, and so C_0 satisfies the same diffusion equation as does G_{0*}, namely:

$$\frac{\sigma_0^2}{2}\frac{\partial^2 C_0}{\partial x_0^2} + \mu_0 \frac{\partial C_0}{\partial x_0} + \frac{\partial C_0}{\partial t_0} - r_0 C_0 = 0 \qquad (42.80)$$

Construct the usual portfolio V of N_S shares and an option. The change dV of the portfolio over time dt_0 is $V(S_1,t_1) - V(S_0,t_0)$. Here the initial portfolio is $V(S_0,t_0) = N_S S_0 + C_0$. Also, $V(S_1,t_1) = N_S S_1 + C_1$ is final portfolio is with S_1 the stochastic variable given by its stochastic equation.

The option value $C_1 = C(S_1,t_1)$ is the expectation integral using the Green function $G_{1*} = G(x_1, x^*; t_1, t^*)$. Here G_{1*} is G_{0*} with the initial step from t_0 to t_1 removed. We have

---

income markets. If scenarios for the real behavior of interest rates in the future differ markedly from the expected behavior of interest rates produced by no-arbitrage, people will do transactions to try to capitalize on their views.

[50] **No-Arbitrage at Other Times:** Similar replacements of the μ_j are made at t_j to ensure no-arbitrage at other times.

[51] **Which Diffusion Equation?** The equation is with respect to the initial variables S_0, t_0 and so is actually the backward Kolmogorov equation.

Path Integrals and Options I: Introduction

$$C_1 \equiv C(S_1, t_1) = \int_{-\infty}^{\infty} G(x_1, x^*; t_1, t^*) C(S^*, t^*) dx^* \qquad (42.81)$$

We get the relation between the Green functions G_{1*} and G_{0*} good to $O(dt_0)$ by straightforward algebra:

$$G(x_1, x^*; t_1, t^*) = \left\{ 1 + (S_1 - S_0) \partial_{S_0} + dt_0 \partial_{t_0} + \tfrac{1}{2}(S_1 - S_0)^2 \partial^2_{S_0 S_0} \right\} G(x_0, x^*; t_0, t^*)$$

$$(42.82)$$

We need to interpret $(S_1 - S_0)^2$ as its expected value, namely $(S_1 - S_0)^2 = S_0^2 \sigma_0^2 \eta_0^2 dt_0^2 = S_0^2 \sigma_0^2 dt_0$. To see this, we simply use the expectation value of η_0^2 over its Gaussian measure with width $1/dt_0$, obtaining[52] $\langle \eta_0^2 \rangle = 1/dt_0$. This is equivalent to the result from the Brownian motion of all paths from (S_0, t_0) to (S_1, t_1), using an arbitrarily fine partition inside the small interval dt_0.

We get the change dV over time dt_0 as

$$\frac{dV}{dt_0} = \left[N_S + \frac{\partial C_0}{\partial S_0} \right] \frac{(S_1 - S_0)}{dt_0} - S_0 \frac{\partial C_0}{\partial S_0} \left(\mu_0 + \tfrac{1}{2}\sigma_0^2 \right) + r_0 C_0 \qquad (42.83)$$

The hedging relation $N_S = -\partial C_0 / \partial S_0$ makes the first bracket disappear and eliminates the dependence of V on the stochastic behavior of the stock price S_1 at t_1. Using the constraint $\mu_0 = r_0 - \tfrac{1}{2}\sigma_0^2$ produces $r_0 N_S S_0$ for the second term. The standard no-arbitrage condition emerges, i.e. the return of the portfolio V of the option and the proper stock hedge is the risk-free rate, viz

$$\frac{dV}{dt_0} = r_0 V \qquad (42.84)$$

---

[52] **Relation to Ito's Lemma:** This statement is equivalent to Ito's lemma. The variable η used here is related to the usual Wiener stochastic variable dz with $dz^2 = dt$ by the relation $\eta = dz/dt$, where we discretize first and then let dt $\to$ 0 after all physical quantities are calculated. It is less confusing to write $d\eta$ than to write ddz.

Appendix C: Perturbation Theory, Local Volatility, Skew

This appendix briefly describes formalism with a local volatility $\sigma(x)$ using perturbation theory [53, xix]. We think this may be useful in skew calculations.

Call the velocity $u(x,t) = [x(t+dt) - x(t)]/dt$. Recalling Eqn. (42.20), write the Lagrangian $\mathcal{L}(x,t) = [u(t) + \frac{1}{2}\sigma^2(x) - r]^2 / [2\sigma^2(x)]$, and call the unperturbed Lagrangian $\mathcal{L}_0 = \mathcal{L}[\sigma(x) \to \sigma_0]$ with σ_0 constant. The path integral Green function is $G^{ab} = \int\limits_{\substack{\text{Paths with} \\ x(t_a)=x_a,\, x(t_b)=t_b}} [Dx(t)] \exp\left[\int_{t_a}^{t_b} \mathcal{L}(x,t)dt\right]$.

Next, define the "potential" $V(x,t)$ via $\mathcal{L} = \mathcal{L}_0 - V$. The perturbation expansion is constructed using $\exp\left[\int_{t_a}^{t_b} V(x,t)dt\right] = \sum_{m=0}^{\infty} \frac{1}{m!}\left[\int_{t_a}^{t_b} V(x,t)dt\right]^m$. The Born approximation, given by the $m = 0, 1$ terms, is $G^{ab} = G_f^{ab} + G_{Born}^{ab}$. Here,

$G_f^{ab} = G^{ab}[\mathcal{L}_0]$ and $G_{Born}^{ab} = \int_{t_a}^{t_b} dt_c \int_{-\infty}^{\infty} dx_c\, G_f^{ac} V(x_c) G_f^{cb}$.

If $\sigma^2(x) = \sigma_0^2/[1 + g^2(x-\tilde{x})^2]$, volatility decreases for increasing x. Expansion for small g gives analytic results. Naturally other forms are possible.

The perturbation expansion is singular, because V is velocity dependent, requiring the Schwinger formalism[iii] for the discretization specification. Integrations by parts are needed. There is a "mass renormalization" involving counterterms [54]. All $1/dt$ terms cancel, producing finite results as $dt \to 0$.

Figure Captions for this Chapter

Fig. 1: A path in co-ordinate x-space, running backward from the final point x^* at time t^*, through intermediate steps to the present time t_0 at x_0. Propagation over each infinitesimal time step $\Delta t \approx 0$ is accomplished by convolution with

---

[53] **History:** Andy Davidson had the idea for the local volatility σ(x) in the context of mortgages in 1986. The V(x,t) that resulted was called the "Mortgage Potential".

[54] **Mass Renormalization:** This is the only finance example I know where this somewhat advanced path integral technique had to be employed.

the "free propagator" Green function G_f, a solution to the underlying diffusion equation for infinitesimal Δt. Multiplication of successive propagators for a given path produces the probability density weight for that path. The path integral is obtained formally in the limit $n \to \infty$. Numerical analysis employs finite n.

Fig. 2: The semi-group or reproducing kernel property for the free propagator corresponding to one-dimensional free diffusion. The significance is that the same functional form holds for the propagator between any two times. This property is valid in general for the path integral for an arbitrary diffusion process, and it holds in any number of spatial dimensions.

Fig. 3: Illustration of the semigroup proof for the path integral denoted here by G_D in one dimension.

Fig. 4a illustrates the geometry for a call option schedule. The European sector at each strike date (where the option is not exercised) is separated from the American sector at that strike date (where the option is exercised) by the point on the discrete Atlantic path at that strike date. The strike prices are also shown assuming lognormal diffusion, although the construction itself is more general.

Fig. 4b illustrates the same for a put option schedule. The American option is reached formally in the limit as the number of strike dates becomes infinite. In that limit the Atlantic path becomes continuous.

Fig. 5: The path classes for pricing a call option with a schedule containing L exercise or strike dates. The paths are only generically illustrated; they must lie below the call-Atlantic path points $\hat{x}_\ell^{call}$ at the exercise dates t_ℓ^* on the schedule, but between exercise dates, they are unconstrained.

Fig. 6: The path classes for pricing a put option with a schedule. The paths must lie above the put-Atlantic path points at the strike dates on the schedule, but between strike dates, they are unconstrained.

Fig. 7: Illustration of the path integral in two dimensions. See Ch. 45 for the mathematical formalism of the path integral in multiple dimensions.

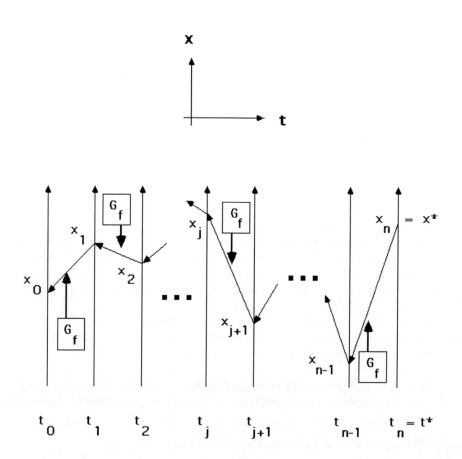

FIG. 1

Path Integrals and Options I: Introduction

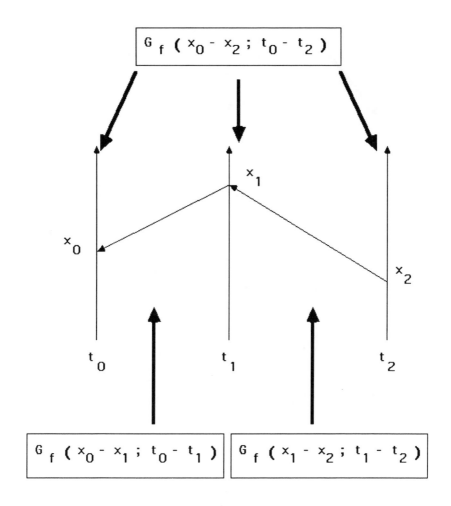

FIG. 2

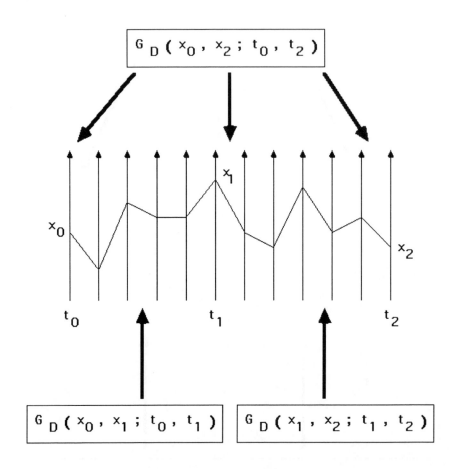

FIG. 3

Path Integrals and Options I: Introduction

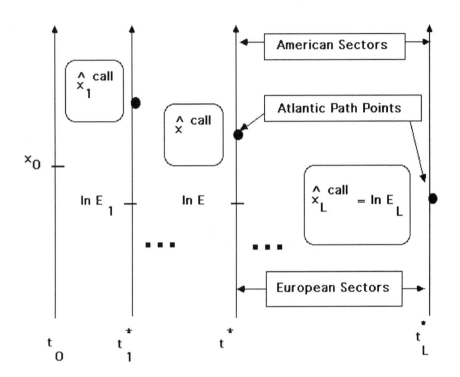

FIG. 4A

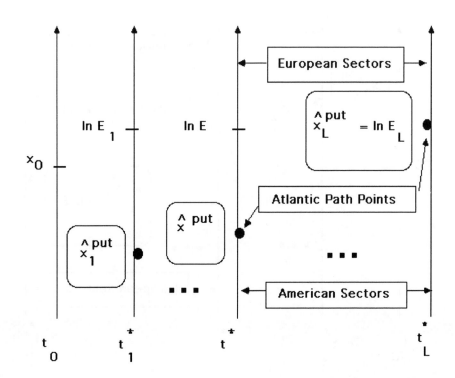

FIG. 4B

Path Integrals and Options I: Introduction 553

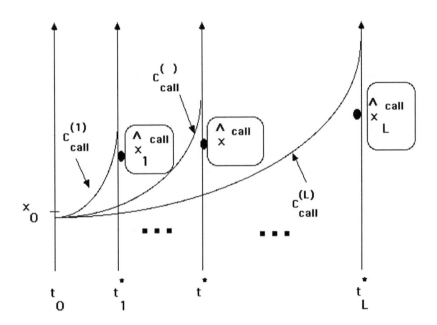

FIG. 5

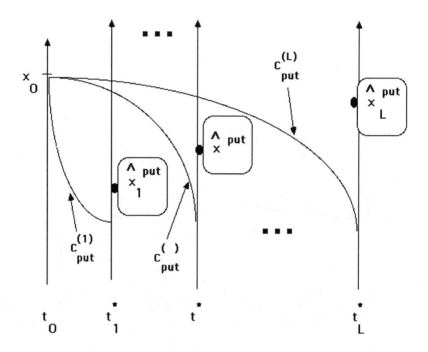

FIG. 6

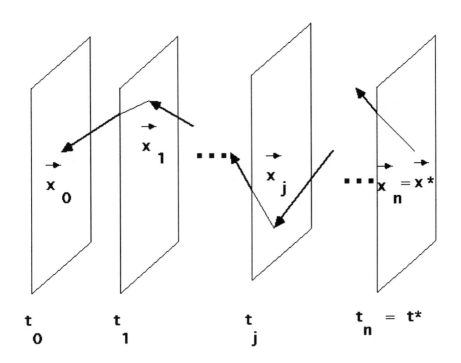

FIG. 7

References

[i] Path Integrals and Finance – Basic References
Dash, J. W., *Path Integrals and Options – I*, Centre de Physique Théorique, CNRS, Marseille preprint CPT-88/PE.2206, 1988.
Dash, J. W., *Path Integrals and Options II: One-Factor Term-Structure Models*, Centre de Physique Théorique, CNRS, Marseille preprint CPT-89/PE.2333, 1989.

[ii] More Path Integrals and Finance References
Linetsky, V., *The Path Integral Approach to Financial Modeling and Options Pricing*. Univ. Michigan College of Engineering report 96-7, 1996.
Otto, M., *Using Path Integrals to Price Interest Rate Derivatives*. U. Göttingen working paper, 1999.
Baaquie, B. E., *A path integral approach to option pricing with stochastic volatility: some exact results*. J. Phys. I France, Vol 7, 1997. Pp. 1733-1753.
Makivic, M. S., *Path integral Monte Carlo method for valuation of deriative securities: algorithm and parallel implemenation*. NPAC Tech. Report SCCS 650, 1995.
Eydeland, A., *A fast algorithm for computing integrals in function spaces: Financial applications*. Comp. Economics Vol 7, 1994. Pp. 277-407.
Chiarella, C. and El-Hassan, N., *Evaluation of derivative security prices in the Heath-Jarrow-Morton framework as path integrals using frast fourier transform techniques*. J. Fin. Engineering, Vol 6 (2), 1997.
Chiarella, C., El-Hassan, N., and Kucera, A., *Evaluation of American option prices in a path integral framework using Fourier-Hermite series expansions*. J. Economic Dynamics & Control, 1999.
Chiarella, C., El-Hassan, N., and Kucera, A., *Evaluation of point parrier options in a path integral framework*. U. Technology, Sydney working paper, 1999.
Benneti, E., Rosa-Clot, M. and Taddei, S., *A Path Integral Approach to Derivative Security Pricing: I. Formalism and Analytical Results*. U. Firenze, 1999 (arXiv: cond-mat/9901277); Int. J. Theor. App. Fin., Vol 2, 1999. Pp. 381-407.
Rosa-Clot, M. and Taddei, S., *A Path Integral Approach to Derivative Security Pricing: II. Numerical methods*, arXiv: cond-mat/9901279, 1999.
Matacz, A., *Path Dependent Option Pricing: the path integral partial averaging method*. U. Sydney working paper, 1999.

[iii] Path Integrals - Basic References
Feynman, R. P., *Space-Time Approach to Non-Relativistic Quantum Mechanics*. Reviews of Modern Physics, Vol. 20, 1948. Pp. 367-387.
Feynman, R. P. and Hibbs, A. R., *Quantum Mechanics and Path Integrals*. McGraw-Hill, 1965. See P. 94 and also Sect. 7-3.

[iv] Path Integrals and Eigenfunction Decompositions
Kac, M., *On Distributions of Certain Wiener Functionals*. Trans. Amer. Math. Soc., Vol. 65, 1949. Pp. 1-13.
Kac, M., *On Some Connections Between Probability Theory and Differential and Integral Equations*, Proc. 2nd Berkeley Symp.on Math. Stat. Prob, Univ. Calif. Press, 1951. Pp. 189-215.

[v] Path Integral Applications - Examples
Feynman, R. P., *Statistical Mechanics*. W. A. Benjamin. 1972.

Dashen, R., *Path Integrals for Waves in Random Media*. J. Math. Phys. 20, 1979. Pp. 894-920.
Williamson, I. P. , *Pulse Broadening due to Multipath Propagation of Radiation*. Proc. R. Soc. London, Vol. A342, 1975. Pp. 131 ff.
Khandekarand, D. C. and Lawande, S. V., *Feynman Path Integrals: Some Exact Results and Applications*. Physics Reports, Vol. 137, North Holland, 1986. Pp. 116-229.
Schulman, *Techniques and Applications of Path Integration*. J. Wiley & Sons, 1981.

[vi] Path Integrals and Diffusion

Einstein, A., *On the Movement of Small Particles Suspended in a Stationary Liquid Demanded By the Molecular Kinetic Theory of Heat*. Annalen der Physik., vol 17, 1905. P. 549; *Investigations on the Theory of the Brownian Movement*, Dover 1956.
Langouche, F. , Roekaerts, D. and Tirapegni, E., *Short Derivation of Feynman Lagrangian for General Diffusion Processes*. J. Phys. A, Vol. 13, 1980. Pp. 449-452.
Morse, P. M. and Feshbach, H. *Methods of Theoretical Physics,*. I. McGraw-Hill. 1953.

[vii] Wiener Integrals

Wiener, N., Proc. Nat. Acad. Sci. Vol 7 (1952). Pp. 253 ff, Pp. 294 ff.
McKean, H. P. , *Stochastic Integrals*, Sect. 1.2, Academic Press. 1969.

[viii] Mean Reversion, Brownian Motion

Uhlenbeck, G. E. and Ornstein, L. S., *On the Theory of Brownian Motion*. Physical Review, Vol. 36, 1930. Pp. 823-841.

[ix] Path Integral Techniques Anticipated

Geske, R. and Johnson, H. E., *The American Put Option Valued Analytically*. J. Finance, Vol. 39, 1984. Pp. 1511-1524 .

[x] Some Classic Options References

Black, F. and Scholes, M. , *The Pricing of Options and Corporate Liabilities*. J. Political Economy, 3, 1973. Pp. 637-654.
Merton, R. C., *Theory of Rational Option Pricing*. Bell J. Econ. Management Science,Vol. 4, 1973. Pp. 141-183.
Cox, J. C. and Rubenstein, M. , *Options Markets*, Prentice-Hall. 1985.
Jarrow, R. A. and Rudd, A., *Option Pricing*, Dow Jones-Irwin. 1983.
Smith, C. W., *Option Pricing: A Review*, J. Fin. Econ., Vol. 3, 1976. Pp. 3-51.
Cox, J. C. and Ross, S. A., *The Valuation of Options for Alternative Stochastic Processes*. J. Financial Economics, Vol. 3, 1976. Pp. 145-166.
Courtadon, G., *The Pricing of Options on Default-Free Bonds*. J. Financial and Quantitative Analysis, Vol. 17, 1982. Pp. 75-100.
Brennan, M. J. and Schwartz, E. S., *Alternative Methods for Valuing Debt Options*, Finance, Vol. 4, 1983. Pp. 120-133.

[xi] Quantum Electrodynamics

Itzykson, C. and Zuber, J. B., *Quantum Field Theory*. McGraw-Hill. 1980.

[xii] Monte Carlo Example

Creutz, M., Jacobs, L. and Rebbi, C., *Monte Carlo Computations in Lattice Gauge Theories*. Physics Reports, Vol. 95, North Holland Pub. Co., 1983. Pp. 203-282.

[xiii] **Perspective on Rigor and Practicality**
Gell-Mann, M., Goldberger, M. L., Low, F.E., Marx, E., and Zachariasen, F., *Elementary Particles of Conventional Field Theory as Regge Poles III*. Phys. Rev. 133, B145, 1964.

[xiv] **Rigorous Path Integrals in the Absence of Oscillations**
Glimm, J. and Jaffe, A., *Quantum Physics – A Functional Integral Point of View*. 2nd Edition. Springer Verlag, 1987. See p. 44

[xv] **Stochastic Calculus, Girsanov's Theorem**
Friedman, A., *Stochastic Differential Equations & Applications I*. Academic Press. 1975.
Harrison, J. M. and Kreps, D. M., *Martingales and Arbitrage in Multiperiod Securities Markets*. J. Economic Theory, Vol. 20, 1979. Pp. 381-408.

[xvi] **Fourier Transforms**
Matthews, J. and Walker, R. L., *Mathematical Methods of Physics*, 2nd Ed., Benjamin. 1970. Sect. 4.2.

[xvii] **Distribution Theory**
Lighthill, M. J., *Introduction to Fourier Analysis and Generalized Functions*. Cambridge U. Press. 1964.

[xviii] **Practical Stochastic Calculus**
Maybeck, P. S., *Stochastic Models, Estimation, and Control*. Vol. 2, Academic Press. 1982. P.177.

[xix] **Perturbation Theory, Local Volatility, Skew**
Dash, J. W., *Path Integrals and Options*. Invited talk, Mathematics of Finance section, SIAM Conference, Philadelphia, July 1993.
Dash, J. W., *A Path Integral Approach to Diffusion with Volatility Dependence on Bond Price - The "Mortgage Potential"*. Merrill Lynch working paper, unpublished, 1986.

43. Path Integrals and Options II: Interest-Rates (Tech. Index 8/10)

Summary of this Chapter

We present the general path-integral framework for one-factor (short-term interest rate), term-structure-constrained models[1]. These include Gaussian, mean-reverting Gaussian (MRG), arbitrary rate-dependent volatilities, and memory effects. It is shown how the stochastic equations for rate dynamics are built directly into the path integral. Analytic results are derived by evaluating standard calculus integrals. No previous knowledge of either the models or path integrals is assumed in this chapter. Those familiar with models may regard this chapter as continuing the pedagogical introduction to path integrals. Those familiar with path integrals will benefit by this straightforward presentation of the models.

This chapter is based on my 1989 paper[i]. My derivation of the Mean-Reverting Gaussian (MRG) analytic model reported in this chapter was done independently and roughly concurrently with other authors [2, 3, ii]. The path integral method for me was merely a tool used to derive solutions to models.

---

[1] **Acknowledgements:** I thank A. Davidson, F. Jamshidian, G. Herman, T. Graham, and A. Nairay for helpful discussions. I especially thank R. Jarnagin for conversations on this and many other topics. I thank Bloomberg L.P. for some support in 1989. Finally, I thank Ben Forest for his wizardry in retrieving the files of my CNRS papers from my old Mac.

[2] **History - Some Earlier MRG Calculations:** Some of my MRG calculations (before the MRG model in the text was developed) are reported in my Merrill working paper *Path Integral Continuous Gaussian Bond Option Model* (10/86, unpublished). Interest rates fluctuated with mean reversion around a classical rate path, determined numerically. Bond drifts μ_{bond} and bond volatilities σ_{bond}, including coupons, were calculated. A Green function to calculate options was constructed directly in bond-price space, using bond drifts and volatilities. An active numerical program existed at Merrill for this hybrid model. The analytic solution for the term-structure constraints of the MRG model in the text, including convexity corrections, is in my Merrill working paper *A Path Integral Gaussian Bond Option Model, Discounting, and Put-Call Parity* (4/87, unpublished).

[3] **History - The MRG Model in the Text, the Hull-White Model, Jamshidian's Work:** Hull and White are given precedence for the MRG model in the text, now known as the *Hull-White Model*. Jamshidian, then a member of Merrill's Quantitative Analysis Group that I managed, independently derived the same model. My own independent calculations of the same model, building on my work in Footnote 2, were finished by early 1989, somewhat after Jamshidian, and form the basis of this chapter and Ref. i.

Introduction to this Chapter

The valuation of fixed-income derivative products, many of which we have discussed in previous chapters, is needed for pricing and risk management. A basic issue is to incorporate information about other financial instruments, notably zero-coupon bond prices (term-structure constraints), into the models. This chapter presents some results of a research program that was carried out in 1986-1989. The approach used (Feynman path integrals) is a general method for solving options problems relevant for practical financial applications.

In the previous chapter, path-integral techniques[iii] were introduced. As a pedagogical introduction, solutions to some standard options models were presented, along with some extensions. This chapter is devoted to a discussion of interest-rate options. We discuss some one-factor short-term interest-rate models, emphasizing consistency with initial, or what here are called "static", term-structure constraints. Some references are at the end of this chapter[iv] and Ch. 42.

The simplest special case that has no mean reversion is the constant-volatility continuous Gaussian limit of the binomial model of Ho and Lee [v].

The next simplest case is the mean-reverting Gaussian (MRG). The mean reversion occurs in fluctuations around a classical path of rates, containing a volatility-dependent term[4].

Some general results for arbitrary rate-dependent volatilities are also given. A simple case that is not Gaussian is the standard lognormal model.

The most general Gaussian model[vi] has a structure that includes mean reversion as a special case[vii], and it can incorporate memory effects. Memory means that the interest rate at one time is correlated to the rate at another time in a fashion that can be specified through parameters in the model.

Other Term-Structure Interest-Rate Models

Many interest-rate term-structure models exist[viii]. This book cannot attempt to deal with a catalog of models, which would require a large volume. We only make a short list with few remarks, and give references for the interested reader.

- The single-factor BDT (Black-Derman-Toy) model[ix] not only incorporates term-structure rate constraints, but also takes a significant further step to fit bond-yield volatilities through the term structure of the short rate volatility[5].

---

[4] **Why Use the Name Classical Path?** It is natural to call the path around which rate fluctuations occur the classical rate path because in physics the state around which stochastic fluctuations occurs generally corresponds to the classical limit. In finance, this classical path is related to the forward interest rates, up to a convexity correction.

[5] **Describing the Dynamical Statistics of the Yield Curve:** The Macro-Micro model (Ch. 47-51), which as a multifactor model tackles the even more complex description of the volatilities of yield-curve *shapes*, has a philosophy similar to the BDT model's

- The CIR (Cox-Ingersoll-Ross) model [x] contains a square-root of the short-term rate in the volatility. The CIR model has an analytic modified Bessel-function solution for the Green function. See the footnote[6] for the connection with path integrals.
- Important developments were made by Heath, Jarrow, and Morton (HJM) [xi], who generalized single-factor models to multiple factors, including analytic solution of the term-structure constraints in Gaussian forward-rate models. Ch. 45 considers dynamics in multiple dimensions in the language of path integrals.
- The HJM work was extended to "market models"[xii] by Brace, Gatarek, and Musiela. See also the work of Jamshidian, and Andersen and Andreasen.
- Duffie and collaborators [xiii] constructed a general class of "affine models".
- Flesaker and Hughston specified processes for discount bonds, with guaranteed positive interest rates[xiv].
- Hughston constructed a general differential geometry framework [xv].

I. Path Integrals: Review

For those readers starting the book with this chapter, we repeat a quick review of path integrals as applied to finance. As was stated in Ch. 41, path integration has many attractive features. First, the picturesque idea of future interest-rate paths is an explicit feature of the approach. Second, the fundamental idea of a contingent claim being equal to the expectation (with respect to the appropriate probability density function) of the discounted terminal value consistent with boundary constraints, is used throughout. Third, a complicated mathematical formalism is avoided in practice: one only has to perform multiple Gaussian integrations for those problems that have closed-form solutions. Sophisticated measure theory issues are incorporated in a straightforward manner using distributions (Dirac delta functions)[xvi], thus connecting the stochastic calculus with path integration. Fourth, the no-arbitrage conditions are implemented simply by specifying drift parameters in the path integral through external constraints. Consistency is obtained with the standard no-arbitrage hedging recipes. Fifth, the connections of

---

recognition of the importance of describing yield dynamical statistics, in addition to the usual term-structure constraints.

[6] **CIR Model and Path Integrals:** The solution of the CIR model can be found in a straightforward manner using path integrals by temporarily introducing artificial extra "angular dimensions" for the interest rate, and using existing path-integral results in polar co-ordinates (see Ref. x). It is much harder to work directly without these extra dimensions since a variety of technical difficulties appear. However I derived results without extra dimensions in 1993-94, starting with the Gaussian measure in the CIR model and after direct calculations ending with a small dt approximation to the Bessel Green function. In order to get nonleading terms in the asymptotic expansion of the Bessel function and also avoid singular behavior at r = 0, a "counter term" is needed. It would take us too far afield to give the details.

the path integral approach with binomial (or multinomial) discretizations, grid discretizations, and Monte-Carlo simulation methods are straightforward. Moreover, discretized path integration can produce efficient numerical techniques. Finally, path integration is easy to generalize to several dimensions to treat multi-factor models.

We emphasize again that path integrals form a general methodology that is consistent with and explicitly uses stochastic equations. The connection of path integrals with stochastic equations is direct. This demonstration has been given in several previous chapters, and it is repeated here for interest rates in App. A.

The Rest of this Chapter

The rest of this chapter is organized as follows. Section II contains a discussion of the Green function and static term-structure constraints for discrete-time Gaussian models with a time-dependent volatility. Section III exhibits the continuous-time limit of the constant-volatility Gaussian model, which is the continuous limit of the Ho-Lee model. Section IV treats mean-reverting Gaussian models with static term-structure constraints. Results for embedded options and caps are given. Section V treats the most general Gaussian model, containing all others as special cases, and including the capability of dealing with memory. Section VI has a summary and outlook. Appendix A treats details of the mean-reverting Gaussian model along with the relation of the path integral and stochastic equation formalisms. The inclusion of coupons is also given. Appendix B treats general volatility models. Appendix C contains a description of the general Gaussian model with memory.

II. The Green Function; Discretized Gaussian Models

We begin with the Green function $G(r_a, r_b; t_a, t_b)$ for propagation back from time t_b where the short-term interest rate is called r_b, to an earlier time t_a with short-term interest rate r_a, and we set $T_{ab} = t_b - t_a > 0$. As described in Ref. i and in Appendix A, the time interval is discretized into intervals of length $dt = -\Delta t > 0$, thus defining times $\{t_j\}$. The short-term interest rate at each t_j is called $r_j = r(t_j)$. The formalism in this section will contain a time-dependent volatility $\sigma_j = \sigma(t_j)$ and drift $\mu_j = \mu(t_j)$. We assume that the interest-rate process is normal, or Gaussian.

Appendix B deals with the case of a general rate-dependent volatility function $\sigma[r(t), t]$.

As shown explicitly in Appendix A, the general expression for the Green function is just the convolution of products of Gaussian conditional probability

Path Integrals and Options II: Interest Rates

density functions times discount factors, one for each interval. The result is the conditional probability density function with the appropriate discounting necessary to obtain the expected discounted value for contingent claims[7],

$$G(r_a, r_b; t_a, t_b) =$$
$$\prod_{j=a+1}^{b-1} \int_{-\infty}^{+\infty} dr_j \prod_{j=a}^{b-1} \left[-2\pi\sigma_j^2 \Delta t\right]^{-\frac{1}{2}} \exp\left\{r_j \Delta t - \frac{[r_j - r_{j+1} - \mu_j \Delta t]^2}{-2\sigma_j^2 \Delta t}\right\} \quad (43.1)$$

The reader will note that the exponent is a quadratic form in each of the integration variables r_j. In order to evaluate the expression for G, we merely complete the square in each r_j variable successively and then perform the corresponding integration. This is done by repeatedly using the following identity

$$\exp\left\{kx \Delta t - \frac{[x-x'-\mu \Delta t]^2}{-2\sigma^2 \Delta t}\right\} =$$
$$\exp\left[kx' \Delta t + k(\Delta t)^2 \left(\mu - \tfrac{k}{2}\sigma^2 \Delta t\right)\right]\exp\left\{-\frac{[x-x'-\tilde{\mu} \Delta t]^2}{-2\sigma^2 \Delta t}\right\} \quad (43.2)$$

Here, $\tilde{\mu} = \mu - k\sigma^2 \Delta t$.

Carrying out the algebra produces the desired result,

$$G(r_a, r_b; t_a, t_b) =$$
$$\left[2\pi \tilde{\sigma}_{ab}^2 T_{ab}\right]^{-\frac{1}{2}} \exp\{\zeta_{ab} - r_b T_{ab}\}\exp\left\{-\frac{[r_a - r_b - \tilde{\mu}_{ab} T_{ab}]^2}{2 \tilde{\sigma}_{ab}^2 T_{ab}}\right\} \quad (43.3)$$

Here, the quantities $\tilde{\sigma}_{ab}^2$, $\tilde{\mu}_{ab}$, and ζ_{ab} are defined by

$$\tilde{\sigma}_{ab}^2 T_{ab} = -\sum_{j=a}^{b-1} \sigma_j^2 \Delta t > 0 \quad (43.4)$$

---

[7] **Notation for Labels and Indices in Path Integrals – please README!** As explained in Ch. 42, the extension of a dummy index labeled **j** in the first product is intended to extend only locally in the formula only up to the second product with a separate dummy index which can be labeled by the same letter **j**.

$$\tilde{\mu}_{ab} T_{ab} = \sum_{j=a}^{b-1} \left[\mu_j \Delta t - (j+1) \sigma_j^2 (\Delta t)^2 \right] - a \tilde{\sigma}_{ab}^2 T_{ab} \Delta t \qquad (43.5)$$

$$\zeta_{ab} = \sum_{j=a}^{b-1} (j+1) \Delta t \left[\mu_j \Delta t - \tfrac{1}{2}(j+1) \sigma_j^2 (\Delta t)^2 \right] \qquad (43.6)$$

Quantities like $\tilde{\sigma}_{0a}^2 T_{0a}$ which will appear are defined by taking $t_a \to \hat{t}_0$ (today) and $t_b \to t_a$ in Eqns. (43.4), (43.5), and (43.6).

We now evaluate the expectation of the discounted terminal value for a zero coupon bond $P^{(t_b)}(r_a, t_a)$ at time t_a where again the short-term interest rate is denoted as r_a. The zero matures at date t_b where its value $P^{(t_b)}(r_b, t_b) = 1$, independent of the short-term rate r_b. The result is

$$P^{(t_b)}(r_a, t_a) = \int_{-\infty}^{+\infty} dr_b \, G(r_a, r_b; t_a, t_b) \, P^{(t_b)}(r_b, t_b)$$

$$= \exp\left\{ \zeta_{ab} - \left[r_a - \tilde{\mu}_{ab} T_{ab} \right] T_{ab} + \tfrac{1}{2} \tilde{\sigma}_{ab}^2 T_{ab}^3 \right\} \qquad (43.7)$$

Now the only unknowns are the drifts μ_j once the volatilities σ_j are given. In order to obtain them, we shall employ the static term-structure constraints. The word "static" is there because (as we shall see in a moment) there are exactly enough drifts in order to constrain the theory to produce the zero-coupon bond prices exactly at one given time that we can call "today", $\hat{t}_0$. The caret is there in order to indicate that this is a special time[8]. The interest rate at $\hat{t}_0$ will be called $\hat{r}_0$. Note that once this procedure of drift-determination is carried out for given volatilities, the theory is specified since there are no more parameters. Zero-coupon bonds at all times $t > \hat{t}_0$ are then given by the above formula.

The static term-structure constraints are obtained as illustrated in Fig. 1 at the end of this chapter. The maturity T-axis is partitioned into intervals of the same length $dt = t_{j+1} - t_j = -\Delta t$ as the length of the intervals into which the future time axis is partitioned. Denote the logarithm of today's zero-coupon bond price $\ln P^{(t_j)}(\hat{r}_0, \hat{t}_0)$ by L_j, i.e.

$$L_j = L^{(t_j)}(\hat{r}_0, \hat{t}_0) = \ln P^{(t_j)}(\hat{r}_0, \hat{t}_0) \qquad (43.8)$$

---

[8] **Notation:** In other parts of this book, "today" is just denoted as t_0 without the caret. The caret corresponds to the notation of the original paper.

In the first time interval ($\hat{t}_0, t_1$), $\hat{r}_0$ is the short-term interest rate and μ_0 is the drift. The shortest-term bond maturing at date t_1 is specified by $\hat{r}_0$ as $L_1 = \hat{r}_0 \Delta t$. The drift μ_0 over the first time interval is actually determined from the second bond $P^{(t_2)}(\hat{r}_0, \hat{t}_0)$ maturing at date t_2 by $L_2 = -\mu_0 (\Delta t)^2 + 2\hat{r}_0 \Delta t - \frac{1}{2} \sigma_0^2 (\Delta t)^3$. The drift μ_1 over the second time interval cancels out in this expression due to the second bond maturity condition. Instead, μ_1 is determined from the third bond $P^{(t_3)}(\hat{r}_0, \hat{t}_0)$ which involves both μ_1 and μ_0, the latter already being determined. Proceeding successively in this way, the general solution for the drift μ_j is obtained as

$$\mu_j = \frac{-1}{(\Delta t)^2} \left[L_{j+2} - 2 L_{j+1} + L_j \right] - \sum_{m=0}^{j-1} \sigma_m^2 \Delta t - \tfrac{1}{2} \sigma_j^2 \Delta t \qquad (43.9)$$

Note that the first term is the negative of the second finite-difference numerical derivative of $L^{(t)}(\hat{r}_0, \hat{t}_0) = \ln P^{(t)}(\hat{r}_0, \hat{t}_0)$ with respect to maturity t, evaluated at $t = t_j$. Formally, we let the time interval shrink to zero to obtain the complete drift function

With the drifts determined, the general result for the zero-coupon bond at time t_a in the discretized theory is

$$P^{(t_b)}(r_a, t_a) = \exp\left[-r_a T_{ab} + (L_b - L_a) + T_{ab} \frac{(L_{a+1} - L_a)}{\Delta t} - \tfrac{1}{2} \tilde{\sigma}_{0a}^2 T_{0a} T_{ab}^2 \right] \qquad (43.10)$$

Note that only the volatility at times between $\hat{t}_0$ and t_a occur in $\tilde{\sigma}_{0a}^2$.

The Green function $G(r_a, r_b; t_a, t_b)$ can be written in terms of the zero-coupon bond $P^{(t_b)}(r_a, t_a)$. The result is

$$G(r_a, r_b; t_a, t_b) = \frac{P^{(t_b)}(r_a, t_a)}{\left[2\pi \tilde{\sigma}_{ab}^2 T_{ab} \right]^{1/2}} \exp\left\{ -\frac{\left[r_a - r_b - \hat{\mu}_{ab} T_{ab} \right]^2}{2 \tilde{\sigma}_{ab}^2 T_{ab}} \right\} \qquad (43.11)$$

Here, $\hat{\mu}_{ab} T_{ab} = \tilde{\mu}_{ab} T_{ab} - \tilde{\sigma}_{ab}^2 T_{ab}^2$, and $\tilde{\sigma}_{ab}^2$ is given above.

Given the Green function $G(r_a, r_b; t_a, t_b)$, a European contingent claim C can be obtained by an expectation integration of the Green function times the terminal intrinsic value. Many examples have been treated earlier in the book, and some further examples are considered in later sections of this chapter.

III. The Continuous-Time Gaussian Limit

In this section, we take the continuous limit of the constant-volatility special case of the model developed in Section II. The result is the continuous limit of the Ho-Lee binomial model that (with an appropriate constraint among their parameters $\pi, \delta, \Delta t$) is Gaussian.

The continuous limit is straightforward to carry out in the path integral formalism. One merely lets Δt approach zero and then drop all terms that go to zero in that limit. Nothing is singular. The limit of the logarithm of the bond price at $\hat{t}_0$ is related to the integral of the forward interest rate $f(t')$, determined at the initial time, for t' between $\hat{t}_0$ and t, namely

$$L^{(t)}(\hat{r}_0, \hat{t}_0) = \ln P^{(t)}(\hat{r}_0, \hat{t}_0) = -\int_{\hat{t}_0}^{t} f(t')\,dt' \qquad (43.12)$$

The forward rate at $\hat{t}_0$ is just $\hat{r}_0$. The drift function $\mu(t)$ is then

$$\mu(t) = -\frac{\partial^2}{\partial t^2}\left[\ln P^{(t)}(\hat{r}_0, \hat{t}_0)\right] + \int_{\hat{t}_0}^{t} \sigma^2(t')\,dt'$$

$$= \frac{\partial f(t)}{\partial t} + \int_{\hat{t}_0}^{t} \sigma^2(t')\,dt' \qquad (43.13)$$

If the volatility is assumed constant, this becomes simply

$$\mu(t) = \frac{\partial f(t)}{\partial t} + \sigma^2(t - \hat{t}_0) \qquad (43.14)$$

The Green function becomes

$$G(r_a, r_b; t_a, t_b) = \frac{P^{(t_b)}(r_a, t_a)}{\left[2\pi\sigma^2 T_{ab}\right]^{1/2}} \exp\left\{-\frac{\chi_{ab}^2}{2\sigma^2 T_{ab}}\right\} \qquad (43.15)$$

Here defining $T_{0a} = t_a - \hat{t}_0$,

$$\chi^2_{ab} = \left[(r_b - r_a) - (f_b - f_a) - \sigma^2 T_{ab} T_{0a} \right]^2 \quad (43.16)$$

The zero-coupon bond price in the continuous limit for constant volatility is

$$P^{(t_b)}(r_a, t_a) =$$
$$\exp\left\{ -\int_{t_a}^{t_b} f(t')\,dt' - T_{ab}(r_a - f_a) - \tfrac{1}{2}\sigma^2 T_{0a} T^2_{ab} \right\} \quad (43.17)$$

Here, the continuous-limit integral of the forward rates is just the difference of the logarithms of initial bond prices,

$$-\int_{t_a}^{t_b} f(t')\,dt' = L_b - L_a \quad (43.18)$$

European Bond Option

As an example, a European call option at time t^* written on a zero-coupon bond maturing at time T is given by

$$C(\hat{r}_0, \hat{t}_0) = \int_{-\infty}^{\infty} dr^* \, G(\hat{r}_0, r^*; \hat{t}_0, t^*) \left[P^{(T)}(r^*, t^*) - E \right]_+ \quad (43.19)$$

Here as usual $[P - E]_+$ is defined as $P - E$ if that is positive (in the money), and otherwise is zero. Since the integral has a Gaussian integrand with finite limits, the result is expressible in terms of the usual normal integral; the answer is

$$C(\hat{r}_0, \hat{t}_0) = P^{(T)}(\hat{r}_0, \hat{t}_0) \, N[\Phi_1] - E \, P^{(t^*)}(\hat{r}_0, \hat{t}_0) \, N[\Phi_2] \quad (43.20)$$

Here, defining $T_{0*} = t^* - \hat{t}_0$ and $T_{*T} = T - t^*$,

$$\Xi_{\pm} = \ln\left[\frac{P^{(T)}(\hat{r}_0, \hat{t}_0)}{E\,P^{(t^*)}(\hat{r}_0, \hat{t}_0)}\right] \pm \sigma^2 T_{0*} T^2_{*T}$$

$$\Phi_{1,2} = \frac{1}{\sigma T_{0*}^{1/2} T_{*T}} \Xi_{\pm} \qquad (43.21)$$

Here, the subscripts 1,2 on $\Phi_{1,2}$ correspond to the +,- subscripts on $\Xi_{\pm}$.

Embedded Call Options with Schedules

Embedded call options have multiple calls. A multiple call schedule is handled as described in the previous chapter on path integrals and options. Classes of paths are defined, each class corresponding to the possibility of calling the bond at each corresponding call date. The critical or "Atlantic path" has to be determined. These details are further examined below and in Ch. 44 on numerical applications of path integrals.

An American option is the continuous limit of a call schedule between the times when the option can be exercised. For example, a two-call schedule on a zero-coupon bond with exercise times t_1^* and t_2^* has option value $C = C^{(1)} + C^{(2)}$. Here $C^{(1)}$ corresponds to the possibility of calling the bond at time t_1^* and $C^{(2)}$ corresponds to the possibility of calling the bond at time t_2^*. Explicitly,

$$C^{(2)}(\hat{r}_0, \hat{t}_0) = \int_{-\infty}^{\hat{r}_2} dr_2^* \int_{\hat{r}_1}^{+\infty} dr_1^* G(\hat{r}_0, r_1^*; \hat{t}_0, t_1^*) G(r_1^*, r_2^*; t_1^*, t_2^*) \left[P^{(T)}(r_2^*, t_2^*) - E_2 \right]_+$$

(43.22)

$$C^{(1)}(\hat{r}_0, \hat{t}_0) = \int_{-\infty}^{\hat{r}_1} dr_1^* G(\hat{r}_0, r_1^*; \hat{t}_0, t_1^*) \left[P^{(T)}(r_1^*, t_1^*) - E_1 \right]_+ \qquad (43.23)$$

Here, $\hat{r}_2$ and $\hat{r}_1$ are the points of the discrete Atlantic path. The point $\hat{r}_2$ is given by the intrinsic value being zero at the second strike or exercise date:

$$P^{(T)}(\hat{r}_2, t_2^*) = E_2 \qquad (43.24)$$

The equality of the first intrinsic value and the option to exercise at the second strike date, all evaluated at the first strike or exercise date, gives $\hat{r}_1$,

Path Integrals and Options II: Interest Rates

$$\int_{-\infty}^{\hat{r}_2} dr_2^* G(\hat{r}_1, r_2^*; t_1^*, t_2^*) \left[P^{(T)}(r_2^*, t_2^*) - E_2 \right]_+ = P^{(T)}(\hat{r}_1, t_1^*) - E_1$$

(43.25)

So far, the formalism has been for zero-coupon bonds. Adding coupons is straightforward provided the coupons do not depend on the interest rate. One simply takes the appropriate linear sum of zero-coupon bonds weighted by the coupons to obtain the forward bonds $\{P^{(T)}(r^*, t^*)\}$ for European options; $P^{(T)}(r_1^*, t_1^*)$ and $P^{(T)}(r_2^*, t_2^*)$ for the two-option problem. The coupons then appear in the formulas for the Atlantic-path points. Appendix A gives the details.

IV. Mean-Reverting Gaussian Models

In this section we show how mean reversion [vii], can be incorporated into static term-structure option models. A review of the mean reversion formalism is given in Appendix A. The mean reversion function $\omega(t)$ is discretized by $\omega_j = \omega(t_j)$. Actually, the case of most practical interest would be the simplest example of constant ω. The limit $\omega \to 0$ gives back the Gaussian Ho-Lee model described in the last section. We shall derive the mean-reverting model directly; the comments at the end of Appendix A can also be used to derive it.

As usual, the main task is to derive the Green function. This is simplified by defining at time t_j the variable $x_j = x(t_j)$ describing fluctuations of $r_j = r(t_j)$ around the classical interest-rate path $r_j^{(CL)} = r^{(CL)}(t_j)$ at time t_j as introduced in Ref. [vi],

$$x_j = r_j - r_j^{(CL)} \quad (43.26)$$

Here, $x(\hat{t}_0) = 0$. The discretized classical path at time t_j is given by

$$r_j^{(CL)} = \hat{r}_0 - \sum_{j'=0}^{j-1} \mu_{j'} \Delta t \quad (43.27)$$

(with no sum at $j = 0$). In the continuous limit this becomes

$$r^{(CL)}(t) = \hat{r}_0 + \int_{\hat{t}_0}^{t} \mu(t') dt' \quad (43.28)$$

The drift function $\mu(t)$ is determined using the procedure discussed in previous sections.

Some Useful Formulae
We have the following useful formulae

$$P^{(t_b)}\left(r_a^{(cl)}, t_a\right) = \exp\left\{-\int_{t_a}^{t_b} r^{(cl)}(t)\,dt + \frac{\sigma^2}{6}T_{ab}^3 \mathcal{J}(\omega T_{ab})\right\} \quad (43.29)$$

$$\int_{t_a}^{t_b} r^{(cl)}(t)\,dt = \int_{t_a}^{t_b} f(t)\,dt + \frac{\sigma^2}{6}\left[T_{0b}^3 \mathcal{J}(\omega T_{0b}) - T_{0a}^3 \mathcal{J}(\omega T_{0a})\right] \quad (43.30)$$

Here $T_{ij} = t_j - t_i$ and

$$\mathcal{J}(\omega\tau) = \frac{3}{(\omega\tau)^3}\left[\omega\tau + 2\left(e^{-\omega\tau} - 1\right) - \frac{1}{2}\left(e^{-2\omega\tau} - 1\right)\right] \quad (43.31)$$

Note that $\mathcal{J}(0) = 1$.

The Green Function
The result for the Green function $G(x_a, x_b; t_a, t_b)$ is given by evaluating the path integral. For the discretized case it is given by (again $\Delta t < 0$):

$$G(x_a, x_b; t_a, t_b) = \prod_{j=a+1}^{b-1}\int_{-\infty}^{+\infty} dx_j \prod_{j=a}^{b-1}\left[-2\pi\sigma_j^2 \Delta t\right]^{-\frac{1}{2}} \exp\left\{r_j \Delta t - \frac{\left[x_{j+1} - (1 + \omega_j \Delta t)x_j\right]^2}{-2\sigma_j^2 \Delta t}\right\} \quad (43.32)$$

Note that to first order in Δt we have $(1 + \omega_j \Delta t) \approx e^{\omega_j \Delta t}$. From a path integral perspective, this is the origin of the exponential factor of the mean-reversion described in Appendix A.

Evaluation of the Green Function
For simplicity in the following discussion, we use constant mean reversion and constant volatility. The general case can also be treated using similar techniques.

Path Integrals and Options II: Interest Rates

The Green function can be evaluated in at least two ways. The first is, as before, to use a recursive method. We just complete the square using

$$\exp\left\{kx\,\Delta t - \frac{[\xi x - x' - \mu\,\Delta t]^2}{-2\sigma^2\,\Delta t}\right\} =$$

$$\exp\left[\frac{k}{\xi}x'\,\Delta t + \frac{k}{\xi}(\Delta t)^2\left(\mu - \frac{k}{2\xi}\sigma^2\,\Delta t\right)\right]\exp\left\{-\frac{[\xi x - x' - \tilde{\mu}\,\Delta t]^2}{-2\sigma^2\,\Delta t}\right\}$$

(43.33)

Here we have defined $\tilde{\mu} = \mu - \dfrac{k}{\xi}\sigma^2\,\Delta t$.

Simple Harmonic Oscillators and the MRG Model

Another method was described by Feynman Ref. [iii]. The problem faced here is formally the same as that of forced harmonic oscillator motion with the replacement of ω to $\sqrt{-1}\cdot\omega$. Feynman's formulae can then be used along with the addition of "surface terms" needed for the correct normalization[9]. The surface terms are the cross terms in the expansion of the action density $\left[\dfrac{dx}{dt} + \omega x\right]^2$.

The result of carrying out either procedure is

$$G(x_a, x_b; t_a, t_b) = \frac{P^{(t_b)}(r_a, t_a)}{\left[2\pi\,\tilde{\sigma}^2_{ab}\,T_{ab}\right]^{1/2}}\exp\left\{-\frac{\chi^2_{ab}}{2\,\tilde{\sigma}^2_{ab}\,T_{ab}}\right\} \quad (43.34)$$

Here,

$$\chi^2_{ab} = \left[x_b - e^{-\omega T_{ab}}x_a + \frac{\sigma^2}{2\omega^2}(1 - e^{-\omega T_{ab}})^2\right]^2 \quad (43.35)$$

$$\tilde{\sigma}^2_{ab}\,T_{ab} = \frac{\sigma^2}{2\omega}(1 - e^{-2\omega T_{ab}}) \quad (43.36)$$

The Zero-Coupon Bond

The zero-coupon bond is given by

---

[9] **Surface Terms:** Watch out. Surface terms are usually dropped as being irrelevant, but they are needed here.

$$P^{(t_b)}(r_a, t_a) =$$

$$\exp\left\{-\int_{t_a}^{t_b} f(t')\,dt' - \frac{1}{\omega}\left[1 - e^{-\omega T_{ab}}\right](r_a - f_a) - \frac{\tilde{\sigma}_{0a}^2 T_{0a}}{2\omega^2}\left[1 - e^{-\omega T_{ab}}\right]^2\right\}$$

(43.37)

Equivalently we can write

$$P^{(t_b)}(r_a, t_a) = \exp\left[-\frac{x_a}{\omega}\left(1 - e^{-\omega T_{ab}}\right)\right] P^{(t_b)}\left(r_a^{(cl)}, t_a\right) \quad (43.38)$$

Here,

$$P^{(t_b)}\left(r_a^{(cl)}, t_a\right) = \exp\left\{-\int_{t_a}^{t_b} r^{(cl)}(t)\,dt + \frac{\sigma^2}{6} T_{ab}^3 \mathcal{J}(\omega T_{ab})\right\} \quad (43.39)$$

The Drift Function
The result for the drift is

$$\mu(t) = \frac{\partial f(t)}{\partial t} + \frac{\sigma^2}{\omega} e^{-\omega(t - \hat{t}_0)}\left[1 - e^{-\omega(t - \hat{t}_0)}\right] \quad (43.40)$$

The Classical Path of Forward Rates with a Correction Term
The classical path at time t is the forward rate at maturity t determined at time $\hat{t}_0$ with a volatility-dependent correction term[10],

$$r^{(CL)}(t) = f(t) + \frac{\sigma^2}{2\omega^2}\left[1 - e^{-\omega(t - \hat{t}_0)}\right]^2 \quad (43.41)$$

This gives $x(t)$ which can be inserted into the above expression for $G(x_a, x_b; t_a, t_b)$ in order to re-express the results in terms of $r(t)$. We call the result $G(r_a, r_b; t_a, t_b)$, keeping the same name G for simplicity.

---

[10] **Futures vs. Forwards Convexity Term:** This is the origin of the convexity correction for this mean-reverting Gaussian model.

European Bond Option in the MRG Model

Armed with this result for the Green function we are in a position to calculate other contingent claims. The European call option is defined as usual by

$$C(\hat{r}_0, \hat{t}_0) = \int_{-\infty}^{+\infty} dr^* \, G(\hat{r}_0, r^*; \hat{t}_0, t^*) \left[P^{(T)}(r^*, t^*) - E \right]_+ \quad (43.42)$$

C is given by the same form as before,

$$C(\hat{r}_0, \hat{t}_0) = P^{(T)}(\hat{r}_0, \hat{t}_0) \, N[\Phi_1] - E \, P^{(t^*)}(\hat{r}_0, \hat{t}_0) \, N[\Phi_2] \quad (43.43)$$

The mean-reverting forms for the arguments of the normal integral are

$$\Phi_{1,2} = \frac{\Xi_\pm}{\frac{\tilde{\sigma}_{0*} \sqrt{T_{0*}}}{\omega} \left[1 - \exp(-\omega T_{*T})\right]} \quad (43.44)$$

Here, the subscripts 1, 2 on Φ correspond to the $\pm$ subscripts on Ξ, and

$$\Xi_\pm = \ln\left[\frac{P^{(T)}(\hat{r}_0, \hat{t}_0)}{E \, P^{(t^*)}(\hat{r}_0, \hat{t}_0)}\right] \pm \frac{\tilde{\sigma}_{0*}^2 T_{0*}}{2\omega^2} \left[1 - \exp(-\omega T_{*T})\right]^2 \quad (43.45)$$

Finally, $\tilde{\sigma}_{0*}^2$ is $\tilde{\sigma}_{ab}^2$ with t_a, t_b replaced by $\hat{t}_0, t^*$. As before, $T_{0*} = t^* - \hat{t}_0$, and $T_{*T} = T - t^*$.

Caplet in the MRG Model

As another example, consider a caplet with threshold strike rate E. The caplet is a single option, whose value at the strike date t^* is defined as given by the difference of the interest rate and E, provided that is positive. Thus, in this model,

$$\text{Caplet}(\hat{r}_0, \hat{t}_0) = \int_{-\infty}^{+\infty} dr^* \, G(\hat{r}_0, r^*; \hat{t}_0, t^*) \left[r^* - E \right]_+ =$$

$$P^{(t^*)}(\hat{r}_0, \hat{t}_0) \left\{ \frac{\tilde{\sigma}_{0*} \sqrt{T_{0*}}}{\sqrt{2\pi}} \exp\left[-\frac{(E - f^*)^2}{2 \tilde{\sigma}_{0*}^2 T_{0*}}\right] + (f^* - E) N\left[\frac{f^* - E}{\tilde{\sigma}_{0*} \sqrt{T_{0*}}}\right] \right\}$$

$$(43.46)$$

Here, $f^* = f(t^*)$ is the forward rate at t^* and the other quantities are as above. The actual caplet as quoted in the market will be this result multiplied by a time interval between resets Dt_{reset}. In addition, other details involve the cap paid "in arrears" at time Dt_{reset} later than t*. A market cap is actually a sum of caplets with different reset dates. There is also a notional principal amount. Caps are usually priced with lognormal rate assumptions, but sometimes they are priced with Gaussian models and sometimes with a mix of lognormal and Gaussian. Practical information and more details on caps can be found in Ch. 10.

Negative Rates in Gaussian Models

In the specific case of Gaussian or mean-reverting Gaussian rate models, the rates can become negative. This is unphysical. For this reason, Gaussian rate models have often been criticized. The probability of negative rates depends on the length of time from the starting time, the volatility, and the starting rate level.

For example, the value of a floor with a strike of zero with positive rates should be exactly zero[11]. This is because the floor only pays off if the rate becomes lower than the strike. On the other hand, floors with zero strikes are not without value in Gaussian rate models due to negative rates.

The extent to which such negative-rate contamination influences the pricing of other securities is not easy to extract.

V. The Most General Model with Memory

In this section, we present the most general one-factor model possible. A Gaussian version Ref. [vi] was originally proposed, and we discuss this first. All other Gaussian models discussed in this paper are special cases. In this general model, the exponent in the integrand for the Green function contains correlations with separate (zero or non-zero) coefficients between the interest rates at any two times, with arbitrary weights[12]. The discretized Green function can be written as

$$G(x_a, x_b; t_a, t_b) = \Omega_{ab} \prod_{j=a+1}^{b-1} \int_{-\infty}^{+\infty} dx_j \exp\left\{ \sum_{j=a}^{b-1} r_j \Delta t - \sum_{k,m=a}^{b} x_k A_{km} x_m (\Delta t)^2 \right\} \quad (43.47)$$

---

[11] **Zero-Strike Floor:** The example of the zero-strike floor was first made by C. Rogers.

[12] **Constraint:** The matrix (A_{km}) needs to be symmetric positive-definite.

Here, A_{km} is a number for given indices k,m. Also, Ω_{ab} is the normalization factor, given by

$$1 = \Omega_{ab} \prod_{j=a+1}^{b} \int_{-\infty}^{+\infty} dx_j \exp\left\{ -\sum_{k,m=a}^{b} x_k A_{km} x_m (\Delta t)^2 \right\} \tag{43.48}$$

With this definition, the Green function still represents the conditional probability density function multiplied by the discount factors needed to produce expected discounted values for contingent claims. The factor $(\Delta t)^2$ is written for convenience; the matrix A is then dimensionless. As before, $x(t)$ is the difference between $r(t)$ and the classical path $r^{(CL)}(t)$.

Applications of Models with Memory

Some attempt was made in 1986-1987 to determine parameters in this Gaussian model to fit term-structure constraints and other market data including both the matrix A and the classical path. Because the number of parameters was not sufficiently restricted, the general form of the model proved to be difficult to implement in practice.

We feel the wide range of market phenomena potentially describable may well justify further effort. In particular, an "effective" memory can be created by stochastic variables that are not normally, included in options pricing (see Ref. [iii], and the remarks at the end of this section). Notice that the matrix element A_{km} indeed connects the rates r_k and r_m at times t_k and t_m. This incorporates memory effects of the interest rate process with itself.

Special Cases

The special cases treated in previous sections are given explicitly by specific choices of the matrix A. For the Gaussian model, the connection is

$$A_{km}^{(Gaussian)} (\Delta t)^2 = \frac{1}{2\sigma^2 \Delta t} \left[\delta_{k,m+1} + \delta_{k,m-1} - 2\delta_{k,m} \right] \tag{43.49}$$

Here, δ_{rs} is the Kronecker delta, equal to one if $r = s$ and zero otherwise. Thus, only nearest neighbor times are connected by the matrix A. For the constant mean-reverting Gaussian (MRG) model, the result is

$$A_{km}^{(MRG)} (\Delta t)^2 = \frac{1}{2\sigma^2 \Delta t} \left[\delta_{k,m+1} + \delta_{k,m-1} - 2\delta_{k,m} \right] - \frac{\omega^2}{\sigma^2 \Delta t} \delta_{k,m} \tag{43.50}$$

This contains the extra term at k = m.

Contingent claims can now evaluated by the usual procedure of taking expectation values. Again, for those problems reducible to iterated Gaussian integrals, closed form solutions can be obtained. We exhibit the calculation of a zero-coupon bond in Appendix C.

As mentioned, A_{km} was assumed a constant in Ref. [vi]. However, a-priori A_{km} can be a function of time and rates at various times,

$$A_{km} = A_{km}\left[\{r_j, t_j\}\right] \tag{43.51}$$

This provides the most general one-factor model, including memory. Special cases include the general-volatility models in Appendix B.

Connection of the Path Integral with the Stochastic Equations

We now give a description of the connection to the stochastic Langevin-Ito equation, using the techniques described in Appendix A. To motivate the discussion, write the finite-difference stochastic equation for the Gaussian model,

$$\left[\frac{dx(t)}{dt}\right]_{t_j} = \frac{x_{j+1} - x_j}{dt} = \sigma_j \eta_j \tag{43.52}$$

This produces (with $dt = -\Delta t$ again),

$$\eta_j = -\frac{1}{\sigma_j \Delta t}\left[\delta_{j+1,k} - \delta_{j,k}\right] x_k = \sqrt{-2\Delta t} \; C_{jk}^{(\text{Gaussian})} x_k \tag{43.53}$$

This defines the matrix $C_{jk}^{(\text{Gaussian})}$. In Eqn. (43.53) we use the summation convention for the repeated index k.

Now we *could* simply write, for a *general* matrix C, the equation

$$\eta_j = \sqrt{-2\Delta t} \; C_{jk} \, x_k \tag{43.54}$$

Defining the matrix $A = C^T C$ as the transpose of C times C we find that the probability density function for the η_j variable becomes (with the summation convention for indices k, m)

Path Integrals and Options II: Interest Rates

$$\exp\left[\frac{\eta_j^2 \, \Delta t}{2}\right] = \exp\left[-C_{jk} \, x_k \, C_{jm} \, x_m \, (\Delta t)^2\right]$$

$$= \exp\left[-x_k \, A_{km} \, x_m \, (\Delta t)^2\right] \qquad (43.55)$$

This is exactly what is required to produce the general Gaussian model.

Stochastic Equations with Memory

Note that if we write out the general Langevin-Ito equation explicitly (this time without the summation convention) we obtain[13]

$$\frac{\eta_j}{\sqrt{2\, dt}} = C_{jj} \, x_j + \left[C_{j,j+1} \, x_{j+1} + C_{j,j-1} \, x_{j-1}\right] + \sum_{i \leq -2} C_{j,j+i} \, x_{j+i} \qquad (43.56)$$

The first term contains mean reversion while the "nearest-neighbor" terms in the bracket are in the Gaussian model Eqn. (43.53). The sum contains potential memory terms, not in the previous models, and corresponding to effects of the noise at time t_j on the variables $x(t)$ at other times.

Physical Intuition for Memory Effects

Physically it is easy to see how memory can occur. Imagine a particle diffusing in a medium that is kicked at some time by an "invisible gremlin". The future trajectory of the particle will remember this kick at all later times; this is the memory effect. In the case of the currency option with stochastic interest rates, for example, the kicks are due to fluctuations in the interest rate processes and the variable left in the description is the exchange rate.

It is easy to invent a simple 2-factor model[14] to exhibit how this effective memory is generated. We start with two random variables, x and q, satisfying

$$\frac{dx(t)}{dt} = A\, x(t) + B\, q(t) + \sigma\, \eta(t) \qquad (43.57)$$

$$\frac{dq(t)}{dt} = C\, q(t) + E\, x(t) + \alpha\, \zeta(t) \qquad (43.58)$$

---

[13] **ARIMA Models:** The formalism is related to ARIMA models. See Ch. 52.

[14] **Acknowledgements:** This example was applied long ago in different contexts by Lindenberg and G. West. It is a special case of the "heat-bath", clearly exposed in Feynman and Hibbs (cf. Ref. iii, Pp. 68 ff.)

Here, $\sigma(t), \zeta(t)$ are two correlated Gaussian random variables. In this two-dimensional formalism where both x and q are explicit, there are no memory effects. Now, however, integrate out $q(t)$ by using the solution to Eqn. (43.58) with initial condition $q_0 = q(0)$,

$$q(t) = e^{Ct} q_0 + \int_0^t dt' \, e^{C(t-t')} \left[E \, x(t') + \alpha \zeta(t') \right] \qquad (43.59)$$

Plugging that into Eqn. (43.57) we find

$$\frac{dx(t)}{dt} = A \, x(t) + \sigma \, \eta(t) + B \, e^{Ct} q_0 + B \int_0^t dt' \, e^{C(t-t')} \left[E \, x(t') + \alpha \zeta(t') \right] \quad (43.60)$$

Explicit memory effects *are* present in Eqn. (43.60) for $x(t)$ with $q(t)$ removed (i.e. the "gremlin" variable q has been made "invisible"). That is, x at time t depends on x at previous times due to the mutual interaction of the x and q variables in the original equations (43.57), (43.58).

VI. Wrap-Up for this Chapter

This chapter has dealt with some term-structure one-factor models. A variety of models was considered, including mean reversion and even memory effects.

What can be improved in the current generation of one-factor options models? First, let us step back and ask what has actually been accomplished. The static term-structure constraints have been included. Still, improvement is possible. For example, there is no guarantee at all that the term-structure time-averaged *statistical* properties of the yield curve, including its fluctuations in shape, will be correctly produced in accord with market observations[15].

As a concrete example of the importance of this remark, mortgage-backed securities require the correct spread statistics between short and long term rates in order to obtain correct prepayment model input. Simple one-factor models do not generate realistic spread statistics, and are therefore not optimal. On the other hand, for embedded options in corporate bonds that are "weak probes" of the actual interest rate process, these models can be quite useful as a parameter-dependent characterization of market data at a certain time from which small

---

[15] **Remark:** Although this point was made in 1989, it is still true today. See the chapters on the Macro-Micro model (Ch. 48-51).

Path Integrals and Options II: Interest Rates

perturbations are made close to that time to price bonds in normal trading and sales activities.

Large effects, e.g. market "crashes", are excluded from all most models constructed so far. We have always believed that a description of crashes requires non-linear phase-transition physics between multiple equilibria. See Ch. 46.

Appendix A: MRG Formalism, Stochastic Equations, Etc.

A.1 Relation of the MRG Path Integral to Stochastic Equations

We first present the formalism of mean reversion in Gaussian models using the path-integral language appropriate to this paper, and we connect it to the Langevin-Ito stochastic equation formalism.

We begin with the Langevin-Ito equation defining the MRG model[16],

$$\frac{dx(t)}{dt} = -\omega(t)x(t) + \sigma(t)\eta(t) \qquad (43.61)$$

We write as before $x(t) = r(t) - r^{(CL)}(t)$ to get the fluctuations of the interest rate about the classical path.

As described in Ch. 42, $\sigma(t)\eta(t)$ has the interpretation of the random slope of the path from time t to time $t+dt$, given $x(t)$ at time t, and accounting for the mean reversion between times t and $t+dt$. The idea is illustrated in Figure 2 at the end of this chapter.

Now $\eta(t)$ is assumed a Gaussian random variable with zero mean and width $\left\langle \eta^2(t) \right\rangle^{1/2} = 1/\sqrt{dt}$, where the expectation value is with respect to the probability density for $\eta(t)$,

$$\wp[\eta(t)] d\eta(t) = \frac{d\eta(t)}{\sqrt{2\pi/dt}} \exp\left[-\tfrac{1}{2}\eta^2(t)\, dt\right] \qquad (43.62)$$

The time interval (t_a, t_b) is discretized into $n+1$ points t_j with $j = [a, a+1, ..., (a+n=b)]$ and interval $dt = t_{j+1} - t_j = -\Delta t > 0$. For notational convenience, we also define an index $i = j - a = [0, 1, ..., n]$. The

---

[16] **Notation:** The left hand side of this equation is [x(t+dt) − x(t)]/dt, which is the same as what we called $d_t x(t)/dt$ in other parts of the book.

variable $x(t)$ is discretized to x_j with the endpoints $x_a = x_{[i=0]}$ and $x_b = x_{[i=n]}$ fixed. The total probability conditioned on the endpoint constraints is the product of the probability densities at times $i = [0,...,n-1]$. However, with the two endpoints constrained, only variable slopes at $i = [0,...,n-2]$ are integrated over; the last slope with $i = n-1$ is then specified once $x_{[i=n-1]}$ and $x_{[i=n]}$ are given.

Using this discretization and again recognizing that the endpoints are fixed we obtain[17]

$$\int_{-\infty}^{+\infty} \wp_{tot}\{\eta\} D\{\eta\} \equiv \wp(\eta_{[i=n-1]}) \cdot \prod_{i=j-a=0}^{n-2} \int_{-\infty}^{+\infty} \wp(\eta_j) d\eta_j$$

$$= \wp(\eta_{[i=n-1]}) \cdot \prod_{i=j-a=0}^{n-2} \int_{-\infty}^{+\infty} \frac{d\eta_j}{\sqrt{2\pi/dt}} \exp\left[-\tfrac{1}{2}\eta_j^2 \, dt\right] \quad (43.63)$$

We multiply this equation by the Dirac delta-function identity

$$1 = \prod_{i=j-a=0}^{n-2} \int_{-\infty}^{+\infty} dx_{j+1} \, \delta\left[x_{j+1} - x_j(1-\omega_j \, dt) - \sigma_j \eta_j dt\right] \quad (43.64)$$

Because the last point is fixed at x_b, we need to fix the last slope as $\eta_{[i=n-1]} = [x_b - x_{b-1}(1-\omega_{b-1} dt)]/(\sigma_{b-1} dt)$. Now in fact we are not interested in the η_j variables since we want the paths to be specified by the x_j variables. The Dirac δ-functions arrange this by killing the η_j integrals and substituting the expressions for the η_j found in the δ-function arguments. We insert the discount product factor $\prod_{i=j-a=0}^{n-1} e^{-r_j dt}$ in the integrals, as is necessary for the expected discounted expression. We need to use the formula (Ref. [xvi])

$$\delta[f(y)] = \frac{\delta(y - \tilde{y})}{\left|\dfrac{df}{dy}\right|_{\tilde{y}}} \quad (43.65)$$

---

[17] **Comment**: We could proceed by including integrals over all slopes η_j up to $j = n-1$, inserting one more delta function and then take away the last dx_b integral to account for the fixed point x_b. I thank Andrew Kavalov for a clarifying discussion on this point.

Here, $f(\tilde{y}) = 0$. Taking the mean reversion as a constant then yields the formula for the Green function for the mean-reverting Gaussian model, Eqn. (43.32).

The zero-coupon bond price $P^{(t_b)}(r_a, t_a)$ with maturity date t_b, evaluated at time t_a with interest rate fixed at one possible value r_a, can be obtained by Gaussian integration using the formula, all with r_a fixed,

$$\left\langle \exp\left[-\int_{t_a}^{t_b} r(t)\, dt\right] \right\rangle = \exp\left[-\int_{t_a}^{t_b} \langle r(t) \rangle\, dt + \tfrac{1}{2} \int_{t_a}^{t_b} dt \int_{t_a}^{t_b} dt\, '\langle r(t)\, r(t\,')\rangle_c \right] \quad (43.66)$$

This generalizes $\left\langle \exp[-r(t)] \right\rangle = \exp\left[-\langle r(t)\rangle + \tfrac{1}{2}\langle [r(t)]^2 \rangle_c\right]$ at fixed time. Here the second-order correlation function is defined as usual,

$$\langle y_1 y_2 \rangle_c = \langle y_1 y_2 \rangle - \langle y_1 \rangle \langle y_2 \rangle \quad (43.67)$$

The expectation values are with respect to the Gaussian measure without the discount factors. The average $\langle r(t) \rangle = \langle x(t) \rangle + r^{(CL)}(t)$ along with $\langle r(t)\, r(t\,') \rangle_c = \langle x(t)\, x(t\,') \rangle_c$, all with r_a fixed, are evaluated from the solution of the Langevin-Ito equation for constant mean reversion,

$$\frac{dx(t)}{dt} = -\omega x(t) + \sigma(t) \eta(t) \quad (43.68)$$

Given x_a at t_a we have

$$x(t) = x_a\, e^{-\omega(t - t_a)} + \int_{t_a}^{t} e^{\omega(\xi - t)}\, \sigma(\xi)\, \eta(\xi)\, d\xi \quad (43.69)$$

Hence

$$\langle x(t) \rangle = x_a\, e^{-\omega(t - t_a)} \quad (43.70)$$

If further $\sigma(\xi) = \sigma$ is taken constant for simplicity we can use the identities $\langle \eta(t) \rangle = 0$, $\langle \eta(t_a)\eta(t_b) \rangle = \delta(t_a - t_b)$ to obtain the results

$$\langle x(t_1) x(t_2) \rangle_c = \frac{\sigma^2}{2\omega} \left\{ e^{-\omega|t_1 - t_2|} - e^{-\omega(t_1+t_2-2t_a)} \right\} \quad (43.71)$$

Also, with $\mathcal{Z}$ given by Eqn. (43.31), we have

$$\int_{t_a}^{t_b} dt \int_{t_a}^{t_b} dt' \langle x(t) x(t') \rangle_c = \frac{\sigma^2}{3} T_{ab}^3 \mathcal{Z}(\omega T_{ab}) \quad (43.72)$$

Some algebra then leads to the results in the text.

For general time-dependent mean reversion, we define a modified volatility function $\sigma_\omega(t)$ by

$$\sigma_\omega(t) = \exp\left[\int_{t_0}^{t} \omega(t') dt' \right] \sigma(t) \quad (43.73)$$

Then the variable

$$x_\omega(t) = \exp\left[\int_{t_0}^{t} \omega(t') dt' \right] x(t) \quad (43.74)$$

after a little algebra is seen to satisfy the equation without mean reversion,

$$\frac{dx_\omega(t)}{dt} = \sigma_\omega(t) \eta(t) \quad (43.75)$$

The time-dependent volatility $\sigma_\omega(t)$ can be handled as described in the text.

A.2 The MRG Diffusion Equation

The diffusion equation solved by the Green function can be obtained by Fourier transform methods Ref. [iv]. From Eqn. (43.32),

$$G(x_a, x_b; t_a, t_b) =$$

$$\prod_{j=a+1}^{b-1} \int_{-\infty}^{+\infty} dx_j \prod_{j=a}^{b-1} \left[-2\pi\sigma_j^2 \Delta t \right]^{-\frac{1}{2}} \exp\left\{ r_j \Delta t - \frac{\left[x_{j+1} - (1+\omega_j \Delta t) x_j \right]^2}{-2 \sigma_j^2 \Delta t} \right\} \quad (43.76)$$

We rewrite the exponential at fixed co-ordinates and again set $dt = -\Delta t > 0$. Introducing the Fourier transform variable k_j at each partition time produces

$$G(x_a, x_b; t_a, t_b) = \prod_{j=a+1}^{b-1} \int_{-\infty}^{+\infty} dx_j \prod_{j=a}^{b-1} \int_{-\infty}^{+\infty} \frac{dk_j}{2\pi}$$

$$\cdot \exp\left\{-dt\left[\omega_j x_j(ik_j) + \tfrac{1}{2}\sigma_j^2 k_j^2 + r_j\right] - ik_j(x_{j+1} - x_j)\right\} \quad (43.77)$$

In Eqn. (43.77), $i = \sqrt{-1}$. The coefficient of dt in the exponent is the infinitesimal time generator $\frac{\partial}{\partial t}\big|_{\text{fixed } x_j}$ for a transition backward from time t_{j+1} at position x_{j+1} to time t_j at position x_j. We replace $\sqrt{-1}\,k_j$ by $\frac{\partial}{\partial x(t)}$ evaluated at t_j according to Fourier prescription. Dropping the subscript j and setting $r(t) = x(t) + r^{(CL)}(t)$, the diffusion operator identity at any time t is:

$$\frac{\partial}{\partial t}\bigg|_{\text{fixed } x(t)} = \omega(t)\, x(t) \frac{\partial}{\partial x(t)} - \tfrac{1}{2}\sigma^2(t) \frac{\partial^2}{\partial x(t)^2} + x(t) + r^{(CL)}(t) \quad (43.78)$$

This derivation is actually more general. The same equation holds if the volatility and mean reversion become functions of $x(t)$ and t. The ordering of the spatial derivatives conforms to the backward Kolmogoroff equation; this point is a little tricky to see. Simplification occurs by using an integrating factor to remove the classical path in the equation propagating back from time t_b to t_a,

$$D^{(CL)}(t_a, t_b) = \exp\left\{-\int_{t_a}^{t_b} r^{(CL)}(t')\, dt'\right\} \quad (43.79)$$

$D^{(CL)}$ is the discount factor produced by the classical path and is very intuitive: it is just the discounting produced by the average interest rate path about which the fluctuations occur. We set

$$G[x(t), x_b; t, t_b] = D^{(CL)}(t_a, t_b)\, \Gamma[x(t), x_b; t, t_b] \quad (43.80)$$

Equivalently,

$$\Gamma[x(t), x_b; t, t_b] = G[x(t), x_b; t, t_b]\big|_{r^{(CL)}=0} \quad (43.81)$$

Then we have

$$\frac{\partial}{\partial t} G[\,x(t)\,,x_b\,;t\,,t_b\,]\,\Big|_{\text{fixed }x(t)} =$$

$$D^{(CL)}\,(t_a\,,t_b\,)\left\{\frac{\partial}{\partial t}\Big|_{\text{fixed }x(t)} - r^{(CL)}(t)\right\}\Gamma[\,x(t)\,,x_b\,;t\,,t_b\,] \quad (43.82)$$

The classical path drops out of the equation for Γ, which is then solved for. Since only the integral of $r^{(CL)}$ enters in G, the forward-rate dependence is eliminated in favor of zero-coupon bond prices. To illustrate, this intermediate step of removing the classical path results in the European call option

$$C(\hat{r}_0,\hat{t}_0) = D^{(CL)}\,(\hat{t}_0,t^*)\int_{-\infty}^{+\infty}\Gamma[\,\hat{x}_0\,,x^*;\hat{t}_0\,,t^*\,]\cdot$$

$$\left\{D^{(CL)}\,(t^*,T)\,\Re^{(T)}(x^*,t^*) - E\right\}_+ dx^* \quad (43.83)$$

Here,

$$\Re^{(T)}(x^*,t^*) = P^{(T)}(x^*,t^*)\Big|_{r^{(CL)}=0} \quad (43.84)$$

Equivalently, defining $\aleph = E/D^{(CL)}(t^*,T)$, we have

$$C(\hat{r}_0,\hat{t}_0) = D^{(CL)}(\hat{t}_0,T)\int_{-\infty}^{+\infty}\Gamma[\,\hat{x}_0,x^*;\hat{t}_0,t^*]\left\{\Re^{(T)}(x^*,t^*) - \aleph\right\}_+ dx^*$$

$$(43.85)$$

We can rewrite the equation in terms of $r(t)$ using $x(t) = r(t) - r^{(CL)}(t)$. Since the classical path at fixed time is fixed, the partial spatial derivatives satisfy $\frac{\partial}{\partial x(t)}\Big|_{\text{fixed }t} = \frac{\partial}{\partial r(t)}\Big|_{\text{fixed }t}$. The partial time derivative picks up an extra term, however, since fixing $x(t)$ is not the same as fixing $r(t)$. We have

$$\frac{\partial}{\partial t}\Big|_{\text{fixed }r(t)} = \frac{\partial}{\partial t}\Big|_{\text{fixed }x(t)} - \frac{dr^{(CL)}(t)}{dt}\frac{\partial}{\partial x(t)}\Big|_{\text{fixed }t} \quad (43.86)$$

This relation can be seen by applying this identity to a function of $x(t)$ and t expanded as a double power series in $x(t)$ and t while using $x(t) = r(t) - r^{(CL)}(t)$.

Path Integrals and Options II: Interest Rates   585

The operator equation expressed in terms of $r(t)$ is then

$$\left.\frac{\partial}{\partial t}\right|_{\text{fixed } r(t)} = -\left\{\frac{dr^{(CL)}(t)}{dt} - \omega(t)\left[r(t) - r^{(CL)}(t)\right]\right\}\frac{\partial}{\partial r(t)}$$

$$-\frac{1}{2}\sigma^2(t)\frac{\partial^2}{\partial r(t)^2} + r(t) \qquad (43.87)$$

The derivative of the classical path involves differentiating the forward-rate function $f(t)$ that in practice is somewhat unstable. Therefore, Eqn. (43.78) can be preferable since it involves the classical path itself; the classical discounting factor is also useful.

This completes the discussion of the diffusion equation that is satisfied by the Green function. Any European contingent claim also satisfies this equation, since it is obtained by convoluting the Green function with its terminal value, consistent with the spatial boundary conditions.

A.3. Inclusion of Coupons for Bond Options in the MRG Model

To close this appendix, we give the details of how to include coupons for bond options in the MRG formalism. The limit of zero mean reversion gives the Gaussian model. A European call option on a coupon bond is given by the convolution of the Green function with the intrinsic value of the option on the forward bond at the exercise date, with strike price E,

$$C(\hat{r}_0, \hat{t}_0) = \int_{-\infty}^{+\infty} dr^* \, G(\hat{r}_0, r^*; \hat{t}_0, t^*) \left[\sum_{t_k > t^*}^{T} c_k \, P^{(t_k)}(r^*, t^*) - E\right]_+ \qquad (43.88)$$

In Eqn. (43.88), all coupons c_k at times t_k after the exercise date t^* to maturity date T are included in the forward bond[18] on which the option is written. At maturity T, we also need to include the par amount. The first "coupon" after t^* is the actual coupon reduced by a fraction equal to the time to that coupon payment from t^* divided by the time between coupon payments.

Now call r_c^* the interest rate at time t^* where the call intrinsic value is zero:

---

[18] **Forward Bond:** This is the value of the bond at the forward exercise time, dependent on whatever rates occur at the forward time. The coupons in the forward bond are the only coupons included, since the previous coupons have been paid before exercise. The idea is a little like the forward stock price that has previous dividend payments removed. See Ch. 42. The forward bonds are also used for forward CMT rates. See Ch. 10.

$$\sum_{t_k > t^*}^{T} c_k \, P^{(t_k)}(r_c^*, t^*) = E \qquad (43.89)$$

In general, this equation must be solved numerically for r_c^*. For $r^* \in (-\infty, r_c^*)$, the intrinsic value is positive since $c_k > 0$. Integration of Eqn. (43.88) then produces the European call option on a coupon bond as

$$C(\hat{r}_0, \hat{t}_0) =$$
$$\sum_{t_k > t^*}^{T} c_k \, P^{(t_k)}(\hat{r}_0, \hat{t}_0) N\left[\Phi_1^{(k)}(r_c^*)\right] - E P^{(t^*)}(\hat{r}_0, \hat{t}_0) N\left[\Phi_2(r_c^*)\right] \qquad (43.90)$$

Setting $T_{*k} = t_k - t^*$, $T_{0*} = t^* - \hat{t}_0$, and $f^* = f(t^*)$ we have

$$\Phi_1^{(k)}(r_c^*) = \frac{r_c^* - f^* + \frac{1}{\omega} \tilde{\sigma}_{0*}^2 T_{0*}\left[1 - e^{-\omega T_{*k}}\right]}{\tilde{\sigma}_{0*} \sqrt{T_{0*}}}$$

$$\Phi_2(r_c^*) = \frac{r_c^* - f^*}{\tilde{\sigma}_{0*} \sqrt{T_{0*}}} \qquad (43.91)$$

The zero-coupon bonds $P^{(t_k)}(\hat{r}_0, \hat{t}_0)$ in Eqn. (43.90) are given by the term structure data at the current time $\hat{t}_0$ as usual. If the coupons are removed, the equation (43.89) for r_c^* can be solved analytically, and the zero-coupon bond European option results of the text are reproduced.

Bermuda options with call schedules and American options are treated as mentioned in the text with coupons included for each forward bond as above. The numerical back-chaining algorithm must be employed.

Appendix B: Rate-Dependent Volatility Models

In this appendix, we deal with one-factor models with general volatility functions, but without memory effects. The lognormal interest-rate model is a special case, as well as the Gaussian models of the text. Combinations of lognormal and Gaussian models can also be incorporated[19]. In order to motivate the ideas, consider the Langevin-Ito equation for a function y of the interest rate $r(t)$, which we write as [20]

$$\frac{dy[r(t)]}{dt} = \mu_y(t) + \sigma_y(t)\eta(t) \qquad (43.92)$$

We can rewrite Eqn. (43.92) using the formula

$$\frac{dy[r(t)]}{dt} = y'[r(t)]\frac{dr}{dt} + \tfrac{1}{2} y''[r(t)]\frac{(dr)^2}{dt} \qquad (43.93)$$

Here, y' and y'' are the first and second derivatives of y with respect to r, evaluated at $r(t)$. Now we define

$$\sigma[r(t),t] = \sigma_y(t) / y'[r(t)] \qquad (43.94)$$

$$\mu[r(t),t] = \left[\mu_y(t) - \tfrac{1}{2} y''[r(t)]\frac{(dr)^2}{dt}\right] \bigg/ y'[r(t)] \qquad (43.95)$$

This produces the equivalent Langevin-Ito equation for $r(t)$ as

$$\frac{dr(t)}{dt} = \mu[r(t),t] + \sigma[r(t),t]\eta(t) \qquad (43.96)$$

Using this we find an alternative expression for μ,

$$\mu[r(t),t] = \frac{1}{y'[r(t)]}\left[\mu_y(t) - \tfrac{1}{2}\sigma_y^2(t)\frac{y''[r(t)]}{\{y'[r(t)]\}^2}\right] \qquad (43.97)$$

---

[19] **Lognorm Models:** Recently linear combinations of Gaussian and lognormal processes have been used. They have the imaginative name "Lognorm" models.

[20] **Acknowledgements:** F. Jamshidian recognized early the importance of this sort of transformation. I thank him for a helpful conversation on the topic.

Eqn. (43.96) gives the equivalence between the Langevin-Ito equation (43.92) for $y[r(t)]$ and that for a general volatility function in the Langevin-Ito equation for $r(t)$. The r-drift is specified once the volatility is given along with the y-drift.

The lognormal model is defined by $y(t) = F[r(t)] = \ln[\, r(t)\, T_s\,]$. Here T_s is a time scale (e.g. 1 year for annualized interest rates). We get $\sigma[r(t),t] = r(t)\,\sigma_{LN}(t)$ and $\mu[r(t),t] = r(t)[\mu_{LN}(t) + \tfrac{1}{2}\sigma^2_{LN}(t)]$, the usual results. In this special case, $y''/(y')^2 = -1$ is independent of r.

The path integral for the Green function for general volatility expressed in terms of the interest rate is given, as explained in Ref. [iv] and in this chapter, by the Langevin-Ito equation (43.96) and the Dirac delta-function procedure given in Appendix A. The Langevin-Ito equation is a constraint in the double integral at each time of the interest rate and the Langevin-Ito variable. The result is

$$G_F(r_a, r_b; t_a, t_b) = \prod_{j=a+1}^{b-1} \int_{r_L}^{r_U} dr_j \prod_{j=a}^{b-1} N_j \exp\left\{-r_j\,dt - \frac{dt}{2\sigma_j^2}\left[\left(\frac{dr(t)}{dt}\right)_{t_j} - \mu_j\right]^2\right\} \quad (43.98)$$

Here $N_j = \left[2\pi\,\sigma_j^2\,dt\right]^{-1/2}$ is the appropriate normalization factor, and $\mu_j = \mu[r(t_j), t_j]$, $\sigma_j = \sigma[r(t_j), t_j]$. The limits on the integral will depend on y. For example, a lognormal distribution has positive rates only.

We can also write the expression for the Green function in terms of the function y. This can be done in either of two ways: (1) rewriting the path integral Eqn. (43.98) through a change of variables, or (2) directly using Appendix A, starting with the Langevin-Ito equation for y, Eqn. (43.92). We write the discretization $y_j = y[r_j] = y[r(t_j)]$, with $r_j = r(y_j)$. The lognormal model has $r(y) = \exp(y)/T_s$. The Gaussian model has $r(y) = y$.

The Green Function for General Volatilities as a Path Integral
We obtain

$$G_y(r_a, r_b; t_a, t_b) = \prod_{j=a+1}^{b-1} \int_{R[y]} dy_j \cdot \prod_{\ell=a}^{b-1} N_{y,\ell} \exp\left\{-r(y_\ell)\,dt - \frac{dt}{2\sigma_y^2(t_\ell)}\left[\left(\frac{dy}{dt}\right)_{t_\ell} - \mu_y(t_\ell)\right]^2\right\}$$

$$(43.99)$$

In the change of variables in the path integral, the Jacobian enters to preserve the unit normalization for the probability density function. The normalization factors are $N_{y,\ell} = \left[2\pi \sigma_y^2(t_\ell) dt \right]^{-1/2}$.

This equation is just the convolution of the probability densities expressed in terms of y, with the discounting factors also expressed in terms of y. The limits on the integrals indicated by $R[y]$ are defined by the lower and upper limits on the possible numerical values of y. The function y can be chosen to limit the integrations over the interest rate by an infinite derivative y' forcing the volatility $\sigma_j = \sigma[\, r(t_j), t_j\,]$ to become zero, and thus setting barriers to the interest-rate process at the desired limits (r_L, r_U) in Eqn. (43.98). For the lognormal process, the lower bound occurs at zero interest rate.

The static term-structure constraints can be imposed using Eqn. (43.99) by choosing the y-drift parameters $\mu_y(t_\ell)$ one at a time using the procedure in Sect. II. This can be done since the y-drift depends (by assumption) only on the time. This would have to be done numerically as analytic expressions for the zero-coupon bond prices do not exist in closed form except for certain cases (e.g. Gaussian models). Even for the simple lognormal model, the term-structure constraints must be carried out numerically using an iterative technique.

Appendix C: The General Gaussian Model With Memory

In this appendix, we exhibit the calculation of a zero-coupon bond in the general model of Section V potentially including memory effects, and with constant A_{km} matrix elements as introduced in Ref. [vi]. We show two things. First, the term-structure constraints may be imposed as for the simpler Gaussian models. Second, the logarithm of the bond price is linear in the interest rate, with quadratic terms in the interest rate canceling out.

The zero-coupon bond is defined as usual,

$$P^{(t_b)}(r_a, t_a) = \int_{-\infty}^{+\infty} dr_b \, G(x_a, x_b; t_a, t_b) \, P^{(t_b)}(r_b, t_b) \qquad (43.100)$$

The terminal maturity condition is as usual, $P^{(t_b)}(r_b, t_b) = 1$. The calculation of the integrals is facilitated by organizing the integrals into those with indices $k, m = (a+1), \ldots, (b-1)$. We then perform the last integral over r_b. We define a sub-matrix B of the matrix A,

$$B_{km} = A_{km} \quad \text{with} \quad [k, m = (a+1), ..., (b-1)] \tag{43.101}$$

We also define the vector ζ with components

$$\zeta_k = 1 \quad \text{with} \quad [k = (a+1), ..., (b-1)] \tag{43.102}$$

We also define vectors J_a, J_b with components

$$J_{ak} = A_{ak}\,\Delta t \,,\; J_{bk} = A_{bk}\,\Delta t \quad \text{with} \quad [k = (a+1)...(b-1)] \tag{43.103}$$

Finally, we define $\alpha, \beta, \psi_a, \psi_b, \psi_c$ by the relations

$$\begin{aligned}
\alpha &= A_{bb}\,(\Delta t)^2 - J_b \cdot B^{-1} \cdot J_b \\
\beta &= -A_{ab}\,(\Delta t)^2 + J_a \cdot B^{-1} \cdot J_b \\
\psi_a &= \zeta \cdot B^{-1} \cdot J_a \\
\psi_b &= \zeta \cdot B^{-1} \cdot J_b \\
\psi_c &= \zeta \cdot B^{-1} \cdot \zeta
\end{aligned} \tag{43.104}$$

Here B^{-1} is the matrix inverse of B, and the sums on the indices of the vectors and of B^{-1} from $(a+1)$ to $(b-1)$ are implied by the dots.

After Gaussian integration, the discretized Green function is found to be (again $\Delta t < 0$ for backward propagation),

$$G(x_a, x_b; t_a, t_b) = \exp\left\{\sum_{j=a}^{b-1} r_j^{(CL)}\Delta t - \psi_a x_a + \tfrac{1}{4}\psi_c\right\} \frac{\exp\left[-\alpha x_b^2 + (2\beta x_a - \psi_b)x_b\right]}{\int_{-\infty}^{+\infty} dx_b \exp\left[-\alpha x_b^2 + 2\beta x_a x_b\right]} \tag{43.105}$$

Performing the final integration, the zero-coupon bond is

$$P^{(t_b)}(r_a, t_a) = \exp\left\{\sum_{j=a}^{b-1} r_j^{(CL)}\Delta t - x_a\left[\psi_a + \frac{\beta \psi_b}{\alpha}\right] + \left[\tfrac{1}{4}\psi_c + \frac{\psi_b^2}{4\alpha}\right]\right\} \tag{43.106}$$

Hence, the logarithm of the bond price is indeed linear in the co-ordinate $x_a = r_a - r_a^{(CL)}$. Moreover, the drift factor enters in the canonical fashion through the classical path; the procedure outlined in Section II for determination of the drifts therefore holds.

Given the general Gaussian form for the Green function Eqn. (43.105), contingent claims like options can be found in a straightforward manner, consistent with the static term-structure constraints.

Figure Captions for This Chapter

Figure 1: Term-Structure Constraints. Determination of the drift function in the static term-structure-constrained models. The vertical maturity T axis at the initial time $\hat{t}_0$ is partitioned in the same way as in the future-time axis. One by one, as shown by the arrows, each drift $\mu_j = \mu(t_j)$ between future times (t_j, t_{j+1}) is determined by the bond at the initial time $\hat{t}_0$ with maturity date $T = t_{j+2}$.

Figure 2: Stochastic Equation Variables. This is a graphical illustration of the connection of the Langevin-Ito variable $\eta_j = \eta(t_j)$ and the co-ordinate $x_j = x(t_j)$ at time t_j for a given discretization into n time intervals. As described in the text, this variable (multiplied by the volatility and corrected for mean reversion) describes the slope of the straight line drawn between x_j and x_{j+1} in one realization. The random slopes produce random paths between the fixed endpoints x_a at time t_a and $x_b = x_{a+n}$ at time $t_b = t_{a+n}$.

Figure 3: The Classical Path. A path (or realization of the random process) for the interest rate $r(t)$ as time varies. The random paths fluctuate about the classical interest rate path labeled by $r^{(CL)}(t)$. The process between times (t_a, t_b) is discretized into n points labeled as $r_1, r_2, \ldots, r_n$ at times $t_1, t_2, \ldots, t_n$. This drawing is taken from Ref. [vi].

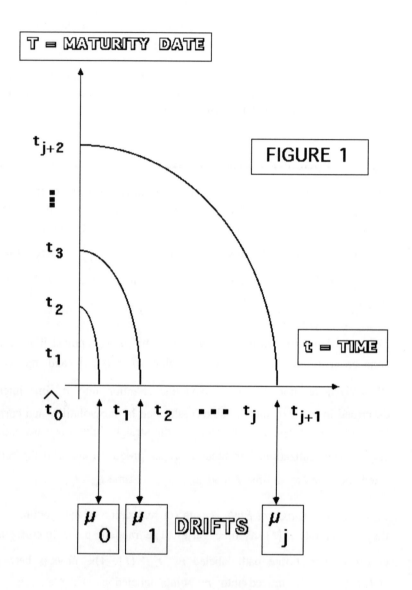

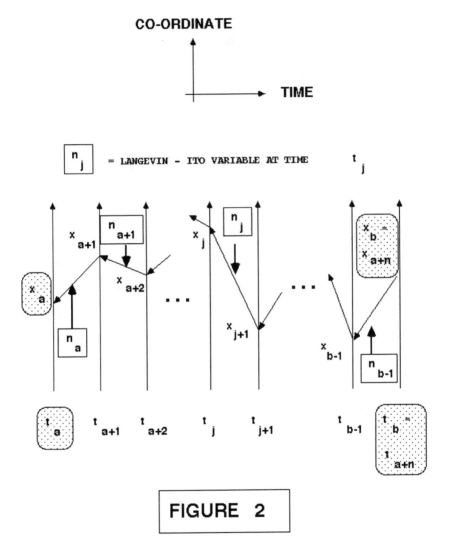

FIGURE 2

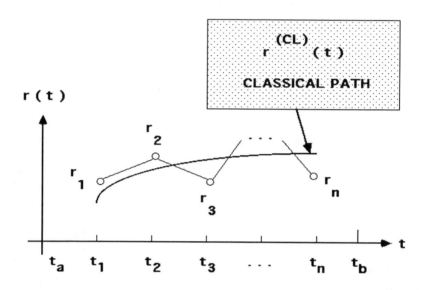

Figure 3

References

[i] **Path Integrals and Term Structure Interest-Rate Modeling**
Dash, J. W., *Path Integrals and Options II: One-Factor Term-Structure Models*, Centre de Physique Théorique, CNRS, Marseille preprint CPT-89/PE.2333, 1989.

[ii] **Hull-White, Jamshidian**
Hull, J. and White, A., *Pricing Interest Rate Derivative Securities*. Review of Financial Studies (3,4) 1990. Pp. 573-92, and references therein.
Jamshidian, F., *An Exact Bond Option Pricing Formula*, Journal of Finance, Vol 44, 1989, Pp. 205-209.
Jamshidian, F., *Pricing of Contingent Claims in the One-Factor Term Structure Model*, Merrill Lynch, 1987.
Jamshidian, F., *Closed-Form Solutions for American Options on Coupon Bonds in the General Gaussian Interest Rate Model*, Merrill Lynch 1989.

[iii] Feynman's Path Integral
Feynman, R. P., *Space-Time Approach to Non-Relativistic Quantum Mechanics*, Reviews of Modern Physics, Vol. 20, 1948. Pp. 367-387.
Feynman, R., Hibbs, A., *Quantum Mechanics and Path Integrals*, McGraw-Hill. 1965.

[iv] Path Integrals in Finance – Other References (cf. also Ch. 42)
Dash, J. W., *Path Integrals and Options – I*. Centre de Physique Théorique, CNRS Marseille preprint CPT-88/PE. 2206, 1988.
Geske, R. and Johnson, H. E., *The American Put Option Valued Analytically*, J. Finance, Vol. 39, 1984. Pp. 1511-1524.
Garman, M. B., *Toward a Semigroup Pricing Theory*, J. Finance, 40, 1985. Pp. 847-861.
Linetsky, V., *The Path Integral Approach to Financial Modeling and Options Pricing*. Univ. Michigan College of Engineering report 96-7, 1996.

[v] Ho-Lee Model
Ho, T. and Lee, S., *Term Structure Movements and Pricing Interest Rates Contingent Claims*, Journal of Finance 41(5), 1986. Pp. 1011-1029.

[vi] Generalized Gaussian Model
Dash, J. W., *Functional Path Integrals and the Average Bond Price*, Centre de Physique Théorique, CNRS, Marseille preprint CPT-87/PE.2077, 1986.

[vii] Mean-Reverting Gaussian Model – Original References
Uhlenbeck, G. E. and Ornstein, L. S., *On the Theory of Brownian Motion*, Physical Review, Vol. 36., 1930. Pp. 823-841.
Vasicek, O. A., *An Equilibrium Characterization of the Term Structure*, Journal of Financial Economics 5, 1977. Pp. 177-188.

[viii] Reviews
Pliska, S., *A Mathematical Analysis of Interest Rate Derivatives Pricing Models*, Risk Seminar, Advanced Mathematics for Derivatives, 1994.
Berger, E., *Modeling Future Interest Rates: Taming the Unknowable*. Bloomberg Magazine, September 1994. PP. 33-39.

[ix] Black, Derman, Toy (BDT) Model
Black, F., Derman, E., and Toy, W., *A One-Factor Model of Interest Rates and Its Application to Treasury Bond Options*. Goldman Sachs working paper, 1987.

[x] Cox, Ingersoll, Ross (CIR) Model and Related Topics
Cos, J. C., Ingersoll, J. E., and Ross, S.A., *A Theory of the Term Structure of Interest Rates*. Econometrica, Vol 54, 1985. Pp. 385-407.
Jamshidian, F., *A Simple Class of Square –Root Interest Rate Models*. Fuji International Finance working paper, 1992.
Peak, D. and Inomata, A., *Summation over Feynman Histories in Polar Co-ordinates*. Journal of Mathematical Physics, Vol. 10, 1969. Pp. 1422-1428.
Edwards, S. F. and Gulyaev, Y. V., *Path integrals in polar co-ordinates*. Proc. Roy. Soc. Lond., Vol A279, 1964. Pp. 220-235.
Langouche, F., Doekaerts, D., and Tirapegui, E., *Short derivation of Feynman Lagrangian for general diffusion processes*. J. Phys. A, Vol. 13, 1980. PP. 449-452.

Khandekar, D. C. and Lwande, S. V., *Feynman Path Integrals: Some Exact Results and Applications*. Physics Reports, Vol 137, 1986. See sect. 6.1.1.

Inomata, A., *Reckoning of the Besselian Path Integral*. Contribution to *On Klauder's Path*, G. Emch et. al. Editors, World Scientific, 1994. Pp. 99-106.

Kleinert, H., *Path Integrals in Quantum Mechanicsw, Statistics, and Polymer Physics*, 2nd Ed. World Scientific, 1995. See Chs. 8, 12.

Benneti, E., Rosa-Clot, M. and Taddei, S., *A Path Integral Approach to Derivative Security Pricing: I. Formalism and Analytical Results*. U. Firenze, 1999 (arXiv: cond-mat/9901277); Int. J. Theor. App. Fin., Vol 2, 1999. Pp. 381-407. See App. C.

[xi] Heath, Jarrow, Morton (HJM) Model

Heath, D., Jarrow, R., and Morton, A., *Bond Pricing and the Term Structure of Interest Rates: A New Methodology*, Cornell U. 1987, rev. 1988.

Heath, D., Jarrow, R., and Morton, A., *Bond Pricing and the Term Structure of Interest Rates: A Discrete Time Approximation*, Cornell U. 1988.

[xii] Market Models

Brace, A., Gatarek, D., and Musiela, M., *The Market Model of Interest Rate Dynamics*. Mathematical Finance, Vol 7 No. 2, 1997. Pp. 127-155.

Jamshidian, F., *LIBOR and swap market models and measures*. Finance and Stochastics, Vol. 1, 1997. Pp. 293-330.

Anderson, L. and Andreasen, J., *Volatility Skews and Extensions of the Libor Market Model*. General Re Financial Products working paper, 1998.

Reed, N., *If the Cap Fits....* Risk Magazine, Vol. 8, August 1995.

[xiii] Affine Models

Duffie, D. and Kan, R., *A Yield-Factor Model of Interest Rates*. Stanford U. working paper, 1993.

[xiv] Bond-Price Model

Flesaker, B. and Hughston, L., *Positive Interest*. Risk Magazine, Vol. 9, No. 1, 1996.

[xv] Differential Geometry Approach

Hughston, L. P., *Stochastic Differential Geometry, Financial Modelling, and Arbitrage-Free Pricing*. Merrill Lynch working paper, 1994.

[xvi] Dirac δ-functions and Distribution Theory

Friedman, B., *Principles and Techniques of Applied Mathematics*, John Wiley, 1956.

Dirac, P. A. M., *Principles of Quantum Mechanics*, Clarendon Press, 1947.

Schwartz, L., *Theorie des distributions*, Actualites scientifiques et industrielles, No. 1091 and 1122, Hermann & Cie, 1950, 1951.

44. Path Integrals and Options III: Numerical (Tech. Index 6/10)

Summary of this Chapter

This chapter presents some aspects of numerical methods for options based on path-integral techniques. We have already emphasized the connection between the binomial algorithm (or any lattice method), Monte-Carlo simulations, and path integrals. A major topic here is the Castresansa-Hogan method for discretizing path integrals[1]. Some simplifying approximations are discussed. An iterative procedure based on "call filtering" for Bermuda options leads to a "quasi-European" approximation. The idea of "geometric volatility" is introduced. We also present an approximation to lognormal dynamics using a mean-reverting Gaussian designed to speed up calculations[2].

Some other aspects of numerical analysis have been treated in other chapters in this book, to which we refer the reader.

Introduction to this Chapter

In previous chapters, the Feynman/Wiener path integral was applied to options in a variety of examples. This chapter examines some aspects of numerical techniques using path integrals as a base. The practicality of direct path-integral discretization was solved by Castresana and Hogan. This approach will be discussed in some detail. All the usual numerical techniques (binomial, multinomial, grid, Monte Carlo) are approximations to the path integral. Because standard texts treat these topics in detail[i], we restrict ourselves here to some interesting aspects of numerical methods motivated by the explicit path integral.

One advantage of using the explicit path-integral formalism involves the elimination of most of the discretization as a grid or lattice in those cases where the Green function (i.e. propagator) can be computed in closed form. Then the

---

[1] **Castresana-Hogan Method:** Juan Castresana implemented the Castresana-Hogan method before 1990. Juan says that Marge Hogan had the basic idea while working in the aircraft industry.

[2] **History:** This chapter started as a sequel for the path-integral paper series in 1989. My work here was performed mostly during 1987-97.

exact solution can be used over those time intervals where free diffusion occurs. Such intervals are not supposed to contain any exercise date or other date when cash flows need to be computed. This eliminates convergence problems between exercise dates (since the exact propagator is used). Further, since the time intervals can be chosen arbitrarily in the path integral, the awkwardness of dates that are not exactly at grid or lattice points in the standard construction is eliminated.

In pioneering work, Geske and Johnson derived a special case of path-integral methods, for multiple-put options. However, the numerical algorithm they chose was slow[3].

Path Integrals and Common Numerical Methods

To start the discussion and for motivation, we consider the connection between the path integral, the binomial approximation, and Monte Carlo simulation.

Path Integrals, MC Simulations, and the Binomial Approximation

Often a binomial numerical recipe is used for evaluating options. Essentially this means that the random numbers generating the paths are replaced by fixed numbers allowing only stylized "up" and "down" movements. The binomial geometry is defined as having one node containing two outgoing legs "up" and "down" as time increases. The binomial lattice is supposed to recombine at each time t_j. The number of points of the binomial lattice in x_j at a given t_j is then $j+1$, not 2^j. Smaller bin sizes, i.e. more points in x_j at t_j are therefore connected with smaller values of Δt and thus more time steps.

See the footnote for a story [4].

---

[3] **Acknowledgements:** R. Geske, private communication.

[4] **Story: Binning the Paths, Risk Talks, and Business Trips:** This connection between Monte Carlo simulation and the binomial approximation using bins is an old idea. The picture (see next page) is from a talk I used to give when I worked for Eurobrokers in 1990. Eurobrokers would rent conference rooms in nice hotels in various cities in Europe. The talk was on risk management for interest-rate derivatives, attended by analysts and traders. The talk was followed by a lot of good food. Then the next day, we would go around to various banks in the city to drum up business. It was fun and it even worked. I thank Don Marshall for his managerial congeniality and acumen in setting all this up.

The Binomial Approximation to a Monte Carlo Simulation

We now consider how the binomial approximation can be connected to Monte Carlo simulation of the path integral for finite time steps and finite sized bins. The following figure shows the idea for three steps.

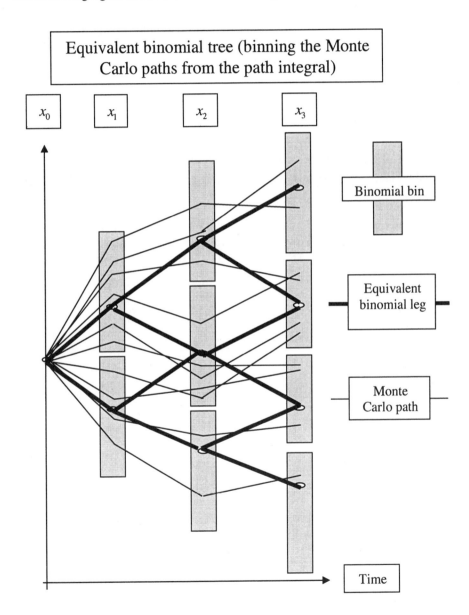

The bins in x_j can be defined through integrated values of the measure from the initial time up to time t_j such that the binomial model geometry appears. Equivalently, we define bins just by counting MC paths (in a large simulation) to reproduce binomial probabilities: e.g. $\left(\frac{1}{8}, \frac{3}{8}, \frac{3}{8}, \frac{1}{8}\right)$ at the third step (x_3, t_3). Therefore, for a given MC simulation we can draw both the MC paths and the equivalent binomial tree with legs running between the bin centers[5]. We discuss the subject in more detail later in the chapter.

Basic Numerical Procedure using Path Integrals

The possible paths from the path integral "fan out". As opposed to the binomial algorithm, the path-integral discretization is allowed to be non-uniform in the spatial variable $x(t)$. This is useful in valuing complex options using importance sampling with increased refinement in regions of sensitivity in $x(t)$.

In addition, the times can be chosen non-uniformly corresponding, for example, to cash-flow payments. For the case of free propagation exhibited below, the free propagator G_f can be used over finite times. In the general case, the diagram still holds but the full path-integral must be used for the propagator.

The next figure shows the complete discretization for the first two steps in a possible numerical approximation using the path integral.

---

[5] **Is the Binomial Approximation a "model"?** Often this binomial algorithm is called a "model", as in the salesman asking, "Do you guys use a binomial model to price options?" This common appellation is inappropriate. The definition of a model should include the assumptions and the parameters, not just the numerical algorithm. Calling a numerical recipe as a model leads to misleading statements about model accuracy by only focusing on numerical convergence and not on the (possibly much greater) uncertainties of the assumptions and parameters. See Ch. 33 on model quality assurance. Nonetheless, the language is so common that everybody lapses into it.

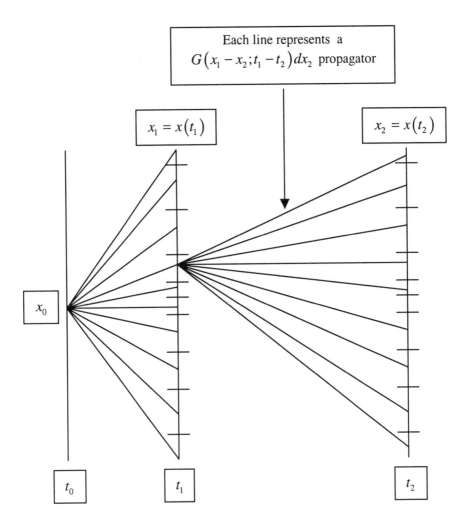

Bins and Path Integrals

Consider the following figure:

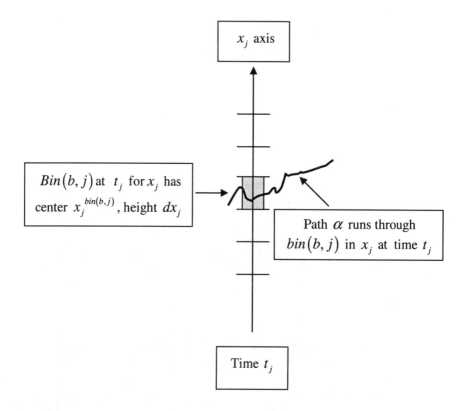

We already discussed binning paths when we talked about the connection with the binomial approximation. In general, paths between the starting point x_0 and end point $x*$ run through all allowed values of $\{x_j\}$ for $j = 1,...,n-1$. We consider these values to lie inside "bins". The bins are vertical intervals with zero widths and heights $\{dx_j\}$ on the $\{x_j\}$ axes, centered around points $\{x_j^{bin(b,j)}\}$. Note that dx_j is an ordinary calculus differential, not a time difference.

The bins[6] serve as intervals for a numerical discretization of the $\{x_j\}$ ordinary integrals. Thus, a value of the variable x_j at time t_j on a given path α

---

[6] **More on Bins and Numerical Approximations:** These bins are important. First, as we describe in the next section, all paths passing through a bin b at time tⱼ can be

called $x_j^{\{path\ \alpha\}}$ lies inside a bin interval for a specific $bin(b,j)$, namely $x_j^{\{path\ \alpha\}} \in \left(x_j^{bin(b,j)} - dx_j/2 \ , \ x_j^{bin(b,j)} + dx_j/2 \right)$.

The path-integral probability measure weight for the set of points $\{x_j\}$ inside bins of size $\{dx_j\}$ is the product of $n-1$ measures $G(x_j - x_{j+1}; t_j - t_{j+1})dx_j$ for successive propagation through $x_1, x_2, ..., x_{n-1}$ from initial time t_0 through intermediate times $t_1, t_2, ..., t_{n-1}$ multiplied by a final free propagator $G(x_{n-1} - x^*; t_{n-1} - t^*)$ to propagate from time t_{n-1} to the final time t*. All possibilities of the values of $\{x_j\}$, or equivalently all paths, are then summed over by the integrals over all intermediate x_j values from $-\infty$ to ∞.

The paths themselves can be produced by randomly choosing the values of the $\{x_j\}$ using the measure in the path integral. This is just Monte Carlo simulation. We discuss the relation between path integrals and Monte Carlo simulation in a little more detail below.

The Castresana-Hogan Path-Integral Discretization

We now arrive at the most important section for this chapter. Juan Castresana and Marge Hogan, with brilliant insight, numerically discretized the path integral in the 1980's. This discretized path-integral method has been used in production[7].

The issues for numerical evaluation of complex interest rate derivatives involve obtaining numerical efficiency and speed. The term structure constraints in addition to any boundary conditions specific to the problem must be satisfied. The main innovation here is to use a "nominal interest-rate grid" that simplifies the calculations. This nominal grid is mapped into the real interest rate grid as determined through the term-structure constraints.

---

approximately collapsed to the central point of the bin. Thus, a lattice model can be constructed from the Monte-Carlo approximation to the path integral. This approximation becomes more accurate for smaller bin sizes. Note also that the bin sizes dx_j can depend on the bin b for importance sampling using the overall probability measure for a given starting point, or equivalently for the density of paths. The bin size can also depend on t_j which can be important if, for example, we are near a possible exercise time or other time of interest in which more or less accuracy is desired. Again, note that dx_j is an ordinary integration measure here, not a time-differenced coordinate.

[7] **Acknowledgements:** I thank Juan Castresana for much hard and dedicated work performing many numerical calculations over the years in two of my quant groups, including using his path-integral discretization method.

The Castresana-Hogan Uniform Nominal Interest Rate Grid

We use the language of interest rates, although the method is general and has been applied to equity products, etc. The main idea starts with a nominal interest rate grid. This grid is *uniform*. The nominal interest rate at time t_j takes on values $R_j^{(\beta)}$ with $\beta = 1...M$ an index. The picture gives the idea.

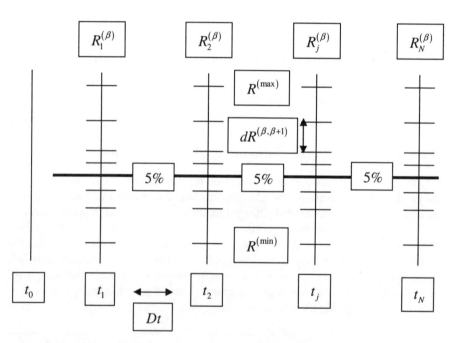

The main point is that the value $R_j^{(\beta)}$ at a given β is *independent* of j, so $R_j^{(\beta)} = R^{(\beta)}$. In the picture, at every time in the grid, the fourth point up is, for example, 5%. Thus, in the drawing, $R^{(\beta=4)} = 5\%$. This is what is meant by "uniformity", namely time-independence.

Note carefully, however, that the spacing between the nominal rates at fixed time t_j, namely $dR^{(\beta,\beta+1)} \equiv R^{(\beta+1)} - R^{(\beta)}$, does *not* need to be constant, as indicated in the figure[8]. This can be used to advantage. The density of points can be increased in regions where more accuracy is desired, for example near a decision rate for cash flow determination or option exercise logic.

The Reason for the Uniform Grid

The reason that the grid points are made time-independent is that the transition probabilities are defined directly on the nominal grid. *Because* the nominal grid is uniform, if the time interval and volatility are fixed, then the transition probabilities $P_{j-1,j}^{(\beta',\beta)}$ between nodes at neighboring times on the grid will be independent of time. For that reason the transition probabilities can be calculated once and then reused. This results in a considerable saving of computer time.

If a volatility term structure is used some time savings can still occur. This is because the volatility term structure is typically determined in practice by a fewer number of constraints (caps, swaptions) than time points in the partition. Hence, if a step-function approximation to the volatility term structure is used, several intervals in the grid can still wind up using the same volatility. For example, if there is a 2-year cap and a 4-year cap used for volatility constraints but no other constraint between 2 and 4 years, then the same forward volatility can be used for all transitions between 2 and 4 years and the probabilities reused in that interval.

The Transition Probabilities on the Uniform Grid

The transition probabilities, as mentioned above, are defined directly on the nominal grid. For lognormal dynamics, we get a familiar-looking form for $P_{j-1,j}^{(\beta',\beta)}$ from node $R^{(\beta')}$ at time t_{j-1} to node $R^{(\beta)}$ at time t_j with volatility σ_{j-1} and finite time interval $Dt = t_j - t_{j-1} > 0$. We lump points between $R^{(\beta-1)}$ and $R^{(\beta)}$ together in a bin. Then $P_{j-1,j}^{(\beta',\beta)}$ is given by the difference of two normal integrals,

$$P_{j-1,j}^{(\beta',\beta)} = N\left[\frac{\ln\left(R^{(\beta)}/R^{(\beta')}\right) + \sigma_{j-1}^2 Dt/2}{\sigma_{j-1}\sqrt{Dt}}\right] - N\left[\frac{\ln\left(R^{(\beta-1)}/R^{(\beta')}\right) + \sigma_{j-1}^2 Dt/2}{\sigma_{j-1}\sqrt{Dt}}\right]$$

(44.1)

---

[8] **Notation:** Careful. Note that this differential $dR^{(\beta,\beta+1)}$ is the difference between neighboring nominal rates at a *fixed* time, *not* a change in rates at neighboring times.

The Number of Points at Fixed Time

The number M of nominal grid points at fixed time should be regarded in the sense of an ordinary integration partition. For example, we can set $M = 100$. There are M transitions on the grid from any given point to the various points at the neighboring time. However, considerable time is saved by cutting off the number of transitions if the probability of transition $P_{j-1,j}^{(\beta',\beta)}$ is below a small value.

The Rectangular Grid and "Pruning the Tree"

The rectangular geometry of the nominal grid is important. It automatically cuts off rates that otherwise would become arbitrarily large. This cuts down calculation time and avoids unphysical rates. Binomial tree implementations also often use such spatial cutoffs, sometimes called "pruning the tree"[9].

Minimum and Maximum Nominal Rates

The minimum and maximum nominal rates are set up to cover the range of all physically reasonable rates. For example, for the nominal grid for US treasury rates we could write $R^{(min)} = 50$ bp and $R^{(max)} = 20\%$. These are to be regarded as minimum and maximum cutoff points for ordinary integrals. If they need to be extended (for example dealing with credit products with spread), they can simply be redefined.

The First Time Interval

The first time interval is a special case because the time to the first cash flow from the value date is arbitrary. Although this is not required, often the rest of the grid points are taken as spaced equally in time. The grid times are placed at cash-flow points. A typical value of the spacing Dt is 6 months, since coupons in bonds are often semiannual. However, Dt can be any value, and it can depend on t_j.

The Physical Interest Rates and the Term Structure Constraints

The interest rates $\left\{R^{(\beta)}\right\}$ in the nominal grid are *NOT* the physical interest rates. The physical interest rates $\left\{r_j^{(\beta)}\right\}$ are obtained through a mapping of the nominal

---

[9] **Acknowledgements:** I first heard about "pruning the tree" from David Haan at Merrill Lynch in 1986.

interest rates. The mapping introduces parameters $\{\alpha_j\}$, one for each time t_j in the partition[10], as follows:

$$r_j^{(\beta)} = \alpha_j R^{(\beta)} \qquad (44.2)$$

This mapping is a special case of the transformation of interest rates discussed in Appendix B in the last chapter. It is set up so that with a lognormal assumption the $\{\alpha_j\}$ parameters are directly related to the drift. However, the mapping can be used with any probability dynamical assumption.

Determination of the Alpha Parameters via Term-Structure Constraints

The $\{\alpha_j\}$ parameters are determined one at a time from the term structure constraints. That is, we evaluate discount factors $P^{(J_T)} = \left\langle \prod_{j=1}^{J_T} \dfrac{1}{\left(1 + r_j^{(\beta)} \cdot Dt\right)} \right\rangle$

for various maturities $T = J_T \cdot Dt$ using various values of J. Here, the bracket $\langle \ \rangle$ indicates the expectation with respect to the probability measure. This expectation involves the spatial index β. The time between partition points $Dt = t_j - t_{j-1}$ is typically independent of j.

Using the mapping Eqn. (44.2) we get

$$P^{(J_T)} = \left\langle \prod_{j=1}^{J_T} \dfrac{1}{\left(1 + \alpha_j R^{(\beta)} \cdot Dt\right)} \right\rangle \qquad (44.3)$$

We then determine the various $\{\alpha_j\}$ parameters from equating these discount factors to known zero-coupon bond prices.

Path Integral Discretization and Stochastic Equations

We have emphasized repeatedly that the path integral formalism is not only consistent with the stochastic equations, but the path integral is explicitly constructed using stochastic equations as constraints. In this section we re-

---

[10] **Notation:** Please do not confuse the Castresana $\{\alpha_j\}$ parameters with the Monte-Carlo path label α or the Laplace transform parameter α_0 (see definition below).

emphasize this fact and show how the uniform grid is connected with stochastic equations. To do this, we assume lognormal dynamics. Write

$$y_j^{(\beta)} = \ln r_j^{(\beta)} = \ln \alpha_j + \ln R^{(\beta)} \qquad (44.4)$$

Consider the transition from some node $y_{j-1}^{(\beta')}$ at time t_{j-1} to the node $y_j^{(\beta)}$ at time $t_j = t_{j-1} + Dt$. Then the time rate of change $d_t y_{j-1}^{(\beta')} = y_j^{(\beta)} - y_{j-1}^{(\beta')}$ over time interval Dt is $d_t y_{j-1}^{(\beta')} = \mu_{j-1} Dt + \sigma_{j-1} dz_{j-1,j}^{(\beta',\beta)} \sqrt{Dt}$ where $dz_{j-1,j}^{(\beta',\beta)}$ is a Gaussian random variable with zero mean and width one.

Identifying the drift term from the stochastic equation with the drift term from Eqn. (44.4) produces $\mu_{j-1} Dt = \ln(\alpha_j / \alpha_{j-1})$. This establishes the connection between the drift and the alpha parameters for the lognormal model.

This ends the description of the Castresana-Hogan method.

Some Numerical Topics Related to Path Integrals

In this section, we give some results for numerical analysis in the path integral framework.

The Path Integral, MC Simulation, and Lattice Approximations

So far, the path integral may sound different from the paths generated in a cone by a binomial algorithm, or other lattice algorithm. However, as the time interval vanishes $\Delta t_j = t_j - t_{j+1} \to 0$ the results are the same. This occurs because paths with large $x_j - x_{j+1}$ for small Δt are numerically suppressed by the Gaussian damping in G. Moreover, a direct connection between MC simulations and lattice models can be constructed.

As briefly discussed above, in an approximate sense all paths passing through a given bin, $bin(b, j)$ at time t_j, can be collapsed to the central point of the bin and forming a point of a lattice. Thus, a lattice model can be constructed from the path integral, and this approximation becomes increasingly accurate for smaller bin sizes. Such an approximation can be useful, because logic often needs to be performed comparing different quantities at each intermediate time for complex options. Reducing the number of points at which calculations are performed facilitates this task.

Consider the following picture that illustrates the ideas:

Path Integrals and Options III: Numerical

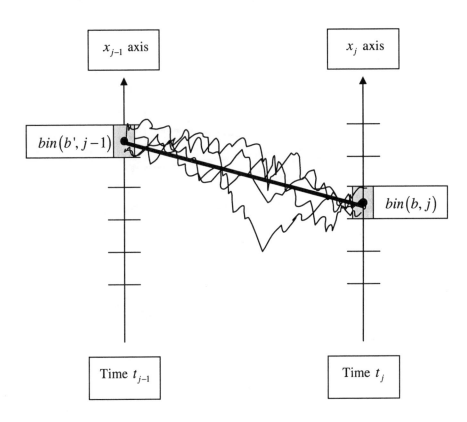

All $N_{j-1,j}^{b',b}$ paths from $bin(b', j-1)$ at t_{j-1} in x_{j-1} to $bin(b, j)$ at t_j in x_j are collapsed to one effective path with weight equal to $N_{j-1,j}^{b',b}$

Monte Carlo Simulation Using Path Integrals

Monte Carlo simulation can be done in the standard brute-force method simply by using the stochastic equations. However, a better idea is to use the binning procedure. We generate paths between bin centers, as shown in the figure above.

Consider the following picture.

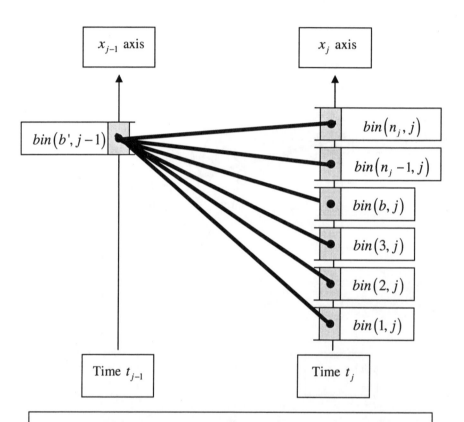

The probability of a path from $bin(b', j-1)$ at t_{j-1} in x_{j-1} to a given $bin(b, j)$ at t_j in x_j is determined by the appropriate propagator.

Denote n_j as the number of bins at time t_j in the x_j variable. The bins need not have the same heights; this provides useful flexibility. Call h_b the height of the $bin(b, j)$. Then $G\left[x_{j-1}^{bin(b', j-1)}, x_j^{bin(b, j)}; t_{j-1}, t_j\right] \cdot h_b$ (not including discounting) is approximately the probability of starting at a point in

$bin(b', j-1)$ and ending somewhere in $bin(b, j)$. Clearly, we have $\sum_{b=1}^{n_j} G\left[x_{j-1}^{bin(b',j-1)}, x_j^{bin(b,j)}; t_{j-1}, t_j\right] \cdot h_b = 1$. We segment the unit interval into partitions with lengths $G\left[x_{j-1}^{bin(b',j-1)}, x_j^{bin(b,j)}; t_{j-1}, t_j\right] \cdot h_b$. Every bin has a unique position corresponding to the integrated probability. Now we generate random numbers $\{R_\alpha^\#\}$ from the uniform distribution $U(0,1)$. Given a particular random number $R_\alpha^\#$, one of the various bins at t_j is chosen for $Path_\alpha$ from $bin(b', j-1)$ just corresponding into which segment of the partitioned unit interval $R_\alpha^\#$ falls. An example of the idea is in the picture below:

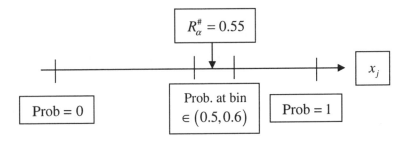

Applications of the Hybrid Monte-Carlo/Lattice Approach

One advantage of this hybrid lattice Monte-Carlo method is that standard lattice back-chaining methods can be used as appropriate for the problem. Then, Monte Carlo simulation can be used following the paths generated as above. In this way, Monte-Carlo simulation can be applied to problems that would otherwise be intractable. One example is Bermuda or American options, discussed below[ii].

Langsam, Broadie, Glasserman, Boyle, Longstaff, and others have contributed to Monte-Carlo evaluation of Bermuda/American options.

Another example is index-amortizing swaps (IAS), where the payoffs are path dependent. We discussed IAS in Ch. 16.

The Standard Back-Chaining Bermuda Algorithm

This section gives details of backward diffusion for a bond with an embedded Bermuda call using the path integral language. The idea is ancient, but we include it for completeness here.

Do We Need the Critical Path?

We show the steps either using the full bond P or else the "bullet" B (the bond with the call removed). If we deal directly with the full bond P, we do *not* need the critical/Atlantic path describing the boundary for option exercise. If we use the bullet bond B, so that the embedded option is considered separately, we do need the critical path. For more information on the critical path, see the Quasi-European approximate solution below.

Steps in the Back-chaining Algorithm

Step 1. Start from last call date t_L^*. Call $P_L^{(+)}$ the full bond price one second after that date; it is the bullet B_L with all coupons after t_L^* and evaluated at t_L^* (recall bullet bonds contain no embedded options). We need to know the bond price $P_L^{(-)}$ one second before t_L^*. Denote the full bond strike price at t_L^* by E_L and the bullet critical path point at t_L^* by $\hat{P}_L$. If $P_L^{(+)} > E_L$ or $B_L > \hat{P}_L$, then $P_L^{(-)} = E_L$ and the bond is called. However, if $P_L^{(+)} < E_L$ or $B_L < \hat{P}_L$, then $P_L^{(-)} = P_L^{(+)}$, and the bond is not called. We have $\hat{P}_L = E_L$. Alternatively, we can write $P_L^{(-)} = \min \left\{ P_L^{(+)}, E_L \right\}$.

Step 2: Propagate backwards from time t_L^* to time t_{L-1}^*, using the exact propagator. We get the bond price $P_{L-1}^{(+)}$ one second after t_{L-1}^* as

$$P_{L-1}^{(+)}(r_{L-1}^*, t_{L-1}^*) = \sum_{\text{bins } b(r_L^*)} G(r_{L-1}^*, r_L^*; t_{L-1}^*, t_L^*) \, P_L^{(-)}(r_L^*, t_L^*) h_{b(r_L^*)}$$

$$+ \text{ Bullet at } t_{L-1}^* \text{ for all coupons between } t_{L-1}^* \text{ and } t_L^* \qquad (44.5)$$

Step 3: The bond price $P_{L-1}^{(-)}$ one second before t_{L-1}^* includes the possibility of calling the bond at t_{L-1}^* with strike price E_{L-1}, where the bullet critical path point is $\hat{P}_{L-1}$. If $P_{L-1}^{(+)} > E_{L-1}$ or $B_{L-1} > \hat{P}_{L-1}$, then $P_{L-1}^{(-)} = E_{L-1}$, and the bond is called. However, if $P_{L-1}^{(+)} < E_{L-1}$ or $B_{L-1} < \hat{P}_{L-1}$, then $P_{L-1}^{(-)} = P_{L-1}^{(+)}$, and the bond is not called. For low enough rates r^*_{L-1}, we know the bond will be called. Thus, we start at large values of r^*_{L-1} where the bond is not called. Then stepping values down for r^*_{L-1} we stop calculating $P^{(+)}_{L-1}$ or B_{L-1} in step #2 the first time either inequality $P^{(+)}_{L-1} > E_{L-1}$ or $B_{L-1} > \hat{P}_{L-1}$ is satisfied.

Step 4: Continue steps #2 and #3 by recursion, replacing $L-1$ by λ and L by $\lambda+1$. The recursion is carried out back for each λ to $\lambda=1$ in order to obtain $P^{(-)}_\lambda$ from $P^{(+)}_\lambda$. The recursion is carried out back for each λ to $\lambda=0$ to obtain $P^{(+)}_\lambda$ from $P^{(-)}_{\lambda+1}$.

Comments for the Back-chaining Bermuda Algorithm

Comment for Step #2. The propagator $G(r^*_{L-1}, r^*_L; t^*_{L-1}, t^*_L)$ is the Green function, including discounting. It is exact, with no grid needed for the backward-diffusion propagation between exercise dates. For Gaussian and mean-reverting Gaussian cases, the propagator can be written down analytically. The second term is there to pick up all coupons between t^*_{L-1} and t^*_L. This term is part of the total bullet bond evaluated at t^*_{L-1} (including all coupons past t^*_{L-1}); we call this B_{L-1}. In fact, $P^{(+)}_{L-1}$ equals B_{L-1} less $C^{(L)}_{L-1}$ (the t^*_L call evaluated at t^*_{L-1}). When $P^{(+)}_{L-1} = E_{L-1}$ at the critical rate, $B_{L-1} = \hat{P}_{L-1}$, and $\hat{P}_{L-1} - E_{L-1} = C^{(L)}_{L-1}$.

Comment for Step #3. Note that $P^{(-)}_{L-1} = \min\{P^{(+)}_{L-1}, E_{L-1}\}$ for the transition. This is because the call option at time t^*_{L-1} is written on the bond $P^{(+)}_{L-1}$. The bondholder is short this call option which has intrinsic value $\max\{[P^{(+)}_{L-1} - E_{L-1}], 0\}$. The bond $P^{(+)}_{L-1}$ with this option subtracted is just $P^{(-)}_{L-1}$ by definition. Note that $\min\{P^{(+)}_{L-1}, E_{L-1}\} = P^{(+)}_{L-1} - \max\{[P^{(+)}_{L-1} - E_{L-1}], 0\}$.

American Option Backchaining

For an American option, the algorithm is applied at each step in the time partition because the bond can be called at any time.

Backchaining for Bonds with Both Puts and Calls

A simple modification allows the valuation of bonds with puts as well as calls. If the option at date t^*_λ is a put the $<$ and $>$ symbols are interchanged for the determination of $P^{(-)}_\lambda, P^{(+)}_\lambda$; and "min" is replaced by "max". Arbitrary sequences of puts and calls can be handled in this manner.

A Few Aspects of Numerical Errors

In this section, we describe some aspects of numerical errors with numerical codes. The topic is very large and a full discussion would require a large volume. We focus here on a few issues that arise in practice.

Sociology and Numerical Errors

The first issue to discuss is sociological. The word "error" can be interpreted very differently by different people. Experts in numerical analysis know that any numerical method carries inherent uncertainties that are called errors. On the other hand, non-technical people sometimes think of "errors" as human mistakes that occur because not enough care was taken, producing "wrong" answers.

The sensitivity of people to the numerical errors depends on the people and the situation. Sometimes, microscopic noise becomes a center of attention to nervous people. Other times, real code errors go unmonitored.

Quantification of Numerical Errors for Pricing

The second issue is the quantification of numerical errors. There are textbook results that give generic guidelines. However, these are often not specific enough. Consider a price C generated by a code with numerical error on the order of $\pm \Delta_{\text{NumErr}}^{(C)}$. This means that the "true" price roughly is somewhere in the interval $C_{True} \in \left(C - \Delta_{\text{NumErr}}^{(C)}, C + \Delta_{\text{NumErr}}^{(C)} \right)$. Unfortunately, in practice the magnitude of the numerical error $\Delta_{\text{NumErr}}^{(C)}$ depends on the grimy details of the problem. The real errors often can only be discovered through painful empirical investigation with a number of specific cases.

Variable time steps help in reducing numerical errors. Small time steps are often needed in the short-time region.

Numerical errors can also be reduced by increasing the number of grid nodes, especially in regions where cash flows exist or logical decisions need to be made.

Oscillations in the Price as the Number of Nodes Increases

Often an oscillatory behavior is observed for the price C as a function of the number of nodes $C(N_{nodes})$ in the calculation. For a binomial model, increasing the number of nodes means adding more time steps with smaller time interval. For the path integral discretization, besides adding more time steps we can also increase the number of nodes at a given time.

Oscillatory behavior can often be characterized by Laplace transform methods[11, iii]. The generic example is provided by a contour integral $\mathscr{I}(s;g,b)$ in a complex j plane with

$$\mathscr{I}(s;g,b) = \int_{c-i\infty}^{c+i\infty} \frac{dj}{2\pi i} s^j \frac{e^{-bj}}{j - \alpha_0 - g^2 e^{-bj}} \qquad (44.6)$$

Here g, $b > 0$, and α_0 are parameters, while c is to the right of all singularities of the integrand in the complex j plane. The variable s is taken here as a monotonically increasing function of the number of nodes in the calculation N_{nodes}. We are interested in the behavior at large N_{nodes}, i.e. $s \to \infty$. Then the integral $\mathscr{I}$ is controlled by the leading zero of the denominator, $D(j) = j - \alpha_0 - g^2 e^{-bj} = 0$ at $j = \alpha_R$. We move the contour to the left in the complex j plane and pick up this leading pole. For finite values of s, the description is more complicated, and non-leading terms enter. In general, we get

$$\mathscr{I}(s;g,b) = \frac{1}{D'(\alpha_R)} s^{\alpha_R} e^{-b\alpha_R} + \text{Non-Leading Terms} \qquad (44.7)$$

In order to make $\mathscr{I}$ independent of N_{nodes} as $N_{nodes} \to \infty$, we choose $\alpha_R = 0$. This implies $\alpha_0 = -g^2 < 0$.

It is equivalent and convenient to expand the integrand in a series in powers of g^2 and then perform the integrations term by term. We get

$$\mathscr{I}(s;g,b) = \sum_{m=0}^{M(s)} \frac{g^{2m}}{m!} \left[\ln(s) - (m+1)b\right]^m s^{\alpha_0} e^{-(m+1)b\alpha_0} \qquad (44.8)$$

Here, $M(s)$ is the largest integer less than $-1 + \ln(s)/b$. Successive terms in this series enter as s increases, producing oscillations in $\mathscr{I}$ as a function of s. Using the analogy, as the number of nodes increases, oscillations in the price are observed.

In practice, oscillations take a variety of forms. The oscillations can be highly damped, or less damped. The results can oscillate around the exact value, or

---

[11] **Oscillation Example:** This formalism comes from studies of the effects on high-energy total cross sections of successive thresholds for production of increasingly massive and different types of particles.

about a curve approaching the exact value. All these behaviors have been seen in practical examples.

Of course to characterize these oscillations quantitatively we would need to know the Laplace transform of the price as produced by the code. Since this is unavailable, we actually need to rely on numerical empirical studies.

The graph below gives an example taking $N_{nodes} = \ln(s)/b$ with $b = 1$, $g^2 = 0.5$, and an overall normalization of 1000.

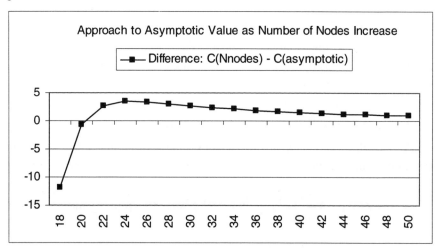

The curve first overshoots and then decreases close to the exact asymptotic value as the number of nodes increases up to 50.

Anomalous Noisy Fluctuations in Option Model Output

An apparently anomalous fluctuating noise-like behavior from day to day in option calculations can result from oscillations, even at a fixed number of nodes. First, the oscillations are "non-universal". This means that the nature of the oscillations depends on market parameters. Hence, with different set of market parameters, the numerical error changes, even for fixed number of nodes. This jumpy behavior, while unpleasant, is unavoidable.

Risk Management and Numerical Errors

The third issue for numerical errors is risk management. Risk management presents a more difficult situation than pricing. Risk is concerned with differences in prices $\delta C = C^{(2)} - C^{(1)}$ under some change of parameters $\delta x = x^{(2)} - x^{(1)}$. Numerical errors for differences are magnified relative to pricing. The relative pricing error $\Delta_{NumErr}^{(C)}/C_{True}$ can be small but the risk error can be large. Suppose $C^{(1)} = C^{(1)}_{True} - \Delta^{(C)}_{NumErr}/2$ and $C^{(2)} = C^{(2)}_{True} + \Delta^{(C)}_{NumErr}/2$. Then the code produces $\delta C = \delta C_{True} + \Delta^{(C)}_{NumErr}$. The true risk is δC_{True}. If $\Delta^{(C)}_{NumErr} \approx O(\delta C_{True})$ with the same sign, the code risk δC will be off by a factor of two from the true risk, viz $\delta C = 2 \cdot \delta C_{True}$. On the other hand, with the opposite sign for $\Delta^{(C)}_{NumErr}$, the code risk will be smaller than the true risk.

The Unobserved P&L

The unobserved P&L is a monitor. This means that the models are used to calculate the code-generated price changes $\delta C = C^{(2)} - C^{(1)}$ and this is compared with the market change δC_{Market} if that is available. The difference $P\&L_{Unobserved} = \delta C_{Market} - \delta C$ gives a handle on the risk errors in the code.

Risk Anomalies in Interest-Rate Risk Ladders

In Ch. 8, 11 we described interest-rate delta ladders used for risk management. These ladders are labeled by the changes in rates for discrete and successively increasing maturities. The type of rate moved (forward rates, zero-coupon rates, swap rates etc.) determines the type of ladder. We emphasized in Ch. 11 that moving the swap rates independently can lead to large fluctuations in individual forward rates. Because codes work with the forward rates, unusual sensitivities can occur if cash flows or decision logic occurs in the region where forward rates are moving substantially.

Some anomalies can occur in the ladders. These can be magnified under unusual market conditions. For example, ladder buckets for short maturities can exhibit instabilities if an inverted cash curve is present[12].

In Ch. 7 we also discussed the construction of the forward rate curve needed as input for pricing interest-rate derivatives. Some discontinuous behaviors across certain transition points can result in the output curve, including the futures/swap boundary, swap maturity points, and the cash/futures boundary. Sometimes, the ladders will exhibit instabilities that can be traced to these discontinuities.

A Few Numerical Methods in the Literature

This book is not a treatise on numerical methods, and no systematic search of the literature has been performed. However, a few innovative methods will be mentioned briefly.

---

[12] **Inverted Curve:** An inverted curve means that a longer-term rate is less than a shorter-term rate. Inverted curves are rare but do occur from time to time.

The Makivic Path-Integral Monte Carlo Approach

Miloje Makivic[iv] has written a paper regarding an efficient path-integral Monte-Carlo simulator for options. He uses the Metropolis algorithm. He has also parallelized the code using High-Performance Fortran.

Moment Methods for Arbitrary Processes

Jarrow and Rudd[v] have set up the formalism for evaluation of options using moments, for arbitrary stochastic underlying processes.

Parametric Analysis of Derivatives Pricing

Bossaerts and Hillion[vi] have formulated a method for derivatives pricing in incomplete markets, fitting hedge ratios locally and using parameter estimation.

Some Miscellaneous Approximation Methods

This section describes some approximation methods. Understanding the ideas behind the approximations can increase intuition.

A "Call-Filtering" Iterative Method

The call-filtering iteration is based on the numerical observation that the first exercise date after the time under consideration is often the most important one. Still, the options associated with the other exercise dates have non-zero value. The first approximation in the iteration is to assume that the value of the option for the purposes of valuing the critical/Atlantic path at a given time is obtained by considering that option as European. Successive corrections to that approximation can then be envisioned which take into account the potentially multiple-option characteristics.

The idea works best if the option is in the money. If so, it will probably be called at the first possible call date; this is the filtering effect in action. If the option is not called at the first call date and stays in the money, it will probably be called at the next (second) call date. If the option is out of the money, its value is small, and the neglect of the rest of the call schedule is therefore small, though perhaps comparable. If the option is at the money, the full complexity of the call schedule arises, but even here, the filtering effect is still operative to some extent. Iteration becomes more important in this case.

Quasi-European Approximation for Bermudas – Some Details

The "quasi-European approximation" for a Bermuda option with a schedule (or its limit, the American option), is based on the filtering effect. The filtering approximation says either the option has small value or else it will be called soon.

An option holder at exercise date t_λ^* needs, in principle, to compare the remaining compound option at all future exercise dates with the intrinsic value of the option at t_λ^* in order to determine the price $\hat{P}_\lambda$ above which he/she will exercise the option. The idea of filtering is to assume first that the most probable option exercise after t_λ^* will happen the next time the option can be exercised, i.e. at $t_{\lambda+1}^*$. To obtain $\hat{P}_\lambda$ at exercise date t_λ^* in this approximation means to evaluate the remaining option as if it were a European option (along with a digital option) at the next exercise date $t_{\lambda+1}^*$. This is very fast numerically.

The second iteration assumes that the determination of the critical path point at each exercise date t_λ^* contains the European option with exercise date $t_{\lambda+1}^*$ as well as the double or "class-2" option with exercise date $t_{\lambda+2}^*$. This can be computed and compared with the results of the quasi-European first iteration. If the convergence is sufficient, the iteration stops. Otherwise, "class-3" options are added with exercise date $t_{\lambda+3}^*$ and so on. By definition, this iteration converges to the correct exact result. In the best case, the QE approximation suffices.

The Critical Atlantic Path in the Quasi-European (QE) Approximation

This algorithm gives the QE approximation to the critical Atlantic path for a Bermuda call embedded in a bond. Coupons can be included as indicated before.

Step 1: Start from last call date t_L^*. The critical price at that date $\hat{P}_L$ is just the bond strike price E_L.

Step 2: At the next-to-last call date t_{L-1}^*, the bond strike price is E_{L-1}. Calculate the European call $C_{\text{Eur. Call}}^{(L)}(P_{L-1}^*, t_{L-1}^*)$ at t_{L-1}^* with exercise date t_L^* and strike E_L. Iterate $\hat{P}_{L-1} - E_{L-1} = C_{\text{Eur. Call}}^{(L)}(\hat{P}_{L-1}, t_{L-1}^*)$ to get $\hat{P}_{L-1}$ at t_{L-1}^*. This is exact.

Step 3: By recursion, obtain the approximate QE critical price $\hat{P}_\lambda^{(QE)}$ at any previous call date t_λ^* back to t_1^*. Get the European call $C_{\text{Eur. Call}}^{(\lambda+1)}(P_\lambda^*, t_\lambda^*)$ at t_λ^*, with exercise date $t_{\lambda+1}^*$ and strike $\hat{P}_{\lambda+1}^{(QE)}$. We also need the digital option $\left(\hat{P}_{\lambda+1}^{(QE)} - E_{\lambda+1}\right) C_{\text{Unit Digital}}^{(\lambda+1)}(P_\lambda^*, t_\lambda^*)$ with strike $\hat{P}_{\lambda+1}^{(QE)}$, since $P_{\lambda+1}^* > \hat{P}_{\lambda+1}^{(QE)} > E_{\lambda+1}$.

We have $\hat{P}_\lambda^{(QE)} - E_\lambda = \int_{\hat{P}_{\lambda+1}^{(QE)}}^{\infty} \left(P_{\lambda+1}^* - E_{\lambda+1}\right) G_{\lambda,\lambda+1} dP_{\lambda+1}^* \Big|_{P_\lambda^* = \hat{P}_\lambda^{(QE)}}$. To get $\hat{P}_\lambda^{(QE)}$, we therefore iterate the resulting equation:

$$\hat{P}_\lambda^{(QE)} - E_\lambda = C_{\text{Eur. Call}}^{(\lambda+1)}(\hat{P}_\lambda^{(QE)}, t_\lambda^*) + \left(\hat{P}_{\lambda+1}^{(QE)} - E_{\lambda+1}\right) C_{\text{Unit Digital}}^{(\lambda+1)}(\hat{P}_\lambda^{(QE)}, t_\lambda^*) \quad (44.9)$$

Shielded "Geometrical Volatility" for Bermudas

The shielded geometrical volatility gives an intuitive feel for Bermudas. The paths for successive possibilities of exercise of a Bermuda option are classified into sets, as described in the Ch. 42. The paths corresponding to exercise at a given time t_λ^* in the schedule must remain below the critical path in price space for all exercises at times $t_{\lambda-k}^* < t_\lambda^*$ before t_λ^*, and then cross the critical path during the interval $\left(t_{\lambda-1}^*, t_\lambda^*\right)$ in order for this exercise at t_λ^* to occur. The probability $\mathcal{P}_\lambda^{(\text{Exercise})}$ of this exercise is just the fraction of the number of paths crossing the critical path in $\left(t_{\lambda-1}^*, t_\lambda^*\right)$ (cf. Ch. 4). The idea is in the picture below:

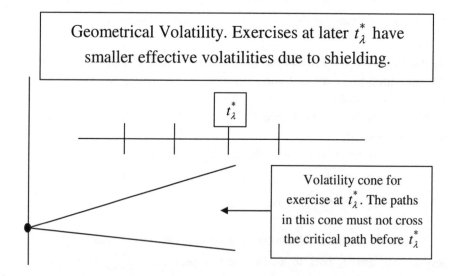

For simplicity, assume that the interest-rate volatility is a constant σ. We introduce a "geometrical volatility" $\sigma_\lambda^{(\text{Geometrical})} < \sigma$ as corresponding to an approximate volatility that would lead to the set of paths $\{\text{Paths}_\alpha\}_\lambda^{(\text{Exercise})}$

producing the exercise at t_λ^*. The geometrical appellation is because the paths are constrained by the geometry of the problem. Then, $\sigma_\lambda^{(\text{Geometrical})}$ can be thought of as the volatility shielded and reduced by the previous exercises at $\{t_{\lambda-k}^*\}$. The idea is to construct a cone of transverse dimension $2\sigma_\lambda^{(\text{Geometrical})}\sqrt{\tau}$ at time τ from today, such that the desired paths $\{\text{Paths}_\alpha\}_\lambda^{(\text{Exercise})}$ are approximately realized, to 1SD.

Given the geometrical volatilities, the resulting sets of paths have the appropriate constraints of previous exercises. Therefore, each exercise at t_λ^* can be treated approximately as a European option with volatility $\sigma_\lambda^{(\text{Geometrical})}$.

Mean-Reverting Gaussian Approximation to Lognormal Model

This section describes a tractable approximation to a lognormal process in terms of an approximately equivalent mean-reverting Gaussian (MRG) process.

The essential analytic problem with a lognormal model in the interest rate is to express the discount factor of products of terms $\exp[-r(t)\,dt]$ in a tractable form in terms of the logarithm of r. This is because then everything can be written in terms of the same variables, allowing analytic progress to be made.

Call $y(t) = \ln[\,r(t)\,T_s\,]$ where T_s is a scale (e.g. 1 year). Then we want to express the discount factor in terms of $y(t)$ in a simple way. To this end, an approximate form for $r(t)$ is found by truncating the sum for $\exp[\,y(t)\,]$,

$$r(t)\,T_s = \exp[\,y(t)\,] \cong \exp[\,y^{(cl)}(t)\,] \sum_{k=0}^{K_{\max}} \frac{1}{k!}[\,y(t) - y^{(cl)}(t)\,]^k \qquad (44.10)$$

We choose $y^{(cl)}(t) = \ln[\,r^{(cl)}(t)\,T_s\,]$ as a convenient expansion point. Here $r^{(cl)}(t)$ is the classical interest-rate path. Up to convexity corrections, the Gaussian or mean-reverting Gaussian result is $r^{(cl)}(t) \cong f(t)$ with $f(t)$ the forward rate for maturity t. Here we are just concerned with some reasonable expansion point; the term structure constraints are implemented later.

The maximum value of the index k has to be discussed. For stability in the discount factor near zero rates, k must be even. We shall stop at the quadratic term, but multiply it by a factor, which approximates effects of the rest of the sum, determined empirically to give good results.

The approximation, with $r^{(cl)}(t) \cong f(t)$ and λ a parameter, is

$$r(t)_{approx} \equiv f(t)\left[1 + \ln\left(\frac{r(t)}{f(t)}\right) + \frac{1}{2\lambda}\ln^2\left(\frac{r(t)}{f(t)}\right)\right] \quad (44.11)$$

This would be used for discounting purposes. Using a Gaussian volatility $[d_t r(t)]^2 = \sigma^2\, dt$ gives the time difference $d_t r(t)$ approximation,

$$d_t r(t)_{approx} \equiv d_t r(t)\frac{f(t)}{r(t)}\left[1 + \frac{1}{\lambda}\ln\left(\frac{r(t)}{f(t)}\right)\right]$$
$$- \frac{dt}{\lambda}\left[\frac{\sigma^2}{2\,r^2(t)} + f'(t)\right]\ln\left(\frac{r(t)}{f(t)}\right) - \frac{\sigma^2\, dt}{2\,r^2(t)}\left[1 - \frac{1}{\lambda}\right] \quad (44.12)$$

Here $f'(t)$ is the time derivative of the forward rate.

Because the quadratic term enters with a positive sign in the sum for $r(t)_{approx}$ regardless of the value of $r(t)$, it acts as a mean-reverting term, canceling out the effects of the first term when that term becomes negative. The quadratic mean-reverting term tends to imitate the barrier at $r = 0$ for the lognormal process, keeping rates positive. However, numerically it actually provides a stronger barrier at zero rates than does the lognormal process. At zero rate, the discount factor is one, but the quadratic approximation above produces a discount factor equal to zero at zero rate. Thus, the quadratic approximation dampens out all paths near zero more than the damping provided by the lognormal model. To compensate, the parameter λ is chosen as greater than one to soften the effects of the quadratic term.

We can test the approximation by inserting $r(t)$ and $d_t r(t)$ on the right hand sides of Eqn. (44.11) and (44.12), and testing the outputs for consistency. The problem for the approximation occurs at small values of $r(t)$ where the logarithm is singular. As an example, $\lambda \approx 1.4$ keeps rates at around 5% reasonably consistently. If the starting forward rate is 10% with a slope of 3% for 30 years and the Gaussian interest rate volatility is $\sigma = 100\, bp/(yr)^{3/2}$ then the approximate $r(t)$ is 4.7% (compared to 5%). The approximate $d_t r(t)$ is [4.4% - 6.4%]. This is calculated using $\pm\,\sigma$ as values for $d_t r(t)$ in the right-hand side of the formula, compared to [4% - 6%] for the exact $\pm\,\sigma$ range.

Put together with the lognormal Green function (which is Gaussian or normal in the logarithm of the interest rate), the above approximation forms an effective mean-reverting Gaussian model, where the variable is the logarithm of the interest rate. Note that this is different from the usual mean-reverting Gaussian model where the variable is the interest rate itself.

An Exponential Interpolation Approximation

Path integrals have exponentials that are sometimes time consuming to evaluate. Here is a robust fast algorithm to approximate interpolated exponentials that generalizes others [vii]. Start with a table of exponentials $E_i = \exp(x_i)$ at points $\{x_i\}$. To find $\exp(\zeta)$ at an interpolated point $\zeta = \lambda x_i + (1-\lambda) x_{i+1}$ with $dx = x_{i+1} - x_i$, a good approximation is

$$\exp(\zeta) \cong \lambda E_i + (1-\lambda) E_{i+1} - \tfrac{1}{2} \lambda (1-\lambda)(E_{i+1} - E_i) dx \qquad (44.13)$$

This approximation is exact at the endpoints and agrees with the quadratic approximation to the expansions about the endpoints with small corrections. For example if $\lambda = \tfrac{1}{2}$ and $dx = 1$, the approximation is off only by 0.2% for $x_i \in (-100, 100)$, and it gets better if dx is smaller.

Canonical Normal Integral Uniform Approximation

Here is the venerable normal-integral approximation [vii] for $x \geq 0$. Setting $t_{AS} = 1/(1 + px)$ with $p = 0.2316419$, up to uniform error with bound $|\varepsilon(x)| < 7.5 \times 10^{-8}$, we have

$$N(x)_{AS\ (26.2.17)} \cong 1 - \frac{1}{\sqrt{2\pi}} \exp(-x^2/2) \sum_{k=1}^{5} b_k t_{AS}^k \qquad (44.14)$$

Here we have $b_1 = 0.31938153$, $b_2 = -0.356563782$, $b_3 = 1.781477937$, $b_4 = -1.821255978$, $b_5 = 1.330274429$. For $x < 0$, use $N(-x) = 1 - N(x)$.

Possible Framework for Multivariate Integral Approximations

It is a challenge to derive a uniform approximation to multivariate normal integrals generalizing Eqn. (44.14). The payoff for a reliable and fast approximation for multivariate integrals would be high, since many models wind up with multivariate integrals. For bivariate integrals, a variety of useful methods exists (cf. Ch. 19 refs.), but even here, none are sufficiently uniform. Different methods work best for different regions of the two variables. Few methods exist for multivariate integrals.

In the late 1980's, I attempted to use the cluster decomposition methods described in Ch. 49 as a framework for calculating approximations to multivariate integrals. The idea was to construct a uniform approximation to each term in the cluster decomposition. Problems due to singular behaviors at boundaries were handled by the cluster decomposition, which has explicit subtractions of factorized pieces. While initial results were promising, there was never time to complete the work, which the enterprising reader might pursue.

References

[i] Standard Numerical Methods for Options
Willmott, P., Dewynne, J., and Howison, S., *Option Pricing – Mathematical models and computation.* Oxford Financial Press, 1993.
Jarrow, R. A. and Rudd, A., *Option Pricing*, Dow Jones-Irwin, 1983.

[ii] Monte-Carlo and Other Numerical Methods for American Options
Langsam, J., *American Exercise and Monte Carlo Techniques.* Morgan Stanley Fixed Income Research working paper, 1994.
Broadie, M. and Glasserman, P., *Pricing American-Style Securities Using Simulation.* Columbia U. working paper, 1995;
Broadie, M. and Glasserman, P., *A Pruned and Bootstrapped American Option Simulator*, 1995 Winter Simulation Conference (Ed. C. Alexopoulos et. al.).
Boyle, P., Broadie, M., Glasserman, P., *Monte Carlo methods for security pricing.* Journal of Economic Dynamics and Control, 21, 1997. Pp. 1267-1321.
Glasserman, P., Heidelberger, P., Shahabuddin, P., *Importance Sampling in the Heath-Jarrow-Morton Framework.* The Journal of Derivatives, Fall 1999. Pp. 32-50.
Li, A., Ritchken, P., and Sankarasubramanian, L., *Lattice Models for Pricing American Interest Rate Claims.* The Journal of Finance, Vol L, 1995. Pp. 719-737.
Longstaff, F. and Schwartz, E., *Valuing American Options by Simulation: A Simple Least-Squares Approach.* Anderson School, UCLA working paper, #25-98, 1998.

[iii] Laplace Transforms and Threshold Effects
Dash, J. W. and Koplik, J., *Energy Scales and Diffraction Scattering.* Physical Review D12, 1975. Pp. 785-791.

[iv] Makivic's Monte-Carlo using Path Integrals
Makivic, M., *Numerical Pricing of Derivative Claims: Path Integral Monte Carlo Approach.* NPAC Technical Report SCCS 650, 1994.

[v] Moment Methods for Arbitrary Processes
Jarrow, R. and Rudd, A., *Approximate Option Valuation for Arbitrary Stochastic Processes.* Journal of Financial Economics 10, 1982. Pp. 347-368.

[vi] Parametric Hedging Formalism
Bossaerts, P. and Hillion, P., *Local Parametric Analysis of Derivatives Pricing.* Cal Tech Social Science working paper, 1997.

[vii] General Numerical Reference
Abramowitz, M., Stegun, I., *Handbook of Mathematical Functions.* National Bureau of Standards, Applied Mathematics Series #55, 1964. See Sect. 26.2.17.

45. Path Integrals and Options IV: Multiple Factors (Tech. Index 9/10)

In this chapter, we describe multi-factor path integrals in an arbitrary number of internal dimensions[1,i]. This is needed to describe models and correlated risk with many factors, multi-factor yield curve models, baskets of equities or FX rates, or any other problem containing multiple variables. The figure below gives the idea.

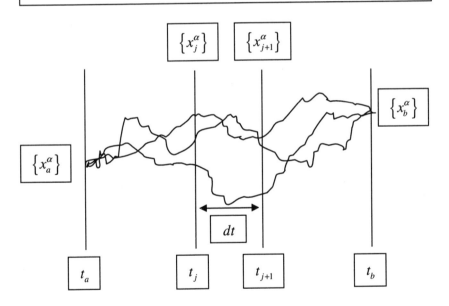

Multi-dimensional path integral. Internal index $\alpha = 1...n_v$

We call n_v the number of variables. The formalism is a straightforward generalization of the results[2] for two variables, say S and P. The variables at

---

[1] **History:** The work in this chapter is in my SIAM talk in 1993 (Ref i). The two-dimensional case with $n_v = 2$ was treated in my 1988 path-integral paper. Path-integral calculations of multivariate yield-curve mean-reverting Gaussian models began in 1988.

[2] **Background:** This chapter assumes you have read the preceding chapters on path integrals and interest rate options, or else you already know about path integrals.

time t_j are denoted as $\{x_j^\alpha\}$ with $\alpha = 1...n_v$, or with another Greek letter β, γ as required by indexing. You have to think of each vertical fiber in the figure as being an n_v-dimensional plane, in which the set of variables $\{x_j^\alpha\}$ is a point, all at fixed t_j. The case $n_V = 2$ was illustrated in Fig. 7, Ch. 42.

The $\{x_j^\alpha\}$ variables will be assumed for simplicity to be Gaussian. As explained in the pedagogical Ch. 42 on path integrals, x_j^α can itself be a function, so lognormal or other non-Gaussian processes, mean reversion, etc. can be included.

We begin with the discretized matrix stochastic equation for x_j^α,

$$d_t x_j^\alpha \equiv x_{j+1}^\alpha - x_j^\alpha = \mu_j^\alpha dt + \sigma_j^\alpha \sum_{\beta=1}^{n_v} R_j^{\alpha\beta} \eta_j^\beta dt \tag{45.1}$$

Here $R_j^{\alpha\beta}$ is the square root of the correlation matrix $\rho_j^{\alpha\beta}$ (c.f. Ch. 28, 48). We have included the possibility of time dependence in the correlations. Each η_j^β is a Gaussian random variable. We have the matrix equation in the internal indices for each t_j:

$$\rho_j = R_j \cdot R_j^T \tag{45.2}$$

We also have the discretized Gaussian orthogonality conditions between the various $\{\eta_j^\alpha\}$, namely

$$\langle \eta_j^\alpha \eta_{j'}^\beta \rangle = \delta_{\alpha\beta} \delta_{jj'}/dt \tag{45.3}$$

The probability measure for the Gaussian variables $\{\eta_j^\alpha\}$ is derived from

$$\mathcal{P} = \prod_{\alpha=1}^{n_V} \prod_j \left\{ \int_{-\infty}^{\infty} d\eta_j^\alpha \frac{1}{\sqrt{2\pi/dt}} \exp\left[-\tfrac{1}{2}(\eta_j^\alpha)^2\right] \right\} \tag{45.4}$$

In this expression, we have suppressed the time indexing for simplicity, as in the chapters treating one-variable path integral discretization. We need to fix the last $j = N-1$ variable η_{N-1}^α by the constraint that we specify x_N^α in the Green function. Alternatively, we just remove the last $\int dx_N^\alpha$ integral once we introduce the constraints[3]. Just as in the one-dimension path integral formalism, we use the stochastic equations as constraints in order to change variables from the $\{\eta_j^\alpha\}$ variables to the $\{x_j^\alpha\}$ variables. So we write the Dirac delta function identity:

$$1 = \prod_{\alpha=1}^{n_V}\prod_j \left\{ \int_{-\infty}^{\infty} dx_{j+1}^\alpha \delta\left[x_{j+1}^\alpha - x_j^\alpha - \mu_j^\alpha dt - \sigma_j^\alpha (R_j \cdot \eta_j)^\alpha dt \right] \right\} \quad (45.5)$$

Here, we introduce the matrix notation $(R_j \cdot \eta_j)^\alpha = \sum_{\beta=1}^{n_V} R_j^{\alpha\beta} \eta_j^\beta$. Now we have the delta function matrix identity for any matrix ς_j,

$$\prod_\alpha \delta\left[\varsigma_j^\alpha - (R_j \cdot \eta_j)^\alpha \right] = \frac{1}{\sqrt{\det \rho_j}} \prod_\alpha \delta\left[\eta_j^\alpha - (R_j^{-1} \cdot \varsigma_j)^\alpha \right] \quad (45.6)$$

Here, we used $\det R_j = \sqrt{\det \rho_j}$. Also, R_j^{-1} is the inverse matrix.

The path integral is then obtained exactly as in the one-dimensional case by letting the delta functions kill the integrals over the $\{\eta_j^\alpha\}$, replacing them by the $\{x_j^\alpha\}$ variables. We denote by $G\left[\{x_a^\alpha\},\{x_b^\beta\};t_a,t_b\right]$ the resulting path-integral Green function from initial time t_a with variables $\{x_a^\alpha\}$ to final time t_b with variables $\{x_b^\beta\}$, without discounting yet. We obtain (restoring the time indexing),

$$G\left[\{x_a^\alpha\},\{x_b^\beta\};t_a,t_b\right] = \left\{ \prod_{j=a}^{b-2}\prod_{\alpha=1}^{n_V} \int_{-\infty}^{\infty} dx_{j+1}^\alpha \right\} \prod_{j=a}^{b-1} \frac{1}{\sqrt{\det \rho_j}} \prod_{\alpha=1}^{n_V} \frac{1}{\sqrt{2\pi(\sigma_j^\alpha)^2 dt}}$$

$$\cdot \prod_{j=a}^{b-1} \exp\left\{ -\sum_{\beta,\gamma=1}^{n_V} \frac{1}{2\sigma_j^\beta \sigma_j^\gamma dt} \left[x_{j+1}^\beta - x_j^\beta - \mu_j^\beta dt \right] (\rho_j^{-1})^{\beta\gamma} \left[x_{j+1}^\gamma - x_j^\gamma - \mu_j^\gamma dt \right] \right\}$$

(45.7)

---

[3] **Last variable:** This has to be done for each α of course. I thank Andrew Kavalov for helping to clarify this point.

We also need to introduce the discount factor, which we left off this expression because we have not specified which of the variables correspond to the variables in the discount factor.

Calculating Options with Multidimensional Path Integrals

Aside from the scary looking Greek alphabet soup, there is really no difference here from what we had in the one-dimensional case. Before calculating anything, the discount factors have to be included. Exactly as in one dimension, the integrations $\prod \int dx_b^\beta$ over the final $\{x_b^\beta\}$ variables at final time t_b have to be reinstated. This is in order to calculate discounted expected values of payoffs $C^*\left[\{x_b^\beta\}, t_b\right]$ specified at time t_b. In general, we may calculate expected discounted cash flows by inserting the cash flows at the appropriate time points.

Analytic results can be obtained as usual, when it is possible to get them, by completing the square, and then performing ordinary calculus integrals.

Boundary conditions can be introduced by restricting appropriately the integration limits over the various $\{x_j^\alpha\}$. In this way, barrier options in several dimensions can be formulated. In Ch. 19, we discussed the calculation of two-dimensional "hybrid" barrier options.

Numerical methods can be formulated as usual. Monte Carlo simulation methods evaluate the multivariable path integral just as in one dimension. Path integral discretization along the $\{x_j^\alpha\}$ fibers for each time can be done using standard numerical integration techniques. Other discretization methods can also be used.

For only one time step, once the integration of the final variables is reinstated, there is one multidimensional integration. For example, a one time step Monte-Carlo simulation of is done for VAR calculations. In Ch. 27, we described the Stressed VAR. The correlation matrix is consistently stressed. The volatilities are taken as fat-tailed vols. The final states $\{x_b^\beta\}$ are characterized statistically and risk measures are used to assess individual risks of the various variables (the stressed CVARs) and the total risk (Stressed VAR) of all correlated variables. More ambitious calculations with several time steps can also be done, and the risk expressed statistically.

Principal-Component Path Integrals

One application deserves special mention. We can choose the variables $\{x_j^\alpha\}$ to be principal components[4]. We can reformulate calculations using path integrals with the principal components as the variables.

Because we want to use α to designate a principal component, we need to use another index, say $m = 1...M$, to designate underlying physical variables $\{r_j^m\}$, of which a linear combination gives x_j^α at fixed t_j. We need some data set in order to define the principal components (or some other ansatz). Therefore, write $F_{mj} = \left(r_{j+1}^m - r_j^m\right)_{\text{Data Set}}$, and define the positive definite variance matrix R by the quadratic form

$$R_{mm'} = \frac{1}{N}\sum_{j=1}^{N} F_{mj}F_{m'j} - \frac{1}{N^2}\sum_{j=1}^{N} F_{mj}\sum_{j=1}^{N} F_{m'j} \qquad (45.8)$$

We look for the R-eigenfunctions $\{\Psi^\alpha\}$ with components $\{\Psi_m^\alpha\}$, and eigenvalues $\{\lambda^\alpha\}$, viz

$$\sum_{m'=1}^{M} R_{mm'}\Psi_{m'}^\alpha = \lambda^\alpha \Psi_m^\alpha \qquad (45.9)$$

The eigenfunctions give linear components for the principal component variables,

$$x_j^\alpha = \sum_{m=1}^{M} \Psi_m^\alpha r_j^m \qquad (45.10)$$

For example, the $\alpha = \text{Flex}$ eigenfunction $\Psi^{\alpha=Flex} = \frac{1}{\sqrt{6}}(1,-2,1)$ for rates with maturities $\{m\} = (2\text{ yr}, 5\text{ yr}, 10\text{ yr})$ produces the flex principal component $x_j^{\alpha=Flex} = \frac{1}{\sqrt{6}}\left(r_j^{2\text{ yr}}, -2r_j^{5\text{ yr}}, r_j^{10\text{ yr}}\right)$.

We can specify stochastic equations applied to the principal components, and then invert Eqn. (45.10) to get information for the physical variables $\{r_j^m\}$.

---

[4] **Principal Components:** See Ch. 48 for a more detailed discussion and some history. There, the variables r_j^m are interest rates of various maturities at different times. We might want to renormalize the eigenfunctions by dividing out the r-volatility, as is done in the flex example below Eqn. 45.10.

Alternatively, starting with the path integral starting using the physical variables $\{r_j^m\}$ gives the dynamics of the principal components $\{x_j^\alpha\}$.

In any case, the eigenfunctions and eigenvalues defining the principal components are assumed fixed in the calculations.

References

[i] **Path Integrals in Several Dimensions**
Dash, J. W., *Path Integrals and Options*. Invited talk, Mathematics of Finance section, SIAM Conference, Philadelphia, July 1993.
Dash, J. W., *Path Integrals and Options – I*, Centre de Physique Théorique, CNRS, Marseille preprint CPT-88/PE.2206, 1988.
Linetsky, V., *The Path Integral Approach to Financial Modeling and Options Pricing*. Univ. Michigan College of Engineering report 96-7, 1996.

46. The Reggeon Field Theory, Fat Tails, Chaos (Tech. Index 10/10)

Introduction to the Reggeon Field Theory (RFT)

This chapter outlines what I have believed for many years[1] would be a fruitful area of research for mathematical finance using the theory of non-linear diffusion called the Reggeon Field Theory (RFT)[i]. One potential use for the RFT lies in the calculation of fat tails in a more fundamental fashion than a phenomenological description using fat-tail volatilities and ordinary diffusion.

The RFT produces critical exponents. Similar exponents are now known in finance as Hurst exponents[ii].

The RFT produces not only scaling exponents but also the scaling laws for the Green function. When the nonlinearities are set to zero, the RFT reduces to the free-diffusion $\sqrt{time}$ scaling, and produces a Gaussian for the Green function, thus reproducing standard financial theory. The nonlinearities in the RFT are non-trivial and modify the standard theory.

The critical exponents and scaling Green functions can be calculated ab-initio in the RFT given some starting assumptions. The starting assumptions involve the types of nonlinearities in the nonlinear diffusion, along with technical points regarding the renormalization group. The calculations are difficult, sometimes involving a hundred pages of dense equations.

I do not know if the results of RFT calculations when they are done will be applicable to finance in a numerical sense. Regardless, lessons learned from the calculations could aid in the description of financial risk.

---

[1] **History, Stories of RFT, Chaos:** I worked on the RFT as a high-energy theoretical physicist, calculating scaling exponents and Green functions, including assessing relevance to experimental data. I used to run a quantitative Options Seminar at Merrill. At the first meeting in 1987, I said that the free diffusion in the Black-Scholes model was probably too simple, and more general scaling laws could be applicable. Unfortunately I never had the time to do the calculations.

Around 1988 I started a program involving the investigation of chaos time-series techniques. After a promising start, the management asked me if I would bet my bonus on the success of the program. I said no. The program was canceled.

So this chapter is really intended to introduce the RFT and leave it as a big exercise for some ambitious readers to actually do the calculations and the empirical comparisons[2,3].

Chaos

Chaos theory[iii], vigorously promoted by B. Mandelbrot, W. Brock, B. Savit, and others, also has critical exponents. The RFT has similarities to chaos theory, but the two theories are not identical[4].

Summary of the RFT in Physics

This section contains a summary of the Reggeon Field Theory convenient for our purposes here[5]. The RFT aims to describe high-energy diffractive scattering. In the language of elementary-particle physics, the RFT is a theory for the Pomeron. The RFT without any interactions is trivial free-field theory equivalent to free diffusion. The interactions generate nonlinearities and non-linear diffusion.

It is important to note that the nonlinearities of the RFT have nothing to do with other "nonlinear" ideas, e.g. a nonlinear dependence of volatility on the underlying variable or a nonlinear transformation of the underlying variable. Rather, nonlinearity here refers to a nonlinear dependence on the Green function itself in the relevant equations. In fact, different Green functions exist, which are coupled in a nonlinear fashion.

The RFT was motivated by perturbation theory in the interactions. The most common interaction assumed is the imaginary triple-Pomeron PPP coupling, commonly denoted as ir with $i = \sqrt{-1}$, shown below:

---

[2] **Kavalov's RFT Calculations:** Andrew Kavalov has performed some RFT calculations for finance using the nonlinear quartic $\psi^{+2}\psi^2$ interaction at D = 1 directly in (x,t) space.

[3] **Acknowledgements:** I thank Andrew Kavalov for stimulating conversations regarding the potential application of the Reggeon Field Theory to finance.

[4] **Acknowledgements:** I thank Bob Savit for informative discussions on chaos techniques and their possible relationship to finance.

[5] **Reader Background Assumed:** In this chapter, no punches are pulled since there is no space to provide the background. The reader is assumed to be familiar with some aspects of scattering theory, the renormalization group, field theory, second-order phase transitions, critical exponents, dimensional regularization, the Wilson ε-expansion, scaling behavior, irrelevant variables etc. It would also be helpful to know something about the Reggeon Field Theory itself.

RFT Nonlinear Diffusion: Triple Coupling Interaction

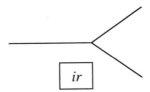

The spatial variables in the RFT in physics correspond to the transverse dimensions perpendicular to the incoming beam of particles in accelerator experiments. The time variable in the RFT is a theoretical construct. It is Fourier-conjugate to an "energy" variable that is really $E = 1 - j$ where j is a "cross-channel" angular momentum. The RFT is of interest in the infrared $E \approx 0$ or $j \approx 1$ region, which governs high-energy, low momentum transfer scattering.

The RFT may be relevant for describing high-energy diffraction scattering, provided that the energies are well above thresholds corresponding to finite-energy heavy-particle production effects.

The RFT Lagrangian

The RFT non-linear diffusion with the PPP coupling has the nonlinear Lagrangian L:

$$L = \frac{1}{2} i \left[\psi^+ \frac{\partial}{\partial t} \psi - h.c. \right] - \alpha'_0 \nabla_x \psi^+ \cdot \nabla_x \psi - \frac{1}{2} i r_0 \left(\psi^+ \psi^2 + h.c. \right) \quad (46.1)$$

Bold type is being used for vector notation. The field $\psi(x,t)$ depends on the spatial variable x and the "time" t, while ψ^+ is its hermitian conjugate (h.c.). The "bare" PPP coupling is denoted as r_0. Correspondingly, the interaction Lagrangian is seen to involve the nonlinear product of three fields. The "bare slope" α'_0 is relevant to the first-order description of the high-energy scattering away from the forward direction.

Non-interacting Free Diffusion Limit

If the interaction is set to zero, the Lagrangian just has a free form. The corresponding dynamical equation is the same as the free diffusion equation, but with imaginary time. The Green function is just a Gaussian with free-field $\sqrt{\Delta t}$ scaling. Up to normalization it reads:

$$G_{free}(\Delta x, \Delta t) = N \exp\left[i \frac{(\Delta x)^2}{4\alpha'_0 \Delta t}\right] \quad (46.2)$$

The interactions change the form of the free Green function and also the scaling behavior. Feynman rules for diagram calculations exist that summarize perturbation theory and are consistent with unitarity. However the applications of interest here are non-perturbative. That is, the results for the Green functions of the RFT are of infinite order in perturbation theory.

Calculations in the RFT

The non-perturbative RFT calculations proceed using renormalization group equations. The critical dimension D_{crit}, in which the coupling constant becomes dimensionless, is relevant for the possible occurrence of a second-order phase transition. Such a second-order phase transition is associated with scaling behavior and an infinite correlation length. However, the critical dimension is not the dimension D_{phys} of physical interest. With a PPP coupling, the critical dimension is $D_{crit} = 4$. However, the physical dimension is $D_{phys} = 2$, corresponding to the two real transverse dimensions perpendicular to the beam direction. We need to calculate for arbitrary dimension D, i.e. at nonzero Wilson variable $\varepsilon = D_{crit} - D$. At the end, $\varepsilon_{phys} = D_{crit} - D_{phys}$ is fixed.

In order to test for the existence of a phase transition, it is necessary to calculate the Gell-Mann-Low (GML) β-function. This function is calculated in perturbation theory[6]. If the β-function has a zero at a non-zero value of the coupling constant, a phase transition can occur. Results are then obtained for the critical exponents as a power series in ε. Scaling behavior for the propagator (the lowest-order Green function) can be calculated depending on ε and the critical exponents. This is done by explicitly solving the renormalization-group equation for the propagator. The vertex function Green function, generalizing the simple bare interaction pictured above, must also be calculated.

---

[6] **Consistency:** There is no contradiction between the perturbative expansion of the GML function and the nonperturbative aspects of the final results for the Green function.

If the PPP coupling is present, higher-order couplings are irrelevant variables and do not change the scaling behavior (cf. Bardeen et. al., ref). However, if the PPP coupling is not present and irreducible higher-order primitive interactions occur, the RFT leads to different results.

Results of the RFT in Physics

The most complete discussion of the results assuming a PPP interaction are given in Baig et. al. (Ref. i). They are summarized as follows. The imaginary part of the amplitude, $\mathrm{Im}\,T_{ab}$ for elastic scattering $ab \to ab$ at Mandelstam variables s, t that results from the calculation is[7]

$$\mathrm{Im}\,T_{ab}(s,t) = \beta_a(t)\beta_b(t)s(\ln s)^{-\gamma} T^{(1,1)}\left[\frac{-\alpha'_0 t}{k_2}(\ln s)^{1-\zeta/\alpha'}\right] \quad (46.3)$$

Here γ and ζ/α' are the critical exponents. These critical exponents were calculated by Abarbanel and Bronzan to $O(\varepsilon)$, and independently by Baker and by Bronzan and Dash to $O(\varepsilon^2)$. They are given by

$$-\gamma = \frac{\varepsilon}{12} + \left(\frac{\varepsilon}{12}\right)^2\left[\frac{161}{12}\ln\frac{4}{3} + \frac{37}{24}\right] \quad (46.4)$$

$$-\zeta/\alpha' = \frac{\varepsilon}{24} + \left(\frac{\varepsilon}{12}\right)^2\left[\frac{59}{24}\ln\frac{4}{3} + \frac{79}{48}\right] \quad (46.5)$$

For the real world, $\varepsilon = 2$. The quantity $T^{(1,1)}$ is the Green-function propagator. The form of the RFT result for $T^{(1,1)}$ is not easy to describe. For a complete discussion including the definitions of all the symbols in Eqn. (46.3), the interested reader is referred to Baig et. al. (Ref. i). Evaluation of $T^{(1,1)}$ must be done numerically.

If the PPP coupling is set to zero, the critical exponents are simply zero. The propagator $T^{(1,1)}$ reduces to the non-interacting free-diffusion trivial result, both for its functional form and for the form of the argument.

---

[7] **Notation: Mandelstam variables s, t:** These variables were introduced by S. Mandelstam and describe the kinematics of elastic scattering. For high-energy diffractive scattering, we are interested in large s and small t. Unfortunately, by convention, the same letter t is also used for the RFT "time", which is really ln(s).

RFT Scaling Expressed in More Standard Language

Here is a translation into more recognizable language. The translation to finance variables from Eqn. (46.3) is $\ln s \to (\Delta t)$, $t \to 1/(\Delta x)^2$. Hence, $t(\ln s)^{1-\zeta/\alpha'} \to (\Delta t)^{1-\zeta/\alpha'}/(\Delta x)^2$, so the relevant scaling variable is of the form $\sqrt{(\Delta x)^2}/(\Delta t)^{0.5(1-\zeta/\alpha')}$. The anomalous dimension, i.e. the difference in the power from square-root scaling, is therefore given by $-0.5\, \zeta/\alpha'$, which is actually a positive number. The total power and the anomalous dimension for different dimensions are given in the table below.

Power of Δt for Different Dimensions Given by the Reggeon Field Theory		
Dimension	Total Power	Anomalous
1	0.64	0.14
2	0.57	0.07
3	0.53	0.03
4	0.50	0.00

The equivalent statement for the variance is $(\Delta x)^2 \propto (\Delta t)^{(1-\zeta/\alpha')}$. For example, at dimension $D=1$ we get $(\Delta x)^2 \propto (\Delta t)^{1.27}$ instead of the Brownian or Gaussian free-field result $(\Delta x)^2 \propto (\Delta t)^1$. Therefore, there is extra variance coming from the interactions in the RFT.

The other critical exponent γ gives a violation of scaling. For our purposes here, this is less interesting than the scaling property.

We note that the RFT calculations above only makes sense for $D \leq 4$.

What should we remember about all this?

The two important things to retain at this point are:

- Definite expressions are calculable in the presence of nonlinear diffusion.
- The limit as the interaction vanishes is just the free diffusion.

Aspects of Applications of the RFT to Finance

The starting point of the potential application of the RFT to finance is to see which assumptions make sense, translate the variables, etc. There is no fundamental finance theory[8]. Essentially, the idea is just to postulate a mapping, calculate the consequences, and see how the results work in practice. Without interactions, the RFT reduces to standard model assumptions in finance, corresponding to free diffusion. With interactions, changes from the standard models will occur.

The main idea, of course, is that critical exponents will emerge to describe deviations from "square-root time" scaling characteristic of the standard Brownian assumption used in finance. That is, we expect to get functions of $\sqrt{(\Delta x)^2}/(\Delta t)^{0.5+\nu}$, where ν gives the deviation from $\sqrt{time}$ due to nonlinear interactions.

Mapping of the RFT to Finance

We want the mapping to give the usual finance results when the interaction is turned off. The mapping could be as follows. The spatial variable x maps to the underlying variable under consideration. For example, for single-stock diffusion we would just take $x = \ln S$ to correspond to lognormal diffusion in the zero interaction case. In this case the physical dimension corresponds to a single variable, so $D_{phys} = 1$. The RFT time (*with the factor i removed*) corresponds to real time. The analog triple PPP interaction term would be taken as being real, not imaginary. Other types of interactions (quartic, etc.) could be envisioned.

It would be assumed that the RFT Feynman rules apply to the calculations in finance. Calculations involving the renormalization group and the Wilson ε-expansion could be carried out. Hopefully the coupling at the first zero of the GML function will be non-trivial. Calculations could also be carried out in D_{phys} dimensions directly provided that the calculations are finite.

Jumps and Fat Tails

The distribution in movements of the underlying variables is critical for risk management, as emphasized many times in this book. The nonlinear interactions may generate interesting fat-tail jump events. If the RFT in physics is any guide, extra variance like that shown in the table may result, producing fat-tail events.

---

[8] **Fundamental Theorem of the Theory of Finance?** Ready? Here it is: *"There Is No Fundamental Theory Of Finance"*. This statement might be elevated into a Third Fundamental Theorem, complementing the two Fundamental Theorems for Data and for Systems. However, it would be much less profound than these two heavy-duty real-world theorems, and so probably only qualifies as some sort of Heretical Remark.

Lessons from Critical Exponents

Numerical values of theoretical critical exponents are probably useful only in a limited sense. First, critical exponents[9] from data vary by underlying, and also vary according to the data time window. Second, critical nonlinear theories are probably restricted to only a subset of behavior over time, so the notion of a critical exponent for an entire time series is only a phenomenological approximation. Third, the notion of the RFT dimension is only valid for $D \leq 4$ ($\varepsilon \geq 0$), so association of an index would need to be made as a composite object.

Still useful information may be uncovered from nonlinear critical theories like the RFT independent of the numerical values of the critical exponents.

References

[i] Reggeon Field Theory in Physics

Abarbanel, H.D.I.A. and Bronzan, J.B., *Structure of the Pomeranchuk Singularity in Reggeon field theory*. Physical Review D9, 1974, Pp. 2397-2410.; with Sugar, R. and White, A., *Reggeon Field Theory*.... Physics Reports 21, 1975, Pp. 119-182.

Bronzan, J. B. and Dash, J. W., *Higher-order ε terms in Reggeon field theory*. Phys. Lett. 51B, 1974, Pp.496-498; Phys. Rev. D10, 1974, Pp.4208-4217; D12, 1975, p.1850.

Baker, M., *The ε Expansion of Pomeron Amplitudes*. Physics Letters 51B, 1974, P 158 ff; Nuclear Physics B80, 1974, Pp. 62-76.

Bardeen, W. A., Dash, J. W., Pinsky, S. S., Rabl, V., *Infrared behavior of the Reggeon field theory for the Pomeron*. Phys. Rev. D12, 1975, Pp. 1820-1828.

Baig, M., Bartels, J. and Dash, J. W., *The Complete $O(\varepsilon^2)$ Reggeon Field Theory Scaling Law for $d\sigma_{el}/dt$*.... Nuclear Physics B237, 1984. Pp. 502-524.

[ii] Hurst Exponents

Hurst, H. E., *The Long-Term Storage Capacity of Reservoirs*. Transactions of the American Society of Civil Engineers 116, 1951.

[iii] Chaos and Finance

Bass, T., *The Predictors – How a Band of Maverick Physicists Used Chaos Theory to Trade Their Way to a Fortune on Wall Street*. Henry Holt and Company, 1999.

Peters, E., *Fractal Market Analysis – Applying Chaos Theory*. John Wiley & Sons, Inc. 1994; *Chaos and Order in the Capital Markets*. 2nd Ed. John Wiley & Sons. 1996.

Mandelbrot, B., *The Fractal Geometry of Nature*. W. H. Freeman, 1982; *Fractals and Scaling in Finance; Discontinuity, Concentration, Risk*. W. H. Freeman, 1997.

Savit, R., *When Random is Not Random: An Introduction to Chaos in Market Prices*. Journal of Futures Markets Vol 8, p. 271, 1988.

Brock, W. A., *Distinguishing Random and Deterministic Systems*. Journal of Economic Theory 40, 1986, Pp. 168-195; *Nonlinearity and Complex Dynamics in Economics and Finance*. Univ. Wisconsin working paper, 1987.

---

[9] **Critical Exponents:** The Bloomberg, L.P. system calls these KAOS exponents, and can calculate them for various time series of different lengths.

PART VI

THE MACRO-MICRO MODEL (A RESEARCH TOPIC)

47. The Macro-Micro Model: Overview (Tech. Index 4/10)

Explicit Time Scales Separating Dynamical Regions

The Macro-Micro model was developed in a systematic program in the late 1980's. The first goal was to model phenomenologically the real-world statistical behavior of yield curves using multifactor models. The idea of different dynamics in different regimes is implied by real-world data. The second goal was to be able to provide a framework to price contingent claims.

The Macro-Micro model explicitly includes time scales. Standard model assumptions of the movements of financial variables do not include time scales that separate different dynamical regions. On the other hand, the behavior of these variables in the real world clearly exhibits a variety of time scales. Interest rates, FX rates, stock prices etc. behave very differently at long time scales (e.g. months) relative to their behavior at short time scales (e.g. days). However, the absence of time scales in standard models makes the description of these different behaviors at different time scales difficult or perhaps impossible to understand.

It is true that mean reversion parameters with units 1/time can be used. We make extensive use of mean reversion, in fact very strong mean reversion. However, the yield-curve data coupled with the pricing of derivatives require more than mean reversion.

The Macro-Micro model is quite intuitive. The Macro component with long time scales is associated with macroeconomic behavior, providing a connection to economics. The Micro component with short time scales is connected with trading. Although first formulated for interest rates, it now appears that the idea is more general and may apply to the FX and equity markets as well.

There are three parts to this chapter. The first part summarizes the original Macro-Micro multifactor yield-curve investigation in Ch. 48 - 50. The second part deals with further Macro-Micro developments in Ch. 51. The third part in Ch. 52 deals with a possibly related topic that is called a "function toolkit".

I. The Macro-Micro Yield-Curve Model

The Macro-Micro model originated from a program at Merrill Lynch in the late 1980's to describe yield-curve movements using a multifactor model[1]. The most important point to retain is that the data for yield-curve movements are consistent with long-term Macro quasi-random "quasi-equilibrium" smooth behavior, around which rapid but small strong Micro mean-reverting fluctuations take place[2]. This is very different from the behavior produced by standard models.

The quasi-random quasi-equilibrium Macro behavior has an associated macro cutoff time scale below which changes in the Macro component do not happen. This is very different from any Brownian model (with or without mean reversion). Thus, a sort of spectral time decomposition is implied with different properties at short and long time scales, governed by different dynamics.

The different dynamics of the Macro and Micro components of the underlying variable movements has a natural and attractive physical interpretation. The Macro component can be associated with a moving quasi-equilibrium governed by the response of the markets to long-term macroeconomic considerations. The Micro component can be regarded as the result of market fluctuations following the macro trends. The Micro component also contains occasional fat-tail jumps. In this way, the Macro-Micro model fits in both with macroeconomics and with trading activity. The model is summarized in the next three chapters, presented in historical order.

Ch. 48. *The Multifactor Lognormal Model and Yield-Curve Kinks*

The first chapter[i] considers the multivariate generalization of the single-variate lognormal (LN) interest-rate model with eleven factors, using input historical data for volatilities and correlations. This is a natural model to consider. Historically, it was one of the first multifactor yield-curve models. Today, similar models are still used in some risk management contexts.

A variety of statistical techniques was used to analyze the data and the model, including one of the first uses of principal components on the Street.

---

[1] **Acknowledgement:** I thank Les Seigel, former Senior Adviser and Manager, Financial Technical Assistance in the Treasury Group at the World Bank, for very lively and insightful conversations on this and many other topics.

[2] **"Quasi-Random Quasi-Equilibrium Yield Curve Path"**, **"Historical Quasi-Equilibrium Yield Curve Path"**: The nomenclature is meant to invoke several ideas. "Equilibrium" signifies that mean-reverting fluctuations exist around an equilibrium path, stable on time scales long with respect to these fluctuation times. "Quasi-Equilibrium" signifies that the equilibrium path changes slowly with time. "Quasi-random" means that such future slow changes are drawn from a random distribution that does not scale down to small times. Historically, it is assumed that one particular realization of this quasi-random behavior took place to form the "Historical quasi-equilibrium yield curve path".

The multifactor LN model produces simulated yield curves containing unphysical yield-curve shape kinks. We believe that this is a general property of models without strong mean reversion.

To drive the point home, here is a picture:

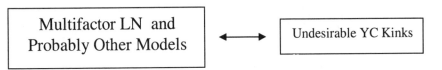

Yield-curve kinks present a major issue. Usually the problem is ignored. The avoidance of undesirable yield-curve kinks constitutes an important challenge to any putative multifactor yield-curve model.

Ch. 49. The Strong Mean-Reverting Yield-Curve Multifactor Model

The second chapter in this series involves the construction of a multifactor strong mean-reverting multifactor model that replaces the unsuccessful LN model[ii]. This strong mean-reverting multifactor model does successfully describe yield-curve statistical data, *and without the presence of undesirable yield-curve kinks.*

A statistical tool called Cluster Decomposition Analysis (CDA) is described that was used to discover the need for strong mean reversion. CDA uses third-order correlation Green functions. Using this sharp probe, the yield-curve data are observed to imply the existence of very strong mean reversion and small, rapid fluctuations about an historically determined slowly varying quasi-equilibrium yield curve.

Here is a picture outlining the process determining the multifactor strong mean reverting Micro Model[3].

---

[3] **Statistical Methods:** These methods were also used for probing the multifactor LN model. In the figure, Principal Components (ef, ev) means that both eigenfunctions and eigenvalues were used to compare model output with the data.

```
┌─────────────────────────────────────────────────┐
│   Determination of Strong Mean-Reverting Micro  │
│          Multifactor Yield Curve Model          │
└─────────────────────────────────────────────────┘

┌──────────────────────────┐  ┌──────────────────────────┐
│       Micro Input        │  │       Macro Input        │
│ Historical Vols,         │  │ Quasi-equilibrium yield  │
│ Correlations             │  │ curve                    │
└──────────────────────────┘  └──────────────────────────┘

        ┌──────────────────────────────────┐
        │       Statistical Methods        │
        │  Cluster Decomposition Analysis  │
        │  Third Order Green Functions     │
        │  Principal Components (ef, ev)   │
        │  Standard Statistical Probes     │
        └──────────────────────────────────┘

┌──────────────────────┐          ┌──────────────────────────┐
│     Micro Model      │  ◄────►  │    Output Quantitities   │
│ Strong Mean Reversion│          │  Curve Shape Volatilities│
│                      │          │  Avoid Yield-Curve Kinks │
│                      │          │  Output vols, correlations│
└──────────────────────┘          └──────────────────────────┘
```

Ch. 50. The Macro-Micro Yield-Curve Model

The third chapter contains the construction of the Macro-Micro yield-curve model for the generation of future yield curves needed to price options and perform risk analysis[iii]. There are two generic time scales, in which the dynamics are very different.

Here is a diagram of the construction of the Macro-Micro model:

Macro-Micro Multifactor Yield Curve Model for Pricing Options

Macro Input
Quasi-random quasi-equilibrium yield curves
Implied Macro volatilities, correlations
Macro interpretation via macroeconomics
Long Time Scales

Micro Input
Strong Mean Reverting Multifactor YC Model
Occasional Jumps
Micro interpretation via Trading
Short Time Scales

No-Arbitrage Constraints
Yield-Curve term-structure constraints
Forward stock price, FX for other markets

Output Quantitities
Realistic Yield-Curve Shapes, Statistics
Paths do spread out in time
Option Prices

For long Macro times, Macro quasi-stochastic variables produce quasi-random means, thus allowing future yield-curve paths to spread or fan out in a smooth fashion over long times.

For short Micro times, the multifactor strong mean reversion model is used. This maintains consistency with the successful description of historical yield-curve data without kinks. The fluctuations from the data and the micro model that fits the data imply interest rate paths that do not fan out much due to the strong mean reversion.

A third component is due to the occasional fat-tail jumps. This third component is added on to the Micro component separately. It is possible that these fat-tail jumps are associated with nonlinear diffusion or chaos, as described in Ch. 46. It should be emphasized that these effects, while important, are additional in this framework.

In a sense, the Macro component acts as a WKB approximation to the full set of fluctuations. The Micro fluctuations are the small fluctuations around the Macro WKB approximation.

Interpretation of the Macro Component

We propose an interpretation of the Macro slowly varying quasi-equilibrium yield curves as due to correspondingly slowly varying macroeconomic trends (Fed. policy, inflation ...).

Interpretation of the Micro Component

The rapid Micro fluctuations are proposed due to trading activities following the smooth Macro trends and reacting to market events. Occasional fat-tail jumps are due to exceptional market movements. However, the dominant Micro dynamics are strongly mean reverting.

II. Further Developments in the Macro-Micro Model

Some further developments in the Macro-Micro Model are in Ch. 51. They include the following topics.

The Micro-Macro Model Applied to the FX and Equity Markets

We summarize some preliminary analyses that indicate the relevance of the Macro-Micro idea to FX and equities. These include:

- Strong Mean Reversion and Cluster Decomposition Analysis
- Probability Analyses for FX and the Macro-Micro Model

Models in the Economics Literature Resembling the Macro-Micro Model

We summarize some references to models that resemble the Macro-Micro model in the literature, especially for FX.

Formal Developments in the Macro-Micro Model

We deal with some formal developments in the Macro-Micro model. This includes a discussion of hedging, consistency with forward quantities, term-structure constraints, and no-arbitrage. A general class of parameters $\left\{\lambda_\beta\right\}$ is introduced to parameterize the Macro dynamics. Included here are subsections on the following topics:

- The Green Function with General Quasi-Random Drift
- Averaging the Green Function over the Macro Parameters
- Option Pricing with the Macro Parameter - Averaged Green Function
- The Macro Parameter - Averaged Diffusion Equation

No Arbitrage and the Macro Micro Model

We discuss various aspects of no arbitrage for the Macro-Micro model. As mentioned in Ch. 50, the basic idea is to have a sufficient number of parameters that can be fixed from no-arbitrage considerations. Included are the following topics:

- The Macro-Micro Model, Hedging, and No Arbitrage for Equity Options
- The Macro-Micro Model and the Satisfaction of the Term-Structure Constraints for Interest-Rate Dynamics

Other Topics

Other topics treated in Ch. 51 are:
- Derman's Equity Regimes and the Macro-Micro Model
- Seigel's Nonequilibrium Dynamics and the MM Model
- Macroeconomics and Fat Tails (Currency Crises)
- Some Remarks on Chaos and the Macro-Micro Model
- Technical Analysis and the Macro-Micro Model
- The Macro-Micro Model and Interest-Rate Data 1950-1996
- Data, Models, and Rate Distribution Histograms
- Negative Forwards in Multivariate Zero-Rate Simulations

III. A Function Toolkit

In this last chapter (Ch. 52), a toolkit of functions is presented potentially useful both at long and at short time scales. The functions were originally used in describing some phenomena in high-energy physics and also in engineering.

Possible Additional Macro Component for Cycles

In this chapter, the function toolkit is introduced and suggested for analyzing business cycles operating over long time scales. Specific topics are:

- Time Thresholds; Time and Frequency; Oscillations

- Relation of the Function Toolkit to Other Approaches
- The Full Macro: Quasi-Random Trends + Toolkit Cycles

The Micro Component (Trading)

In this section, we suggest that the function toolkit might find some applications in trading.

Technical Analysis and the Macro-Micro Model

We briefly describe trading technical analysis, and propose a qualitative connection with the Macro-Micro model.

References

[i] **Multifactor Lognormal Yield Curve Micro Model**
Beilis, A. and Dash, J. W., *A Multivariate Lognormal Model*, Centre de Physique Théorique CPT-89/PE.2334 preprint, CNRS Marseille, France, 1989.

[ii] **Multifactor Strong Mean Reversion Yield Curve Micro Model**
Dash, J. W. and Beilis, A., *A Strong Mean-Reverting Yield-Curve Model*. Centre de Physique Théorique CPT-89/PE.2337 preprint, CNRS Marseille, France, 1989.

[iii] **Multifactor Macro-Micro Yield Curve Model**
Dash, J. W. and Beilis, A., *A Macro-Micro Yield-Curve Simulator*. CNRS, Centre de Physique Théorique CPT-89/P.E.2336, 1989.

48. A Multivariate Yield-Curve Lognormal Model (Tech. Index 6/10)

Summary of this Chapter

This chapter is the first of four in this book dealing with the Macro-Micro Model. The motivating idea was to construct a multivariate yield-curve model that can successfully describe the statistics and dynamics of the yield curve as it moves in time. To this end, putative models were compared to yield-curve data using a battery of statistical tools. This chapter[1,i] contains the first attempt, a multivariate lognormal (LN) interest rate model. This model is the natural generalization of the popular single-factor LN model.

While not the final product, the examination of the multifactor LN model is useful if only because it highlights the problems in describing yield-curve statistics. The next chapter, Ch. 49, presents a successful description of yield-curve dynamics in terms of strong mean reversion about a slowly varying quasi-equilibrium yield curve. Occasional fat-tail jumps are also present. This model then forms the Micro component of the Macro-Micro model, described in Ch. 50.

Because the goal is to fit historical yield-curve statistical data, no-arbitrage drifts are irrelevant at this stage. Instead, an historically based quasi-equilibrium yield curve is introduced around which fluctuations occur. In the Macro-Micro Model of Ch. 50, a set of quasi-equilibrium curves is generated by quasi-random variables to form the theoretical Macro component. The determination of an overall average drift through no-arbitrage considerations becomes relevant for pricing contingent claims. The no-arbitrage properties of the Macro-Micro model in various markets and recent developments of the model are examined in Ch. 51.

---

[1] **History and Acknowledgements:** This chapter is mostly taken from Ref. i. The work was done in 1988-89 in collaboration with Alan Beilis. I thank Alan for his collegial and dedicated work over many years with me. Alan had the seminal idea of the multifactor lognormal model in early 1988. This was among the first multifactor models that directly modeled the yield curve. It may have been the first yield-curve model to have a separate factor for each maturity.

The Problem of Kinks in Yield Curves for Models

Some yield-curve data properties are extremely difficult for models to reproduce. The most difficult is to include the large magnitudes of non-parallel movements without anomalous large local inversions (kinks), as described by the statistical properties of spreads of neighboring maturities.

Important anomalies of the multivariate lognormal model with respect to the data are discovered. Namely, the multivariate lognormal model yield curves are not smooth enough and do contain kinks, even when data are used for volatilities and correlations.

We believe that avoiding undesirable kinks in yield curves presents a major challenge for all putative yield curve models[2]. Usually the problem is overlooked. Note that one or two factor models cannot generate kinks, since three points on the yield curve are required to make a kink. The idea is illustrated below.

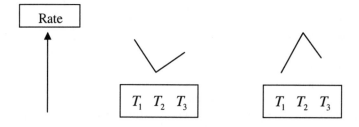

I. Introduction to this Chapter

The single-factor lognormal model is often used for interest-rate-risk analysis and contingent-claims. The single factor is taken as the short interest rate. This model can be constrained to be arbitrage-free using initial yield-curve constraints[ii].

The point of view in this chapter is that if the short rate is assumed lognormal, then logically speaking, rates corresponding to other maturities could be assumed lognormal as well. This is a generalization of a two-factor lognormal model sometimes used in pricing mortgage-backed securities[iii].

Ideally for risk management purposes, it is desirable to have a model that provides reasonable yield curves moving forward in time that possess statistical

---

[2] **Kinky Yield-Curve Challenge:** *It is a challenge for any multifactor model purporting to describe yield-curve dynamics to avoid kinks in forward-time yield curves.* I have met other quants who told me that their multifactor models also contain yield-curve kinks. I believe that these anomalies are generic to all models without very strong mean reversion. A strong mean-reverting model without yield-curve kinks is described in the next chapter.

A Multivariate Yield-Curve Lognormal Model

properties in agreement with actual data. For example, this would insure that the correct spread statistics between the 10-year rate driving a prepayment model and the short-term rate used for discounting cash flows in a mortgage-backed-securities framework would be incorporated.

This chapter presents the first part of the results of a research program carried out at Merrill Lynch around 1988-89. The work was dedicated to the construction of a multivariate yield-curve simulator with two properties: (1) The simulator agrees with the statistical properties of yield-curve data, and (2) The simulator can be used for interest-rate-risk analysis and contingent claims valuation. The lognormal multivariate model is the first step along this direction. The next two chapters present steps toward a more successful model.

Statistical Tests: Brief Description

The first statistical probe tests for the presence of local yield-curve inversions, or kinks (we distinguish between smooth yield-curve inversions and kinks). These kinks or local inversions are relatively rare and of small magnitude in actual data, and so the test is a sharp discriminator for models. The rest of the statistical probes are the means and volatilities of neighboring-maturity yield-shifts and finally the usual interest rate volatilities for different maturities, as well as the interest rate correlation matrix between different maturities.

The second part of the analysis, EOF Analysis or Principal Component Analysis, breaks down the yield curve into its component movements[iv]. Roughly, but not exactly, these movements correspond to parallel shift, tilt, flex, and behaviors that are more complex[3].

Data Used in the Analysis

We applied these statistical probes to characterize weekly U.S. Treasury yield data over various time windows, namely 2, 5, and 10 years[4]. We focus here on the data window of 5 years.

The yield curve was taken as consisting of eleven maturities between 3 months and 30 years. The notation is shown in the figure below:

---

[3] **Standardization of Principal Components:** Today, after over fifteen years of the writing of the original paper (ref. i), principal components are completely standard. When this work was done, the technique was almost unknown in finance. For this reason a description of the framework was included in the paper, which is retained in the book.

[4] **Acknowledgements:** The yield-curve data were current as of the writing of the original paper (ref. i) in 1989. The data are US treasury yields, including coupons. These data have been presented in numerous talks, and are in the CPT-CNRS preprints (refs. i and v). I thank Merrill Lynch for the use of the statistics and graphs from these data.

> **Multifactor Yield Curve Notation**
>
> Maturity indices: $1, 2, 3, ..., 11$
>
> Maturities: 3M, 6M, 1Y, 2Y, 3Y, 4Y, 5Y, 7Y, 10Y, 20Y, 30Y

The same statistical probes were used to analyze the output yield curves generated by the multifactor lognormal model.

The historical volatilities and correlation matrix, averaged over suitable time windows, were used as input to the model.

The Historical Quasi-Equilibrium Yield-Curve Path

Since we were interested in the statistical properties of the yield curves, a simple model for the historical trends of the yield curve was used as input. This gives the model the best chance at describing the yield-curve fluctuation data. The trend model just consists of several long-time straight-line trend segments. We call this the historical "quasi-equilibrium yield-curve path".

Summary of the Results

The most striking way of presenting the results is simple visual examination[5]. Fig. 1A exhibits a three-dimensional plot of weekly yield-curve data over the period 1983-1988. Fig. 1B shows one typical yield-curve path in time (of all maturities) from the Monte-Carlo simulation for the multivariate LN model, starting from the initial data yield curve. The reader can see that the lognormal model yield-curve path exhibits a variety of unphysical features, including unacceptable local yield curve inversions or kinks[6]. Again, we believe that this problem constitutes an important challenge for any realistic multifactor yield-curve simulator.

The Rest of the Chapter

The remainder of the chapter is organized as follows. In Sections IIA, IIB the results from the statistical tests are discussed, including the problem of yield-curve kinks. Section III presents the analysis of the data and model yield-curve

---

[5] **Figures:** All numbered figures are at the end of the chapter, and are taken from Ref. i.

[6] **See the Next Chapter:** These yield-curve kink problems are largely eliminated with the introduction of strong mean reversion.

movements using (EOF) principal components. Section IV describes the reduction of the eleven-variate lognormal model into a simpler model with fewer variates. Section V summarizes the paper and briefly describes the next two papers in the series[v]. The appendices contain details and equations complementing the discussion of the text. Appendix A defines various quantities and contains a description of stochastic dynamics for the multivariate lognormal model. Appendix B reviews the methodology for the decomposition of the yield-curve movements into EOF or principal components.

IIA. Statistical Probes, Data, Quasi-Equilibrium Drift

In this section, we describe the statistical probes used to test the multivariate lognormal model with data. These include mean shifts and fluctuations between rates of adjacent maturities along the yield curve, means, volatilities, and correlations. A very important measure of the yield-curve shape is the amount of local inversions, or "kinks", in which the yield of a longer maturity is lower than the yield of a shorter maturity but where the yield curve itself is not inverted. Large kinks as opposed to smooth inversions are unphysical, usually representing uncharacteristic shapes not seen in the U.S. Treasury data. Avoiding kinks turns out to constitute a sharp test that is very difficult for models to pass. Even if the average time-series statistics (e.g. volatilities, inter-maturity correlations) are correct, the way the yield curves actually look can be wrong.

All equations are given in the appendices. Statistical quantities are defined for the data, as usual, over a given time window. For the simulator, quantities are calculated for each yield-curve path and then averaged over all paths. We distinguish between the "short end" of the yield curve, which we denote by 3 months to 1 year and the "long end", 2 years to 30 years.

Yield-Curve Data

Fig. 1A shows weekly Treasury yield-curve data from 5 years (1983-88). Fig. 2A shows the time-averaged statistical properties. These data are the mean shifts $\mu_{shift}(T;\Delta T)$ between rates of adjacent maturities, the standard deviations $\sigma_{shift}(T;\Delta T)$ of the shifts between rates of adjacent maturities, rate volatilities $\sigma_x(T)$, correlations $\rho(3M,T)$ of the 3-month rate with rates of other maturities, and correlations $\rho(10yr,T)$ of the 10-year rate with rates of other maturities. Again, the maturities are labeled by the maturity index.

The results for these data show that the mean shift between the 3 and 6-month rates is about 35 bp/yr while the standard deviation of this shift is around 20 bp/yr. Other results can be read off the graphs. One feature was the 30-year rate that was consistently below the 20-year rate.

Input Volatilities and Yield-Curve Correlations

The short-end rate volatilities are around $0.15/\sqrt{yr}$ while the long-end rate volatilities are around $0.13/\sqrt{yr}$. We have chosen the convention of calculating volatilities for yield movements normalized by the initial yield at the beginning of the time window. The units are appropriate for lognormal models.

The historical correlation matrix between yield movements of different maturities breaks up into several distinct parts. There is a clear division between the short and long ends. The short end (by itself) is highly correlated. Short rate - short rate correlation coefficients are $\rho \geq 0.75$. The long end (by itself) is also highly correlated with long rate - long rate correlation coefficients $\rho \geq 0.9$. However, the short end and the long end are only loosely correlated, with short rate - long rate correlation coefficients around $\rho \cong 0.4$.

The Input Historical "Quasi-Equilibrium" Yield-Curve Path for the Model

A critical input to the Monte-Carlo simulators is the drift for each yield, defined by the corresponding mean in the stochastic equation (cf. App. A). The drift defines the Historical Quasi-Equilibrium Yield Curve Path, denoted as $r_{\text{Quasi-Equil. Path}}(t,T)$. While the definition of these trend regions is subjective, it is also extremely useful. The details of the description we will obtain depend on the precise nature of these definitions; the overall features will not. We write[7]

$$d_t r_{\text{Quasi-Equil. Path}}^{\text{Historical}}(t,T) = \sum_{\kappa=1}^{3} \lambda_\kappa^T Dt_\kappa \left[\theta(t - t_{\kappa-1}) - \theta(t - t_\kappa) \right] \qquad (48.1)$$

We describe the trends of these 5 years of data as three regions $\kappa = 1, 2, 3$ of time lengths $\{Dt_\kappa\}$ between transition points $\{t_\kappa\}$, with $Dt_\kappa = t_\kappa - t_{\kappa-1}$, during which interest rates first trended upward, then downward, and again upward. The reader can see from Fig. 1A that this is a reasonable general description of the interest rate trends in the data. The parameters $\{\lambda_\kappa^T, Dt_\kappa\}$ in Eqn. (48.1) were chosen to reflect these trends[8].

---

[7] **Notation:** Please do not confuse these quasi-equilibrium drift parameters λ_κ^T with the eigenvalues of the principal components, discussed later. As usual, $\Theta = 1$ for positive argument, 0 otherwise.

[8] **Generalization of the Quasi-Equilibrium Path in the Macro-Micro Model:** In the Macro Micro model, described in Ch. 50 and 51, these drifts are taken as quasi-random. They then define quasi-random quasi-equilibrium yield-curve paths.

IIB. Yield-Curve Kinks: Bête Noire of Yield Curve Models

Consider a yield curve rising in maturity at a given fixed time. If there is a local inversion in maturity, the yield curve is not monotonic in maturity, and contains a kink.

The kinks can be characterized qualitatively by two numbers: (1) The percentage of cases in which there was a kink; (2) The average size of the kink. A kink between the 3 month and 1 year rate occurred in the data 7% of the time. This means that the 6-month rate was below the 3-month rate 7% of the time, with a small average kink size, the difference between the 3 and 6-month rates, when there was a local inversion, of around 0.8 bp/yr. Other maturities yield similar results. Except for the 30-year rate that was consistently below the 20-year rate, average local inversion sizes are all less than 7 bp/yr, and most are on the order of 1-2 bp/yr.

Fig. 1B shows one path (#100) from the multivariate lognormal simulation[9]. Other paths exhibit similar characteristics. Large kinks are produced by this model. In particular, typical average kink sizes were around 20 bp/yr. Kinks occurred 30-50% of the time in this model averaged over paths. *These numbers are far in excess of the small and rare kinks in the data.*

Fig. 2B shows the results for the time and path averaged statistics from the multivariate lognormal model. Most of them are reasonably in accord with data, including the output correlations, consistent with the input correlations. The output short-maturity vols are a little high. However, the most noteworthy point is that a large discrepancy exists for $\sigma_{shift}(T;\Delta T)$, the standard deviation of the mean shift between adjacent maturities, which is far too large in the LN yield-curve model. This is a reflection of the large kinks generated by the model.

Example of Consequence of Kinks for Risk Management

An important significance of kinks is incorrect spread statistics between different maturities. For example, the long rate used in Mortgage-Backed Securities calculations to drive a prepayment calculation for fixed-rate mortgages will take incorrect jumps with respect to the short rate used in the discounting procedure, resulting in anomalous prepayments for a given path in the MC simulation.

Kinks and Arbitrage

Formal arbitrage possibilities can be created if the kinks are large enough to produce negative forward rates. Negative forward rates imply that zero-coupon bonds with longer maturity have prices above zeros with shorter maturity.

To illustrate, a kink originating from a local inversion of 92 bp would produce a higher price of a zero-coupon bond with a maturity of 11 years

---

[9] **Why Path Number 100?** We simply chose this path number at random for presenting results. There was nothing special about it, and other paths gave similar conclusions.

compared to a zero with maturity of 10 years with a spot rate of 10%. A local inversion of 340 bp would produce a higher price of a zero with a maturity of 15 years compared to the same 10-year zero.

Histogram for Fluctuations of a Given Rate

A further characterization of the model is found in the statistical distribution or histogram corresponding to a given rate. Figs. 3A and 3B show the histograms for the fluctuations of the 10-year rate $\delta r_{fluctuations}$ $(t, T = 10 \, yr)$ and those of the 3 month rate, respectively. It is important to note that these fluctuations are measured with respect to the historical quasi-equilibrium path $r_{\text{Quasi-Equil. Path}}^{\text{Historical}}(t,T)$ defined above. The distributions of the data and of the lognormal model are shown. The lognormal model generates fluctuations that are much too wide compared to the data. This is because the paths generated by the model fluctuate away from the quasi-equilibrium path. For one path, the width of the model distribution is not far from the data width. However, the model distributions are generally displaced from the center of the data distribution, the total model distribution over all paths is very wide. Thus, this discrepancy by itself does not imply that the model is wrong, but it is a telling observation. We discuss the point further in the next two chapters.

Attempted Modifications Without Strong Mean Reversion

We attempted a variety of modifications to try to improve the multivariate LN model. One modification included memory effects to attempt to reduce the kinks and fluctuations by recall of the initial smooth yield curve; others were smoothing recipes. In spite of our determined effort over a rather long period, none of these efforts (without strong mean reversion) was successful.

We also examined a multivariate Gaussian model. Similar conclusions were obtained.

III. EOF / Principal Component Analysis

A useful characterization of yield curve movements is given by the empirical orthogonal function (EOF) analysis, also called principle component analysis (and other names)[10]. This analysis has long been standard in applied physics and engineering. It was, to the best of our knowledge, first applied to yield-curve analysis by Garbade in 1986, and independently around the same time by

---

[10] **Notation:** EOF analysis is usually referred to as principal component analysis. Jerry Herman (and also David Haan), who introduced this technique at Merrill Lynch, used the name EOF analysis.

Herman[11, 12, iv]. The movements of the yield curve are decomposed into independent movements; technically these movements are orthogonal. These movements correspond to reasonably down-to-earth descriptions.

The leading movement usually corresponds to a parallel shift of the yield curve (though not exactly, and occasionally not even qualitatively). The next most important movement can roughly be described as a tilt of the yield curve about some maturity as a fulcrum. Following this is a series of movements of decreasing importance.

The physical picture is analogous to the decomposition of the motions of a violin string into simple harmonic motions or modes, each with its own frequency, making up the rich sound. The movements are described by eigenfunctions and their importance by eigenvalues. See Appendix B.

Principal Components of the Yield-Curve Data

Fig. 4A shows the first three eigenfunctions of the yield curve movements in the data, labeled parallel shift, tilt, and flex. Indeed, the "parallel shift" eigenfunction does resemble a parallel shift. The "tilt" eigenfunction resembles a twist of the yield curve, and the flex is a more complex motion. There are (for the 11 variates used in the multifactor modeling here) eight more eigenfunctions representing movements that are even more complicated. The sum, with the correct weighting, gives back the average yield-curve movements over the time window of 5 years on which we focused.

The relative importance or weighting of the various contributions is given by the eigenvalues. An eigenvalue measures the amount of yield-curve fluctuations characterized by the corresponding eigenfunction. The first three eigenvalues for the data are in the approximate ratio 36/8/1. Higher-order modes have smaller eigenvalues and characterize the fine details of the fluctuations.

Figs. 4B show the same graphs for the lognormal simulation.

A Principal-Component Simulator

Another interesting Monte-Carlo model, originally proposed by Garbade [iv], is to assume a lognormal model for yield-curve time changes along each of the EOF

---

[11] **Acknowledgements:** Jerry Herman was the manager of Merrill's Financial Strategies Group. The PhD Quantitative Analysis Group that I managed in 1987-89 was a subgroup. Jerry was an unusual and excellent manager who understood quantitative issues well. I thank him for many discussions and assistance.

[12] **History:** Very early, at a seminar I was giving at Merrill in 1986, Jerry Herman correctly identified the incorporation of yield curve dynamics as a major unsolved issue in models, and he developed a yield-curve model in 1987. Jerry's work was the lynch pin that led to the models described here.

eigenfunction directions[13] rather than using the maturity-correlation matrix technique used in the text and described in Appendix A. We have investigated this and found similar results. In particular, the multivariate lognormal model based on this modified technique still produces anomalous yield-curve kinks.

IV. Simpler Lognormal Model with Three Variates

We have dealt in this chapter with an eleven-variate lognormal model. The choice of 11 variates is not mandated. For applications, a simpler model composed of, e.g. two or three variates, is preferable. Such a model can now be constructed from the eleven-variate model.

As an example, we have examined a three-factor reduction (i.e. three maturities: long, medium, short). As mentioned in the text, the three-factor model is the simplest one that is still capable of being tested for kinks. *Again, note that by definition a two-factor model cannot have kinks. A kink requires three points on the yield curve to exist at all.*

One can follow several procedures for this reduction. First, one can extract the paths for these three maturities from the paths generated by the model for all 11 maturities. Second, one can start the process over and model the three maturities directly. Third, one can by average over variates (e.g. long-end averaging over, say, the 7 year to 20 year rates to produce a proxy long rate driving prepayment models). The results for these various procedures are statistically similar when averaged over paths, but of course will not be equivalent path by path for a given maturity.

Kinks in the 3-Variate LN Model

The problems of the eleven-variate lognormal model do not disappear when this reduction is made. One of the most serious problems, as mentioned many times already, is the presence of kinks. Kinks large enough to produce negative forward rates were in fact generated in some cases of the three-variate lognormal model[14].

---

[13] **Acknowledgement:** I thank Sheldon Epstein for conversations on principal component simulators.

[14] **Generality:** Negative forwards are a potential problem in any multifactor model whose underlying variables are composite and not the forwards themselves, unless strong mean reversion is present. Another model based on zero-coupon rates (also without strong mean reversion) that generates negative forwards is described in Ch. 51.

V. Wrap-Up and Preview of the Next Chapters

We have presented a methodology for examining and characterizing yield-curve dynamics and have obtained what we feel is important insight into the stochastic properties of yield-curve movements. The various statistical tests and the EOF or principal component yield-curve analysis have led to the conclusion that the multivariate lognormal model, while a straightforward generalization of the popular single-variate lognormal model, fails in some important regards.

First Improvement: Using Strong Mean Reversion (Ch. 49)

The next chapter Ch. 49 describes an improved model using strong mean-reversion for fluctuations, which removes the problematic yield-curve kink anomalies. As a result, we believe that interest rates, as exhibited in historical data, do not exhibit the statistical properties of a stochastic process that spreads out in time. The same slowly varying quasi-equilibrium historical yield-curve path is used, around which the fluctuations occur.

Further Improvement: The Macro-Micro Model (Ch. 50, 51)

Ch. 50, 51 contain the description of the Macro-Micro model. This model can be used for pricing contingent claims and sophisticated interest-rate risk analysis. The basic innovation is that the historical quasi-equilibrium yield curve path is generalized to possess a special quasi-random nature. The Macro-Micro model contains explicit time scales. An interpretation of slowly varying quasi equilibrium paths is given in terms of long-term trending quasi-random macroeconomic effects on interest rates. The rapid fluctuations about a quasi-equilibrium yield curve path is given in terms of trading activities.

Appendix A: Definitions and Stochastic Equations

In this appendix, we give some definitions of quantities used in the text and present the stochastic equations defining the multivariate lognormal model. Define the T-year maturity rate at time t for path α as $r_\alpha(t,T)$. Data will be denoted using the same symbol without the path label, $r(t,T)$. For some function x of the rates and the time, the differential change over infinitesimal time dt is written as

$$d_t x\,[r_\alpha(t,T),t] = \mu_x[r_\alpha(t,T),t]\,dt + \sigma_x[r_\alpha(t,T),t]\,dZ_\alpha(t,T) \quad (48.2)$$

Here, $\mu_x[r_\alpha(t,T),t]$ and $\sigma_x[r_\alpha(t,T),t]$ are the (possibly) rate-dependent mean or drift and volatility for maturity T at time t along path α. The Gauss-

Wiener process is defined as usual as having zero mean when time-series is averaged over an infinite time window for path α,

$$\langle dZ_\alpha (t,T) \rangle_t = 0 \qquad (48.3)$$

The cross-correlations among maturities give the correlation matrix, which for an infinite time window is independent of the path,

$$\langle dZ_\alpha (t,T_1) dZ_\alpha (t,T_2) \rangle_t = \rho(T_1,T_2) dt \qquad (48.4)$$

This is solved in terms of independent normal distributions by[15]

$$dZ_\alpha (t,T) = \sum_{T'} \Lambda(T,T') N_\alpha^{(0,1)}(t,T') \sqrt{dt} \qquad (48.5)$$

Here

$$\rho(T,T') = \sum_{T''} \Lambda(T,T'') \Lambda(T',T'') \qquad (48.6)$$

Also, $N_\alpha^{(0,1)}(t,T')$ is the random number drawn from an $N(0,1)$ normal distribution for the yield with maturity T', at time t on path α. This equation can be checked by squaring it and using the independence property

$$\langle N_\alpha^{(0,1)}(t,T_1') N_\alpha^{(0,1)}(t,T_2') \rangle_t = \delta_{T_1',T_2'} \qquad (48.7)$$

Correlation Formalism in Two and Three Dimensions

Call the correlations[16] $c_{ij} = \rho_{ij}$, with $s_{ij} = \sqrt{1-\rho_{ij}^2}$ (where maturities are labeled by i, j). Then if $d_t x_i = \mu_i dt + \sigma_i dz_i$ with $\langle dz_i\, dz_j \rangle = \rho_{ij}\, dt$ we can define independent measures $\{dw_i\}$. These measures satisfy the orthogonal condition $\langle dw_i\, dw_j \rangle = \delta_{ij}\, dt$. We get $d_t x_i = \mu_i dt + \sigma_i \sum_j \Lambda_{ij}\, dw_j$, where the correlation matrix is $\rho = \Lambda \Lambda^{transpose}$. This can be solved directly by orienting

---

[15] **Real-World Issues:** See Ch. 24 for a description of the problems of getting Λ in practice from the data.

[16] **Correlation Formalism:** Correlations were discussed in detail in Ch. 22-25. We retain the discussion given in the original paper here in order to make the discussion self-contained in this chapter.

the $d_t x_i$ variables simply with respect to the $\{dw_i\}$ variables, taking successive products of two $d_t x_i$ variables to get the correct correlations, and then insuring that the normalizations of each of them is correct. We get

$$d_t x_1 = \mu_1 dt + \sigma_1 \, dw_1$$
$$d_t x_2 = \mu_2 dt + \sigma_2 \left[c_{12} \, dw_1 + s_{12} \, dw_2 \right]$$
$$d_t x_3 = \mu_3 dt + \sigma_3 \left[c_{13} \, dw_1 + \left(\frac{c_{23} - c_{12} c_{13}}{s_{12}} \right) dw_2 + \Lambda_{33} \, dw_3 \right] \quad (48.8)$$

Here the "triangle function" is

$$\Lambda_{33} = \frac{1}{s_{12}} \left[1 - c_{12}^2 - c_{13}^2 - c_{23}^2 + 2 c_{12} c_{13} c_{23} \right]^{1/2} \quad (48.9)$$

Means and Volatilities

For a given time window and a given set of paths, we write the time-averaged and path-averaged drift and squared volatility as

$$\begin{aligned}\mu_x(T) &= \langle \mu_x \left[r_\alpha (t,T), t \right] \rangle_{t;\alpha} \\ \sigma_x^2(T) &= \langle \sigma_x^2 \left[r_\alpha (t,T), t \right] \rangle_{t;\alpha} \end{aligned} \quad (48.10)$$

Averages for the data are defined the same way except, of course, there is no path averaging. For the LN case, the time change $d_t x$ up to a convexity correction is

$$d_t x \left[r_\alpha (t,T), t \right] = d_t r_\alpha (t,T) / r_\alpha (t,T) \quad (48.11)$$

Spread Shifts between Rates of Adjacent Maturities

The shift or spread between yields of adjacent maturities T and $T + \Delta T$ on the yield curve, at fixed time on a given path, is defined as

$$\Delta_T \, r_\alpha (t,T; \Delta T) \equiv r_\alpha (t, T + \Delta T) - r_\alpha (t,T) \quad (48.12)$$

Here, $\Delta T = \Delta T(T)$ is the maturity increment between defined nearest-neighbor maturities on the yield curve (e.g. 10 years between the 20 year and 30 year rates). Averaging $\Delta_T r_\alpha(t,T;\Delta T)$ over time and paths, with averaging notated by $\langle \cdot \rangle_{t;\alpha}$, gives the mean spread or shift,

$$\mu_{shift}(T;\Delta T) = \langle \Delta_T r_\alpha(t,T;\Delta T) \rangle_{t;\alpha} \qquad (48.13)$$

The spread or shift squared volatility is given by

$$\sigma^2_{shift}(T;\Delta T) = \left\langle \left[\Delta_T r_\alpha(t,T;\Delta T) \right]^2 \right\rangle_{t;\alpha} - \mu^2_{shift}(T;\Delta T) \qquad (48.14)$$

The Kink Definition

We define a yield-curve "kink" when a yield of higher maturity is lower than a yield of lower maturity. While this does not yield much useful information in the case of an overall-inverted yield curve, it certainly is useful in discriminating between models when the yield curve is normal and increasing. As described in the text, kinks play a major role in formulating sharp tests between various models.

The mean shifts and volatilities for kinks are defined exactly as before, with the additional filter that the yield difference between neighboring maturities is negative, thus producing the kink in the first place, i.e.

$$\Delta_T r_\alpha(t,T;\Delta T) \leq 0 \qquad (48.15)$$

Appendix B: EOF or Principal-Component Formalism

In this appendix we briefly describe the Empirical Orthogonal Function (EOF) or principal component expansion analysis used in the text [iv]. This expansion has other names, including the Karhunen-Loeve expansion, modal analysis, etc. We first construct the matrix F consisting of N observations in time (labeled by n) of the time changes of each of M different rates (labeled by m). In this work, we had eleven variates, so $M = 11$. We therefore write

$$F_{mn} = r(t_{n+1}, T_m) - r(t_n, T_m) \qquad (48.16)$$

The MxM matrix R is then constructed as

$$R_{mm'} = \frac{1}{N} \sum_{n=1}^{N} F_{mn} F_{m'n} - \frac{1}{N^2} \sum_{n=1}^{N} F_{mn} \sum_{n=1}^{N} F_{m'n} \qquad (48.17)$$

Up to normalization, R is the correlation matrix ρ; it is actually the covariance matrix with matrix elements

$$R_{mm'} = \sigma_m \sigma_{m'} \rho_{mm'}$$ (48.18)

We now look for the M solutions labeled by β to the eigenvalue equation for R, namely

$$\sum_{m'=1}^{M} R_{mm'} \psi_{m'}^{(\beta)} = \lambda^{(\beta)} \psi_{m}^{(\beta)}$$ (48.19)

The quantities $\psi_m^{(\beta)}$ for a given β can be thought of as the components of an M-vector; this vector is called an eigenvector. The number $\lambda^{(\beta)}$ is called the eigenvalue associated with this eigenvector. The eigenfunctions for the three largest eigenvalues for the data and for the lognormal model are plotted against the maturity index m in Figs. 4A and 4B, respectively.

We have the completeness and orthogonality relations for the eigenfunctions,

$$\sum_{\beta} \psi_m^{(\beta)} \psi_{m'}^{(\beta)} = \delta_{mm'} \;,\quad \sum_{m} \psi_m^{(\beta)} \psi_m^{(\beta')} = \delta_{\beta\beta'}$$ (48.20)

We can decompose the individual rates as mentioned in Ch. 45,

$$r(t_n, T_m) = \sum_{\beta} \psi_m^{(\beta)} Y_n^{(\beta)}$$ (48.21)

Here, $Y_n^{(\beta)}$ is given by the inverse formula,

$$Y_n^{(\beta)} = \sum_{m=1}^{M} \psi_m^{(\beta)} r(t_n, T_m)$$ (48.22)

We can also define, for each time,

$$Y_{mn}^{(\beta)} = Y_m^{(\beta)}(t_n) = \psi_m^{(\beta)} \sum_{m'=1}^{M} \psi_{m'}^{(\beta)} r(t_n, T_{m'}) = \psi_m^{(\beta)} Y_n^{(\beta)}$$ (48.23)

We write the time-averaged m^{th} rate $\langle r \rangle_m \equiv \langle r(t_n, T_m) \rangle_{n(time)-averaged}$. Then,

$$\langle r \rangle_m = \sum_{m=1}^{M} \langle Y^{(\beta)} \rangle \psi_m^{(\beta)} \quad , \quad \langle Y^{(\beta)} \rangle = \sum_{m=1}^{M} \langle r \rangle_m \psi_m^{(\beta)} \qquad (48.24)$$

We can reconstruct the amount of movement of the yield curve "along" the "direction" of each eigenfunction. Write $E_{m\beta} = \psi_m^{(\beta)}$ as the components of an $M \times M$ matrix. This matrix is an orthogonal matrix, namely its inverse is its transpose. Now define the matrix with elements $C_{\beta n}$, depending on the eigenfunction label β and the time index n, as

$$C_{\beta n} = \sum_{m=1}^{M} [E^{-1}]_{\beta m} F_{mn} = \sum_{m=1}^{M} \psi_m^{(\beta)} F_{mn} \qquad (48.25)$$

Then we can define

$$f_{mn}^{(\beta)} = C_{\beta n} \psi_m^{(\beta)} = \psi_m^{(\beta)} \sum_{m'=1}^{M} \psi_{m'}^{(\beta)} F_{m'n} \qquad (48.26)$$

This gives the component of the measurement of the T_m maturity yield change at time t_n along the βth eigenvector $\psi_m^{(\beta)}$. The time average of the differenced rate at fixed maturity is $\langle D_t r \rangle_m = \langle F_{mn} \rangle_{n(time)-averaged}$. We can then write the various time-averaged components of the yield shifts as

$$\langle D_t Y_m^{(\beta)} \rangle = \frac{1}{N} \sum_{n=1}^{N} f_{mn}^{(\beta)} = \psi_m^{(\beta)} \sum_{m'=1}^{M} \psi_{m'}^{(\beta)} \langle D_t r \rangle_{m'} \qquad (48.27)$$

These equations can also be written using the projection operator $\mathcal{P}_{mm'}^{(\beta)}$ which is

$$\mathcal{P}_{mm'}^{(\beta)} = \psi_m^{(\beta)} \psi_{m'}^{(\beta)} \qquad (48.28)$$

It is useful to define the scalar quantity $\langle D_t Y^{(\beta)} \rangle$ using

$$\langle D_t Y^{(\beta)} \rangle \equiv \sum_{m=1}^{M} \psi_m^{(\beta)} \langle D_t r \rangle_m \qquad (48.29)$$

With this definition, we have

$$\langle D_t Y_m^{(\beta)} \rangle = \psi_m^{(\beta)} \langle D_t Y^{(\beta)} \rangle \tag{48.30}$$

It is useful to get the composition of the average rate move for a given maturity in terms of the principal component average moves with different β. This is

$$\langle D_t r \rangle_m = \sum_{\beta=1}^{M} \langle D_t Y_m^{(\beta)} \rangle \tag{48.31}$$

Simpler Three Principal Component Example

All this formalism may seem unintuitive. As the canonical simple example[17], we take three principal EOF components. These are generally called "parallel", "tilt" and "flex". Three components suffice for most yield curve movements most of the time[18]. Say there are three points on the yield curve $m = 1, 2, 3$ corresponding to 2, 5, 10 years. The eigenfunctions and scalar principal components are

$$\psi^{(\beta=ParallelShift)} = \frac{1}{\sqrt{3}} \begin{pmatrix} 1 \\ 1 \\ 1 \end{pmatrix}, \quad \psi^{(\beta=Tilt)} = \frac{1}{\sqrt{2}} \begin{pmatrix} 1 \\ 0 \\ -1 \end{pmatrix}, \quad \psi^{(\beta=Flex)} = \frac{1}{\sqrt{6}} \begin{pmatrix} 1 \\ -2 \\ 1 \end{pmatrix} \tag{48.32}$$

$$\langle D_t Y^{(\beta=ParallelShift)} \rangle = \frac{1}{\sqrt{3}} \left[\langle D_t r \rangle_{2-yr} + \langle D_t r \rangle_{5-yr} + \langle D_t r \rangle_{10-yr} \right]$$

$$\langle D_t Y^{(\beta=Tilt)} \rangle = \frac{1}{\sqrt{2}} \left[\langle D_t r \rangle_{2-yr} - \langle D_t r \rangle_{10-yr} \right]$$

$$\langle D_t Y^{(\beta=Flex)} \rangle = \frac{1}{\sqrt{6}} \left[\langle D_t r \rangle_{2-yr} - 2\langle D_t r \rangle_{5-yr} + \langle D_t r \rangle_{10-yr} \right] \tag{48.33}$$

For simplicity, we set $Dr_m = \langle D_t r \rangle_m$, $DY^{(\beta)} = \langle D_t Y^{(\beta)} \rangle$. The vector quantities (in maturity) are given by equations in which the projection matrices $\mathcal{P}_{mm'}^{(\beta)}$ enter:

---

[17] **Canonical Butterfly Example:** As a physics colleague Henry Abarbanel once wisely remarked, "One Example is Worth Two Theorems". This is the canonical principal component example. The flex component is called a "butterfly". The parallel component is an exactly parallel shift, and the tilt component is the "Twos-Tens" tilt.

[18] **Are just three EOF principal components really enough?** Usually, but not always. For example, if some desk has some complicated strategy with a messy yield-curve composition, three components are not enough. Therefore, if you structure your risk management system to include only three principal components you may miss some important risk on the desk that is invisible to your measuring probes.

$$DY^{(\beta=ParallelShift)} = \frac{1}{3}\begin{pmatrix} 1 & 1 & 1 \\ 1 & 1 & 1 \\ 1 & 1 & 1 \end{pmatrix}\begin{pmatrix} Dr_2 \\ Dr_5 \\ Dr_{10} \end{pmatrix} = \frac{1}{3}\begin{pmatrix} Dr_2 + Dr_5 + Dr_{10} \\ Dr_2 + Dr_5 + Dr_{10} \\ Dr_2 + Dr_5 + Dr_{10} \end{pmatrix} \text{ for } \begin{pmatrix} m=1 \\ m=2 \\ m=3 \end{pmatrix},$$

$$DY^{(\beta=Tilt)} = \frac{1}{2}\begin{pmatrix} 1 & 0 & -1 \\ 0 & 0 & 0 \\ -1 & 0 & 1 \end{pmatrix}\begin{pmatrix} Dr_2 \\ Dr_5 \\ Dr_{10} \end{pmatrix} = \frac{1}{2}\begin{pmatrix} Dr_2 - Dr_{10} \\ 0 \\ -Dr_2 + Dr_{10} \end{pmatrix} \text{ for } \begin{pmatrix} m=1 \\ m=2 \\ m=3 \end{pmatrix} \quad (48.34)$$

$$DY^{(\beta=Flex)} = \frac{1}{6}\begin{pmatrix} 1 & -2 & 1 \\ -2 & 4 & -2 \\ 1 & -2 & 1 \end{pmatrix}\begin{pmatrix} Dr_2 \\ Dr_5 \\ Dr_{10} \end{pmatrix} = \frac{1}{6}\begin{pmatrix} Dr_2 - 2Dr_5 + Dr_{10} \\ -2Dr_2 + 4Dr_5 - 2Dr_{10} \\ Dr_2 - 2Dr_5 + Dr_{10} \end{pmatrix} \text{ for } \begin{pmatrix} m=1 \\ m=2 \\ m=3 \end{pmatrix}$$

The 2-year rate time averages difference, for example, can be reconstructed as

$$\langle D_t r \rangle_{2-yr} = \langle D_t Y_{2-yr}^{(\beta=ParallelShift)} \rangle + \langle D_t Y_{2-yr}^{(\beta=Tilt)} \rangle + \langle D_t Y_{2-yr}^{(\beta=Flex)} \rangle$$
$$= \frac{1}{3}[Dr_2 + Dr_5 + Dr_{10}] + \frac{1}{2}[Dr_2 - Dr_{10}] + \frac{1}{6}[Dr_2 - 2Dr_5 + Dr_{10}] \quad (48.35)$$
$$= Dr_2$$

Real-Life Principal Components

In general, we will not get these simple forms from the data, but they will be reminiscent of the canonical example. For example, in 1992 the flex component was

$$\langle D_t Y^{(\beta=Flex)} \rangle = \left[0.16 \langle D_t r \rangle_{2-yr} - 0.81 \langle D_t r \rangle_{5-yr} + 0.57 \langle D_t r \rangle_{10-yr} \right] \quad (48.36)$$

The First 3 Principal Components for the Data

Fig. 5A shows the initial and final yield curves for the period 5/88 to 3/89. Fig. 5B shows the first three $\langle D_t Y^{(\beta=ParallelShift)} \rangle$, $\langle D_t Y^{(\beta=Tilt)} \rangle$, $\langle D_t Y^{(\beta=Flex)} \rangle$ time-averaged components of the yield shifts over this period. Roughly, as can be seen from these graphs, yield movements over this period did behave as a parallel shift along with a clockwise tilt around the middle of the yield curve. A small amount of flexing was also present.

Anomalous Principal Component Behavior

Occasional anomalous behavior sometimes does occur where the first eigenfunction that we label with "parallel" does *not* in fact correspond to a parallel movement.

To illustrate, Fig. 6A shows a breakdown in small intervals of data between the dates 2/19/88 and 1/30/89. Fig. 6B shows the components of the yield shifts over the period 2/19/88 to 5/13/88. The leading component labeled "parallel shift" is not at all parallel but rather tilted opposite to the second component labeled "tilt". The reason for the anomalous behavior was connected to negative correlations between the short and long ends of the yield curve during this period.

Figures: Multivariate Lognormal Yield-Curve Model

Fig. [1]. Fig. 1A shows the three-dimensional plot of weekly U.S. Treasury yield-curve data over the period 1983-1988. Fig. 1B exhibits a yield-curve path (consisting of eleven maturities) from the Monte-Carlo simulation for the multivariate lognormal LN model, starting from the same initial yield curve as in Fig. 1A. Note that the LN model yield curves have kinks not seen in the data.

Fig. [2]. Statistical measures used to characterize the yield-curve time series are shown. The maturity index 1 to 11, represents the 3M, 6M, and 1, 2, 3, 4, 5, 7, 10, 20, 30-year rates, respectively. Figs. 2A show data for spread shifts between neighboring maturities. These are the shift means $\mu_{shift}(T;\Delta T)$ and standard deviations $\sigma_{shift}(T;\Delta T)$ at the top. The yield volatilities $\sigma_x(T)$ are in the center. At the bottom are the correlations $\rho(3M,T)$ for the 3M rate with other rates, and the correlations $\rho(10yr,T)$ of the 10-year rate. Figs. 2B show the same graphs for the time and path-averaged lognormal simulation. Note that the vols of the shift spreads are too large in the LN model as compared with the data.

Fig. [3]. Figs. 3 show histograms of the fluctuations of rates, measured around the slowly varying quasi-equilibrium yield curve discussed in the text. Figs. 3A and 3B show the fluctuations of the 10-year and of the 3-month rates for the data compared to the lognormal simulation. Note that the LN fluctuations are too wide compared to the data.

Fig. [4]. Principal component EOF analysis (see Appendix B). Figs. 4A show data for the first three eigenfunctions plotted against maturity, labeled "parallel shift", "tilt", and "flex". Note that the "parallel shift" is not exactly parallel. The eigenvalues are also listed. Figs. 4B show the same graphs for the LN simulation.

Fig. [5A]. Treasury yield curves in 5/88 and in 3/89.

Fig. [5B]. The first three principal components of the movements of the yield curve between 5/88 and 3/89 (see App. B), labeled parallel, tilt, and flex.

Fig. [6A]. Five treasury yield curves showing anomalous behavior.

Fig. [6B]. The principal components of the data between 2/19/88 and 5/13/88. Note that the "parallel shift" component is not parallel. See text.

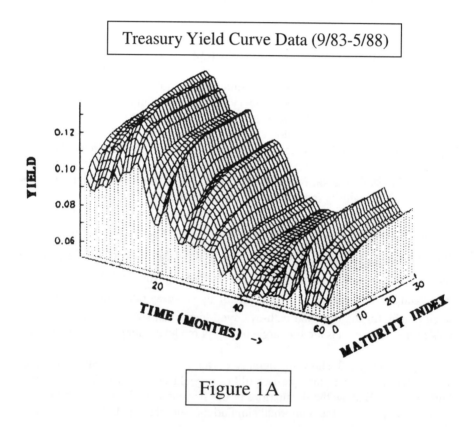

Figure 1A

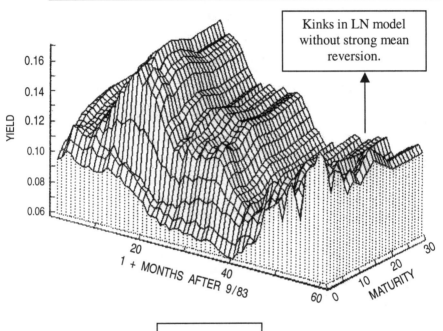

Figure 1B

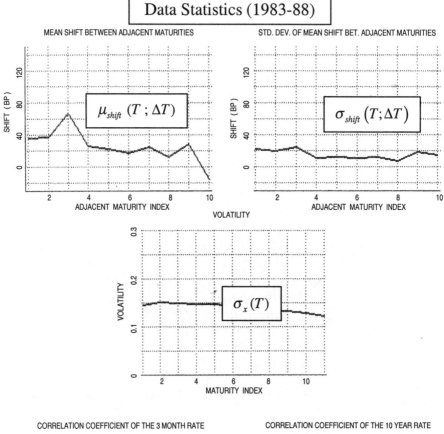

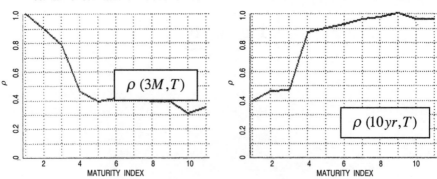

Figure 2A

A Multivariate Yield-Curve Lognormal Model

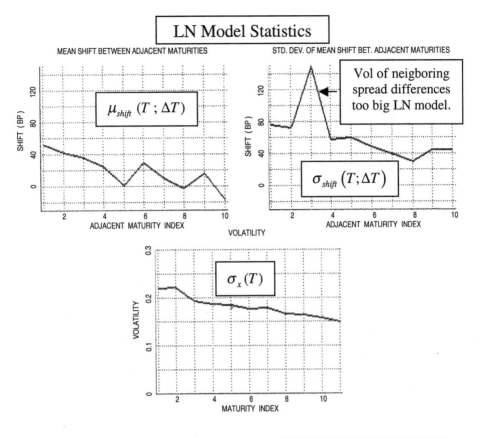

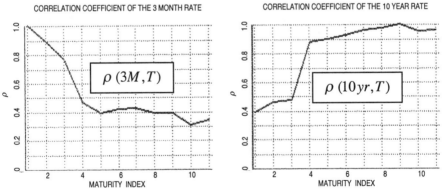

Figure 2B

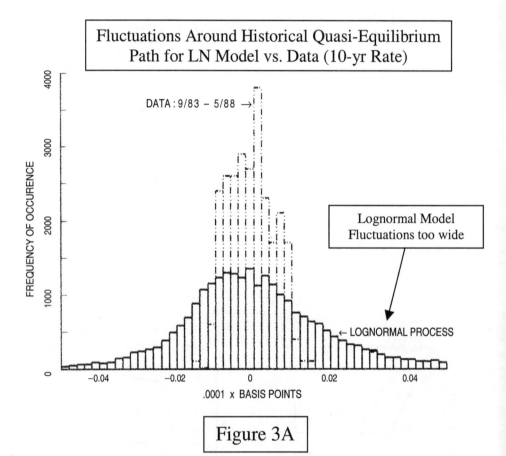

Figure 3A

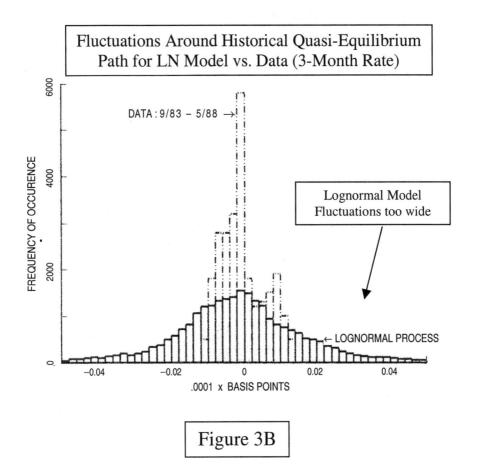

Figure 3B

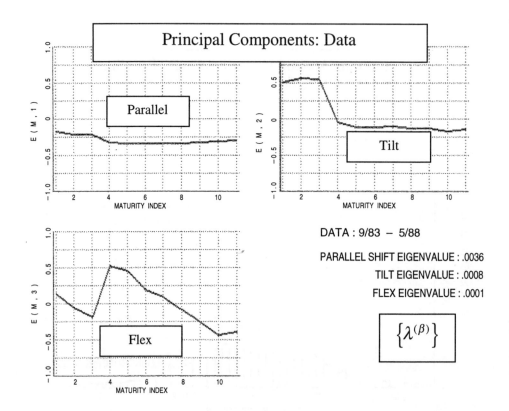

Figure 4A

A Multivariate Yield-Curve Lognormal Model

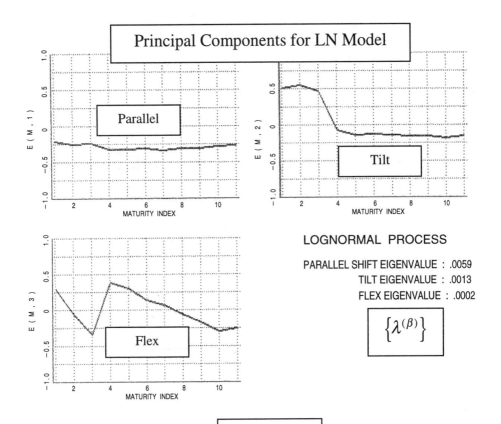

Figure 4B

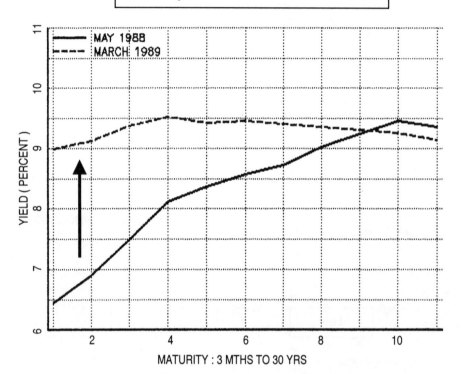

Figure 5A

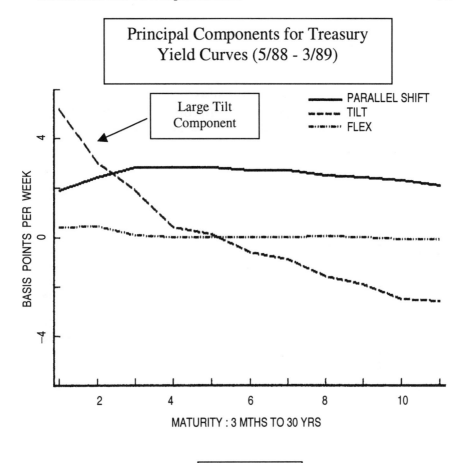

Figure 5B

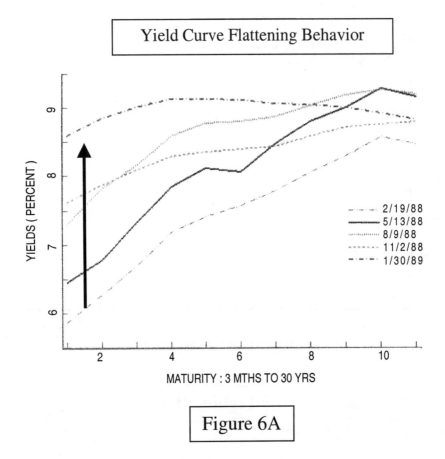

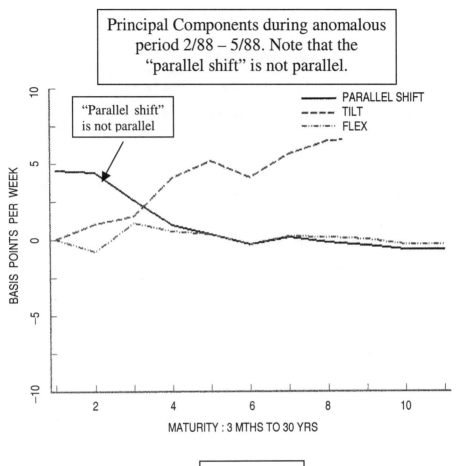

Figure 6B

References

[i] **Multifactor Lognormal Yield Curve Model**
Beilis, A. and Dash, J. W., *A Multivariate Lognormal Model*, Centre de Physique Théorique CPT-89/PE.2334 preprint, CNRS Marseille, France, 1989.

[ii] **Term Structure Single Factor Lognormal Rate Model**
Dattatreya, R. E. and Fabozzi. F. J., *Active Total Return Management of Fixed Income Portfolios*, Probus Publishing Co., 1989. Chapter 11.

[iii] **Two-Factor Models for MBS**
Jacob, D. P. and Toevs, A. L.. *An Analysis of the New Valuation, Duration and Convexity Models for Mortgage-Backed Securities*, in Fabozzi, F. J., *The Handbook of Mortgage-Backed Securities*, Revised Edition, 1988. P. 687.
Hayre, L. (Ed), *Guide to mortgage-backed and asset-backed securities*. Salomon Smith Barney, John Wiley & Sons, 2001.

[iv] **Principal Components for the Yield Curve**
Garbade, K. D., *Modes of Fluctuation in Bond Yields: An Analysis of Principal Components*, Topics in Money and Securities Markets, Banker's Trust, 1986.
Herman, G. F., *Modeling Yield Curve Variations*, Merrill Lynch Capital Markets Working Paper, 1987.

[v] **Strong Mean Reversion Micro Model, the Macro-Micro Model**
Dash, J. W. and Beilis, A., *A Strong Mean-Reverting Yield-Curve Model.* Centre de Physique Théorique CPT-89/PE.2337 preprint, CNRS Marseille, France, 1989.
Dash, J. W. and Beilis, A., *A Macro-Micro Yield-Curve Simulator.* CNRS, Centre de Physique Théorique CPT-89/P.E.2336, 1989.
Dash, J. W., *Evaluating Correlations and Models.* RISK Conference talk, 11/94.

49. Strong Mean-Reverting Multifactor YC Model (Tech. Index 7/10)

Summary of this Chapter

This chapter[1] is the second in the series in this book on modeling the dynamics of the yield curve. You do not have to read the previous chapters to understand this chapter. Necessary background will be summarized.

A strong mean-reverting multivariate model is constructed that agrees well with yield-curve data[i]. This dramatically improves on the multivariate lognormal yield-curve model introduced in Ch. 48. Many statistical probes are used in the analysis. It should be mentioned that the data for the yield volatilities and correlations are input to all the multifactor models we examine.

One very useful probe, Cluster-Decomposition Analysis (CDA), is introduced. It is based on third-order correlation functions in a manner taken from theoretical physics, generalizing skewness. The difficult data properties to reproduce are the third-order correlation measurements along with the absence of kinks or anomalous yield curve shape changes, as described by the statistical properties of spreads of neighboring maturities.

An economical description is achieved by identifying a smooth, slowly varying historical quasi-equilibrium yield curve, about which small, rapid fluctuations occur with very strong mean reversion. Models without strong mean reversion fail to describe yield-curve shape-change statistical data.

The next chapter describes a generalization (the Macro-Micro Model) that is appropriate for valuing contingent claims and for interest-rate-risk analysis. The main idea is the generalization to quasi-random quasi-equilibrium yield curves.

---

[1] **History:** This chapter is based on Ref i. The work was done in collaboration with Alan Beilis. The strong mean-reverting approach is, I believe, the first and perhaps the only multifactor model that successfully reproduces the statistics of the movements and shapes of the yield curve data, and at the same time avoids kinks. The important thing to keep in mind is that the strong nature of the mean reversion is not an arbitrary assumption, but seems to be implied by the data.

I. Introduction to this Chapter

This is the second chapter in a series summarizing an effort aimed at producing a realistic yield-curve multivariate model. Models usually have one factor, although sometimes two-factor models have been used[ii]. Our philosophy, described in the previous chapter, is to describe the yields directly with a multivariate model[2].

What Should a Realistic Multifactor Yield-Curve Model Accomplish?

We want to demand that a yield-curve model possess two desirable attributes. The main point is that we want the model to reproduce the essential statistical characteristics of yield-curve data. We eventually also want the model to be capable of pricing contingent claims. The satisfaction of these two requirements is highly non-trivial. In the previous chapter, we showed [iii] that the multivariate lognormal model fails to describe important yield-curve shape statistical properties and in particular generates kinks. On the other hand, an extension of the model in this chapter will be needed to enable the pricing of options.

The Cluster-Decomposition Analysis (CDA)

A sharp statistical probe is introduced, called Cluster-Decomposition Analysis or (CDA) from theoretical physics. This analysis, although unfortunately sharing the same name has nothing to do with the cluster decomposition in statistical regression. The CDA technique uses third-order correlation functions to uncover properties of the stochastic processes underlying the yield-curve data.

The CDA is used as a probe to produce a new model. This new model is based on exceptionally strong mean-reversion with rapid fluctuations about a moving quasi-equilibrium slowly varying yield curve. While mean reversion is an old idea, the way we use mean reversion is new. First, a quasi-equilibrium yield curve is defined by the long-time behavior of the yield-curve data. Second, the mean reversion is very strong. The mean reversion needed to describe the yield-curve statistical data is in fact so strong that the fluctuations are tightly bounded around the quasi-equilibrium curve.

Problematic Yield-Curve Features Described by the SMRG Model

The strong mean reversion is built into a multivariate Gaussian model; we call it the Strong Mean-Reverting Gaussian (SMRG) model[3]. The SMRG model passes

---

[2] **Yield-Curve Simulator:** The yields for different maturities are modeled as correlated stochastic processes.

[3] **What About Strong Mean-Reverting Models other than Gaussian?** The mean reversion implied by the yield-curve data is so strong that other models would also no doubt work, for example a strong mean-reverting lognormal model. See below.

the CDA tests. Models without strong mean reversion fail the CDA tests. This failure occurs for *both* the single-factor *and* for the multi-factor models without strong mean reversion.

The multivariate SMRG model passes other statistical tests described in the previous chapter that models without strong mean reversion fail. These include: (1) the absence of large kinks and consistency with the spread statistics between yields of neighboring maturities; and (2) the eigenfunctions and eigenvalues of the Empirical Orthogonal Function (EOF) or principal component analysis as compared to data. Other more common statistical probes are also included, such as the yield correlation matrix and yield volatilities.

Generalization of the SMRG Model to the Macro-Micro Model

The multivariate SMRG model described in this chapter is not appropriate for a simulator to price contingent claims such as options. This is because the strong mean reversion about the quasi-equilibrium yield curve needed to describe data when placed in a simulator produces tightly bunched paths. The paths do not "fan out", and therefore do not sample the range of future interest rates present in standard options models that do describe market options prices.

It is important to understand that the bunching up of paths in yield-curve space is not an assumption but rather an apparently unavoidable consequence of describing the details of historical yield-curve data.

In the next chapter[iv,v] we describe how the SMRG model can be extended to produce the Macro-Micro model that can price contingent claims while maintaining consistency with the absence of kinks and other statistical properties of yield-curve data. The main idea is the generalization of the historical quasi-equilibrium yield curve to quasi-random quasi-equilibrium yield curves.

WKB: Small Fluctuations from a Quasi-Equilibrium Yield-Curve Path

The SMRG model achieves simplicity in a manner well known from physics. The basic concept is to look for a quasi-equilibrium state that varies smoothly and from which small, rapid fluctuations occur. An example is the WKB method[vi].

Our fundamental point is that the same idea works for the description of yield-curve data as well. We believe that this fact is both attractive and important.

Yield-Curve Data Used for the Analysis

The same data were used as in Ch. 48: weekly U.S. Treasury yield data over 2, 5, and 10 year periods, and for eleven maturities between 3 months and 30 years. Mostly the analysis focused on the period 1983-88. Input data included the yield volatilities and the yield correlation matrix.

Results of the Numerical Analysis

The most striking way of presenting the results is simple visual examination. Figure 1 exhibits a three-dimensional plot of weekly yield-curve data over the period 1983-1988 as well as a path (of all maturities) from the Monte-Carlo simulation for the strong mean-reverting Gaussian model, starting from the same initial yield curve. The two plots are intentionally unlabeled. The reader is invited to distinguish between them[4]. It is clear that there is statistical similarity and the yield curves generated in the model are visually similar to the data. Other SMRG model paths yield similar results.

In contrast, paths from the lognormal or Gaussian models without mean reversion exhibit a variety of unphysical features, including unacceptably large local yield-curve inversions or kinks, as discussed in Ch. 48 and Ref. iii.

Short and Long Time Scales

Our main conclusion is that, barring exceptional events like the 1987 crash, the yield-curve data can be described in a simple and economical fashion using the following construction. There are two time scales, "long" and "short". On long time-scales, e.g. months, interest rates drift along slowly time-varying quasi-equilibrium means. On short time-scales (e.g. days), fluctuations about these means are mainly observed to take place in a bounded region in the multidimensional multi-maturity interest-rate space.

That is, the data do not act as if they correspond to a stochastic process that spreads out in time according to standard models. We attribute the failure of traditional models to satisfy the CDA and other statistical tests to this property.

On the contrary, the strong mean-reversion produces an effective lower and upper bounding characteristic about the means, with very low probability of the rates exceeding these bounds. In particular, no negative or zero model interest rate values are ever observed in practice in the SMRG model we construct in this paper containing this large mean reversion. We believe that this holds more generally as independent of the presence or absence of any barriers at zero interest rate provided the quasi-equilibrium yield curve stays away from zero rates (a lower bound of about 3% is enough in practice[5]).

---

[4] **Fifty-Fifty Votes: "Can You Tell the Difference Between the SMRG Model and the Data"?** Since writing the original paper in Ref. i, I have given a number of talks on this work. I first show the yield-curve data. Then later in the talk I show an unlabeled plot containing both the data and the SMRG result for one path. I ask the audience to vote on which plot contains the data. The results are on the average 50-50.

The human eye is an excellent pattern recognition instrument. The fact that trained analysts are generally unable to distinguish the difference between the data and the SMRG model is a powerful statement that the SMRG model indeed provides a reasonable phenomenological description of the data. This is backed up by the successful statistical yield-curve analysis.

[5] **Very Low Rates:** This statement should be re-examined for recent very low rates.

Remainder of this Chapter

The remainder of the chapter is organized as follows. In Section II, we describe the Cluster Decomposition Analysis. The main results of the paper are also given in this section. Section III presents the results from the rest of the statistical tests. Section IV summarizes the paper and introduces the next chapter. Appendix A reviews the stochastic dynamics for the various models considered. Appendix B contains a short description of the cluster-decomposition analysis formalism.

II. Cluster Decomposition Analysis and the SMRG Model

In this section, we describe the Cluster Decomposition Analysis applied to the yield-curve data and the models. We find that the SMRG model, i.e. the multivariate Gaussian model with strong mean reversion about a "quasi-equilibrium" slowly varying mean yield-curve path, is preferred by the CDA tests. Multivariate Gaussian or lognormal models without mean reversion essentially fail the CDA tests.

The method of Cluster Decomposition Analysis is based on the following observation. Consider first a simple Gaussian or normal probability density function. It is specified by its first two moments—the mean and width, or volatility. All higher moments (skewness, etc.) factor into a sum of products of powers of these first two moments. This idea generalizes to a time series $x(t)$, as explained in Appendix B. The mean is just the time average over a given window.

The second-order correlation functions generalizing the second moments are defined as follows. Take one photo-copy of a given time series, displace it by a lag time with respect to the original, multiply the points of the series and the displaced series together over some time window where both series exist, and average over time. Divide this by the product of the means of the original and displaced series over the same time window. We call the result the "total two-point function" M_2. It is convenient to define the "connected two-point function" or auto-correlation function, schematically $C_2 = M_2 - 1$.

Third-Order Correlation Functions

The new probes we use are third-order correlation functions that are the generalizations of the "skewness". Take two photo-copies of the given time series, displace them each by a lag time with respect to the original, multiply the three points of the series and the two displaced series together over some time window where all series exist, and average over time. Dividing this by the product of the three means, we obtain the "total three-point function" M_3. Now subtract from M_3 the quantity that M_3 would equal if the process $x(t)$ were

Gaussian. The result is the "connected three-point function" C_3. This function is the analog of the skewness of the distribution for ordinary statistics. For a Gaussian stochastic process, the identity $C_3 = 0$ holds exactly. That is, C_3 is zero for all time lags of the two photocopies.

The same construction can be used with different time series instead of identical time series; for a multivariate Gaussian process defining these time series, the connected three-point function (with any of the variates taken for any time series) is identically zero.

Higher-point functions can be defined using more time series; we began to look at these but found it unnecessary since it turns out that the three-point function suffices for sharp tests.

How the Third-Order Correlation Function is Used in Practice

We use these third-order or three-point correlation functions to test for underlying stochastic property of data in the following way. First, make a hypothesis regarding the stochastic property. Formulate the hypothesis in such a way that the variable used to define a time series is in fact Gaussian by definition. Then, for this time series, if the hypothesis is in fact correct, C_3 will be identically zero by definition. We call this the "zero-C_3 test". If the hypothesis is "approximately" correct, C_3 will be small compared to M_3. If the hypothesis "fails", C_3 and M_3 will have comparable magnitude. While (as in all phenomenology) subjectivity is involved in the words "approximately" and "fails", we believe that the results we obtain are striking enough to speak for themselves.

Mathematical Example

In order for the reader to better understand the ideas, we first consider one realization of a purely lognormal process, $L = \exp\left[\sigma N(0,1)\right]$ using the exponent of a Gaussian random-number generator. We plot one-dimensional graphs showing one variable; hence, the times for the starts of photocopies are fixed. We fixed these times at 10 weeks from the end of the Gaussian series for convenience. The start time for the original time series $x(t)$ is the variable t (in weeks from the end of the series).

The results for the lognormal random-number generator L are shown in Figs. 2A. When $t = 10$, we expect and indeed we see peaks in M_2, C_2, and M_3 because then the original and xerox-copied time series "line up". For a discretized Gaussian process, these peaks decay rapidly with the time lag (infinitely rapid decay as the time window becomes infinitely long); this demonstrates the "short-range correlations" of the Gaussian process. However, as mentioned above, C_3 cannot have any peak for a Gaussian process and in fact should be zero. Because we must choose a finite-sized time window, this is only be approximately realized in practice; indeed $C_3 \approx 0$ is seen to be true.

LN Model Analysis Using 3rd-Order or Three-Point Correlation Function

We now consider actual data viewed as a lognormal process. If we hypothesize that the T-year yield follows a lognormal process, we would construct from the data the time series of the temporal shift in this T-year yield divided by the T-year yield, i.e. $d_t x(t) = d_t r(t;T) / r(t;T)$. According to the hypothesis we just made, $x(t)$ is supposed to follow a Gaussian process. To check the hypothesis, we merely construct the quantities M_2, C_2, M_3, and C_3 for this putative Gaussian process. If they look like the results in Figs. 2A generated by a Gaussian process, then there is evidence that the T-year rate follows a lognormal process. If on the other hand C_3/M_3 is not small, the hypothesis of a lognormal process for $r(t;T)$ is not supported.

Figs. 2B shows the result of applying the lognormal hypothesis to the 10-year rate, where the time (in weeks) goes from 5/88, back towards 9/83. The two-point functions M_2 and C_2 do peak at ten weeks. This means that the short-time correlations characteristic of the hypothesized Gaussian time series seem to be present. However, the ratio of the three-point functions C_3/M_3 is not at all small as it should be if the hypothesis of lognormality of the 10-year rate were true. We regard this as a failure of the 10-year rate generated by the lognormal model to satisfy the zero-C_3 cluster-decomposition analysis test.

The same CDA tests were constructed for other maturity yields, and for cross-maturity correlations. Similar results were obtained. In particular, the zero-C_3 test fails in every case for the multivariate lognormal model.

We could also hypothesize that the yield curve movements are purely normal or Gaussian. Here, $x(t)$ is just assumed to be the short-term rate, or as proxy, a short-term yield. However, similar results to the lognormal case are obtained here for all functions M_2, C_2, M_3, and C_3 across maturities. Again, the zero-C_3 test fails, with C_3 again only slightly smaller than M_3.

The Historical Quasi-Equilibrium Data Yield Curve and Mean Reversion

Another possible stochastic process involves mean reversion. The mean reversion can be taken about the historical quasi-equilibrium yield curve. As mentioned many times, the historical quasi-equilibrium yield curve is the path of drifts for each yield over long time regions. Our purpose here is to decipher the historical statistical properties of the underlying yield-curve dynamics. This complicated task is facilitated by inserting a simple description of the quasi-equilibrium yield curve historically and then examining the fluctuations carefully.

While the details of the description we will obtain depend on the precise nature of these definitions, the overall features will not. However, not all definitions are equivalent. Choosing a drift which does not "reasonably" reproduce data trends will lead to a definition of fluctuations that will be more complicated than that given here, since some of what "should" be described as trend is mixed in with what "should" be described as fluctuation.

The historical quasi-equilibrium yield curve is not the no-arbitrage yield curve defined by the drifts producing consistency with bond price data. We want to perform a comparison of the SMRG model with yield-curve data. The moving averages of the data were what they were, and were naturally not bound by any considerations of today's yield curve used to produce future yield curves through no-arbitrage. No-arbitrage term-structure constraints will be discussed in Ch. 51.

Details: The Historical Quasi-Equilibrium Yield Curve Path for Data

The goal is neither to produce many descriptions, nor to prove that some description is unique (since it is not). We want a simple and useful description. This goal is achieved by describing the trends of the five years of data as being composed of three regions $\kappa = 1, 2, 3$ during which interest rates first trended upward, then downward, and again upward.

As in Ch. 48, we introduce rate slope parameters $\{\lambda_\kappa^T\}$ over macroscopic time intervals $\{Dt_\kappa = t_\kappa - t_{\kappa-1}\}$ between transitition points $\{t_\kappa\}$, and write

$$d_t r_{\text{Quasi-Equil. Path}}^{\text{Historical}}(t,T) = \sum_{\kappa=1}^{3} \lambda_\kappa^T Dt_\kappa \left[\theta(t - t_{\kappa-1}) - \theta(t - t_\kappa) \right] \quad (49.1)$$

The reader can see from Figs. 1 that this is indeed a reasonable general description of interest rate trends. The trends do not capture the fine details of the fluctuations; that task will be accomplished by the SMRG model. The drifts of the yields in these three regions are included in the Monte-Carlo models as inputs, thus guaranteeing that the overall trends will be correct.

Conversely, if "unreasonable" drifts not reproducing trends are input, the performance of every simulator we tried deteriorated.

SMRG Model Analysis Using 3rd-Order Correlation Function

The mean-reverting Gaussian model can be turned into a form that can be examined by the above methodology. See Appendix A. One defines $d_t x(t)$ by adding to the temporal shift of a T-year yield a mean-reverting term defined by multiplying a mean-reversion parameter by the difference of the T-year yield and its mean defined above. If the hypothesis of a mean-reverting Gaussian process for the T-year yield is correct, the presumed Gaussian process in the stochastic equation will indeed satisfy the mathematical properties of a Gaussian process.

Figs. 2C shows the correlation functions under the assumption of Gaussian mean reversion for the 10-year rate. As opposed to the previous cases, with a sufficiently large value for the mean-reversion parameter, C_3 is now seen to be quite small, relative to M_3. Indeed, all functions now look quite similar to the pure Gaussian process of Figs. 2A; the zero-C_3 test now seems quite well satisfied. We take this as evidence that the 10-year rate closely follows a strongly mean-reverting process.

The CDA tests were performed over time windows of 2, 5, and 10 years and across the yield curve, all with the same conclusion.

Large Value of the Mean Reversion

The value of the mean-reverting parameter ω does appear to be very large. For values of ω less than about 1/week, failure for the zero-C_3 test is obtained. A transition occurs at around $\omega = 3$ / week, which leads to positive results (Figs. 2C), so we use this value. Values up to $\omega = 5$/week may be acceptable as described in the next section[6].

An interesting result is that, although it might have been true that the mean-reversion parameter could depend on maturity, similar values seem to hold for all maturities. Moreover, similar parameters hold for cross-maturity correlation functions.

Strong Mean Reversion with Other Models

We have examined a strong mean-reverting Gaussian process. The mean reversion can be combined with any volatility assumption (Gaussian, lognormal, CIR model[vii] etc.). We believe that similar results would hold for other models. This is because barriers at zero interest rate are largely irrelevant for stron mean-reverting processes, if the quasi-equilibrium yield curve path is sufficiently away from zero rates. We find that the data do satisfy this criterion for the quasi-equilibrium yield curve. Large mean-reverting processes in practice only very rarely lead to interest rate paths anywhere near zero; hence the details of the presence or absence of a barrier make little difference.

---

[6] **Mean Reversion Effects on Interest-Rate Fluctuations:** We have checked the effect on interest-rate fluctuations due to the strong mean reversion. A mean-reversion parameter of 3-5/week typically produces mean-reversion-induced changes in rates per time period on the order of 25% of the changes due to random fluctuations.

The SMRG Model is Quite Different from the Vasicek Model

Mean reversion about the quasi-equilibrium yield curve is quite different from the original application of Vasicek [viii] that had mean reversion about one fixed rate (at infinite time).

Fat Tails Producing Jumps/Gaps are an Extra Component

It is interesting to speculate on the non-zero part of C_3, forming a correction to the SMRG process we have been considering. We believe that this correction is connected to "crashes" like that of October 1987 producing long tails on the interest-rate process. While the idea of fat tails is not new (Fama [ix] discussed the problem in 1965), our version is different. The tails in our view are just a small remnant of the total, with most of the dynamics controlled by the strongly mean-reverting process about the quasi-equilibrium yield curve. The tails may or may not give rise to a small component of infinite-variance Pareto statistics in the dynamics; this is in any case difficult to examine with only a few crashes. We note that for the October '87 crash, the interest-rate fluctuation was less than 2%. We consider the point a bit further in the next section[7].

In Ch. 46, we intodroduced the Reggeon Field Theory as a possible mechanism for generating such fat tails.

Troublesome Aspects of Standard Approaches to Interest Rate Dynamics

Our results imply that interest rates, as exhibited in historical data, do *not* behave as corresponding to a stochastic process that spreads out much in time. Such processes include the standard zero (or small) mean-reversion models in common use. Such models fail the zero-C_3 test, as we have seen. It is a challenge to any putative multifactor model to come to grips with the statistics of yield curve movements.

Physical Picture of the Way Interest Rates Really Seem to Behave

From our analysis, interest rates seem to behave as if there were a mean "quasi-equilibrium" moving-average yield-curve path, which changes smoothly and slowly. The quasi-equilibrium yield-curve path as defined by the yield-curve data does move with time and can lead to high or low values for rates, thus producing high or low trends in a smooth way. High or low values only succeed each other over relatively long time scales.

---

[7] **Fat Tails are Important but are Not the Dominant Feature for most Yield Curve Movements:** This topic is treated in some detail in Ch. 21. Note that the fat tails do not affect the main conclusion, that most of the dynamics corresponds to small strongly mean-reverting fluctuations around a quasi-equilibrium slowly moving yield curve. The fat tails produce jumps from time to time that constitute an extra component.

We believe that these results are a-priori reasonable. Interest-rate fluctuations are in fact tightly constrained about a moving mean, as a glance at the data in the bottom part of Fig. 1 will show. The magnitude of these rapid fluctuations can be estimated by eye from the data to lie in a region of about $\pm 1\%$ about the smooth, slowly varying quasi-equilibrium rates. There are almost never large, sudden interest-rate fluctuations over small time scales. That is, the fluctuations in the data do seem essentially bounded. The essentially bounded property is consistent with and is implied by the large mean reversion we find.

We believe that the words "smooth", "slow", and "long" can in practice be defined with respect to time scales of, e.g., several months. Actually, we believe that an infinite number of time scales exist. We are merely dividing the spectrum up into two parts, a low frequency part corresponding to the slow variation of the quasi-equilibrium means and a high-frequency part corresponding to the small rapid fluctuations.

III. Other Statistical Tests and the SMRG Model

In this section, we briefly describe other statistical probes used to test the models and data. They are the same as in Ch. 48 and in Ref iii. These include mean shifts and fluctuations between rates of adjacent maturities along the yield curve, means, volatilities, and correlations. As mentioned in Ch. 48, a critical measure of the yield-curve shape is the amount of local inversions, or kinks. These occur when the yield of a longer maturity is lower than the yield of a shorter maturity but where the yield curve itself is not inverted. Large kinks are unphysical, and the need to avoid kinks constitutes a sharp test that is very difficult for models to pass.

Equations are given in Appendix A. Statistical quantities are defined for the data, as usual, over a given time window. For the simulator, they are calculated for each yield-curve path and then averaged over all paths.

Figs. 3A shows the time-averaged statistical properties of weekly Treasury yield-curve data (9/83 to 5/88), repeated from Ch. 48. These are the mean shifts $\mu_{shift}(T; \Delta T)$ between rates of adjacent maturities, standard deviations $\sigma_{shift}(T; \Delta T)$ of the mean shifts between rates of adjacent maturities, rate volatilities $\sigma_x(T)$, correlations $\rho(3M, T)$ of the 3-month rate with other rates, and correlations $\rho(10yr, T)$ of the 10-year rate with other rates.

Repeat: Problems with Kinks in Models without Strong Mean Reversion

The local inversions or kinks can be characterized qualitatively by the percentage of cases in which there was a kink and the average size of the kink. A kink between the 3-month and 1-year rates occurred in the data only 7% of the time.

Other maturities yield similar results. Except for the 30-year rate that was consistently below the 20-year rate, average local inversion sizes are all less than 7 bp/yr, and most are on the order of 1-2 bp/yr.

The multivariate lognormal model described in Ch. 48 produced anomalous and large kinks not seen in the data. The spread statistics were also not in agreement with data. In particular, the spread-shift volatilities $\sigma_{shift}(T;\Delta T)$ were too large. The aspect that is being measured is the same that leads the LN model to fail the CDA tests; the fluctuations are not tightly constrained.

The multivariate Gaussian model without mean reversion gives results similar to the multivariate lognormal model. Again, the fluctuations are not tightly constrained. Strong mean reversion appears to be critical.

Results for the SMRG Model Description of Yield Curve Data

The Strong Mean-Reverting Gaussian (SMRG) Model exhibits much better properties. One path, taken arbitrarily as #100 but typical, is the top yield curve in the unlabelled Fig. 1 along with the data. The sequence of random numbers generating this path is the same sequence that generated the lognormal path shown in a similar graph in Ch. 48. The statistics corresponding to the data are shown in Figs. 3B. The situation is vastly improved over the non mean-reverting models described in Ch. 48. The model kink sizes are now typically on the order of (and only slightly larger than) those of the data. The spread statistics, along with the other statistics, are in quite reasonable agreement with the data.

Figs. 4A, 4B show histograms of the fluctuations $\delta r_{fluctuations}(t, T = 10yr)$ of the 10-year rate generated by the SMRG model, the fluctuations of the 3-month rate $\delta r_{fluctuations}(t, T = 3M)$, and the fluctuations of the corresponding data. All fluctuations are measured away from the historical quasi-equilibrium yield-curve path $r_{\text{Quasi-Equil. Path}}^{\text{Historical}}(t,T)$. In contrast to the models without strong mean reversion, the agreement between the SMRG model and the data is good. This means that the fluctuations produced by the SMRG model are realistic.

Results Including a Fat Tail Gaussian Component

As mentioned before, evidence exists in the data for fat tails. Fat tails form an important, though temporally occasional, phenomenon. Fat tails are by definition not described by the SMRG model. Evidently, we need to add fat-tail effects into the description. In the language of the Macro-Micro Model, fat tails need to be added on as a correction to the Micro component.

To get an idea of how that might work phenomenologically, we investigated including a small Gaussian amplitude $d_t r_{\text{Fat Tails}}^{\text{Large Gaussian}}$ for rate changes, with a large width, representing around 10% of the fluctuations.

We first considered a modified SMRG model with an even larger mean reversion $\omega = 5/\text{week}$ representing 90% of the fluctuations. The distribution of the 10-year fluctuations of this modified SMRG model shown in Fig. 4C is now somewhat too narrow compared to that of the data.

Fig. 4D shows the combination of the fat-tail Gaussian added to the modified SMRG model. The agreement with the data is improved with this composite description.

How Many Free Parameters Are We Using to Describe the Yield Curve Fluctuations?

The description we use is actually quite parsimonious. *We emphasize that, aside from modeling the fat tails, there is really only one free parameter in the SMRG model, namely the value of the strong mean reversion.* All other parameters that characterize the rapid fluctuations are fixed by yield-curve data.

It is striking that the same value of the mean-reversion parameter (around 3/week) found in the cluster-decomposition analysis also serves, without further adjustment, to produce model statistical properties in agreement with data. If this parameter is reduced, both the cluster decomposition analysis and the statistical tests deteriorate. As the mean reversion becomes small, the problematic results reappear. Indeed, the mean reversion was determined by increasing it from zero and choosing that value such that the CDA tests began to succeed.

Maturity and Time Independence of the Mean-Reversion Parameter

It is striking that the mean-reversion parameter seems roughly independent of maturity. We have also done the analysis with different time windows with similar results, so the mean-reversion parameter does not seem to depend strongly on the time either.

Other Attempts to Fit the Data without Strong Mean Reversion

As mentioned in Ch. 48, we attempted models without the strong mean reversion but including other effects. This included memory effects to attempt to reduce kinks by recall of the initial smooth yield curve, smoothing recipes, etc.

In spite of a determined effort over a rather long period, no effort to replace strong mean reversion was successful in describing the statistical properties of the yield curve, passing the CDA tests, and avoiding kinks.

We therefore believe that strong mean reversion is essential.

IV. Principal Components (EOF) and the SMRG Model

As described in Ch. 48, a useful characterization of yield curve movements is given by the EOF or principal component analysis[x]. Principal component analysis decomposes yield curve movements into orthogonal components along eigenfunctions of the covariance matrix with associated eigenvalues.

Numerical Results

Figs. 5A shows the first three eigenfunctions of the yield curve movements in the data, labeled parallel shift, tilt, and flex. The first three eigenvalues for the data are in the approximate ratio 36/8/1.

The results of the EOF analysis for the multivariate SMRG model are given in Figs. 5B. Both the eigenfunctions and eigenvalues are in good agreement with the data.

V. Wrap-Up for this Chapter

We have continued and extended in this chapter our methodology for examining and characterizing yield-curve dynamics. The Cluster Decomposition Analysis (CDA) leads to sharp tests for models. Standard models fail these tests. Strong mean reversion about a slowly varying "quasi-equilibrium yield-curve" mean seems to be a preferred dynamical mechanism. We have considered a multivariate strongly mean-reverting Gaussian model. We believe that similar results would hold for other models as long as they include strong mean reversion (e.g. strong mean-reverting lognormal, strong mean-reverting square-root, etc.).

An additional, but small, component is connected to crashes or jumps like that of October 1987, producing fat tails on the interest-rate process.

Summary of the Physical Behavior of Interest Rates over Long Times

Recap: Our results imply that interest rates, as exhibited in historical yield-curve data, do not behave as if they correspond to a stochastic process that spreads out in time according to the statistics of zero (or small) mean reversion models in common use.

Instead, interest rates behave as if there were a mean quasi-equilibrium moving-average yield-curve path, which changes smoothly and slowly. The quasi-equilibrium yield-curve path defined by the data does move slowly with time and can lead to high (e.g. 20%) or low (e.g. 3%) values for rates, as well as inverted yield curves, all in a smooth fashion. We believe that the words "smooth" and "slow" mean with respect to time scales of, say several months. About this smooth quasi-equilibrium path, the actual rates fluctuate with small rapid movements.

Strong Mean-Reverting Multifactor YC Model 695

The concept of small fluctuations about a slowly varying or quasi-equilibrium state is one of the most pervasive and useful ideas in the physical sciences. We find it appealing that this idea also seems to be relevant to financial data.

Options Pricing Needs Generalization to the Macro-Micro Model

The multivariate Monte-Carlo simulator we have presented in this chapter, while agreeing with historical data, is not useful for pricing contingent claims. Option pricing requires future interest-rate paths to spread out in time. It is in fact possible to obtain a simulator that both agrees with historical yield-curve data and has future paths that "spread out".

The essential ingredient of this extended model, called the Macro-Micro Model, is to regard the quasi-equilibrium slowly varying yield curve of the data as one realization of a special kind of quasi-stochastic Macro variable effect[8]. We suggest an interpretation of these quasi-random variables as being due to long time-scale effects of macroeconomic forces (e.g. Fed. policy). One of these quasi-equilibrium paths is realized historically, $r_{\text{Quasi-Equil. Path}}^{\text{Historical}}(t,T)$. An interpretation of the rapid Micro SMRG fluctuations is regarded as being due to trading activity, reacting to individual market events, while following closely the overall macro economically produced slow trends.

The large mean reversion of the Micro SMRG model, coupled with the spreading out of future paths due to the Macro effects, can lead to a small total "effective" mean reversion. This is similar to mean reversion sometimes used in current one or two-factor models.

However, the big difference is retaining the agreement with yield-curve shape statistics not enjoyed by standard models.

The standard no-arbitrage requirements can be incorporated in the Macro-Micro model by finding an appropriate yield-curve path $r_{\text{No-Arbitrage}}^{\text{Term-Structure}}(t,T)$ about which fluctuations (both Macro and Micro) occur. This is explained in Ch. 51.

Appendix A: Definitions and Stochastic Equations

In this appendix, we repeat some definitions of quantities used in the text along with some details of the stochastic equations defining the various models. Further details are in Ref. iii. Define the T-year rate at time t for path α as $r_\alpha(t,T)$. Data will be denoted as the same symbol without the path label. For some function x of the rates and the time, the differential change $d_t x$ over infinitesimal time dt is taken as

---

[8] **Quasi?** That adjective is used to instill the idea that the Macro variables will not be ordinary Brownian variables that scale down to arbitrarily short times.

$$d_t x[r_\alpha(t,T),t] = \mu_x[r_\alpha(t,T),t]dt + \sigma_x[r_\alpha(t,T),t]dZ_\alpha(t,T) \tag{49.2}$$

The correlations between rates of different maturities give the correlation matrix. For the three cases of lognormal, Gaussian, and mean-reverting Gaussian (MRG) the function change $d_t x$ is given in terms of the rates by

$$d_t x(t,T)_{\text{Lognormal}} = d_t r(t,T)/r(t,T)$$
$$d_t x(t,T)_{\text{Gaussian}} = d_t r(t,T)$$
$$d_t x(t,T)_{\text{Mean-Rev. Gaussian}} = d_t r(t,T) + \omega \left[r(t,T) - r_{\text{Quasi-Equil. Path}}^{\text{Historical}}(t,T) \right] dt \tag{49.3}$$

Here $\omega(t,T)$ is the mean-reversion parameter. A-priori it can be a function of both time and maturity, although in practice as mentioned in the text, data are consistent with this quantity roughly independent of both time and maturity. For the MRG model the drift can be written

$$\mu_{MRG}[r_\alpha(t,T),t] = d_t r_{\text{Quasi-Equil. Path}}^{\text{Historical}}(t,T) + \omega(t,T) r_{\text{Quasi-Equil. Path}}^{\text{Historical}}(t,T) \tag{49.4}$$

Here $r_{\text{Quasi-Equil. Path}}^{\text{Historical}}(t,T)$ is the slowly varying quasi-equilibrium path defined by the data as described in the text, around which the yield-curve fluctuations occur. It is similar to $r_{classical}(t,T)$, the "classical path" described earlier in the book. The fluctuations are defined with respect to the quasi-equilibrium path:

$$\delta r_{fluctuations}(t,T) = r(t,T) - r_{\text{Quasi-Equil. Path}}^{\text{Historical}}(t,T) \tag{49.5}$$

The solution to the MRG stochastic equation, as can be checked by differentiating, is

$$r(t,T) = r_{\text{Quasi-Equil. Path}}^{\text{Historical}}(t,T)$$
$$+ \int_{t_0}^{t} \exp\left[-\int_{t'}^{t} \omega(t',T)dt'\right] \sigma_{MRG}(t',T) dZ_\alpha(t',T) dt' \tag{49.6}$$

The shifts between rates of adjacent maturities on the yield curve, as in Ch. 48, are denoted by

$$\Delta_T r_\alpha(t,T;\Delta T) \equiv r_\alpha(t,T+\Delta T) - r_\alpha(t,T) \tag{49.7}$$

Here, ΔT is the maturity increment between defined nearest-neighbor maturities on the yield curve. The spread-shift volatility $\sigma_{shift}(T;\Delta T)$ is averaged over time and paths.

As in Ch. 48, we define a yield-curve "kink" when a yield of higher maturity is lower than a yield of lower maturity. The mean shifts and volatilities for kinks are defined as before, with the additional filter that the yield spread is negative producing the kink in the first place.

Appendix B: The Cluster-Decomposition Analysis (CDA)

In this appendix, we describe the cluster-decomposition analysis CDA, as defined in theoretical particle physics (Ref. xi). This CDA has nothing to do with "cluster decomposition" as sometimes used in statistics.

Consider Γ time series $x_\gamma(t)$, $[\gamma = 1,2,3,...,\Gamma]$, where some or all series may be the same. Call the Γth - order correlation (actually covariance) function of these series $M_\Gamma^{(123...\Gamma)}\left[\{\tau_\gamma\},\{x_\gamma\}\right]$, or $M_\Gamma^{(123...\Gamma)}$ for short. Define $\{\tau_\gamma\}$ time lags. Then $M_\Gamma^{(123...\Gamma)}$ is the point-by-point sum of products of lagged variables[9]:

$$M_\Gamma^{(123...\Gamma)} \equiv \frac{1}{N}\sum_{n=1}^{N} x_1(\tau_1 + n\,\Delta t)\ldots x_\Gamma(\tau_\Gamma + n\,\Delta t) \qquad (49.8)$$

The mean of a time series x_γ with lag τ_γ is $M_1^{(\gamma)} = M_1\left[\tau_\gamma;x_\gamma\right]$. The second-order and third-order connected correlation (actually covariance) functions called $C_2^{(12)} \equiv C_2\left[\tau_1,\tau_2;x_1,x_2\right]$ and $C_3^{(123)} \equiv C_3\left[\tau_1,\tau_2,\tau_3;x_1,x_2,x_3\right]$ are standard looking expressions:

$$C_2^{(12)} \equiv M_2^{(12)} - M_1^{(1)}M_1^{(2)} \qquad (49.9)$$

$$C_3^{(123)} \equiv M_3^{(123)} - \sum_{perm\,(abc)} C_2^{(ab)}M_1^{(c)} - \prod_{\gamma=1}^{3} M_1^{(\gamma)} \qquad (49.10)$$

The sum in Eqn. (49.10) runs over permutations $(abc) = (123), (231), (312)$. It should be emphasized again that these are dynamic Green functions, not simple static statistical measures.

---

[9] **Notation:** The time interval Δt is arbitrary. We assume N>>1, so 1/(N-1) ≈ 1/N etc.

Using Time-Differenced Series in the CDA

The same definitions given above can be made with each $x(t)$ replaced by $d_t x(t) = x(t+dt) - x(t)$. The difference is one of application. The use of $d_t x$ instead of $x(t)$ is that with $d_t x$ correlations, time expectations of products of Wiener measures are dealt with. With $x(t)$ correlations, time integrals of these expectations are encountered. The $d_t x$ correlation provides a "strong" pointwise test in the time lags, whereas the x-correlations would only provide a "weak" integrated test.

We can also use $(d_t x - \mu_x)$ time series, or $(d_t x - \mu_x)/\sigma_x$ time series. The latter measures the number of standard deviations.

Checking Model Hypotheses with the CDA

The main point is that generally for Gaussian processes, and also for lognormal processes with distinct indices, the third-order connected correlation function is identically zero, i.e.

$$C_3^{(123)} \equiv C_3\left[\tau_1, \tau_2, \tau_3; x_1, x_2, x_3\right] = 0 \tag{49.11}$$

The third order function $M_3^{(123)}$ is not equal to zero for Gaussian processes if the means are nonzero, and not equal to zero for lognormal processes, viz

$$M_3^{(123)} \neq 0 \tag{49.12}$$

Therefore, an acid test for the validity of models is to check the relationship

$$\frac{C_3^{(123)}}{M_3^{(123)}} \stackrel{?}{=} 0 \tag{49.13}$$

Formal Proof of $C_3 = 0$ for Gaussian Processes

The proof of this statement[10] as well as the general case for higher-order correlation functions follows from generating functional formalism. In outline, the argument goes as follows. Define the generating functional Z of "currents"

---

[10] **Proof with Time Differences:** The proof with time-differenced variables goes through in the same way. Only the form of the matrix A changes.

$J_\gamma(t_n)$ for a general Gaussian probability density function (exponentiated quadratic function of the x's) by

$$Z(\{J\}) = \prod_{\gamma,n} \left\{ \int_{-\infty}^{+\infty} dx_\gamma(t_n) \right\} \exp\left\{ -X \cdot A \cdot X + J \cdot X \right\} \quad (49.14)$$

Here, A is the matrix defining the quadratic form,

$$X \cdot A \cdot X = \sum_{\gamma,\gamma';n,n'} x_\gamma(t_n) A_{\gamma\gamma'}(t_n, t_{n'}) x_{\gamma'}(t_{n'}) \quad (49.15)$$

Also,

$$J \cdot X = \sum_{\gamma;n} J_\gamma(t_n) x_\gamma(t_n) \quad (49.16)$$

Averages are then defined by

$$\langle \Omega[x_a(t_a), \ldots, x_\zeta(t_\zeta)] \rangle =$$
$$\frac{1}{Z(\{0\})} \prod_{\gamma,n} \left\{ \int_{-\infty}^{+\infty} dx_\gamma(t_n) \right\} \Omega[x_a(t_a), \ldots, x_\zeta(t_\zeta)] \exp\left\{ -X \cdot A \cdot X \right\} \quad (49.17)$$

Notice that Z can be evaluated by Gaussian integration as

$$Z(\{J\}) = Z(\{0\}) \exp\left\{ \tfrac{1}{4} J \cdot A^{-1} \cdot J \right\} \quad (49.18)$$

Here,

$$J \cdot A^{-1} \cdot J = \sum_{\gamma,\gamma';n,n'} J_\gamma(t_n) [A^{-1}]_{\gamma\gamma'}(t_n, t_{n'}) J_{\gamma'}(t_{n'}) \quad (49.19)$$

Also, notice that a factor $x_\gamma(t_n)$ can be included in an average by differentiation of Z with respect to the associated current $J_\gamma(t_n)$, and then setting all currents to zero, by definition. Thus averages of products of x's can be obtained by differentiating the evaluated form of Z by each of the associated currents to each x in the product. This serves to evaluate the correlation functions defined in the text.

From the above remarks along with the fact that the integral of an odd function over symmetric limits is zero, the reader may convince himself after some algebra that the connected three-point function is identically zero for a Gaussian probability density function. The cluster decomposition formulae for higher-order correlation functions follow from similar manipulations.

More About the Cluster Decomposition Analysis

A convenient "bubble" diagram notation[xi] for the third order cluster decomposition equation is shown below[11]:

3rd Order Cluster Decomposition Diagram

To read this, imagine that lines to the left represent "particles" coming "in" and lines to the right represent "particles" going "out". The symbol C in a bubble means that those particles attached to that bubble interact, and M means that particles may or may not interact. One line through an M bubble means that the particle goes in and out untouched. The terms on the right hand side of the equation represent all possible ways that the particles can interact or not interact. The equation can also be written with C on the left hand side, as in Eqn. (49.10).

For Gaussian noise, we have both $M_3^{(123)} = 0$ and $C_3^{(123)} = 0$ (as the number of data points $N \to \infty$). For Gaussian noise plus a deterministic drift, $M_3^{(123)}$ is no longer zero, but $C_3^{(123)} = 0$ still holds.

For a lognormal model, an exponentiated form is needed for the rate difference $d_t r$. We need the replacement

---

[11] **"Bubble Notation" for Green function equations:** This notation was invented by the physicist David Olive. Similar results hold for any arbitrary number of lines in or out. It is a great mnemonic to remember the formulae, which otherwise have to be derived using the functional techniques, and it always works.

$$\sigma N(0,1) \to \exp[\sigma N(0,1)] \qquad (49.20)$$

We wind up with exponentiated Gaussian noise, for which $M_3^{(123)} \neq 0$.

Specifically, for lognormal noise with $L_a = \exp[\sigma_a N_a(0,1)]$ for index a (and similarly for indices b, c), the Green functions through third order are[12]:

$$M_1^{(a)} = \langle \exp[\sigma_a N_a(0,1)] \rangle = \exp[\sigma_a^2/2]$$
$$M_2^{(ab)} = \langle \exp[\sigma_a N_a(0,1) + \sigma_b N_b(0,1)] \rangle = \exp[(\sigma_a^2 + \sigma_b^2 + 2\delta_{ab}\sigma_a\sigma_b)/2]$$
$$C_2^{(ab)} = \exp[(\sigma_a^2 + \sigma_b^2)/2][\exp(\delta_{ab}\sigma_a\sigma_b) - 1]$$
$$(49.21)$$

$$M_3^{(abc)} = \langle \exp[\sigma_a N_a(0,1) + \sigma_b N_b(0,1) + \sigma_c N_c(0,1)] \rangle$$
$$= \exp[(\sigma_a^2 + \sigma_b^2 + \sigma_c^2 + 2\delta_{ab}\sigma_a\sigma_b + 2\delta_{ac}\sigma_a\sigma_c + 2\delta_{bc}\sigma_b\sigma_c)/2]$$
$$C_3^{(abc)} = e^{(\sigma_a^2 + \sigma_b^2 + \sigma_c^2)/2} \left\{ e^{\delta_{ab}\sigma_a\sigma_b} + 2\,perm - \left[e^{\delta_{ab}\sigma_a\sigma_b} + 2\,perm\right] + 2 \right\}$$
$$(49.22)$$

Note that if a, b, c are all distinct then $C_3^{(abc)} = 0$ since the curly bracket is $\{1 - 3 + 2\} = 0$. If $a = b$ and $a \neq c$, we get $C_3^{(abc)} = 0$ for the lognormal case.

Figures: Strong Mean-Reverting Multifactor Yield-Curve Model

Fig. [1]. 3-D plot of weekly yield-curve data over the period 1983-1988 as well as a path (of all maturities) from the Monte-Carlo simulation for the strong mean-reverting Gaussian (SMRG) model, starting from the same initial yield curve. The two plots are intentionally unlabeled. Can you tell which is which?[13]

Fig. [2]. Results from the cluster decomposition analysis. Figs. 2A show the correlation functions using a set of exponentiated Gaussian random numbers. Note that $C_3/M_3 \approx 0$. Figs. 2B show the same analysis using data along with

---

[12] **Notation:** $\delta_{ab} = 1$ if a = b and 0 otherwise.

[13] **Answer:** The top figure contains the model and the bottom figure contains the data.

the hypothesis that the fluctuations in the data result from a zero-mean reversion lognormal process. If this lognormal hypothesis were correct, C_3/M_3 should be small. However, C_3/M_3 is *not* small, showing that the lognormal assumption is incorrect. Figs. 2C show the same analysis using data along with the hypothesis that the fluctuations in the data result from the SMRG model. Note that C_3/M_3 is small, so the SMRG model passes this test.

Fig. [3]. Statistical measures of the yield-curve time series. Figs. 3A show data for the shifts between neighboring maturities. These are the shift means $\mu_{shift}(T;\Delta T)$, standard deviations $\sigma_{shift}(T;\Delta T)$, volatilities $\sigma_x(T)$, correlations between the 3 month rate and other rates $\rho(3M,T)$, and correlations between the 10 year rate and other rates $\rho(10yr,T)$. The maturity index runs from 1 to 11, representing the 3-month, 6 month, and the 1, 2, 3, 4, 5, 7, 10, 20, and 30-year rates, respectively. Figs. 3B show the same graphs for the time and path-averaged SMRG simulation. The agreement of the SMRG model and the data is quite reasonable.

Fig. [4]. Figs. 4 show histograms of fluctuations of rates about the historical "quasi-equilibrium yield-curve path", the slowly varying moving-trend yield curve discussed in the text. Figs. 4A and 4B show the fluctuations of the 10 year and 3 month rates for the data, compared to the SMRG model. Fig. 4C shows the fluctuations of the 10-year rate for the data compared to a modified SMRG model with larger mean reversion. Fig. 4D shows the fluctuations of the 10 year rate for the data compared to a composite model consisting of a 90% component of the modified MRG model in Fig. 4C along with a 10% component of a wide Gaussian distribution, phenomenologically accounting for the fat tails.

Fig. [5]. Principal Component EOF analysis. Figs. 5A show data for the first three eigenfunctions plotted against maturity, labeled parallel, tilt, and flex. The parallel shift is not exactly parallel. The eigenvalues are also shown. Figs. 5B show the same graphs for the SMRG model.

Data and Strong Mean-Reverting Gaussian Model
Figures intentionally not labelled.

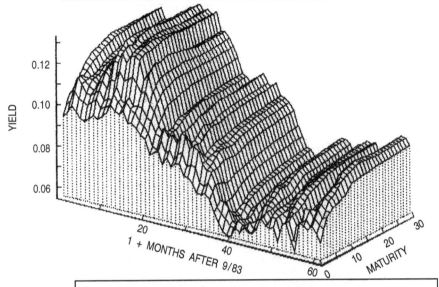

Which graph has the data?
Which graph has the path from the SMRG model?

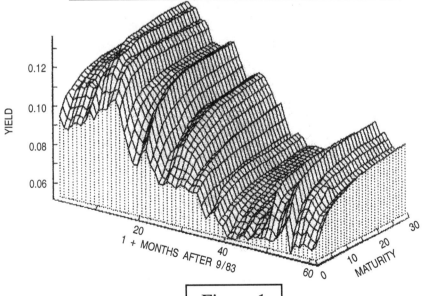

Figure 1

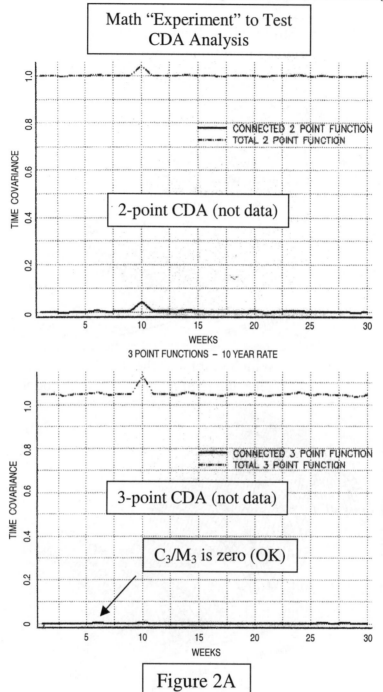

Figure 2A

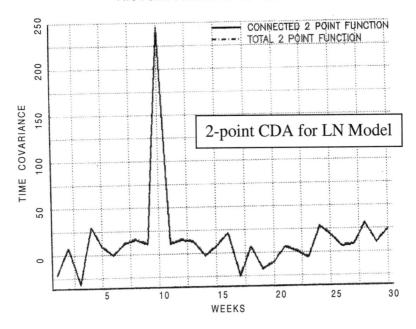

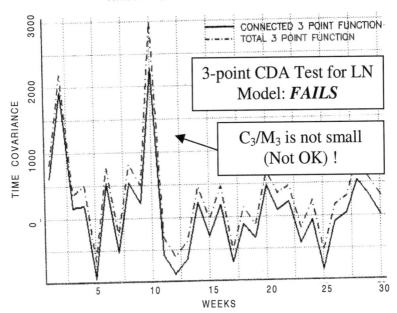

Figure 2B

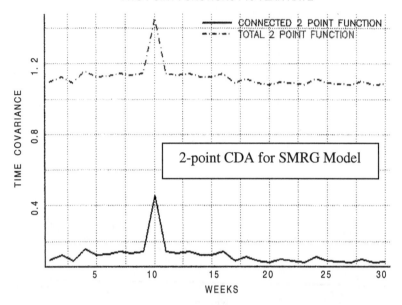
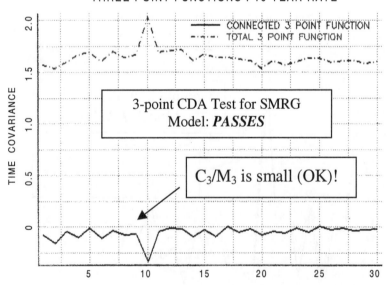

Figure 2C

Strong Mean-Reverting Multifactor YC Model

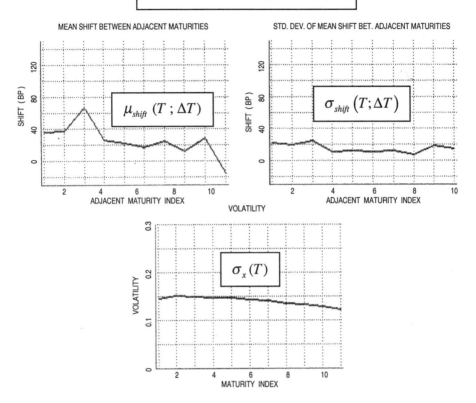

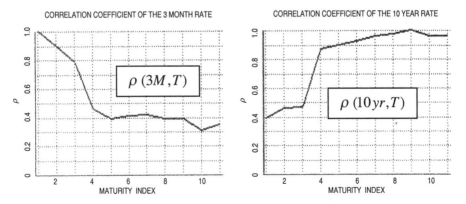

Figure 3A

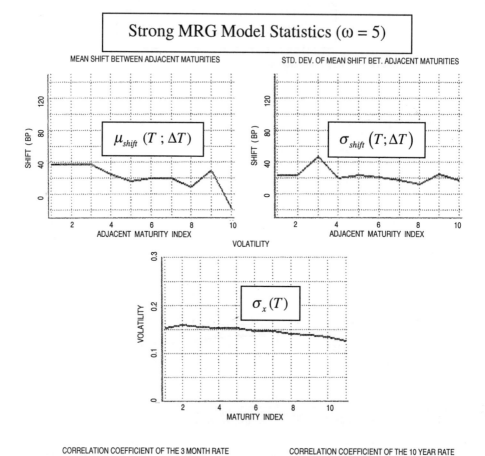

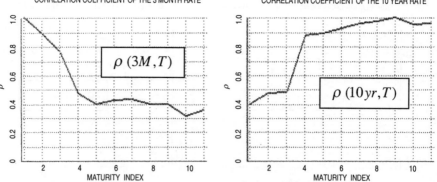

Figure 3B

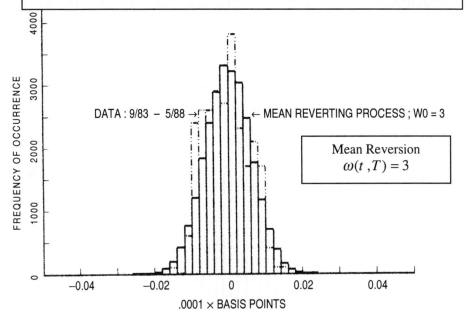

Figure 4A

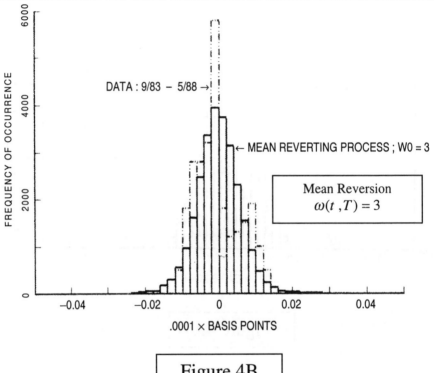

Figure 4B

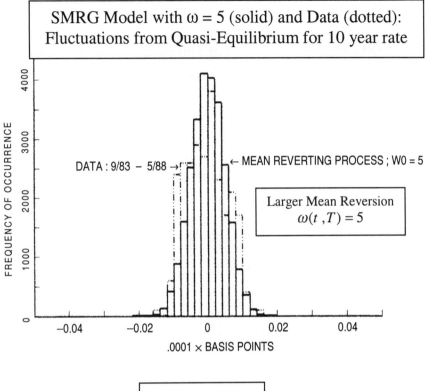

Figure 4C

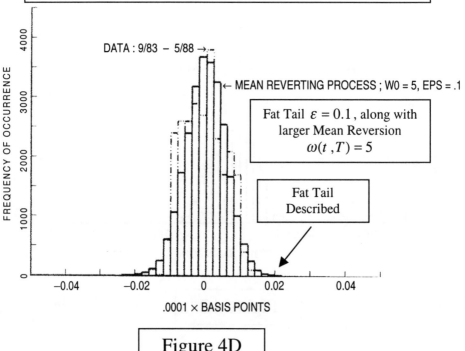

Figure 4D

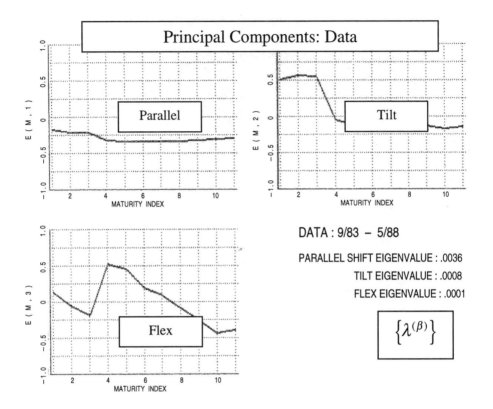

Figure 5A

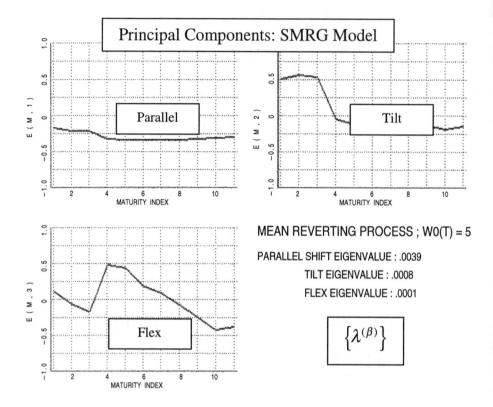

Figure 5B

References

[i] **Strong Mean Reversion Multifactor Yield-Curve Model**
Dash, J. W. and Beilis, A., *A Strong Mean-Reverting Yield-Curve Model*. Centre de Physique Théorique CPT-89/PE.2337 preprint, CNRS Marseille, France, 1989.

[ii] **Two-Factor Yield-Curve Modeling for Mortgage Backed Securities**
D. P. Jacob and A. L. Toevs, *An Analysis of the New Valuation, Duration and Convexity Models for Mortgage-Backed Securities*, in Fabozzi, F. J., *The Handbook of Mortgage-Backed Securities*, Revised Edition, 1988, p. 687.
Hayre, L. (Ed), *Guide to mortgage-backed and asset-backed securities*. Salomon Smith Barney, John Wiley & Sons, 2001.

[iii] **Lognormal Multifactor Yield-Curve Model**
Beilis, A. and Dash, J. W., *A Multivariate Lognormal Model*, Centre de Physique Théorique CPT-89/PE.2334 preprint, CNRS Marseille, France, 1989.

[iv] **Macro-Micro Yield-Curve Model**
Dash, J. W., and Beilis, A. 1989, *A Macro-Micro Yield-Curve Simulator*, Centre de Physique Théorique CPT-89/PE.2336 preprint, CNRS Marseille, France, 1989.

[v] **Fifty-Fifty Votes: Yield-Curve Data vs. Macro-Micro Model**
Dash, J. W., *Evaluating Correlations and Models*. Talk, RISK Conference, 11/94

[vi] **WKB Method**
Morse, P. M. and Feshbach, H., *Methods of Theoretical Physics*. McGraw-Hill Book Company, 1953. See Part II, pp. 1092-1106.

[vii] **CIR Model**
Cox, J. C., Ingersoll, J. E. Jr. and Ross, S. A., *A Theory of the Term Structure of Interest Rates*, Econometrica. Vol. 53, 1985, Pp. 385-407.

[viii] **Vasicek Model**
Vasicek, O. A, *An Equilibrium Characterization of the Term Structure*, Journal of Financial Economics 5, 1977. Pp. 177-188.

[ix] **Fat Tails**
Fama, E. F., *The Behavior of Stock Market Prices*, Journal of Business, Vol. 38, p.34-105, 1965, and references therein.

[x] **Principal Components (EOF) Applied to Yield Curve Movements**
Garbade, K. D., *Modes of Fluctuation in Bond Yields—An Analysis of Principal Components*, Topics in Money and Securities Markets, Banker's Trust, 1986.
Herman, G. F., *Modeling Yield Curve Variations*, Merrill Lynch Capital Markets Working Paper (unpublished), 1987

[xi] **Green Functions and High-Order Correlation Functions; Bubble Notation**
Eden, R. J., Landshoff, P.V., Olive, D.I., Polkinghorne, J.C., *The Analytic S-Matrix*. Cambridge University Press, 1966. *P.190.*

50. The Macro-Micro Yield-Curve Model (Tech. Index 5/10)

Summary of this Chapter

This chapter contains a description of the Macro-Micro multi-time scale, multifactor yield-curve model [1,2,i]. The reader can understand this chapter without reading previous chapters; necessary material will be summarized. In Ch. 49, we showed that yield-curve statistical data are consistent with small rapid fluctuations implying strong mean reversion, with fluctuations about an historical "quasi-equilibrium" yield curve $r^{Historical}_{Quasi-Equil.\ Path}(t,T)$, parameterized from the data[ii]. Two time scale regimes are envisioned: short (the Micro component) and long (the Macro component). For short Micro times, the strong mean reversion model is used, because this agrees with the data fluctuations. A fat-tail component is also present[3].

For the long Macro times, we need an additional ingredient. The basic concept is the quasi-equilibrium yield curve. In this chapter, the historical quasi-equilibrium yield curve used in Ch. 48, 49 is generalized. For the long Macro times, quasi-stochastic variables produce quasi-random quasi-equilibrium yield curves. This allows future interest-rate paths to spread out or fan out to achieve high or low values. The Macro-Micro model not only is in accord with the historical yield-curve dynamics, but it satisfies no-arbitrage properties that we investigate in Ch. 51.

---

[1] **History:** This chapter is based on work done with Alan Beilis in Ref. i. Recent developments in the Macro-Micro model are covered in Ch. 51. The work was described in various talks, and in my CIFEr tutorial.

[2] **Acknowledgements:** I thank Eric Slighton and Alain Nairay for helpful conversations.

[3] **Fat Tails Are an Extra Micro Component:** Besides the Micro strong mean reverting effects, there are occasional large, quick jumps producing fat tails. In Ch. 49, we followed Ref ii, employing a phenomenological form for fat tails using a large-width Gaussian. See Ch. 21 that continues this idea. See also Ch. 46, which describes the nonlinear diffusion Reggeon Field Theory as a possible dynamical mechanism for fat tails.

In a sense, the statistical properties of the yield curve form physical constraints that are just as real as the universally accepted idea that zero-coupon bond prices should be reproduced by models.

We propose an interpretation of the Macro slowly varying quasi-equilibrium paths. The interpretation is that the slow variation of the Macro paths is due to macroeconomic effects (Fed. actions, etc.). The small, rapid Micro fluctuations are proposed to be due to trading activities. These small Micro fluctuations closely follow the smooth Macro trends, reacting to market events.

I. Introduction to this Chapter

In the two previous chapters, empirical studies of yield curve movements using a number of statistical techniques were described. Several multivariate yield-curve models were examined, and two crucial points were made regarding models and data. First, the importance was emphasized of a slowly varying quasi-equilibrium yield-curve path in time defining the mean, (i.e. drift, trend, moving average) of the data over time scales of months to a few years. Second, yield-curve data were observed to behave as rapid movements tightly constrained by strong mean reversion about this slowly moving quasi-equilibrium yield curve.

Specifically, a multivariate strongly mean-reverting Gaussian (SMRG) model was shown to provide good agreement with the statistical properties of the fluctuations of the data about a reasonably defined slowly-varying quasi-equilibrium mean yield-curve path. Phenomenologically, the quasi-equilibrium historical yield curve path was taken to be composed of simple line segments, defined by the data trends. The statistical probes included the usual correlations and volatilities, the absence of anomalous yield-curve shapes (kinks), spread volatilities, a sharp probe involving third-order correlation functions, and principal component (EOF) decompositions of yield-curve movements.

Standard models without strong mean reversion, which produce interest rates that "fan out", surprisingly do not appear to be in accord with important features of data. The most telling disagreement is the presence of large kinks in yield curves produced by multifactor models without strong mean reversion. Other related problems with statistical measures are present with standard models.

The success of this description of data using a multivariate strong mean-reverting model was, however, tempered with the realization that the use of such a model to price contingent claims was problematic. Contingent claims are currently priced mostly using one or two-factor interest rate diffusion models that do not have strong mean reversion (or any mean reversion), and therefore do indeed have interest rate paths that fan out. Hence, the strong mean-reversion model needed to fit past yield-curve data needs modification in order to describe future interest rate movements that can reasonably price options.

We therefore want to modify the strong MRG model with two goals in mind:

1. Maintain the agreement with yield-curve dynamical data
2. Price options with parameters that can be roughly reinterpreted as parameters of standard models.

This chapter constructs a model that achieves these goals; we call it the Macro-Micro yield-curve model, realized as a Monte-Carlo simulator. The interpretation of the Macro-Micro model involves both macroeconomic effects and trading activities.

Spectral Decomposition and Time Scales

Essentially, we are breaking the spectrum of fluctuations up into low and high frequency regimes. The low frequencies are assumed due to macroeconomic effects; we call these "macro fluctuations". The high frequencies are assumed due to trading; we call these "micro fluctuations". Actually, there are probably an infinite number of time scales. The micro component is hardwired and determined to agree with yield-curve historical data. The micro parameters are therefore fixed.

The idea of fundamental time scales is a new ingredient not present in the standard models used in pricing derivatives. We repeat that we are forced into this description by the yield-curve data, especially the dynamical shape statistics with the absence of kinks, as described in the last chapter. In our view, the usual models without explicit time scales that price options only provide approximate parameterizations of market data. The Macro-Micro model can reduce approximately to simpler models.

The Macro Component and Macroeconomics

The basic idea is that the slowly varying quasi-equilibrium means, around which fluctuations occur, are themselves stochastic, *but only on long time scales*. This stochastic process does not scale down to short time scales, and therefore is *not* Brownian. We call this behavior "quasi stochastic". In this model, a stochastic property with average time scales on the order of months to years is attributed to the smooth long time-scale effects on interest rates of these macroeconomic variables.

We do not attempt to model these macroeconomic effects directly, as this is far beyond our abilities. Rather, we describe these effects in a phenomenological way, containing parameters. One of the parameters is a "macro-volatility", providing the quasi randomness, and accounting for the uncertainties in the long-term trends due to macroeconomic effects. This macro-volatility can be determined by (1) taking a view on the overall volatility of interest rates or (2) using implied techniques, i.e. demanding that the model prices and market prices agree for options.

In this model, slowly varying long-time-scale effects are due to inflation, Fed policy, third-world debt, etc., which produce trend-related interest-rate effects and are responsible for the smoothly varying quasi-equilibrium yield-curve path actually realized historically in the real world. Such effects would exist even in the absence of trading activities[4]. These trends can lead very high (20%) and low (1%) rates, as well as inverted yield curves, etc.. In our model, these characteristics are attained in a smooth fashion acting on time scales of months or greater.

A given realization of these macroeconomic events leads to a given macro-path. Thus, in historical data, one realization existed to form the historical mean quasi-equilibrium yield curve path of the mean trends of rates.

Figure 1 (at the end of the chapter) shows a collection of these macro-paths for one yield produced by the macro simulator. The macro simulator is described in Section II.

The Micro Component and Short-Term Trading Activity

Short time-scale fluctuations on the order of minutes to weeks are proposed as due to trading activities that react to individual market events while attempting to follow overall trends. These fluctuations mostly are rapid and small; in fact, traders do an excellent job of following the trends. As they say, "the trend is your friend". Data show that fluctuations are less than 100 bp on either side of a given trend, and this fact is roughly independent of time. That is, traders act the same now as they acted five years ago; the fluctuations are relatively stationary. The fluctuations about a given trend, or macro-path, constitute what we call "micro-paths".

Figure 2 (at the end of the chapter) illustrates the micro-paths about one macro-path for one of the eleven yields produced by the micro multifactor simulator. The micro part of the Macro-Micro model was described in Ch. 49 [ii]. There, the fluctuations were described by a strong mean reversion about the quasi-equilibrium trends of the data [i.e. the macro path as realized historically], which did describe the yield-curve data.

Prototype: Prime (Macro) and Libor (Macro + Micro)

Consider Fig. 3 showing Prime rate and 6M Libor rate data[5]. The Prime rate, which is not traded, is a Macro variable. The Libor rates can be thought of as being composed of small rapid fluctuations (the Micro component) around a

---

[4] **Traders in the Closet?** A picturesque way to describe the macro paths is to imagine an alien world that eliminates trading activities by putting all the traders in a closet. The remaining rate changes would only be due to macro effects. A proxy for this perhaps rather interesting but impractical action is to look at a rate that is not traded, such as the prime cash rate. See Fig. 3 at the end of the chapter.

[5] **Acknowledgements:** I thank Citigroup for the use of these data, which are the same as those shown in my CIFEr tutorial.

Macro path (the Prime rate minus a rather stable spread). The Prime rate is not unique; we could have exhibited the Fed Funds rate for example.

Remainder of this Chapter

The rest of this chapter is organized as follows. In Section II, we present details of the Macro-Micro yield-curve model. Section III contains a wrap-up. Appendix A contains some remarks on no-arbitrage conditions, scenario analysis, and yield-curve dynamics. The no-arbitrage discussion is continued and amplified in the next chapter, Ch. 51.

II. Details of the Macro-Micro Yield-Curve Model

In this section, we give details of the Macro-Micro model simulator. As described in the preceding section, this simulator pictorially consists of random tubes of yields. The construction of a given tube surrounding a given realized mean yield-curve path was shown in Ch. 49 using a strong mean-reverting Gaussian model, producing agreement with the statistics of yield-curve data. We retain this construction here. That is, the micro component is fixed. The part of the simulator left is the quasi-random nature of the means, the macro-path simulator.

Our aim is pragmatic; we wish to construct a "reasonable" macro-path simulator without getting bogged down in an extremely difficult attempt to "derive" its properties from macroeconomic data. This new Macro component of the Macro-Micro model will therefore be phenomenological and contain parameters. These parameters can be constrained both by historical trend data of interest rates and/or by obtaining "implied" parameters obtained by equating model prices of contingent claims to market prices.

The Importance of the Time Step in the Macro Simulator

The basic feature of the macro-simulator is that the time-step, usually an irrelevant variable that in practice is made as small as possible, must instead play a central role. This is because we do not want to allow the macro-fluctuations to scale down to the same time scales as the micro-fluctuations. If that were the case, nothing new would have been accomplished. The macro-fluctuations are supposed, after all, to describe the smooth trends or quasi-equilibrium yield-curve paths around which the micro-fluctuations occur.

The macro simulator is based on the observations that a reasonable parameterization of trends consists of simple straight-line fits over different time periods that exhibit upward or downward movements of interest rates. These time steps have different lengths, and the slopes of the trend lines are different from one period to the next.

While in principle this description could be maturity-dependent, we choose to take it independent of maturity for simplicity.

Model for the Random Macro Time Step Dynamics

We model the (positive) macro time step Δt_{macro}, over which a given straight-line trend occurs, as a random variable. The probability density function $\wp\left[\,\Delta t_{macro}\,\right]$ for macro time steps is taken as a cutoff, displaced lognormal distribution. The distribution has a cutoff τ_{cutoff} below which time-step values are not permitted, on the order of one month. This marks the transition into the Micro time regime. There is a central value τ_{macro}. The macro time volatility $\sigma_{macro}^{(\tau)}$ describes the uncertainty in the macro time intervals.

Explicitly, for $\Delta t_{macro} > \tau_{cutoff} > 0$, we write

$$\wp\left[\,\Delta t_{macro}\,\right] = \frac{1}{\sigma_{macro}^{(\tau)}\sqrt{2\pi}}\exp\left[-\frac{\ln^2\left(\dfrac{\Delta t_{macro}-\tau_{cutoff}}{\tau_{macro}-\tau_{cutoff}}\right)}{2\,[\,\sigma_{macro}^{(\tau)}\,]^2}\right] \quad (50.1)$$

Dynamical Model for the Random Macro Slope or Drift

Over a given time interval with length Δt_{macro} expressed in months, we choose a slope for the interest rate trends, which we call λ. This parameter has units of bp/yr per month, so $\lambda \Delta t_{macro}$ measures a macro change in rates over time Δt_{macro}.

A simple Gaussian probability density function $\wp\left[\,\lambda\,\right]$ is both rational[6] and useful. We write

$$\wp\left[\,\lambda\,\right] = \frac{1}{\sigma_\lambda\sqrt{2\pi}}\exp\left[-(\lambda-\lambda_{av})^2\big/(2\sigma_\lambda^2)\right] \quad (50.2)$$

---

[6] **Brownian Limit and the Gaussian Slope Distribution:** In the limit as the cutoff time goes to zero $\tau_{cutoff}\to 0$, the Gaussian assumption for the slope distribution reproduces Brownian motion, as explained in previous chapters on path integrals. Therefore, the Gaussian slope distribution is a natural assumption with finite time scales.

Here, λ_{av} is an average slope parameter and σ_λ the slope volatility. λ_{av} will play a significant role in no-arbitrage considerations, as we discuss in Ch. 51. Note trend slopes can be either positive or negative[7].

The Macro Volatilities for Times and Slopes

The two overall macro-volatilities used in the model are $\sigma_{macro}^{(\tau)}$ for macro time intervals, and σ_λ for interest-rate slopes. These parameters can be derived by considering the shape of the envelope of macro-paths for a given rate, and approximately fitting the width of the envelope to that given by a standard model.

For example, in Fig. 1 the effective Gaussian macro-volatility corresponding to the macro-paths generated by the simulator at ten years is approximately $\sigma_{eff.Gaussian} \approx 100\, bp\, /\, (yr)^{3/2}$. This is on the order of implied-volatility values needed to correctly price options using a simple Gaussian rate model with no mean reversion.

Weak Mean Reversion and the Connection to Standard Models

The macro envelope for a given rate can also be fit with a macro weakly mean-reverting one-factor model.

The reader should carefully note that the macro weak mean reversion we are now talking about is completely different from and independent of the strong mean reversion for the micro component. The effective mean reversion for the macro envelope will be much smaller than the very strong mean reversion needed to constrain the tightly fluctuating micro-paths about each macro-path.

In this sense, the strong mean reversion needed to describe yield curve data coupled with the macro quasi-random component can be made consistent with the weak (or zero) mean reversion used in standard models.

The Overall Average Macro Drift and No Arbitrage

The set of paths can include an overall average yield-curve path $R_{Avg}(t,T)$ about which all macro and micro fluctuations occur. If this is done, the average slope parameter λ_{av} is set to zero.

For no arbitrage, $R_{Avg}(t,T) = R_{NoArb}(t,T)$ is composed of the forward yields $f(t,T;t_0)$ determined by today's t_0 yield curve, up to convexity

---

[7] **Avoiding Negative Rates:** In practice, we first pick the macro time step and then pick the macro slope. If the macro slope is negative enough so that the macro rate happens to approach zero, we replace the macro time step by a smaller number to avoid negative rates. This constraint makes the macro time-step distribution more complicated than the formula in the text.

corrections[8]. These convexity corrections were first determined for specific model assumptions for multifactor models by Heath, Jarrow and Morton[iii]. Further discussion is in Appendix A and in the next chapter.

Option Pricing Using the Macro-Micro Model

European options, for example depending on the short rate, can be priced. The discounting can be done either using the mean model spot-rate curve at time t or else discounting back along the short-rate paths one at a time; the latter method is the conventional approach.

We have checked that European option prices using the Macro-Micro simulator, with appropriate parameters, are in reasonable accord with option prices produced by conventional one-factor models.

Developments in the Macro-Micro Model

Various developments in the Macro-Micro Model are described in the next chapter.

III. Wrap-Up of this Chapter

We have presented a new multifactor yield-curve "Macro-Micro" model. This model can be used for pricing and risk-management purposes, where the need for consistency with the data for the statistics of the shape of the yield curve can make a significant difference.

Two very different sorts of dynamical variables are included. Yield-curve Macro-paths are generated by a quasi-stochastic process that does *not* scale down to small time scales, but rather generates smooth trends in interest rates over months to years, the quasi-equilibrium macro-path yield curves. About these smooth yield-curve macro paths, the yield-curve micro paths exhibit small rapid fluctuations. This description is consistent with historical yield-curve statistics, as shown in the last chapter.

We proposed a physical interpretation of the origins of the macro and micro variables. Macroeconomic effects (themselves uncertain) produce the smooth long-term trends. Trading activities, responding to individual market events and following a given realized macro trend, generate the rapid fluctuations. The macro component contains several parameters that can be determined by historical data, or as implied parameters by pricing options.

---

[8] **Coupons:** Actually, the phenomenology described in this part of the book was done directly with coupon bonds, for which we had the data . The more usual description is in terms of zero-coupon bonds.

All fluctuations (slow macro and rapid micro) occur around $R_{Avg}(t,T)$ that can be specified by no-arbitrage arguments for pricing options. The trading activities of the Micro component can be described as attempting to profit by arbitrage while moving the market toward a local state of no-arbitrage. On the other hand, the Macro component due to macroeconomic effects on interest rates would seem to have nothing to do with trading arbitrage or the lack of it. Still, we are able to incorporate no-arbitrage constraints, as discussed in Ch. 51.

Because standard models serve well in market situations to price options, we do not suggest that our complicated simulator be used for this purpose. On the other hand, the Macro-Micro can be re-expressed approximately in terms of standard models for pricing options. For example, the macro-micro multivariate yield-curve simulator, or a projected version of it using only a short and long rate, can be used to price mortgage-backed securities products. The micro-fluctuations are guaranteed to produce accurate short-rate/long-rate spread statistics. Thus, the long-rate input into a prepayment model will have a historically-consistent relation with the short discounting rate.

Still, contingent claims are just weak probes of the real underlying interest-rate process. A model agreeing with the dynamics of yield curves provides a better tool for interest-rate-risk analysis.

Appendix A. No Arbitrage and Yield-Curve Dynamics

In this appendix, we make some general remarks. Formal properties of no-arbitrage, yield-curve dynamical properties, and the Macro-Micro Model are derived in Ch. 51.

- No-arbitrage constraints are universally used in pricing contingent claims. Still no guarantee is given that the historical *statistical* properties of yields will be reproduced, even with historical volatility and correlation input.
- The statistical properties of the yield curve form physical constraints that are just as real as the universally accepted idea that zero-coupon bond prices should be reproduced by models.
- The statistics of the shapes of yield curves predicted by standard models without very strong mean reversion do not appear to reproduce the statistics of yield-curve data.
- Disagreement with yield-curve dynamics is important since that can lead to model predictions of statistical arbitrage that are not present in the market. This is because statistical properties of yields in the data but not in the model can lead to portfolios that the model will statistically misprice.
- As an example, we have models that produce kinks in the yield curve that are not observed. Such models will misprice options that are sensitive to yield-curve shape.

- Standard models produce market prices for options by adjustment of implied volatility. However, options are only "weak probes" of the underlying interest-rate process; i.e. the details of the actual interest-rate process are not tested by options.
- No-arbitrage constraints are imposed by determination of the yield-curve path $R_{Avg}(t,T)$, about which all fluctuations occur. For the short rate, this function is chosen such that the average discount factors over all short-rate paths agree with initial yield-curve data. In general, $R_{Avg}(t,T)$ up to a convexity correction is the forward yield $f(t,T;t_0)$ of maturity T at forward time t, as obtained from today's yield curve.
- No-arbitrage implies the absence of a higher total return of one portfolio of bonds over another. Even without imposing the no-arbitrage drifts, we have checked portfolio returns are roughly constant. That is, no arbitrage is approximately true even for the Monte-Carlo paths used in the analysis of the historical data, i.e. $R_{Avg}(t,T)$ as determined historically. This is consistent with the historical yield-curve data not containing much arbitrage.

Scenario Analysis

Scenario analysis is a different issue. Instead of following the forward rates moving forward in time, portfolio managers often want to know what will happen to their portfolios under scenario viewpoints on yield-curve changes. In this case $R_{Avg}(t,T) = R_{Scenario}(t,T)$ can be chosen as the preferred yield-curve scenario. The strong mean-reverting fluctuations of the Micro component would exist around this preferred scenario. Since risk managers in general do not like to postulate scenarios far out into the future, the scenario can be used out to a horizon time, after which time the no-arbitrage yield curve can be used. The maturity T- dependence can be chosen to produce inverted or non-inverted average yield curves at future time t.

Figures: Macro-Micro Model

Fig. [1]. Macro quasi-equilibrium yield-curve paths in time for fixed maturity T produced by the macro simulator relative to the initial value, $r_{macro}(t,T) - r(0,T)$. The quasi-random fluctuations in the macro paths exist over long times. They are imagined due to uncertainties in macroeconomic variables. These macro fluctuations can be defined around an overall average path $R_{Avg}(t,T)$ chosen to satisfy no-arbitrage term-structure constraints. This is taken as $r(0,T)$ for simplicity in the figure. In the model, one macro-path realization is picked out of all possibilities as the smoothly varying quasi-equilibrium path of historical data.

Fig. [2]. Micro paths in time for a rate $r(t,T)$ at fixed T forming a "tube" of small rapid highly mean-reverting fluctuations around a macro path. The dynamics for micro-path fluctuations are described by the strong mean-reverting Gaussian process. These restricted fluctuations agree with historical data for yield-curve fluctuations.

Fig. [3]. Data for the Prime rate and 6-month Libor. The interpretation is that the Prime rate is macroeconomically determined, and varies along a quasi-equilibrium path. The Libor rate contains both Macro effects and short-time fluctuating Micro effects. The difference between the two rates is roughly constant, up to the small Micro fluctuations.

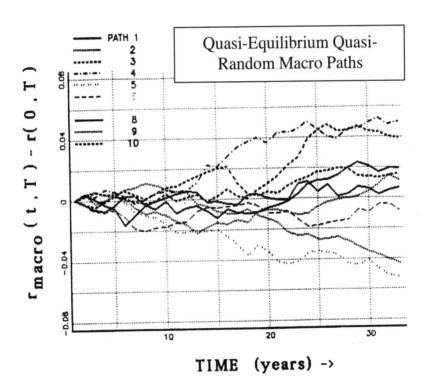

Figure 1

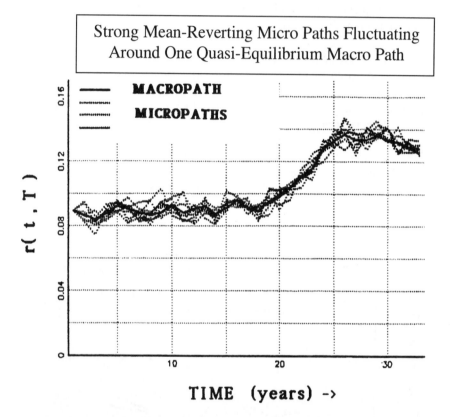

Figure 2

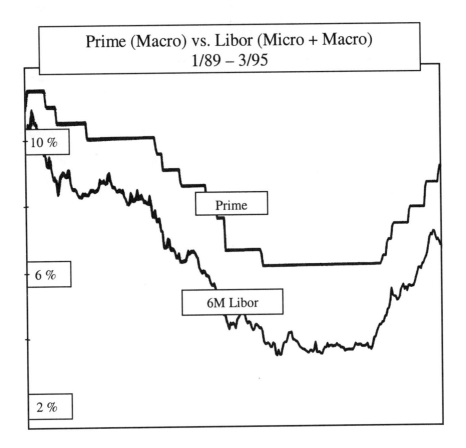

Figure 3

References

[i] Macro-Micro Model
Dash, J. W. and Beilis, A., *A Macro-Micro Yield-Curve Simulator*. CNRS, Centre de Physique Théorique CPT-89/P.E.2336, 1989.
Dash, J. W., *Evaluating Correlations and Models*. Talk, RISK Conference, 11/94

[ii] Strong Mean-Reverting Yield-Curve Multifactor Micro Model
Dash, J. W. and Beilis, A., *A Strong Mean-Reverting Yield-Curve Model*. Centre de Physique Théorique CPT-89/PE.2337 preprint, CNRS Marseille, France, 1989.

[iii] Heath, Jarrow, Morton
Heath, D., Jarrow, R. and Morton, A., *Contingent Claim Valuation with a Random Evolution of Interest Rates*. The Review of Futures Markets 9 (1), 1990.
Heath, D., Jarrow, R. and Morton, A., *Bond Pricing and the Term Structure of Interest Rates*: A New Methodology. Econometrica 60, 1 (1992).

51. Macro-Micro Model: Further Developments (Tech. Index 6/10)

Summary of This Chapter

In this chapter, we consider various developments in the Macro-Micro model since the original 1989 paper [i]. We think that these results are quite encouraging and urge others to consider performing research and analysis.

The Micro-Macro Model Applied to the FX and Equity Markets
We summarize some preliminary analyses that indicate the relevance of the Macro-Micro idea to FX and equities. These include:

- Strong Mean Reversion and Cluster Decomposition Analysis
- Probability Analyses for FX and the Macro-Micro Model

Models in the Economics Literature Resembling the Macro-Micro Model

- We give some references to work in papers in the economics literature that the Macro-Micro model resembles, both for interest rates and for FX.

Formal Developments in the Macro-Micro Model
We deal with some formal developments in the Macro-Micro model. This includes a discussion of hedging, consistency with forward quantities, term-structure constraints, and no-arbitrage. A general class of parameters $\{\lambda_\beta\}$ is introduced to parameterize the Macro dynamics. Included here are subsections on the following topics:

- The Green Function with General Quasi-Random Drift
- Averaging the Green Function over the Macro Parameters
- Option Pricing with the Macro Parameter - Averaged Green Function
- The Macro Parameter - Averaged Diffusion Equation

- For equities, we show that the standard forward stock price formula can be obtained. A modified Black-Scholes equity option formula is derived. Standard local hedging is retained.

No Arbitrage and the Macro Micro Model

We discuss various aspects of no arbitrage for the Macro-Micro model. As mentioned in Ch. 50, the basic idea is to have a sufficient number of parameters that can be fixed from no-arbitrage considerations. Included are the following topics:

- An extra term is shown to be present for the usual equity option + stock portfolio no-arbitrage argument, which shows up at long Macro times.
- No arbitrage for interest rates is discussed. It is shown that the standard term-structure constraints can be satisfied in the presence of a Macro component.

Other Topics

- Derman's Equity Regimes and the Macro-Micro Model
- Seigel's Nonequilibrium Dynamics and the Macro-Micro Model
- Macroeconomics, Fat Tails, Imitating Chaos and the Macro-Micro Model. We describe work possibly connecting the Macro component with fat tails[1]. First, recent work by economists shows that good predictive results for currency crises or breakdowns can be obtained from macroeconomic variables and signal-extraction techniques. Second, anomalous dimensions, deviating from sqrt time scaling, can appear numerically using the Macro-Micro model.
- Technical Analysis and the Macro-Micro Model
- The Macro-Micro Model and Interest-Rate Data over long times (1950-1996).
- Data, Models, and Rate Distribution Histograms: We present a clarifying numerical example for historical vs. theoretical rate distribution histograms, as discussed in Ch. 48, 49.
- We present some further evidence that mean reversion is needed for the Micro component. Negative forward-forward rates violating no-arbitrage are generated in a multifactor lognormal MC simulation for zero-coupon rates, without mean reversion.
- A dynamical model for correlation instability using the Macro component was already presented in Ch. 25.

---

[1] **Fat Tails:** Up to now, we have assumed that fat tails are added into the Micro component. Here we investigate the possibility that the Macro component could also be associated with fat tails.

The Macro-Micro Model for the FX and Equity Markets

The Macro-Micro model was originally formulated for interest rates, as described in preceding chapters. The origin of the model was the desire to describe the statistics of yield-curve data including the absence of kinks, and to be able to price contingent claims.

Review: The CDA Test and the Yield Curve

As described in Ch. 50, a sharp statistical tool enabling us to understand yield-curve dynamics is the 3rd-order generalized skewness Green function in the Cluster Decomposition Analysis (CDA). This quantity vanishes for a Gaussian measure, i.e. $C_3^{(Gaussian)} = 0$. The application of the CDA proceeds in two steps. First, using algebra we isolate the assumed Gaussian measure $dz(t)$ occurring in the stochastic equation of a putative model. Then we use the data to test whether or not $C_3/M_3 \approx 0$ for this supposed Gaussian $dz(t)$, and thereby test the model's validity directly using the data. Here M_3 is used for normalization purposes, and is nonzero for nonzero drift (which always occurs in data). If the requirement $C_3/M_3 \approx 0$ is satisfied, this test is consistent with the data. If this requirement is not satisfied, e.g. if $C_3/M_3 >> 0$ for $dz(t)$, then the model is inconsistent with the data.

For the yield curve, we found that if strong mean reversion about a quasi-equilibrium moving average is used, then $C_3/M_3 \approx 0$ is satisfied. This enabled us to develop a strong mean-reverting MC simulator that produced yield curves that look like real data, in addition to passing a battery of statistical tests.

Equities, FX, Strong Mean Reversion, and the CDA Test

We have carried out similar preliminary analyses for both the FX and equity markets, and find results similar to the yield-curve case. The $C_3/M_3 \approx 0$ requirement is satisfied only if strong mean reversion is assumed for FX and equities, similar to the interest rate case. This provides some evidence that strong mean reverting fluctuations about a quasi-equilibrium mean also operates in the FX and equity markets.

Probability Analyses for FX and the Macro-Micro Model

Even a casual look at FX data over many years presents striking behavior of long successive periods of time during which distinct long-term trends were present, and about which small oscillations occur. Such behavior is natural and expected with the Macro-Micro model with quasi-random slopes and strong mean-reverting oscillations. Conversely, this observed behavior is unnatural and highly unlikely in standard models. We now consider some details explicitly.

In my CIFER tutorial[ii], some probability calculations were presented that indirectly imply the existence of the Macro-Micro model for FX. These calculations were based on a simple analysis of the standard no-arbitrage FX model.

The standard no-arbitrage FX model was described in Ch. 5. It says that relative changes of an exchange rate fluctuate about a drift given by interest-rate parity. Contrary to the equity market where the no-arbitrage drift is determined by portfolio arguments and not supposed to have much to do with the actual behavior of equities, the FX drift is a physically motivated quantity directly tied to FX forwards, and these are transacted in the market. To the extent that the actual spot FX time dependence behavior is correlated with that predicted by FX forwards, the market spot FX time behavior should at least approximately exhibit the behavior of the no-arbitrage FX model. Therefore, we can use the no-arbitrage drift of FX to gain some physical insight.

The calculation performed was a straightforward check of how probable the observed market behavior on the average over the long term (1972-1999) could be described by the no-arbitrage FX model. A variety of time series was considered, all with similar results. These are described next.

Long-Term FX Data: Probability Standard Model Holds is 10^(-8)

For the case of $\eta[DEM/USD]$, four distinct regions occurred for the average behavior from 1972-1999. Qualitatively, these regions exhibited: (I) *Decrease* 1972-1980, (II) *Increase* 1980-1985, (III) *Decrease* 1985-1988, and (IV) *Flat* 1988-1999. The total amounts of movement in the first three regions were very large, making it quite improbable that the observed behavior could be explained by the standard model.

The probability that the behavior observed in the first three regions can be explained by the standard no-arbitrage FX model can be evaluated using barrier probabilities[2]. The probability that the observed decrease occurred in region I is a Down-In probability. Numerically, this probability is 5%, $P_{\text{Region I}}^{(\text{Down-In})} = 0.050$. Region II involves an Up-In probability of 0.01%, $P_{\text{Region II}}^{(\text{Up-In})} = 0.0001$. Region III involves another Down-In probability of 0.1%, $P_{\text{Region III}}^{(\text{Down-In})} = 0.001$. All these probabilities are small.

The composite probability is the product of these three probabilities, $P_{\text{All Regions}}^{(\text{Total})} = O(10^{-8})$. This small value is the probability that long-term FX rate

---

[2] **Parameters:** We used volatilities and rates current as of the time of the data to evaluate the probabilities. For a discussion of barrier probabilities, see Ch. 17.

behavior can be explained by the standard FX no-arbitrage drift plus Brownian noise model. To put this in human terms, 10^{-8} is the probability of correctly picking out one pre-specified second of time in three years.

Macro-Micro-Related Models in the Economics Literature

This section presents some references to models in the economics literature. These models all relate to some aspect of the Macro-Micro idea[3].

The development of the Macro-Micro model took place in the world of finance. The association of macroeconomics with the Macro component was an hypothesis that came at the end of the investigation.

It is quite gratifying to see after the fact that the Macro-Micro idea fits in with strikingly parallel ideas developed in the world of macroeconomics.

Indeed, it is perhaps high time that the economists and the finance gurus started talking to each other regarding the underlying dynamics of variables over different time scales.

Related Models for Interest Rates in the Literature

The Fed and the Macro Component

The fact that the average long-time-scale behavior of the short-maturity sector of the yield curve is driven by the Fed is obvious. Innumerable articles and papers exist describing and commenting on this situation. Fed policy is a large subject, but all we need here is that the Fed actions are driven and determined one way or the other by macroeconomics. Thus, the average behavior of the short end of the yield curve is driven by macroeconomics. To the extent that the long-maturity end of the curve is correlated with the short end, the average long-time-scale behavior of the whole yield curve is driven by macroeconomics.

This idea was in fact a strong part of the motivation behind the original construction of the Macro-Micro model in 1988.

TIPS: Bertonazzi and Maloney

In Ch. 16, we discussed TIPS. In that chapter, we mentioned an analysis of Bertonazzi and Maloney[iii]. They stated: *"The implication is that inflation is the biggest component of variance in the yield on government bonds"*. This is consistent with the Macro-Micro model. The most straightforward consequence

---

[3] **Partial Literature Search:** We have not performed a systematic search of the literature.

of the Macro-Micro model is that most interest-rate variation results from the Macro component with macroeconomic causes, with small fluctuations due to the Micro component. Bertonazzi and Maloney identified inflation as the main macroeconomic driver for government bonds.

Related Models for FX in the Literature

Feiger and Jacquillat

A striking and straightforward description of the FX market containing separate dynamics for short-time-scale and long-time-scale descriptions is found in the work of Feiger and Jacquillat [iv, 4]. The description of these dynamics seems qualitatively very similar to the Macro-Micro model.

These authors state in Chapter 5: Exchange Rate Behavior in the Long Run: *"First, it will become apparent that exchange rate behavior can be understood only in the context of a global macroeconomic perspective.... Understanding is better for the material dealt with in this chapter and not very good for short-run exchange rate behavior analyzed in Chapter 6"*. This is very similar to the statement of the Macro long-term component.

These authors state in Chapter 6: Short-Run Exchange Rate Movements in a Free Market: *"Thus we propose that, in the short-term, the spot rate is simply pulled around by the actions of speculators in the forward exchange markets and by trading on the capital markets"*. This is quite similar to the statement of the Micro short-term component.

Blomberg

S. B. Blomberg[iv, 5] mentions work in the literature that ascribes different dynamics along the lines of the Macro-Micro model. For short time scales (the Micro component region) he states: *Starting with the influential work of Meese & Rogoff (1983) a large body of research has documented that the random walk model out-performs a wide class of structural and time series (univariate and multivariate) models in short run out-of-sample prediction.*

For long time scales (the Macro component region) he states: *However, Taylor (1995) suggests models that rely on rich dynamic specifications can beat*

---

[4] **Acknowledgement:** I thank Alain Nairay for pointing out the work of Feiger and Jacquillat to me.

[5] **Short Time Scales:** Blomberg's focus is on alternative descriptions at short time scales. He presents evidence that short-run exchange fluctuations can be described using "narrative measures" involving certain dummy variables of monetary policy.

a random walk over the long horizon, see Mark (1995) and Chin and Meese (1995).

Chin

M. Chin[iv] also distinguishes short and long time scales with different dynamics in the FX market. He states: *In the short run, nominal exchange rates depend primarily on financial market variables and expectations* .... At long time scales, he states: *I then present results of empirical tests of the theory, which suggest that, in many instances in East Asia, enhanced productivity growth in the tradable goods sector is associated with a long-run strengthening of the real exchange rate*[6]. Again, a division of dynamics between short and long time scales is present in the Macro-Micro approach.

Formal Developments in the Macro-Micro Model

In this section, we deal with some formal developments in the Macro-Micro model. This includes a discussion of hedging, consistency with forward quantities, term-structure constraints, and no-arbitrage.

In general, we will have a collection of macro parameters that we generically call $\{\lambda_\beta\}$. In Ch. 50, we considered quasi-random slopes or trends over random times above a cutoff time. In the next chapter, we will introduce a set of "Toolkit" functions with several parameters that might serve to describe quasi-random cycles. For purposes of this chapter, we keep the idea general.

It is important first to consider fixed specific values of the parameters $\{\lambda_\beta\}$ and then to perform the averaging over their various possible values. A specific deterministic function (i.e. *not* random) will be used to enforce constraints.

The Green Function with a Specific Quasi-Random Drift

For fixed specific values of the macro parameters $\{\lambda_\beta\}$, the quasi-equilibrium quasi-random drift $\mu(t,\{\lambda_\beta\})$ is deterministic. Therefore, for fixed $\{\lambda_\beta\}$, the usual convolution theorems hold for Gaussian propagators in $x(t)$. We therefore get the Green function $G_{0*}(\{\lambda_\beta\})$ over time $T_{0*} = t^* - t_0$ from x_0 to x^* with the fixed parameters $\{\lambda_\beta\}$ directly in the usual form. Here the subscripts on

---

[6] **Nominal and Real FX Rates:** The "nominal" FX rate is the actual traded rate. The "real" exchange rate is an inflation-adjusted exchange rate; the price at which goods and services produced at home can be exchanged for those produced abroad.

$G_{0*}(\{\lambda_\beta\})$ indicate the spatial and time dependences of the fully notated Green function $G(x_0, x^*; t_0, t^*; \{\lambda_\beta\})$. We obtain (including discounting)

$$G_{0*}(\{\lambda_\beta\}) dx^* = \frac{\Theta(T_{0*}) e^{-r_0 \cdot T_{0*}}}{\sqrt{2\pi \sigma_{0*}^2 T_{0*}}} \exp\left[-\Phi_{0*}(\{\lambda_\beta\})\right] dx^* \tag{51.1}$$

Here $\Phi_{0*}(\{\lambda_\beta\}) = \left[x^* - x_0 - \mu_{0*}(\{\lambda_\beta\}) T_{0*}\right]^2 / \left[2\sigma_{0*}^2 T_{0*}\right]$. The total macro drift $\mu_{0*}(\{\lambda_\beta\})$ for fixed $\{\lambda_\beta\}$ is the usual time-averaged expression,

$$\mu_{0*}(\{\lambda_\beta\}) = \frac{1}{T_{0*}} \int_{t_0}^{t^*} \mu(t, \{\lambda_\beta\}) dt \tag{51.2}$$

Averaging the Green Function over Macro Parameters $\{\lambda_\beta\}$

To describe *future* paths, we naturally do not know which macro parameters $\{\lambda_\beta\}$ to use. Call $\wp[\{\lambda_\beta\}]$ the probability density function of the $\{\lambda_\beta\}$. This generalizes the $\wp[\lambda]$ probability function of the slope in Ch. 50. The associated measure of $\wp[\{\lambda_\beta\}]$ is $DM_\wp = \wp[\{\lambda_\beta\}] \prod_\beta d\lambda_\beta$, where the normalization is $\int_{\{\lambda_\beta\} \subset Set_\mu} \wp[\{\lambda_\beta\}] \prod_\beta d\lambda_\beta = 1$. Here Set_μ is the set of possible values of the various parameters $\{\lambda_\beta\}$. We keep Set_μ independent of time for simplicity, although the theory can be generalized to make Set_μ depend on time.

For option pricing, we need the expectation over all variables; hence, we want to average over the macro parameters $\{\lambda_\beta\}$. Therefore we multiply $G_{0*}(\{\lambda_\beta\})$ by $\wp[\{\lambda_\beta\}]$ and integrate over the possible values of the $\{\lambda_\beta\}$. We call the resulting expression $G_{0*}^{Avg\{\lambda\}}$ where

$$G_{0*}^{Avg\{\lambda\}} = \int_{\{\lambda_\beta\} \subset Set_\mu} G_{0*}(\{\lambda_\beta\}) \cdot \wp[\{\lambda_\beta\}] \prod_\beta d\lambda_\beta \tag{51.3}$$

Options and the Macro Parameter - Averaged Green Function

The quantity $G_{0*}^{Avg\{\lambda\}}$ averaged over the $\{\lambda_\beta\}$ parameters as just defined is the Green function needed for performing discounted expected values of cash flows in pricing for equities, FX, and commodities options. For a European option valued today t_0 with payout $C(S^*, t^*)$ at expiration t^*, we merely commute the integrals[7] over $\{\lambda_\beta\}$ and x^*. We obtain

$$C_0^{Avg\{\lambda\}} \equiv C^{Avg\{\lambda\}}(S_0, t_0) = \int_{-\infty}^{\infty} G_{0*}^{Avg\{\lambda\}} \cdot C(S^*, t^*) dx^* \tag{51.4}$$

To repeat, this is just the usual option expression averaged over the $\{\lambda_\beta\}$ macro parameters.

No Arbitrage and the Macro-Micro Model: Formal Aspects

This section discusses the formal properties of no-arbitrage conditions in the presence of quasi-equilibrium drifts with quasi-random movements including minimum time scales. We discuss both the equity market and the interest-rate market.

We will see that forward quantities can be matched. An important result is the proof that short-term hedging arguments remain intact.

Naturally, some changes occur in some quantities, as we can expect, since we have extra dynamical variables relative to standard models. These include an extra term in an effective diffusion equation and an extra term in the no-arbitrage formalism.

First, we will define the general expectation $\langle X \rangle_\lambda$ of any quantity with respect to $\{\lambda_\beta\}$,

$$\langle X \rangle_\lambda \equiv \int_{\{\lambda_\beta\} \subset Set_\mu} X(\{\lambda_\beta\}) \wp[\{\lambda_\beta\}] \prod_\beta d\lambda_\beta \tag{51.5}$$

For example, writing $\mu_a(\{\lambda_\beta\}) = \mu(t_a, \{\lambda_\beta\})$ at time t_a, we have

---

[7] **Commutative Restrictions.** As mentioned in the section on stochastic volatility in Ch. 6, App. B, the integrals may not commute unless appropriate restrictions are placed on the parameters.

$$\mu_a^{Avg\{\lambda\}} = \langle \mu(t_a) \rangle_\lambda = \int\limits_{\{\lambda_\beta\} \subset Set_\mu} \mu_a(\{\lambda_\beta\}) \cdot \wp[\{\lambda_\beta\}] \prod_\beta d\lambda_\beta \quad (51.6)$$

Also, note that $G_{0*}^{Avg\{\lambda\}} = \langle G_{0*} \rangle_\lambda$ as above.

The (Drift, Green Function) Correlation $\langle \mu_a G_{ab} \rangle_{\lambda;C}$

The correlation function $\langle \mu_a G_{ab} \rangle_{\lambda;C}$ will turn out to be important. This quantity is defined as

$$\langle \mu_a G_{ab} \rangle_{\lambda;C} \equiv \langle \mu_a G_{ab} \rangle_\lambda - \langle \mu_a \rangle_\lambda \langle G_{ab} \rangle_\lambda \quad (51.7)$$

The subscripts a, b are shorthand for (x_a, t_a) and (x_b, t_b). We have

$$\langle \mu_a G_{ab} \rangle_{\lambda;C} = \int\limits_{\{\lambda_\beta\} \subset Set_\mu} \left[\mu_a(\{\lambda_\beta\}) - \mu_a^{Avg\{\lambda\}} \right] \cdot G_{ab}(\{\lambda_\beta\}) \wp[\{\lambda_\beta\}] \prod_\beta d\lambda_\beta$$

$$(51.8)$$

The Macro Parameter - Averaged Diffusion Equation

Again, assume that the variable x is Gaussian. With a fixed set of parameters $\{\lambda_\beta\}$, the Green function $G_{ab}(\{\lambda_\beta\}) = G(x_a, x_b; t_a, t_b; \{\lambda_\beta\})$ satisfies the standard diffusion equation with drift $\mu(t_a, \{\lambda_\beta\})$. Here the (backward) diffusion operator is with respect to (x_a, t_a). This is because $G_{ab}(\{\lambda_\beta\})$ has the canonical Gaussian form for fixed $\{\lambda_\beta\}$.

We average this diffusion equation with respect to the $\{\lambda_\beta\}$. We obtain

$$\left[\frac{\partial}{\partial t_a} + \frac{1}{2}\sigma^2(t_a) \frac{\partial^2}{\partial x_a^2} + \mu_a^{Avg\{\lambda\}} \frac{\partial}{\partial x_a} - r(t_a) \right] G_{ab}^{Avg\{\lambda\}} = -\frac{\partial}{\partial x_a} \langle \mu_a G_{ab} \rangle_{\lambda;C}$$

$$(51.9)$$

The term on the right-hand side is the x_a-derivative of the correlation function $\langle \mu_a G_{ab} \rangle_{\lambda;C}$. Again, the average is over the macro parameters $\{\lambda_\beta\}$. This is an extra term not present in the standard formulation

Hedging, Forward Prices, No Arbitrage, Options (Equities)

In this section, we deal with hedging, forward prices, and no arbitrage for equity (and FX, commodity etc.) options in the presence of the quasi-random drifts specified by variation of the $\{\lambda_\beta\}$ parameters. The standard delta hedging procedure goes through. We introduce a deterministic average slope function $\lambda_{av}(t)$ that can be used to enforce the forward stock price. The no-arbitrage equation picks up an extra term containing the correlation $\langle \mu_a G_{ab} \rangle_{\lambda;C}$.

Hedging at Fixed $\{\lambda_\beta\}$ Parameters

The usual hedging arguments go straight through for fixed parameters $\{\lambda_\beta\}$. This is because the drift for fixed $\{\lambda_\beta\}$ is deterministic. We consider the usual portfolio $V(S,t) = N_S S + C(S,t)$ of one option plus stock. The portfolio value V_0 at time t_0 changes to V_1 at time $t_1 = t_0 + dt_0$. Calling the drift $\mu_0(\{\lambda_\beta\}) = \mu(t_0, \{\lambda_\beta\})$, we get the standard result at fixed $\{\lambda_\beta\}$,

$$V_1 - V_0 = \left[N_S + \frac{\partial C_0}{\partial S_0} \right] (S_1 - S_0) - \left[\mu_0(\{\lambda_\beta\}) + \tfrac{1}{2}\sigma_0^2 \right] S_0 \frac{\partial C_0}{\partial S_0} dt_0 + r_0 C_0 dt_0$$

(51.10)

The standard hedging prescription $-N_S = \Delta = \partial C_0 / \partial S_0$ eliminates the stochastic term proportional to $(S_1 - S_0)$ at fixed $\{\lambda_\beta\}$.

Hedging Including Averaging over the $\{\lambda_\beta\}$ Parameters

Call $V^{Avg\{\lambda\}}(S,t) = N_S S + C^{Avg\{\lambda\}}(S,t)$, the $\{\lambda_\beta\}$-averaged portfolio at time t. Call $\mu_0^{Avg\{\lambda\}}$ the averaged drift at time t_0. We get directly

$$V_1^{Avg\{\lambda\}} - V_0^{Avg\{\lambda\}} = \left[N_S + \frac{\partial C_0^{Avg\{\lambda\}}}{\partial S_0}\right](S_1 - S_0)$$

$$-\left[\mu_0^{Avg\{\lambda\}} + \tfrac{1}{2}\sigma_0^2\right]S_0 \frac{\partial C_0^{Avg\{\lambda\}}}{\partial S_0} dt_0 + r_0 C_0^{Avg\{\lambda\}} dt_0 \qquad (51.11)$$

$$-\frac{\partial}{\partial x_0}\int_{-\infty}^{\infty}\langle \mu_0 G_{0*}\rangle_{\lambda;C} C(S^*,t^*)dx^*$$

The hedging prescription that eliminates the stochastic term proportional to $(S_1 - S_0)$ is just $\Delta^{Avg\{\lambda\}}$, the $\{\lambda_\beta\}$-averaged expression for Δ, namely

$$-N_S = \Delta^{Avg\{\lambda\}} = \frac{\partial C_0^{Avg\{\lambda\}}}{\partial S_0} \qquad (51.12)$$

This produces the usual hedging prescription exactly, since $C_0^{Avg\{\lambda\}}$ is the final model option value at t_0 after averaging.

The bottom line is that the standard hedging prescription goes through in the presence of $\{\lambda_\beta\}$ averaging.

The Forward Stock Price is the Standard Expression

The forward stock price at t^* is obtained by writing down the standard expression at fixed $\{\lambda_\beta\}$ and then averaging over $\{\lambda_\beta\}$. We take the prescription fixing the averaged-$\{\lambda_\beta\}$ drift for $T_{0*} = t^* - t_0$ as

$$\langle \exp[\mu_{0*}(\{\lambda_\beta\})T_{0*}]\rangle_\lambda = \exp[(r_{0*} - \tfrac{1}{2}\sigma_{0*}^2)T_{0*}] \qquad (51.13)$$

This can be done using a point-by-point specification of the deterministic function $\lambda_{av}(t)$ discussed before. The above condition reduces to $\mu_0^{Avg\{\lambda\}} = r_0 - \sigma_0^2/2$ when $T_{0*} \to dt_0$ becomes infinitesimal.

We then obtain the usual expression for the $\{\lambda_\beta\}$-averaged forward stock price $S_{fwd}^{Avg\{\lambda\}}(t^*)$ at time t^*, viz (ignoring dividends),

$$S_{fwd}^{Avg\{\lambda\}}\left(t^*\right) = S_0 \exp\left[\int_{t_0}^{t^*} r(t)\,dt\right] \tag{51.14}$$

Modified No Arbitrage Condition for Equity Options

The no-arbitrage condition $\dfrac{dV}{dt_0} = r_0 V$ could have been produced at fixed $\{\lambda_\beta\}$ by setting $\mu_0\left(\{\lambda_\beta\}\right)$ equal to the usual expression $r_0 - \sigma_0^2/2$. However, we do not want to do this because we would have to impose this constraint for every set $\{\lambda_\beta\}$. Thus, we would lose the $\{\lambda_\beta\}$ variability of the drift, which is the whole point of the Macro component. For this reason, we consider no-arbitrage only after averaging over $\{\lambda_\beta\}$ by writing $\mu_0^{Avg\{\lambda\}} = r_0 - \sigma_0^2/2$. Because options in this framework are averaged over $\{\lambda_\beta\}$, this approach is consistent.

We need to specify the drift a bit further. We refer the reader to Ch. 50 on the Macro-Micro model, where a distribution of slopes was used to describe the drifts. The distribution of slopes has an average slope. We take this average slope $\lambda_{av}(t)$ as a deterministic function of time. The averaging over quasi-random variations of the $\{\lambda_\beta\}$ parameters will *not* include variation of this deterministic function $\lambda_{av}(t)$. We then write the drift $\mu\left(t,\{\lambda_\beta\}\right)$ in the form

$$\mu\left(t,\{\lambda_\beta\}\right) = \lambda_{av}(t) + \delta\mu\left(t,\{\lambda_\beta\}\right) \tag{51.15}$$

We define $\delta\mu^{Avg\{\lambda\}}(t_0)$ via $\{\lambda_\beta\}$-averaging of $\delta\mu\left(t,\{\lambda_\beta\}\right)$ at time t_0. We then choose $\lambda_{av}(t_0) = -\delta\mu^{Avg\{\lambda\}}(t_0) + \left(r_0 - \sigma_0^2/2\right)$ to enforce $\mu_0^{Avg\{\lambda\}} = r_0 - \sigma_0^2/2$. With this done, we get

$$V_1^{Avg\{\lambda\}} - V_0^{Avg\{\lambda\}} = r_0 V_0^{Avg\{\lambda\}} dt_0 - \frac{\partial}{\partial x_0}\int_{-\infty}^{\infty}\langle\mu_0 G_{0^*}\rangle_{\lambda;C}\, C\left(S^*,t^*\right)dx^* \tag{51.16}$$

This is as close as we can get to the standard no-arbitrage condition. The second term containing the integral of the correlation function $\langle\mu_0 G_{0^*}\rangle_{\lambda;C}$ is an extra piece.

The Modified Black-Scholes Equity Option Formula

Consider a European call option. We first carry out the usual integral of the payoff over the terminal co-ordinate x^* and afterward perform the $\{\lambda_\beta\}$ averaging. Define $\varsigma(\{\lambda_\beta\}) = \dfrac{\exp[\mu_{0^*}(\{\lambda_\beta\})T_{0^*}]}{\langle \exp[\mu_{0^*}(\{\lambda\})T_{0^*}] \rangle_\lambda}$. Standard manipulations lead to a modified Black-Scholes formula with $\{\lambda_\beta\}$ averaging,

$$C_0^{Avg\{\lambda\}} = \langle S_0 \cdot \varsigma(\{\lambda\}) \cdot N[d_+\{\lambda\}] - E e^{-r_0 \cdot T_{0^*}} \cdot N[d_-\{\lambda\}] \rangle_\lambda \quad (51.17)$$

Here $d_\pm(\{\lambda_\beta\})$ are the usual expressions but containing the drift $\mu_{0^*}(\{\lambda_\beta\})$.

Satisfying the Interest-Rate Term-Structure Constraints

For term-structure constraints for interest rate dynamics, we need consistency with the zero-coupon bond prices given by the data at time t_0. The zero-coupon bond at t_0 paying one at maturity date t^* (time to maturity T_{0^*}) must equal the x^*-spatial integral at t^* of the Green function $G_{0^*}^{Avg\{\lambda\}}$ including discounting, viz

$$P_{ZC}^{(t^*)}(t_0) = \int_{-\infty}^{\infty} G_{0^*}^{Avg\{\lambda\}} dx^* \quad (51.18)$$

As in the above sections on equity options, we utilize the deterministic slope function $\lambda_{av}(t)$, writing the drift as $\mu(t,\{\lambda_\beta\}) = \lambda_{av}(t) + \delta\mu(t,\{\lambda_\beta\})$. We specify $\lambda_{av}(t)$ to satisfy the term-structure constraints in the presence of $\{\lambda_\beta\}$ averaging over the rest of the quasi-random drift function. This idea was already present in the original Macro-Micro paper in 1989.

The idea is to define $R_{\lambda_{av}}(t) = \int_{t_0}^{t} \lambda_{av}(t')dt'$, which acts like an interest rate. We then determine $\exp\left[-\int_{t_0}^{t^*} R_{\lambda_{av}}(t)dt\right]$. This quantity is proportional to the t^*-maturity zero-coupon bond. The proportionality constant depends on the model.

We then get $\lambda_{av}(t) = dR_{\lambda_{av}}(t)/dt$. There are enough degrees of freedom in this arbitrary function (actually an infinite number) so that arbitrary term structure constraints can be incorporated.

For the case of the mean-reverting Gaussian model[8], for example, the proportionality constant is $1/\mathcal{X}^{(T_{0*})Avg\{\lambda\}}(r_0, t_0)$ where

$$\mathcal{X}^{(T_{0*})Avg\{\lambda\}}(r_0, t_0) \equiv K_{0*} e^{-r_0 T_{0*}} \left\langle \exp\left[-\int_{t_0}^{t^*} dt \int_{t_0}^{t} \delta\mu(t', \{\lambda_\beta\}) dt' \right] \right\rangle_\lambda \quad (51.19)$$

Calling ω the mean reversion, $K_{0*} = \exp\left[-\frac{x_0}{\omega}(1 - e^{-\omega T_{0*}}) + \frac{\sigma^2}{6} \mathcal{Z}(\omega T_{0*}) \right]$.

Here, $\mathcal{Z}(\omega\tau) = \frac{3}{(\omega\tau)^3}\left[\omega\tau + 2(e^{-\omega\tau} - 1) - \frac{1}{2}(e^{-2\omega\tau} - 1) \right]$. In addition, the starting value is $x_0 = 0$.

Other Developments in the Macro-Micro Model

In this last part of the chapter, we present other significant developments.

Derman's Equity Regimes and the Macro-Micro Model

In the chapter on volatility skew, we discussed various regimes for equities as characterized by Derman (trending, range-bound, jumpy). These regimes seem to correspond to the characteristics of the Macro-Micro model. In the Macro-Micro model, we have quasi-random drifts or slopes corresponding to trending (with positive or negative drifts) and range-bound behavior (small drift), all with limited volatility. Jumps play a supplemental role in the micro component.

Seigel's Nonequilibrium Dynamics and the MM Model

Les Seigel[v] has proposed an innovative and interesting time-dependent dynamical framework. He writes a relation between the stock price S and stock index level I as $S = S(I, t)$, allowing for a specific time dependence. Assuming lognormal

---

[8] **Mean-Reverting Gaussian Model:** See Ch. 43 on path integrals and options.

dynamics for the index produces a diffusion equation for the stock price, using the usual rules. Solutions of the diffusion equation produce relaxation time scales for non-equilibrium behavior for stocks relative to the index.

Seigel finds long characteristic times, on the order of years. These are representative of Macro times in the Macro-Micro Model.

Macroeconomics and Fat Tails (Currency Crises)

So far, we have considered the Macro and Micro components of the Macro-Micro model to be independent. We now quote some evidence that these two components may be connected in the FX market. In particular, the improbable-event fat-tail jumps that form a critical part of the Micro component may be directly related to macroeconomics[1].

It has long been a goal of economics to predict currency crises. Using the language of this book, the topic is the attempt to establish a connection of long time scale macroeconomics with fat tails in the FX market. A body of work that has made significant progress along these lines has emerged in the economics literature. This work was pioneered by Kaminsky and Reinhart (KR). The field has recently been significantly extended by Omarova[9], and we follow this work as described in Mills and Omarova (Ref. vi).

Currency crises are first defined historically using a "currency pressure index" defined by KR that involves a normalized sum of relative changes of the exchange rate[10] and relative changes of FX reserves. When this currency pressure index drops enough from its average, a currency crisis is defined.

Next, threshold levels are defined for various economic variables[11]. A "crisis window" before one of the historical crises is defined during which a positive alarm signal is considered to have been successful and relevant for predicting that crisis. Threshold levels for each economic variable are defined by minimizing combinations of type-I errors (missed crises) and type-II errors (false alarms). Two threshold levels ("stress" and "critical") are defined.

The next step constructs the alarm signals. Alarm signals are defined by indicator functions of economic variables that go above their respective threshold levels. A composite signal-alarm indicator function is constructed for each country. If this alarm goes above a certain level, a currency crisis is signaled.

---

[9] **Acknowledgements:** I thank Nora Omarova for very helpful discussions of her work (Ref. vi).

[10] **REER:** Actually, the "Real Effective Exchange Rate" is used by Mills and Omarova. For a given country, the REER is a trade-weighted combination of inflation-adjusted ("real") bilateral exchange rates for that country with its top trading partners.

[11] **Economic Variables:** Mills and Omarova use eight variables. They are the REER, Short Term Capital Inflows (% GDP), Current Account Balance (% GDP), Short Term Debt/Reserves, Equity Prices, Industrial Production, Exports, and M2/Reserves.

Backtesting and out-of-sampling testing show that the results are successful. The Omarova model produces 79% correctly-called crises over many countries, with over 40% probability of crisis given an alarm.

Some Remarks on Chaos and the Macro-Micro Model

In Ch. 46, we consider chaos-like models, including the Reggeon Field Theory. These are scaling models without any explicit time scales, and so they are very different from the Macro-Micro model that has explicit time scales. While we are convinced that chaos-like models have a place in describing rare fat-tail jumps, we are equally convinced that the bulk of the data requires the presence of time scales not present in chaos models.

Data, the Macro-Micro Model, and Chaos

As we saw in detail in Ch. 49, the description of yield-curve statistical data was possible with a strong mean-reverting model. Conversely, the data were problematic for other models to describe. The presence of distinct time scales seemed unavoidable. It is a challenge to chaos models without distinct time scales to fit the details of the yield-curve data.

The Econometric Model for FX Crises and Chaos

The significance of the econometric KR/Omarova model just described above has important significance for chaos models. Explicit long time scales of one to two years are used in the analysis for the crisis windows. Such distinguished time scales are not present in scaling theories that by definition are supposed to work at all time scales.

Anomalous Dimensions, Chaos, and the Macro-Micro Model

The Macro-Micro model can numerically mimic anomalous dimensions. Anomalous dimensions represent deviations from square-root time scaling $\sqrt{t}$, i.e. t^ς with. $\zeta^{\text{Square Root}} = 0.5$.

Deviation from $\sqrt{t}$ scaling is often associated with chaos-like behavior. The point here is that the very different underlying dynamics of the Macro-Micro model can mimic this chaos-like behavior without any such scaling properties in the model. Contrariwise, the Macro-Micro model contains explicit time scales.

We can explicitly construct an "effective" time-dependent volatility for the Macro-Micro model with a mean-reverting Micro component and the quasi-random slope Macro component. For simplicity, we model one variable.

Specifically, we can write $\sigma_\kappa^{MicroMacro}$ as the effective total volatility in the κth macro time interval of length $\tau_{\kappa,\kappa+1}$ due to both macro and micro effects. Denote the micro volatility by σ_κ^{Micro} and the macro slope volatility by $\sigma_\lambda^{MacroSlope}$. Using standard convolution arguments, we get the total variance in the κth period as

$$\left(\sigma_\kappa^{MicroMacro}\right)^2 \tau_{\kappa,\kappa+1} = \left(\sigma_\kappa^{Micro}\right)^2 \tau_{\kappa,\kappa+1} + \left(\sigma_\lambda^{MacroSlope}\right)^2 \tau_{\kappa,\kappa+1}^2 \qquad (51.20)$$

The second term is due to the uncertainty of the macro slope and induces a time dependence in $\sigma_\kappa^{MicroMacro}$.

When this effective time-dependent volatility $\sigma_\kappa^{MicroMacro}$ is fit with a power law t^ζ in time, we find that the power can appear to be different from one half, e.g. $\zeta = 0.6$.

The first figure below shows an example. Here we took the micro volatility as given by a mean-reverting Gaussian model and took equal step lengths $\tau_{\kappa,\kappa+1}$.

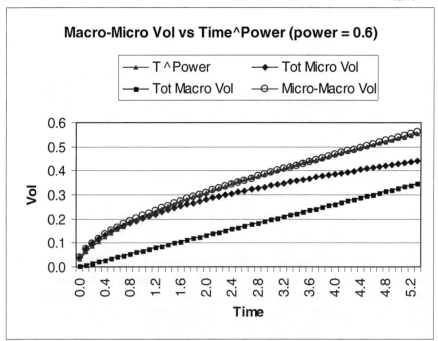

To show the versatility of the Macro-Micro model in producing anomalous dimensions, consider the second figure below. It shows a two-component structure for the macro vol, but this time with no micro vol. There are two step lengths chosen at random with equal probabilities. The example has been constructed to give the same anomalous power behavior, $\zeta \approx 0.6$.

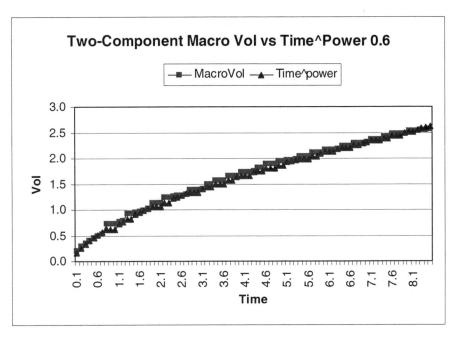

Technical Analysis and the Macro-Micro Model

For convenience of the reader, we give a lightning summary of technical analysis[vii,12]. Technical analysis looks for patterns directly in the time-series data and employs a variety of measuring probes. The most basic are support and resistance levels. Support and resistance levels are defined using the closing price level plus or minus a volatility estimate based on the recent trading price range. Patterns in historical price data are classified with various pictorial expressions like "head and shoulders". Trend indicators, for example different moving averages, are employed. Different moving averages are used and compared with each other, with special attention to crossing points. Time-differenced series are used to generate "momentum" indicators. Cyclic behavior is also considered, from seasonal behavior with a physical basis (e.g. weather) to abstract schemes (e.g. "Elliott Waves" involving Fibonacci numbers). Besides prices, other

---

[12] **Acknowledgement:** I thank Rick Sheffield and John Pisanchik for informative conversations regarding technical analysis.

variables like volume and open interest are considered. Concordance between indications of some subset of these probes is used as input to trading decisions.

The usefulness of technical analysis is a subject of disagreement. Some technical indicators are simple versions of volatility measures used in standard risk management. The relation to trading patterns to technical analysis is not clear. Regardless, many traders and systems use technical analysis. Perhaps technical analysis acquires some numerical validity in a self-fulfilling prophecy through feedback of trader activities using technical analysis to the market.

Connection of Technical Analysis with the Macro-Micro Model

The obvious conjecture for a connection of technical analysis with the Macro-Micro model is simple and physical:

- Moving averages in technical analysis correspond to the Macro component, when taken over macro time scales.
- The short-term fluctuations around the moving averages in technical analysis correspond to the Micro component.

The Macro-Micro Model and Interest-Rate Data 1950-1996

We present here some further indications for long time scales. We used the interest rate data in Ref. viii between 1950 and 1996, i.e. 46 years. Without doing any fits, we generated MC paths using the Macro model. *That is, in distinction to the previous chapters, we did not fix the Macro drifts using data.*

The general behavior of the data can in fact be qualitatively reproduced by the model over a wide range of rates and over long time periods. This shows that the Macro-Micro model is capable of providing realistic results on long time scales for interest rates, much longer than the time period presented in previous chapters.

Consider the following graph giving illustrative results for the 3-month treasury rate. The path shown was selected by hand, but again no fit was performed.

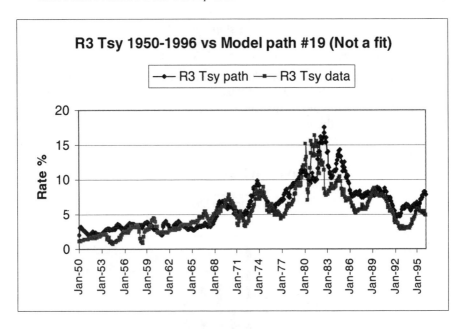

Data, Models, and Rate Distribution Histograms

In the discussion of data and models in Ch. 48, 49 we presented histograms of distributions of rates for data and also for a set of Monte-Carlo model paths. All models received as input the data volatilities. While the strong mean-reverting Gaussian (SMRG) model distribution was in accord with the data, the distribution histograms did not agree for the lognormal (LN) model.

In Ch. 48, we mentioned that, path by path, the model output will produce a distribution of width on the order of the output volatility, which is (within statistical noise) the same as the input data volatility. The problem arises if we sum over the distributions of paths. Each path produces a distribution of approximately the right width. However, without strong mean reversion, the center of the model distribution for a path is displaced randomly away from the center of the data distribution. When summed over paths, the model distribution without strong mean reversion will appear spread out, even though the output and input volatilities are consistent.

To illustrate, we use a simple model that generates "data" using a Gaussian process with a set of fixed random numbers. We also use a Gaussian model, with the same input volatility, as the "data".

The deviations are measured with respect to long-time-scale trends, defining the quasi-equilibrium historical path, as explained in previous chapters. For the toy model, we used a straight line with slope $\mu^{QE\ \text{Drift}} = \left(x_{avg}^{data} - x_0\right)/N$, where x_{avg}^{data} was the average of the "data" points and N the number of time intervals.

The graph below shows a sample run of the Gaussian model along with the "data". The widths of the model and "data" distributions are roughly equal (as they should be), but the model distribution is displaced relative to the "data" (in this particular case, to the left).

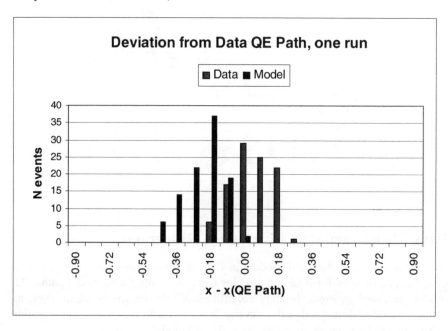

In other runs, the model distribution will be displaced to the right. This phenomenon generates a wider distribution of the model over many runs with respect to the data, even though each run has (within statistical error) the same width as the data.

Negative Forwards in Multivariate Zero-Rate Simulations

In this section, a simple lognormal simulation for zero-coupon rates without mean reversion is shown to generate negative forward rates. This generalizes Ch. 48, where coupon yields were used. Our conclusion remains: multivariate models using composite rates without strong mean reversion may be problematic[13].

---

[13] **Non-Composite Forward Rate Simulations:** If the forward rates themselves are simulated directly and not derived from composite rates, then a zero-rate barrier imposed by lognormal or other dynamics can be imposed to prevent negative rates.

We use a pedagogical 5-dimensional simulation for 21 unit time steps. The simulation is for term forward zero-coupon rates $f^{(t)}(t;t_0;T)$, in the notation of Ch. 7. This rate, observed at time t, applies between dates $(t,t+T)$. We start at time t_0 with an upward sloping forward curve. Correlations and volatilities are typical values $\rho^{(T,T')} \in (0.55,0.9)$, $\sigma^{(T)} \in (0.1,0.2)$. No drift was included. No-arbitrage drift corrections are quadratic in the volatility and generally small.

Now consider the forward rate relabeled $f^{(T,T+1)}(t) = f^{(t+T)}(t;t_0;\Delta T = 1)$, observed at time t, which applies between dates $(t+T,t+T+1)$.

The graph below shows the results for one run that produced negative values $f^{(3,4)}(t) < 0$ for $t \geq 15$. Other forward rates were observed to go negative in other runs.

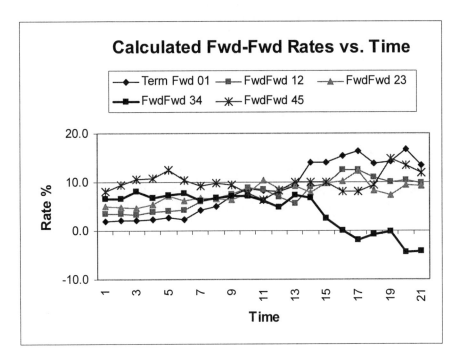

Negative forward rates are forbidden by no arbitrage considerations. Again, we believe that this sort of problem may be a general property of multifactor yield-curve models for composite rates, without strong mean reversion.

References

[i] Macro-Micro Model
Dash, J. W. and Beilis, A., *A Macro-Micro Yield-Curve Simulator*. CNRS, Centre de Physique Théorique CPT-89/P.E.2336, 1989.

[ii] Probability Calculations for FX Trends and Standard FX Model
Dash, J., *Risk Management*. Tutorial, IEEE/IAFE Conference on Computational Intelligence for Financial Engineering (CIFEr). 1999, 2000.

[iii] Interest Rates and the Macro-Micro Model
Bertonazzi, E. and Maloney, M. T., *Sorting the Pieces to the Equity Risk Premium Puzzle*. Clemson U. working paper, 2002. See P. 3.

[iv] FX Market Articles Related to the Macro-Micro Model
Feiger, G. and Jacquillat, B., *International Finance*, Allyn and Bacon, 1982. Ch. 5, 6.
Blomberg, S. B., *Dumb and Dumber Explanations for Exchange Rate Dynamics*. Journal of Applied Economics, Vol. IV, 2001. Pp. 187-216.
Chin, M. and Meese, R., *Banking on Currency Forecasts: Is the Change in Money Predictable?* Journal of International Economics, 1995. Pp. 161-178.
Mark, N., *Exchange Rates and Fundamentals: Evidence on Long Horizon Predictability and Overshooting*. American Economic Review, 1995. Pp. 201-218.
Meese, R. and Rogoff, K., *Empirical Exchange Rate Models of the Seventies: Do They Fit Out of Sample?* Journal of International Economics, 1983. Pp. 3-24.
Taylor, M. P., *The Economics of Exchange Rates*. Journal of Economic Literature, Vol 33, 1995. Pp. 12-47.
Chin, M. D., *Long-run Determinants of East Asian Real Exchange Rates*. www.sf.frb.org/econrsrch/wklyltr/wklyltr98/el98-11.html . Federal Reserve Bank of San Francisco Economic Letter 98-11, Pacific Basin Notes, 1998.

[v] Seigel's Non-Equilibrium Dynamics and Relaxation Time Scales
Seigel, L. , *Toward a Nonequilibrium Theory of Stock Market Behavior*. World Bank Working Paper; Seminar, Georgetown U., 4/16/91.

[vi] Economics and Currency Breakdown
Kaminsky, G. L. and Reinhart, C. M., *The Twin Crises: The Causes of Banking and Balance-of-Payment Problems"*, American Economic Review, 89 (3), 1999. Pp. 473-500.
Kaminsky, G. L., Lizondo, S., Reinhart, C., *Leading Indicators of Curency Crises*, IMF Staff Papers, Vol. 45, No. 1, 1998. Pp. 1-48.
Mills, C. and Omarova, E., *Predicting Currency Crises: A Practical Application for Risk Managers*, Citigroup working paper July 2003, to be published in the National Association for Business Economics.

[vii] Technical Analysis
Fink, R. and Feduniak, R., *Futures Trading: Concepts and Strategies*. New York Institute of Finance, 1988. Ch. 17.

[viii] Economic Modeling and Data
Pindyck, R. S. and Rubinfeld, D. L., *Economic Models and Economic Forecasts (4th Ed)*. McGraw-Hill, 1998.

52. A Function Toolkit (Tech. Index 6/10)

In the final chapter of this book, a "toolkit" of functions potentially useful for analyzing business cycles is presented. These functions could then form part of the macro component of financial markets operating over long time scales [i,1]. These functions may also be useful on shorter time scales for trading, as described at the end of this chapter.

The functions were originally used in describing some threshold phenomena in high-energy physics [ii,2] and subsequently in engineering [iii,3]. Applications using simpler forms of these functions have been used in finance for a long time. The functions can be used to specify a general form of a drift.

The first part of the macro component is composed of quasi-random trends, discussed in Ch. 50. The connection of the formalism with derivative pricing was shown. A general description of the macro component would be a combination of these quasi-random trends and the function toolkit cycles [iv,4].

In this chapter, the signals $x(t)$ are first assumed deterministic without a stochastic random component. While the macro component as used in this book does have a postulated quasi-random behavior, this behavior does not scale down to small times. Hence, a given realization of a macro path essentially acts as if it

---

[1] **Cyclical Analysis and the Macro Component of Finance:** We believe that the cyclical analysis approach to financial time-series analysis, with long-standing proponents, is valid for part of the macro component. Our purpose here is to introduce a set of functions that could be useful in such analyses.

[2] **Connection of the Function Toolkit with Thresholds in High-Energy Physics:** The application was the description of data in high-energy diffractive scattering. The translation of the formalism: the function f(t) is the imaginary part of a quantum-mechanical amplitude, time is the logarithm of the beam energy, and the time thresholds correspond to the successive production in energy of particles with different quantum numbers (like strangeness and charm), having successively increasing masses.

[3] **Connection of the Function Toolkit with Time Thresholds in Engineering:** The application was the description of the measured response of some equipment in spatially separated locations to input transient electromagnetic fields. The time-threshold behavior occurred because different parts of the equipment reacted to the electromagnetic fields at different times. The example presented in this chapter comes from this work.

[4] **Quasi-Random Trends and the Macro Component of Finance:** Trends over random but macro-length time periods ("quasi-random") are the first macro component. We describe a macro model for quasi-random secular trends in Ch. 50.

were deterministic. We are now faced with a classic signal-processing problem. One basic problem in signal analysis is the choice of reasonable functions to describe the data. The function toolkit sets out to address this problem.

The practical applications so far have involved performing a least-squares fit of a given signal with a limited number of functions, in either the time domain or the frequency domain. This is parametric signal estimation. We are not after "exact" results. As mentioned above, the method has already proved to be practical and useful for good characterizations of various data.

At the end of the chapter, we consider the relation of the toolkit functions to other techniques, including wavelets.

Time Thresholds; Time and Frequency; Oscillations

Many signals have an arrival onset-time threshold nature[5]. In finance, these time thresholds may have a clear significance—something definite happens at a given time, providing an excitation to the financial system. There is usually some inertia in the system, so the reaction of the variable x to the excitation may have a continuous lagged response, with some time to ramp up. The reaction naturally has some intrinsic magnitude. In addition, the initial reaction may be either positive or negative, so there is an initial phase. The reaction will damp down over some time period, depending on the excitation and situation. Various successive excitations can occur in a given macro time period.

Signals that are due to given excitations tend to be limited in time. Signals also tend to be band-limited in frequency ω. There may be a tendency to oscillate as markets go higher and lower in response to an excitation. First the market may over-react going (say) down, then may over-react in a rebound fashion going up in some characteristic up-down time τ_α. Continuation of this behavior forms oscillations with frequency $\omega_\alpha = 1/\tau_\alpha$. However, things are not so definite because successive up-down movements may have reaction times influenced by the environment. For a given excitation α, the characteristic up-down time τ_α occurs with some uncertainty $\Delta\tau_\alpha$. This determine the bandwidth of the oscillations $\Delta\omega_\alpha$ if we use the standard uncertainty relation $\Delta\omega_\alpha = 1/\Delta t_\alpha$. Choosing oscillations with various weights inside this bandwidth produces a variable period of oscillations in the time series $x(t)$. However, the nature of the variable period may be even more complicated than indicated by this standard uncertainty relation.

---

[5] **Thresholds:** Please note that here we mean by threshold a point in time characterized by a time delay, not a minimum value for the signal.

Summary of Desirable Properties of Toolkit Functions

The description above indicates that functions with the following seven properties should be useful in describing business-cycle behavior in finance:

1. Time-onset threshold $t_{threshold}$
2. Ramp-up time $\tau_{ramp-up}$
3. Damping time $\tau_{damping}$
4. Oscillation time τ_{osc}
5. Variable period in oscillations
6. Intrinsic magnitude
7. Initial phase

With functions of this sort, the goal is to provide a physically motivated and economical description of time series. Each function has seven possible parameters.

*The main point is that **because** each function looks like a physically possible signal, the **number** of functions required can turn out to be small.*

What about Gaps or Jumps?

We could put in gaps or jumps over short times as a separate requirement. However, our interest here is in behaviors over macro time periods (weeks to years). We can in fact model a gap behavior by specific choices of parameters, as we shall see. Moreover, in so doing, the gap behavior will have a natural associated time scale. However, a more natural explanation for gaps or jumps is perhaps nonlinear diffusion, as described in Ch. 46.

Construction of the Toolkit Functions

A set of functions $f(t)$ that has all seven of the features is defined as follows[6].

$$f\left(t;\{\lambda_\beta\}\right) = \mathrm{Re}\left[|C|e^{i\psi}\Theta(Y)Y^z e^{\alpha Y}\right] \qquad (52.1)$$

We have indicated generically the parameters by $\{\lambda_\beta\}$. These parameters should be understood when we write $f(t)$. We now describe the parameters individually.

---

[6] **Practical Functions:** Examples using simpler functions that are subsets of the f(t) toolkit functions have been used on the Street and are in the finance literature.

Time Thresholds

An explicit threshold time $t_{threshold}$ is built in, and Y is the time difference from that time:

$$Y = t - t_{threshold} \qquad (52.2)$$

The usual Heaviside function makes $f(t) = 0$ below $t_{threshold}$,

$$\Theta(Y) = \begin{cases} 0 & (t < t_{threshold}) \\ 1 & (t > t_{threshold}) \end{cases} \qquad (52.3)$$

Power-Law Takeoff

Write the complex parameter z as $z = z_{Re} + iz_{Im}$ with $z_{Re} \geq 0$. If $z_{Re} > 0$, a power-law increase prevents $f(t)$ from taking off suddenly from threshold. The ramp-up time $\tau_{ramp\text{-}up}$ is provided by the real part of Y^z. Setting $|Y^z| = e$, we get

$$\tau_{ramp-up} = \exp(1/z_{Re}) \qquad (52.4)$$

Damping

The parameter $\alpha = \alpha_{Re} + i\alpha_{Im}$ is complex, and the real part $\alpha_{Re} < 0$ provides the characteristic damping of the function in time $\tau_{damping}$. Setting $|e^{\alpha Y}| = 1/e$ we get

$$\tau_{damping} = -1/\alpha_{Re} \qquad (52.5)$$

Oscillations—Standard and Nonstandard

Oscillations ω of a standard nature are provided by the imaginary part $\omega = \alpha_{Im}$ that is conjugate to Y. If we take a half-period as characterizing an oscillation, we have the characteristic half-period oscillation time $\tau_{osc}^{(\alpha_{Im})}$ given by

$$\tau_{osc}^{(\alpha_{Im})} = \pi/\alpha_{Im} \qquad (52.6)$$

A Function Toolkit

Variability in the oscillations can be taken into account with a weighted sum inside a bandwidth, but there is another possibility provided by the factor Y^z using the imaginary part z_{Im}. This variable is conjugate to $\ln(Y)$ instead of Y. If $\alpha_{\text{Im}} = 0$, $f(t)$ oscillates with characteristic time for the first half oscillation as

$$\tau_{osc}^{(z_{\text{Im}})} = \exp(\pi/z_{\text{Im}}) \tag{52.7}$$

Since this expression is nonlinear, the oscillations are not uniform. The second half of the first oscillation takes place over the time $\left[\exp(2\pi/z_{\text{Im}}) - \exp(\pi/z_{\text{Im}})\right]$ that is different from the time for the first half of the first oscillation.

Now if both α_{Im} and z_{Im} effects are present, we can for small z_{Im} find the first half-oscillation time perturbatively. Write $\tau_{osc} = \tau_{osc}^{(\alpha_{\text{Im}})} + \delta\tau_{osc}$ and expand $z_{\text{Im}} \ln \tau_{osc}$ to first order to get $\delta\tau_{osc}$. We get the approximation for τ_{osc} as

$$\tau_{osc} \approx \tau_{osc}^{(\alpha_{\text{Im}})} \left[1 - \frac{z_{\text{Im}} \ln\left(\tau_{osc}^{(\alpha_{\text{Im}})}\right)}{(z_{\text{Im}} + \pi)}\right] \tag{52.8}$$

Magnitude and Phase
Parameters $|C|$ and ψ specify the magnitude and phase of $f(t)$.

Example of a Function Toolkit Application
To illustrate the richness of the phenomena that can be described with the toolkit, here is a graph of a series $x(t)$ built up by some functions $f(t)$:

Example of a Signal Described by the Function Toolkit

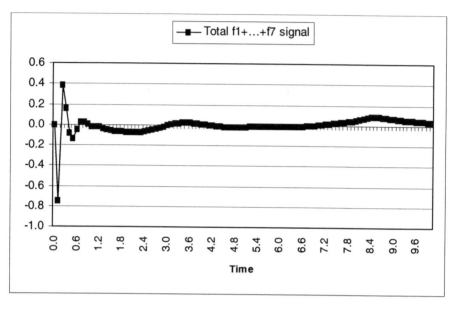

There is at first a sharp drop and rebound, followed by a sequence of some quasi-oscillatory behaviors. Two terms f_1, f_2 are used to produce the drop/rebound:

Functions f_1, f_2 Describing the Drop and Rebound

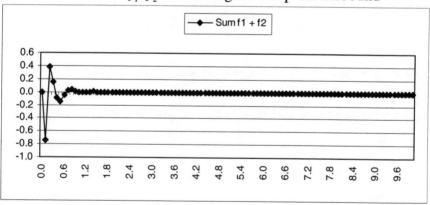

Then the rest of the seven $f(t)$ functions $f_3...f_7$ describe the rest of the series:

A Function Toolkit

Function $f_3...f_7$ past the Initial Drop/Rebound

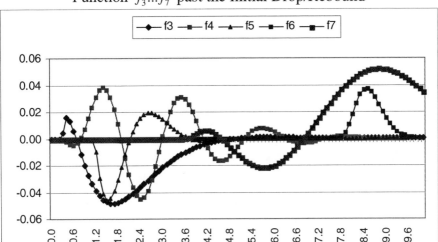

All seven features described above can be seen in these functions, including the time thresholds, ramp-up, damping, oscillations (sometimes with anomalous features) and phase. Their appearances are qualitatively different. The parameters for these functions follow[7]:

Parameters for the Signal Components

Term	t_{Thresh}	α_{Re}	α_{Im}	$|C|$	ψ	z_{Re}	z_{Im}
1	0.022	-7.586	8.783	4.456	3.159	0.469	0.070
2	0.160	-11.038	8.599	1.041	0.276	0.000	-0.020
3	0.266	-1.055	0.505	0.199	2.611	0.897	0.400
4	0.201	-1.113	2.962	0.102	2.975	2.023	-0.020
5	1.215	-2.046	1.934	0.269	2.640	0.992	0.000
6	7.090	-7.222	1.102	60.662	-2.101	7.465	0.000
7	4.390	-0.719	-1.066	0.054	2.748	2.828	2.133

Laplace Transform of a Function $f(t)$ in the Toolkit

The Laplace transform of any function $f(t)$ is defined as usual as

---

[7] **What is this Example?** This is the seven-term fit of the response to an electric field in ref iii with somewhat different parameters. This example exhibits many of the characteristics that we have seen in macro financial time series.

$$L(J;\{\lambda_\beta\}) = \int_0^\infty e^{-Jt} f(t;\{\lambda_\beta\}) dt \qquad (52.9)$$

The inverse is

$$f(t;\{\lambda_\beta\}) = \int_{c-i\infty}^{c+i\infty} e^{Jt} L(J;\{\lambda_\beta\}) \frac{dJ}{2\pi i} \qquad (52.10)$$

Again the parameters λ_β are those described above. In Eqn. (52.10), c is taken to the right of all singularities of $L(J;\{\lambda_\beta\})$ in the complex J-plane. The connection with the usual Fourier transform variable is $J = -i\omega$.

Since we have thresholds in time, they must be reproducible by the Laplace transform. This is the case; for $Y < 0$, we close the contour in the right-half J-plane, producing zero for $f(t)$ below $t_{threshold}$ as desired. For reference, the Laplace transform of $f(t)$ is:

$$L(-i\omega;\{\lambda_\beta\}) = \frac{1}{2}|A|e^{i\omega t_{threshold}}\left[\frac{e^{i\phi}}{(-i\omega-\alpha)^{z+1}} + \frac{e^{-i\phi}}{(-i\omega-\alpha*)^{z*+1}}\right] \qquad (52.11)$$

Here,

$$|A|e^{i\phi} = |C|e^{i\psi}\Gamma(z+1) \qquad (52.12)$$

If z is an integer, there are multiple poles in Eqn. (52.11). If z is not an integer there are branch cuts; these are treated by standard techniques. Also $\Gamma(z+1)$ is the usual gamma function; if $z = n$ is integral, $\Gamma(z+1) = n!$.

Relation of the Function Toolkit to Other Approaches

We now compare the function toolkit approach for describing part of the macro component of time series to several other approaches for time series analysis:

- Fourier Series [v]
- Prony Analysis [vi]
- Wavelets [vii]
- ARIMA Models [viii]

Function Toolkit Relation to Fourier Transforms

A term in a Fourier series is a sine wave, and therefore is a special case of $f(t)$ with no thresholds, ramp-up, or damping. Although a general method, Fourier series have two drawbacks. The first is that a sine wave does not look much like any signal in practice. Second, to describe real signals even approximately, a large number of terms are needed. Time thresholds are problematic to describe because sine waves start infinitely far in the past and go on forever.

Function Toolkit Relation to Prony Analysis

Prony analysis[8] was invented in 1795. Prony analysis forms a special case of the analysis described here, lacking the time thresholds[9]. The variable for Prony analysis is damping with α_{Re}, with $z = 0$. Prony analysis is applicable for radar, where an object has apparent dimensions small compared to the distance to the object. Then, a returning signal from the object arrives back at essentially at a single time t_0. An infinite number of Prony terms starting at t_0 are needed to reproduce a time threshold past t_0.

Function Toolkit Relation to Wavelets

Wavelets have produced an explosion in numerical analysis[10]. The spirit of wavelets is very close to ours; indeed our $f(t)$ functions can be considered in some sense as wavelets with more complex parameterizations than usual.

Perhaps the main advance for wavelets is that expansions can now be carried out in systematic fashion. Wavelets can have different time thresholds with a limited time extent, so they look more like signals than Fourier sine waves.

Still, from our perspective a disadvantage of the wavelets used in practice is that an individual wavelet is chosen to have a simple shape with only a few parameters, and so still does not look too much like a real signal. Therefore, a

---

[8] **Acknowledgements:** I thank Santa Federico for bringing Prony analysis to my attention. We had some good times at Bell Labs applying Prony analysis and generalizations along the lines of the functions described here.

[9] **Who was Prony?** The Baron de Prony (1755 - 1839) was a French engineer and a contemporary of the famous mathematician Fourier (1768 - 1830). It is interesting that Prony, besides inventing Prony analysis, also collaborated in measuring the speed of sound, supervised the construction of logarithmic tables to 19 decimals, invented a type of brake, and modernized several ports of Europe.

[10] **Acknowledgement:** I thank Alex Grossmann for informative discussions on the mathematical properties of wavelets, and also for lots of fun playing music.

large number of wavelets are needed (though fewer than for Fourier series) to describe a time series to some accuracy.

Again our philosophy here is to obtain a reasonable description in terms of only a very few $f(t)$ functions that individually generically resemble real-world signals to begin with. The functions $f(t)$ have the properties of both "time-scale" wavelets and "time-frequency" wavelets.

The Appendix discusses wavelets a little further.

Function Toolkit Relation to ARIMA(p,d,q) Models

ARIMA(p,d,q) models[11,12] describe a time series in terms of itself and input Gaussian noises. ARIMA models are complementary to the method we are proposing. We are proposing to describe at least part of the macro long-time-scale behavior of time series using the function toolkit. This does not include noise. Noise is supposed to be assigned to the "micro" component, appropriate for the description of short time scales, and ARIMA models can play an important role in describing micro component. ARIMA models include memory effects. Memory effects have been investigated, as described in Ch. 43 (Section V and App. B) [ix].

Example of Standard Micro "Noise" Plus Macro "Signal"

We now illustrate the incorporation of micro noise along with the macro toolkit functions. So far, the series $x(t)$ has been deterministic, built up by the "signal" $x^{Signal}(t)$ of the toolkit $f(t)$ functions. Based on empirical studies, the appropriate micro noise $x^{Noise}(t)$ seems to be highly mean reverting, and we use this property. The total is $x(t) = x^{Signal}(t) + x^{Noise}(t)$. The noise changes are given by $d_t x^{Noise}(t) = -\omega x^{Noise}(t) dt + \sigma \sqrt{dt} N(0,1)$ for the time evolution with

---

[11] **ARIMA(p,d,q) Models – Summary Description:** Several stages are combined to get an ARIMA(p,d,q) model. The first is a moving average MA(q) and the second is an autoregressive AR(p) that together form an ARMA(p,q) description. The MA(q) model assumes that each x(t) is generated by a weighted average of q independent Gaussian noise terms G(t – jdt) with j = 1,...,q. The AR(p) model assumes that x(t) can be written as a weighted average of p prior x(t – kdt) values with k = 1,...,p. If the series x(t) needs to be time-differenced d times in order to be described by an ARMA(p,q) process, then x(t) is itself an integrated ARMA(p,q) process. This is called an ARIMA(p,d,q) process.

[12] **Joke:** Even though ARIMA models are described by the parameters *(p,d,q)* there is no evidence that Prof. Peter Schickele had any influence on the topic. See P.D.Q. Bach's opera video, *The Abduction of Figaro*, 1984. Get it?

mean reversion ω. An illustrative $x(t)$ path with the noise specified by $\omega dt = 0.5$ and $\sigma\sqrt{dt} = 0.03$ is plotted below:

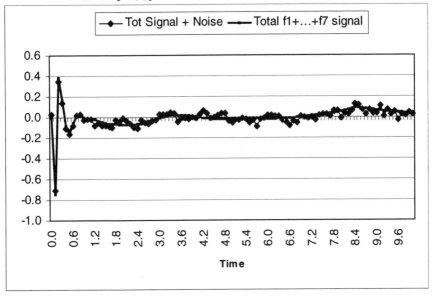

The sharp drop and rebound produced by f_1, f_2 are not affected by the noise. Past this point, the functions $f_3 ... f_7$ take over. The noise, being mean reverting, bounces around the signal function sum.

Connected 3-Point Green Function / Auto-Correlation Analysis

In Ch. 49, we emphasized the importance of the third order time-dependent Green functions or correlation functions. These were employed and provided a sharp test in the analyses of interest-rate data that led to the Macro-Micro model in 1989. We anticipate that this type of analysis will continue to be a useful probe in obtaining workable models in the present context. For those who skipped Ch. 49, we recall the formalism. Define $\tau_k = k\Delta t$ and $x^{(k)}(t) = x(t - \tau_k)$ with k lags of time step Δt. We work here with time-differenced series $d_t x(t) \equiv x(t+dt) - x(t)$. Define $M_1^{(k)} = M(\tau_k)$ as the time-average of $d_t x^{(k)}(t)$ and call $C_2^{(ij)} = C_2(\tau_i, \tau_j)$ the $\left[d_t x^{(k_i)}(t), d_t x^{(k_j)}(t)\right]$ covariance function. Then define $M_3^{(123)}$ (for N observation points) by

$$M_3^{(123)} = \frac{1}{N}\sum_{l=1}^{N} d_t x^{(k_1)}(t_l) d_t x^{(k_2)}(t_l) d_t x^{(k_3)}(t_l)$$. The connected 3-point function $C_3^{(123)} = C_3(\tau_1, \tau_2, \tau_3)$ is

$$C_3^{(123)} = M_3^{(123)} - \sum_{CyclicPerm(ijk)} C_2^{(ij)} M_1^{(k)} - \prod_{k=1}^{3} M_1^{(k)} \quad (52.13)$$

We use the exponentiated version of the formalism just described. For illustration, we calculate with lags (actually forward moves) of (0,0), (0,10), and (4,10) measured in units of $\Delta t = 1$. The results for the signal (taken as the sum of the functions f_3 to f_7) and for the noise are plotted below[13]:

Plots of $C_3^{(123)} / M_3^{(123)}$ for the Sample Path

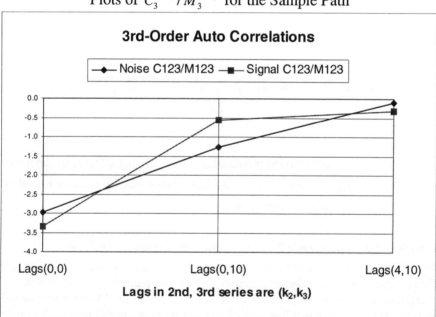

We see that for this illustrative example, both the signal and the noise satisfy $C_3^{(123)} / M_3^{(123)} \approx 0$ as the lags become nonzero. Again, for a real application we would use the constraint $C_3^{(123)} / M_3^{(123)} \approx 0$ as a probe to indicate a reasonable

---

[13] **Parameters:** The signal had n = 100 points and the noise had n = 543 points. Also, Δt was taken as one unit (arbitrary). The results for the noise can be calculated analytically, but we present the Monte Carlo results as illustrative of what happens in practice.

A Function Toolkit

model and a result like $C_3^{(123)} \approx M_3^{(123)}$ to indicate a failure. Thus, for example, if we have data $r_{data}(t)$ and we believe that a lognormal noise model is relevant, we would write the putative stochastic equation involving a "data-defined" Wiener noise term $dz_{data}(t)$ as follows:

$$\frac{d_t r_{data}(t)}{r_{data}(t)} = \mu(t)dt + \sigma_{LN} dz_{data}(t) \qquad (52.14)$$

We solve for $dz_{data}(t)$ as

$$dz_{data}(t) = \left[\frac{d_t r_{data}(t)}{r_{data}(t)} - \mu(t)dt\right]/\sigma_{LN} \qquad (52.15)$$

We then check to see if $C_3^{(123)} / M_3^{(123)} \approx 0$ for $dz_{data}(t)$ or for its exponentiated form $\exp(dz_{data}(t))$. If so, the model passes the test. If not, we need to change the model.

The Total Macro: Quasi-Random Trends + Toolkit Cycles

We have presented the macro-micro model with its different time scales using two complementary models for the macro component. The first, introduced in 1989 and described in Ch. 50, was a quasi-random trend or drift model, involving random slopes over random time intervals, with a cutoff lower bound time. The second, introduced here, is a function toolkit basis for cycles of general shape. Economists use both trends and cycles[14], and we will generically follow the idea in this section[x].

The most general macro model would be a combination. Calling QRT = quasi-random trends and TC = toolkit cycles for short,

$$f_{Macro\_Total}\left(t;\{\lambda_\beta\}\right) = f_{QRT}\left(t;\{\lambda_\beta^{(QRT)}\}\right) + f_{TC}\left(t;\{\lambda_\beta^{(TC)}\}\right) \qquad (52.16)$$

---

[14] **Random Trends and Stationary Cycles?** We differ from those economists who assume that random trends are stochastic on a monthly basis and cycles are stationary statistical quantities. For us, trends are *quasi-random* with explicit time scales, as described in a previous chapter. In addition, cycles here are *not* stationary, but are constructed from the function toolkit.

In the absence of a theory of macroeconomics connecting with our formalism, we would perform numerical phenomenology to fit parameters with historical data.

Forward pricing with the macro-micro model would use implied macro parameters that would specify the future macro paths. Similarly to the quasi-random drift macro model, the cyclic-toolkit macro component would be added.

To illustrate, take the lognormal equation for an interest rate $r(t)$ with $x = \ln(r)$ and a label "Macro" placed on the drift μ_{Macro}, viz

$$d_t x(t) = \mu_{Macro}(t)dt + \sigma_{LN} dz(t) \tag{52.17}$$

We want to use the quasi-random drifts and toolkit cycle functions $f\left(t;\{\lambda_\beta\}\right)$ in the drift μ_{Macro}, and we have purposely exhibited the parametric dependence $\{\lambda_\beta\}$. So we write

$$\mu_{Macro}(t) = \sum_{\{\lambda_\beta\} \subset Set_\mu} \mu_{Macro}\left(t;\{\lambda_\beta\}\right) \tag{52.18}$$

We have already described the probability function for the macro quasi-random-trend component.

For the toolkit cycle component, we have less intuition. For example, we could use a coherent-state prescription involving a Gaussian in the variable ω around a central frequency ω_0 and width σ_ω:

$$\wp[\omega] = \frac{1}{\sqrt{2\pi\sigma_\omega^2}} \exp\left[\frac{-1}{2\sigma_\omega^2}(\omega - \omega_0)^2\right] \tag{52.19}$$

The parameters ω_0 and σ_ω would be implied from long-term option data or taken from historical analysis. Other parameters would be similarly defined. We would want to respect the use of macro time scales above a cutoff τ_{Cutoff} as discussed in the quasi-random trend Macro component.

Short-Time Micro Regime, Trading, and the Function Toolkit

Although the main purpose of this part of the book is the description of long-time-scale macro behavior, we digress briefly to discuss short time-scale

A Function Toolkit 769

extrapolations in the micro time regime where most trading occurs. We will only discuss a few issues qualitatively[15].

Short-term micro dynamics are different from the long-term macro dynamics. In particular, gap jump behavior at short time scales is extremely important. Intermittent trading at some time scale occurs for all products[16], producing gaps between times when trading does occur. Sometimes these jumps are large. Psychology and quick response to news is a driver at short time scales [xi,17].

The efficacy of techniques depends on the quality of real-time data feeds, as well as historical data for backtesting.

A description of various trading applications is in the references[xii].

Trading and the $f(t)$ Function Toolkit

The main point of this short section is to suggest that the function toolkit may be useful in trading applications. The idea here would be to do real-time fast fits to recent data from feeds using functions from the toolkit. These fits would be projected into the future for a short time.

Algorithms using simple functions have been successful[18], so we think it is likely that the more general $f(t)$ functions could provide additional benefit[19].

Appendix: Wavelets, Completeness, and the Function Toolkit

Examples of Wavelet Variables

The set of wavelet variables can, for example, be time translation and damping (dilation). These are "time-scale" wavelets. The wavelets suggested by

---

[15] **Proprietary Short Term Methods:** Most short-term extrapolation methods used in trading are proprietary, so only generalities can be given here.

[16] **Intermittent Trading:** Trades only occur at finite time intervals, small for liquid products and long for illiquid products. The description of prices between times when trading occurs is just an assumption that cannot be tested since those prices don't exist.

[17] **Vol "News" Models:** Some short-term volatility models try to estimate the differential impacts of specific future news events, e.g. an FOMC meeting.

[18] **Analyses Using f(t) Prescriptions:** Explicit fits to historical data have been used involving a few sine waves along with a trend, as input to some trading strategies. We consider this as an existence proof that the f(t) function toolkit, which contains these fits as a subset, will be useful.

[19] **Crash Oscillations:** An excellent example of damped oscillations occurred in day trading for a few days after the 1987 stock market crash. The market "rang like a bell".

Grossmann and Morlet (Ref. vii) were time-translated Gaussians with different widths. Physically this has the attractive picture that a bump in the signal at some time t is approximated by a Gaussian centered at t with an appropriate width. Another possibility is the "time-frequency" wavelets first introduced by Gabor and related to Gaussian coherent states. Here variables are chosen as frequency and time translation with time decay. This is analogous to a note with some frequency or pitch ω - say middle C - in the kth measure of a piece of music, lasting one beat. The time-frequency uncertainty relation is not violated since the note will have a frequency uncertainty (think of vibrato). Enough of these functions are used to cover the entire time-frequency plane, although the choice of exactly how the covering is made is not unique.

Completeness and a Plea to the Mathematicians

One major issue is to find a series expansion of a suitable given function in terms of a complete set of wavelets. Completeness means that a unique set of coefficients can be determined for the expansion of a function in terms of a set of orthonormal basis functions. Orthonormality means that two basis functions when multiplied together and suitably integrated give zero if they are different and one if they are the same. For certain appropriate sets of wavelets, completeness theorems hold.

Our $f(t)$ toolkit functions have more parameters than the usual wavelets. This makes the expansion theoretical analysis more difficult, and no completeness theorems have been proven. The set $\{f(t)\}$ for all possible parameters is actually "over-complete". That means that there are "too many possible $f(t)$ functions". We do not want to just eliminate some parameters, because they are all useful, as we have seen. We do not know how to restrict the parameter values to get a complete set.

We avoid this theoretical topic in practice by simply choosing a small number of functions and performing a least-squares fit to a time series to fix the parameters of these few functions. For this brute-force approach, we essentially do not care if there are potentially "too many" choices of such functions for all possible values of the parameters.

Indeed, the attractiveness of the approach is supposed to be that because the toolkit functions start out looking like real signals, only a small number of functions is needed for a good approximation.

Still, it would be nice (*this is a plea*) if some of the powerful wavelet mathematicians would figure out how to specify a subset of these $f(t)$ functions that is just complete.

References

[i] **Cyclical Analysis, Secular Trends, and the Macro Component of Finance**
Herbst, A. F., *Commodity Futures – Markets, Methods of Analysis, and Management of Risk*. See pp. 127-130. John Wiley & Sons, 1986.
Kaufman, H., *Interest Rates, the Markets, and the New Financial World.*, Ch. 11. Times Books, 1986.
Hildebrand, G., *Business Cycle Indicators and Measures*. Probus Publishing, 1992.
Makridakis, S. and Wheelwright, S. C., *Forecasting – Methods and Applications*. See Ch. 4 and 12. John Wiley & Sons, 1978.
The Conference Board, *The Cyclical Indicator Approach*. Wall Street Journal Online, 9/23/02. Online.wsj.com/documents/bblead.htm
Crossen, C., *Wave Theory Wins Robert Prechter Title of Wall Street Guru*. Wall Street Journal, 3/18/97.

[ii] **Function Toolkit Application in High-Energy Physics**
Dash, J. W. and Jones, S. T., *Flavoring, RFT and $ln^2 s$ Physics at the SPS Collider*. Physics Letters, Vol 157B, p.229, 1985.

[iii] **Function Toolkit Application in Engineering**
Dash, J. W., *Signal Analysis with Time-Delay Onset-Arrival Thresholds*, IEEE Transactions on Electromagnetic Compatibility, Vol. EMC-28, p. 61, 1986.

[iv] **Macro Quasi-Random Trends + Micro Noise Model for Interest Rates**
Dash, J. W. and Beilis, A., *A Macro-Micro Yield-Curve Simulator*. CNRS, Centre de Physique Théorique CPT-89/P.E.2336, 1989.

[v] **Fourier Series and Transforms**
Oppenheim, A. V. and Schafer, R. W., *Digital Signal Processing*. Prentice-Hall, 1975.

[vi] **Prony Analysis**
Prony, R. (Baron de), *Essai expérimental et analytique sur les lois de la dilatabilité de fluides élastiques et sur celle de la force expansive de la vapeur d'alcool, à differentes températures*. Journal de l'Ecole Polytechnique, Paris vol.1, p24, 1795.
Prony, R. Fr.encyclopedia.yahoo.com, *Gaspard Marie Riche, baron de Prony*.
Van Blaricum, M. L., *Historical Summary of Resonance Extraction in General and Prony's Method in Particular*. Advanced Technologies Division, General Research Corp. paper, 1986.

[vii] **Wavelets**
Grossmann, A. and Morlet, J., *Decomposition of Hardy functions into square integrable wavelets of constant shape*, SIAM J. Math. Anal, vol. 15, p. 723, 1984.
Aslaksen, E. W. and Klauder, J.R., *Continuous representation theory using the affine group*. J. Math. Phys. Vol. 10, p. 2269. 1969.
Gabor, D., *Theory of Communications*. J. IEE, 93, p. 429, 1946.
Meyer, Y. (Translation Ryan, R.), *Wavelets – Algorithms & Applications*, SIAM 1993.
Daubechies, I., *Ten Lectures on Wavelets*, SIAM 1992.
Emch, G. G., Hegerfeldt, G. C. and Streit, L. (Editors), *On Klauder's Path*. World Scientific, 1994.

viii ARIMA Models
Pindyck, R. S. and Rubinfeld, D. L., *Economic Models and Economic Forecasts (4th Ed)*. McGraw-Hill, 1998.

Spanos, A., *Statistical foundations of econometric modelling*. Cambridge University Press, 1987.

ix Memory in Financial Stochastic Processes
Dash, J. W., *Path Integrals and Options II: One-Factor Term-Structure Models*, Centre de Physique Théorique, CNRS, Marseille CPT-89/PE.2333, 1989. See Section V.

x Economic Trends and Cycles
Makridakis, S. and Wheelwright, S., *Forecasting: Methods and Applications*. John Wiley & Sons, 1978. See Chapter 4.

xi Psychology and the Markets
Kahneman, D. and Tversky, A., *Choices, Values, and Frames*. Cambridge University Press, 2000.

xii Trading References
Nicolar, J. G., *Market-Neutral Investing: Long/Short Hedge Fund Strategies*. Bloomberg Professional Library, 2000.

Jaeger, L., *Managing Risk in Alternative Investment Strategies: Successful Investing in Hedge Funds and Managed Futures*. Pearson Education Ltd., 2002.

Index

30/360 day count 82
Abarbanel, H. 635, 665
Accrued Interest 119
Acknowledgementsxix, 27, 71, 85, 142, 164,
　　176, 201, 210, 232, 265, 311,
　　323, 325, 343, 393, 408, 421,
　　437, 441, 453, 457, 461, 471,
　　481, 484, 485, 489, 505, 559,
　　577, 587, 598, 603, 606, 632,
　　642, 649, 651, 657, 658, 717,
　　720, 736, 746, 749, 763
Advance Notice 148
Affine Models 561
Aged Inventory 391
Aggregation 387
Algorithms 424
Allocation and CVARs 482
Allocation and Standalone Risk 482
American Options 30, 537
American Swaption 141
Andersen, L. 69, 561
Andreasen, J. 69, 561
Angle Volatilities 315
Angle-Angle Correlations 315
Arbs 164
ARIMA Models 577, 764
ARM Caps 231
Atlantic Path 531, 547
Average Macro Drift 723
Averaged Diffusion Equation 740
Averaged Volatility 277
Average-Rate Caplet 275
Average-Rate Digital 275
Average-Target Correlation 309
Averaging 162, 175
Avoid Double Counting 407
Azimuthal Angles 297
Back of the Envelope 197
Back Office 272
Back-Chaining 165, 175, 612
Backflip Option 23, 494
Backtesting 347

Backward Diffusion 529
Baker, M. 635
Bali, N. 287
Barrier Options 56, 243, 248, 265
Barrier Potential Theory 253
Barrier Step 254
Basis Point 20, 74
Basis Risk 148, 224
Basis Swaps 91
Basket Options 31
Baskets 325, 338, 625
BDT Model 560
Beauty Contest 146
Beilis, A. ....xix, 421, 649, 681
Berger, E. 278
Bermuda Options 530
Bertonazzi, E. 735
Best-of Option 183
Bhansali, V. 210
Bid Side Swap 96
Binomial Approximation 598, 600
Bins 602
Bivariate Integral 173, 269
Bivariate Model 335
Black Formula 125
Black Hole Data Lemma 290
Black Option Model 223
Black, F. 560
Black-Scholes Formula 248
Black-Scholes Model 222, 509
Blomberg, S. 736
Bloomberg L. P. 559
Bond Conventions 119
Bond Futures 96
Bond Issuance 115
Bond Math 118
Bond Trading 116
Bonds 111
Bonds with Puts and Calls 613
Bookstaber, R. 305
Born Approximation 546
Bossaerts, P. 618

Boyle, P. 611
Brace, A. 561
Break-Even Rate 101
Broadie, M. 611
Brock, W. 632
Bronzan, J. 635
Bullet Bonds 106, 612
Bushnell, D. 489
Business Trips 598
Buyback 180
C 439
C++ 128, 229, 438
Calculators 431
Call Filtering 618
Callable Bonds 112
Callable DEC 185
Calling a Bond 144
Caplet 148
Caps 123
Carr, P. 60
Carry 237
Cash Rates 74, 79
Castresana, J.... xix, 176, 320, 421, 508, 597, 603
CDA 685, 697, 733
Centre de Physique Théorique xix
Chameleon Bond 163
Chan, L. P. 421
Changeover Point 79
Chaos 632
Chaos and Macro-Micro Model 747
Charm 40
Chin, M. 737
Cholesky Decomposition 303, 325, 327, 346
CIFEr 3
CIR Model 561
Citibank 4, 85
Citigroup .. xix, 4, 311, 323, 393, 408, 453, 457, 720
Classical Path 560, 591, 696
Clients 201, 210
Cliquets 226
Close of Business 73
Cluster-Decomposition Analysis 623, 682, 685, 694, 697
CMT Caps 135
CMT Rates 132
CMT Swaps 135
CNRS .... xix, 505
Code Comments 22, 423
Color 40

Color Movies 13, 365
Color Printer 157
Commodity Options 214
Communication Issues .. 16, 21, 23, 157, 433
Completeness 770
Compounding Rates 82
Computing Nodes 444
Concentration 402
Confidence Levels.. 286, 289, 346, 350, 401, 480
Consistency Check 198
Constan, A. 489
Contingent Caps 201
Continuum Limit 543, 566
Conventions 43, 96
Convergence 259, 427
Convergence Trades 117, 308
Convertible Issuance 239
Convertibles 113, 186
Convexity 119, 343, 353
Corporate VAR 461
Corporate-Level VAR 387
Correlation Degeneracy 457
Correlation Dependence on Volatility 335
Correlation Dynamics 329
Correlation Instability 330, 453
Correlation Signs 50
Correlation Time Intervals 333
Correlations 50, 220, 297
Correlations and # Data Points 456
Correlations and Data 453
Correlations and Jumps 455
Correlations and the N- Sphere 304
Correlations for Baskets 338
Correlations in High Dimensions 301
Correlations Least-Squares Fitting 321
Cost Cutting 487
Counterparty Credit Risk 107
Counterparty Default 107
Counterparty Risk 394
Coupons 585
Cox, J. C. 561
Creativity 15
Credit Correlations 402
Credit Rating Criteria 475
Credit Ratings 111, 393
Credit Risk 12
Credit Spreads 186
Critical American Path 612
Critical Exponents 631
Cross-Currency Swaps 92

Index

Cross-Market Knock-Out Cap.................212
Crystal Ball 64
Currency Pressure Index 746
Curve Smoothing 80
Cushions 167
Cutoffs 356
CVAR Uncertainty Analogy 348
CVAR Volatility 341, 369, 376, 381, 385
CVARs 341, 389
CVR Option 161
Cycles 647
Daily Trading Volume 240
Damping 758
Dangerous Pronouns 17
Dash, J. W. 635
Data 342, 451
Data Completeness............................450
Data Consistency.............................449
Data Groups 451
Data Preparation.............................451
Data Reliability 450
Data Sentences 449
Data Topics 11
Data Vendors..............................450
Davidson, A. .... xix, 60, 505, 546, 559
Day-Count Conventions...........................74
DEC 183
Delta 39, 142
Delta Ladders.............................97, 143
Delta Strangles 43
Derman, E. 63, 65, 68, 560, 745
Desk-Level Model QA............................421
Difference Option 224
Diffusion and Drunks...........................46
Diffusion Equation.................................582
Digital Options..................205, 261
Dirac Delta Function 273, 371, 462, 518, 542, 580, 627
Disclaimer 167, 342, 423
Discount Factor..........................118
Discounting Rate...........................186
Discrete Barrier 252
Diversification......................... 117, 330, 351
Dividends 523, 524, 527, 530
DKO option..............................260
Documentation..............................423
Double Barrier Option.....................257, 260
Driscoll, M. 441
Duffie, D. 561
Duration 119
DV01 96

Economic Capital............................358, 475
Economic Capital Analogy 476
Economic Capital Cost....................483
Economic-Capital Utility Function 484
Economics and Macro-Micro Model.....646, 731
Exercise 19
Effective Number of SD.................376
Eighty-Twenty Rule 15
Elliptic Functions 259
Enhanced/Stressed VAR 341, 357
Epperlein, E.............................325
Epstein, S. 658
Ergodic Statement 480
Ergodic Trick 349
Errors 197
Estimated Correlation Probability 465
ES-VAR 341, 357
Eurobrokers.............................4, 85, 598
European Swaptions.....................137, 148
Excel 19, 24, 285
Exchangeables...........................184
Expected Losses....................344, 485, 486
Exponential Interpolation.......................623
Exposure 343
Exposure Basket..........................494
Exposure Changes............................403
Exposure Double-Up............................364
Exposure Ending Level 363
Exposure Fractions...........................491
Exposure Reduction 361, 364
Exposure Scenarios..........................496
Exposure Starting Level 362
Extension 162
Extension Option..................198
Fama, E. F 690
FAS 123 215
Fat Tails 284, 690, 717, 746
Fat-Tail Example............................284
Fat-Tail Gaussians............................288
Fat-Tail Volatility........... 286, 287, 358, 377
Favorite Options Model 415
FCMC 85
Fed and the Macro Component 718, 735
Federico, S. ....xix, 484, 505, 763
Feedback 424
Feiger, G. 736
Feynman, R. 501, 508, 571, 597
Fibers 297
Fifty-Fifty Votes..........................684

776   *Quantitative Finance and Risk Management*

Figures
    CDA Test 701
    Correlations 667, 702
    Eigenfunctions 667, 702
    Eigenvalues 667, 702
    Eleven Maturities 667
    Fat Tails 702
    Libor (Macro + Micro) 727
    Macro Paths 726
    Micro Paths 727
    Prime Rate (Macro) 727
    Quasi-Equilibrium Yield Curve 667, 702, 726
    Shift Means 667, 702
    SMRG YC Simulation 702
    Statistical Measures 667, 702
    Yield Volatilities 667, 702
    YC Principal Components 667, 702
    YC Principal Components SMRG Model 702
    Yield Fluctuations 702
    Yield-Curve Data 667, 701
    Yield-Curve SMRG Model 701
First Time Interval 606
Fisher's Transform 459, 461
Fixed Rate 111
Flesaker, B. 561
Flight to Quality 117, 330
Floating Rate 112
Floors 127
Footnotes 4, 19
Forest, B. 559
Fortran 90 438
Forward - Future Correction 78, 572
Forward Bond 585
Forward Rates 77, 143
Forward Stock Price 742
Forward Swap 148
Fourier Transform 273, 371, 763
Fox, B. 201
FRAs 94
FRN 106
Fudge Factor for Systems 432
Fuller, J. 393, 481
Function Toolkit 647
Function Toolkit and Trading 648
Function Toolkit Example 759
Function Toolkit in Engineering 755
Function Toolkit in Physics 755
Fundamental Theorem of Data 19, 449
Fundamental Theorem of Systems 432

Fundamental Theory of Finance? 637
Futures 19, 74, 75, 78, 93
FX Forwards 35, 36
FX Options 35
FX Triangles 298
FX Volatility Skew 41
Gabor, D. 770
Gadgets 65
Gamma 40
Gamma for a Swap 97
Gamma Ladders 104
Gamma Matrix 104
Gantt and Pert Charts 434
Gaps 417
Garbade, K. 657
Garman, M. 196
Gatarek, D. 561
Gaussian 283
Gaussian Model With Memory 589
Gell-Mann, M. 508, 634
General Model With Memory 574
Generalized Kurtosis 733
Generating Function 371
Geometrical Volatility 620
Geske, R. 506, 598
Graphs 22
Girsanov's Theorem and Path Integrals . 509, 538, 540
Gladd, T. 58, 421, 471
Glasserman, P 611
Goat Language 21
Goldberg, G. 21
Graham, T. 559
Greeks 9, 39, 90
Green Function 189, 274, 562, 570, 627, 631
Green Function with Quasi-Random Drift 737
Grossmann, A 763, 770
Haan, D. 606, 656
Heath, D. 365, 561
Hedges 9, 152
Hedging . 27, 39, 43, 93, 96, 98, 128, 129, 130, 140, 176, 207, 223, 238, 308, 339, 544, 741
Heisenbugs 439
Heretical Remark 637
Herman, G. 559, 656, 657
Herskovitz, M. 505
Heston, S. 60
High-Performance Fortran 445
Hill, J. 441, 457

Index

Hillion, P. 618
Hiscano, D. 164
Historical Data 451
Historical Quasi-Equilibrium YC Path 688
Historical Scenarios 10, 478
Historical VAR 343
History   3, 4, 47, 51, 56, 60, 85, 105, 136,
            180, 201, 209, 211, 214, 219,
            236, 243, 252, 253, 257, 265,
            271, 287, 295, 320, 325, 341,
            369, 393, 421, 441, 461, 475,
            489, 505, 541, 546, 559, 597,
            625, 629, 631, 649, 657, 681,
            717
HJM 77
Hogan, M. 508, 597
Ho-Lee Model 560
Horizon Time 392
Hughston, L. 561
Hull, J. 60, 559
Hull-White Model 559
Humphreys, D. 343
Hurst Exponents 631
Hybrid 2-D Barrier Options 211
Hybrid Lattice Monte-Carlo 611
Hybrid Options 265
Hyperbolic Geometry 297
Iacono, F. 410
Idiosyncratic Risks 121, 342, 359
Illiquidity 130, 391
Illiquidity Penalty 360
Image Green function 259
Images 245, 258
IMM Dates 19
IMM Swaps 93
Implied Volatilities 126, 130
Improved Plain-Vanilla VAR 341, 353
Independent Models 422
Index-Amortizing Swaps 232
Indicative Pricing 192
Ingersoll, J. E. 561
Integrated VAR Measures 366
Interest-Rate Risk Ladders 617
Intuition 51, 193, 221
IPV-VAR 341, 353
ISDA 95
It (The Most Dangerous Word) ... 17, 22, 432
Ito prescription 512
Ito's Lemma 545
Jacquillat, B. 736
Jamshidian, F. 138, 344, 559, 587

Jargon 75, 96, 213
Jarnagin, R. 559
Jarrow, R. 561, 618
Jian, K. 325
Job Interview Question 154
Johnson, C. 71
Johnson, H. 506, 598
Jumps 293, 417, 523
Kaminsky, G. 746
KAOS Exponents 638
Kavalov, A. 580, 627, 632
Kink Definition 662
Kinks 650, 658, 667, 697
Knock-In Options 249
Knock-Out Options 245
Lagrangian 633
Langsam, J. 611
Laplace Transform 761
Lattice Approximations 608
Least Squares 322
Lee, R. xix
Legal 180
Legal Correlation Matrix 319
Levy, E. 278
Libor 19, 74, 128
Libor Swaps 75
Limit Exceptions 491
Linear Parallelization 445
Linetsky, V. 254
Liquidation Penalty 392
Liquidation Time 392
Liquidity 148, 165, 355, 403
Local Volatility 62
Lognorm model 149
Lognormal Multifactor Model 642
Longstaff, F. 611
Lost Opportunity Spread 483
Lotus 439
LTCM 366
Macro Component 417, 719
Macro Simulator 721
Macro Time Step Dynamics 722
Macroeconomics 719, 727, 746
Macro-Micro Correlation Model 331
Macro-Micro Model . 69, 560, 642, 644, 683,
            721, 724, 733
Macro-Micro Model and Derman's Equity
            Regimes 745
Macro-Micro Model and No Arbitrage ... 737
Macro-Micro Model for FX 733

Macro-Micro Model Formal Developments 737
Magic Dust 326
Maloney, M. 735
Managers .... xix, 21
Mandel, R. 142
Mandelbrot, B. 632
Mandelstam, S. 635
Margin Account 78
Marinovich, N. 410
Marker, J. 393, 481
Market Input Rates 73
Market Models 561
Market Risk 11, 406
Market Risk Managers 489
Marshall, D. 598
Mass Renormalization 546
Matrix Pricing 121
MaxVol 294
Mean Reversion 292, 417
Mean-Reverting Gaussian Model 276, 569
Merrill Lynch 4, 441, 505, 559, 606, 651
Methodology 424
Micro Component 720
Micro Fluctuations 727
Milevsky, M. 278
Model Assumptions 424
Model for Unused Limit Risk 491
Model Limitations 417
Model QA Documentation (Quants) 425
Model QA Documentation (Systems) 425
Model QA Documentation (Users) 425
Model Quality Assurance 419, 421
Model Reserves 418
Model Risk 81, 415, 416, 418
Model Testing Environment 424
Model Topics 11
Model Validation 419, 421
Models and Aristotle 149
Models and Parameters 419
Models and Psychological Attitudes 418
Models and Reporting 124
Modified Black-Scholes Formula 744
Modified No Arbitrage Condition 743
Monet, C. 410
Moneyness 68
Monte Carlo Simulation 196, 599
Monte Carlo Simulation Using Path Integrals 609
Monte-Carlo CVARs 347
Monte-Carlo Effective Paths 216
Monte-Carlo VAR 346
Moody's Ratings 111
Morlet, J. 770
Mortgage Potential 546
Mortgage Servicing 235
MPI (Message Passing Interface) 446
MRG Model 559
Multidimensional Path Integrals 628
Muni Derivatives 221
Muni Futures 222
Muni Options 222
Munis 114
Musiela, M. 561
N Sphere 319
Nairay, A. 559, 717, 736
Nathan, A. 27
Negative Forward Rates 753
No Arbitrage 510, 543, 723, 739, 741
No Arbitrage and Yield-Curve Dynamics 725
No-Arbitrage, Hedging and Path Integrals 541
Nominal Interest Rate Grid 604
Noncallable Bonds 106
Nonlinear Diffusion 417
Non-Technical Managers 431
Normal Integral Approximation 623
Notation 190, 248, 258, 259, 265, 271,
 333, 341, 385, 465, 507, 513,
 514, 522, 524, 564, 605, 607,
 635, 656
Notation for Path Integrals 516, 563
Nozari, A. 461
Number of Parameters 693
Numerical Approximations 602
Numerical Code 51, 136, 142, 304
Numerical Errors 616
Numerical Instabilities 120
Numerical Noise 142
O(2,1) 298
OAS 121
Olive, D. 700
Omarova, N. 746
One-Sided Markets 338
Operational Risk 12, 415
Optimal PD Stressed Matrix 321
Option Extension 167
Option Replication 65
Option-Adjusted Spread 121
Oscillations 614, 615, 758
Outliers 284

Index

Overcompleteness 770
P&L Correlations 390
Pairs Trading 308
Parallel Processing 437, 442
Parallel Supercomputer 444
Parameters 419
Parisi, G. 462
Path Integral Discretization....................603
Path Integrals 273
Path Integrals - Basics............................502
Path Integrals - Review 561
Path Integrals and Finance 501
Path Integrals and Green Functions.........542
Path Integrals and Physics......................501
Path Integrals and Stochastic Equations..579
Path Integrals and VAR 378
Path Integrals Introduction 506
Payments in Arrears 92
PD Correlation Matrix............................319
Perelberg, A. 232
Periodic Caps 231
Personnel Risk.........................23, 423, 433
Perturbation Theory 63
Perturbative Skew 56
Phase Transition.............................117, 634
Phi 40
Physical Picture for Rates 690
Picoult, E. 109, 421, 485
Pisanchik, J. 749
Pit Options 130, 238
Plain-Vanilla VAR 341
Portfolios 33, 151
Positive-Definite Correlation Matrix......319, 331
Posner, S. 278
Potential Losses.....................................107
Power Options.......................................230
Presentations 22
Prices 152
Pricing . 27, 85, 164, 169, 173, 260, 267, 391, 614
Prime Caps..131
Prime Rate 720
Prime Rate and Macro-Micro Model 131
Principal-Component Analysis................656
Principal-Component Formalism 662
Principal-Component Options.................209
Principal-Component Path Integrals........629
Principal-Component Risk Measures 210
Principal-Component Simulation............657

Principal-Component Stressed Correlations 313
Principal-Component VAR 342
Programming Difficulties.......................434
Prony Analysis 763
Prony, Baron de.....................................763
Proprietary Models.................................423
Proprietary Trading 769
Prototyping....................................416, 435
Pruning the Tree 606
Put Spread 162
Put-Call Parity................................54, 139
Quadratic Plain-Vanilla VAR 344
Quant-Group Managers............................16
Quantitative Analysis Groups 4
Quanto Options 50
Quants ....xix, 15, 421
Quasi-Equilibrium Yield Curve Path 642, 654, 683
Quasi-European Approximation..............619
Quasi-Random Correlation Slope............332
Quasi-Random Trends............................755
Random Correlation Matrices 313
Random Macro Slope.............................722
Random Trends 767
Rapaport, G. 437
Rate Conventions 81
Rate Interpolation....................................83
Rate-Dependent Volatility Models..........586
Realized Volatilities 130
Rebates 48, 243, 262, 263
Rebates at Touch....................................263
Recovery Rate Uncertainty 401
REER 746
Reggeon Field Theory............................631
Reinhart, C. 746
Reload Option Exercise..........................215
Reload Options......................................214
Renormalization 326
Renormalization Group 634
Replicating Portfolio 58
Repo 95, 134, 236
Repo Rate 222
Reporting 20, 399
Reserves 344
Reset Rates..272
Resettable Options 226
Restrictions on Sale................................216
RFT 631
RFT Application to Finance 637
RFT in Physics......................................632

Rho 40
Rich-Cheap Analysis................................164
Rigor 32, 132, 508
Risk Aggregation 96, 150
Risk Classification 479
RISK Conference 410, 460, 715, 730
Risk Limits...489
Risk Reversals...43
Risk-Free Rate.......................................510
Rodriguez, M. 311, 320, 323
Rogers, C. 574
Rosen, D. 410
Ross, S. 506
Ross, S. A. 561
Rubenstein, M. 243
Rubin, L. 27
Rudd, A. 618
Rules of Thumb...63
S&P Ratings...111
Salomon Smith Barney 4, 85
Sample Documentation 424
Sampling Error.......................................461
Sanity Checks..................................51, 229
Savit, R. 632
Scenario Simulator..................................157
Scenarios ...10, 33, 90, 153, 155, 167, 178,
    204, 205, 238, 307, 344, 726
Schickele, P..764
Schwartz Distributions............................518
Schwartz, T. 481
Schwinger Formalism 546
Science and Finance 7
Seigel, L. ....xix, 642, 745
Selvaggio, B. 410
Semi Group 247, 515, 547
Separating Market and Credit Risk 407
Serial Correlation 290
Shapiro, H. 441
Sharpe Ratios 484
Sheffield, R. 749
Short Positions 401
Simple Harmonic Oscillator....................277
Singular Value Decomposition 326
Skew 53, 546
Skew-Implied Probability.........................63
Sliding Down the Forward Curve 157
Sliding Down the Yield Curve 76
Slighton, E...717
Smith Barney 4, 85
SMRG Model.................................682, 692
Sociology 188, 236, 614

Software Development Problems............440
Sparks, N. 410
Speed 40
Spherical Representation........................303
Spread 120
Spread Uncertainty.................................401
Spreads Between Adjacent Maturities......661
Stand-Alone Risks..................................389
Static Replication 58
Statistical Measures................................478
Statistical Scenario 10
Statistical Tests 651, 691
Sticky Delta...68
Sticky Moneyness 68
Sticky Strike...68
Stochastic Equations 591, 607, 626, 659, 695
Stochastic Equations and Path Integrals ..579
Stochastic Transition/Default Matrices ...398
Stochastic Volatility 60, 358
Story 113, 128, 214, 284, 421, 424,
    438, 439, 441, 524, 598
Stratanovich Prescription (not used) 512
Strategy 360
Stress Testing...477
Stressed Correlations.......................325, 359
Stressed Transition/Default Matrices 396
Stressed VAR 358
Strong Mean Reversion....................689, 733
Stuckey, R. 393
Surface terms..................................277, 571
SVD 325, 346
Swaplets 88
Swaps 79, 85
Swaption Deal 146
Swaptions 32, 137
Swaptions and Risk Management 144
Synthetic Convertibles 183
Systems Solution....................................440
Systems Topics 11
Systems Traps 432
Taleb, N. 56
Tech. Index (Definition).............................3
Technical Analysis 749
Technology, Strategy and Change...........446
Term Sheet..92
Term-Structure Constraints.............591, 744
The Bloomberg 119
Theta 40, 150
Third-Order Correlation Function ..685, 687,
    688
Three Principal Component Example 665

Index

Three Strikes Rule 491
Three-Fives Systems Criteria 431
Three-Point Correlation Analysis 765
Tilt Delta 210
Tilt Gamma 210
Time Scales 416, 485, 641
Time Step 721
Time Thresholds 756
TIPS 219, 735
Toolkit Functions 757
Total Return 204
Toy, W. 560
Tracy, T. 410
Trader on a Stick 124
Traders . xix, 27, 292, 330, 356, 400, 421
Traders and Risk Management 13
Traders in the Closet 720
Trading 769
Traditional Measures of Capital 487
Transition Probabilities 605
Transition/Default Probability Matrices .. 394
Transition/Default Probability Uncertainties
.... 400
Treasuries 75
Treasury-Bond Option Models 222
Trivariate Integral 173
Turetsky, J. 265
Turnbull, S. 278, 410
Tutorial 3
Tutorial (CIFEr) 734
Two-Country Paradox 48
Two-Dimensional Options 265
Uncertainties 8, 197
Unified Credit + Market Risk 410
Unitarity 529
Units 74, 178, 377, 484
Unused Limit Economic Capital 497
Unused-Limit Risk 489
Useful Integrals 251, 268, 274, 515
VAR Aggregation 388
VAR and Subadditivity 365
VAR Calculations 343
VAR Scaling-Factors 358
VAR Synopsis 381

VAR Uncertainty 357
Variable Strike 193
Vasicek, O. 690
VCR Option 161
Vectorization 445
Vega 40, 131, 140, 150
Vega Ladders 129
Vega Paradox 140
Velocity 546
Vendor Systems 436
Vol of Vol 60
Volatility 139, 510
Volatility Estimate 462
Volatility Regimes 68, 69
Volatility Surface 55
Wakeman, L. M. 278
Warrant 164
Watson, E. 95
Wavelets 763
Weights 322
What-if Scenario (WIS) 10
Whipsaw Analysis 207
White, A. 60, 559
Window Size 454
Windows 290, 454
WIS Numerical 477
WIS and Entropy 478
Wishart Distribution 464
Wishart- Fourier Transform 468
Wishart Generating Function 470
Wishart Theorem 461
WKB Approximation 467, 508, 646
Workstation Networks 444
Workstations 441
World Scientific xix
Yield 118
Yield Conventions 118
Yield-Curve Correlations 654
Yield-Curve Kinks 642, 655, 691
Yield-Curve Shape Options 209
Zero Correlations 329
Zero-Coupon Bond 112, 560, 567
Zhu, Y. 344

Errata

The following corrections have been made in this 2005 reprint of *Quantitative Finance and Risk Management: A Physicist's Approach.*

Equations: Notation is (Chapter C, Equation Eq, Page P)

(C17, Eq15, P251) for v_{max}, v_{min}: Switched A, B; changed $\lambda\sigma\sqrt{\tau^*} \to \lambda\sigma^2\tau^*$.

Corrected: $v_{max} = \dfrac{1}{\sigma\sqrt{\tau^*}}\left[x_a - \ln A + \mu\tau^* + \lambda\sigma^2\tau^*\right]$, $v_{min} = v_{max}(A \to B)$

(C25, Eq2, P333), 2nd paragraph below for $\Im$: Take the square root.

Corrected: $\Im\left[\rho_{\alpha\beta}(t)\right] = \sqrt{1 - \rho_{\alpha\beta}^2(t)}$

(C26, Eq7, P348): Dropped subscript $\mathcal{R}$ to get power $\aleph(k_{CL})$.

Corrected: $\left(^\$CVAR_\alpha\right)^{\aleph(k_{CL})} = {}^\$\mathcal{E}_\alpha(d_t x_\alpha)^{\aleph(k_{CL})}$

(C49, Eq13, P698): Interchanged numerator, denominator.

Corrected: $\dfrac{C_3^{(123)}}{M_3^{(123)}} \stackrel{?}{=} 0$

Text: Notation is (Chapter C, Page P, Paragraph Pr, Lines L)

(C6, P64, Pr2, L1-4): Switched labels "1", "2" and switched "bigger", "smaller"

(C49, P681, Pr3, L3), (C51, P733, Pr2, L2): Replaced "kurtosis" by "skewness"

(C49, P686, Pr1, L2): Deleted "kurtosis, or"